Insect Behavior

Insect Behavior

From mechanisms to ecological and evolutionary consequences

EDITED BY

Alex Córdoba-Aguilar
Universidad Nacional Autónoma de México, México

Daniel González-Tokman
CONACYT, Instituto de Ecología A. C., México

Isaac González-Santoyo
Universidad Nacional Autónoma de México, México

OXFORD
UNIVERSITY PRESS

Great Clarendon Street, Oxford, OX2 6DP,
United Kingdom

Oxford University Press is a department of the University of Oxford.
It furthers the University's objective of excellence in research, scholarship,
and education by publishing worldwide. Oxford is a registered trade mark of
Oxford University Press in the UK and in certain other countries

First Edition published in 2018
Impression: 1

British Library Cataloguing in Publication Data
Data available

Library of Congress Control Number: 2018932146

ISBN 978–0–19–879750–0 (hbk.)
ISBN 978–0–19–879751–7 (pbk.)

DOI: 10.1093/oso/9780198797500.001.0001

Printed in Great Britain by
Bell & Bain Ltd., Glasgow

Foreword

John Alcock

Emeritus Regents' Professor, School of Life Sciences, Arizona State University

Niko Tinbergen (1963) wrote a now famous paper in which he proposed that a complete study of animal behavior required research into the development of behavior, the physiological control of behavior, the adaptive value of behavior, and the evolutionary history of behavior. Given the broad range of the disciplines needed for a total picture of the causes of behavior, ranging from genetics to evolutionary biology, it is not surprising that most previous books on the subject of insect behavior have been largely limited to some portion of the four areas of research. So, for example, classic books by Vincent Dethier (1976) and Kenneth Roeder (1963) dealt with the physiology of behavior in certain insects while Choe and Crespi (1997) edited a book on the evolution of social behavior in insects. Evolutionary adaptations were the focus of a book that Thornhill and Alcock (1983) wrote, a book that was updated recently by Shuker and Simmons (2014). The second edition of the book on insect behavior by Matthews and Matthews (2010) did discuss both proximate and ultimate aspects of insect behavior, but almost a decade has passed since it was published and, moreover, the authors intended to reach undergraduates, rather than a more advanced audience. Therefore, previous books, whether edited or written entirely by one or two persons, left room for a modern survey of both proximate (developmental and physiological) and ultimate (adaptive and historical) causes of insect behavior.

The current edited compendium fills the need for a complete survey of the causes of insect behavior by taking advantage of the ability of specialists in all facets on insect behavior, including the relationship between behavior and pest control, as well as insect conservation, to communicate with readers about the most recent developments in their specialty, whether they be primarily proximate or ultimate in content. Graduate students in behavior and entomology will be the main beneficiaries inasmuch as many of the authors provide suggestions for additional research in their field. So, for example, Hunt and co-authors point out in this volume that, although genetic effects on the reproductive behavior of insects have been well documented, the relationship between natural selection and genes for elements of reproductive behavior requires much more work because many genes contribute both to reproduction and to the development of other important attributes. Sherratt and Kang suggest that use of the comparative method, a key tool for tracing the evolutionary history of attributes of interest, could help explain why, in groups of related species, some but not all exhibit certain characteristics, such as the brightly coloured underwings of certain *Catocala* moths. Olzer and her colleagues note that cryptic female choice in which females choose mates on the basis of their ability to manipulate stored sperm remains controversial and poorly studied. Vale and his co-authors examine the fascinating subject of parasites that change the behavior of infected insects, while cautioning that it is difficult to show that infected insects are preyed upon by the appropriate hosts of the parasites. Many additional examples of the kinds of useful future research are provided by the authors of this book's chapters providing interesting challenges for readers.

Although graduate students could clearly gain by reading this book, all behavioral biologists and entomologists would do well to peruse the book's chapters. Insects, of which there are more than a millions species, are not only extremely diverse behaviorally, but the ever increasing number of first rate research reports means that the task of keeping abreast of new developments related to insect behavior is ever more difficult. This book will do much to help in this regard particularly since one of the recurrent themes of the book is the importance and utility of investigating the connection between proximate mechanisms and the evolution of behavior, a still imperfectly studied phenomenon. So despite the fact that much has been done with insects, as this book documents, much more remains for inspired researchers to examine. The authors of this collection help us identify what still needs to be done if we are to more fully understand the behavior of the small-brained, but behaviorally complex inhabitants of our world.

References

Choe, J., and B. Crespi (Eds) (1997). *The Evolution of the Social Behavior of Insects and Arachnids*. Cambridge University Press, Cambridge.

Dethier, V. (1976). *The Hungry Fly*. Harvard University Press, Cambridge, MA.

Matthews, R.W., and Matthews, J.R. (2010). *The Behavior of Insects*. Springer, New York, NY.

Roeder, K. (1963). *Nerve Cells and Insect Behavior*. Harvard University Press, Cambridge, MA. [Revised edition 1998].

Shuker, D., and Simmons, L.W. (2014). *The Evolution of Insect Mating Systems*. Oxford University Press, Oxford.

Thornhill, R., and Alcock, J. (1983). *The Evolution of Insect Mating Systems*. Harvard University Press, Cambridge, MA.

Tinbergen, N. (1963). On aims and methods of ethology. *Zeitschrift für Tierpsychologie*, **20**, 410.

Acknowledgements

Many coincidences shape our life and decisions. For the case of this book, one coincidence is that we all editors happen to be good friends. When you can trust another person in shared tasks (such as producing a book), your life becomes much easier and this is why we not only made this adventure possible but also had fun during the process. Moreover, a huge task such as reviewing biological aspects of the most diverse animal group can become a hopeless adventure had not we had the help of quite a few fellows. We would therefore like to thank all people that contributed, including reviewers of chapter drafts and people who made suggestions of varying nature: John Alcock, Wolf Blanckenhorn, Bruno Buzatto, Adolfo Cordero-Rivera, Federico Escobar, Thomas Flatt, James Gilbert, Michael Greenfield, Johnattan Hernández-Cumplido, Héctor Méndez-Maldonado, Salvador Hernández-Martínez, Humberto Lanz-Mendoza, Zenobia Lewis, Germán Octavio López-Riquelme, Pierre-Olivier Montiglio, Miguel Moreno-García, Roberto Munguía-Steyer, Fernando Noriega, Diana Pérez-Staples, Bernard Roitberg, Gianandrea Salerno, Marla Sokolowski, Robert Srygley, Kate Umbers and Maren Wellenreuther. A big "thank you" for devoting their time and knowledge. Our acknowledgement also goes to Ian Sherman and Bethany Kershaw from Oxford University Press for your encouragement and patience.

Our granting agencies also deserve a place for indirectly supporting this book. We would therefore like to thank the following funding sources: PAPIIT UNAM IN203115 and IN206618 to AC-A, Project CONACYT Ciencia Básica 257894 to DG-T, PAPIIT UNAM IA209416 and Project CONACYT Ciencia Básica 241744 to IG-S.

Finally, AC-A wishes to thank his family: Ana, Rodrigo and Santiago.

Alex Córdoba-Aguilar
Daniel González-Tokman
Isaac González-Santoyo

Contents

List of contributors   xi

1. **Introduction**   1
   Daniel González-Tokman, Isaac González-Santoyo, and Alex Córdoba-Aguilar

2. **The genetics of reproductive behavior**   3
   John Hunt, James Rapkin, and Clarissa House

3. **Neurobiology**   32
   Anne C. von Philipsborn

4. **The role of hormones**   49
   H. Frederik Nijhout and Emily Laub

5. **Phenotypic plasticity**   63
   Karen D. Williams and Marla B. Sokolowski

6. **Habitat selection and territoriality**   80
   Darrell J. Kemp

7. **Long-range migration and orientation behavior**   98
   Don R. Reynolds and Jason W. Chapman

8. **Feeding behavior**   116
   Stephen J. Simpson, Carlos Ribeiro, and Daniel González-Tokman

9. **Anti-predator behavior**   130
   Thomas N. Sherratt and Changku Kang

10. **Chemical communication**   145
    Bernard D. Roitberg

11. **Visual communication**   158
    James C. O'Hanlon, Thomas E. White, and Kate D.L. Umbers

12. **Acoustic communication**   174
    Heiner Römer

13. **Reproductive behavior**   189
Rachel Olzer, Rebecca L. Ehrlich, Justa L. Heinen-Kay, Jessie Tanner, and Marlene Zuk

14. **Parental care**   203
Glauco Machado and Stephen T. Trumbo

15. **Sociality**   219
Jennifer Fewell and Patrick Abbot

16. **Personality and behavioral syndromes in insects and spiders**   236
Carl N. Keiser, James L.L. Lichtenstein, Colin M. Wright, Gregory T. Chism,
and Jonathan N. Pruitt

17. **Cognition and learning**   257
Reuven Dukas

18. **The influence of parasites**   273
Pedro F. Vale, Jonathon A. Siva-Jothy, André Morrill, and Mark R. Forbes

19. **Behavioral, plastic, and evolutionary responses to a changing world**   292
Wolf U. Blanckenhorn

20. **Behavior-based control of insect crop pests**   309
Sandra A. Allan

21. **Behavior-based control of arthropod vectors: the case of mosquitoes,
ticks, and Chagasic bugs**   332
Ana E. Gutiérrez-Cabrera, Giovanni Benelli, Thomas Walker, José Antonio De Fuentes-Vicente,
and Alex Córdoba-Aguilar

22. **Insect behavior in conservation**   348
Tim R. New

General Glossary   363
Index   375

List of contributors

Patrick Abbot Department of Biological Sciences, Vanderbilt University, Nashville, TN, USA

Sandra A. Allan Center for Medical, Agricultural and Veterinary Entomology, ARS/USDA, Gainesville, FL, USA

Giovanni Benelli Department of Agriculture, Food and Environment, University of Pisa, Pisa, Italy

Wolf U. Blanckenhorn Evolutionary Biology & Environmental Studies, University of Zürich-Irchel, Zürich, Switzerland

Jason W. Chapman Centre for Ecology and Conservation, and Environment and Sustainability Institute, University of Exeter, Penryn, Cornwall, UK

Gregory T. Chism Entomology and Insect Science Program, University of Arizona, Tuscon, AZ, USA

Alex Córdoba-Aguilar Departamento de Ecología Evolutiva, Instituto de Ecología, Universidad Nacional Autónoma de México, CDMX, México

Reuven Dukas Department of Psychology, Neuroscience and Behaviour, McMaster University, Hamilton, Ontario, Canada

Rebecca L. Ehrlich Department of Ecology, Evolution, and Behavior, University of Minnesota, Twin Cities, St. Paul, Minnesota, USA

Jennifer Fewell School of Life Sciences, Arizona State University, Tempe, AZ, USA

Mark R. Forbes 209 Nesbitt Bldg, Carleton University, 1125 Colonel By Drive, Ottawa ON K1S 5B7 Canada

José Antonio De Fuentes-Vicente Universidad Pablo Guardado Chávez, Tuxtla Gutiérrez, Chiapas, Mexico

Isaac González-Santoyo Departamento de Psicobiología y Neurociencias, Facultad de Psicología, Universidad Nacional Autónoma de México, CDMX, México

Daniel González-Tokman CONACYT. Red de Ecoetología, Instituto de Ecología, A. C. Xalapa, México

Ana E. Gutiérrez-Cabrera CONACYT-Centro de Investigación sobre Enfermedades Infecciosas, Instituto Nacional de Salud Pública, Cuernavaca, Morelos, México

Justa L. Heinen-Kay Department of Ecology, Evolution, and Behavior, University of Minnesota, Twin Cities, St. Paul, Minnesota

Clarissa House School of Science and Health and the Hawkesbury Institute for the Environment, Western Sydney University, Penrith, NSW, Australia

John Hunt Centre for Ecology and Conservation, University of Exeter, Cornwall, UK. School of Science and Health and the Hawkesbury Institute for the Environment, Western Sydney University, Penrith, NSW, Australia

Changku Kang Department of Biosciences, Mokpo National University, Muan, Republic of Korea

Carl N. Keiser Department of Biology, University of Florida, Gainesville, Florida, United States

Darrell J. Kemp Department of Biological Sciences, Macquarie University, North Ryde, New South Wales, Australia

Emily Laub Department of Biology, Duke University, Durham, NC, USA

James L.L. Lichtenstein Department of Ecology, Evolution and Marine Biology, University of California at Santa Barbara, Santa Barbara, CA, USA

Glauco Machado LAGE do Departamento de Ecologia, Instituto de Biociências, Universidade de São Paulo, São Paulo, SP, Brazil

André Morrill Department of Biology, Carleton University, Ottawa, Canada

Tim R. New Department of Ecology, Environment and Evolution, La Trobe University, Victoria, Australia

H. Frederik Nijhout Department of Biology, Duke University, Durham, NC, USA

James C. O'Hanlon School of Environmental and Rural Science, University of New England, Armidale, Australia

Rachel Olzer Department of Ecology, Evolution, and Behavior, University of Minnesota, Twin Cities, St. Paul, Minnesota

Anne C. von Philipsborn Danish Research Institute of Translational Neuroscience (DANDRITE), Aarhus University, Aarhus C, Denmark

Jonathan N. Pruitt Department of Psychology, Neuroscience and Behaviour, McMaster University, Hamilton, Ontario, Canada

James Rapkin Centre for Ecology and Conservation, University of Exeter, Cornwall, UK

Don R. Reynolds Natural Resources Institute, University of Greenwich, Chatham, Kent, UK

Carlos Ribeiro Champalimaud Research, Champalimaud Centre for the Unknown, Lisbon, Portugal

Bernard D. Roitberg Biology, Simon Fraser University, Burnaby, Canada

Heiner Römer Zoology, University of Graz, Austria

Thomas N. Sherratt Department of Biology, Carleton University, Ottawa, Canada

Stephen J. Simpson Charles Perkins Centre, The University of Sydney, NSW, Australia

Jonathon A. Siva-Jothy Institute of Evolutionary Biology, School of Biological Sciences, University of Edinburgh, Scotland, UK

Marla B. Sokolowski Department of Ecology and Evolutionary Biology, University of Toronto, Canada

Jessie Tanner Department of Ecology, Evolution, and Behavior, University of Minnesota, Twin Cities, St. Paul, Minnesota, USA

Stephen T. Trumbo Department of Ecology and Evolutionary Biology, University of Connecticut, Waterbury, CT, USA

Kate D.L. Umbers School of Science and Health Western Sydney University, Hawkesbury, Richmond, NSW, Australia

Hawkesbury Institute for the Environment, Western Sydney University, Hawkesbury, Richmond, NSW, Australia

Pedro F. Vale Institute of Evolutionary Biology, School of Biological Sciences, University of Edinburgh, Scotland, UK

Thomas Walker Department of Disease Control, London School of Hygiene and Tropical Medicine, London, UK

Thomas E. White School of Life and Environmental Sciences, The University of Sydney, Sydney, New South Wales, 2006, Australia

Karen D. Williams Department of Ecology and Evolutionary Biology, University of Toronto, Canada

Colin M. Wright Department of Ecology, Evolution and Marine Biology, University of California at Santa Barbara, Santa Barbara, CA, USA

Marlene Zuk Department of Ecology, Evolution, and Behavior, University of Minnesota, Twin Cities, St. Paul, Minnesota, USA

CHAPTER 1

Introduction

Daniel González-Tokman[1], Isaac González-Santoyo[2], and Alex Córdoba-Aguilar[3]

[1] CONACYT, Red de Ecoetología, Instituto de Ecología, A. C. Xalapa, México
[2] Departamento de Psicobiología y Neurociencias, Facultad de Psicología, Universidad Nacional Autónoma de México. CdMx México
[3] Departamento de Ecología Evolutiva, Instituto de Ecología, Universidad Nacional Autónoma de México. CdMx, México

1.1 Introduction

With over 1 million described species and 4–6 million species hypothesized to exist (Schowalter 2016), insects are the most diverse group of animals in the world, and such diversity is reflected in their behavior. Such vast diversity has propelled studies to put forward explanations of its basis, patterns, and consequences. Due to this, the study of insect behavior has attracted not only geneticists, physiologists, ecologists, evolutionary biologists, entomologists, and agronomists, who may be directly interested in the causes and consequences of insect behavior, for academic reasons or simple personal fascination, but also psychologists, nutritionists, economists, and mathematicians that have based or are currently basing their own research on the knowledge generated with insects. This attraction has found a fertile ground for another reason—insects can be great study subjects as they allow a fairly easy manipulation of key variables to investigate their behavior. This practical property has permitted insect behavior to be studied at all levels of analysis, from proximal causes, such as physiology, genetic regulatory mechanisms, and development, to the ultimate consequences, such as evolution and ecology. This is the reason why studies using several insect systems have championed our understanding of biological phenomena. To put a simple example that illustrates such 'insect strength', the 2017 Nobel prize was given to Jeffrey C. Hall, Michael Rosbash, and Michael W. Young, for elucidating the molecular mechanisms underlying circadian rhythms, a research fundamentally carried out in *Drosophila* flies.

Scientific knowledge of insect biology expands every day and this urgently needs updated reviews that facilitate our access to such information, especially for 'newcomers' in the insect behavior discipline. This feeling of an empty niche emerged several years ago, when we taught different graduate and postgraduate courses. These made use of different sources that never really captured an updated and/or summarized version of insect behavior at all levels. With this in mind, the aim of this book was to create a textbook of insect behavior for both students (mainly) and researchers that provides the key classical and modern concepts and approaches to understanding insect behavior, all

González-Tokman, D., González-Santoyo, I., and Córdoba-Aguilar, A., *Introduction*. In: *Insect Behavior: From mechanisms to ecological and evolutionary consequences*. Edited by Alex Córdoba-Aguilar, Daniel González-Tokman, and Isaac González-Santoyo: Oxford University Press (2018). © Oxford University Press.
DOI: 10.1093/oso/9780198797500.003.0001

set within a multidimensional framework—from genes to ultimate evolutionary and ecological consequences. As it was highly likely that the target may be missed if this book were to include topics outside our study fields, it was decided to produce a multi-authored work, where experts in each aspect of insect behavior could provide specialist views of the field.

As it is usual in science, this book did not start from scratch, but uses Matthews and Matthews' (2009) and Alcock's (2013) magnificent books as fundamental starting points. Thus, several specialists were asked to provide a clear and concise state-of-the-art review of their fields directed to new generations in areas that are or have become fruitful grounds for research, including traditional (e.g. genetics, hormonal control) and topical (e.g. personality, parasite-induced insect behavior), or even fields that are usually included in other areas, such as global change, and pest and vector management. The authors are fully aware that this treatise is by no means complete. On one hand, there are issues that this book does not cover, but that are fortunately found in other textbooks, such as the history of insect behavior (Matthews and Matthews 2009), multitrophic interactions (e.g. herbivory, predation, and pollination; Rosenthal and Berenbaum 1992; Price et al. 2011), and thermoregulation behavior (Matthews and Matthews 2009) to quote a few. On the other hand, there are issues that could not be covered for reasons of space limitation, but that remain pending for future treatises, such as oviposition behavior and behavioral adaptations of insects as predators (a counterpart of Chapter 9). Other topics included in the selection were written with some explicit limitations, as the field was too vast to be reviewed in fewer than 7000 words [e.g. Chapter 2 used only reproductive behavior, rather than all behaviors to illustrate the genetic basis or Chapter 21 where only three vectors (two of which are insects) are explained].

Besides this introductory chapter, the book includes further 21 chapters, which are divided in three main sections. The first section includes four chapters about the interacting mechanisms controlling behavior—genes, hormones, and the nervous system. The second section, which is the core of the book, includes thirteen chapters about the diversity of behaviors, and their ecological and evolutionary consequences, incorporating emerging topics that have traditionally been studied in other animal groups, such as learning, cognition, and animal personality. The final section of four chapters comprises the application of insect behavior, including the importance of climate change on insect behavior, management of crop pests and disease vectors, and the importance of behavior in insect conservation. Since this book was conceived essentially for students, readers will find a glossary section at the end of the book, where concepts used throughout all chapters have been defined by contributing authors. These concept terms can be found in bold the first time they are mentioned in the text.

Finally, the aim of this book will not be achieved if the reader does not find this book an indispensable part of their library. Thus, the authors are open to feedback in case anyone wants to reach them. Meanwhile, the authors wholeheartedly expect their readers to enjoy each chapter in the same fashion as they did as they paved their way to publication.

References

Alcock, J. (2013). *Animal behavior*, 10th edn. Oxford University Press, Oxford.

Matthews, R.W., and Matthews, J.R. (2009). *Insect behavior*, 2nd edn. Springer Science & Business Media, Berlin.

Price, P.W., Denno, R.F., Eubanks, M.D., Finke, D.L., and Kaplan, I. (2011). *Insect ecology: behavior, populations and communities*. Cambridge University Press, Cambridge.

Rosenthal, G.A., and Berenbaum, M.R. (1992). *Herbivores: their interactions with secondary plant metabolites: ecological and evolutionary processes*. Academic Press, San Diego.

Schowalter, T.D. (2016). *Insect ecology: an ecosystem approach*. Academic Press, Cambridge, MA.

CHAPTER 2

The genetics of reproductive behavior

John Hunt[1,2], James Rapkin[1], and Clarissa House[2]

[1] *Centre for Ecology and Conservation, University of Exeter, Cornwall Campus, Penryn, TR10 9EZ, UK*
[2] *School of Science and Health and the Hawkesbury Institute for the Environment, Western Sydney University, Hawkesbury Campus, Locked Bag 1797, Penrith, NSW, Australia*

2.1 Introduction

Understanding the relative contribution of genes and the environment to observed variation in behavior has been the central aim of behavioral geneticists for nearly six decades. This is clearly an important endeavour as behavior must have a genetic basis if it is to evolve and drive key evolutionary processes such as adaptation and speciation (Boake 1994). However, after decades of empirical research on this topic, it is safe to say that the majority of researchers would agree that most (if not all) behaviors have a genetic basis, but are also influenced, to some degree, by the environment. Consequently, the question is no longer whether behavior is under genetic control, but rather what is the distribution of genetic effects for behavior (many genes with a small effect or few genes with a large effect), how do these genes interact with each other, with genes for other traits, and with the environment, and what are the wider implications of this complex genetic architecture to the evolutionary process?

Insects have played a key role in the understanding of how genes influence behavior. This is for three main reasons. First, many behaviors in male and female insects are highly stereotyped, meaning they are performed the same way each time. This enables behavior to be easily and accurately quantified for a large number of individuals. Secondly, the short generation times and high fecundity of many insect species (relative to vertebrates) makes them well suited to a variety of quantitative genetic breeding designs that span a few (parent–offspring regression, full and half-sibling designs) or multiple generations (i.e. artificial selection). Furthermore, many insect species can be inbred without a substantial decline in fitness, enabling inbred and iso-female lines to be easily created and used in various crossing designs (e.g. diallel) to estimate **non-additive genetic variance** and to create mapping populations for genomic analysis. Finally, many insects have a relatively simple genome (compared with vertebrates) that is well-annotated, notable examples include *Drosophila melanogaster*, the silk moth (*Bombyx mori*), and honey bees (*Apis mellifera*). This increases the ease and effectiveness of genomic studies investigating the specific gene(s) that regulates

Hunt, J., Rapkin, J., and House, C., *The genetics of reproductive behavior.* In: *Insect Behavior: From mechanisms to ecological and evolutionary consequences.* Edited by Alex Córdoba-Aguilar, Daniel González-Tokman, and Isaac González-Santoyo: Oxford University Press (2018). © Oxford University Press.
DOI: 10.1093/oso/9780198797500.003.0002

behavior in insects (e.g. Mackay et al. 2005). Given these features, it is not surprising that the genetic basis of a large variety of different behaviors have been investigated in insects (e.g. foraging, Page et al. 1995; personality, Løvlie et al. 2014; courtship and mating, Gaertner et al. 2015; division of labour, Smith et al. 2008; learning, Dunlap and Stephens 2014; and grooming, Hamiduzzaman et al. 2017), making this topic an incredibly broad one.

In an attempt to help narrow this extensive list, this chapter will focus exclusively on the genetics of insect reproductive behavior. It takes a broad view of 'behavior' as 'the response of an individual to a particular stimulus' and 'reproductive behavior' as 'any behavior that influences reproduction in either sex'. The view presented here includes obvious reproductive behaviors, such as choice of mate, courtship and mating displays, and oviposition preference, as well as a range of traits in males and females that are not traditionally viewed as behaviors. This includes fecundity that can be increased by females in response to a variation in male quality (e.g. Kotiaho et al. 2003) or with impending mortality (e.g. Staudacher et al. 2015). Likewise, it includes several sexual traits in males, such as cuticular hydrocarbons (e.g. Kent et al. 2008) and acoustic signals (e.g. Kasumovic et al. 2012), that can be rapidly altered in response to changes in the social environment.

We cover a range of topics in this chapter that are considered to be fundamental to the understanding of the genetics of insect reproductive behavior. In the first section, it is argued that the majority of insect reproductive behaviors are governed by many genes (i.e. polygenic), that each have a small effect, and use empirical evidence from quantitative genetic studies and genomic approaches to support this argument. Section 2.2 examines the exception to this general polygenic rule, where insect reproductive behavior is determined by a small number of genes of major effect. Although empirical support for genes of major effect is currently weak for reproductive behavior and limited to species with well-annotated genomes, it is suspected that this view may change as the number of genomic studies increase. Section 3 shows that genes for insect reproductive behavior are often associated with genes for a diversity of other important traits, including those involving morphology and life-history. This suggests that reproductive behaviors are unlikely to be free to evolve independently. Section 4 examines the importance of non-additive genetic effects (dominance and epistasis) to insect reproductive behavior. While greatly under-studied, it is likely that dominance and epistasis make important contributions to the observed variation in insect reproductive behaviors. Section 5 shows that the genes for reproductive behavior in many insect species interact with both the abiotic and social environment. This indicates that the influence of genes on reproductive behavior is likely to be highly context-dependent, with **genotype-by-environment (GEI)** and **genotype-by-social environment** interactions both complicating the link between genotype and phenotype. The final section focuses on the wider implications that genetic architecture has for the evolution of insect reproductive behavior, as well as outline some future research directions that the authors view as exciting and deserving of more attention.

2.2 Reproductive behaviors in insects are polygenic and each gene has a small effect

Most behaviors in animals, especially those associated with reproduction, are complex quantitative traits that show considerable variation along a continuous (and normal) distribution of phenotypes. Early theoretical models for the inheritance of quantitative traits assume that they are controlled by an infinite number of loci, each having an infinitely small effect; the so-called 'infinitesimal' model (Fisher 1918; Bulmer 1980; Barton et al. 2016). Under this model, the genome is treated as a 'black box' with genetic effects described through statistical parameters (such as variances and covariances), rather than focusing on the effects of individual loci, which are considered to be small and unmeasurable (Falconer and Mackay 1996; Barton and Keightley 2002; Conner and Hartl 2004).

2.2.1 Exploring genetic variation in insect reproductive behavior using quantitative genetics

Before we can discuss how **genetic variances** and covariances can be used to describe the importance of genetic effects on quantitative traits, we must first provide a brief overview of the basic principles of quantitative genetics. Quantitative genetics posits that the phenotype, P, of an individual is the sum of the effects of genes, G, and the environment, E (Falconer and Mackay 1996; Lynch and Walsh 1998):

$$P = G + E \qquad [2.1]$$

Quantitative geneticists, however, typically focus on partitioning phenotypic variation within a population to genetic and environmental sources, rather than focusing on the phenotype of specific individuals. Therefore, Eqn [2.1] can be expressed in terms of population variances as:

$$V_P = V_G + V_E \qquad [2.2]$$

where V_P is the **phenotypic variance**, V_G is the genetic variance, and V_E is the **environmental variance** in the population. Eqn [2.2] represents the simplest way that V_P can be partitioned into genetic and environmental sources, and is most useful when considering clonal (or highly self-fertilizing) organisms because the parental diploid genotypes are replicated in the offspring. It is less useful in sexually reproducing organisms where novel genotypes are created in each offspring by a random combination of one allele from each parent at each locus. In these species, we need to further partition V_G:

$$V_G = V_A + V_D + V_I \qquad [2.3]$$

where V_A is the **additive genetic variance**, V_D is the **dominance variance**, and V_I is the epistatic variance. V_A is the most important form of genetic variation for sexually reproducing organisms because only the additive effects of genes are transmitted directly from parents to offspring and, therefore, contribute to changes in phenotype across generations. V_D and V_I are collectively referred to as non-additive genetic variance. Unlike V_A, V_D, and V_I are not directly transmitted from parents to offspring.

Historically, the environment was considered to have a 'random' (and non-genetic) effect on phenotype and is viewed as a source of variation that reduces the resemblance between parents and offspring. When the environment influences phenotype in this way it generates **general environmental variance** (V_{Eg}). However, in many cases, the environment is provided by other individuals in the population. This social environment is often experienced non-randomly by a given individual, for example, as occurs when a parent provisions their offspring. Collectively, the effect of the social environment on phenotype is referred to as **special environmental variance** (V_{Es}) and we discuss this source of variance further in section 6. Just like V_G, V_E can therefore also be further partitioned as:

$$V_E = V_{Eg} + V_{Es} \qquad [2.4]$$

Finally, Eqns [2.1] and [2.2] assume that genes and the environment have independent effects on phenotype, which is unlikely to ever be the case. GEIs exist whenever genotypes respond differently to environmental variation and this can also represent an important source of variance in phenotype (V_{GEI}).

Eqn [2.2] can now be extended to include all of the previously mentioned sources of phenotypic variation in the population:

$$V_P = V_A + V_D + V_I + V_{Eg} + V_{Es} + V_{GEI} \qquad [2.5]$$

2.2.2 Estimating genetic variance and heritability for phenotypic traits

The central aim of quantitative genetics is to estimate the variance components outlined in the above equations, especially V_G and V_A, and a variety of different breeding designs that use individuals of known relatedness are used to achieve this aim. A commonly used metric to describe the importance of genes to phenotypic variation is the **heritability**. Heritability is simply the ratio of genetic variance to total phenotypic variance and, therefore, estimates theoretically range from 0 to 1. Importantly, however, it can be measured in two ways: as a broad-sense estimate (H^2) or a narrow-sense estimate (h^2). H^2 is estimated as V_G/V_P, whereas h^2 is estimated as V_A/V_P. Consequently, h^2 is more informative as an agent of evolutionary change than H^2. The benefit of both metrics, however, is that dividing V_A and V_G by V_P means that the relative importance of genes can be compared across different traits and studies. However, it is important to remember that because H^2 and h^2 are ratios, changes in both the numerator (V_A or V_G) and the denominator (V_P) can influence the magnitude of these parameters. Thus, it is possible that differences in H^2 and h^2 may also reflect differences in V_D, V_I, V_E and/or V_{GEI}.

Next, some of the commonly used laboratory approaches to estimate the genetic contribution to phenotypic variation are outlined, placing particular emphasis on their strengths and weakness.

- *Common garden experiment*: This is the simplest way to demonstrate that a phenotypic trait has a genetic basis. Individuals from different populations that exhibit natural variation in a given phenotypic trait are collected and reared under the same environmental conditions in the laboratory, and the divergence in phenotype is again assessed across populations. If the populations are still divergent then the phenotypic trait has a genetic basis, whereas if differences are no longer apparent then this initial divergence is due to the environment. While this approach is relatively simple to implement, it does not allow key genetic parameters (such as h^2) to be estimated. Furthermore, at least two generations of **common garden experiment** rearing are needed to remove any population differences due to maternal effects.
- *Parent–offspring regression*: Each male is mated to a single female in the parental generation and the phenotypic trait of interest is measured for one (or both) parent(s). Offspring are reared under the same environmental conditions and the trait measured at the same age (or developmental stage) as their parents. The average of the trait in offspring for each family is then regressed against the parental value(s)—either one parent or the average of both—using linear regression. The slope of this regression line can be used to estimate the H^2 for the trait of interest. If the average phenotypic value of both parents is used in the regression, the slope equals H^2 and if the phenotypic value of only one parent is used, H^2 is twice the slope. While being one of the simpler breeding designs to execute, it is possible for estimates of H^2 to be biased if maternal and/or paternal or dominance effects are large, and if the environments experienced by parents and offspring are dramatically different.
- *Full-sibling analysis*: A full-sibling design is identical to the parent–offspring regression, with the exception that the trait of interest is *not* measured in the parents. Instead the among-family variance is estimated using a one-way analysis of

variance (ANOVA). Just like the parent–offspring regression, estimates of H^2 can be biased by maternal and/or paternal effects and dominance variance, and because V_G is extracted from a 'family' term in the ANOVA, it does not allow the effects of mothers and fathers to be statistically separated. However, as the phenotypic trait of interest is not measured in both parents and offspring, the issue of differences in the environment experienced by parents and offspring is no longer a concern when estimating H^2.
- *Half-sibling analysis*: In this approach a series of males (sires) are each mated to a unique set of randomly chosen females (dams) and the trait of interest is measured for a number of offspring for each dam. Consequently, this design differs from the full-sibling design in that each sire is mated to multiple females, meaning that the design contains full-siblings (same mother and father), half-siblings (different mother, same father) and unrelated offspring (different mother and father). A nested ANOVA (with dams nested within sires) can then be used to determine the independent effects of males and females on offspring phenotype. This design is considered the 'gold standard' because V_A (and ultimately h^2) can be estimated through sires and, therefore, is free from maternal effects and dominance variance.
- *Inbred and iso-female lines*: Both of these approaches are based on the same principle—fixing a series of genotypes from the general population using inbreeding. In the case of iso- female lines, a series of gravid females are collected from the field, and their offspring isolated and subjected to brother–sister mating to provide individuals for subsequent generations. As females are collected from the field, it can be argued that the genotypes contained in these iso-female lines captures the genetic architecture present in the field, including **linkage disequilibrium**. In the case of inbred lines, male–female pairs are isolated from a laboratory population, mated to produce offspring, and brother–sister mating used to populate subsequent generations. After twenty generations, the inbred lines will be homozygous at 99.98 per cent of loci and will largely be identical by descent. The phenotypic trait of interest can then be measured in individuals from the

different lines and differences across lines assessed using ANOVA. Significant divergence in the trait across lines indicates a genetic variance for the trait, but is not possible to determine whether this genetic basis is due to additive and non-additive gene effects. Furthermore, as only certain genotypes in the population are likely to survive the inbreeding process, it has been argued that iso-female and inbred lines may upwardly bias genetic estimates (David et al. 2005).

- *Crosses*: There are various types of crossing designs available to estimate the various forms of genetic variance contained in Eqn [5]. For example, in a *diallel* cross, male and female parents (typically taken from iso-female or inbred lines) are crossed to produce hybrid offspring in which the trait of interest is measured. The crossing design is usually 'complete' (all possible parental combinations), but partial, reciprocal, or pooled reciprocal designs also exist. A two-factor ANOVA, including male and female genotype plus their interaction as terms in the model, can then be used to analyse the variation in offspring phenotype. The male term can be used to estimate V_A for the trait being examined and a significant interaction term indicates that non-additive genetic variance also contributes to the variation in this trait. More complex, cross-classified designs, such as the North Carolina Design III and the triple test cross, can be then used to estimate V_D and V_I, respectively (Lynch and Walsh 1998). While crosses provide a powerful way to estimate additive and non-additive genetic variance for phenotypic traits, it is not possible for many sexually reproducing organisms where replicate individuals of a single genotype are not available.
- *Artificial selection*: The evolutionary response (*R*) of a given phenotypic trait to selection (*S*) can be predicted by the univariate breeder's equation as $R = h^2 S$. Artificial selection experiments that enforce a known regime of selection on a trait and measure the evolutionary response across generations can therefore be used to estimate the h^2 of the trait by rearranging this equation to $h^2 = R/S$. Importantly, this estimate is referred to as a *realized* h^2 because it has been measured after an evolutionary response has already been observed.

2.2.3 Empirical evidence from quantitative genetics for the polygenic control of reproductive behaviors in insects

Studies using quantitative genetics have been instrumental in demonstrating the genetic contributions to differences in insect reproductive behavior (Ewing and Manning 1967; Krebs et al. 1993; Pianka 1999), and this is especially true for reproductive behaviors in insects (Thornhill and Alcock 1983; Arnholt and Mackay 2004; Markow and O'Grady 2005; Shuker and Simmons 2014). Table 2.1 provides a modest collection of quantitative genetic studies that have examined insect reproductive behavior using a variety of different breeding designs. Three clear patterns are apparent in these examples. First, there is a taxonomic bias in the quantitative genetics of reproductive behaviors in insects with the greater majority of studies being conducted on *Drosophila* and a number of field cricket species. This probably reflects the fact that these species are easier to breed in large numbers in the laboratory. Secondly, within these commonly used taxa, the quantitative genetics of reproductive behavior has been examined using a greater variety of breeding designs in *Drosophila*. The fast generation times in *Drosophila* and their suitability to inbreeding makes artificial selection and breeding designs based on inbred and iso-female lines possible. These approaches are much more difficult and time-consuming in other insect species. Finally, there does not appear to be any consistent differences in h^2 estimates across insect species, the sexes, the different reproductive behaviors examined, or the different breeding designs used. h^2 estimates vary from 0.03 to 0.90, although most are upwards of 0.20. This clearly shows that reproductive behavior has a strong genetic basis in insects and is likely to be under the control of many genes (polygenic).

2.2.4 Revealing specific gene effects through quantitative trait loci mapping and 'omics' approaches

A key development in the genetic analysis of quantitative traits has been the establishment of a large collection of molecular markers that have been used to construct genetic maps for a number of insect

Table 2.1 Some examples of quantitative genetic studies of reproductive behavior in insects.

Insect Order	Species name	Common name	Experimental design	Sex	Behavior examined	Heritability estimate	Reference
Coleoptera	*Callosobruchus maculatus*	Cowpea seed beetle	Parent–offspring regression	Female	Oviposition preference	0.35–0.88[A]	[1]
	Onthophagus taurus	Dung beetle	Half-sib	Female	Offspring provisioning	0.13 ± 0.09	[2]
Diptera	*Drosophila melanogaster*	Fruit fly	Inbred lines	Male	Courtship and mating behaviors	0.03–0.09[B]	[3]
			Iso-female lines	Female	Early life mating frequency	0.63 ± 0.18	[4]
			Artificial selection	Male	Courtship song structure	0.26 ± 0.03	[5]
				Female	Fecundity	0.28 ± 0.11	[6]
	Drosophila serrata	Fruit fly	Parent–offspring regression	Male	Cuticular hydrocarbons	0.06–0.73[C]	[7]
	Drosophila simulans	Fruit fly	Artificial selection	Female	Preference for ebony males	0.26 ± 0.11	[8]
Hymenoptera	*Nasonia vitripennis*	Parasitoid wasp	Half-sib	Female	Polyandry	0.03–0.82[D]	[9]
Lepidoptera	*Achroia grisella*	Pyramid moth	Half-sib	Female	Preference for male song	0.21 ± 0.13	[10]
	Euphydryas editha	Edith's checkerspot butterfly	Parent–offspring regression	Female	Oviposition preference	0.90	[11]
Mescoptera	*Panorpa vulgaris*	Scorpion fly	Half-sib	Male	Fighting ability[E]	1.07 ± 0.44	[12]
Orthoptera	*Gryllus bimaculatus*	Field cricket	Artificial selection	Male	Sperm length	0.52 ± 0.06	[13]
	Gryllodes sigillatus	Decorated cricket	Parent–offspring regression	Male	Spermatophylax investment	0.47 ± 0.21	[14]
	Gryllus firmus	Sand cricket	Half-sib	Male	Courtship song components	0.10–0.35[F]	[15]
	Gryllus firmus	Sand cricket	Full-sib	Female	Oviposition behavior	0.17–0.45[G]	[16]
	Teleogryllus commodus	Black field cricket	Half-sib	Male	Advertisement call structure	0.17–0.72[H]	[17]
	Teleogryllus oceanicus	Polynesian field cricket	Half-sib	Male	Courtship call components	0.06–0.60[I]	[18]

References: [1] Fox (); [2] Hunt and Simmons (2002); [3] Jang and Greenfield (2000); [4] Travers et al. (2015); [5] Ritchie and Kyriacou (1996); [6] Rose (1984); [7] Hine et al. (2004); [8] Sharma et al. (2010); [9] Shuker et al 2007; [10] Jang and Greenfield (2000); [11] Singer and Thomas (1988); [12] Thornhill and Sauer (1992); [13] Morrow and Gage (2001); [14] Sakaluk and Smith (1988); [15] Webb and Roff (1992); [16] Réale and Roff 2002; [17] Hunt et al. (2007); [18] Simmons et al. (2010).

Notes:

[A] Heritabilites calculated as the regression of family average preference on parental preference in two different populations (Bay Area and Davis). Courtship and mating behaviors include movement patterns—orientating towards female, approaching female, wing vibration, genital licking, attempted copulation, and copulation.

[C] Measured a cocktail of four different CHC components. [D] Behaviors measured include courtship duration and copulation duration.

[F] Song components include pulses per chirp, pulse length, pulse rate, chirp length, and frequency. [G] Oviposition behaviors measured include digging depth, egg depth, egg distribution, and fecundity.

[H] Call components measured include chirp pulse number, chirp inter-pulse duration, trill number, inter-call duration, and dominant frequency.

[I] Call components measured include chirp length, chirp pulse length, chirp pulse interval, chirp-trill interval, trill length, trill pulse interval, trill pulse length, and pulses per trill.

species (e.g. Hill 2012). These markers are the foundation for **quantitative trait loci** (QTL) mapping approaches, which includes techniques such as single-marker, interval, and multiple trait mapping. QTL mapping allows for the statistical analysis of associations between phenotype and genotype, and the dissection of the regions of the genome that significantly contribute to the variation of quantitative traits (Hill 2012). It aims to open the 'black box' of quantitative genetics by locating and identifying the genomic regions responsible for quantitative genetic variation.

However, even when significant associations between quantitative traits and molecular markers are identified, studies have found that these genomic regions are often too large (and too expensive) to identify the specific genes that contribute to genetic variation (Doerge 2002; Hill 2012). Fortunately, the rapid development (and reduction in cost) of a high throughput 'omic' methods, such as **genome-wide association studies** (GWAS), has provided an opportunity to identify some of these genes. Genome sequences have the potential to provide a comprehensive list of genes in an organism and functional **genomics** approaches can then be used to generate information about gene functions, and about genetic interactions between gene complexes and the environment.

2.2.5 Identifying QTLs and candidate genes of interest for insect reproductive behavior

Initial QTL approaches were linkage-based analysis, which used related individuals (specifically F_1 individuals originating from inbred lines) to provide an observable number of loci to identify segregating genetic markers. These F_1 individuals were then crossed and the segregation of genetic markers and QTLs in the F_2 generation statistically modelled. However, further developments have accommodated the use of composite mapping, multiple loci and family analysis in random-mating populations (Doerge 2002; Mackay et al. 2009; Hill 2012). This has led to the use of a number of linkage-based methodologies to detect QTLs associated with a quantitative trait of interest. These include:

- *Single-marker analysis*: Single-marker tests using *t*-tests, ANOVA, or simple linear regression, assess the segregation of a phenotype with respect to a marker genotype. These tests ascertain which markers are associated with the quantitative trait of interest and suggest the existence of QTLs. Studies using single-marker analyses deal primarily with detecting individual markers, rather than genomic regions and are useful for screening a large population for specific traits (Hill 2012).

- *Genetic-linkage maps*: Single-marker analyses investigate individual genetic markers without any reference to their position, order on the chromosome, or relative distances between these genetic markers. Additional genetic information can be gained about the interactions between these markers by placing them in map order. A genetic-linkage map, therefore, provides a genetic representation of the chromosome on which the markers and QTL reside (Mackay et al. 2009; Hill 2012).

- *Interval mapping*: Uses an estimated genetic map as the framework for determining the location of QTLs. Interval mapping statistically tests for a single QTL at each increment across the ordered genetic markers in the genome (Lander and Botstein 1989).

- *Multiple QTL*: Statistical approaches for locating multiple QTLs are more powerful than locating single QTLs because they can potentially differentiate between linked and/or interacting QTLs. However, locating multiple QTLs is a more complicated approach due to the large number of potential QTLs and their interactions. A number of methods have been developed to test for multiple QTLs. A simple technique is to first identify a single QTL, then to build a statistical model with these QTLs and their interactions, and then search in one dimension for significant interactions. However, one-dimensional searches can be challenged by the multiplicity of the effects of QTL interactions. An alternative approach is to split the search for interactions between QTLs into two parts—the relationship between QTLs and the quantitative trait, and the location of the QTLs (Mackay et al. 2009).

Linkage-based analyses have proved highly successful in identifying QTLs associated with variation in quantitative traits. However, limitations such as the inability to do finer scale mapping has seen linkage-based analyses being replaced with 'association

Table 2.2 Some examples of QTL-based studies showing the location and number of genes for reproductive behavior in insects.

Insect Order	Species name	Common name	Sex	Behavior examined	Location of genes/loci controlling behavior	Number of genes/loci controlling behavior	Reference
Diptera	*Drosophila elegans/gunungcola*	Fruit fly	Male	Wing spot	Ch 3 and X	4 QTL[A]	[1]
				Courtship wing display			
	Drosophila melanogaster	Fruit fly	Male	Courtship song	Ch 2, 3, 4, 5, and X	21 QTL[B]	[2] [3]
			Male and female	Cuticular hydrocarbon production	Ch 2, 3, and X	15–25 QTL	[4]
			Male	Courtship and copulation occurrence and latency	Ch 2, 3, and X	4 QTL[C]	[5]
			Male and female	Aggression	Ch 2 and 3	5 QTL[D]	[6]
			Male and female	Aggressive behavior	—	10 genes identified[E]	[7]
	Drosophila simulans/sechellia	Fruit fly	Male	Courtship song	Ch 2, 3, and 4	1–20 QTL[F]	[8]
	Drosophila virilise	Fruit fly	Male	Courtship song	Ch 2, 3, 4, and X	8–13 QTL[G]	[9]
Hymenoptera	*Nasonia giraulti/oneida*	Jewel wasp	Male and female	Male pheromone production	Ch 1, 2, 3, and 4	1–3 QTL[H]	[10]
				Male courtship behavior			
				Female mate discrimination			
Lepidoptera	*Achroia grisella*	Lesser wax moth	Male and female	Male courtship song	-	20–25 QTL[I]	[11]
				Female preference			
Orthoptera	*Laupala paranigra/kohalensis*	Hawaiian cricket	Male	Courtship song	Ch 1, 3, 4, 5, and X	5 QTL	[12]

References: [1] Yeh et al. (2006); [2] Etges et al. (2007); [3] Etges et al. (2007); [4] Foley et al. (2007); [5] Moehring and Mackay (2004); [6] Edwards and Mackay (2009); [7] Zwarts et al. (2011); [8] Gleason and Ritchie (2004); [9] Huttunen et al. (2004); [10] Diao et al. (2016); [11] Limousin et al. (2012); [12] Shaw et al. (2007).

Notes:

[A] Two pairs of loci, *y* and *Moe*, on the X chromosome, and *e* and *Tfl/A-L* on the third chromosome right arm.

[B] Six loci (2_2868a, 2_6540c, 2010, 2030, 2_1603a, 2200) on chromosome 2, three loci (3030, 3101, 3100) on chromosome 3, five loci (4010, 4050, 4300, 4301, 4302) on chromosome 4, three (5_1232a, 5100, 5200b) on chromosome 5 and four loci (X010, X030, X090, X110) on X chromosome.

[C] One on chromosome 2, two on chromosome 3 and one on X chromosomes. [D] Two on chromosome 2 and three on chromosome 3.

[E] Genes encompassed many biological and molecular processes—a transcription factor (*mbl*), protein kinases (*Doa*), a guanine exchange factor (*siz*), an NMDA receptor subunit (*Nmdar1*), a UDP-glucose transferase (*sgl*), an extracellular matrix protein (*LanA*), a cell adhesion molecule (*ed*), two Notch signalling regulation genes (*neur* and *Gp150*).

[F] Eight found on X chromosome, sixteen found on chromosome 2, twenty found on chromosome 3, one found on chromosome 4.

[G] Eight significant loci markers affecting variation in pulse train length (one on chromosome 2, six on chromosome 3 and one on chromosome 3) and thirteen significant loci markers affecting variation in pulse train (four on chromosome 2, six on chromosome 3 and two on chromosome 4).

[H] Three found for male pheromone quantity (one on chromosome 1, one on chromosome 4, one on chromosome 5), one found on chromosome 4 for male courtship behavior and one loci on chromosome 3 for female mate discrimination. Ten candidate genes on chromosome 1 were associated with copulation success and five associated with copulation success, six candidate genes on chromosome 3 were associated with copulation success, five candidate genes on chromosome 4 were associated with pheromone quantity and three candidate genes on chromosome 5 were associated with pheromone quantity.

[I] Between two brood groups twenty QTLs found in brood *Xt7* and 25 QTLs in Xt19. Most QTLs were distributed among thirty linkage groups in the *A.grisella* genome, but the authors did not find any obvious cluster of QTLs in certain chromosomes.

mapping', which uses individuals from natural populations that experience linkage disequilibrium (Mackay et al. 2009; Ott et al. 2011). Linkage disequilibrium is a sensitive indicator of the population genetic forces that structure the genome (Falconer and Mackay 1996; Lynch and Walsh 1998) and the resultant strong association between markers and QTLs it generates allows for much finer mapping and potentially uncovers the specific genes or mutations that are responsible for quantitative genetic variation (Ott et al. 2011; Hill 2012).

Association studies are routinely conducted across the entire genome using GWAS. GWAS are performed by genotyping many thousands of **single-nucleotide polymorphisms** (SNPs), which are single base-pair changes occurring at a high frequency in a DNA sequence and are used as genetic markers in GWAS (Ott et al. 2011; Bush and Moore 2012). GWAS utilize many of the same basic methodologies as linkage-based QTL studies and have become highly successful in identifying QTLs for quantitative traits of interest in a variety of species (e.g. Stranger et al. 2011; Bush and Moore 2012). Furthermore, because many GWAS use samples from the entire population, they potentially reflect natural genetic variation in quantitative traits, allowing for more accurate predictions about the underlying genetic architecture (Mackay et al. 2009; Ott et al. 2011). However, there are limitations to GWAS approaches, namely the low number of insect species that have a library of fully sequenced genomic data available, which can affect the availability of SNP markers and make detecting QTLs with strong environmental effects harder to detect (Hill 2012). Furthermore, methods for incorporating GEIs in GWAS studies are currently lacking.

2.2.6 Empirical evidence from QTL-based studies examining the polygenic control of reproductive behaviors in insects

Table 2.2 provides some examples of empirical studies utilizing QTL analyses to examine the polygenic control of reproductive behaviors in a number of insect species. Similar to the examples provided in Table 2.1, there is a strong taxonomic bias in studies using QTL-based approaches to locate the position and number of genes (or QTL regions)

responsible for insect reproductive behavior. By far the greatest numbers of studies have used *Drosophila* (especially *D. melanogaster*) as a model and this probably reflects the availability of a fully mapped genome for a number of *Drosophila* species, which makes identifying genes of interest much more efficient and easier than in other insect species (Markow and O'Grady 2005). The examples provided show a simple pattern. In most cases, multiple QTLs have been identified, providing further support for the polygenic control of reproductive behaviors in insects. Furthermore, in those instances where the location of QTLs was identified, they appear to be spread across the entire genome. Perhaps not surprisingly, a large number of these studies (six of the nine studies where the location of QTLs was identified) have shown that QTLs for insect reproductive behavior occur on the X chromosome. This supports the more widespread view that the X chromosome is a 'hot spot' for genomic evolution (e.g. Bailey et al. 2004). Finally, other than the increased occurrence of QTLs on the X chromosome, there does not appear to be any consistent patterns in the location or number of QTLs across the different reproductive behaviors. For example, the production of a courtship song is influenced by a large number of QTLs in *Drosophila* (typically more than ten QTLs), but is only influenced by five QTLs in *Laupala*. Whether this represents a more widespread taxonomic difference between Diptera and Orthoptera, however, will require more empirical testing.

2.3 Genes that have a major effect on insect reproductive behavior: the exception to the polygenic rule

In Section 2.2, we make the argument that insect reproductive behavior is polygenic and provides empirical support from quantitative genetic and QTL-based studies to support this argument. We would be negligent, however, if we did not mention the obvious exception to this argument—when single genes (or a small number of genes) have a major effect on reproductive behavior. Genes of major effect have been shown to be important for a range of non-reproductive behaviors in insects, including stinging behavior in the honey bee (*Apis mellifera*; Hunt et al. 1998) and feeding behavior in the pea aphid (*Acyrthosiphon pisum*; Caillaud and Via 2012). There is also evidence to suggest that genes that have a major effect on non-reproductive behavior in insects may also have a conserved function in other taxonomic groups. For example, the *for* gene and its associated orthologs have been found to be responsible for variation in foraging behavior in a number of insect species, including *Drosophila melanogaster* (e.g. Allen et al. 2017), honey bees (Ben-Shahar et al. 2002) and ants (e.g. Malé et al. 2017).

Unfortunately, there are only a limited number of studies that have examined genes of major effect on insect reproductive behavior. Table 2.3 provides an overview of existing empirical studies that have used a variety of QTL mapping and GWAS analysis to locate genes that have a major effect. The majority of studies have identified genes of major effect for reproductive behaviors in *Drosophila*. Work on this genus has identified genes having a major effect for male courtship behaviors, especially elements of the courtship song (Table 2.3). For example, both the *fruitless* (*fru*) and *doublesex* (*dsx*) sex determining genes in *D. melanogaster* are located in close proximity on the right arm of the third chromosome and play a key role in courtship song production (Rideout et al. 2007). Similarly, genes having a major effect on male courtship song have also been documented in the brown planthopper (*Nilaparvata lugens*, Butlin 1996) and the Australian field cricket (*Teleogryllus oceanicus*, Tinghitella 2008), although in the latter species this relationship is driven by a wing mutation (*flatwing*) at a single loci, which results in males lacking the wing apparatus needed to produce a courtship song. Finally, genes having a major effect on **pheromone** production have been identified in females of two lepidopteran species (*Heliothis subflexa* and *Ostrinia nubilalis*; Lassance et al. 2010; Groot et al. 2013). Both studies have been facilitated by the availability of a well-annotated reference genome for the moth, *Bombyx mori*.

Table 2.3 Examples of genes of major effect on reproductive behavior in insects.

Insect Order	Species name	Common name	Sex	Reproductive behavior examined	Number of genes involved	Reference
Diptera	*Drosophila melanogaster*	Fruit Fly	Male	Courtship behavior	Fruitless '*fru*' gene[A]	[1]
			Male	Courtship Song	Doublesex '*dsx*' gene[A], and '*fru*' gene	[2]
			Male	Interpulse interval courtship song	3 loci	[3]
	Drosophila virilis/littoralis		Male	Courtship song	2-6 loci	[4]
	Drosophila elegans/gunungcola		Male	Wing pigmentation and display	'Few' loci	[5]
Homoptera	*Nilaparvata lugens*	Brown planthopper	Male and female	Courtship song and female response	1.5–5 loci	[6]
Lepidoptera	*Heliothis subflexa*	Noctuid moth	Female	Pheromone production	KAIKOGA052256 BGIBMGA013924 BGIBMGA013740	[7]
	Ostrinia nubilalis	European corn borer	Female	Pheromone production	pgFAR	[8]
Orthoptera	*Teleogryllus oceanicus*	Polynesian field cricket	Male	Presence/absence of courtship song	1 loci	[9]

References: [1] Ryner et al. (1996); [2] Rideout et al. (2007); [3] Gleason et al. (2002); [4] Hoikkala et al. (2000); [5] Yeh et al. (2006); [6] Butlin (1996); [7] Groot et al. (2013); [8] Lassance et al. (2010); [9] Tinghitella (2008).

Notes:

[A] Both the '*fru*' gene and '*dsx*' gene are located on the right arm of chromosome 3.

2.4 Genes for reproductive behavior are often linked to other traits

We have so far limited our discussion to individual insect reproductive behaviors. Organisms, however, are not simply collections of independent phenotypic traits, but rather these traits are often interconnected at the genetic level due to shared functional, developmental, and/or physiological pathways (Falconer and Mackay 1996; Conner and Hartl 2004). This genetic association means that phenotypic traits are seldom free to evolve independently in the population, with a change in one trait influencing the expression of any other traits genetically associated with it. Quantitative genetic theory posits that the strength of the genetic association between two traits can be quantified through the sign and magnitude of the **genetic correlation** (Falconer and Mackay 1996; Lynch and Walsh 1998). As with phenotypic correlations, values for genetic correlations range from −1 to +1. The closer the genetic correlation is to these limits, the stronger the association is between the genes for the two traits,whereas the sign indicates whether the genes that increase one trait are linked to genes that increase (a positive correlation) or decrease (negative correlation) the second trait.

A genetic correlation between two traits can be generated in two ways: pleiotropy and linkage disequilibrium (Falconer and Mackay 1996). **Pleiotropy** generates a genetic correlation between two traits when a locus has a casual effect on both traits. In contrast, linkage disequilibrium will generate a genetic correlation between two traits when the alleles at two or more loci for the traits are associated with a higher or lower degree than would be expected through random association (Falconer and Mackay 1996). Genetic correlations generated through pleiotropy are expected to evolve, either through adaptation or by genetic drift, and are produced by common functional mechanism(s) that underlie the production of these correlated traits. Correlations through pleiotropy are not expected to break down in the population through neutral processes, such as random genetic drift. In contrast, genetic correlations generated by linkage disequilibrium are expected to be temporary, contributing very little to evolutionary change, and are expected to be eroded over time through recombination

(Falconer and Mackay 1996; Lynch and Walsh 1998; Saltz et al. 2017).

It is well established in quantitative genetic theory that selection rarely targets single phenotypic traits in isolation and that traits are often genetically correlated (Lande 1979; Lande and Arnold 1983). Furthermore, it has been known for well over three decades that this pattern of complex selection and the genetic variance in and covariance between traits can be used to predict the phenotypic evolution of traits across generations with the multivariate breeder's equation:

$$\Delta \bar{z} = \beta \mathbf{G} \qquad [2.6]$$

where $\Delta \bar{z}$ is the vector of phenotypic responses of traits across generations, β is the vector of linear selection gradients targeting those traits and $\mathbf{G}$ is a matrix of genetic variances in, and covariances between, these traits (Lande 1979). The key outcome of this equation is that the evolution of a given phenotypic trait is not only due to selection *directly* targeting the genetic variance in the trait, but also *indirectly* due to selection targeting other, genetically correlated traits. Consequently, to understand how a reproductive behavior evolves, it is necessary to know both the genetic variance in this behavior and how this behavior is genetically correlated with other important traits under selection.

2.4.1 Estimating genetic correlations between traits using quantitative genetics

The quantitative genetic breeding designs outlined in Section 2.2 can be used to estimate the genetic correlation between different traits (Falconer and Mackay 1996; Lynch and Walsh 1998). The key difference is that because the genetic relationship between two traits is now being examined, it is necessary to estimate the genetic variance in both traits, as well as the genetic covariance between the traits. As was discussed for heritability estimates in Section 2.2, however, genetic correlations have different meaning when estimated from these different breeding designs. That is, genetic correlations can be derived from the additive genetic (co)variance between traits (r_A) and represent a narrow-sense estimate or from the total genetic (co)variance between traits (r_G) and represent a broad-sense estimate (i.e. includes variance due to dominance and/or epistasis). As with heritability estimates, r_A provides a better estimate than r_G in how the genetic association between different traits directs phenotypic evolution. The following equations outline how r_A and r_G can be estimated from the breeding designs outlined in Section 2.2.

Using a parent–offspring regression, r_G can be calculated by dividing the covariances between different traits X and Y (cov_{XY}) in parents and offspring with the square root product of the covariances between the same traits (cov_{XX} and cov_{YY}, respectively) in parents and offspring (Falconer and Mackay 1996; Lynch and Walsh 1998):

$$r_G = \frac{cov_{XY}}{\sqrt{cov_{XX}cov_{YY}}} \quad [2.7]$$

As there are two possible products of cov_{XY}, there are two estimates of r_G (r_{G1} and r_{G2}) and the arithmetic mean of both estimates is generally provided. When using a full-sibling analysis, r_G is simply calculated as the covariance between mean of the two traits across full-sibling families using a regression (Falconer and Mackay 1996; Lynch and Walsh 1998). Likewise, this approach can also be used to calculate r_G when using inbred or iso-females lines, with the exception that line means for the two traits are used.

Using a half-sibling design, the additive genetic covariance between the two traits can be calculated at the sire level (and, thus, should largely be free from the effects of dominance and epistasis) using a nested analysis of covariance. r_A can then simply be calculated by dividing the additive genetic covariance between the two traits (cov_{XY}) by the square root product of the additive genetic variance in each trait (var_X and var_Y, respectively; Falconer and Mackay 1996; Lynch and Walsh 1998):

$$r_A = \frac{cov_{XY}}{\sqrt{var_X var_Y}}[6] \quad [2.8]$$

Finally, r_G can be measured in two ways when using artificial selection. First, r_G can be measured indirectly through the correlated response to selection. That is, if a given trait (X) is subject to artificial selection and shows an evolutionary response across generations, then a second trait (Y) can also be measured in the terminal generation. As Y has not been selected directly, any response to the selection regimes would indicate that that X and Y are genetically correlated. This commonly used approach, however, does not measure the strength of r_G, only the sign. If X and Y respond in the same direction, r_G is positive; if X and Y respond in opposite directions, r_G is negative. Secondly, a *double* selection experiment (where X is selected in one line and Y in another) can be used to measure both the direct (R_X and R_Y) and correlated responses (CR_X and CR_Y) of both traits. A joint estimate of r_G can then be obtained as (Falconer and Mackay 1996):

$$r_G = \sqrt{\frac{CR_X}{R_X}\frac{CR_Y}{R_Y}}[7] \quad [2.9]$$

This approach is not often used, however, given that it is twice the work of normal artificial selection experiment.

2.4.2 Empirical examples of genetic correlations between reproductive behavior and other traits in insects

In Table 2.4 we provide some examples of genetic correlations between reproductive behavior and other important phenotypic traits in insects. These examples clearly illustrate that insect reproductive behavior is genetically correlated with a range of other important **life-history traits**. Most available data has examined two important genetic correlations—between lifespan and reproductive behavior, and between immunity and reproductive behavior—that are often collectively viewed as 'costs of reproduction'. For females in the majority of species examined, there is a negative genetic correlation between lifespan and reproductive behavior (most commonly fecundity) and this appears to be independent of the particular quantitative genetic design used. The notable exception to this is the study by Khazaeli and Curtsinger (2010) that found a strong and positive genetic correlation between these traits in female *Drosophila melanogaster* when using inbred lines. The genetic correlation between lifespan and reproductive behavior is less clear in males, being negative in some species (Hunt et al. 2006; Brown et al. 2009) and positive in others (Brandt and Greenfield 2004). Studies on this relationship in males has largely been restricted to

Table 2.4 Examples of empirical studies showing that genes for reproductive behavior in insects are associated with genes for other important phenotypic traits. Standard errors or 95 per cent confidence intervals (in brackets) are provided for estimates of r_A or r_G. In studies using artificial selection, the sign (+ve or −ve) of the genetic correlation is provided.

Insect Order	Species name	Common name	Experimental design	Sex	Reproductive behavior examined	Linked trait	r_A or r_G	Reference
Coleoptera	Gnatocerus cornutus	Broad horned flour beetle	Artificial selection	Male	Fighting behavior	Mandible length	+ve	[1]
	Nicrophorus vespilloides	Burying beetle	Artificial selection	Male	Mating rate	Genital shape	+ve	[2]
	Callosobruchus maculatus	Seed beetle	Half-sib	Male	Copulation duration	Lifespan	-0.16 ± 0.20	[3]
	Callosobruchus chinensis	Azuki bean weevil	Half-sib	Female	Fecundity	Lifespan	-0.89 ± 0.12	[4]
Diptera	Drosophila melanogaster	Fruit fly	Half-sib	Female	Fecundity	Lifespan	-0.71	[5]
			Inbred lines	Female	Fecundity	Lifespan	0.75	[6]
	Drosophila nigrospiracula		Artificial selection	Female	Fecundity	Immunity to ectoparasitic mite	−ve	[7]
	Bactrocera cucurbitae	Melon fly	Artificial selection	Female	Fecundity	Lifespan	−ve	[8]
Hemiptera	Bactericera cockerelli	Potato psyllid	Inbred lines	Female	Fecundity	Lso infection[A]	-0.40	[9]
Hymenoptera	Apis melifera	Honey bee	Artificial selection	Male	Worker reproduction	Age at foraging	−ve	[10]
Lepidoptera	Achroia grisella	Acoustic moth	Half-sib	Male	Attractiveness	Lifespan	0.64 ± 0.09	[11]
	Pieris napi	White butterfly	Half-sib	Male	Spermatophore weight	Body size	0.35 ± 0.30	[12]
Orthoptera	Allonemobius socius	Ground cricket	Full-sib	Female	Fecundity	Presence of wings	-0.53 ± 0.15	[13]
	Gryllodes sigillatus	Decorated cricket	Inbred line	Male	Spermatophylax weight	Encapsulation ability	0.76 ± 0.03	[14]
				Male	Early-life calling effort	Rate of ageing	0.44 ± 0.17	[15]
				Female	Early-life fecundity	Rate of ageing	0.97 ± 0.06	[15]
	Gryllus firmus	Sand cricket	Artificial selection	Female	Fecundity	Proportion of winged morph	−ve	[16]
			Half-sib	Female	Fecundity	Wing morph	-0.86 ± 0.17	[17]
	Teleogryllus oceanicus	Polynesian field cricket	Half-sib	Male	Amount of trill in the courtship song	Encapsulation ability	-0.47 $(-0.49, -0.45)$	[18]

(Continued)

Table 2.4 Continued

Insect Order	Species name	Common name	Experimental design	Sex	Reproductive behavior examined	Linked trait	r_A or r_G	Reference
					Amount of trill in the courtship song	Haemocyte load	−0.48 (−0.50, −0.46)	[18]
	Teleogryllus commodus	Black field cricket	Artificial selection	Male	Time spent calling	Lifespan	−ve	[19]
			Half-sib	Female	Fecundity	Lifespan	−0.63 ± 0.27	[20]

References: [1] Okada and Miyatake (2009); [2] Hopwood et al. (2016); [3] Brown et al. (2009); [4] Nomura and Yonezawa (1990); [5] Rose and Charlesworth (1980); [6] Khazaeli and Curtsinger (2010); [7] Luong and Polak (2007); [8] Miyatake (1998); [9] Nachappa et al. (2014); [10] Oldroyd and Beekman (2008); [11] Brandt and Greenfield (2004); [12] Wedell (2006); [13] Roff and Bradford (1996); [14] Gershman et al (2010), [15] Archer et al. (2012); [16] Roff et al. (1999); [17] Roff et al. (1997); [18] Simmons et al. (2010); [19] Hunt et al. (2006); [20] Zajitschek et al. (2007).

Notes:

^A Also infection refers to infection by the bacterium *Candidatus Liberibacter solanacearum*.

insect species where males produce an acoustic signal (field crickets and an acoustic moth), as this provides a much easier way to assess reproductive effort (Hunt et al. 2006) and attractiveness (Brandt and Greenfield 2004) than in species lacking this form of **communication**. In the decorated cricket (*Gryllodes sigillatus*) both sexes show a strong positive genetic correlation between reproductive behaviors early-in-life and the rate of ageing, which further supports the view that reproduction is costly in insects and also demonstrates that reproductive behavior in the sexes (especially females) has important implications for the evolution of lifespan and ageing (Archer et al. 2012).

The examples presented in Table 2.4 also show that a negative genetic correlation between reproductive behavior and immune function is common in insects. Again, this genetic relationship appears to be largely consistent in both males (e.g. Gershman et al. 2010; Simmons et al. 2010) and females (e.g. Luong and Polak 2007; Nachappa et al. 2014) and also appears to be independent of the specific measure of immunity used (i.e. encapsulation ability, haemocyte load, immunity to bacterial or ectoparasite challenge) and the type of breeding design used. In females, fecundity was the most commonly studied reproductive behavior, whereas a much broader range of reproductive behaviors were examined in males, including the production of a large **nuptial gift** (spermatophylax, Gershman et al. 2010) and a more elaborate courtship song (Simmons et al. 2010). In both examples, the negative genetic correlation between immunity and these aspects of reproductive behavior is likely to have important implications for the operation of **sexual selection** in these species, challenging the view that females gain 'good genes' for immune function by mating with males with more elaborate sexual traits.

Reproductive behavior also appears to be genetically correlated with a range of important morphological traits in insects. For example, artificial selection has been used to show that there is a positive genetic correlation between fighting behavior and mandible length in the broad-horned flour beetle (*Gnatocerus cornutus*, Okada and Miyatake 2009), and between mating rate and genital shape in the burying beetle (*Nicrophorus vespilloides*, Hopwood

et al. 2016). Furthermore, a positive genetic correlation between spermatophore weight and body size was also shown in the white butterfly (*Pieris napi,* Wedell 2006) using a half-sibling design. Mandible length, genital shape, and body size are all known to be important determinants of male reproductive success in these, as well as other species of insects. Perhaps one of the best known examples of genetic correlations between reproductive behavior and morphology occurs in females of a number of cricket species—the negative genetic correlation between fecundity and the development of long wings (known as macroptery). This relationship has been documented using both artificial selection and sibling designs in two different cricket species, *Allonemobius socius* (Roff and Bradford 1996) and *Gryllus firmus* (Roff et al. 1997, 1999), although considerably more work has been done in the latter species. Although macroptery is an important determinant of flight capability and, therefore, the capacity for **dispersal**, these studies clearly show that the genes for this trait have a negative effect on those for reproduction; a finding that is also supported in male *G. firmus*, where a negative genetic correlation between testis mass and macroptery has also been documented (Saglam et al. 2008).

In the examples provided in Table 2.4, we only examine the genetic correlation between reproductive behavior and other important phenotypic traits using breeding designs. It is also possible to use molecular approaches to quantify the genetic association between reproductive behavior and other traits. For example, Kronforst et al. (2006) used QTL genetic linkage mapping to show an association between male mate preference and female forewing colour in two species of *Heliconius* (*H. cydo* and *H. pachinus*) butterflies. More specifically, mapping places the preference locus in the same genomic region as the locus determining forewing colour, which itself is linked to the wing patterning candidate gene, *wingless*. This suggests that wing colour and colour preference are either controlled by loci that are located in an inversion or result from the pleiotropic effect of a single locus (Kronforst et al. 2006). This tight genetic association between preference and wing colour patterns is likely to have played a key role in the high degree of speciation in the *Heliconius* genus.

2.5 Genes can have non-additive effects on reproductive behavior

As discussed in Section 2.2, additive genetic effects are the most important type of gene action in sexually reproducing species because they are directly transmitted across the generations and, therefore, contribute to evolution in a relatively straightforward manner. However, as noted in Eqn [2.5], non-additive gene effects also contribute to an individual's phenotype. Despite this, the majority of studies that investigate the genetic basis of reproductive behavior (and animal behavior more generally) tend to ignore dominance and epistasis, either because of the difficulty in estimating their effects or because their effects are considered to be unimportant (Meffert et al. 2002; Roff and Emerson 2006). This is an unfortunate trend given that the effects of dominance and epistasis on important phenotypic traits appear to be large, especially for traits that are more closely related to fitness (Roff and Emerson 2006). For example, a review of additive and non-additive genetic effects on morphological and life-history traits found that epistatic effects were detected more often in life-history than in morphological traits (79 versus 67 per cent, respectively), whereas dominance effects were reported for 95 per cent of traits, irrespective of trait type (Roff and Emerson 2006). Furthermore, for both dominance and epistasis, the ratio of non-additive to additive effects in life-history traits is approximately twice as large as for morphological traits (Roff and Emerson 2006). Given the close link to fitness, it is likely that non-additive genetic effects will also be important for insect reproductive behavior.

2.5.1 Estimating the effects of dominance and epistasis on phenotype using quantitative genetics

In Section 2.2, we discuss how a significant interaction between male and female genotypes in a *diallel* breeding design indicates that non-additive genetic effects have an important influence on the phenotypic trait being examined. A number of additional approaches have also been used to show the importance of non-additive genetic effects. The first approach is to examine the difference in the estimates of V_G from a parent-son and a parent–daughter regression. If genes are additive (and autosomal) in effect, half should be inherited from each parent and these regression coefficients should be the same. Any deviance from this (especially when the parent–daughter coefficient exceeds the parent–son coefficient) has been taken as evidence for non-additive genetic effects. The second approach is to compare the genetic variance explained by sires and dams in a nested half-sibling design. Again, any asymmetry in these variance estimates (especially if the dam variance exceeds the sire variance) is often taken as evidence of non-additive genetic effects. The main issue with this approach, however, is a bias in the dam variance, which can be caused by non-genetic maternal effects. It is important to note that none of the above approaches allow dominance or epistatic variance to be directly estimated, just that one or both is likely to be important in determining phenotypic variation.

A large number of different breeding designs are available to directly estimate the contribution of dominance and epistasis to any observed non-additive genetic effects. We do not attempt to cover all of these designs here, but instead provide two commonly used breeding designs to estimate dominance and epistatic variance in phenotypic traits. The first is a line-cross technique (known as the *North Carolina Design III)* that can be used to estimate the degree of dominance (Lynch and Walsh 1998). This approach crosses two parental genotypes (inbred lines, different populations or species are commonly used) to produce an F_1 generation that is then subject to random breeding to generate an F_2 generation. Random members of the F_2 generation are then backcrossed to each of the parental lines and the phenotype of these backcrossed families is measured. If $\bar{z}_1$ and $\bar{z}_2$ denote the mean phenotypes of progeny derived from the F_2 individuals backcrossed to parental line 1 and 2, respectively, and the sum of families is $S = \bar{z}_1 + \bar{z}_2$ and the family differences is $\Delta = \bar{z}_1 - \bar{z}_2$, then a one-way ANOVA can be used to estimate the variances of the family sums $[\sigma^2(S)]$ and differences $[\sigma^2(\Delta)]$. In the absence of epistasis and gametic phase disequilibrium, $\sigma^2(S)$ is equivalent to the V_A in the F_2 backcrossed population, while $\sigma^2(\Delta)$ is equivalent to twice the V_D. This approach will lead to inflated estimates of V_D, however, when gene frequencies are not equal, but when this assumption applies, the benefit of this approach is that it estimates V_A and V_D with nearly equal precision (Lynch and Walsh 1998).

The second approach, known as the *triple test cross*, is specifically designed to test the importance of epistatic variance (Lynch and Walsh 1998). This approach is very similar to the *North Carolina Design III*, with the major exception that F_2 individuals are backcrossed to both the parental lines and the F_1 population (not just the parentals). The logic behind this test is that F_1 individuals produce recombinant gametes, whose average gene expression will deviate from that of the mean of the parental line gametes if epistatic interactions are significant (Lynch and Walsh 1998). If $\bar{z}_3$ represents the mean phenotype of progeny from a backcross between F_1 and F_2 individuals, then $\bar{z}_1 + \bar{z}_2 - 2\bar{z}_3$ will have an expectation of zero in the absence of epistasis. A one-way ANOVA can again be used to test the significance of epistatic variance by evaluating whether then variance among the observed family values of $\bar{z}_1 + \bar{z}_2 - 2\bar{z}_3$ is greater than expected from sampling error. More complex analyses can also be used to estimate additive and dominance effects, as well as partitioning the different forms of epistatic variance (additive × additive epistasis, additive × dominance epistasis, dominance × dominance epistasis; Lynch and Walsh 1998). Furthermore, if reciprocal crosses and backcrosses are included in this design, additive genetic maternal variance, dominance genetic maternal variance, cytoplasmic variance, and Y chromosome variance can also be estimated.

2.5.2 Empirical examples of non-additive genetic effects for insect reproductive behavior

Unfortunately, there are only a handful of empirical studies that have investigated the role of non-additive genetic effects for insect reproductive behavior and we provide an overview of these studies in Table 2.5. The most compelling evidence

Table 2.5 Examples of empirical studies documenting non-additive genetic effects (dominance and/or epistasis) for reproductive behavior in insects.

Insect Order	Species name	Common name	Experimental design	Sex	Behavior examined	Reference
Coleoptera	*Callosobruchus maculatus*	Seed beetle	Reciprocal backcrosses	Female	Egg dispersion behavior[A]	[1]
			Reciprocal backcrosses	Female	Egg dispersion behavior[A]	[2]
	Acanthos celides obtectus	Seed beetle	Reciprocal backcrosses	Female	Oviposition site preference[B]	[3]
Diptera	*Drosophila tripunctata*	Fruit fly	Reciprocal backcrosses	Female	Oviposition-site preference[B]	[4]
	Musca domestica	Housefly	P–O regression	Male and female	Courtship behavior[C]	[5]
			Reciprocal backcrosses	Male and female	Courtship behavior[C]	[5]
	Eurosta soligaginis	Tephritid fly	Reciprocal backcrosses		Oviposition-site preference[D]	[6]
Hymenoptera	*Nasonia vitripennis*	Parasitoid wasp	Half-sib	Female	Polyandry[E]	[7]

References: [1] Fox et al. (2004); [2] Fox et al. (2009); [3] Tucić and Šešlija (2007); [4] Jaenike (1987); [5] Meffert et al. (2002); [6] Craig et al. (2001); [7] Shuker et al. (2007).

Notes:

[A]Egg dispersion behavior describes how uniformly females disperse eggs across seeds.

[B]Oviposition site preference was measured as the number of eggs laid on each host.

[C]A total of eight courtship behaviors were measured in males (mount, close, creep, touch, buzz, lunge, hold, and lift) and two courtship behaviors in females (female and wing out).

[D]Oviposition site preference was measured as the amount of ovipunctures on each host plant.

[E]Polyandry was measured as the product of four female behaviors (receptivity at first courtship (R_1), receptivity after 10 minutes (R_{10}), courtship duration, and copulation duration).

for non-additive effects on reproductive behavior is for egg dispersion behavior and oviposition-site preference in female insects. Egg dispersion behavior has been studied exclusively in female seed beetles (*Callosobruchus maculatus*) and describes how uniformly females disperse their eggs across seeds (Fox et al. 2004, 2009). Using a series of reciprocal backcrosses, Fox et al. (2004) showed that dominance, additive × additive epistasis, and dominance × dominance epistasis all significantly influenced egg dispersion behavior when females were reared on cowpea seeds, but that only the latter two forms of epistatic variance influenced this behavior when females were reared on mung bean seeds. A subsequent study, however, found that additive genetic, dominance, and additive–additive epistasis all influenced female egg dispersion in this species, irrespective of whether females were reared on cowpea or mung bean seeds (Fox et al. 2009).

Oviposition site preference describes the behavior that females exhibit when deciding which host to lay their eggs on when given the choice and the genetics of this behavior has been examined in a more taxonomically diverse range of insect species. In the seed beetle *Acanthos celides obtectus*, additive × additive epistasis, dominance × dominance epistasis and additive × dominance epistasis all influence oviposition site preference when females are reared on bean seeds, whereas additive genetic, dominance, and additive × additive and dominance × dominance epistasis influence this behavior when females are reared on chickpea seeds (Tucic and Seslija 2007). This contrasts work on the seed beetle *C. maculatus* where oviposition site preference in females was best described by an additive model, irrespective of whether females were reared on mung bean or cowpea seeds (Fox et al. 2004). Oviposition site preference has also been examined in a number of dipteran species. In *Drosophila tripunctata*, both dominance and epistasis were shown to be important sources of variation in female oviposition site preference for mushrooms or tomatoes, although the more explicit forms of the epistasis were not examined (Jaenike 1987). In the tephritid fly (*Eurosta soligaginis*), however, dominance, but not epistasis appears to regulate oviposition-site preference in females for two species of goldenrod (*Solidaginis gigantea* and *S. altissima*; Craig et al. 2001).

The importance of non-additive genetic variation to courtship and mating behavior in insects has also been examined. In the housefly (*Musca domesticus*), average heritability estimates for a range of courtship behaviors in males and females were significantly higher from parent–daughter analysis than parent–son analysis suggesting that non-additive genetic variance is likely to contribute to these reproductive behaviors (Meffert et al. 2002). A more detailed analysis using reciprocal backcrosses verified that both dominance and epistasis have important effects on courtship behavior in male and female houseflies but that the exact nature of these genetic effects varied for the different behaviors (Meffert et al. 2002). For example, only additive genetic effects were present for the 'buzz' courtship behavior in females, whereas dominance, dominance × additive epistasis, and dominance × dominance epistasis were important for the 'lunge', 'hold', 'lift wing', and 'wing out' courtship behaviors (Meffert et al. 2002). Furthermore, Shuker et al. (2007) used a nested half-sibling design to show that dam heritability estimates were, on average, seven times greater than sire heritability estimates for four reproductive behaviors linked to **polyandry** (receptivity at first courtship (R_1), receptivity after 10 minutes (R_{10}), courtship duration, and copulation duration) in female **parasitoid** wasps (*Nasonia vitripennis*). While this was taken as evidence that non-additive genetic effects were important for these behaviors, it is important to note that this asymmetry in sire and dam variances could also be driven by non-genetic maternal effects. Thus, further experimentation is needed to verify the contribution of non-additive genetic effects to these reproductive behaviors in *N. vitripennus*.

2.6 Genes for reproductive behavior frequently interact with the environment

Genotype-by-environment interactions (GEIs) exist whenever genotypes respond differently to environmental variation and are often illustrated using reactions norms where the phenotypic value of a trait in each environment is plotted separately for different genotypes (see Chapter 5). Most empirical studies investigating GEIs have focused on the influence of abiotic environments (Hunt and

Hosken 2014). This is especially true for insects, where many studies have documented GEIs involving a wide range of abiotic factors (e.g. Rodríguez and Greenfield 2003; Danielson-Francois et al. 2005; Weddle et al. 2012).

A number of studies have also started to consider GEIs that include the biotic environment, especially the presence of conspecifics or competitors (e.g. Saltz 2013; Pascoal et al. 2016b). Collectively, the environment provided by others in the population is referred to as the 'social' environment and as we outline in Section 2.2, this represents an important source of phenotypic variation referred to as *special environmental variance* (V_{Es}). V_{Es} differs from sources of general environmental variance (V_{Eg}), such as diet and temperature, because the social environment is provided by other individuals in the population. This means that genotype-by-social environment interactions (GSEIs) can have very different effects on the evolutionary dynamics of phenotypic traits compared with GEIs that involve the abiotic environment (Wolf et al. 2014). For example, traits that are influenced by GSEIs are expected to be much more labile and more evolutionarily dynamic than those subject to GEIs, especially in viscous populations with little re-assortment of individuals between environments (Wolf et al. 2014). This is due to the fact that both the focal trait and the social environments evolve simultaneously, making it easier to build and lose genotype-environment combinations compared with GEIs (Wolf et al. 2014).

2.6.1 Estimating GEIs and GSEIs for phenotypic traits using quantitative genetics

Most of the breeding designs outlined in Section 2.2 can be easily modified to estimate the variance in phenotype explained by GEIs or GSEIs. In all cases, this modification involves splitting each genotype across alternate environments, an approach known as a 'split brood' design (Lynch and Walsh 1998). However, the breeding designs in Section 2.2 will differ in the strength of support they provide for a GEI (or GSEI). In the case of common garden and artificial selection experiments, populations and selection lines can be split across environments, and any significant interactions between population or selection line, and the environment taken as evidence

of a GEI. These approaches, however, should be interpreted with caution as both populations and selection lines will have a range of different genotypes, rather than a single, fixed genotype. Furthermore, these approaches do not allow the direct estimation of V_{GEI} and, therefore, should only be taken as evidence that GEIs are likely to exist (not as definitive evidence that they do).

In the case of inbred and iso-female lines, as well as full- and half-sibling designs, V_{GEI} can be estimated directly when discrete genotypes are split across environments, although it is only in the latter design that additive × environment interactions can be estimated. In each design, the statistical models outlined in Section 2.2 can easily be extended to include an 'environment' term as a fixed effect, and V_{GEI} estimated using an ANOVA-based approach (Lynch and Walsh 1998). A variety of statistical approaches also exist where the environment can be included as a continuous variable and also where the genetic basis of the **reaction norm** across environments can be estimated (Lynch and Walsh 1998). Parent–offspring analysis can also be conducted in different environments and any difference in the slope of the regression lines in the different environments assessed using analysis of covariance (ANCOVA), and used to estimate V_{GEI}, although this approach is rarely used.

2.6.2 Empirical examples of GEIs for insect reproductive behavior

Table 2.6 provides some empirical examples of GEIs and GSEIs for insect reproductive behavior. Although this list is not exhaustive, a number of clear patterns exist. First, inbred and iso-female lines are the most commonly used designs to empirically measure GEIs and GSEIs for reproductive behavior. This is not surprising, given that GEIs and GSEIs are far simpler to estimate using these designs. The use of inbred or iso-female lines, however, is restricted to those insect species where the effects of inbreeding depression are minor, including numerous *Drosophila* species, the lesser wax moth (*Achroia grisella*) and the decorated cricket (*Gryllodes sigillatus*). For species where inbred or iso-female lines are not feasible, full- and half-sibling designs have been used to quantify both GEIs (Rodríguez and Greenfield 2003; Lewis

Table 2.6 Some examples of empirical studies documenting genotype-by-environment (GEI) and genotype-by-social environment effects on reproductive traits in insects.

Insect Order	Species name	Common name	Experimental design	Environment	Sex	Reproductive behavior	Reference
Coleoptera	*Tribolium castaneum*	Flour beetle	Half-sib	Diet	Male	Mating rate	[1]
Diptera	*Drosophila melanogaster*	Fruit fly	Iso-female lines	Cold stress[A]	Female	Mate choice[B]	[2]
		Fruit fly	Iso- female lines	Social environment[C]	Male	Courtship display	[3]
					Female	Mating frequency	[3]
	Drosophila simulans	Fruit fly	Inbred lines	Social environment[D]	Male	Aggression	[4]
		Fruit fly	Iso-female lines	Diet	Male	Cuticular hydrocarbon expression	[5]
				Temperature	Male	Cuticular hydrocarbon expression	[5,6]
			Iso-female lines	Temperature	Female	Mate choice[E]	[6]
Hemiptera	*Enchenopa binotata*	Treehopper	Full-sib	Social environment[F]	Female	Mate choice[G]	[7]
			Full-sib	Host plant species	Male	Courtship song	[8]
Lepidoptera	*Achroia grisella*	Lesser wax moth	Full-sib	Temperature	Female	Mate choice[H]	[9]
			Inbred lines	Diet	Male	Courtship song	[10]
			Inbred lines	Social environment[I]	Male	Courtship song	[11]
Orthoptera	*Gryllodes sigillatus*	Decorated cricket	Inbred lines	Diet	Male	Cuticular hydrocarbon expression	[12]
	Teleogryllus oceanicus	Polynesian field cricket	Common garden[J]	Social environment[K]	Male	Cuticular hydrocarbons expression	[13]
					Male	Advertisement song	[14]
	Teleogryllus commodus	Australian field crickets	Common garden[L]	Social environment[M]	Female	Mate choice[N]	[15]

References: [1] Lewis et al. (2011); [2] Narraway et al. (2010); [3] Higgins et al. (2005); [4] Saltz (2013); [5] Ingleby et al. (2013a); [6] Ingleby et al. (2013b); [7] Rebar and Rodriguez (2013); [8] Rodriguez and Al-Watchiqui (2012); [9] Rodriguez and Greenfield (2003); [10] Danielson-Francois et al. (2005); [11] Danielson-Francois et al. (2009); [12] Weddle et al. (2012); [13] Pascoal et al. (2016b); [14] Pascoal et al. (2017); [15] Bailey and Macleod (2014).

Notes

[A] Cold stress was measured as cold shock (4°C for 15 minutes every day for 10 days versus non-stress (maintained at 25°C constantly).

[B] Female preference was measured in two ways—mate acceptance and mating latency.

[C] Social behavior was examined in small populations consisting of five male and five female competitors.

[D] The social environment consisted of a focal individual, plus two other males that were always the same genotype. [E] Mate choice was measured as female preference and choosiness.

[F] Social environment was manipulated by placing different full-sib families together on the same host plant. [G] Mate choice was measured as female preference (with preference functions generated).

[H] Mate choice was measured as the time taken to respond to a song playback. [I] Social environment was manipulated by rearing different larvae of known genotype together.

[J] A total of seven different populations were examined after common garden rearing.

[K] The social environment was manipulated by either playing or not playing acoustic signals to focal individuals from each population during development. [L] Two different populations were examined after common garden rearing.

[M] Social environment was manipulated by either playing the song of other males (from *T. commodus* or *T. oceanicus*) or not during development.

[N] Mate choice was measured as female responsiveness and preference for songs produced by males from the same species or from *T. oceanicus* males et al. 2011, Rodríguez and Al-Wathiqui 2012) and GSEIs (Rebar and Rodríguez 2013), and a common garden approach has been used to demonstrate the potential for GSEIs in the field crickets, *Teleogryllus oceanicus* (Pascoal et al. 2016b, 2017) and *T. commodus* (Bailey and Macleod 2014).

Secondly, the most common abiotic environments examined in GEI studies of insect reproductive behavior are diet and temperature. This reflects the biological importance of these environmental factors to the life-history and fitness of most insect species. Importantly, most GEI studies have treated diet and temperature as discrete (Table 2.6), rather than as continuous variables. While sufficient to demonstrate the existence of GEIs, this approach focuses exclusively on linear reaction norms and, therefore, is likely to seriously underestimate the complexity of how different genotypes respond to the abiotic environment.

Thirdly, there has been a recent increase in the number of empirical studies examining GSEIs for insect reproductive behavior. These studies have focused on a diverse range of reproductive behaviors that are fundamental to sexual interactions in many insect species (Table 2.6). This range, however, appears far greater in males than females, where reproductive behaviors examined include courtship displays (Higgins et al. 2005) and signals (Danielson-François et al. 2009), aggressive behavior (Saltz 2013), and the expression of sexual traits that are known to be socially flexible and enhance mating success (Pascoal et al. 2016b, 2017). In contrast, GSEIs for reproductive behavior in female insects have focused on mating frequency (Higgins et al. 2005) and mate choice, with support for the latter being demonstrated in treehoppers (Rebar and Rodríguez 2013), wax moths (Rodríguez and Greenfield 2003), and field crickets (Bailey and Macleod 2014). The way that the social environment has been manipulated in these studies also varies, taking two major forms. The first approach alters the social environment by rearing or housing focal individuals of known genotype in different social groups (Higgins et al. 2005; Danielson-François et al. 2009; Rebar and Rodríguez 2013; Saltz 2013). The second uses social cues to manipulate a focal individual's **perception** of the social environment (Pascoal et al. 2016b, 2017). This latter approach, however, requires that important social cues are known and can be manipulated in a reliable manner. Consequently, this approach may not be possible for many insect species.

Finally, we note that all of the empirical examples provided in Table 2.6 are based on quantitative genetic data, which constitutes most of the support for GEIs and GSEIs. It is, however, possible to use genomic approaches to study GEIs and GSEIs, and this approach is becoming increasingly common. For example, Etges et al. (2007) used QTL analysis on two divergent populations of *Drosophila mojavensis* to show significant GEIs for mating success and a number of different courtship song parameters (especially inter-burst interval, number of bursts), when flies were reared on two different host cacti species. Interestingly, four QTLs showing GEIs were located for mating success, and two each for interburst interval and the number of bursts, making GEI effects as common as main effects, and likely to play a key role in incipient speciation in *D. mojavensis*. We expect that genomic studies of this nature will become even more common as the price of sequencing continues to decrease.

2.7 Wider evolutionary implications and areas for future research on the genetic architecture of insect reproductive behavior

This chapter has covered what are considered to be key topics on the genetics of insect reproductive behavior. It shows that many (if not most) insect reproductive behaviors are polygenic, with genes each having a small effect (although we acknowledge that genes having a major effect do exist) and that reproductive behavior is often genetically correlated with other important traits. Furthermore, it has been shown that non-additive genetic effects and interactions between genes, and the biotic (GEIs) and social (GSEIs) environments are likely to make important contributions to insect reproductive behaviors. While this is undoubtedly important information to have, it can be argued that more interesting questions arise when considering how this complex genetic architecture influences the evolutionary process. Simply put, a shift is needed away from studies that focus exclusively on characterizing the genetic architecture of insect reproductive behavior, towards those that also examine the wider evolutionary implications of this genetic architecture.

The evolutionary response of a given reproductive behavior is the product of selection acting on this behavior and the amount of additive genetic variance regulating this behavior: this is the core of the univariate breeder's equation ($R = h^2S$). Since many reproductive behaviors are known to be under strong selection in insects (e.g. Brooks et al. 2005; Bentsen et al. 2006; Steiger et al. 2013) and the examples presented in Tables 2.1 and 2.2 suggest a likely abundance of additive genetic variation, it appears that reproductive behaviors have the core ingredients necessary for rapid evolution. However, phenotypic evolution is far more complex than this because many reproductive behaviors consist of multiple components, which are genetically correlated and often targeted differentially by selection. Consequently, the mean response ($\Delta \bar{z}$) of a complex behavior is determined by the genetic architecture (characterized by the **additive genetic variance-covariance matrix** or **G**) and the pattern of selection targeting these specific components (characterized by the vector of linear selection gradients or β), and can be predicted by the multivariate breeder's equation ($\Delta \bar{z} = \beta \mathbf{G}$; Lande 1979). Exactly how **G** is aligned with β will determine whether **G** facilitates or constrains the evolution of reproductive behavior, and it has been argued that this alignment has been fundamental in population divergence and possibly even speciation (Schluter 2000). Few studies, however, have examined how **G** interacts with β to influence these processes, especially for reproductive behaviors in insects. The notable exception to this is work on cuticular hydrocarbons (CHCs) expression in the fruit fly *Drosophila serrata*. Like most insect species, CHCs in this species play a key role in desiccation resistance and are also the target of female mate choice. *D. serrata* shows a pronounced **latitudinal cline** in CHC expression along the east coast of Australia, which correlates with temperature and moisture differences across populations (Frentiu and Chenoweth 2010). However, differences in female mate choice for CHCs across populations only weakly predicted the observed

divergence in male CHCs. This relationship was greatly improved when population estimates of **G** were included in the statistical models (Chenoweth et al. 2010), demonstrating that **G** has biased the evolutionary trajectories of CHCs in these populations (Chenoweth et al. 2010).

The examples presented in Table 2.4 show that insect reproductive behaviors are often genetically correlated with other important morphological and life-history traits. This suggests that the potential for reproductive behavior to also constrain or facilitate the evolution of such traits. Unfortunately, there are currently few clear empirical examples documenting this process, although numerous studies manipulating the degree of polyandry in experimental populations of insects have shown that evolved changes in this reproductive behavior are associated with changes in a range of non-reproductive traits, such as lifespan (e.g. Martin and Hosken 2003) and immunity (e.g. McNamara et al. 2013). An obvious exception to this is recent work in the rapid evolution of a 'flatwing' mutation in the field cricket *Teleogryllus oceanicus* (Zuk et al. 2006). *T. oceanicus* has a wide geographic distribution spanning northern Australia, Polynesia, and three Hawaiian islands (Oahu, Hawaii, and Kauai), where there is overlap with the acoustically orientating parasitoid fly, *Ormia ochracea* (Zuk et al. 2006). This fly finds its host using the calling song and the fly larvae burrow into the male cricket and develop inside, killing the host on emergence. Due to this intense selection, more than 90 per cent of male crickets on Kauai island have a wing mutation (flatwing) where the normal stridulatory apparatus required for sound production (the file and scraper) is missing, rendering them silent (Zuk et al. 2006). Crosses of laboratory populations have shown that the flatwing phenotype is inherited as a sex-linked single gene (Tinghitella 2008), and RNA-seq analysis has shown that most differentially expressed transcripts in flatwing versus wild-type males were down-regulated (625 up versus 1716 down), with differences between morphs not restricted to a single pathway (Pascoal et al. 2016a). Genomic analysis (using RAD-seq) of the genetic divergence of Oahu compared with Kauai populations has shown that of the 7226 flatwing-associated SNP markers, only 0.30 per cent were shared between the two islands (Pascoal et al. 2014), which is consistent with independent mutational events. This demonstrates the powerful effects that reproductive behavior can have on morphology and how rapidly convergent evolution can occur.

Although not covered in this chapter, it is also likely that many shared reproductive behaviors will be positively genetically correlated between the sexes because males and females share most of their genome. Under contrasting selection on shared behaviors, intralocus **sexual conflict** (ISC) will exist and prevent the independent evolution of shared reproductive behaviors (Bonduriansky and Chenoweth 2009). Ultimately, ISC should oppose the evolution of sexual dimorphism in any shared reproductive behaviors, yet this phenomenon remains pervasive in nature (Bonduriansky and Chenoweth 2009). For example, CHC expression in *D. serrata* is sexually dimorphic, the magnitude of which is known to vary across populations in eastern Australia (Chenoweth and Blows 2008), despite opposing selection on CHCs between the sexes (Chenoweth and Blows 2004), and strong positive genetic correlations between CHC components in the sexes (Chenoweth and Blows 2008). Various mechanisms are known to help resolve ISC, and in the case of *D. serrata* many CHC components are X-linked, which reduces the intersexual genetic correlations (Chenoweth and Blows 2008). However, we still know very little about the operation of other mechanisms (such as genomic imprinting and gene duplication) that help resolve ISC for insect reproductive behaviors and more genomic studies are desperately needed.

Non-additive genetic effects occur due to interactions between alleles, either at the same locus (dominance) or different loci (epistasis). In general, there are far fewer studies documenting the contribution of non-additive genetic effects to phenotypic traits compared with additive gene effects, particularly for reproductive behaviors. This is surprising, given the important role that non-additive genetic effects are predicted to have for the evolution of phenotypic traits. Theoretical models show that dominance and epistatic variance can be converted into additive genetic variance by a number of different processes (Hansen and Wagner 2001) and can influence the response to selection through the build-up of linkage disequilibrium, as parents not

only transmit half of the additive effects to offspring, but also a quarter of pairwise epistatic effects and smaller fractions of high-order interactions (Lynch and Walsh 1998). This suggests that some of the linkage disequilibrium built by gene interactions can be converted into response to selection and that non-additive genetic effects are likely to have important long-term evolutionary consequences. While the limited examples in Table 2.5 demonstrate the importance of non-additive genetic effects on insect reproductive behaviors, there are currently no studies available that have examined the wider consequences of non-additive genetic effects to the evolution of these behaviors.

The empirical examples we provide in Table 2.6 suggest that GEIs are important contributors to the genetic architecture of reproductive behavior. Most GEI studies, however, have focused on interactions involving diet and temperature, and have taken a dichotomous approach (e.g. high versus low temperature). Clearly, this approach does not encompass the full complexity of environments experienced by most insect species, especially in nature, and therefore limits biological interpretation. Consequently, more studies are needed that cover both a broader range of environmental factors and more levels within a given environmental factor. As outlined in Section 6, the latter will enable the genetic basis of reaction norms to be estimated using random regression-based approaches. Theoretically, GEIs are predicted to have a number of important evolutionary consequences, including the maintenance of genetic variation in a population, facilitating the evolutionary response to a changing environment and promoting population divergence (e.g. Via and Lande 1985, 1987). Unfortunately, there are no empirical studies that have examined how GEIs for reproductive behaviors influence the above processes, making this a priority for future research.

The examples in Table 2.6 also show that GSEIs represent an important source of variation in insect reproductive behavior. As GSEIs involve the social environment they are predicted to be more labile and evolutionary dynamic compared with GEIs involving the abiotic environment (Wolf et al. 2014). It is also possible for the social environment to have a genetic basis, although **indirect genetic effects** (IGEs), meaning that the social environment can itself evolve (Wolf et al. 1998). Like GSEIs, IGEs are expected to have important effects on the evolution of phenotypic traits, including altering the rate and direction of evolution, promoting evolutionary time-lags and permitting traits to evolve that lack a genetic basis (Wolf et al. 1998). IGEs have been shown for a range of reproductive behaviors in insects, including parental care (Agrawal et al. 2001; Hunt and Simmons 2002; Head et al. 2012), CHC production (Petfield et al. 2005), and female mate choice (Rebar and Rodríguez 2013). Unfortunately, there are currently no empirical tests of the long-term evolutionary consequences of either GSEIs or IGEs for reproductive behaviors in insects. This should be a priority for future research, especially since studies on livestock and poultry have shown that IGEs can significantly alter the evolutionary response of traits in artificial selection experiments spanning multiple generations (e.g. Muir 2005, Camerlink et al. 2015).

References

Agrawal, A. (2001). Parent–offspring coadaptation and the dual genetic control of maternal care. *Science*, **292**, 1710.

Agrawal, A.F., Brodie, E.D., and Brown, J. (2001). Parent-offspring coadaptation and the dual genetic control of maternal care. *Science*, **292**(5522), 1710–12.

Alcok, J. (1983). *Animal behavior and evolution approach.* Sinauer Associates, Sunderland, MA.

Allen, A.M., Anreiter, I., Neville, M.C., and Sokolowski, M. B. (2017). Feeding-related traits are affected by dosage of the foraging gene in *Drosophila melanogaster*. *Genetics*, **205**, 761.

Archer, C.R., Zajitschek, F., Sakaluk, S.K., Royle, N.J., and Hunt, J. (2012). Sexual selection affects the evolution of lifespan and ageing in the decorated cricket. *Gryllodes sigillatus*. *Evolution*, **66**, 3088.

Arnholt, R.R.H. and Mackay, T.F.C. (2004). Quantitative genetic analyses of complex behaviours in Drosophila. *Nature Review Genetics*, **5**, 838.

Bailey, J.A., Baertsch, R., Kent, W.J., Haussler, D., and Eichler, E.E. (2004). Hotspots of mammalian chromosomal evolution. *Genome Biology*, **5**, R23.

Bailey, N.W., and Macleod, E. (2014). Socially flexible female choice and premating isolation in field crickets (*Teleogryllus* spp.). *Journal of Evolutionary Biology*, **27**, 170.

Barton, N.H., Etheridge, A.M., and Véber, A. (2016). The infinitesimal model. *bioRxiv*, 39768.

Barton, N.H., and Keightley, P.D. (2002). Understanding quantitative genetic variation. *Nature Review Genetics*, **3**, 11.

Ben-Shahar, Y., Robichon, A., Sokolowski, M.B., and Robinson, G.E. (2002). Influence of gene action across different time scales on behavior. *Science*, **296**, 741.

Bentsen, C.L., Hunt, J., Jennions, M.D., and Brooks, R. (2006). Complex multivariate sexual selection on male acoustic signaling in a wild population of *Teleogryllus commodus*. *American Naturalist*, **167**, 102.

Boake, C.R.B. (1994). *Quantitative genetic studies of behavioral evolution*. University of Chicago Press, Chicago, IL.

Bonduriansky, R., and Chenoweth, S.F. (2009). Intralocus sexual conflict. *Trends in Ecology and Evolution*, **24**, 280.

Brandt, L.S.E., and Greenfield, M.D. (2004). Condition-dependent traits and the capture of genetic variance in male advertisement song. *Journal of Evolutionary Biology*, **17**, 821.

Brooks, R., Hunt, J., Blows, M.W., *et al.* (2005). Experimental evidence for multivariate stabilizing sexual selection. *Evolution*, **59**, 871.

Brown, E.A., Gay, L., Vasudev, R., *et al.* (2009). Negative phenotypic and genetic associations between copulation duration and longevity in male seed beetles. *Heredity*, **103**, 340.

Bulmer, M.G. (1980). *The mathematical theory of quantitative genetics*. Oxford University Press, Oxford.

Bush, W.S., and Moore, J.H. (2012). Genome-wide association studies. *PLoS Computational Biology*, **8**, e1002822.

Butlin, R.K. (1996). Co-ordination of the sexual signalling system and the genetic basis of differentiation between populations in the brown planthopper, *Nilaparvata lugens*. *Heredity*, **77**, 369.

Caillaud, M.C., and Via, S. (2012). Quantitative genetics of feeding behavior in two ecological races of the pea aphid, *Acyrthosiphon pisum*. *Heredity*, **108**, 211.

Camerlink, I., Ursinus, W.W., Bijma, P., Kemp, B., and Bolhuis, J.E. (2015). Indirect genetic effects for growth rate in domestic pigs alter aggressive and manipulative biting behaviour. *Behavioural Genetics*, 45, 117.

Chenoweth, S.F., and Blows, M. (2008). QST meets the G matrix: the dimensionality of adaptive divergence in multiple correlated quantitative traits. *Evolution*, **62**, 1437.

Chenoweth, S.F., and Blows, M.W. (2004). Contrasting mutual sexual selection on homologous signal traits in *Drosophila serrata*. *American Naturalist*, **165**, 281.

Chenoweth, S.F., Rundle, H.D., and Blows, M.W. (2010). The contribution of selection and genetic constraints to phenotypic divergence. *American Naturalist*, **175**, 186.

Conner, J.K., and Hartl, D.L. (2004). *A primer of ecological genetics*. Sinauer Associates Incorporated, Sunderland, MA.

Craig, T.P., Horner, J.D., and Itami, J.K. (2001). Genetics, experience, and host-plant preference in *Eurosta solidaginis*: implications for host shifts and speciation. *Evolution*, **55**, 773.

Danielson-Francois, A.M., Kelly, J.K., and Greenfield, M.D. (2005). Genotype × environment interaction for male attractiveness in an acoustic moth: evidence for plasticity and canalization. *Journal of Evolutionary Biology*, **19**, 532.

Danielson-François, A.M., Zhou, Y., and Greenfield, M.D. (2009). Indirect genetic effects and the lek paradox: inter-genotypic competition may strengthen genotype × environment interactions and conserve genetic variance. *Genetica*, **136**, 27.

David, J.R., Gibert, P., Legout, H., *et al.* (2005). Isofemale lines in *Drosophila*: an empirical approach to quantitative trait analysis in natural populations. *Heredity*, **94**, 3.

Diao, W., Mousset, M., Horsburgh, G.J., et al. (2016). Quantitative trait locus analysis of mating behavior and male sex pheromones in *Nasonia* wasps. *G3. Genes Genomes Genetics*, **6**, 1549.

Doerge, R.W. (2002). Mapping and analysis of quantitative trait loci in experimental populations. *Nature Review Genetics*, **3**, 43.

Dunlap, A.S., and Stephens, D.W. (2014). Experimental evolution of prepared learning. *Proceedings of the National Academy of Sciences USA*, **111**, 11750.

Edwards, A.C., and Mackay, T.F.C. (2009). Quantitative trait loci for aggressive behavior in *Drosophila melanogaster*. *Genetics*, **182**, 889.

Etges, W.J., De Oliveira, C.C., Gragg, E., *et al.* (2007). Genetics of incipient speciation in *Drosophila mojavensis*. I. Male courtship song, mating success, and genotype × environment interactions. *Evolution*, **61**, 1106.

Etges, W.J., Over, K.F., De Oliveira, C.C., and Ritchie, M.G. (2006). Inheritance of courtship song variation among geographically isolated populations of *Drosophila mojavensis*. *Animal Behaviour*, **71**, 1205.

Ewing, A.W., and Manning, A. (1967). The evolution and genetics of insect behaviour. *Annual Review of Entomology*, **12**, 471.

Falconer, D.S., and Mackay, T.F. (1996). *Introduction to Quantitative Genetics*, 4th edn. Longman, Harlow.

Fisher, R.A. (1918). The correlation between relatives on the supposition of Mendelian inheritance. *Transactions of the Royal Society of Edinburgh*, **52**, 399.

Foley, B., Chenoweth, S.F., Nuzhdin, S.V., and Blows, M.W. (2007). Natural genetic variation in cuticular hydrocarbon expression in male and female *Drosophila melanogaster*. *Genetics*, **175**, 1465.

Fox, C.W. (1993). A quantitative genetic analysis of oviposition preference and larval performance on two hosts in the bruchid beetle, *Callosobruchus maculatus*. *Evolution*, **47**, 166.

Fox, C.W., Stillwell, R.C., Amarillo-S, A.R., Czesak, M.E., and Messina, F.J. (2004). Genetic architecture of population differences in oviposition behaviour of the seed beetle *Callosobruchus maculatus*. *Journal of Evolutionary Biology*, **17**, 1141

Fox, C.W., Wagner, J.D., Cline, S., Thomas, F.A., and Messina, F.J. (2009). Genetic architecture underlying convergent evolution of egg-laying behavior in a seed-feeding beetle. *Genetica*, **136**, 179.

Frentiu, F.D., and Chenoweth, S.F. (2010). Clines in cuticular hydrocarbons in two *Drosophila* species with independent population histories. *Evolution*, **64**, 1784.

Gaertner, B.E., Ruedi, E.A., McCoy, L.J., *et al.* (2015). Heritable variation in courtship patterns in *Drosophila melanogaster*. *G3 Genes Genomes Genetics*, **5**, 531.

Gershman, S.N., Barnett, C.A., Pettinger, A.M., *et al.* (2010). Give 'til it hurts: trade-offs between immunity and male reproductive effort in the decorated cricket, *Gryllodes sigillatus*. *Journal of Evolutionary Biology*, **23**, 829.

Gleason, J.M., Nuzhdin, S.V., and Ritchie, M.G. (2002). Quantitative trait loci affecting a courtship signal in *Drosophila melanogaster*. *Heredity*, **89**, 1.

Gleason, J.M., and Ritchie, M.G. (2004). Do quantitative trait loci (QTL) for a courtship song difference between *Drosophila simulans* and *D. sechellia* coincide with candidate genes and intraspecific QTL? *Genetics*, **166**, 1303.

Groot, A.T., Staudacher, H., Barthel, A., et al. (2013). One quantitative trait locus for intra- and interspecific variation in a sex pheromone. *Molecular Ecology*, **22**, 1065.

Hamiduzzaman, M.M., Emsen, B., Hunt, G.J., et al. (2017). Differential gene expression associated with honey bee grooming behavior in response to varroa mites. *Behavioural Genetics*, **47**, 335.

Hansen, T.F., and Wagner, G.P. (2001). Modeling genetic architecture: a multilinear theory of gene interaction. *Theoretical Population Biology*, **59**, 61.

Head, M.L., Berry, L.K., Royle, N.J., and Moore, A.J. (2012). Paternal care: direct and indirect genetic effects of fathers on offspring performance. *Evolution*, **66**, 3570.

Higgins, L.A., Jones, K.M., and Wayne, M.L. (2005). Quantitative genetics of natural variation of behavior in *Drosophila melanogaster*: the possible role of the social environment on creating persistent patterns of group activity. *Evolution*, **59**, 1529.

Hill, W.G. (2012). Quantitative genetics in the genomics era. *Current Genomics*, **13**, 196.

Hine, E., Chenoweth, S.F., and Blows, M.W. (2004). Multivariate quantitative genetics and the lek paradox: genetic variance in male sexually selected traits of *Drosophila serrata* under field conditions. *Evolution*, **58**, 2754.

Hoikkala, A., Päällysaho, S., Aspi, J., and Lumme, J. (2000). Localization of genes affecting species differences in male courtship song between *Drosophila virilis* and *D. littoralis*. *Genetic Research*, **75**, 37.

Hopwood, P.E., Head, M.L., Jordan, E.J., et al. (2016). Selection on an antagonistic behavioral trait can drive rapid genital coevolution in the burying beetle, *Nicrophorus vespilloides*. *Evolution*, **70**, 1180.

Hunt, G.J., Guzmán-Novoa, E., Fondrk, M.K., and Page, R.E. (1998). Quantitative trait loci for honey bee stinging behavior and body size. *Genetics*, **148**, 1203.

Hunt, J., Blows, M.W., Zajitschek, F., Jennions, M.D., and Brooks, R. (2007). Reconciling strong stabilizing selection with the maintenance of genetic variation in a natural population of black field crickets (*Teleogryllus commodus*). *Genetics*, **177**, 875.

Hunt, J., and Hosken, D.J. (2014). *Genotype-by-environment interactions and sexual selection*. John Wiley and Sons, Hoboken, NJ.

Hunt, J., Jennions, M.D., Spyrou, N., and Brooks, R. (2006). Artificial selection on male longevity influences age-dependent reproductive effort in the black field cricket *Teleogryllus commodus*. *American Naturalist*, **168**, 72.

Hunt, J., and Simmons, L.W. (2002). The genetics of maternal care: direct and indirect genetic effects on phenotype in the dung beetle *Onthophagus taurus*. *Proceedings of the National Academy of Sciences USA*, **99**, 6828.

Huttunen, S., Aspi, J., Hoikkala, A., and Schlötterer, C. (2004). QTL analysis of variation in male courtship song characters in *Drosophila virilis*. *Heredity*, **92**, 263.

Ingleby, F.C., Hunt, J., and Hosken, D.J. (2013a). Genotype-by-environment interactions for female mate choice of male cuticular hydrocarbons in *Drosophila simulans*. *PLoS One*, 8, e67623.

Ingleby, F.C., Hunt, J., and Hosken, D.J. (2013b). Heritability of male attractiveness persists despite evidence for unreliable sexual signals in *Drosophila simulans*. *Journal of Evolutionary Biology*, **26**, 311.

Jaenike, J. (1987). Genetics of oviposition-site preference in *Drosophila tripunctata*. *Heredity*, **59**, 363.

Jang, Y., and Greenfield, M.D. (2000). Quantitative genetics of female choice in an ultrasonic pyralid moth, *Achroia grisella*: variation and evolvability of preference along multiple dimensions of the male advertisement signal. *Heredity*, **84**, 73.

Kasumovic, M.M., Hall, M.D., and Brooks, R.C. (2012). The juvenile social environment introduces variation in the choice and expression of sexually selected traits. *Ecology and Evolution*, **2**, 1036.

Kent, C., Azanchi, R., Smith, B., Formosa, A., and Levine, J.D. (2008). Social context influences chemical communication in *D. melanogaster* males. *Current Biology*, **18**, 1384.

Khazaeli, A.A., and Curtsinger, J.W. (2010). Life history variation in an artificially selected population of *Drosophila melanogaster*: pleiotropy, superflies and age-specific adaptation. *Evolution*, **64**, 3409.

Kotiaho, J.S., Simmons, L.W., Hunt, J., and Tomkins, J.L. (2003). Males influence maternal effects that promote sexual selection: a quantitative genetic experiment with dung beetles *Onthophagus taurus*. *American Naturalist*, **161**, 852.

Krebs, J.R., Davies, N.B., and Parr, J. (1993). *An introduction to behavioural ecology*.: Blackwell Scientific Publications, Cambridge, MA.

Kronforst, M.R., Young, L.G., Kapan, D.D., *et al.* (2006). Linkage of butterfly mate preference and wing color preference cue at the genomic location of wingless. *Proceedings of the National Academy of Sciences USA*, **103**, 6575.

Lande, R. (1979). Quantitative genetic analysis of multivariate evolution, applied to brain: body size allometry. *Evolution*, **33**, 402.

Lande, R., and Arnold, S.J. (1983). The measurement of selection on correlated characters. *Evolution*, **37**, 1210.

Lander, E.S., and Botstein, D. (1989). Mapping mendelian factors underlying quantitative traits using RFLP linkage maps. *Genetics*, **121**, 185.

Lassance, J-M., Groot, A.T., Liénard, M.A., et al. (2010). Allelic variation in a fatty-acyl reductase gene causes divergence in moth sex pheromones. *Nature*, **466**, 486.

Lewis, Z., Wedell, N., and Hunt, J. (2011). Evidence for strong intralocus sexual conflict in the Indian meal moth, *Plodia interpunctella*. *Evolution*, **65**, 2085.

Limousin, D., Streiff, R., Courtois, B., *et al.* (2012). Genetic architecture of sexual selection: QTL mapping of male song and female receiver traits in an acoustic moth. *PLoS One*, **7**, e44554.

Løvlie, H., Immonen, E., Gustavsson, E., Kazancioğlu, E., and Arnqvist, G. (2014). The influence of mitonuclear genetic variation on personality in seed beetles. *Proceedings of the Royal Society B: Biological Sciences*, **281**, 20141039.

Luong, L.T., and Polak, M. (2007). Costs of resistance in the Drosophila–Macrocheles system: a negative genetic correlation between ectoparasite resistance and reproduction. *Evolution*, **61**, 1391.

Lynch, M., and Walsh, B. (1998). *Genetics and analysis of quantitative traits*. Sinauer Associates Incorporated, Sunderland, MA.

Mackay, T.F.C., Heinsohn, S.L., Lyman, R.F., *et al.* (2005). Genetics and genomics of Drosophila mating behavior. *Proceedings of the National Academy of Sciences USA*, **102**, 6622.

Mackay, T.F.C., Stone, E.A., and Ayroles, J.F. (2009). The genetics of quantitative traits: challenges and prospects. *Nature Review Genetics*, **10**, 565.

Malé, P-J.G., Turner, K.M., Doha, M., *et al.* (2017). An ant–plant mutualism through the lens of cGMP-dependent kinase genes. *Proceedings of the Royal Society B: Biological Sciences*, **284**, 20170896.

Markow, T.A., and O'Grady, P.M. (2005). Evolutionary genetics of reproductive behavior in *Drosophila*: connecting the dots. *Annual Review Genetics*, **39**, 263.

Martin, O.Y., and Hosken, D.J. (2003). Costs and benefits of evolving under experimentally enforced polyandry or monogamy. *Evolution*, **57**, 2765.

McNamara, K.B., Wedell, N., and Simmons, L.W. (2013). Experimental evolution reveals trade-offs between mating and immunity. *Biology Letters*, **9**, 20130262.

Meffert, L.M., Hicks, S.K., and Regan, J.L. (2002). Nonadditive genetic effects in animal behavior. *American Naturalist*, **160**, S198–213.

Miyatake, T. (1998). Genetic changes of life history and behavioral traits during mass-rearing in the melon fly, *Bactrocera cucurbitae* (Diptera: Tephritidae). *Researches on Population Ecology*, 40, 301.

Moehring, A.J., and Mackay, T.F.C. (2004). The quantitative genetic basis of male mating behavior in *Drosophila melanogaster*. *Genetics*, **167**, 1249-63.

Morrow, E.H., and Gage, M.J.G. (2001). Artificial selection and heritability of sperm length in *Gryllus bimaculatus*. *Heredity*, **87**, 356.

Muir, W.M. (2005). Incorporation of competitive effects in forest tree or animal breeding programs. *Genetics*, **170**, 1247.

Nachappa, P., Levy, J., Pierson, E., and Tamborindeguy, C. (2014). Correlation between "*Candidatus Liberibacter solanacearum*" infection levels and fecundity in its psyllid vector. *Journal of Invertebrate Pathology*, **115**, 55.

Narraway, C., Hunt, J., Wedell, N., and Hosken, D.J. (2010). Genotype-by-environment interactions for female preference. *Journal of Evolutionary Biology*, **23**, 2550.

Nomura, T., and Yonezawa, K. (1990). Genetic correlations among life history characters of adult females in the azuki bean weevil, *Callosobruchus chinensis* (L.) (Coleoptera: Bruchidae). *Applied Entomology and Zoology*, **25**, 423.

Okada, K., and Miyatake, T. (2009). Genetic correlations between weapons, body shape and fighting behaviour in the horned beetle *Gnatocerus cornutus*. *Animal Behaviour*, **77**, 1057.

Oldroyd, B.P., and Beekman, M. (2008). Effects of selection for honey bee worker reproduction on foraging traits. *PLoS Biology*, **6**, e56.

Ott, J., Kamatani, Y., and Lathrop, M. (2011). Family-based designs for genome-wide association studies. *Nature Review Genetics*, **12**, 465.

Page, R.E., Waddington, K.D., Hunt, G.J., and Fondrk, M.K. (1995). Genetic determinants of honey bee foraging behaviour. *Animal Behaviour*, **50**, 1617.

Pascoal, S., Cezard, T., Eik-Nes, A., et al. (2014). Rapid convergent evolution in wild crickets. *Current Biology*, **24**, 1369.

Pascoal, S., Liu, X., Ly, T., et al. (2016a). Rapid evolution and gene expression: a rapidly evolving Mendelian trait that silences field crickets has widespread effects on mRNA and protein expression. *Journal of Evolutionary Biology*, **29**, 1234.

Pascoal, S., Mendrok, M., Mitchell, C., *et al*. (2016b). Sexual selection and population divergence I: the influence of socially flexible cuticular hydrocarbon expression in male field crickets (*Teleogryllus oceanicus*). *Evolution*, **70**, 82.

Pascoal, S., Mendrok, M., Wilson, A.J., Hunt, J., and Bailey, N.W. (2017). Sexual selection and population divergence II. Divergence in different sexual traits and signal modalities in field crickets (*Teleogryllus oceanicus*). *Evolution*, **71**, 1614.

Petfield, D., Chenoweth, S.F., Rundle, H.D., and Blows, M.W. (2005). Genetic variance in female condition predicts indirect genetic variance in male sexual display traits. *Proceedings of the National Academy of Sciences USA*, **102**, 6045.

Pianka, E.R. (1999). *Evolutionary ecology*, 6th edn. Benjamin Cummings, San Francisco, USA.

Réale, D., and Roff, D.A. (2002). Quantitative genetics of oviposition behaviour and interactions among oviposition traits in the sand cricket. *Animal Behaviour*, **64**, 397.

Rebar, D., and Rodríguez, R.L. (2013). Genetic variation in social influence on mate preferences. *Proceedings of the Royal Society B: Biological Sciences*, **280**, 20130803.

Rideout, E.J., Billeter, J.C., and Goodwin, S.F. (2007). The sex-determination genes fruitless and doublesex specify a neural substrate required for courtship song. *Current Biology*, **17**, 1473.

Ritchie, M.G., and Kyriacou, C.P. (1996). Artificial selection for a courtship signal in *Drosophila melanogaster*. *Animal Behaviour*, **52**, 603.

Rodríguez, R.L., and Al-Wathiqui, N. (2012). Causes of variation in sexual allometry: a case study with the mating signals and genitalia of Enchenopa treehoppers (Hemiptera: Membracidae). *Ethology Ecology and Evolution*, **24**, 187.

Rodríguez, R.L., and Greenfield, M.D. (2003). Genetic variance and phenotypic plasticity in a component of female mate choice in an ultrasonic moth. *Evolution*, **57**, 1304.

Roff, D.A., and Bradford, M.J. (1996). Quantitative genetics of the trade-off between fecundity and wing dimorphism in the cricket *Allonemobius socius*. *Heredity*, **76**, 178.

Roff, D.A., and Emerson, K. (2006). Epistasis and dominance: evidence for differential effects in life-history versus morphological traits. *Evolution*, **60**, 1981.

Roff, D.A., Stirling, G., and Fairbairn, D.J. (1997). The evolution of threshold traits: a quantitative genetic analysis of the physiological and life-history correlates of wing dimorphism in the sand cricket. *Evolution*, **51**, 1910.

Roff, D.A., Tucker, J., Stirling, G., and Fairbairn, D.J. (1999). The evolution of threshold traits: effects of selection on fecundity and correlated response in wing dimorphism in the sand cricket. *Journal of Evolutionary Biology*, **12**, 535.

Rose, M., and Charlesworth, B. (1980). A test of evolutionary theories of senescence. *Nature*, **287**, 141.

Rose, M.R. (1984). Genetic covariation in *Drosophila* life history: untangling the data. *American Naturalist*, **123**, 565.

Ryner, L.C., Goodwin, S.F., Castrillon, D.H., et al. (1996). Control of male sexual behavior and sexual orientation in *Drosophila* by the fruitless gene. *Cell*, **87**, 1079.

Saglam, I.K., Roff, D.A., and Fairbairn, D.J. (2008). Male sand crickets trade-off flight capability for reproductive potential. *Journal of Evolutionary Biology*, **21**, 997.

Sakaluk, S.K., and Smith, R.L. (1988). Inheritance of male parental investment in an insect. *American Naturalist*, **132**, 594–601.

Saltz, J.B. (2013). Genetic composition of social groups influences male aggressive behaviour and fitness in natural genotypes of *Drosophila melanogaster*. *Proceedings of the Royal Society B: Biological Sciences*, **280**, 20131926.

Saltz, J.B., Hessel, F.C., and Kelly, M.W. (2017). Trait correlations in the genomics era. *Trends in Ecology and Evolution*, **32**, 279.

Schluter, D. (2000). *The ecology of adaptive radiation*. Oxford University Press, Oxford.

Sharma, M.D., Tregenza, T., and Hosken, D.J. (2010). Female mate preferences in *Drosophila simulans*: evolution and costs. *Journal of Evolutionary Biology*, **23**, 1672.

Shaw, K.L., Parsons, Y.M., and Lesnick, S.C. (2007). QTL analysis of a rapidly evolving speciation phenotype in the Hawaiian cricket Laupala. *Molecular Ecology*, **16**, 2879.

Shuker, D.M., Phillimore, A.J., Burton-Chellew, M.N., Hodge, S.E., and West, S.A. (2007). The quantitative genetic basis of polyandry in the parasitoid wasp, *Nasonia vitripennis*. *Heredity*, **98**, 69.

Shuker, D.M., and Simmons, L.W. (2014). *The evolution of insect mating systems*. Oxford University Press, Oxford.

Simmons, L.W., Tinghitella, R.M., and Zuk, M. (2010). Quantitative genetic variation in courtship song and its covariation with immune function and sperm quality in the field cricket Teleogryllus oceanicus. *Behavioural Ecology*, **21**, 1330.

Singer, M.C., Ng, D., and Thomas, C.D. (1988). Heritability of oviposition preference and its relationship to offspring performance within a single insect population. *Evolution*, **42**, 977.

Smith, C.R., Toth, A.L., Suarez, A.V., and Robinson, G.E. (2008). Genetic and genomic analyses of the division of labour in insect societies. *Nature Review Genetics*, **9**, 735.

Staudacher, H., Menken, S.B.J., and Groot, A.T. (2015). Effects of immune challenge on the oviposition strategy of a noctuid moth. *Journal of Evolutionary Biology*, **28**, 1568.

Steiger, S., Ower, G.D., Stokl, J., *et al.* (2013). Sexual selection on cuticular hydrocarbons of male sagebrush crickets in the wild. *Proceedings of the Royal Society B: Biological Sciences*, **280**, 20132353.

Stranger, B.E., Stahl, E.A., and Raj, T. (2011). Progress and promise of genome-wide association studies for human complex trait genetics. *Genetics*, **187**, 367.

Thornhill, R., and Alcock, J. (1983). *The evolution of insect mating systems*. Harvard University Press, Cambridge, MA.

Thornhill, R., and Sauer, P. (1992). Genetic sire effects on the fighting ability of sons and daughters and mating success of sons in a scorpionfly. *Animal Behaviour*, **43**, 255.

Tinghitella, R.M. (2008). Rapid evolutionary change in a sexual signal: genetic control of the mutation 'flatwing' that renders male field crickets (*Teleogryllus oceanicus*) mute. *Heredity*, **100**, 261.

Travers, L.M., Garcia-Gonzalez, F., and Simmons, L.W. (2016). Genetic variation but weak genetic covariation between pre- and post-copulatory episodes of sexual selection in *Drosophila melanogaster*. *Journal of Evolutionary Biology*, **29**, 1535.

Tucic, N., and Seslija, D. (2007). Genetic architecture of differences in oviposition preference between ancestral and derived populations of the seed beetle *Acanthoscelides obtectus*. *Heredity*, **98**, 268.

Via, S., and Lande, R. (1985). Genotype-environment interaction and the evolution of phenotypic plasticity. *Evolution*, **39**, 505.

Via, S., and Lande, R. (1987). Evolution of genetic variability in a spatially heterogeneous environment: effects of genotype–environment interaction. *Genetic Research*, **49**, 147.

Webb, K.L., and Roff, D.A. (1992). The quantitative genetics of sound production in *Gryllus firmus*. *Animal Behaviour*, **44**, 823.

Weddle, C.B., Mitchell, C., Bay, S.K., Sakaluk, S.K., and Hunt, J. (2012). Sex-specific genotype-by-environment interactions for cuticular hydrocarbon expression in decorated crickets, *Gryllodes sigillatus*: implications for the evolution of signal reliability. *Journal of Evolutionary Biology*, **25**, 2112.

Wedell, N. (2006). Male genotype affects female fitness in a paternally investing species. *Evolution*, **60**, 1638.

Wolf, J.B., Brodie III, E.D., Cheverud, J.M., Moore, A.J. and Wade, M.J. (1998). Evolutionary consequences of indirect genetic effects. *Trends in Ecology and Evolution*, **13**, 64.

Wolf, J.B., Royle, N.J., and Hunt, J. (2014). In: Hunt, J., and Hosken, D.J. (Eds) *Genotype-by-environment interactions when the social environment contains genes*. In: *Genotype-by-environment interactions and sexual selection*, pp. 63–97. John Wiley and Sons, Hoboken, NJ.

Yeh, S.D., Liou, S.R., and True, J.R. (2006). Genetics of divergence in male wing pigmentation and courtship behavior between *Drosophila elegans* and *D. gunungcola*. *Heredity*, **96**, 383.

Zajitschek, F., Hunt, J., Zajitschek, S.R., Jennions, M.D., and Brooks, R. (2007). No intra-locus sexual conflict over reproductive fitness or ageing in field crickets. *PLoS One*, **2**, e155.

Żuk, M., Rotenberry, J.T., and Tinghitella, R.M. (2006). Silent night: adaptive disappearance of a sexual signal in a parasitized population of field crickets. *Biology Letters*, **2**, 521.

Zwarts, L., Magwire, M.M., Carbone, M.A., et al. (2011). Complex genetic architecture of Drosophila aggressive behavior. *Proceedings of the National Academy of Sciences USA*, **108**, 17070.

CHAPTER 3

Neurobiology

Anne C. von Philipsborn

Danish Research Institute of Translational Neuroscience (DANDRITE),
Aarhus University, Ole Worms Alle 3, Building 1170,
DK-8000 Aarhus C, Denmark

3.1 Methods in insect behavioral neurobiology

Insects not only fascinate by their endless variety of external form, but also by the astounding diversification of their behavior—aerial manoeuvres of flies, foraging of ants, concerted **swarming behavior** of bee colonies, singing of crickets, dramatic fights of two male beetles. Observing the lives of insects cannot but intrigue the curious scientists with the question about *how* these actions are generated by so marvellously small nervous systems.

The nervous system is the proximate cause of behavior. Ecology and evolution of a species can explain why and to what extent behavioral strategies are adaptive. By scrutinizing the mechanistic, circuit level basis for behavior, we can ask how and why adaptive behaviors have been implemented in the particular way we see them, with their stunning effectiveness, as well as their constraints.

The question about how the nervous system controls and shapes behavior is approached on different levels. First, the general structure and the wiring architecture of nervous system can be observed and compared across species (comparative neuroanatomy). Ideally, one can observe the activity of neurons and the information flow through the nervous system while the animal is behaving. On the next level, experimental manipulation of cells and their functional connectivity gives insight as to whether the activity of a given neuron or a circuit is necessary, or sufficient for a behavior, or impacts it in any other way.

Some structures of the insect nervous system, for example, the well-developed mushroom bodies in **eusocial** insects (see Section 3.3.3), can be seen under a microscope without staining or processing of the tissue (Figure 3.1a). In order to analyse the arrangement of neurons, **fibre tracts** and **neuropil** structures, single cells have to be stained, e.g. by the traditional **Golgi method** (Figure 3.1b). Visualization of neuronal architecture can suggest blueprints for information flow and processing.

The advent of **electrophysiology** in the 1960s made it possible to not only look at brains, but to record and manipulate the electric activity of neurons. Arthropod nervous systems are unique in that anatomically and functionally recognizable neurons can be found across different animals and even species (Bullock 1977). The activity pattern of neurons is most easily interpreted when it is closely correlated with muscle contractions or when it encodes a quantifiable sensory stimulus, such as light, sound, or mechanical disturbance. Among the first insect behaviors that were mechanistically probed regarding their underlying neuronal path-

von Philipsborn, A. C., *Neurobiology*. In: *Insect Behavior: From mechanisms to ecological and evolutionary consequences.* Edited by Alex Córdoba-Aguilar, Daniel González-Tokman, and Isaac González-Santoyo: Oxford University Press (2018). © Oxford University Press.
DOI: 10.1093/oso/9780198797500.003.0003

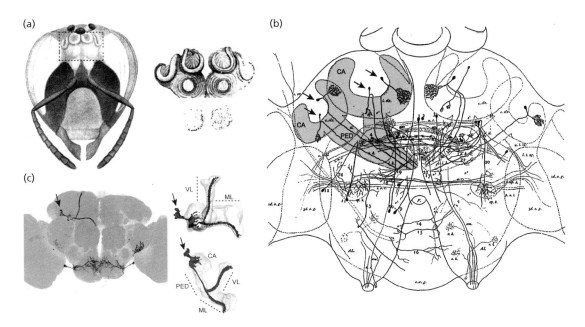

Figure 3.1 History of insect neuroanatomy. (a) Drawing of the mushroom bodies of a bee (to the left, indicated through the head capsule (in the box), to the right in isolation) by the nineteenth century's naturalist Dujardin (modified from Dujardin 1850). (b) Single neurons stained by the Golgi method in the brain of a bee, from the first publication using this method in the brain of an insect. The left mushroom body is highlighted in grey, with arrows pointing at mushroom body intrinsic neurons, later named Kenyon cells, after the author (modified from Kenyon 1896, *Journal of Comparative Neurology*, with permission from John Wiley and Sons). (c) Confocal microscopy image of a transgenic *Drosophila melanogaster* central brain with a small group of Kenyon cells and neurons in the ventral part of the brain expressing green fluorescent protein. To the right, 3D volume reconstruction of the left mushroom body and neurons from the sample, in frontal view and tilted. Arrow points to cell bodies. Note the two cup shaped calyces in the bee and the single, disk-shaped calyx in *Drosophila*.

CA, calyx; PED, pedunculus; VL, vertical lobe; ML, medial lobe.

ways were escape responses, e.g. jumping of locusts evoked by visual stimuli (O'Shea and Rowell 1977), response of moths to bat echolocation cries or startle response in cockroaches (Roeder 1963). These studies pinpoint how sensory systems can effectively filter and encode behaviorally relevant cues, enabling fast evasive behaviors adapted to the specific situation. Importantly, the nervous system is not only responding to sensory stimuli in a simple input–output fashion, but also generates endogenous activity. Motor patterns are often generated by so-called **central pattern generators**, i.e. neuronal circuits that can produce rhythmic output in the absence of rhythmic input or sensory feedback. Early studies on locust flight (Wilson 1968; Burrows 1996), ventilation of dragonfly larvae (Mill 1977) and cricket **stridulation** (Huber 1962) were seminal in demonstrating these general principles and concepts of motor control.

Following the 1990s, the development of sophisticated genetic tools (Venken et al. 2011; Riemensperger et al. 2016) and large transgenic stock libraries (Jenett et al. 2012) made the fruit fly *Drosophila melanogaster* the model organism of choice for understanding not only the genetic, but also neuronal basis of behavior (Yoshihara and Ito 2012; Figure 3.1c, Box. 3.1, Genetically encoded tools for *Drosophila* neurobiology). Over the past years, an abundance of studies has addressed sensory processing and integration, motor control, **learning**, sexual behavior, feeding, and sleep at the level of neurons and circuits, sometimes rediscovering concepts outlined decades earlier in bigger insects such as blowflies, cockroaches, grasshoppers, crickets, and stick insects.

Advances in functional neuroanatomy, as well as the use of genetic techniques in species other than the classical model *Drosophila melanogaster* will

---

Box 3.1 Genetically encoded tools for *Drosophila* neurobiology

The GAL4/UAS transcriptional system is commonly used in transgenic *Drosophila* to visualize or manipulate different sets of neurons in a reproducible way. The components of the system come from yeast; GAL4 is a transcription activator protein and UAS (Upstream Activation Sequence) its cognate enhancer sequence. When the GAL4 protein is expressed, it specifically binds to the UAS DNA sequence and induces expression of the gene located upstream of this UAS sequence. Any of the large number of driver GAL4 lines (expressing the transcriptional activator in different cell types) can be combined by a simple genetic cross with an effector UAS line specifying the tool expressed in the GAL4 positive cells—fluorescent proteins for visualizing neurons or their compartments, calcium indicators, ion channels, mutated pro-

teins or toxins for manipulating neuronal functions, as well as RNAi constructs for knockdown of gene expression. Various extensions, modifications, and intersectional strategies for the GAL4/UAS system have been devised for better temporal and spatial control of effector expression. Fluorescent, genetically encoded calcium indicators allow the monitoring of neuronal activity, which is correlated with a rise in calcium. With calcium imaging, it is possible to monitor many cells and neuropil regions simultaneously and detect activity in neurites, which are too small for electrophysiological recordings. **Optogenetics** provide a possibility to activate or inhibit neurons with light; for this cation or anion channel rhodopsins are expressed, which open upon illumination, depolarizing or hyperpolarizing the neuron.

---

undoubtedly pave new roads for comparative neurobiology. For many species, three-dimensional brain atlases have been established (el Jundi and Heinze in press). Progress in electron microscopy and computational approaches to interpret large anatomical datasets are furthering **connectomics**, the study of connectivity maps of nervous systems (Schlegel et al. 2017). A complete electron microscopy volume, which was first acquired for the *Drosophila* larval nervous system, where large-scale reconstruction of neuronal connections has provided biological insight into circuit function (Ohyama et al. 2015). Data of the adult brain (Zheng et al. 2017) is now available as well. In the future, connectomes from other insect species might provide detailed comparative data on circuit architecture underlying behavior.

CRISPR/Cas9-mediated **genome editing** opens the possibility of genetically manipulating the nervous system of species with sequenced genomes and has already been successfully applied in many insect orders (Sun et al. 2017; Tanaka et al. 2017; Trible et al. 2017; Yan et al. 2017). There is a good prospect that these developments will help in closing the divide between neuroscientists working with *Drosophila melanogaster* (where the focus on genetic tools sometimes can eclipse context from behavioral ecology) and neuroscientists studying other species, leading to a more comprehensive understanding of insect behavioral neurobiology.

3.2 The insect nervous system

3.2.1 General structure

The insect nervous system is a segmented structure with a common ground plan of 1 neuromere per body segment (Figure 3.2a). It consists of the central nervous system (CNS) and the peripheral nervous system (PNS). The CNS is a series of interconnected ganglia. The brain (cerebral ganglia) is located anteriormost, followed by the gnathal ganglia, three thoracic ganglia, and several abdominal ganglia. In some insects (e.g. Diptera and Hymenoptera), cerebral and gnathal ganglia are fused. Thoracic and abdominal ganglia can be fused as well to various degrees, differing not only across species, but also in some cases across sex and social **caste** (Niven et al. 2008) (Figure 3.2b,c). The nervous system consists of **sensory neurons**, **interneurons**, **motor neurons**, and glia cells. Sensory neurons detect stimuli such as light, sound, temperature, mechanical forces, or the presence of airborne or contact chemicals, and transform them into electrical activity. Motor neurons receive stimuli from other neurons and connect to muscles, regulating their contraction and thus orchestrating movement. The vast majority of neurons are interneurons, i.e. they are neither sensory nor motor neurons, but connect with other neurons, forming neuronal circuits. The CNS ganglia contain the cell bodies of interneurons and motor neurons,

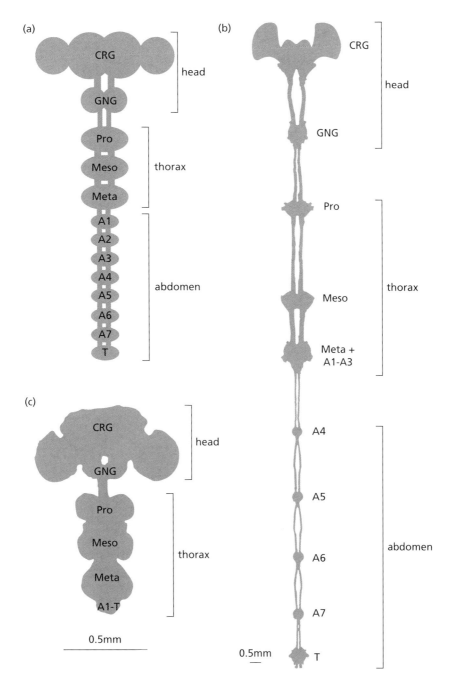

Figure 3.2 General structure of the insect nervous system. (a) Ground plan of the insect nervous system (after Niven et al. 2008). (b) Nervous system of the desert locust, *Schistocerca gregaria*. The last thoracic ganglion is fused to the first three abdominal ganglia and located in the thorax (after Burrows 1996). (c) Nervous system of the fruit fly, *Drosophila melanogaster*. Cerebral and gnathal ganglia are fused, and the three thoracic ganglia form a continuous structure with all abdominal ganglia, located in the thorax.

CRG, cerebral ganglia; GNG, gnathal ganglia; Pro, prothoracic ganglion; Meso, mesothoracic ganglion; Meta, metathoracic ganglion; A1–A7, abdominal ganglia; T, terminal ganglion.

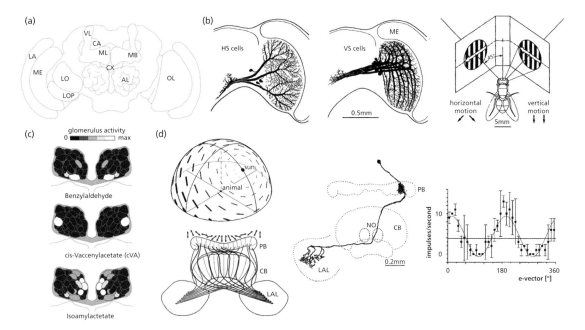

Figure 3.3 Identified brain regions and neurons for behavior. (a) Brain of *Drosophila melanogaster* with main neuropil areas. (b) Motion sensitive neurons (so-called tangential cells) in the lobula plate of the blowfly, *Calliphora erythrocephala*. HS cells respond to horizontal motion, VS cells to vertical motion. Behavioral set-up for playback of motion stimuli to the animal (modified from Hengstenberg 1982, with permission of Springer.). (c) Glomerulus activity pattern in response to three different olfactory stimuli measured by neuropil calcium imaging in the AL of *Drosophila melanogaster* (modified from Grabe and Sachse 2017, with permission from Elsevier). (d) Polarization pattern of the sky and schematic of the sky compass of the desert locust, *Schistocerca gregaria*. The e-vector tuning of the sixteen neuropil slices of the protocerebral bridge is indicated by arrows. To the right, identified CX neuron of *Schistocerca gregaria* with arborization in the protocerebral bridge and its response profile to polarized light, tuned to approximately 13° (modified from Homberg 2015 and Vitzthum et al. 2002, with permission from Society for Neuroscience).

AL, antennal lobe; OL, optic lobes; LA, lamina; ME, medulla; LO, lobula; LOP, lobula plate; MB, mushroom body; CA, mushroom body calyx; ML, mushroom body medial lobe; VL, mushroom body vertical lobe; CX, central complex; PB, protocerebral bridge of the CX; CB, central body of the CX; NO, noduli of the CX; LAL, lateral accessory lobe (nomenclature after Ito et al. 2014).

which form an outer layer surrounding the neuropil, i.e. the intertwining arborizations and synaptic areas of neurons. The PNS contains sensory neurons, which send information via **afferent nerves** into the CNS, as well as the **efferent nerves** originating from the CNS and innervating muscles or sensory organs.

Inside the brain, several conspicuous neuropil regions, as well as landmark fibre bundles can be distinguished. Clearly defined structures, which are easily found across species are the optic lobes (see Section 3.3.1), the antennal lobes (see Section 3.3.2), the mushroom bodies (see Section 3.3.3) and the central complex (see Section 3.3.4; Figure 3.3a; Strausfeld 1976; Wolff and Strausfeld 2015). Many smaller and less demarcated regions have been given different names in different species, and recent efforts by a

working group have been directed to establishing a systematic nomenclature (Ito et al. 2014). The chain of ganglia in the thorax, lying under the digestive system, is called the ventral nerve cord (VNC; Figure 3.2b). The VNC contains leg and wing motor neurons, which mainly arborize in the dorsal part, whereas the ventral neuropil is dominated by sensory afferents (Niven et al. 2008). Local circuits allow for adaptive motor control, which can be further modulated by descending input from the brain (Tuthill and Wilson 2016).

3.2.2 Neurons and glia

How small are insect brains numerically? A *Drosophila* brain consists of an estimated 100,000 neurons

(Chiang et al. 2011). The brain of a bee has around ten times more neurons, approximately 960,000 (Menzel and Giurfa 2001). Extremely small feather-wing beetles of the Ptiliidae family still have 40,000 miniature neurons in their nervous system, which is much larger in relative volume compared with bigger insects (Polilov 2008). Compared with verte-brate counterparts, insect neurons are relatively small (5–50 µm cell body diameter) and mostly unipolar, but have often very elaborate arborizations. In uni-polar cells, a single **neurite** leaves the cell body and branches into axonal (input) and dendritic (output) regions, which are sometimes intermingled. This means that electrical information flowing from input to output site can bypass the cell body, which might not or only weakly depolarize when the neuron is active. Structurally separated arborizations of a large neuron may act as functionally independent compu-tational units (Strausfeld 1976; Meinertzhagen 2017).

Insect neurons use conserved **neurotransmitters**. Neurotransmitters are molecules transmitting infor-mation across chemical synapses. They are released from vesicles in the presynaptic neuron into the syn-aptic cleft, bind to receptors on the post-synaptic neuron and, by this change, its activity. The most widespread CNS excitatory neurotransmitter is acetylcholine, whereas GABA transmits most inhibi-tory signals. Glutamate mediates excitation at the neuromuscular junction. **Biogenic amines** like dopa-mine, serotonin (5-hydroxytryptamine), octopamine, and tyramine (analogues of vertebrate adrenalin and noradrenaline), as well as many different neuropep-tides act as **neuromodulators**. They are produced and released by so-called neuromodulatory neurons. Neuromodulators change neuronal activity by being either locally released at synapses or by spreading more diffusely into a larger neuropil area (volume transmission). The distinction between neurotransmitters and neuromodulators can be a bit blurred, with many neuromodulators acting at synapses in a very similar way as neurotrans-mitters. Whereas neurotransmitters are released at synapses, neuromodulators can be also released at other sites, affect larger groups of neurons or effector cells, and have longer lasting, slower effects (Burrows 1996).

Most neuromodulators affect multiple behaviors. Octopamine, for example, is known to surge during flight. It enables the thoracic interneurons in the locust to produce the patterned activity for driving the wingbeat and modulates input from wing pro-prioceptive neurons, which entrain this central activity (Orchard et al. 1993). During *Drosophila* flight, octopamine increases the response of visual interneurons to motion, adapting fly vision to fast aerial locomotion (Suver et al. 2012). Octopaminergic neurons are also implicated in regulating sleep, aggression, reproductive behaviors, foraging, and learning and **memory** in many different insect spe-cies (Fahrbach and Mesce 2005; Roeder 2005).

Insects have several types of functionally special-ized glia cells. In the *Drosophila* adult nervous sys-tem, glia makes up approximately 10 per cent of the cells and are categorized into three major classes—surface glia, cortex glia around the cell bodies, and neuropil glia (including astrocytes; Kremer et al. 2017). Glia is important for shielding neurons and buffer-ing neurotransmitters released into the extracellular space, and it plays a role in development and repair of the nervous system. Astrocytes can also be active circuit elements. In *Drosophila* larvae, astrocytes respond to neuronally released neuromodula-tors with increased calcium levels. Through their action, larvae adjust startle-induced behaviors and escape responses (Ma et al. 2016). Mammals use similar molecular pathways in astrocytes for gating motor output and arousal, suggesting conserved mechanisms (Bazargani and Attwell 2017).

To summarize, a small nervous system does not necessarily mean a simpler nervous system and less behavioral complexity. Very few or even a single neuron in a small nervous system can effectively perform the computational task accomplished by a group of hundreds of partially redundant neurons in a larger nervous system (Chittka and Niven 2009; Niven and Farris 2012). There is also evidence that neurons in small invertebrate nervous systems are more often multifunctional (Niven and Chittka 2010). Examples are neurons reused during development (see Section 3.4.1).

3.3 Control of behavior by characterized brain regions

Comparative anatomy and functional studies from different species have accumulated detailed information about the neuronal composition, the wiring architecture and the behavioral relevance of the main brain neuropils (Wolff and Strausfeld 2015). To understand how the insect nervous system extracts information from the environment and uses this information to control and shape behavior, it is illuminating to take a closer look at the different regions dedicated to processing and integration of sensory stimuli.

3.3.1 Vision: optic lobes

Insects are excellent fliers, and motion vision for fast and accurate processing of optic flow, and moving objects, such as conspecifics, prey, or predators is of utmost importance. Many species use colour, including the UV range, for detecting and recognizing floral patterns and body markings of con- and heterospecifics (see Chapter 11). Additionally, many insects can extract the polarization pattern of light and use it for navigation (see Chapter 7).

The large compound eyes are composed of repeating ommatidia, housing photoreceptor neurons, which send information to the optic lobes (OLs). OLs can make up more than half of the brain by number of neurons and neuropil volume (Chittka and Niven 2009; Strausfeld 2012). The visual neuropils have a successive distal to proximal arrangement, comprising lamina, medulla, lobula, and lobula plate (Figure 3.3a). Neuronal circuits in the OL are characterized by a very orderly, layered architecture, which contains the same amount of repetitive columns as there are omatidia (Strausfeld 1976; see Chapter 11).

Motion vision has been intensely studied in Diptera. It can be probed by a basic orientation behavior. When exposed to moving stripes, an animal compensates for the motion of the environment by steering in the same direction to stabilize its course (optomotor response). The optomotor response depends on large neurons (so-called tangential cells) in the lobula plate, which respond to motion in a particular direction (Figure 3.3b). When experimentally activated, these neurons elicit a turning response. Ablating them abolishes the optomotor response. Tangential cells connect to neck motor neurons controlling head movements and to descending interneurons, which convey steering commands to motor centres in the VNC. Tangential cells receive input from medulla neurons, which are not direction selective themselves, but compute motion by detecting temporal delays between spatially adjacent channels, which respond to strong luminance changes, i.e. light or dark edges moving across the **retina** (Behnia and Desplan 2015). In the 1950s, a model for such a circuit, the so called 'elementary motion detector', was proposed by Hassenstein and Reichert, who studied the optomotor response of the beetle *Chlorophanus viridis*. Motion computation in the vertebrate retina relies on a remarkably similar neuronal mechanism, illustrating how common circuit motifs for basic computations, perfected by convergent evolution, are found across all animals with nervous systems (Borst and Helmstaedter 2015). On the other hand, motion detection is finely tuned to ecological demands. Lobula plate neurons in insect species with different flight velocities, such as butterflies, bumblebees, hoverflies, and hawk moths, show species-specific temporal frequency tuning matched to the visual behavior of the animal (O'Carroll et al. 1996).

Colour vision requires comparing output of at least two photoreceptor classes with different spectral sensitivity. Different insect species vastly differ in their ability to discriminate colour. Many species can separately detect UV, green, and blue light. Butterflies are known for their photoreceptor diversity, holding the record with fifteen spectrally distinct classes in the common bluebottle, *Graphium sarpedon* (Lebhardt and Desplan 2017). The downstream circuits for colour vision are not yet well understood. A current model is that signals from different photoreceptor classes are compared by so-called colour opponent neurons. Candidates for such neurons have been found in the medulla and lobula of bees and flies (Kelber 2016).

The polarization pattern of light carries information about the position of sun or moon, the presence of water surfaces, and can even serve as a mate-recognition signal when reflected from iridescent butterfly scales. Many insects, e.g. desert ants, crickets, and locusts, have structurally specialized photoreceptors for the detection of polarized light of different orientation. These are located at a small zone of the eye margin, the so-called dorsal rim area. Polarization information is processed in the medulla and further relayed to the central complex in the brain, a major area for orientation and navigational control (Homberg 2015).

Output neurons of the OL can be seen as a bottleneck, conveying presence of behaviorally relevant visual attributes to demarcated neuropil structures (optic glomeruli) in the central brain (Strausfeld and Okamura 2007). In *Drosophila,* over 20 different types of lobula columnar neurons with projections to distinct optic glomeruli have been characterized with genetic tools. Some of them respond to looming stimuli, others specifically to small, moving objects. Optogenetic activation of these lobula columnar neurons induces distinct behaviors such as escape-like jumping, changes in locomotion, movements resembling reaching during gap-climbing or aggressive encounters. This corroborates the view that interneurons projecting from OLs to the central brain signal ready computed salient features of the environment, which are normally met with motor responses by the animal (Wu et al. 2016).

3.3.2 Olfaction: antennal lobes

The sense of smell is pivotal for most insects, guiding foraging, feeding, oviposition, and social interactions (see Chapter 10). Some responses to odorants are independent of experience, whereas others are guided by learning and memory (see chapter 17). The antennal lobes (ALs) are the first brain processing area for olfactory information and receive the afferents from olfactory sensory neurons housed in sen palps. The majority of olfactory sensory neurons express only one of the many different receptor types. The receptor type determines to which set of airborne molecules the neuron responds. Neurons expressing the same type of receptor arborize in the same spatially defined region of the AL, a so called olfactory **glomerulus**. This anatomical organization has been most rigorously characterized in *Drosophila*, but is assumed to exist in many other species as well. The number of glomeruli can thus give a good estimate of olfactory receptor types expressed in a species. *Drosophila* has fifty-four glomeruli, honey bees, and cockroaches have over 100, some ants more than 400. Each odour or odour blend activates different sets of olfactory sensory neurons and leads to a stereotyped combinatorial activation pattern of AL glomeruli (Figure 3.3c). The

AL is innervated by local inhibitory and neuromodulatory neurons, involved in inter glomeruli processing resulting in gain control, sharpening, or broadening of response profiles. Excitatory projection neurons, which in general innervate only one glomerulus, relay activity to the mushroom bodies and a region in the central brain called the lateral horn (Grabe and Sachse 2017; Wolff and Strausfeld 2015). There, olfactory information is further processed, evaluated, and integrated with other sensory stimuli (see Section 3.3.3).

Some glomeruli are very narrowly tuned, responding only to one or very few specific volatiles with crucial behavioral relevance for the insect, e.g. pheromones or cues from host plants. In *Drosophila*, for example, citrus odours guide the choice of oviposition substrates, whereas the microbial compound geosmin triggers avoidance. In both cases, the behavioral adaptation depends on identified receptors and the neurons expressing them, segregating into fixed AL glomeruli (Mansourian and Stensmyr 2015).

In many eusocial insect species, olfaction is very important for intraspecific communication and coordination. This has been dramatically demonstrated in two species of ants, the raider ant *Ooceraea biroi* and *Harpegnathos saltator*. By genetic engineering, most olfactory sensory neurons were rendered nonfunctional, accompanied by massive loss of AL glomeruli. The nearly anosmic animals are unable to forage, cannot interact normally with conspecifics and have a strongly reduced fitness (Trible et al. 2017, Yan et al. 2017).

Gustatory (or contact chemosensory) sensilla in the periphery are, in general, much broader distributed than olfactory sensilla, clustering at mouthparts, inside the pharynx, at the legs, sometimes also ovipositor and anterior wing margins. Their multiple afferent zones, e.g. in the VNC and the gnathal ganglia are less conspicuous than the antenna lobe. In *Drosophila*, a certain afferent mapping logic, e.g. segregation according to taste quality (sweet versus bitter) is observed (Freeman and Dahanukar 2015).

3.3.3 Integration and learning: mushroom bodies

Mushroom bodies (MBs) look indeed like mushrooms, with calyces on the posterior surface of the brain, and pedunculi and lobes extending into the central neuropil (Figure 3.1). The main input to Kenyon cells in the calyces comes from the AL and carries olfactory information. In some species, Kenyon cells also receive major input from the OL (Wolff and Strausfeld 2015). MB pedunculi and lobes constitute local circuits with input and output from different neuropils of the central brain and intrinsic neuromodulatory neurons, as well as inhibitory feedback loops.

MB size dramatically varies across species. It is large in many social species, such as ants, wasps, and bees. However, there is no correlation between MB size and eusociality, since large MBs also occur in solitary species. Rather, it appears that MB size depends on sensory ecology, the presence of visual input to the calyces and a lifestyle requiring spatial learning for foraging (Farris 2016). Anosmic species, such as aquatic beetles or species with non-feeding adults that do not use olfaction for mate location and recognition (cicadas, mayflies), lack antenna lobes, and have missing or greatly reduced calyces. Still, they possess MBs with lobes, suggesting functions independent of olfaction, and involvement in polymodal information processing and memory formation (Strausfeld 2012).

Most functional studies of MBs have focused on their role in the formation of olfactory memory. *Drosophila* can learn to associate a previously neutral odour with attractive (sugar reward) or aversive (foot shock) stimuli, and after training either seek it out or avoid it. It is not clear to what extent these simplified assays reflect natural behavior of the fly during foraging or predator avoidance, but they have been instrumental in demonstrating learning ability and elucidating underlying circuits in the MB. Any one odour leads to the activation of only few of the around 2000 Kenyon cells. Sparse representation of odour identity is thought to increase coding capacity and to make discrimination easier. The current opinion is that learning takes place when odour-specific activity patterns coincide with signals from neuromodulatory neurons encoding valence (punishment or reward), leading to modifications of the synapse between Kenyon cells and MB output neurons (Stopfer 2014; Owald and Waddell 2015). In *Drosophila*, there are only 21 MB output neuron types, which innervate

spatially segregated portions of the MB lobes. Optogenetic activation of these neurons can trigger attraction or aversion in the absence of any other stimulus, and different sets of them are required for various associative memory tasks (Aso et al. 2014). The basic functional organization of the much larger MBs of the honey bee, *Apis mellifera*, is similar to the one of *Drosophila*. The behavioral repertoire of bees comprises configural learning, categorization, and flexible navigation based on topographic, vector-based landscape memory, to an extent not observed in *Drosophila*. It still remains to be elucidated how much these capabilities rely on MB circuits in bees (Menzel and Giurfa 2001).

Memory associating odours with positive or negative outcomes is only useful for the animal, if it can also be extinguished or consolidated dependent on the validity of the prediction (see Chapter 17). Networks of MB output neurons and modulatory dopaminergic neurons mediate this re-evaluation and update learned information (Felsenberg et al. 2017). An encountered odour is also evaluated by the fly as being either familiar or novel. In the latter case, flies are alerted and interrupt their ongoing behavior. Specific MB output neurons, which only respond to novel stimuli, mediate this response. This exemplifies that additional to associative memory, many more aspects of the individual history of an animal are represented in the MB circuits and guide responses to the environment (Hattori et al. 2017).

Olfactory information is not only routed from the AL to the MB, but also to an anatomically less segregated area, the lateral horn. In the lateral horn, connectivity patterns of AL projection neurons are more stereotyped than in the MB. There is indication that lateral horn circuits are crucial for experience-independent responses to ecologically relevant odours, such as pheromones and volatiles from food or predators (Schultzhaus et al. 2017).

3.3.4 Navigation and motor control: central complex

The central complex (CX) is an unpaired structure in the middle of the brain, consisting of several subregions—protocerebral bridge, central body (with the upper division called fan-shaped body and the lower division called ellipsoid body), and paired noduli (Figure 3.3d). Since it does not receive conspicuous direct input from primary sensory neuropil, the CX has since long been recognized as a higher order processing and multimodal integration area, representing space and time for refined motor control, navigation and orientation behavior (Wolff and Strausfeld 2015). The protocerebral bridge is subdivided in sixteen neuropil slices, which are connected by a regular, midline crossing array of columnar neurons to vertical subunits of the central body (Figure 3.3d). Columnar neurons are also the main output neurons of the CX (Homberg 2015).

In migrating species, such as the desert locust, *Schistocerca gregaria*, and the monarch butterfly, *Danaus plexippus*, CX neurons are sensitive to the plane of polarized light, with topographically arranged *E*-vector tuning (Figure 3.3d). The CX in monarch butterflies integrates the sunlight polarization plane with the position of the sun and time of day, serving thus as a time-compensated sky compass for the astonishing long distance orientation across North America (Reppert et al. 2016, see Chapter 7).

In *Drosophila*, neuronal silencing and mutant studies suggest that CX circuits are involved in visual place learning, pattern recognition, and spatial memory (Varga et al. 2017). They are also implicated in coordinating sleep and arousal state (Donlea 2017). The activity of neurons innervating the ellipsoid body reflects the fly's body orientation (head direction) relative to visual landmarks (Seelig and Jayaraman 2015). Importantly, the internal orientation sense can be maintained independently of vision. The neuronal representation of heading persists and is updated in the dark, when the fly moves. Turning (rotational velocity) of the animal results in transient asymmetric activity in the protocerebral bridge, followed by a shift of the head direction representation in the ellipsoid body (Green et al. 2017).

Keeping correctly updated head direction representations is especially important for insects performing daily foraging trips and returning to a central location. They are thought to use a strategy called path integration—direction and distance are registered during the outward journey. From this, a homeward-pointing vector is generated, allowing the animal to return on a straight path. In the tropical bee, *Megalopta genalis*, a candidate circuit for path integration in the CX has been proposed. As in locusts and monarch butterflies, direction information is signalled by characteristic CX neurons sensitive to polarized light. Special speed neurons innervating the paired noduli encode translational optical flow and, in this way, provide a distance signal. The two parameters, direction and distance, are likely to be integrated and combined to a home vector representation by neurons interconnecting the different CX regions (Heinze 2017).

In the tropical cockroach, *Blaberus discoidalis*, electrophysiological recordings show that CX neurons encode head directions and are tuned to preferred angles, like their *Drosophila* counterparts. Apart from visual information, the cockroach CX also responds to mechanical cues, e.g. antennal deflections during exploratory walking and climbing. Behavioral studies underline that the spatio-temporal representation of the environment, which is computed in the CX, is crucial for fine tuning and adjusting goal-directed motor behaviors and coordinated locomotion in complex terrain (Varga et al. 2017).

To summarize, the major behavioral function of the CX seems to depend on the ecology of the species. It serves as a sky compass for migrants, a short-range navigational tool for cockroaches foraging in cluttered leaf litter and a target-tracking device for stalking predators like mantis. Head-direction encoding neurons are also found in mammalian brains, serving as yet another example of how nervous systems use common principles for effective solutions to basic problems (Varga et al. 2017).

3.4 Metamorphosis and nervous system plasticity

3.4.1 Restructuring of the nervous system during metamorphosis

Over 80 per cent of named insect species are **holometabolous** (Grimaldi and Engel 2005) and undergo complete metamorphosis, which restructures almost all morphological structures, including the nervous system. Larvae and adults often occupy very different **habitats**, exploit different food sources, and show different behavioral adaptations, requiring specialized sensory and motor neuronal circuits. In short, the genome of a holometabolous insect has to build two nervous systems, with the additional challenge of recycling the first, while assembling the second during pupal stage, on top of maintaining pupal-specific behavior. Metamorphosis is under the control of steroid hormones, which also mediate major changes of the nervous system (see Chapter 4). Some larval neurons die by programmed cell death (apoptosis), whereas others are remodelled. Many motor neurons, for example, re-innervate new adult muscles, which form around the remnants of larval muscles. Their central arborizations are restructured, which include the pruning of larva-specific dendrites and the outgrowth of adult specific dendrites. Additionally, newly born neurons differentiate and are integrated into adult circuits (Truman 1992; Tissot and Stocker 2000). A striking example is motor neuron MN5; in *Manduca* larvae, it innervates a muscle participating in slow crawling, whereas in the adult, it controls muscles powering the fast wing down-strokes during flight. To undergo this switch in functionality, MN5 changes not only its dendritic morphology, but also expression of membrane channels and receptors, determining its electrical properties and excitability (physiological remodelling) (Consoulas et al. 2000).

Since holometabolous insects integrate many new neurons into their adult nervous system, they not only have an embryonic phase of neurogenesis (like **hemimetabolous** insects), but also a second, post-embryonic phase of neurogenesis. New neurons are generated by neural progenitors, so called **neuroblasts**. Their divisions are under tight spatial and temporal control, following a fixed programme, which not only specifies the final size of a nervous system region, but also the genetically determined diversity of different neuronal types. Post-embryonic neuroblast proliferation and neurogenesis during larval stages and pupation is especially prominent in the OL and the MB of holometabolous insects. Adult specific neurons are predominantly interneurons. These are needed for processing information from new adult sensory organs, and for coordinating new and more complex modes of locomotion (Tissot and Stocker 2000; Yasugi and Nishimura 2016).

It is an intriguing question for behavioral ecology if experience dependent changes of the nervous system made during larval life are retained in adults. In many lepidopterans, there is evidence that larval

feeding experience influences adult oviposition decisions (Hopkins host-selection principle). The mechanisms and neuronal substrates for these plastic changes in behavior remain to be elucidated for the majority of documented cases. Sometimes, chemicals from the larval feeding place are present in pupal case and haemolymph during metamorphosis. When the newly emerging adult is exposed to them, conditioning can take place (chemical legacy; Jones and Agrawal 2017). With this mechanism, no particular neural substrate for the memory is maintained during metamorphosis, but a new memory is formed at adult emergence, using information from the larval stage. Indication that olfactory associative memory, i.e. the learned avoidance of an odour presented repeatedly together with electroshock, might under certain circumstances survive metamorphosis comes from *Drosophila* (Tully et al. 1994) and *Manduca* (Blackiston et al. 2008). Olfactory memories are thought to be stored in the MB, which to a large extent undergoes dramatic morphological reorganization in the pupa. So far, it is not understood if the phenomenon relies on the specific subset of MB neurons and their connections, which are retained during metamorphosis (Lee et al. 1999), but these would be the most likely candidates.

Hemimetabolous insects like crickets, cockroaches, and bugs hatch from eggs as miniature adults without wings and reproductive organs, and undergo fewer changes in behavioral repertoire and nervous system layout during their progressive moults. They have only few dividing neuroblasts during post-embryonic development in the OL and the MB, and none in the ventral nerve cord. Their nervous system grows by strong expansion of neuropil and proliferation of glia (Bentley 1977; Tissot and Stocker 2000). Neuronal circuits are often progressively assembled, with their main layout and basic functionality being present before adult behavior is expressed. This has been, for example, shown for flight networks in crickets (Bentley 1977) or locusts (Stevenson and Kutsch 1988).

3.4.2 Plasticity of the adult nervous system

An important function of the nervous system is to respond adaptively to the surroundings. The prevalence of stereotyped, identifiable neurons can give the mistaken idea that the insect nervous system shows little plasticity apart from the fixed restructuring during metamorphosis. Neuronal plasticity, more precisely, alteration of the nervous system depending on individual environmental circumstances is well documented for many insect species (Meinertzhagen 2001). An example of response to injury comes from locusts. Their wing-beat pattern during flight is influenced by feedback from mechanosensory cells on the hindwing. When these are removed, flight motor patterns are first abnormal, but progressively recover. Electrophysiological recordings show that forewing mechanosensory neurons, which normally do not impact flight patterns, gradually take over the function of the hindwing afferents and restore the behavior (Büschges et al. 1992).

Adult neuropil can change in volume in several brain regions in response to environmental stimuli, as can be most dramatically illustrated by examples from eusocial insects undergoing caste or work task transitions. Honeybees, *Apis mellifera*, show a division of labour between nurses and foragers. Compared with nurses of the same age, forager bees have larger MB neuropils (Withers et al. 1993). Likewise, in the Florida carpenter ant, *Camponotus floridanus*, the relative volume of the MB neuropil depends on the tasks the animal is performing. Nurses have larger neuropils than animals kept artificially idle by separation from queen and brood. Age-matched foragers from the same morphological caste show the largest MB neuropils, suggesting strong growth of neuronal arborizations upon behavioral experience and increased processing demands (Gronenberg et al. 1996). Conversely, in the highly visual ponerine ant, *Harpegnathos*, OL neuropils of hunting foragers decrease markedly in size when they become reproductives and do not leave the dark **nest** any more (Gronenberg and Liebig 1999).

In several species of crickets, beetles, praying mantis, milkweed bugs, and moth adult neurogenesis has been observed in the MB. For the house cricket, *Acheta domesticus*, neurogenesis is stimulated in an environment enriched by olfactory and visual sensory input. Animals that have their adult born MB neurons ablated by irradiation perform worse in learning a specific escape paradigm using olfactory cues (Cayre et al. 2007). *Drosophila melanogaster* was long thought to lack adult neurogenesis. Recently, a low level of neurogenesis was detected in the visual system, which can be further upregulated by acute tissue damage (Fernández-Hernández et al. 2013). Prevalence and general behavioral relevance of adult neurogenesis in comparison to neuropil volume and structure plasticity of existing neurons is not yet well understood (Simões and Rhiner 2017).

3.5 Case study: neurons and circuits for *Drosophila* sexual behavior

Among the most fascinating and multifaceted behaviors displayed by insects are courtship displays and mating strategies (see Chapter 13). For some time, they have attracted interest from evolutionary biologists (Thornhill and Alcock 1983; Shuker and Simmons 2014). Moreover, understanding the neurobiology of insect reproduction is highly relevant for pest and disease **vector** control (see also Chapters 20 and 21).

The neuronal circuits underlying sexual behavior in *Drosophila melanogaster* are well understood, and illustrate the neuronal basis of sensory processing, motor patterning, and central control of arousal. When a sexually mature male fly encounters a female, he shows characteristic courtship behavior. After the initial approach and following, he touches the female with his forelegs, extends one wing and vibrates it, producing an acoustic signal—the courtship song (Figure 3.4a). The song consists of trains of pulses, interspersed by a humming sound, the sine song (Figure 3.4b). Its temporal parameters are species-specific and enhance the receptivity of the female. When the female is willing to mate, she slows down locomotion in response to male displays and allows copulation.

Male courtship only occurs in the presence of females, but its most conspicuous element, wing song, can be triggered in isolated males by artificially activating a population of CNS neurons marked by the male-specific expression of a gene of the sex determination cascade, *fruitless* (Clyne and Miesenböck 2008). More specifically, activation of P1, a specific class of *fruitless* positive neurons, is sufficient and necessary for song behavior (von Philipsborn et al. 2011). The P1 class contains 20–30

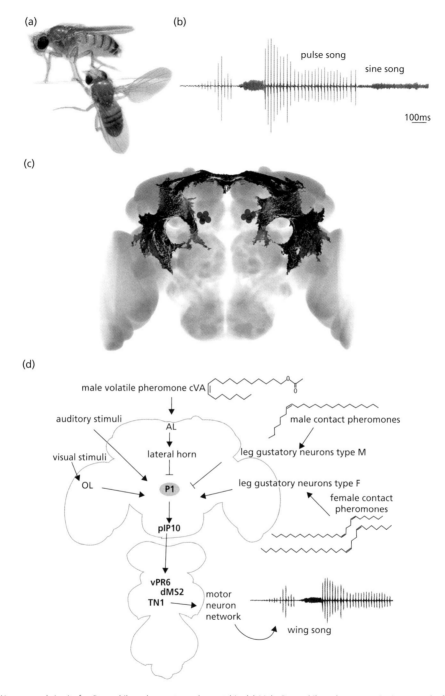

Figure 3.4 Neurons and circuits for *Drosophila melanogaster* male courtship. (a) Male *Drosophila melanogaster* singing toward a female by unilateral wing extension (Photo: Solvin Zankl). (b) Oscillogram of wing courtship song, with pulse and sing song elements. (c) Reconstruction of the P1 neuronal class in a male *Drosophila* brain. Cell bodies are indicated by grey circles. (d) Schematic of neuronal circuit for courtship song production with multimodal input to the central brain integration neuronal class P1 and identified song neuronal classes pIP10, vPR6, vMS2 and TN1 indicated.

neurons, which are only present in male flies, since sex-specific apoptosis targets them during metamorphosis in females (Kimura et al. 2008). The large dendritic and axonal arbours of P1 form ring-like structures around the MB lobes (Figure 3.4c). Calcium imaging shows that P1 is activated when a male fly touches the female abdomen with his foreleg, a step that normally precedes song during courtship (Kohatsu et al. 2011).

How does sensory information reach P1? Gustatory neurons on the male's leg are activated by female-specific cuticular pheromones. The afferents of these neurons synapse onto ascending interneurons, which in turn activate P1. In parallel, the inhibitory neuronal class mAL is activated, which balances the excitation of P1 and might mediate the gain control of the chemosensory signal. Touching a male does not trigger song, because male pheromones activate a different set of gustatory neurons, which only connect to mAL. Courtship toward another male is further prevented by a volatile male-specific pheromone, cis-vaccenyl acetate (cVA). cVA activates specific olfactory neurons on the antenna, which activate an identified AL glomerulus. From there, AL projection neurons carry the signal to the lateral horn and, ultimately, to P1. As a male pheromone, cVA is attractive to females. Sexual differences in the wiring patterns of lateral horn neurons ensure that the cVA signal is processed differently in females, leading to the activation of the female specific receptivity enhancing neuronal class pC1. Neuronal elements processing visual and auditory stimuli, which further guide and enhance male courtship behavior, have also been identified, completing the picture that P1 integrates multimodal sensory information and enables the recognition of conspecific females (Ellendersen and von Philipsborn 2017; Figure 3.4d). How can activity of P1 elicit song behavior? The descending, male-specific interneuron pIP10 connects P1 to different interneurons in the VNC, which are thought to influence song pattering. Activity of the neuronal class vPR6, for example, influences the temporal spacing of song pulses, while activity of the neuronal class dMS2 is required for correct song structure (von Philipsborn et al. 2011). Sine song is elicited by yet another neuronal class, TN1, which directly connects to a motor neuron controlling wing movement (Figure 3.4d). Many of the neuronal com-

ponents of the circuit downstream of P1 are sexually dimorphic, and their development and arborization pattern is controlled by the *fruitless* gene, or a second sex-determination gene, *doublesex* (von Philipsborn et al. 2014; Shirangi et al. 2016).

Male flies can identify females and produce the species specific song without prior experience. Various external and internal factors, however, modulate his innate response and allow for behavioral flexibility (Ellendersen and von Philipsborn 2017). For example, males can learn to specifically suppress courtship to recently mated, unreceptive females, a phenomenon dependent on specific MB neurons (Griffith and Ejima 2009; Keleman et al. 2012). Female reproductive behavior is controlled by distinct neuronal circuits, computing which male to accept, if and when to remate, and where to oviposit (Ellendersen and von Philipsborn 2017).

D. melanogaster courtship is arguably one of the best understood complex behaviors on the level of **identified neurons**, but we are still far from a complete reconstruction of the full circuit and its computations. Reproductive behaviors and sexual communication in the genus *Drosophila* show fascinating species-specific specializations and idiosyncrasies (Markow and O'Grady 2005). In the future, it will be of great interest to analyse conserved and divergent courtship circuit elements in other *Drosophila* species, and elucidate how they were shaped by ecology and evolution. Genome engineering holds great potential for an in-depth comparative approach to the circuit mechanisms underlying behavior. First steps in this direction have been made with *D. subobscura*, where optogenetic activation of *fruitless* expressing neurons also induces species specific courtship behavior (Tanaka et al. 2017).

Acknowledgement

The author would like to thank the editors, an anonymous reviewer, and M. Verzijden, S. Nolte, A. O'Sullivan, P. Kerwin, and B. Ellendersen for helpful comments on the manuscript. I apologize to colleagues whose work was not mentioned due to limited space or ignorance from my side. Research at DANDRITE is funded by the Lundbeck Foundation (Lundbeckfonden grant no. DANDRITE-R248-2016-2518).

References

Aso, Y., Sitaraman, D., Ichinose, T., et al. (2014). Mushroom body output neurons encode valence and guide memory-based action selection in *Drosophila*. *eLife*, **3**, e04580.

Bazargani, N., and Attwell, D. (2017). Amines, astrocytes, and arousal. *Neuron*, **94**, 228.

Behnia, R., and Desplan, C. (2015). Visual circuits in flies: beginning to see the whole picture. *Current Opinion in Neurobiology*, **34**, 125.

Bentley, D.R. (1977). Development of insect nervous systems. In: G. Hoyles (Ed.), *Identified Neurons and Behavior of Arthropods*, pp. 461. Plenum Press, New York.

Blackiston, D.J., Silva Casey, E., and Weiss, M.R. (2008). Retention of memory through metamorphosis: can a moth remember what it learned as a caterpillar? *PLOS One*, **3**, e1736.

Borst, A. and Helmstaedter, M. (2015). Common circuit design in fly and mammalian motion vision. *Nature Neuroscience*, **18**, 1067.

Bullock, T.H. (1977). Some perspectives on comparative neurophysiology. In G. Hoyles (Ed.) *Identified Neurons and Behavior of Arthropods*, pp. 533–8. Plenum Press, New York.

Burrows, M. (1996). *The Neurobiology of an Insect Brain.* Oxford University Press, Oxford.

Büschges, A., Ramirez, J-M., and Pearson, K.G. (1992). Reorganization of sensory regulation of locust flight after partial deafferentation. *Journal of Neurobiology*, **23**, 31.

Cayre, M., Scotto-Lomassese, S., Malaterre, J., Strambi, C., and Strambi, A. (2007). Understanding the regulation and function of adult neurogenesis: contribution from an insect model, the house cricket. *Chemical Senses*, **32**, 385.

Chiang, A-S., Lin, C-Y., Chuang, C-C., et al. (2011). Three-dimensional reconstruction of brain-wide wiring networks in *Drosophila* at single-cell resolution. *Current Biology*, **21**, 1.

Chittka, L., and Niven, J. (2009). Are bigger brains better? *Current Biology*, **19**, R995.

Clyne, J.D., and Miesenböck, G. (2008). Sex-specific control and tuning of the pattern generator for courtship song in *Drosophila*. *Cell*, **133**, 354.

Consoulas, C., Duch, C., Bayline, R.J., and Levine, R.B. (2000). Behavioral transformations during metamorphosis: remodeling of neural and motor systems. *Brain Research Bulletin*, **53**, 571.

Donlea, J.M. (2017). Neuronal and molecular mechanisms of sleep homeostasis. *Current Opinion in Insect Science*, 24: p. 51.

Dujardin, F. (1850). Mémoire sur le système nerveux des insectes. *Annales des Sciences Naturelles*, **14**, 195.

el Jundi, B., and Heinze, S. (2016). Three-dimensional atlases of insect brains. In: R. Pelc, R. Doucette, and W. Walzs (Eds) *Neurohistology and Imaging: Basic Techniques*. Springer, Humana Press, in press.

Ellendersen, B.E., and von Philipsborn, A.C. (2017). Neuronal modulation of *D. melanogaster* sexual behaviour. *Current Opinion in Insect Science*, **24**, 21.

Fahrbach, S.E., and Mesce, K.A. (2005). 'Neuroethoendocrinology': integration of field and laboratory studies in insect neuroendocrinology. *Hormones and Behavior*, **48**, 352.

Farris, S.M. (2016). Insect societies and the social brain. *Current Opinion in Insect Science*, **15**, 1.

Felsenberg, J., Barnstedt, O., Cognigni, P., Lin, S., and Waddell, S. (2017). Re-evaluation of learned information in *Drosophila*. *Nature*, **544**, 240.

Fernández-Hernández, I., Rhiner, C., and Moreno, E. (2013). Adult neurogenesis in *Drosophila*. *Cell Reports*, **3**, 1857.

Freeman, E.G., and Dahanukar, A. (2015). Molecular neurobiology of *Drosophila* taste. *Current Opinion in Neurobiology*, **34**, 140.

Grabe, V., and Sachse, S. (2017). Fundamental principles of the olfactory code. *Biosystems*, **164**, 94–101.

Green, J., Adachi, A., Shah, K.K., et al. (2017). A neural circuit architecture for angular integration in *Drosophila*. *Nature*, **546**, 101.

Griffith, L.C., and Ejima, A. (2009). Courtship learning in *Drosophila melanogaster*: diverse plasticity of a reproductive behavior. *Learning & Memory*, **16**, 743.

Grimaldi, D., and Engel, M.S. (2005). *Evolution of the Insects.* Cambridge University Press, New York.

Gronenberg, W., Heeren, S., and Hölldober, B. (1996). Age-dependent and task-related morphological changes in the brain and the mushroom bodies of the ant *Camponotus floridanus*. *Journal of Experimental Biology*, **199**, 2011.

Gronenberg, W., and Liebig, J. (1999). Smaller brains and optic lobes in reproductive workers of the ant Harpegnathos. *Naturwissenschaften*, **86**, 343.

Hattori, D., Aso, Y., Swartz, K.J., et al. (2017). Representations of novelty and familiarity in a mushroom body compartment. *Cell*, **169**, 956.

Heinze, S. (2017). Unraveling the neural basis of insect navigation. *Current Opinion in Insect Science*, **24**, 58.

Hengstenberg, R. (1982). Common visual response properties of giant vertical cells in the lobula plate of the blowfly Calliphora. *Journal of Comparative Physiology*, **149**, 179.

Homberg, U. (2015). Sky compass orientation in desert locusts—evidence from field and laboratory studies. *Frontiers in Behavioral Neuroscience*, 9.

Huber, F. (1962). Central nervous control of sound production in crickets and some speculations on its evolution. *Evolution*, **16**, 429.

Ito, K., Shinomiya, K., Ito, M., et al. (2014). A systematic nomenclature for the insect brain. *Neuron*, **81**, 755.

Jenett, A., Rubin, G.M., Ngo, T-T.B., et al. (2012). A GAL4-driver line resource for *Drosophila* neurobiology. *Cell Reports*, **2**, 991

Jones, P.L., and Agrawal, A.A. (2017). Learning in insect pollinators and herbivores. *Annual Review of Entomology*, **62**, 53.

Kelber, A. (2016). Colour in the eye of the beholder: receptor sensitivities and neural circuits underlying colour opponency and colour perception. *Current Opinion in Neurobiology*, **41**, 106.

Keleman, K., Vrontou, E., Kruttner, S., et al. (2012). Dopamine neurons modulate pheromone responses in *Drosophila* courtship learning. *Nature*, **489**, 145.

Kenyon, F.C. (1896). The meaning and structure of the so-called "mushroom bodies" of the hexapod brain. *American Naturalist*, **30**, 643.

Kimura, K., Hachiya, T., Koganezawa, M., Tazawa, T., and Yamamoto, D. (2008). Fruitless and doublesex coordinate to generate male-specific neurons that can initiate courtship. *Neuron*, **59**, 759.

Kohatsu, S., Koganezawa, M., and Yamamoto, D. (2011). Female contact activates male-specific interneurons that trigger stereotypic courtship behavior in *Drosophila*. *Neuron*, **69**, 498.

Kremer, M.C., Jung, C., Batelli, S., Rubin, G.M., and Gaul, U. (2017). The glia of the adult *Drosophila* nervous system. *Glia*, **65**, 606.

Lebhardt, F., and Desplan, C. (2017). Retinal perception and ecological significance of color vision in insects. *Current Opinion in Insect Science*, **24**, 75.

Lee, T., Lee, A., and Luo, L. (1999). Development of the *Drosophila* mushroom bodies: sequential generation of three distinct types of neurons from a neuroblast. *Development*, **126**, 4065.

Ma, Z., Stork, T., Bergles, D.E., and Freeman, M.R. (2016). Neuromodulators signal through astrocytes to alter neural circuit activity and behaviour. *Nature*, **539**, 428.

Mansourian, S., and Stensmyr, M.C. (2015). The chemical ecology of the fly. *Current Opinion in Neurobiology*, **34**, 95.

Markow, T.A., and O'Grady, P.M. (2005). Evolutionary genetics of reproductive behavior in *Drosophila*: connecting the dots. *Annual Review of Genetics*, **39**, 263.

Meinertzhagen, I. (2017). Morphology of invertebrate neurons and synapses. In J. H. Byrne (Ed.) *The Oxford Handbook of Invertebrate Neurobiology*. Oxford Handbooks, Oxford.

Meinertzhagen, I.A. (2001). Plasticity in the insect nervous system. In: P. Evans (Ed.) *Advances in Insect Physiology*, pp. 84–167. Academic Press, London.

Menzel, R., and Giurfa, M. (2001). Cognitive architecture of a mini-brain: the honeybee. *Trends in Cognitive Sciences*, **5**, 62.

Mill, P.J. (1977). Ventilation motor mechanisms in the dragonfly and other insects. Identified neurons and behavior of arthropods. In: G. Hoyles (Ed.) *Identified Neurons and Behavior of Arthropods*, pp. 187–208. Plenum Press, New York.

Niven, J.E., and Chittka, L. (2010). Reuse of identified neurons in multiple neural circuits. *Behavioral and Brain Sciences*, **33**, 285.

Niven, J.E., and Farris, S.M. (2012). Miniaturization of nervous systems and neurons. *Current Biology*, **22**, R323.

Niven, J.E., Graham, C.M., and Burrows, M. (2008). Diversity and evolution of the insect ventral nerve cord. *Annual Review of Entomology*, **53**, 253.

O'Carroll, D.C., Bidweii, N.J., Laughlin, S.B., and Warrant, E.J. (1996). Insect motion detectors matched to visual ecology. *Nature* **382**, 63.

O'Shea, M., and Rowell, C.H.F. (1977). Complex neural integration and identified interneurons in the locust. In: G. Hoyles (Ed.) *Identified Neurons and Behavior of Arthropods*, pp. 307–28. Plenum Press, New York.

Ohyama, T., Schneider-Mizell, C.M., Fetter, R.D., et al. (2015). A multilevel multimodal circuit enhances action selection in *Drosophila*. *Nature*, **520**, 633.

Orchard, I., Ramirez, J.M., and Lange, A.B. (1993). A multifunctional role for octopamine in locust flight. *Annual Review of Entomology*, **38**, 227.

Owald, D., and Waddell, S. (2015). Olfactory learning skews mushroom body output pathways to steer behavioral choice in *Drosophila*. *Current Opinion in Neurobiology*, **35**, 178.

Polilov, A.A. (2008). Anatomy of the smallest coleoptera, featherwing beetles of the tribe nanosellini Coleoptera, Ptiliidae, and limits of insect miniaturization. *Entomological Review*, **88**, 26.

Reppert, S.M., Guerra, P.A., and Merlin, C. (2016). Neurobiology of monarch butterfly migration. *Annual Review of Entomology*, **61**, 25.

Riemensperger, T., Kittel, R.J., and Fiala, A. (2016). Optogenetics in *Drosophila* neuroscience. In: A. Kianianmomenis (Ed.) *Optogenetics: Methods and Protocols*, pp. 167–75. Springer, New York.

Roeder, K.D. (1963). *Nerve Cells and Insect Behavior*. Harvard University Press, Cambridge, MA.

Roeder, T. (2005). Tyramine and octopamine: ruling behavior and metabolism. *Annual Review of Entomology*, **50**, 447.

Schlegel, P., Costa, M., and Jefferis, G.S.X.E. (2017). Learning from connectomics on the fly. *Current Opinion in Insect Science*, **24**, 96.

Schultzhaus, J.N., Saleem, S., Iftikhar, H., and Carney, G.E. (2017). The role of the *Drosophila* lateral horn in olfactory information processing and behavioral response. *Journal of Insect Physiology*, **98**, 29.

Seelig, J.D., and Jayaraman, V. (2015). Neural dynamics for landmark orientation and angular path integration. *Nature*, **521**, 186.

Shirangi, T.R., Wong, A.M., Truman, J.W., and Stern, D.L. (2016). Doublesex regulates the connectivity of a neural circuit controlling *Drosophila* male courtship song. *Developmental Cell*, **37**, 533–44.

Shuker, D., and Simmons, L. (2014). *The Evolution of Insect Mating Systems*. Oxford University Press, Oxford.

Simões, A.R., and Rhiner, C. (2017). A cold-blooded view on adult neurogenesis. *Frontiers in Neuroscience*, **11**, 327.

Stevenson, P.A., and Kutsch, W. (1988). Demonstration of functional connectivity of the flight motor system in all stages of the locust. *Journal of Comparative Physiology A* **162**, 247.

Stopfer, M. (2014). Central processing in the mushroom bodies. *Current Opinion in Insect Science*, **6**, 99.

Strausfeld, N.J. (1976). *Atlas of an Insect Brain*. Springer-Verlag, Berlin.

Strausfeld, N.J. (2012). *Arthropod Brains: Evolution, Functional Elegance, and Historical Significance*. Belknap Press of Harvard University Press, Cambridge, MA.

Strausfeld, N.J., and Okamura, J-Y. (2007). Visual system of calliphorid flies: organization of optic glomeruli and their lobula complex efferents. *Journal of Comparative Neurology*, **500**, 166.

Sun, D., Guo, Z., Liu, Y., and Zhang, Y. (2017). Progress and prospects of CRISPR/Cas systems in insects and other arthropods. *Frontiers in Physiology*, 8, 608.

Suver, M.P., Mamiya, A., and Dickinson, M.H. (2012). Octopamine neurons mediate flight-induced modulation of visual processing in *Drosophila*. *Current Biology*, **22**, 2294.

Tanaka, R., Higuchi, T., Kohatsu, S., Sato, K., and Yamamoto, D. (2017). Optogenetic activation of the fruitless-labeled circuitry in *Drosophila subobscura* males induces mating motor acts. *Journal of Neuroscience*, **37**(48), 11662–74.

Thornhill, R., and Alcock, J. (1983). *The Evolution of Insect Mating Systems*. Harvard University Press, Cambridge, MA.

Tissot, M., and Stocker, R.F. (2000). Metamorphosis in *Drosophila* and other insects: the fate of neurons throughout the stages. *Progress in Neurobiology*, **62**, 89.

Trible, W., Olivos-Cisneros, L., Mckenzie, S.K., et al. (2017). Orco mutagenesis causes loss of antennal lobe glomeruli and impaired social behavior. *Cell*, **170**, 727.

Truman, J.W. (1992). Developmental neuroethology of insect metamorphosis. *Journal of Neurobiology*, **23**, 1404.

Tully, T., Cambiazo, V., and Kruse, L. (1994). Memory through metamorphosis in normal and mutant *Drosophila*. *Journal of Neuroscience*, **14**, 68.

Tuthill, J.C., and Wilson, R.I. (2016). Mechanosensation and adaptive motor control in insects. *Current Biology*, **26**, R1022.

Varga, A.G., Kathman, N.D., Martin, J.P., Guo, P., and Ritzmann, R.E. (2017). Spatial navigation and the central complex: sensory acquisition, orientation, and motor control. *Frontiers in Behavioral Neuroscience* 11.

Venken, K.J.T., Simpson, J.H., and Bellen, H.J. (2011). Genetic manipulation of genes and cells in the nervous system of the fruit fly. *Neuron*, **72**, 202.

Vitzthum, H., Müller, M., and Homberg, U. (2002). Neurons of the central complex of the locust *Schistocerca gregaria* are sensitive to polarized light. *Journal of Neuroscience*, **22**, 1114.

von Philipsborn, A.C., Jörchel, S., Tirian, L., et al. (2014). Cellular and behavioral functions of fruitless isoforms in *Drosophila* courtship. *Current Biology*, **24**, 242.

von Philipsborn, A.C., Liu, T., Yu, J.Y., et al. (2011). Neuronal control of *Drosophila* courtship song. *Neuron*, **69**, 509.

Wilson, D.M. (1968). The flight-control system of the locust. *Scientific American*, **218**, 83.

Withers, G.S., Fahrbach, S.E., and Robinson, G.E. (1993). Selective neuroanatomical plasticity and division of labour in the honeybee. *Nature*, **364**, 238.

Wolff, G.H., and Strausfeld, N.J. (2015). The insect brain: a commentated primer. In: A. Schmidt-Rhaesa, S. Harzsch, and G. Purschkes (Eds) *Structure and Evolution of Invertebrate Nervous Systems*, pp. 597–639. Oxford University Press, Oxford.

Wu, M., Nern, A., Williamson, W.R., et al. (2016). Visual projection neurons in the *Drosophila lobula* link feature detection to distinct behavioral programs. *eLife*, **5**, e21022.

Yan, H., Opachaloemphan, C., Mancini, G., et al. (2017). An engineered orco mutation produces aberrant social behavior and defective neural development. *Cell*, **170**, 7367.e9.

Yasugi, T., and Nishimura, T. (2016). Temporal regulation of the generation of neuronal diversity in *Drosophila*. *Development, Growth & Differentiation*, **58**, 73.

Yoshihara, M., and Ito, K. (2012). Acute genetic manipulation of neuronal activity for the functional dissection of neural circuits—a dream come true for the pioneers of behavioral genetics. *Journal of Neurogenetics*, **26**, 43.

Zheng, Z., Lauritzen, J.S., Perlman, E., et al. (2017). A complete electron microscopy volume of the brain of adult *Drosophila melanogaster*. *bioRxiv*, 140905.

CHAPTER 4

The role of hormones

H. Frederik Nijhout and Emily Laub

Department of Biology, Duke University, Durham, NC 27708, USA

4.1 Introduction

Hormones provide a bridge between an environmental stimulus, and the behavioral and physiological states of the organism. Hormones activate response pathways by altering cell function, allowing organisms to quickly adapt to environmental changes. Although insects have fewer known hormones than vertebrates, they are capable of profound hormone-mediated behavioral diversity. In this chapter, we outline how hormones influence insect behavior with particular focus on metamorphosis, **migration**, reproduction, and social behaviors. Furthermore, we will describe the multi-functionality of insect hormones and how relatively few hormones can stimulate diverse responses in adult and juvenile insects.

4.2 Behavior and hormonally-controlled life-stage transitions

As insects metamorphose from larvae to adults both morphology and behavior can change dramatically. This is particularly true in insects with complete metamorphosis where there has been a progressive evolutionary divergence between larval and adult morphology and behavior. Insofar as the transformations during metamorphosis are controlled by hormones, it could be said that hormones also control the new behavior of the adult, at least

indirectly, by controlling the development of the nervous and muscular machinery of the adult. During metamorphosis larva-specific neural circuits degenerate and adult specific neurons arise from neuroblasts (neural stem cells). Likewise, larva-specific muscles atrophy and degenerate, and muscles used for adult behavior are generated de novo, or by hypertrophy and repurposing of larval muscles. In such cases, hormones act in a way as enablers of behavior, but it is not possible to associate any given behavior specifically with one hormone or another.

Hormones also have many direct effects and are used extensively to initiate or modulate a broad diversity of behaviors, including migration, courtship, dominance, and parental care in adults. It turns out that many of these cases happen to involve ecdysteroids and **juvenile hormone**, the same hormones that are involved in controlling moulting and metamorphosis so, although we will not deal with the indirect effects of these hormones, it will be useful to provide context by outlining briefly how moulting and metamorphosis are controlled by hormones.

In insects, as in all arthropods, moulting is controlled by ecdysteroids, primarily 20-hydroxy-ecdysone (20E). 20E is a steroid hormone that acts by binding to a nuclear receptor, which then functions as a transcriptional regulator. The **ecdysone** receptor is a heterodimer of two proteins, usually **ECR** (the receptor 'proper') **and USP** (ultraspiracle). 20E induces a stereotyped sequence of events

Nijhout, H. F., and Laub, E., *The role of hormones*. In: *Insect Behavior: From mechanisms to ecological and evolutionary consequences*. Edited by Alex Córdoba-Aguilar, Daniel González-Tokman, and Isaac González-Santoyo: Oxford University Press (2018). © Oxford University Press.
DOI: 10.1093/oso/9780198797500.003.0004

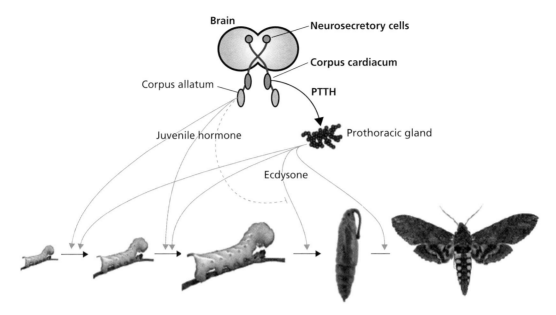

Figure 4.1 Hormonal control of moulting and metamorphosis. The brain controls the moulting cycle via the secretion of the prothoracicotropic hormone (PTTH), which is manufactured by neurosecretory cells in the brain and released from the *corpora cardiaca* into the haemolymph. PTTH then stimulates the prothoracic glands to secrete the moulting hormone, ecdysone. The type of moult is determined by the secretion of juvenile hormone (JH) from the *corpora allata*. When JH levels are high, insects moult to a larger larva. When JH disappears, insects moult to the next developmental stage: a larva moults to a pupa and a pupa moults to an adult, when stimulated by ecdysone. During the last larval instar, JH inhibits the secretion of ecdysone, preventing the moult until all JH has been metabolized. Adult insects do not moult, but the same hormones, JH and ecdysone, are secreted during the adult phase, and control a diversity of physiological functions and associated behaviors.

beginning with apolysis (separation of the epidermis from the cuticle), cell divisions in the epidermis, and the secretion of a new cuticle. At the molecular level, 20E causes a complex and extensive sequence of gene expression. However, the functions of the many genes induced by 20E are mostly unknown.

The timing of ecdysteroid secretion is controlled by the brain. Neurosecretory cells in the brain produce the **prothoracicotropic hormone** (PTTH), which is released from the *corpora cardiaca*, a pair of small neurohaemal organs behind the brain. PTTH then stimulates the prothoracic glands to secrete ecdysone (Figure 4.1). Ecdysone is a relatively inactive pro-hormone that is converted to the active 20E by the fat body and epidermis. During larval life the brain periodically secretes a large pulse of PTTH, which stimulates 20E to initiate a moult.

The prothoracic glands degenerate during metamorphosis, but in many adult insects the follicle cells of the ovaries secrete ecdysone, which controls reproductive functions such as yolk protein synthesis and egg development. Adult insects do not moult

because ecdysteroid receptors are not expressed in the epidermis.

When a larva is fully grown a pulse of ecdysone induces the metamorphic moult to the adult in the Hemimetabola or to the pupal stage in the Holometabola (Table 4.1). In the Holometabola a subsequent pulse of 20E during the pupal stage initiates the moult to the adult. The characteristics of a moult, whether it is to a larger larva or whether it is a metamorphic moult is determined by another hormone, juvenile hormone (JH), which is secreted by the *corpora allata*, a small pair of glands usually closely associated with the *corpora cardiaca*. If a moult occurs in the presence of JH the insect retains the characteristics of the current stage (a larva moults to a larger larva, and if JH is applied to a pupa it moults to another pupa; for this reason JH is often called a *status quo* hormone). A moult that occurs in the absence of JH is a so-called progressive moult, either from a larva to an adult (in Hemimetabola), or from a larva to a pupa, or a pupa to an adult (in Holometabola). The idea is that, in the presence of

Table 4.1 Hormones and life transitions.

Hormones	Larval behaviors
20-Ecdysone (20E)	Apolysis of old cuticle, moulting, wandering
Juvenile hormone (JH)	Determines if moult is to next larval instar or adult ('status quo' hormone)
Prothoracicotropic hormone (PTTH)	Stimulate prothoracic glands to secrete ecdysone

JH, ecdysone induces expression of the current gene set so the animal moults to the same form (larva to larva, or pupa to pupa), whereas in the absence of JH, genes characteristic of the current developmental stage (larva or pupa) are turned off and genes characteristic of the next developmental stage (pupa or adult) are turned on (Nijhout 1994).

Secretion of JH stops in the last larval stage and the JH level in the haemolymph gradually declines. In the last larval stage (of Holometabola at least), JH actually inhibits the secretion of PTTH and ecdysone (Figure 4.1). This is a safety mechanism that ensures that all JH has been eliminated ensuring a clean genetic switchover to the pupal stage. At the end of larval life the first pulse of ecdysone to occur in the absence of JH stimulates a behavior known as 'wandering' in many insects, during which they go in search of a suitable place in which to undergo metamorphosis (Nijhout 1994). Both JH and ecdysone are secreted again in many adult insects, JH by the corpora allata and ecdysone by the ovaries, and these hormones are used for a broad diversity of physiological and behavioral functions, as we will see later.

4.3 Polyphenisms and behavior

Many insects can metamorphose into alternative adult forms, depending on environmental signals received during larval life. This is a kind of **phenotypic plasticity** called **polyphenism**. Best known among these are the distinctive seasonal forms of butterflies, the solitary and migratory forms of locusts, the winged/apterous and sexual/parthenogenetic forms of aphids, and the various castes of social insects. The developmental switch that causes one form or another to develop at metamorphosis is controlled by changes in the temporal pattern of secretion of either ecdysteroids or JH, depending

on the species. In the butterflies *Junonia coenia* and *Araschnia levana*, for instance, there is an ecdysone-sensitive period in the early pupal stage when the seasonal form is determined and alternative forms develop by a shift in the timing of ecdysteroid secretion into or out of this sensitive period (Nijhout 1999, 2003). In the butterfly *Bicyclus anynana*, the sizes of **eyespots** on the wing differ in the wet and dry seasons and this is also controlled by ecdysone (Brakefield et al. 1998). In the ant, *Pheidole bicarinata*, there is a JH-sensitive period in the last larval instar and, if JH is present at that time, the larva will metamorphose into a soldier, whereas if JH is absent the larva develops into a worker (Nijhout and Wheeler 1982; Wheeler 1991; Evans and Wheeler 2000). Likewise, in *Apis mellifera* there is a JH-sensitive period in the middle of larval life that determines whether a larva will develop into a queen or into a worker (Nijhout and Wheeler 1982).

The alternative forms not only look very different from each other, but they also have distinctive behaviors. The dry-season forms of tropical butterflies are more sedentary than the wet-season forms—the autumn form of *Junonia coenia* migrates, whereas the summer form does not. The behavioral repertoires of the worker, soldier, and reproductive castes of social insects are very different, often resulting in a strict **division of labour** in the colony.

4.4 Hormones, receptors, and sensitive periods

The general principle revealed by studies on the hormonal control of metamorphosis and polyphenism is that hormones act during specific sensitive periods to alter developmental trajectories and patterns of gene expression.

It is believed that these are periods during which hormone receptors are expressed in a particular tissue or organ. Expression of the ecdysteroid receptor, for instance, is known to vary over time, and is different in different tissues (Nijhout, 2003).

When hormones trigger or change a behavior they do so by acting on specific receptors in the brain or abdominal ganglia. What happens next depends on the kind of hormone. Steroid hormones, such as ecdysteroids, bind to a receptor that is a transcriptional activator and thus result in a change in gene expression. JH also acts by binding to a transcriptional regulator. Some behaviors are under the regulation of neurosecretory hormones that are produced by the brain or the ganglia of the thorax or abdomen. All neurosecretory hormones are small peptides and act by binding to cell surface receptors. They then trigger an intracellular signalling pathway, which activates a variety of proteins that alter the physiological properties of the recipient cell, tissue, or organ.

Hormones only act on cells and tissues that express the right kind of receptor, and the response of a cell to the hormone depends, in turn, on the kind of signalling pathway that is activated. The fact that cells, in effect, control their own response to a hormone means that a single hormone can have many different effects, depending entirely on the time and tissue on which it acts. As will be seen later, JH is used to initiate or modify an extremely diverse array of behaviors, ranging from migration to aggression and parental care, to name but a few. This diversity of responses to a single hormone is a clear example of the fact that hormones merely serve as triggers, and that the actual behavior initiated by that trigger is a property of the cells and tissues on which the hormone acts.

4.5 Hormonally-induced behaviors associated with moulting and metamorphosis

Insects undergo a series of moulting cycles during their growth and metamorphosis. The last step in the moulting cycle is **ecdysis**, the shedding of the old cuticle. Ecdysis to the adult form is sometimes referred to as eclosion. Ecdysis often occurs at night and its timing is regulated by a biological clock (Truman 1972). The timing of ecdysis is believed to be regulated so that it occurs at a time when the insect is least subject to predation or disturbance. After ecdysis the new exoskeleton still needs to harden and the animal is incapable of significant movement and largely defenceless.

Ecdysis occurs via a stereotyped sequence of behaviors that crack the old cuticle and gradually shed it via a series of peristaltic muscle contractions of the abdomen. Ecdysis behavior is triggered by the **eclosion hormone**, a neurosecretory hormone produced by the brain (Horodyski et al. 1989; Truman 1981; Suzuki et al. 1990; Table 4.2). The entire behavior appears to be hard-wired into the abdominal ganglia of the central nervous system and only requires the eclosion hormone to trigger the entire series of behaviors. Preparations of the eclosion hormone or even crude brain homogenates, will immediately trigger eclosion behavior when injected as much as 24 hours before eclosion would normally have occurred (Truman 1992). That the behavior is encoded in the nervous system and does not require feedback from the animal was shown by dissecting out the ventral nerve cord from ready-to-eclose pupae of the American silk moth, *Hyalophora cecropia*, or the tobacco hornworm, *Manduca sexta* (Truman 1992), incubating it in a saline-containing eclosion hormone, and recording action potentials from the motor nerves leaving each ganglion (Truman 1978; Gammie and Truman 1999). The pattern of action potentials is at first synchronous in the anterior ganglia, consistent with anterior movements designed to rupture the pupal cuticle. This is followed by waves of action potentials moving posterior to anterior from ganglion to ganglion, consistent with peristaltic movements designed to escape from the pupal cuticle.

The secretion and action of eclosion hormone (EH) follows a positive feedback cascade. EH initially stimulates a population of cell associated with the peripheral tracheal system, called the **Inka cells** (Zitnan et al. 1996), to secrete a hormone called the **ecdysis-triggering hormone** (ETH). ETH, in turn, acts back on the brain neurosecretory system to stimulate the secretion of more EH (Ewer et al., 1997; Gammie and Truman 1997). This positive feedback results in a large surge of EH/ETH, which then stimulates a complex reaction in the central nervous system that initiates ecdysis behavior. With some molecular variation, EH, ETH and Inka cells have been found throughout the insects (Zitnan and Adams, 2005). Positive feedback mechanisms are commonly used to rapidly trigger a physiological response, or, as in this case, a critical behavior that must be carefully timed.

Prior to metamorphosis, many insects seek a secluded place in which to pupate. The European lime hawk moth, *Mimas tiliae*, climbs down from

Table 4.2 Hormones and metamorphosis.

Hormones	Larva behavior	Adult behavior
Eclosion hormone (EH)	Eclosion	–
Ecdysis-triggering hormone (ETH)	Stimulates brain to secrete more EH	–
Juvenile hormone (JH)	Geotaxis	Stimulates ovarian development, host-seeking behavior, and mating
Ecdysone	Silk spinning	–

the tree in which it was feeding and burrows underground to pupate. When such larvae receive implanted corpora allata they reverse this behavior and climb back up the tree (Piepho et al. 1960). This suggests that JH somehow controls geotactic behavior. When JH levels are high during the caterpillar stage the larva stays up in the tree, and when JH levels drop prior to metamorphosis the larva becomes positively geotactic and moves down the tree. This behavior is reversible, and supplementation with JH from implanted corpora allata reverses the sign of the geotaxis.

The larvae of many Lepidoptera spin a cocoon before pupation. In many species, the cocoon has distinctive inner and outer envelopes. The inner envelope, which is spun last, is smoother and more tightly spun than the outer envelope, which is spun first. When larvae of the moth *Ephestia küniella* are injected with ecdysone while they are spinning the outer envelope, they immediately switch to spinning the inner envelope (Giebultowicz et al. 1980). Under normal circumstances, the switchover to spinning the inner envelope occurs when the long pulse of ecdysone secretion that initiates the pupal moult begins.

Many insects mate soon after adult emergence, and this behavior appears to be regulated by the rising levels of JH that also initiate ovarian development (Table 4.2). In the yellow fever mosquito, *Aedes aegypti*, both host-seeking and mating behaviors depend on rising levels of JH, and can be inhibited by removing the corpora allata from freshly eclosed females (Lea 1968). In the grasshopper, *Gomphocerus rufus*, receptivity of females to courting males can be inhibited by removing the corpora allata. Indeed,

allatectomized females actively reject courting males. Normal mating receptivity is restored by injections of JH (Loher and Huber 1966).

4.6 Hormones and migration

Many insects perform some kind of migratory behavior, either as a brief dispersal flight, typically right after metamorphosis to the adult form, or as a more prolonged and long-distance flight to a new geographic region (see Chapter 7). Both types of flight are adaptations. Dispersal flights serve to move away from the region where the insect grew up, either to enhance outbreeding or to move to an area with better food resources. Long-distance migration is either seasonal and induced by stimuli such as shortening day lengths that predict the onset of an unfavourable season, or as a response to increased population density or deteriorating food resources. In either case, migration is designed to move an insect population to an area where climatic conditions and the prospect for survival are likely to be more favourable.

Migration and dispersal flights always occur during a non-reproductive period. Typically, these flights take place immediately after the metamorphic moult, as the first act in an insect's adult life. When it occurs later during adult life, as in the monarch butterfly, *Danaus plexippus*, and the greater milkweed bug, *Oncopeltus fasciatus*, migration is associated with the cessation of egg development and reproduction.

Egg development in many insects requires juvenile hormone (JH), either for yolk protein synthesis by the fat body or yolk protein uptake by the oocytes (Nijhout 1994). Because of the association of migration with a reproductive stand-still, it has been thought for a long time that migratory behavior might depend on low levels of JH in the adult. High JH levels would promote reproduction and inhibit migration. This was investigated in *Oncopeltus*, a species whose propensity to migrate can be studied in the laboratory. When an *Oncopeltus* is tethered to a flight mill (a simple counterbalanced lightweight rod balanced on a pivot), it will undertake spontaneous 'flights' that last a few seconds to a minute for animals in a reproductive non-migratory state, but can last for many minutes to hours (until fuel for flight is exhausted), for animals in a non-reproductive migratory state (Caldwell and Rankin 1972). Flight mill experiments showed that migratory flights can be stimulated by short day lengths, cool temperatures, and reduced food quality. These environmental stimuli signal the onset of winter and also cause egg development in the ovaries to stop. Measurements of JH levels in the haemolymph of adult *Oncopeltus* revealed that reproduction was associated with high levels of JH and migration with low to intermediate levels of JH. When JH was absent, metabolism declined and bugs became quiescent with undeveloped ovaries, the hallmarks of **diapause** (Rankin and Riddiford 1977, 1978). When 'migrating' bugs were treated with JH they stopped flying and began to develop eggs. Starved animals also stopped flying because they entered diapause, but when such animals received implants of corpora allata they began making long flights (Rankin 1974).

Experimental work revealed that there are two thresholds of sensitivity to JH. If JH levels are above the upper threshold animals fly little, develop eggs, and are in their reproductive phase. If JH levels are below the lower threshold, animals are in overwintering diapause. When JH levels are between the two thresholds, animals are in the migratory phase, with high flight activity, but undeveloped ovaries (Rankin and Riddiford 1978).

Studies with *Oncopeltus* show how dispersal flight and migration may be linked (Figure 4.2). Freshly eclosed females have undeveloped ovaries and low levels of JH, and this disposes them to take flight. If they encounter an adequate food source, and temperature and **photoperiod** are adequate, their JH levels rise, ovaries begin to develop, and the adults stop flying and settle to reproduce. This would be considered a dispersal flight. If the habitat deteriorates due to senescence of the host plant, or by reduced temperature and/or short photoperiods, JH levels begin to decline, ovaries regress and the adults undertake flight again. This could be a long-term migratory flight if environmental conditions remain poor. Adults will then either end up in a warmer region with improved food quality or, if not, JH levels will drop below the migration threshold and the adult will enter diapause. Then, when temperatures rise and photoperiod lengthens, JH levels begin to rise again, and dispersal flights and ovarian development begin once more.

Migration of the monarch butterfly, *Danaus plexippus*, is also associated with undeveloped ovaries and low levels of JH (Rankin et al. 1986). In the migrating grasshopper *Melanoplus sanguinipes*, JH levels rise after a long flight (Min et al. 2004), suggesting that, after a dispersal flight, females will settle into their reproductive phase. An interesting

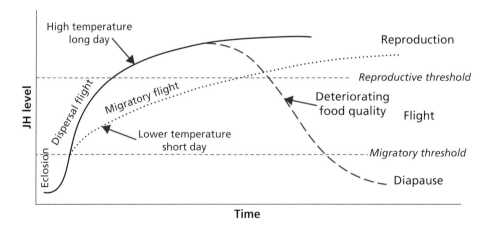

Figure 4.2 Juvenile hormone (JH) profile in the burying beetle, *Nicrophorus orbicollis*. There is a slight rise in JH after discovery of the carcass, which leads to oviposition. This is followed by a much greater increase in JH that stimulates the female to prepare the carcass and feed young larvae. Males tend to have much lower levels of JH (dashed curve) and do not participate in brood rearing, but when the female is removed, the JH level of the remaining 'solitary' male increases (dotted curve) and he takes over the brood care role during the first few days of larval development. After Panaitof et al. (2004) and Scott and Panaitof (2004).

case of hormonal control of flight was discovered in the European cockchafer beetle, *Melolontha melolontha*. This beetle switches between two habitats during its adult life. Adult beetles feed primarily in forests, but lays eggs in open fields. Migration to open fields occurs when ovaries start to develop and the reverse migration, to forest, occurs after the eggs are laid and the beetles begin to feed again. Directionality of migration can be studied in enclosures in the field, where beetles fly in the direction of dark forest or bright fields, depending on their reproductive state. Implantation of *corpora allata*, the glands that produce JH, into feeding beetles causes them to switch their direction of flight from forest to field (Stengel 1974). A similar hormonally controlled change in directionality of locomotion has been described in caterpillars of *Mimas tilieae* (see section on hormones associated with moulting and metamorphosis above).

4.7 Hormonal control of pheromone production and mating activity

In addition to undertaking a dispersal flight, one of the first acts in the life of adult insects is to mate. In some insects, female receptivity to mating is controlled by the rising level of JH soon after eclosion, and allatectomy can abolish a female's receptivity to courtship and mating (Truman and Riddiford 1974; Table 4.3).

In many insects, females attract males by means of a species-specific chemical **sex pheromone** (see Chapter 10). Synthesis and release of sex pheromones is modulated in many species by JH. In the roaches *Byrsotria fumigata* and *Pycnoscelus surinamensis*, and in the beetle *Tenebrio molitor*, allatectomy inhibits pheromone secretion, and re-implantation of corpora allata or treatment with JH re-establishes pheromone production (Barth 1965; Menon, 1970). In some Lepidoptera, by contrast, a neurosecretory hormone from the brain or the sub-oesophageal ganglion is necessary for normal pheromone production. There is actually a diverse family of hormones with this function, collectively called the pheromone biosynthesis-activating neuropeptides (PBANs; Kitamura et al. 1989; Raina et al. 1989). Whether PBANs act in conjunction with JH or

Table 4.3 Hormones and mating behaviors.

Hormone	Adult Behavior
Juvenile hormone (JH)	Pheromone production, egg production
Pheromone biosynthesis activating neuropeptides (PBANs)	Pheromone production,
Matrone	Refractory behavior, host seeking, and feeding,
Sex peptide	Refractory behavior, increase JH, terminate pheromone production

independently is not known. PBANs appear to act directly on the pheromone gland, but interestingly, can only stimulate pheromone synthesis at night, at the time pheromones would normally be produced. Injections of PBAN at any other time are without effect (Christensen et al. 1991). Therefore, it appears that the pheromone gland only expresses hormone receptors, or critical elements of the hormone response pathway, at a particular species-specific time of night.

The females of many insects are **semelparous**, meaning that they only mate once in their life, usually shortly after emerging as adults, and are refractory to any further mating after that. Refractory behavior typically consists of contracting the terminal abdominal segments to make the genitalia unavailable or by actively repelling courting males. In mosquitoes, **monogamy** is induced by a peptide called matrone, secreted by the male's accessory glands and transferred to the female in the seminal fluid (Craig 1967). Matrone appears to act directly in the terminal abdominal ganglion to induce refractory behavior. Although not technically a hormone, matrone in effect acts like a hormone due to its influence on physiology and behavior, and because injections of matrone have the same effect as matrone passed by **insemination**. A molecule with similar effects on female reproductive behavior has been called the sex peptide in *Drosophila* (Kubli 1992). In mosquitoes, injections of matrone also increase host-seeking and biting behavior (Judson 1967). Sex peptide transmitted to female Lepidoptera during mating stimulates JH production and halts pheromone production (Fan et al. 1999), thus increasing egg development and inhibiting further mating.

4.8 Hormones and parental care

Although insect reproductive strategies are generally of low investment in many eggs, many insects produce relatively few offspring and provide parental care (see Chapter 14). Parental care includes provisioning, guarding, and nest building, with larvae initiating interactions with parents through begging. The challenges of raising offspring demand significant behavioral plasticity that is typically mediated by hormones. However, insects use fewer known hormones compared with vertebrates that provide similar care. JH is the primary hormone that modulates parental care in insects, although its role is not consistent across insect species. Parental care has been most extensively documented in burying beetles, (*Nicrophorus sp.*, but it is widespread in most insect orders (see Chapter 14). JH is involved in the control of breeding-pair synchronization, offspring defence, food provisioning, and

larval begging. This broad diversity of JH functions illustrate once more how this one hormone may be co-opted for social, rather than metamorphic or gonadotropic roles. Although parental care in other species will be described, this section begins by focusing on the behaviors of burying beetles of the genus *Nicrophorus*.

Family life for the burying beetle begins by locating a carcass to use as a breeding resource (Scott 1998). The JH titre of female beetles rises steadily as their ovaries develop, although JH will plateau at a low level until a carcass is discovered, after which it rises dramatically (Scott and Panaitof 2004, Trumbo et al., 1995). In females the JH titer can rise over 100 per cent within ten minutes of carcass discovery (Figure 4.3). Burying beetles form mated pairs and, like the female, the male's JH titer also rises with discovery and manipulation of the carcass (Scott and Panaitof, 2004; Trumbo and Robinson, 2008). Following the initial carcass discovery, the JH titres of males and females decline as they bury the carcass and transform it into a suitable nursery (Trumbo et al. 1995; Scott 1998).

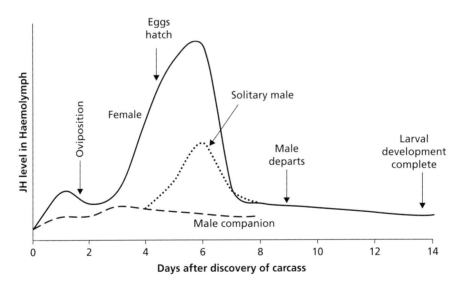

Figure 4.3 Juvenile hormone control of migration, reproduction, and diapause in the milkweed bug, *Oncopeltus fasciatus*. There are two threshold levels of juvenile hormone (JH). When JH is below the lower threshold ovarian development stops and the adult goes into diapause. If JH is between the two thresholds, females are stimulated to take long flights and the ovaries remain undeveloped. If JH is above the higher threshold, flight is inhibited, ovarian development is rapid, and females remain sedentary. Under ideal conditions, JH levels rise rapidly after metamorphosis (solid line), stimulating a brief dispersal flight and leading to maturation of the ovaries and reproduction. Under cool temperatures and short-day conditions JH levels rise more slowly (dotted line), stimulating a prolonged migratory flight. If the environment deteriorates, JH levels begin to decline (dashed line), which stimulates flight again, either to an improved environment or to a diapause site.
After Rankin and Riddiford (1978).

The breeding carcass is a valuable commodity that mated pairs of burying beetles defend from infanticidal interlopers (Trumbo 1997). JH titres rise with competition from conspecifics and increases aggression towards rivals (Scott 2006b). Prior to oviposition, a female's JH titres rises when confronted with an intruder of the same sex, although after oviposition, JH titres rise irrespective of the interloper's sex. Experimental application of **methoprene**, a JH analog, to pairs of burying beetles without a resource to defend increases their aggressiveness towards other beetles (Scott 2006b). Mate recognition also contributes to joint nest defence. Both members of the mated pair emit **(E)-methylgeranate** (MG) during the high JH period of parental care. The amount of emitted MG, a compound structurally similar to JH, is correlated with levels of JH, indicating that it serves as an honest signal of reproductive state. Non-reproductive interlopers treated with MG are tolerated by reproductive beetles (Haberer et al. 2010).

In addition to defending larvae, burying beetles provision their offspring through masticating flesh from the carcass and passing it to the young. Although both parents provision their offspring, female burying beetles provide more care, and stay with the larvae longer than males (Fetherston et al. 1990; Scott 1998). During the first and second larval instars, maternal JH titres remain high, but decline after the third instar (Trumbo et al., 1995; Panaitof et al. 2004). Paired male JH titres are significantly lower than their mates during the first two larval instars. However, when their mate is removed, male JH titres rise to that of the female (Figure 4.3), and they will provide equivalent care to the larvae (Fetherston et al. 1990; Panaitof et al. 2004,). Larvae are instrumental in stimulating their parent's endocrine profiles as shown by the fact that both female and male JH titres remain high when their brood is consistently replaced by first instar larvae (Scott and Panaitof 2004; Trumbo

Table 4.4 Hormones and parental care.

Hormone	Adult behavior	Larva behavior
Juvenile hormone (JH)	Breeding pair synchronization, resource defence, E-methylgerenate secretion, competition, food provisioning, egg dumping	Begging

and Robinson 2008). Larvae solicit feeding from parents through begging by touching the mandibles of the adult with their forelegs and larvae treated with methoprene spend significantly more time begging than controls, suggesting that JH stimulates begging behavior in larval burying beetles (Crook et al. 2008).

Other insect species that provide parental care do not have the same relationship to JH as burying beetles, prompting questions of JH multi-functionality. Earwig females guard the eggs and provision young, but high levels of JH will terminate care (Rankin et al. 1997). Lace bugs guard their eggs or dump them with a conspecific; egg dumping is correlated with high levels of JH, guarding is correlated with low levels of JH (Tallamy et al. 2002). Why is parental care halted by JH in some species, but extended in burying beetles? One hypothesis is that JH titres are uncoupled from mating and vitellogenesis in burying beetles because their reproductive cycle is dependent on a carcass, a limited and unpredictable resource. Although Scott (2006b) argues that JH directly impacts social aggression, similarly to testosterone, Trumbo and Robinson (2008) argue that elevated JH titres during social conflicts and parental care could be due to the increased metabolic demands of these activities, which indirectly elevate JH titres. While JH may directly or indirectly influence parental behavior, it is clear that insect endocrine profiles respond to complex social and non-social cues.

4.9 Dominance and social behavior

Eusocial hymenopteran colonies are characterized by a reproductive dominance hierarchy and division of labour (Robinson and Vargo 1997). Wasps and bumblebees establish reproductive dominance through aggression, where subordinates change their endocrine profile and lower their reproductive capacity (Penick et al. 2014). Honey bees and ants, by contrast, irreversibly assign reproductive and worker castes during larval and pupal development (Nijhout, 1994; Hartfelder and Emlen, 2012). Although there are no morphological castes among worker honey bees, they progress through a sequence of temporal behavioral castes during their lifetime moving from nursing and nest maintenance, to guard and forager duties. The transition between these behavioral tasks is controlled by JH, which increases as workers age (Robinson 1987; Huang and Robinson 1992; Table 4.5). Interestingly, although JH is the primary gonadotropic hormone in most insect species (Nijhout 1994), JH serves as a 'behavioral pace-maker' in worker honey bees and has lost its reproductive function. JH is fully uncoupled from reproduction; queens and egg-laying workers have low JH titres, and experimental application of methoprene will retard ovary development (Robinson et al. 1992). Furthermore, moderating colony task allocation with JH allows hives to respond plastically to environmental demands (Huang and Robinson, 1992). The number of workers performing different tasks is modulated by physical contact with workers of different ages. Without contact with young bees, foragers regress, reduce JH titres and perform nurse tasks (Huang and Robinson 1996). Young bees are prevented from transitioning to foraging by an inhibitory signal secreted by the mandibular glands of older bees (Huang et al. 1998). Although the rise in JH initiates the transition to foraging and guard behaviors, it does not appear to control foraging behavior directly, because removal of the corpora allata does not inhibit foraging behavior, but does delay its onset (Sullivan et al. 2000).

The transition of JH to a social, rather than reproductive role is of great evolutionary interest; two competing hypotheses have emerged to describe its origin (Robinson et al. 1992; West-Eberhard 1996; Robinson and Vargo 1997; Giray et al. 2005). The 'novel-function' hypothesis, proposed by Robinson et al. (1992), suggests that JH would only have adopted a behavioral pacemaker role after losing reproductive function. However, West-Eberhard (1996) proposes the 'split-function' hypothesis, purporting that, in ancestral species, JH would serve reproductive and social roles, depending on nutritional and ovary status. Recent studies with primitively eusocial species have not provided definitive support for either hypothesis. Foragers in the primitive ants *Harpegnathos saltator* have elevated JH titres, while reproductive gamergates do not, and application of JH fails to induce egg-laying, suggesting that JH loses its reproductive function to moderate foraging (Penick et al., 2014). However, Giray et al. (2005) find that JH influences both worker foraging

and queen ovarian development in primitive wasps, supporting the 'split-function' hypothesis.

In place of the temporal castes organization of honey bee hives, wasps and bumblebees establish dominance hierarchies through aggressive contests (Röseler et al. 1985; Brent et al. 2006; Tibbetts and Izzo 2009). The hormonal control of dominance and aggression are best studied in *Polistes* wasps and *Bombus terrestris* bumblebees (Röseler et al., 1985; Bloch et al. 2000b). *Polistes* wasps form colonies in the spring, where foundresses form solitary nests, or join a nest with conspecifics and compete for dominance. The most dominant individual becomes the queen and has reproductive priority, and the workers form a linear dominance hierarchy (Wilson 1971; Tibbetts and Izzo 2009). Wasps with the largest corpora allata and ovaries are most likely to become dominant, and high titres of JH and ecdysteroids are associated with dominance (Röseler et al. 1984, 1985). High JH titres are also associated with visual and chemical signals of dominance and reproductive quality (Tibbetts and Shorter 2009). *Bombus terrestris* colonies, by contrast, are founded by a single queen (Robinson and Vargo 1997). As the queen ages, workers begin to develop functional ovaries and lay eggs, ultimately expelling the queen from the hive. The workers then enter the 'competition phase', where they battle for dominance (Doorn 1989). Dominant bumblebees have higher JH titres, ecdysteroid titres, and ovary development than subordinates (Bloch and Hefetz 1999; Bloch et al. 2000a,b). Application of **precocene**, a JH inhibitor, decreases aggression and the status of dominant bumblebees (Amsalem et al., 2014).

Although there are strong correlations between aggression, dominance, and JH titres, the multifunctionality of JH obscures its role in social conflict. It is hypothesized that JH modulates social aggression similarly to testosterone in vertebrates (Scott 2006a; Tibbetts and Crocker 2014). The 'challenge hypothesis', that JH titres rise during periods of social instability and determine competitive ability, is supported across many insect taxa. JH titres increase during queen loss in both bumblebees and wasps, and the JH titres of burying beetles will elevate when confronted by a rival of the same sex (Scott 2006b; Tibbetts and Huang 2010). Male cockroaches fight for territory, and high JH titres increase the

Table 4.5 Hormones and social behavior.

Hormone	Adult behavior
Juvenile hormone (JH)	Temporal caste transition, dominance, contest ability
Ecdysteroids	Dominance

probability of victory (Kou et al. 2009). Furthermore, JH titres remain elevated in cockroach contest winners, resembling the elevated levels of testosterone experienced by vertebrate victors. Despite these compelling examples, elevated JH titres during aggressive conflict are not consistently observed in insects with a dominance hierarchy (Brent et al. 2006; Kelstrup et al., 2014).

4.10 Conclusions

Insects display astonishing morphological and behavioral diversity. Despite having small brains, they integrate dynamic sensory input, communicate, cooperate, and compete. While utilizing fewer known hormones than vertebrates, the role of insect hormones in behavior remains an intriguing and much understudied area of research. Directions for further studies could include the evolution of hormone signalling pathways, the roles in behavior of hormone-like neurotransmitters like serotonin, dopamine, and octopamine, the mechanisms by which one hormone can have multiple functions in different contexts and the role of hormones in learning. Many examples of hormones in behavior have been studied only in a single species, and we usually do not know how widespread the phenomenon is—comparative studies across genera and families of insects are almost entirely lacking, but could yield novel and important insights. In addition to these questions, understanding the roles of insect hormones in behavior and life history is vital to developing pest management, understanding pollination biology, and preventing vector behavior associated with disease transmission.

References

Amsalem, E., Teal, P., Grozinger, C.M., and Hefetz, A. (2014). Precocene-I inhibits juvenile hormone biosynthesis, ovarian activation, aggression and alters sterility signal production in bumble bee (Bombus terrestris) workers. *Journal of Experimental Biology*, **217**, 3178.

Barth, R.H. (1965). Insect mating behavior: endocrine control of a chemical communication system. *Science*, **149**, 882.

Bloch, G., Borst, D.W., Huang, Z-Y., et al. (2000a). Juvenile hormone titers, juvenile hormone biosynthesis, ovarian development and social environment in *Bombus terrestris*. *Journal of Insect Physiology*, **46**, 47.

Bloch, G. and Hefetz, A. (1999). Regulation of reproduction by dominant workers in bumblebee (*Bombus terrestris*) queenright colonies. *Behavioral Ecology and Sociobiology*, **45**, 125.

Bloch, G., Hefetz, A., and Hartfelder, K. (2000b). Ecdysteroid titer, ovary status, and dominance in adult worker and queen bumble bees (*Bombus terrestris*). *Journal of Insect Physiology*, **46**, 1033.

Brakefield, P., Kesbeke, F., and Koch, P. (1998). The regulation of phenotypic plasticity of eyespots in the butterfly *Bicyclus anynana*. *American Naturalist*, **152**, 853.

Brent, C.S., Peeters, C., Dietmann, V., Crewe, R., and Vargo, E.L. (2006). Hormonal correlates of reproductive status in the queenless Ponerine ant, *Streblognathus peetersi*. *Journal of Comparative Physiology A*, **192**, 315.

Caldwell, R.L. and Rankin, M.A. (1972). Effects of a juvenile hormone mimic on flight in the milkweed bug, *Oncopeltus fasciatus*. *General and Comparative Endocrinology*, **19**, 601.

Christensen, T.A., Itagaki, H., Teal, P.E., et al. (1991). Innervation and neural regulation of the sex pheromone gland in female *Heliothis* moths. *Proceedings of the National Academy of Sciences of the United States of America*, **88**, 4971.

Craig, G.B. (1967). Mosquitoes: female monogamy induced by male accessory gland substance. *Science*, **156**, 1499.

Crook, T.C., Flatt, T., and Smiseth, P.T. (2008). Hormonal modulation of larval begging and growth in the burying beetle *Nicrophorus vespilloides*. *Animal Behaviour*, **75**, 71.

Doorn, A.V. (1989). Factors influencing dominance behaviour in queenless bumblebee workers (*Bombus terrestris*). *Physiological Entomology*, **14**, 211.

Evans, J. and Wheeler, D. (2000). Expression profiles during honeybee caste determination. *Genome Biology*, **2**, 1–6.

Ewer, J., Gammie, S.C., and Truman, J.W. (1997). Control of insect ecdysis by a positive-feedback endocrine system: roles of eclosion hormone and ecdysis triggering hormone. *Journal of Experimental Biology*, **200**, 869.

Fan, Y., Rafaeli, A., Gileadi, C., Kubli, E., and Applebaum, S.W. (1999). *Drosophila melanogaster* sex peptide stimulates juvenile hormone synthesis and depresses sex pheromone production in *Helicoverpa armigera*. *Journal of Insect Physiology*, **45**, 127.

Fetherston, I.A., Scott, M.P., and Traniello, J.F.A. (1990). Parental care in burying beetles: the organization of male and female brood-care behavior. *Ethology*, **85**, 177.

Gammie, S.C., and Truman, J.W. (1997). Neuropeptide hierarchies and the activation of sequential motor behaviors in the hawkmoth, *Manduca sexta*. *Journal of Neuroscience*, **17**, 4389.

Gammie, S.C., and Truman, J.W. (1999). Eclosion hormone provides a link between ecdysis-triggering hormone and crustacean cardioactive peptide in the neuroendocrine cascade that controls ecdysis behavior. *Journal of Experimental Biology*, **202**, 343.

Giebultowicz, J.M., Zdarek, J., and Chróścikowska, U. (1980). Cocoon spinning behaviour in *Ephestia kuehniella*; correlation with endocrine events. *Journal of Insect Physiology*, **26**, 459.

Giray, T., Giovanetti, M., and West-Eberhard, M.J. (2005). Juvenile hormone, reproduction, and worker behavior in the neotropical social wasp *Polistes canadensis*. *Proceedings of the National Academy of Sciences of the United States of America*, **102**, 3330.

Haberer, W., Steiger, S., and Muller, J.K. (2010). (E)-methylgeranate, a chemical signal of juvenile hormone titre and its role in the partner recognition system of burying beetles. *Animal Behaviour*, **79**, 17.

Hartfelder, K., and Emlen, D.J. (2012). Endocrine control of polyphenism. In: L. I. Gilbert (Ed.) *Insect Endocrinology*, pp. 464–522. Academic Press, London.

Horodyski, F.M., Riddiford, L.M., and Truman, J.W. (1989). Isolation and expression of the eclosion hormone gene from the tobacco hornworm, *Manduca sexta*. *Proceedings of the National Academy of Sciences of the United States of America*, **86**, 8123.

Huang, Z.Y., Plettner, E., and Robinson, G.E. (1998). Effects of social environment and worker mandibular glands on endocrine-mediated behavioral development in honey bees. *Journal of Comparative Physiology A*, **183**, 143.

Huang, Z.Y., and Robinson, G.E. (1992). Honeybee colony integration: worker-worker interactions mediate hormonally regulated plasticity in division of labor. *Proceedings of the National Academy of Sciences of the United States of America*, **89**, 11726.

Huang, Z-Y., and Robinson, G.E. (1996). Regulation of honey bee division of labor by colony age demography. *Behavioral Ecology and Sociobiology*, **39**, 147.

Judson, C.L. (1967). Feeding and oviposition behavior in the mosquito *Aedes aegypti* (L.). I. Preliminary studies of physiological control mechanisms. *Biological Bulletin*, **133**, 369.

Kelstrup, H.C., Hartfelder, K., Nascimento, F.S., and Riddiford, L.M. (2014). Reproductive status, endocrine physiology and chemical signaling in the neotropical, swarm-founding eusocial wasp *Polybia micans*. *Journal of Experimental Biology*, **217**, 2399.

Kitamura, A., Nagasawa, H., Kataoka, H., et al. (1989). Amino acid sequence of pheromone-biosynthesis-activating neuropeptide (PBAN) of the silkworm, *Bombyx mori*. *Biochemical and Biophysical Research Communications*, **163**, 520.

Kou, R., Chou, S-Y., Chen, S-C., and Huang, Z.Y. (2009). Juvenile hormone and the ontogeny of cockroach aggression. *Hormones and Behavior*, **56**, 332.

Kubli, E. (1992). My favorite molecule. The sex-peptide. *BioEssays*, **14**, 779.

Lea, A.O. (1968). Mating without insemination in virgin *Aedes aegypti*. *Journal of Insect Physiology*, **14**, 305.

Loher, W., and Huber, F. (1966). Nervous and endocrine control of sexual behavior in a grasshopper (*Gomphocerus rufus* L., Acridinae). *Symposia of the Society for Experimental Biology*, **20**, 381.

Menon, M. (1970). Hormone–pheromone relationships in the beetle, *Tenebrio molitor*. *Journal of Insect Physiology*, **16**, 1123.

Min, K.J., Jones, N., Borst, D.W., and Rankin, M.A. (2004). Increased juvenile hormone levels after long-duration flight in the grasshopper, *Melanoplus sanguinipes*. *Journal of Insect Physiology*, **50**, 531.

Nijhout, H.F. (1994). *Insect Hormones*. Princeton University Press, Princeton, NJ.

Nijhout, H.F. (1999). Control mechanisms of polyphenic development in insects. *BioScience*, **49**, 181.

Nijhout, H.F. (2003). Development and evolution of adaptive polyphenisms. *Evolution and Development*, **5**, 9.

Nijhout, H.F. and Wheeler, D.E. (1982). Juvenile hormone and the physiological basis of insect polymorphisms. *Quarterly Review of Biology*, **57**, 109–33.

Panaitof, S.C., Scott, M.P., and Borst, D.W. (2004). Plasticity in juvenile hormone in male burying beetles during breeding: physiological consequences of the loss of a mate. *Journal of Insect Physiology*, **50**, 715.

Penick, C.A., Brent, C.S., Dolezal, K., and Liebig, J. (2014). Neurohormonal changes associated with ritualized combat and the formation of a reproductive hierarchy in the ant *Harpegnathos saltator*. *Journal of Experimental Biology*, **217**, 1496.

Piepho, H., Böden, E., and Holz, I. (1960). Über die Hormonabhängigkeit des Verhaltens von Schwärmerraupen vor den Häutungen. *Zeitschrift für Tierpsychologie*, **17**, 261.

Raina, A.K., Jaffe, H., Kempe, T.G., *et al.* (1989). Identification of a neuropeptide hormone that regulates sex pheromone production in female moths. *Science*, **244**, 796.

Rankin, M.A. (1974). The hormonal control of flight in the milkweed bug, *Oncopeltus fasciatus*. *In:* L. Barton Browne (Ed.) *Experimental Analysis of Insect Behaviour*, pp. 317–28. Springer, Berlin.

Rankin, M.A., Mcanell, M.L., and Bodenhamer, J.E. (1986). The oogenesis-flight syndrome revisited. In: W. Danthanarayana (Ed.) *Insect Flight: Dispersal and Migration*, pp. 27–48. Springer-Verlag, Berlin.

Rankin, M.A., and Riddiford, L.M. (1977). Hormonal control of migratory flight in *Oncopeltus fasciatus*: The effects of the corpus cardiacum, corpus allatum, and starvation on migration and reproduction. *General and Comparative Endocrinology*, **33**, 309.

Rankin, M.A. and Riddiford, L.M. (1978). Significance of haemolymph juvenile hormone titer changes in timing of migration and reproduction in adult *Oncopeltus fasciatus*. *Journal of Insect Physiology*, **24**, 31.

Rankin, S.M., Chambers, J., and Edwards, J.P. (1997). Juvenile hormone in earwigs: Roles in oogenesis, mating, and maternal behaviors. *Archives of Insect Biochemistry and Physiology*, **35**, 427.

Robinson, G.E. (1987). Regulation of honey bee age polyethism by juvenile hormone. *Behavioral Ecology and Sociobiology*, **20**, 329.

Robinson, G.E., Strambi, C., Strambi, A., and Huang, Z-Y. (1992). Reproduction in worker honey bees is associated with low juvenile hormone titers and rates of biosynthesis. *General and Comparative Endocrinology*, **87**, 471.

Robinson, G.E., and Vargo, E.L. (1997). Juvenile hormone in adult eusocial hymenoptera: gonadotropin and behavioral pacemaker. *Archives of Insect Biochemistry and Physiology*, **35**, 559.

Röseler P-F., Röseler I., and Strambi, A. (1985). Role of ovaries and ecdysteroids in dominance hierarchy establishment among foundresses of the primitively social wasp, *Polistes gallicus*. *Behavioral Ecology and Sociobiology*, **18**, 9.

Röseler, P-F., Röseler, I., Strambi, A., and Augier, R. (1984). Influence of insect hormones on the establishment of dominance hierarchies among foundresses of the paper wasp, *Polistes gallicus*. *Behavioral Ecology and Sociobiology*, **15**, 133.

Scott, M.P. (1998). The ecology and behavior of burying beetles. *Annual Review of Entomology*, **43**, 595.

Scott, M.P. (2006a). Resource defense and juvenile hormone: the 'challenge hypothesis' extended to insects. *Hormones and Behavior*, **49**, 276.

Scott, M.P. (2006b). The role of juvenile hormone in competition. and cooperation by burying beetles. *Journal of Insect Physiology*, **52**, 1005.

Scott, M.P., and Panaitof, S.C. (2004). Social stimuli affect juvenile hormone during breeding in biparental burying beetles (Silphidae: *Nicrophorus*). *Hormones and Behavior*, **45**, 159.

Stengel, M.M.C. (1974). Migratory behaviour of the female of the common cockchafer *Melolontha melolontha* L, and its neuroendocrine regulation. In: L. Barton Browne (Ed.) *Experimental Analysis of Insect Behaviour*, pp. 297–303. Springer, Berlin.

Sullivan, J.P., Jassim, O., Fahrbach, S.E., and Robinson, G.E. (2000). Juvenile hormone paces behavioral development in the adult worker honey bee. *Hormones and Behavior*, **37**, 1.

Suzuki, A., Nagasawa, H., Kono, T., et al. (1990). *Bombyx* eclosion hormone. In: A. B. Borkovec and E. P. Masler (Eds) *Insect Neurochemistry and Neurophysiology—1989*, pp. 211–14. Humana Press, Totowa, NJ.

Tallamy, D.W., Monaco, E.L., and Pesek, J.D. (2002). Hormonal control of egg dumping and guarding in the lace bug, *Gargaphia solani* (Hemiptera: Tingidae). *Journal of Insect Behavior*, **15**, 467.

Tibbetts, E.A., and Crocker, K.C. (2014). The challenge hypothesis across taxa: social modulation of hormone titres in vertebrates and insects. *Animal Behaviour*, **92**, 281.

Tibbetts, E.A. and Huang, Z.Y. (2010). The challenge hypothesis in an insect: juvenile hormone increases during reproductive conflict following queen loss in *Polistes* wasps. *American Naturalist*, **176**, 123.

Tibbetts, E.A., and Izzo, A.S. (2009). Endocrine mediated phenotypic plasticity: condition-dependent effects of juvenile hormone on dominance and fertility of wasp queens. *Hormones and Behavior*, **56**, 527.

Tibbetts, E.A., and Shorter, J.R. (2009). How do fighting ability and nest value influence usurpation contests in *Polistes* wasps? *Behavioral Ecology and Sociobiology*, **63**, 1377.

Truman, J.W. (1972). Physiology of insect rhythms. I. Circadian organization of the endocrine events underlying the moulting cycle of larval tobacco hornworms. *Journal of Experimental Biology*, **57**, 805.

Truman, J.W. (1978). Hormonal release of stereotyped motor programmes from the isolated nervous system of the Cecropia silkmoth. *Journal of Experimental Biology*, **74**, 151.

Truman, J.W. (1981). Interaction between ecdysteroid, eclosion hormone, and bursicon titers in *Manduca sexta*. *American Zoologist*, **21**, 655.

Truman, J.W. (1992). The eclosion hormone system of insects. In: *Progress in Brain Research*, pp. 361–74. Elsevier, Amsterdam.

Truman, J.W., and Riddiford, L.M. (1974). Hormonal mechanisms underlying insect behaviour. In: J. E. Treherne, M. J. Berridge, and V. B. Wigglesworth (Eds) *Advances in Insect Physiology*, pp. 297–352. Academic Press, Cambridge, CA.

Trumbo, S.T. (1997). Juvenile hormone-mediated reproduction in burying beetles: from behavior to physiology. *Archives of Insect Biochemistry and Physiology*, **35**, 479.

Trumbo, S.T., Borst, D.W., and Robinson, G.E. (1995). Rapid elevation of juvenile hormone titer during behavioral assessment of the breeding resource by the burying beetle, *Nicrophorus orbicollis*. *Journal of Insect Physiology*, **41**, 535.

Trumbo, S.T., and Robinson, G.E. (2008). Social and nonsocial stimuli and juvenile hormone titer in a male burying beetle, *Nicrophorus orbicollis*. *Journal of Insect Physiology*, **54**, 630.

West-Eberhard, M.J. (1996). Wasp societies as microcosms for the study of development and evolution. In: S. Turillazzi and M. J. West-Eberhard (Eds) *Natural History and Evolution of Paper Wasps*, pp. 291–317. Oxford University Press, New York.

Wheeler, D.E. (1991). The developmental basis of worker caste polymorphism in ants. *American Naturalist*, **138**, 1218.

Wilson, E. (1971). *The Insect Societies*. Harvard University Press, Cambridge, MA.

Zitnan, D., and Adams, M.E. (2005). Neuroendocrine regulation of insect ecdysis. In: L.I. Gilbert, K. Iatrou, and S.S. Gill (Eds) *Comprehensive Molecular Insect Science*. Elsevier, Boston.

Zitnan, D., Kingan, T.G., Hermesman, J.L., and Adams, M.E. (1996). Identification of ecdysis-triggering hormone from an epitracheal endocrine system. *Science*, **271**, 88.

CHAPTER 5

Phenotypic plasticity

Karen D. Williams and Marla B. Sokolowski

Department of Ecology and Evolutionary Biology, University of Toronto, Canada

5.1 Introduction

5.1.1 Phenotypic plasticity, genes, and environments, and their interaction

This chapter addresses sources of variability in insect behavior with a particular focus on phenotypic plasticity (defined later). A number of contexts (social, nutritional and seasonal) were chosen to explore how environmental variation can give rise to phenotypic plasticity. The examples are limited to insect behavior, but it is worth mentioning that many of the concepts discussed in this chapter are applicable to all organisms including humans. The intent is not to perform an exhaustive review, but rather to use selected illustrative examples from the literature. For completeness, the topic of genetic contributions to behavioral variability is briefly discussed, which is covered more thoroughly in Chapter 2. Discussions of phenotypic plasticity, learning, and memory can be found in Chapters 3 and 17. Smid and Vet (2016) also provided a review of insect learning and memory, and discussed the considerable natural variation between species and within populations in these plastic traits. The extensive literature on the evolution of plasticity is not discussed in this chapter, but readers are referred to Chapter 19 and to several informative reviews (Via et al. 1995; Dewitt et al. 1998; West-Eberhard 2003; Murren et al. 2015).

5.1.2 Sources of behavioral variability

Sources of individual differences in behavior can arise from variation in genes, the environment, or the interactions of these (Figure 5.1). Variation attributable to genes (the DNA sequence) is also known as **heritable variation** and its measurement can inform responses to artificial and natural selection (Falconer and Mackay 1996). An organism's response to the environment measured within the lifetime of the individual is called phenotypic plasticity (Robinson et al. 2008) and can also be referred to as a norm of reaction (Gupta and Lewontin 1982), behavioral flexibility or modifiability. From a geneticist's perspective, when a group of genetically homogenous organisms behave differently in two or more environments, they are said to exhibit phenotypic plasticity (Figures 5.1 and 5.2a,b). This is phenotypic plasticity narrowly defined. However, in the ecological and physiological literature, the term phenotypic plasticity is often used more loosely, for example, as when an organism (or population of organisms) changes its behavior in response to a change in the environment (West-Eberhard 2003).

Williams, K. D. and Sokolowski, M. B., *Phenotypic plasticity*. In: *Insect Behavior: From mechanisms to ecological and evolutionary consequences.* Edited by Alex Córdoba-Aguilar, Daniel González-Tokman, and Isaac González-Santoyo: Oxford University Press (2018). © Oxford University Press.
DOI: 10.1093/oso/9780198797500.003.0005

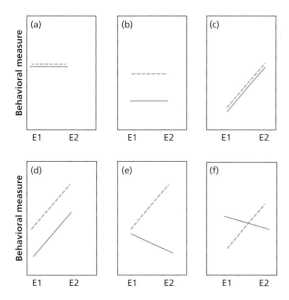

Figure 5.1 Conceptual models of gene, environment and gene-by-environment interactions using two environments (E1 and E2) and two genotypes (G1 solid line and G2 dashed line) in a two-way analysis of variance (ANOVA) framework. (a) In both environments (E1 and E2) the two genotypes exhibit similar behaviors and there is no statistically significant effect of genotype (G = NS), or environment (E = NS) or gene-by-environment interaction (GxE = NS) in the two-way analysis of variance (ANOVA). (NS, non-significant). (b) The genetic variants differ in the level of their response to the two environments, thus the ANOVA G = significant, E = NS, GxE = NS. (c) The response of the genetic variants to the environments does not differ: ANOVA shows G = NS, E = significant, GxE = NS. (d) The genetic variants differ in their level of response to the environment, although they both have the same pattern of response. G = significant, E = significant, GxE = NS. (e). The two genetic variants respond to one environment (E1) in the same way, but differ in their responses to the other environment (E2), thus G = significant, E = significant, GxE = significant. (f). The genetic variants respond to the different environments in different ways, G = significant, E = significant, GxE = significant. If the crossing of the lines is a mirror image, then the output of a two-way ANOVA, will yield significant GxE, while G and E will not be significant. However, in general, significant GxE does not preclude the significance of G or E. Figure 5.1 redrawn from Sokolowski & Levine (2010).

When differences in phenotypic plasticity are exhibited between two genetically different groups, this is a case of gene-by-environment interaction (GxE) where organisms with different genotypes show different behavioral responses to two or more environments (Sokolowski and Wahlsten 2001; Anholt and Mackay 2004; Sokolowski and Levine 2010; Figure 5.1e,f). At the cellular level of organization,

plasticity is usually reflected in changes in gene expression (Robinson et al. 2008). A more recent term, gene–environment interplay describes reciprocal feedback between an organism's genome and its environment (Boyce et al. 2012). In this case, one can use the analogy that the genome or genes are 'listening' to the environment and responding to it via changes in gene expression. gene–environment interplay explores the realm of **epigenetics** also discussed in this chapter.

Changes in behavior can occur on short time scales, within the lifetime of the individual, as well as over long evolutionary time scales, as for example, during artificial and natural selection involving multiple generations (Robinson et al. 2008).

5.2 Polyphenisms

In general, the term 'polyphenism' refers to two or more distinct forms produced in response to alterations in the environment (Nijhout 2003). A more narrow definition of polyphenism refers to the case wherein a single genome gives rise to multiple phenotypes (Nijhout 2003). Insect morphological polyphenisms are well known (West-Eberhard 1989; Via et al. 1995; Nijhout 2003). For example, wing polymorphisms in aphids (Ogawa and Miura 2014), crickets and locusts (Simpson et al. 2011; Wang and Kang 2014) are associated with polyphenisms in dispersal and migratory behavior. A polyphenic trait is one where multiple, discrete phenotypes can arise from a single genotype as a result of varying environmental conditions (Nijhout 2003). Polyphenisms are, therefore, a special case of phenotypic plasticity.

5.2.1 Social context and polyphenisms

Social insects display plasticity in their behaviors depending on the needs of the colony (Villalta et al. 2016). Worker honey bees (*Apis mellifera*) begin life as nurses and, as they mature, they transition to become foragers (Robinson et al. 2005). Foragers search for pollen, nectar, and water, which they bring back to the hive (Michener 2000; Page et al. 2012). They use many different cues in their foraging decisions (Grüter and Leadbeater 2014) and like many other social insects perform different tasks or

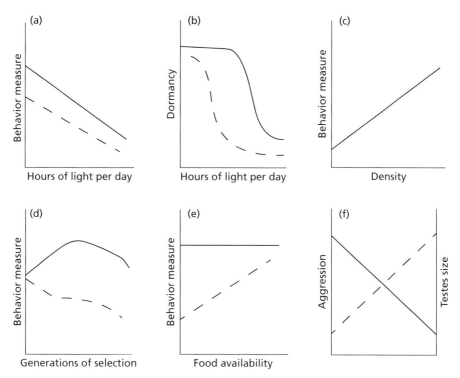

Figure 5.2 Conceptual diagrams of behavioral responses to various environmental factors based on examples from the literature. (a) Diapause and related behaviors vary with the photoperiod (the number of hours of light in a day). Among geographic variants of *Leptinotarsa decemlineata* beetles, burrowing behaviors and diapause are associated; high burrowing occurs at short photoperiods and decreases under long photoperiods (Lehmann et al. 2014) The northern variant (solid line) consistently exhibits more burrowing than the low burrowing southern variant (dashed line; Lehmann et al. 2014). Similarly, the propensity to diapause (measured at low temperature conditions) also varies with the photoperiod. For example, allelic variants (solid and dashed lines) at the *timeless* locus of *Drosophila melanogaster* both respond to photoperiod with higher diapause proportions under wintry short photoperiodic conditions (Tauber et al. 2007). Flies differing only at *tim*, respond to increasing hours of light per day with decreased propensity to diapause: flies bearing the *ls-tim* allele (solid line) show more diapause than genetically identical flies bearing the *s-tim* allele (dashed line). Diagram inspired by Tauber et al. (2007). (b) *Drosophila melanogaster* diapause (dormancy) typically shows a threshold response to the hours of light in the day with the proportion of flies undergoing diapause changing dramatically beyond a certain critical photoperiod. Flies from Maine (solid line) showed a higher proportion in diapause than did the laboratory strain Oregon–R (dashed line). Diagram inspired by Saunders and Gilbert (1990). (c) Environmentally sensitive genotypes respond to changes in the environment. For example, when the density of non-soldier aphids increases, this triggers more soldier aphids and increased aggression (Shibao et al. 2004). Similarly, an increasingly crowded environment triggers aggregation of locusts and further changes in behavior (Anstey et al. 2009). (d) Artificial selection for increases (solid line) or decreases (dashed line) in behavior over many generations demonstrates genetic variation in the starting population for the trait under selection, e.g. pollen hoarding in bees (Page et al. 2012) or mating frequency in burying beetles (Head et al. 2014). Diagram inspired by Head et al. (2014). (e) Genotype affects the response to variation in food availability producing a gene by environment interaction (GxE). For example, *Drosophila melanogaster foraging* allelic variants show different food leaving behaviors in response to food deprivation as adults (Kent et al. 2009)—rover flies (dashed line), sitter flies (solid line). Diagram inspired by Kent et al. (2009). (f) A trade-off is a negative correlation between two traits, such that selection cannot maximize the fitness of both traits in one individual. For example, in *Onthophagus* beetles, aggression is negatively correlated with size and hornless (dashed line) beetles have larger testes than do highly aggressive horned (solid line) beetles (Moczek 2009).

jobs in the hive that are responsive to social conditions (Robinson et al. 2005). In honey bees, whether a larva will become a reproductive (a queen) is determined during development, whereas primitively eusocial wasps become reproductive long after they become adults (Judd et al. 2015).

Changes in gene expression in response to the environment provide a molecular readout of an organism's social experience. At approximately 3 weeks of age, honey bee workers transition from working in the hive as nurses, who look after developing larvae, to foragers who work outside the hive

(Ben Shahar 2005). The transition from nurse to forager is also affected by the social environment in the hive. For example, removal of all foragers from a hive results in young bees becoming foragers precociously; this manipulation is used to experimentally produce same-age nurses and foragers (Ben Shahar et al. 2002; Whitfield et al. 2003; Ben Shahar 2005). The behavioral task performed by bees is associated with the RNA expression pattern (the **transcriptome**) in their brains (Robinson et al. 2005). Gene expression can be measured from the perspective of individual genes or at the genome-wide level. Whitfield et al. (2003, 2006) showed that same-age nurse and forager worker bees differed significantly in their brain transcriptomes.

Although social insects are an excellent model for studies of the relationship between behavior and gene expression, it has been difficult to perform genetic analyses of social behavior in eusocial insects. Recently, however, Trible et al. (2017) successfully introduced a mutation in the clonal raider ant, *Ooceraea biroi*, which affected social organization, fitness, and the development of olfactory glomeruli. Specifically, they used the CRISPR/Cas9 system to generate mutations in the *orco* gene, a gene required for the functioning of the olfactory receptors. This important study opens the door for future studies of how genes interact with the social environment to affect individual and colony level behavioral traits in all social insects.

The social environment also affects gene expression and behavior of the fire ant, *Solenopsis invicta*. When these invasive ants were discovered in the United States, their colonies each had a single queen, and members of the colony were extremely aggressive toward individuals who had migrated from other colonies (reviewed in Gotzek and Ross 2007). However, researchers soon found that certain *S. invicta* colonies had multiple queens and even tended to adopt queens from other colonies. It turned out that different alleles of a single gene called *Gp-9* affect the colony social system of this species (Keller and Ross 1993). *Gp-9* affects whether a *S. invicta* colony will tolerate only one queen (*Gp-9* homozygotes) or accept more than one queen (*Gp-9* heterozygotes).

Gp-9 encodes an odorant-binding protein that has consequences for the number of queens accepted into a colony and probable effects of chemical communication in the colony (Krieger and Ross 2005; Leal and Ishida 2008). Wang et al. (2008) went on to further show that there were differences in gene expression between ants with the same genotype, depending on whether the colony they were part of was a **polygyne** (accepting of multiple queens) or a **monogyne** colony. These results demonstrate that the *Gp-9* variant interacts with the social environment of the colony to affect behavior in this ant species (Wang et al. 2008; Manfredini et al. 2013).

In migratory locusts, changes in odour, touch (physical contact), and food availability affect gene expression, which influences the frequency of the production of solitary versus gregarious foragers (Simpson et al. 2011). This locust phase polyphenism is proximately mediated by physical contact, which increases with the density of the colony (Figure 5.2c). The amount that locusts' touch one another can be thought of as part of the social environment of the organism. Solitary locusts that experimentally received a high degree of leg touching (simulating the high density of a swarm) became gregarious foragers (Anstey et al. 2009). Hind leg touch also stimulated an increase in serotonin, a known neuromodulator of behavior, sufficient to trigger the formation of gregarious behavior (Anstey et al. 2009). Locusts provided with environments of increased density, leg touching or an increase in serotonin change from solitarily feeding on a few types of plants to generalist feeding (reviewed by many including Applebaum and Heifetz 1999; Simpson et al. 2011; Wang and Kang 2014; Simões et al. 2016; and see Chapter 7 on migration and Chapter 8 on feeding).

5.2.2 Nutritional context and polyphenisms

Nutrition is important for insect development and behavioral plasticity (Dethier 1976; Oettler et al. 2015 ; see also Chapter 8). In honey bees, nurse bees feed developing larvae copious amounts of a substance called royal jelly, which then elicits queen-like ovary development resulting in adult queen bees that will mate and produce thousands of progeny (Kamakura 2011; Kamakura et al. 2016). Worker bees, unlike their sisters who are queens, care for the brood, forage, dispose of rubbish, and do not reproduce. Larvae destined to become workers are not fed large amounts of royal jelly throughout their development (Chittka and Chittka 2010).

Nutrition also affects behavioral plasticity. Winged *Cataglyphis* ants capable of reproduction ingest more protein and perform different behaviors to worker ants of this species (Amor et al. 2016). Nutrition can change developmental rate and produces behavioral variation by uncovering genetic variation leading to changes in the developmental trajectory of insects (Gibson and Dworkin 2004; Scofield and Mattila 2015). Food availability can trigger a change in reproductive development either along a foundress or a gyne trajectory in primitively eusocial adult wasps, *Polistes* spp. (Judd et al. 2015). In the mosquito, *Aedes aegypti,* mating induces changes in feeding behavior (Alfonso-Parra et al. 2016). Behavioral plasticity is triggered during bee larval development. Starvation during the larval period results in adult honey bees being more resilient toward starvation (Wang et al. 2016). These adult bees have reduced ovary size, elevated glycogen stores, and juvenile hormone titres, and feed on sugar of lower concentration than honey bees that have not been starved as larvae (Wang et al. 2016). In nutrient-variable environments, greater food-related behavioral plasticity may positively affect lifetime fitness. For further discussion of insect feeding behavior see Chapter 8.

5.2.3 Seasonal context and polyphenisms

Diapause

Many insects respond to short photoperiods by entering a reproductive diapause (Saunders et al. 1989; Denlinger 2002). Diapause is a seasonal polyphenism, induced in response to predictive cues, usually short photoperiod and low temperature; diapause leads to different metabolic and life history responses that often include changes in behavior (Flatt et al. 2013; Lehmann et al. 2014). *Culex pipiens* mosquito females in diapause do not take blood meals (Robich and Denlinger 2005; Denlinger and Armbruster, 2014). Winter imposes selection for diapause-related burrowing behavior, high body mass, and low metabolic rate in *Leptinotarsa decemlineata* beetles (Piiroinen et al. 2011).

Examining the genetic mechanisms at play in the induction and maintenance of diapause provides insight into the regulation of polyphenisms. *Drosophila melanogaster* has been a useful model for the molecular genetic analyses of diapause. In response to short day lengths and low temperatures, young female *D. melanogaster* go into a reproductive diapause. It is quantified as the proportion of ovaries that have remained immature when the flies are kept under short day length and low temperature conditions (Figure 5.2a,b; Saunders et al. 1989). Genetic variation exists in the tendency to enter diapause in natural populations (Williams and Sokolowski 1993; Schmidt et al. 2005). In general, genes that affect diapause in *D. melanogaster* fall into three categories:

- genes that influence the photoperiodic aspect of the clock used to detect the onset of winter conditions via photoperiod induction cues (Sandrelli et al., 2007; Tauber et al. 2007);
- genes that decrease insulin signalling;
- genes that inhibit ovary development (Williams et al., 2006; Schmidt et al. 2008).

For example, *timeless*, involved in biological rhythmicity (Tauber et al. 2007; Sandrelli et al. 2007), *couch potato* (*cpo*), an RNA binding protein (Schmidt et al. 2008; Schmidt 2011) and *dp110,* a phosphatidylinositol 3 kinase involved in the insulin signalling pathway (Williams et al. 2006) are all involved in *Drosophila* diapause. Diapausing *Drosophila* divert resources towards somatic maintenance and away from reproduction (Williams et al. 2010). Insulin signalling is the lynchpin of energy diversion, and both insulin metabolism and *timeless* affect steroid hormone levels (Di Cara and King-Jones 2016; Katewa et al. 2016). Thus, diapause in *D. melanogaster* is a seasonal polyphenism, associated with differences in energy allocation. Interestingly, the insulin signalling pathway is critical for diapause in a number of insect species (Sim and Denlinger 2008; Kubrak et al. 2014; Schiesari et al. 2016).

Young *Polistes metricus* larvae that experience less provisioning enter a different developmental pathway, which bypasses the reproductively dormant state of their more abundantly fed sister wasps (Hunt and Amdam 2005). Such paper wasps have a food-induced, seasonal polyphenism that leads to a difference in foundress behavior in wasps and ultimately to different life history (Hunt et al., 2007, Berens et al., 2015). Larvae that have less food, develop into workers that reproduce, build nests and provide maternal care (Hunt and Amdam, 2005; Hunt et al. 2007; Berens et al. 2015). In contrast, larvae that have more food develop into gynes that delay reproduction until they have completed diapause, and these gynes do not build nests or provide maternal care (Hunt and Amdam 2005; Hunt et al. 2007; Berens et al. 2015). Wasps that exhibit maternal behaviors show a distinct pattern of gene transcription related to their nursing task, and among the genes involved are genes in the insulin pathway (Toth et al. 2007). Genetic pathways underlying seasonal life-history polyphenisms share some common components. They show a threshold-like change in their phenotype and differential transcription in response to changes in environmental cues, with effects that spill over to pathways that alter growth and reproduction (reviewed in Flatt et al. 2013).

Other seasonal polyphenisms affect growth and reproduction and show changes in the levels of key insect hormones. For example, in *Polistes* wasps, diapausing gynes store hexamerin proteins (Hunt et al. 2007), which are also known to alter the levels of juvenile hormone (JH) among termites (Zhou et al. 2006; Martins et al. 2010) and affect mating behavior in other insects (Goto 2013). Intriguingly, the insulin pathway and JH are also involved in reproductive plasticity in aphids. In the aphids species, *Megoura viciae* and *Aphis fabae*, long nights (short days) trigger egg diapause, resulting in granddaughters that can sexually reproduce, whereas short nights (long days) do not trigger diapause, resulting in parthenogenetic granddaughters that reproduce asexually (reviewed in Hardie 2010). In this case there is an egg diapause in the daughters and these affects extend to the granddaughters. Interestingly, there is a critical night length that puts aphids on a developmental trajectory that results in morphological and behavioral changes in addition to changes in reproductive physiology (Hardie 2010), light sensitivity (Hardie 1989) and host-plant choice (Hardie and Glascodine 1990; Hardie 2010). Changes in insulin signalling that occur as a result of the aphids perceiving short day lengths diminishes JH, which in turn affects key enzymes important for development (Le Trionnaire et al. 2009).

Migration

The long trip made by Monarch butterflies, *Danaus plexippus*, during migration to Mexico is initiated by a response to seasonal cues. Butterflies respond to cues from photoperiod, temperature, and the social environment that predict a change in season (Guerra and Reppert 2013, 2015). The variety of cues makes this seasonal polyphenism a particularly interesting example because there are differences in gene expression between summer (non-migrating) and migrating butterflies (Zhu et al. 2008). Monitoring their sensitivity to changing environmental conditions is critical for the future conservation of Monarch butterflies (Zhu et al. 2008).

Monarch butterflies use a time-compensated sun compass during migration (Reppert et al. 2010). Skylight cues are processed through both eyes and the information is likely integrated in the central brain, where the sun compass is thought to reside. The butterfly's antennae house the circadian clock and its molecular mechanisms underlying the time compensation needed to interpret the skylight cues obtained from the sun and polarized light. Monarchs may also use a magnetic compass for navigation (reviewed in Reppert et al. 2016). Current research aims at identifying genes important for navigation during long-distance migration (Reppert et al., 2016). For further discussion of insect migration see Chapter 7.

During their migration and overwintering season, Monarch butterflies are in a reproductive diapause and rarely mate (Reppert 2006; Zhu et al. 2008, 2009). Diapause is dispensable for the directed oriented flight characterizing Monarch migration (Zhu et al. 2009), although diapause may be important for conserving energy resources for the long journeys. The ancestors of Monarchs in North America were apparently already migratory to some extent (Zhan et al. 2014). Perhaps the behavioral plasticity seen in modern North American Monarchs is the result of the genetic factors that underlie both migration and diapause, both of which respond to cues that presumably triggered an energy-saving programme in their evolutionary ancestors.

5.3 Gene-by-environment interactions (GxE)

5.3.1 The *Drosophila* foraging gene model

The rover and sitter allelic variants of the *D. melanogaster foraging* gene were described in laboratory strains and in larvae from an orchard in Toronto, Canada (Sokolowski 1980; Pereira and Sokolowski 1993; Sokolowski et al. 1997). The locomotion of rover and sitter larvae did not differ in the absence of food. However, on a nutritive yeast paste rover larvae had longer foraging trails than sitters. Quantitative genetic analyses of larval foraging trail length revealed that a factor on the second pair of chromosomes accounted for much of the variation in foraging trail length between rover and sitter larvae, with the rover phenotype exhibiting genetic dominance over sitter (de Belle and Sokolowski 1987). (A more minor effect of the X chromosome was also found but not characterized.) The *foraging* gene was later mapped to a region on the second pair of chromosomes (de Belle et al. 1989). Subsequent cloning of the *foraging* gene revealed that it encodes cGMP dependent protein kinase (PKG), a signalling molecule (Osborne et al. 1997). Larvae with short sitter-like foraging trails resulted when the entire gene was deleted using homologous recombination (Allen et al. 2017). When the entire gene was put back into the fly using a transgenic approach larval foraging trail length increased. Manipulation of the dose of the *foraging* gene revealed that more copies of the gene increased larval foraging trail length making them more rover-like. This proved that the PKG gene, *foraging*, accounts for the foraging trail length differences between larval rovers and sitters. The *foraging* gene is a large (35kb) complex gene that is alternatively spliced with four promoters that generate twenty-one transcripts (Allen et al. 2017). *Foraging*'s complex regulation may facilitate both its pleiotropy (one gene with multiple functions) and its environmental sensitivity (further described later).

Rover and sitter larvae also differ in their food intake and fat levels. Sitters move less, eat more, and have higher levels of fat compared with rovers that move more, eat less, and are lean. All of these traits genetically map within the *foraging* gene, demonstrating the pleiotropic effects of the *foraging* gene on larval behavior and metabolism. In *D. melanogaster* it affects phenotypes involved in larval and adult food-related traits, learning and memory, aggression, social behaviors all of which are discussed further in Section 5.3.4.

5.3.2 Phenotypic plasticity and the *Drosophila* foraging gene

Even though allelic variation in the *foraging* gene can be said to predispose larvae to have longer or shorter foraging trails (as described previously), the environment that the larvae experiences can completely modify their behavior. In order to vary larval nutritional environment, food levels have been manipulated chronically throughout larval life (Kaun et al. 2007) or acutely for a number of hours (Kaun et al. 2008). Under well-fed conditions, rovers feed less and have higher larval PKG enzyme activities than sitters. These differences disappear when larvae are chronically food-deprived throughout their lives; rovers become like sitters both from a behavioral (food intake) and molecular (PKG enzyme activity) perspective. Under acute food deprivation, rover food intake increases to a sitter level after 2 hours. Thus, the rover and sitter allelic variants of the *foraging* gene are able to exhibit plastic responses to food deprivation despite their genetic predispositions to behave differently (Figure 5.2e). This example makes the important point that there is not necessarily an inverse relationship between heritability and plasticity (Sokolowski and Wahlsten 2001).

Early nutritional adversity also has a long reach into adulthood. Chronic food deprivation in larval life affects adult movement in an open field (a petri dish). In this case, adult sitters who were deprived of nutrition in their early life become more like rovers, darting into the centre of the arena instead of hugging the edges, something sitters, well fed in their early life, commonly do as adults (Burns et al. 2012). In contrast, manipulations of food availability during the adult stage alone did not affect adult movement patterns in the open field. Chronic food deprivation in the larval stage also affects adult defecation; in this case, rovers are more plastic in their response than sitters (Urquhart-Cronish and Sokolowski 2014). Another investigation of adult plasticity in rovers and sitters showed that sitters could only learn and remember when they were taught and tested with a group of flies, whereas rovers did well when they were taught and tested alone or in groups (Kohn et al. 2013). The latter study suggested that the olfaction-based learning and memory of sitters is more sensitive to the social context than that of rovers. It should be clear from this discussion that environmental modifications result in plastic behavioral responses in larval and adult flies, and that which morph is more plastic, rover or sitter, depends on the phenotype and the environmental context. It is not the case that one morph is always more environmentally sensitive than the other. These different patterns of plasticity of the rover and sitter genetic variants are important examples of GxE.

5.3.3 The foraging gene in eusocial insects

The task or job that social insects perform in their nest is also affected by the expression of the *foraging* gene. Honey bee nurses exhibit lower expression of honey bee *foraging* (*Amfor*) than do foragers. *Amfor* is up-regulated when honey bees reach about 3 weeks in age (Ben Shahar et al. 2002). This regulation happens within the lifetime of the individual, as *Amfor* gene expression is responsive to the social needs of the colony. When the number of foragers decreases in the hive and more are needed, there is an increase in *Amfor* expression in young nurses and precocious foragers are generated. Even though there are hundreds of genes that change their expression during the worker transition from nurse to forager, it is sufficient to pharmacologically manipulate levels of PKG to change bee task (Ben Shahar et al. 2002). Similar relationships between gene expression and nurse/forager tasks are found in ants (Ingram et al. 2005). In the ant, *Pheidole pallidula*, the level of *foraging* expression influences the tendency to forage or defend the nest (Lucas and Sokolowski 2009), and the brain expression pattern differs in major and minor ants in this species. Thus, the *foraging* gene plays a role in phenotypic plasticity within the lifetime of the individual in social insects.

5.3.4 Pleiotropy and the foraging gene

When one gene affects multiple traits it is said to be pleiotropic. A pleiotropic gene can affect the behavior of an insect in profound ways. The *foraging* gene of *Drosophila melanogaster* is a good example of this. Originally described for its effects on the food-related movement of larvae, *foraging* has been shown to affect additional phenotypes including adult fly foraging (Pereira and Sokolowski 1993; Kent et al. 2009; Hughson, Anreiter et al. 2017; Anreiter et al. 2017), adult olfactory responses (Shaver et al. 1998), exploration in an open field (Burns et al., 2012), adult sucrose responses (Scheiner et al. 2004), long- and short-term olfactory-based aversive memory (Mery et al. 2007), retroactive interference in memory (Reaume et al. 2011), a type of working memory (Kuntz et al. 2012), social **cognition** (Kohn et al. 2013), and learning after sleep deprivation (Donlea et al. 2012). Allelic variation in *foraging* also affects larval foraging behavior, food intake, triglyceride levels, appetitive learning, stress responses (Dawson-Scully et al. 2007), Malpighian tubule function (MacPherson et al. 2004), and nutrient-dependent defecation patterns (Urquhart-Cronish and Sokolowski, 2014). Finally, when *foraging* is deleted, the resultant animal dies during the pupal stage of development making *foraging* a vital gene (Allen et al. 2017). Another pleiotropic gene is the *fruitless* gene in *Drosophila melanogaster*, which affects both courtship and fruit fly aggression (Vrontou et al. 2006; Chan and Kravitz 2007; Koganezawa et al. 2016). Each of these genes has many transcripts, each varying in its temporal and spatial expression (Attrill et al. 2016; Allen et al. 2017). This suggests that genes of complex structure, which are developmentally regulated with multiple promoters and transcripts (they have a modular structure), may lend themselves to pleiotropy. How pleiotropy affects behavioral phenotypes is not well understood from both mechanistic and evolutionary perspectives. For example, it is not known whether *foraging* acts as a modifier of the traits mentioned above through its overall effect on metabolism and/or whether the gene regulates each of these traits independently. Future investigations using a wide variety of insects will help us understand the modularity of pleiotropic effects on behavior (Goto, 2013). Overall, the gene is responsive to different environments and acts on different timescales to regulate phenotypic plasticity.

5.3.5 Trade-offs

Trade-offs are negative genetic correlations between traits (Flatt et al. 2005). They can occur when traits affect, or are affected by, the same genetic pathway, or because the genetic factors involved are closely linked (Jandt et al. 2015). Pathways underlying behavior polyphenisms suggest particular trade-offs in aphids (Brisson 2010; Vellichirammal et al. 2016). Changes in insulin signalling and JH that occur when aphids perceive short day lengths result in decreases in key enzymes, which are important for cuticle sclerotization (Le Trionnaire et al. 2009). Female *Nicrophorus vespilloides* burying beetles that care for young incur a cost resulting from changes in JH (Engel et al. 2014), which is linked to mechanisms involved in female *N. vespilloides* anti-aphrodisiac pheromone production, suggesting the possibility of a trade-off (Engel et al. 2016).

Horned and hornless beetles

Dung beetle, *Onthophagus* spp., horns exhibit spectacular variations. Two male morphs exist in these species, horned and hornless (Moczek 2009). Males with long horns are more successful in territorial defence fights and this provides them with more mating opportunities (Karino et al. 2005). However, hornless males have larger testes that contribute to their reproductive success even though they are unable to defend territories and are not aggressive (Simmons and Emlen, 2006; Moczek 2009; Figure 5.2f). Horn development is remarkably plastic and, recently, the molecular basis of this plasticity has been the focus of several studies (Moczek 2010; Schwab and Moczek 2016). Horn growth is sensitive to body condition; in low nutrition conditions during development, males do not grow large horns (Kijimoto and Moczek, 2016). Aggressive beetle behavior and horn weapon shape are correlated, and selection based on weapon shape, results in changes in beetle behavior (Okada and Miyatake 2009).

Research on *Onthophagus* species has identified mechanisms for sensing **nutrient** availability in the environment and these mechanisms involve insulin signalling (Emlen et al. 2012; Snell-Rood and Moczek 2012), JH (Emlen and Nijhout 2001), the *hedgehog* limb developmental pathway (Kijimoto and Moczek 2016),

and *doublesex*, a key player in the sex determination pathway of insects (Kijimoto et al. 2012). The *hedgehog* pathway depresses the production of male horns at low food availability (Kijimoto and Moczek 2016), and the *doublesex* gene increases the growth of male horns in high food availability environments (Kijimoto et al. 2012). The insulin signalling pathway, which is conserved in its function among insects and mammals, has the potential to coordinate resources and energy availability with the ability to grow horns (Emlen et al. 2012). As mentioned previously, JH is a powerful insect hormone with the ability to dramatically change development, life-history (Dingle and Winchell 1997), and behavior in insects. Thus, JH and insulin signalling are probably key players involved in the relationships between developmental trade-offs and behavior (reviewed by Erion and Sehgal 2013). See also Chapter 4 for further discussions of hormones and phenotypic plasticity.

Parasitoids

Most insects, including parasitoid wasps, may express rapid behavioral responses to changes in the environment. Weather affects parasitoid behavior, which has consequences for life expectancy (Roitberg et al. 1993). A trade-off between putting energy into soma versus reproduction is thought to impact upon life-history polyphenisms because immune system components of somatic maintenance are energetically costly (reviewed in Schwenke et al. 2016). In *Plutella xylostella* and *Pseudoplusia includens* moths, their hormonal environment is changed when they are infected by a developing parasitoid wasp (and the polydnavirus it carries); this results in altered larval development of the moth due to changes in ecdysteroids and JH (Pruijssers et al., 2009; Kwon et al. 2010). Parasitoid infections can also affect behavior to augment survival (Milan et al. 2012). *Drosophila* larvae tend to move onto ethanol-containing food when they are infected by their parasitoids—the wasps *Leptopilina boulardi* and *L. heterotoma* (Milan et al. 2012). Ethanol-containing food increases the fitness of parasitized *Drosophila* larvae because it is detrimental to the parasitoid larva inside the host (Milan et al. 2012). Female *D. melanogaster* are thought to medicate their offspring by laying their eggs on a medium supplemented with alcohol after seeing wasp parasitoids (Kacsoh et al. 2013, 2015). Thus, immune challenges can affect environment-specific fitness outcomes in insects.

5.4 Potential molecular mechanisms of plasticity: behavioral epigenetics

In recent years, the study of behavioral variation has included epigenetics as a mechanism by which the environment affects gene expression (Boyce and Kobor 2015). Epigenetics has been thought of as a mechanism for nurture—how our environment gets embedded in our biology. More recent studies suggest that epigenetics may reflect GxE interactions in that the placement of epigenetic marks can be dependent on DNA sequence variation in the genome, and this has consequences for gene expression and behavioral variability (Anreiter et al. 2017). Epigenetic modifications can arise through a number of mechanisms, including DNA methylation, histone modification, and microRNAs (Yan et al. 2014; Maleszka 2016). These mechanisms and others act above the DNA sequence level, and in some cases result in a change in gene expression.

DNA methylation has been most studied from the perspective of behavioral epigenetics in mammals (Boyce and Kobor 2015). It requires specific enzymes also found in many social insects (Bonasio et al. 2010; Bonasio 2015; Yan et al. 2015; Maleszka 2016). Studies of epigenetic modification of behaviors have also been performed using eusocial insects known for their behavioral flexibility (Maleszka 2008; Bonasio et al. 2010; Weiner et al. 2013; Yan et al. 2014, 2015; Holman et al., 2016). Examples include studies showing that DNA methylation is correlated with the development of the eusocial bee larvae from a worker to queen (Kucharski et al. 2008). DNA methyltransferases altered long-term memory associated with odours in honey bees (Biergans et al. 2012). Ants that differ in colony structure differ in DNA methylation (Bonasio et al. 2010, 2012). Finally, nutrition appears to modulate both honey bee behavior and DNA methylation (reviewed in Yan et al. 2014; Cridge et al. 2015).

Epigenetic mechanisms have also been found to change the behavior of primitively eusocial wasps and bumble bees (Weiner et al. 2013; Amarasinghe et al. 2014), as well as solitary insects such as locusts

(Boerjan et al. 2011; Ernst et al. 2015). Histone modifications play a role in the transition between major (defenders of the colony) and minor (foragers of the colony) *Camponotus floridanus* ants (Simola et al. 2016). Histone modifications are also involved in the transition from solitary to gregarious phases of many species of orthopterans known for their density phase polyphenism (Boerjan et al. 2011; Ernst et al. 2015; Guo et al. 2016). *Drosophila melanogaster* may not have DNA methyltransferases (Yan et al. 2015), but they do have enzymes important for histone modification. The euchromatin histone methyl transferase encoded by *G9a/EHMT* is known to be involved in intellectual disability in humans, but is also important for learning and memory in *Drosophila* (Kramer 2016; Kramer et al. 2011). Fruit fly learning and memory are negatively impacted by release from epigenetic control due to the deletion of key histone acetylase enzymes (Tip60 HAT; Xu et al. 2016). Also in fruit flies, restoration of cognitive abilities by environmental enrichment appears to be mediated by the ability of the enriched environment to act to increase in histone acetylation on key environmentally responsive genes (Xu et al. 2016).

From a historical perspective, studies of the effects of environment on behavior were dismissed as ideas echoing the 'ghost of Lamarck' in that they suggested environmentally acquired traits would be inherited (West-Eberhard 2007; Sokolowski and Levine 2010). How and whether epigenetic mechanisms affect the evolution of behavior is a key question for future studies (Charlesworth et al. 2017; Jablonka 2017). Addressing both intergenerational and transgenerational inheritance will provide much needed insights into the mechanisms by which epigenetic marks might affect the evolution of behavior. Intergenerational inheritance has begun to be investigated in insects (Öst et al. 2014), but how and whether epigenetic modifications are mechanisms for intergenerational inheritance is currently not known. Insects will provide excellent models for these investigations with their relatively fast generation time.

5.5 Conclusions

Polyphenisms are a type of phenotypic plasticity where a single genotype produces two or more discrete behavioral variants in response to the environment. We have discussed examples of environmental contexts that can result in polyphenisms, including the social, nutritional, and seasonal environment. An emerging theme from this chapter is that environmental factors influence gene expression and, even for a given genotype, the altered timing or location of gene expression can change the expression of suites of genes important for behavioral plasticity.

Identifying the gene networks that change in response to environmental conditions permits assessment of the impacts of changing environments on insect behavior. The changes in gene networks that mediate the response to changes in the environment confer behavioral plasticity by eliciting changes in development, energy metabolism, and environmental sensitivity. Amid these networks, each gene can have its own effects on development and physiology, and together these effects can additively or antagonistically alter the development, physiology, and behavior of insects. For each genotype, the environmentally induced changes in physiology and gene expression can result in behavioral plasticity.

The mechanisms by which insects inherit the ability to be behaviorally flexible are not well understood. We have discussed examples of behavioral plasticity that appear to co-opt ancestral environmental response mechanisms, such as circadian rhythmicity, JH, and insulin signalling; they are tied to changes in gene expression, energy allocation, and usage. Further research into the mechanisms underlying pleiotropy will lead to a deeper understanding of the ways that behavioral variation and trade-offs might be influenced by environmental change. Similarly, future investigations into epigenetic mechanisms that affect behavioral variation will provide greater insight into the ways by which environmentally induced changes might be written onto the genome of the insect.

Increasingly, individual genes important for a particular behavior will be identified and causally manipulated in many other (and perhaps in all) organisms. This will open the door into investigations of the timing of gene expression, alternate splicing, spatial patterning. and experience-dependent gene expression. The mechanisms for environmentally sensitive changes in gene expression, the complex interdependence of genes and their environments, and the integrated gene networks through which trade-offs are mediated will provide an increased understanding of the phenotypic plasticity of insect behaviors.

Acknowledgements

The authors thank the reviewers Wolf Blanckenhorn and Bernard Roitberg, as well as the editors Daniel González-Tokman, Isaac González-Santoyo and Alex Córdoba-Aguilar, for their generous comments on an earlier version of the manuscript. Funding was from NSERC (grant #RGPIN -2016-06185 to MBS).

References

Alfonso-Parra, C., Ahmed-Braimah, Y.H., Degner, E.C., et al. (2016). Mating-induced transcriptome changes in the reproductive tract of female *Aedes aegypti*. *Plos Neglected Tropical Diseases*, **10**, e0004451.

Allen, A.M., Anreiter, I., Neville, M.C., and Sokolowski, M.B. (2017). Feeding-related traits are affected by dosage of the foraging gene in *Drosophila melanogaster*. *Genetics*, **205**, 761.

Amarasinghe, H.E., Clayton, C.I., and Mallon, E.B. (2014). Methylation and worker reproduction in the bumble-bee (*Bombus terrestris*). *Proceedings of the Royal Society B-Biological Sciences*, **281**, 20132502.

Amor, F., Villalta, I., Doums, C., et al. (2016). Nutritional versus genetic correlates of caste differentiation in a desert ant. *Ecological Entomology*, **41**, 660.

Anholt, R.R.H., and Mackay, T.F.C. (2004). Quantitative genetic analyses of complex behaviours in *Drosophila*. *Nature Reviews Genetics*, **5**, 838.

Anreiter, I. Kramer, J.M., and Sokolowski, M.B. (2017). Epigenetic mechanisms modulate differences in *Drosophila* foraging behavior. *Proceedings of the National Academy of Sciences of the United States of America*, **114**(47), 12518–23.

Anstey, M.L., Rogers, S.M., Ott, S.R., Burrows, M., and Simpson, S.J. (2009). Serotonin mediates behavioral gregarization underlying swarm formation in desert locusts. *Science*, **323**, 627.

Applebaum, S.W., and Heifetz, Y. (1999). Density-dependent physiological phase in insects. *Annual Review of Entomology*, **44**, 317.

Attrill, H., Falls, K., Goodman, J.L., et al. (2016). FlyBase: establishing a Gene Group resource for *Drosophila melanogaster*. *Nucleic Acids Research*, **44**, D786.

Ben-Shahar, Y. (2005). The foraging gene, behavioral plasticity, and honeybee division of labor. *Journal of Comparative Physiology a-Neuroethology Sensory Neural and Behavioral Physiology*, **191**, 987.

Ben-Shahar, Y., Robichon, A., Sokolowski, M.B., and Robinson, G.E. (2002). Influence of gene action across different time scales on behavior. *Science*, **296**, 741.

Berens, A.J., Hunt, J.H., and Toth, A.L. (2015). Nourishment level affects caste-related gene expression in *Polistes* wasps. *BMC Genomics*, **16**, 235.

Biergans, S.D., Jones, J.C., Treiber, N., Galizia, C.G., and Szyszka, P. (2012). DNA methylation mediates the discriminatory power of associative long-term memory in honeybees. *Plos One*, **7**, e39349.

Boerjan, B., Sas, F., Ernst, U.R., et al. (2011). Locust phase polyphenism: does epigenetic precede endocrine regulation? *General and Comparative Endocrinology*, **173**, 120.

Bonasio, R. (2015). The expanding epigenetic landscape of non-model organisms. *Journal of Experimental Biology*, **218**, 114.

Bonasio, R., Li, Q.Y., Lian, J.M., et al. (2012). Genome-wide and caste-specific DNA methylomes of the ants *Camponotus floridanus* and *Harpegnathos saltator*. *Current Biology*, **22**, 1755.

Bonasio, R., Tu, S.J., and Reinberg, D. (2010). Molecular signals of epigenetic states. *Science*, **330**, 612.

Boyce, W.T., and Kobor, M.S. (2015). Development and the epigenome: the 'synapse' of gene-environment interplay. *Developmental Science*, **18**, 1.

Boyce, W.T., Sokolowski, M.B., and Robinson, G.E. (2012). Toward a new biology of social adversity. *Proceedings of the National Academy of Sciences of the United States of America*, **109**, 17143.

Brisson, J.A. (2010). Aphid wing dimorphisms: linking environmental and genetic control of trait variation. *Philosophical Transactions of the Royal Society B-Biological Sciences*, **365**, 605.

Burns, J.G., Svetec, N., Rowe, L., et al. (2012). Gene-environment interplay in *Drosophila melanogaster*: chronic food deprivation in early life affects adult exploratory and fitness traits. *Proceedings of the National Academy of Sciences of the United States of America*, **109**, 17239.

Chan, Y.B., and Kravitz, E.A. (2007). Specific subgroups of Fru(M) neurons control sexually dimorphic patterns of aggression in *Drosophila melanogaster*. *Proceedings of the National Academy of Sciences of the United States of America*, **104**, 19577.

Charlesworth, D., Barton, N.H., and Charlesworth, B. (2017). The sources of adaptive variation. *Proceedings of the Royal Society B-Biological Sciences*, **284**, 20162864.

Chittka, A., and Chittka, L. (2010). Epigenetics of royalty. *Plos Biology*, **8**, e1000532.

Cridge, A.G., Leask, M.P., Duncan, E.J., and Dearden, P.K. (2015). What do studies of insect polyphenisms tell us about nutritionally-triggered epigenomic changes and their consequences? *Nutrients*, **7**, 1787.

Dawson-Scully, K., Armstrong, G.A.B., Kent, C., Robertson, R.M., and Sokolowski, M.B. (2007). Natural variation in the thermotolerance of neural function and behavior due to a cGMP-dependent protein kinase. *Plos One*, **2**, e773.

de Belle, J.S., Hilliker, A.J., and Sokolowski, M.B. (1989). Genetic localization of foraging (for)—a major gene for larval behavior in *Drosophila-melanogaster*. *Genetics*, **123**, 157.

de Belle, J.S., and Sokolowski, M.B. (1987). Heredity of rover sitter—alternative foraging strategies of *Drosophila melanogaster* larvae. *Heredity*, **59**, 73.

Denlinger, D.L. (2002). Regulation of diapause. *Annual Review of Entomology*, **47**, 93.

Denlinger, D.L., and Armbruster, P.A. (2014). Mosquito diapause. *Annual Review of Entomology*, **59**, 73.

Dethier, V.G. (1976). *The Hungry Fly: The Physiological Study of the Behavior Associated with Feeding*. Harvard University Press, Cambridge, MA.

Dewitt, T.J., Sih, A., and Wilson, D.S. (1998). Costs and limits of phenotypic plasticity. *Trends in Ecology and Evolution*, **13**, 77.

Di Cara, F., and King-Jones, K. (2016). The Circadian clock is a key driver of steroid hormone production in *Drosophila*. *Current Biology*, **26**, 2469.

Dingle, H., and Winchell, R. (1997). Juvenile hormone as a mediator of plasticity in insect life histories. *Archives of Insect Biochemistry and Physiology*, **35**, 359.

Donlea, J., Leahy, A., Thimgan, M.S., et al. (2012). Foraging alters resilience/vulnerability to sleep disruption and starvation in *Drosophila*. *Proceedings of the National Academy of Sciences of the United States of America*, **109**, 2613.

Emlen, D.J., and Nijhout, H.F. (2001). Hormonal control of male horn length dimorphism in *Onthophagus taurus* (Coleoptera : Scarabaeidae): a second critical period of sensitivity to juvenile hormone. *Journal of Insect Physiology*, **47**, 1045.

Emlen, D.J., Warren, I.A., Johns, A., Dworkin, I., and Lavine, L.C. (2012). A mechanism of extreme growth and reliable signaling in sexually selected ornaments and weapons. *Science*, **337**, 860.

Engel, K.C., Stokl, J., Schweizer, R., et al. (2016). A hormone-related female anti-aphrodisiac signals temporary infertility and causes sexual abstinence to synchronize parental care. *Nature Communications*, **7**, 11035.

Engel, K.C., Von Hoermann, C., Eggert, A.K., Muller, J.K., and Steiger, S. (2014). When males stop having sex: adaptive insect mating tactics during parental care. *Animal Behaviour*, **90**, 245.

Erion, R., and Sehgal, A. (2013). Regulation of insect behavior via the insulin-signaling pathway. *Frontiers in Physiology*, **4**, 353.

Ernst, U.R., Van Hiel, M.B., Depuydt, G., et al. (2015). Epigenetics and locust life phase transitions. *Journal of Experimental Biology*, **218**, 88.

Falconer, D.S., and Mackay, T.F.C. (1996). *Introduction to Quantitative Genetics*. Longman, Harlow.

Flatt, T., Amdam, G.V., Kirkwood, T.B.L., and Omholt, S.W. (2013). Life-history evolution and the polyphenic regulation of somatic maintenance and survival. *Quarterly Review of Biology*, **88**, 185.

Flatt, T., Tu, M.P., and Tatar, M. (2005). Hormonal pleiotropy and the juvenile hormone regulation of *Drosophila* development and life history. *Bioessays*, **27**, 999.

Gibson, G., and Dworkin, I. (2004). Uncovering cryptic genetic variation. *Nature Reviews Genetics*, **5**, 681.

Goto, S.G. (2013). Roles of circadian clock genes in insect photoperiodism. *Entomological Science*, **16**, 1.

Gotzek, D., and Ross, K.G. (2007). Genetic regulation of colony social organization in fire ants: an integrative overview. *Quarterly Review of Biology*, **82**, 201.

Grüter, C., and Leadbeater, E. (2014). Insights from insects about adaptive social information use. *Trends in Ecology and Evolution*, **29**, 177.

Guerra, P.A., and Reppert, S.M. (2013). Coldness triggers northward flight in remigrant Monarch butterflies. *Current Biology*, **23**, 419.

Guerra, P.A., and Reppert, S.M. (2015). Sensory basis of lepidopteran migration: focus on the monarch butterfly. *Current Opinion in Neurobiology*, **34**, 20.

Guo, S.Y., Jiang, F., Yang, P.C., et al. (2016). Characteristics and expression patterns of histone-modifying enzyme systems in the migratory locust. *Insect Biochemistry and Molecular Biology*, **76**, 18.

Gupta, A.P., and Lewontin, R.C. (1982). A study of reaction norms in natural populations of *Drosophila-pseudoobscura*. *Evolution*, **36**, 934.

Hardie, J. (1989). Spectral specificity for targeted flight in the black bean aphid, *Aphis-fabae. Journal of Insect Physiology*, **35**, 619.

Hardie, J. (2010). Photoperiodism in insects: aphid polyphenism. In: R. J. Nelson, D.L, Denlinger, and D.E. Somers (Eds) *Photoperiodism*, pp. 342–63. Oxford University Press, New York, NY.

Hardie, J., and Glascodine, J. (1990). Polyphenism and host-plant preference in the black bean, *Aphis-fabae* Scop. *Acta Phytopathologica et Entomologica Hungarica*, **25**, 323.

Head, M.L., Hinde, C.A., Moore, A.J., and Royle, N.J. (2014). Correlated evolution in parental care in females but not males in response to selection on paternity assurance behaviour. *Ecology Letters*, **17**, 803.

Holman, L., Trontti, K., and Helantera, H. (2016). Queen pheromones modulate DNA methyltransferase activity in bee and ant workers. *Biology Letters*, **12**, 20151038.

Hughson, B.N., Anreiter, I., Jackson Chornenki, N.L., et al. (2017). The adult foraging assay (AFA) allows detection of strain and food deprivation effects in feeding-related traits of *Drosophila melanogaster. Journal of Insect Physiology*, pii: S0022–1910(17)30108-7.

Hunt, J.H., and Amdam, G.V. (2005). Bivoltinism as an antecedent to eusociality in the paper wasp genus Polistes. *Science*, **308**, 264.

Hunt, J.H., Kensinger, B.J., Kossuth, J.A., et al. (2007). A diapause pathway underlies the gyne phenotype in *Polistes* wasps, revealing an evolutionary route to caste-containing insect societies. *Proceedings of the National Academy of Sciences of the United States of America*, **104**, 14020.

Ingram, K.K., Oefner, P., and Gordon, D.M. (2005). Task-specific expression of the foraging gene in harvester ants. *Molecular Ecology*, **14**, 813–18.

Jablonka, B. (2017). The evolutonary implications of epigenetic inheritance. *Interface Focus*, **7**, 20160135.

Jandt, J.M., Thomson, J.L., Geffre, A.C., and Toth, A.L. (2015). Lab rearing environment perturbs social traits: a case study with *Polistes* wasps. *Behavioral Ecology*, **26**, 1274.

Judd, T.M., Teal, P.E.A., Hernandez, E.J., Choudhury, T., and Hunt, J.H. (2015). Quantitative differences in nourishment affect caste-related physiology and development in the paper wasp *Polistes metricus*. *Plos One*, **10**, e0116199.

Kacsoh, B.Z., Bozler, J., Hodge, S., Ramaswami, M., and Bosco, G. (2015). A novel paradigm for nonassociative long-term memory in *Drosophila*: predator-induced changes in oviposition behavior. *Genetics*, **199**, 1143.

Kacsoh, B.Z., Lynch, Z.R., Mortimer, N.T., and Schlenke, T.A. (2013). Fruit flies medicate offspring after seeing parasites. *Science*, **339**, 947.

Kamakura, M. (2011). Royalactin induces queen differentiation in honeybees. *Nature*, **473**, 478.

Kamakura, M., Ihling, C.H., Pietzsch, M., and Moritz, R.F.A. (2016). Royalactin is not a royal making of a queen. *Nature*, **537**, E10.

Karino, K., Niiyama, H., and Chiba, M. (2005). Horn length is the determining factor in the outcomes of escalated fights among male Japanese horned beetles, *Allomyrina dichotoma* L. (Coleoptera : Scarabaeidae). *Journal of Insect Behavior* 18: 805.

Katewa, S.D., Akagi, K., Bose, N., et al. (2016). Peripheral circadian clocks mediate dietary restriction-dependent changes in lifespan and fat metabolism in *Drosophila*. *Cell Metabolism*, **23**, 143.

Kaun, K.R., Chakaborty-Chatterjee, M., and Sokolowski, M.B. (2008). Natural variation in plasticity of glucose homeostasis and food intake. *Journal of Experimental Biology*, **211**, 3160.

Kaun, K.R., Riedl, C.A.L., Chakaborty-Chatterjee, M., et al. (2007). Natural variation in food acquisition mediated via a *Drosophila* cGMP-dependent protein kinase. *Journal of Experimental Biology*, **210**, 3547.

Keller, L., and Ross, K.G. (1993). Phenotypic basis of reproductive success in a social insect—genetic and social determinants. *Science*, **260**, 1107.

Kent, C.F., Daskalchuk, T., Cook, L., Sokolowski, M.B., and Greenspan, R.J. (2009). The *Drosophila* foraging gene mediates adult plasticity and gene-environment interactions in behaviour, metabolites, and gene expression in response to food deprivation. *Plos Genetics*, **5**, e1000609.

Kijimoto, T., and Moczek, A.P. (2016). Hedgehog signaling enables nutrition-responsive inhibition of an alternative morph in a polyphenic beetle. *Proceedings of the National Academy of Sciences of the United States of America*, **113**, 5982.

Kijimoto, T., Moczek, A.P., and Andrews, J. (2012). Diversification of doublesex function underlies morph-, sex-, and species-specific development of beetle horns. *Proceedings of the National Academy of Sciences of the United States of America*, **109**, 20526–31.

Koganezawa, M., Kimura, K., and Yamamoto, D. (2016). The neural circuitry that functions as a switch for courtship versus aggression in *Drosophila* males. *Current Biology*, **26**, 1395.

Kohn, N.R., Reaume, C.J., Moreno, C., et al. 2013. Social environment influences performance in a cognitive task in natural variants of the foraging gene. *Plos One*, **8**, e81272.

Kramer, J.M. (2016). Regulation of cell differentiation and function by the euchromatin histone methyltransferases G9a and GLP. *Biochemistry and Cell Biology*, **94**, 26.

Kramer, J.M., Kochinke, K., Oortveld, M.A.W., et al. (2011). Epigenetic regulation of learning and memory by *Drosophila* EHMT/G9a. *Plos Biology*, **9**, e1000569.

Krieger, M.J.B., and Ross, K.G. (2005). Molecular evolutionary analyses of the odorant-binding protein gene Gp-9 in fire ants and other *Solenopsis* species. *Molecular Biology and Evolution*, **22**, 2090.

Kubrak, O.I., Kucerova, L., Theopold, U., and Nassel, D.R. (2014). The Sleeping beauty: how reproductive diapause affects hormone signaling, metabolism, immune response and somatic maintenance in *Drosophila melanogaster*. *Plos One*, **9**, e113051.

Kucharski, R., Maleszka, J., Foret, S., and Maleszka, R. (2008). Nutritional control of reproductive status in honeybees via DNA methylation. *Science*, **319**, 1827.

Kuntz, S., Poeck, B., Sokolowski, M.B., and Strauss, R. (2012). The visual orientation memory of *Drosophila* requires foraging (PKG) upstream of Ignorant (RSK2) in ring neurons of the central complex. *Learning and Memory*, **19**, 337.

Kwon, B., Song, S., Choi, J.Y., Je, Y.H., and Kim, Y. (2010). Transient expression of specific Cotesia plutellae bracoviral segments induces prolonged larval development of the diamondback moth, *Plutella xylostella*. *Journal of Insect Physiology*, **56**, 650.

Le Trionnaire, G., Francis, F., Jaubert-Possamai, S., et al. (2009). Transcriptomic and proteomic analyses of seasonal photoperiodism in the pea aphid. *BMC Genomics*, **10**, 456.

Leal, W.S., and Ishida, Y. (2008). GP-9s are ubiquitous proteins unlikely involved in olfactory mediation of social organization in the red imported fire ant, *Solenopsis invicta*. *Plos One*, **3**, e3762.

Lehmann, P., Lyytinen, A., Piiroinen, S., and Lindstrom, L. (2014). Northward range expansion requires synchronization of both overwintering behaviour and physiology with photoperiod in the invasive Colorado potato beetle (*Leptinotarsa decemlineata*). *Oecologia* **176**, 57.

Lucas, C., and Sokolowski, M.B. (2009). Molecular basis for changes in behavioral state in ant social behaviors. *Proceedings of the National Academy of Sciences of the United States of America*, **106**, 6351.

Macpherson, M.R., Broderick, K.E., Graham, S., et al. (2004). The dg2 (for) gene confers a renal phenotype in *Drosophila* by modulation of cGMP-specific phosphodiesterase. *Journal of Experimental Biology*, **207**, 2769.

Maleszka, R. (2008). Epigenetic integration of environmental and genomic signals in honey bees. *Epigenetics*, **3**, 188.

Maleszka, R. (2016). Epigenetic code and insect behavioural plasticity. *Current Opinion in Insect Science*, **15**, 45.

Manfredini, F., Riba-Grognuz, O., Wurm, Y., et al. (2013). Sociogenomics of cooperation and conflict during colony founding in the fire ant *Solenopsis invicta*. *Plos Genetics*, **9**, e1003633.

Martins, J.R., Nunes, F.M.F., Cristino, A.S., Simoes, Z.P., and Bitondi, M.M.G. (2010). The four hexamerin genes in the honey bee: structure, molecular evolution and function deduced from expression patterns in queens, workers and drones. *BMC Molecular Biology*, **11**, 23.

Mery, F., Belay, A.T., So, A.K.C., Sokolowski, M.B., and Kawecki, T.J. (2007). Natural polymorphism affecting learning and memory in *Drosophila*. *Proceedings of the National Academy of Sciences of the United States of America*, **104**, 13051.

Michener, C.D. (2000). *The Bees of the World*, 2nd edn. The Johns Hopkins University Press, Baltimore, MD.

Milan, N.F., Kacsoh, B.Z., and Schlenke, T.A. (2012). Alcohol consumption as self-medication against blood-borne parasites in the fruit fly. *Current Biology*, **22**, 488.

Moczek, A.P. (2009). The origin and diversification of complex traits through micro- and macroevolution of development: Insights from horned beetles. *Current Topics in Developmental Biology*, **86**, 135.

Moczek, A.P. (2010). Phenotypic plasticity and diversity in insects. *Philosophical Transactions of the Royal Society B-Biological Sciences*, **365**, 593.

Murren, C.J., Auld, J.R., Callahan, H., et al. (2015). Constraints on the evolution of phenotypic plasticity: limits and costs of phenotype and plasticity. *Heredity*, **115**, 293.

Nijhout, H.F. (2003). Development and evolution of adaptive polyphenisms. *Evolution and Development*, **5**, 9.

Oettler, J., Nachtigal, A.L., and Schrader, L. (2015). Expression of the foraging gene is associated with age polyethism, not task preference, in the ant *Cardiocondyla obscurior*. *Plos One*, **10**, e0144699.

Ogawa, K., and Miura, T. (2014). Aphid polyphenisms: trans-generational developmental regulation through viviparity. *Frontiers in Physiology*, **5**, 1.

Okada, K., and Miyatake, T. (2009). Genetic correlations between weapons, body shape and fighting behaviour in the horned beetle *Gnatocerus cornutus*. *Animal Behaviour*, **77**, 1057.

Osborne, K.A., Robichon, A., Burgess, E., et al. (1997). Natural behavior polymorphism due to a cGMP-dependent protein kinase of *Drosophila*. *Science*, **277**, 834.

Öst, A., Lempradl, A., Casas, E., et al. (2014). Paternal diet defines offspring chromatin state and intergenerational obesity. *Cell*, **159**, 1352.

Page, R.E., Rueppell, O., and Amdam, G.V. (2012). Genetics of reproduction and regulation of honeybee (*Apis mellifera* L.) social behavior. *Annual Review of Genetics*, **46**, 97.

Pereira, H.S., and Sokolowski, M.B. (1993). Mutations in the larval foraging gene affect adult locomotory behavior after feeding in *Drosophila melanogaster*. *Proceedings of the National Academy of Sciences of the United States of America*, **90**, 5044.

Piiroinen, S., Ketola, T., Lyytinen, A., and Lindstrom, L. (2011). Energy use, diapause behaviour and northern range expansion potential in the invasive Colorado potato beetle. *Functional Ecology*, **25**, 527.

Pruijssers, A.J., Falabella, P., Eum, J.H., et al. (2009). Infection by a symbiotic polydnavirus induces wasting and inhibits metamorphosis of the moth *Pseudoplusia includens*. *Journal of Experimental Biology*, **212**, 2998.

Reaume, C.J., Sokolowski, M.B., and Mery, F. (2011). A natural genetic polymorphism affects retroactive interference in *Drosophila melanogaster*. *Proceedings of the Royal Society B-Biological Sciences*, **278**, 91.

Reppert, S.M. (2006). A colorful model of the circadian clock. *Cell*, **124**, 233.

Reppert, S.M., Gegear, R.J., and Merlin, C. (2010). Navigational mechanisms of migrating monarch butterflies. *Trends in Neurosciences*, **33**, 399.

Reppert, S.M., Guerra, P.A., and Merlin, C. (2016). Neurobiology of Monarch butterfly migration. *Annual Review of Entomology*, **61**, 25.

Robich, R.M., and Denlinger, D.L. (2005). Diapause in the mosquito *Culex pipiens* evokes a metabolic switch from blood feeding to sugar gluttony. *Proceedings of the National Academy of Sciences of the United States of America*, **102**, 15912.

Robinson, G.E., Fernald, R.D., and Clayton, D.F. (2008). Genes and social behavior. *Science*, **322**, 896.

Robinson, G.E., Grozinger, C.M., and Whitfield, C.W. (2005). Sociogenomics: social life in molecular terms. *Nature Reviews Genetics*, **6**, 257.

Roitberg, B.D., Sircom, J., Roitberg, C.A., Vanalphen, J.J.M., and Mangel, M. (1993). Life expectancy and reproduction. *Nature*, **364**, 108.

Sandrelli, F., Tauber, E., Pegoraro, M., et al. (2007). A molecular basis for natural selection at the timeless locus in *Drosophila melanogaster*. *Science*, **316**, 1898.

Saunders, D.S., and Gilbert, L.I. (1990). Regulation of ovarian diapause in *Drosophila-melanogaster* by photoperiod and moderately low-temperature. *Journal of Insect Physiology*, **36**, 195.

Saunders, D.S., Henrich, V.C., and Gilbert, L.I. (1989). Induction of diapause in *Drosophila-melanogaster*—photoperiodic regulation and the impact of arrythmic clock mutations on time measurement. *Proceedings of the National Academy of Sciences of the United States of America*, **86**, 3748.

Scheiner, R., Page, R.E., and Erber, J. (2004). Sucrose responsiveness and behavioral plasticity in honey bees (*Apis mellifera*). *Apidologie*, **35**, 133.

Schiesari, L., Andreatta, G., Kyriacou, C.P., O'connor, M.B., and Costa, R. (2016). The insulin-like proteins dILPs-2/5 determine diapause inducibility in *Drosophila*. *Plos One*, **11**, e0163680.

Schmidt, P.S. (2011). Evolution and mechanisms of insect reproductive diapause: A plastic and pleiotropic life history syndrome. In: T. Flatt, and A. Heyland (Eds) *Mechanisms of Life History Evolution: the Genetics and Physiology of Life History Traits and Trade-offs*, pp. 230–42. Oxford University Press, Oxford.

Schmidt, P.S., Matzkin, L., Ippolito, M., and Eanes, W.F. (2005). Geographic variation in diapause incidence, life-history traits, and climatic adaptation in *Drosophila melanogaster*. *Evolution*, **59**, 1721.

Schmidt, P.S., Zhu, C.T., Das, J., et al. (2008). An amino acid polymorphism in the couch potato gene forms the basis for climatic adaptation in *Drosophila melanogaster*. *Proceedings of the National Academy of Sciences of the United States of America*, **105**, 16207.

Schwab, D.B., and Moczek, A.P. (2016). Nutrient stress during ontogeny alters patterns of resource allocation in two species of horned beetles. *Journal of Experimental Zoology Part a-Ecological Genetics and Physiology*, **325**, 481.

Schwenke, R.A., Lazzaro, B.P., and Wolfner, M.F. (2016). Reproduction-immunity trade-offs in insects. *Annual Review of Entomology*, **61**, 239.

Scofield, H.N., and Mattila, H.R. (2015). Honey bee workers that are pollen stressed as larvae become poor foragers and waggle dancers as adults. *Plos One*, **10**, e0121731.

Shaver, S.A., Varnam, C.J., Hilliker, A.J., and Sokolowski, M.B. (1998). The foraging gene affects adult but not larval olfactory-related behavior in *Drosophila melanogaster*. *Behavioural Brain Research*, **95**, 23.

Shibao, H., Kutsukake, M., and Fukatsu, T. (2004). Density-dependent induction and suppression of soldier differentiation in an aphid social system. *Journal of Insect Physiology*, **50**, 995.

Sim, C., and Denlinger, D.L. (2008). Insulin signaling and FOXO regulate the overwintering diapause of the mosquito *Culex pipiens*. *Proceedings of the National Academy of Sciences of the United States of America*, **105**, 6777.

Simmons, L.W., and Emlen, D.J. (2006). Evolutionary trade-off between weapons and testes. *Proceedings of the National Academy of Sciences of the United States of America*, **103**, 16346.

Simões, P.M.V., Ott, S.R., and Niven, J.E. (2016). Environmental adaptation, phenotypic plasticity, and associative learning in insects: the desert locust as a case study. *Integrative and Comparative Biology*, **56**, 914.

Simola, D.F., Graham, R.J., Brady, C.M., et al. (2016). Epigenetic (re)programming of caste-specific behavior in the ant *Camponotus floridanus*. *Science*, **351**, 42.

Simpson, S.J., Sword, G.A., and Lo, N. (2011). Polyphenism in insects. *Current Biology*, **21**, R738–49.

Smid, H.M., and Vet, L.E.M. (2016). The complexity of learning, memory and neural processes in an evolutionary ecological context. *Current Opinion in Insect Science*, **15**, 61.

Snell-Rood, E.C., and Moczek, A.P. (2012). Insulin signaling as a mechanism underlying developmental plasticity: the role of FOXO in a nutritional polyphenism. *Plos One*, **7**, e34857.

Sokolowski, M.B. (1980). Foraging strategies of *Drosophila melanogaster*—a chromosomal analysis. *Behavior Genetics*, **10**, 291.

Sokolowski, M.B., and Levine, J.D. (2010). Nature–nurture interactions. In: T. Szekely, A. J. Moore, J. Komdeur (Eds) *Social Behaviour, Genes, Ecology and Evolution*, pp. 11–25. Cambridge University Press, Cambridge.

Sokolowski, M.B., Pereira, H.S., and Hughes, K. (1997). Evolution of foraging behavior in Drosophila by density-dependent selection. *Proceedings of the National Academy of Sciences of the United States of America*, **94**, 7373.

Sokolowski, M.B., and Wahlsten, D. (2001). Gene–environment interaction and complex behaviour. In: M.O. Moldin (Ed.) *Methods in Genomic Neuroscience*, pp. 3–28. CRC Press, Boca raton, FL.

Tauber, E., Zordan, M., Sandrelli, F., et al. (2007). Natural selection favors a newly derived timeless allele in *Drosophila melanogaster*. *Science*, **316**, 1895.

Toth, A.L., Varala, K., Newman, T.C., et al. (2007). Wasp gene expression supports an evolutionary link between maternal behavior and eusociality. *Science*, **318**, 441.

Trible, W., Chang, N-C., Matthews, B.J., et al. (2017). *orco* Mutagenesis causes loss of antennal lobe glomeruli and impaired social behavior in ants. *Cell*, **170**(4), 727–35.e10.

Urquhart-Cronish, M., and Sokolowski, M.B. (2014). Gene-environment interplay in *Drosophila melanogaster*:

chronic nutritional deprivation in larval life affects adult fecal output. *Journal of Insect Physiology*, **69**, 95.

Vellichirammal, N.N., Madayiputhiya, N., and Brisson, J.A. (2016). The genomewide transcriptional response underlying the pea aphid wing polyphenism. *Molecular Ecology*, **25**, 4146.

Via, S., Gomulkiewicz, R., Dejong, G., et al. (1995). Adaptive phenotypic plasticity—consensus and controversy. *Trends in Ecology and Evolution*, **10**, 212.

Villalta, I., Blight, O., Angulo, E., Cerda, X., and Boulay, R. (2016). Early developmental processes limit socially mediated phenotypic plasticity in an ant. *Behavioral Ecology and Sociobiology*, **70**, 285.

Vrontou, E., Nilsen, S.P., Demir, E., Kravitz, E.A., and Dickson, B.J. (2006). Fruitless regulates aggression and dominance in *Drosophila*. *Nature Neuroscience*, **9**, 1469.

Wang, J., Ross, K.G., and Keller, L. (2008). Genome-wide expression patterns and the genetic architecture of a fundamental social trait. *Plos Genetics*, **4**, e1000127.

Wang, X.H., and Kang, L. (2014). Molecular mechanisms of phase change in locusts. *Annual Review of Entomology*, **59**, 225.

Wang, Y., Kaftanoglu, O., Brent, C.S., Page, R.E., and Amdam, G.V. (2016). Starvation stress during larval development facilitates an adaptive response in adult worker honey bees (*Apis mellifera* L.). *Journal of Experimental Biology*, **219**, 949.

Weiner, S.A., Galbraith, D.A., Adams, D.C., et al. (2013). A survey of DNA methylation across social insect species, life stages, and castes reveals abundant and caste-associated methylation in a primitively social wasp. *Naturwissenchaften*, **100**, 795.

West-Eberhard, M.J. (2003). *Developmental Plasticity and Evolution*, Oxford University Press, Oxford.

West-Eberhard, M.J. (2007). Dancing with DNA and flirting with the ghost of Lamarck. *Biology and Philosophy*, **22**, 439.

West-Eberhard, M.J. (1989). Phenotypic plasticity and the origins of diversity. *Annual Review of Ecology and Systematics*, **20**, 249.

Whitfield, C.W., Ben-Shahar, Y., Brillet, C., et al. (2006). Genomic dissection of behavioral maturation in the honey bee. *Proceedings of the National Academy of Sciences of the United States of America*, **103**, 16068.

Whitfield, C.W., Cziko, A.M., and Robinson, G.E. (2003). Gene expression profiles in the brain predict behavior in individual honey bees. *Science*, **302**, 296.

Williams, K.D., Busto, M., Suster, M.L., et al. (2006). Natural variation in *Drosophila melanogaster* diapause due to the insulin-regulated PI3-kinase. *Proceedings of the National Academy of Sciences of the United States of America*, **103**, 15911.

Williams, K.D., Schmidt, P.S., and Sokolowski, M.B. (2010). Photoperiodism in insects: Molecular basis and consequences of diapause. In: R. J. Nelson, D. L, Denlinger, and D. E. Somers (Eds) *Photoperiodism*, pp. 287–317. Oxford University Press, New York, NY.

Williams, K.D., and Sokolowski, M.B. (1993). Diapause in *Drosophila melanogaster* females—a genetic-analysis. *Heredity*, **71**, 312.

Xu, S.J., Panikker, P., Iqbal, S., and Elefant, F. (2016). Tip60 HAT action mediates environmental enrichment induced cognitive restoration. *Plos One*, **11**, e0159623.

Yan, H., Bonasio, R., Simola, D.F., et al. (2015). DNA methylation in social insects: how epigenetics can control behavior and longevity. *Annual Review of Entomology*, **60**, 435.

Yan, H., Simola, D.F., Bonasio, R., et al. (2014). Eusocial insects as emerging models for behavioural epigenetics. *Nature Reviews Genetics*, **15**, 677.

Zhan, S., Zhang, W., Niitepold, K., et al. (2014). The genetics of monarch butterfly migration and warning colouration. *Nature*, **514**, 317.

Zhou, X.G., Oi, F.M., and Scharf, M.E. (2006). Social exploitation of hexamerin: RNAi reveals a major caste-regulatory factor in termites. *Proceedings of the National Academy of Sciences of the United States of America*, **103**, 4499.

Zhu, H.S., Casselman, A., and Reppert, S.M. (2008). Chasing migration genes: a brain expressed sequence tag resource for summer and migratory Monarch butterflies (*Danaus plexippus*). *Plos One*, **3**, e1345.

Zhu, H.S., Gegear, R.J., Casselman, A., Kanginakudru, S., and Reppert, S.M. (2009). Defining behavioral and molecular differences between summer and migratory monarch butterflies. *BMC Biology*, **7**, 14.

CHAPTER 6

Habitat selection and territoriality

Darrell J. Kemp

*Department of Biological Sciences, Macquarie University,
North Ryde, New South Wales, Australia*

6.1 Introduction

The vast evolutionary radiation of insects has seen them exploit virtually every habitat type on earth. Insects dominate the terrestrial domain, where they pose the lion's share of metazoan biodiversity across land and freshwater habitats, and the skies above. Species are often found in highly specific habitats and individuals may seek to occupy different habitats depending upon their life-history stage. In the adult stage, favoured habitats often vary on both seasonal and diel time frames, and routinely diverge between the sexes.

Most broadly, habitat choice represents the outcome of a sensory/behavioral interaction between individuals and their environment. This is driven by abiotic features, such as thermal microclimate (etc.), but also biotic factors such as the occurrence of conspecific mates or rivals, and heterospecific competitors and/or predators. Individuals are expected to assort among habitats in order to achieve the greatest fitness outcome from the task at hand. In broad terms, the goal of immature insects is to optimize their rate of resource acquisition. Juvenile habitat selection is, therefore, guided by perceived nutritional gradients, and often shaped by the maternal provisioning and/or oviposition decisions (Lancaster et al. 2011). Upon reaching adulthood, primary goals then shift towards dispersal and reproduction. The different fitness priorities of adult males versus females means that optimal habitats often vary, yet they must converge at some point for mating to occur. Habitat selection across life stages is expected to remain sensitive to predation and competition, and seated within phenological strategies, such as migration, quiescence, and diapause (i.e. escape in space versus time).

This chapter deals with habitat selection in reproductively-active adults. Focus is placed upon the selection/defence of mating sites by males, which has offered a fertile testing ground for evolutionary theory. Indeed, the heritage of such work traces from the very first tests of **evolutionary game theory** (Box 6.1), and its progression since offers a narrative of intellectual development in biology.

6.2 Adult sex roles and behavior

A general theme among sexually-reproductive, diploid insects is that males seek to maximize their rate of copulation, whereas females seek to maximize offspring provisioning (ova production, oviposition, and sometimes parental care; see also Chapter 13). Bateman's (1948) principle explains these divergent objectives in terms of investments tracing to the gametic level (i.e. **anisogamy**). Males, by definition, produce gametes consisting of little

Kemp, D. J., *Habitat selection and territoriality*. In: *Insect Behavior: From mechanisms to ecological and evolutionary consequences*. Edited by Alex Córdoba-Aguilar, Daniel González-Tokman, and Isaac González-Santoyo: Oxford University Press (2018). © Oxford University Press.
DOI: 10.1093/oso/9780198797500.003.0006

Box 6.1 Game theory and Insect behavior: an historic perspective

Evolutionary game theory seeks to model the optimization of individual fitness when the costs and benefits of particular tactics depend upon those adopted by competitors. It recognizes that outcomes are 'zero sum,' meaning that one individual's gain must come at the expense of other individuals' losses. The logic of game theory is rooted in applied mathematics, with tenets such as the Nash equilibrium developed for human economics. It was first levelled at biological adaptation in a systematic manner by John Maynard Smith and colleagues (Maynard Smith and Price 1973, Maynard Smith 1982). Evolutionary game theory has proven enormously valuable, not in the least as a heuristic for emphasizing the frequency-dependent context of *individual-level* tactics. As a quantitative framework, it has outperformed classic applications of game theory because evolutionary adaption provides a more reliable cavass for cost-benefit optimization than human decision-making (Gintis 2009).

Game theoretic models seek to define evolutionarily stable strategies (ESSs). These are population-level strategies incapable of being invaded by potential (biologically-realistic) alternatives. The earliest conception of evolutionary game theory considered pairwise (dyadic) contests (Maynard Smith 1972, 1974; Maynard Smith and Price 1973; Maynard Smith and Parker 1976). Here, the 'hawk–dove' game indicated how aggressive and non-aggressive tactics could co-exist as ESSs at equilibrium provided that fighting costs (C) remain sufficiently high relative to the payoffs of winning (V). A simple elaboration of this game then informed a rather profound heuristic: that a strategy based purely upon convention (akin to 'tossing a coin') could under some circumstances trump all (Maynard Smith and Parker 1976). As for a coin toss, the potential stability of this convention demands that roles are unequivocally perceived, assigned at random, and unrelated to success via any other means. Behaviorists acted quickly to test this idea in relation to the asymmetry offered by the resident-intruder roles of territorial male insects (Figure 6.4). Despite initially exciting results, nearly four decades of empirical study has revealed that purely conventional strategies of this nature are rare (if not non-existent) in nature. This is not to say that animal contests do not lack

conventional components. Indeed, both the accumulation of data and progression of theory (e.g. Mesterton-Gibbons et al. 2016b; Mesterton-Gibbons and Sherratt 2016) support a complex interaction between convention and biology in shaping real-world contest behavior (see main text).

The hawk–dove game was presented almost simultaneously with the 'war of attrition' (Maynard Smith 1974). This model, subsequently generalized by Bishop and Cannings (1978a,b) and (Mesterton Gibbons et al. 1996), emphasizes situations where success is determined by persistence. Hence, whereas the hawk–dove game parameterized tactical decisions in discrete terms (i.e. the escalation to costly fighting), the war of attrition recognized an underlying continuous basis to such decisions (i.e. when to 'give-up'[1]). For symmetric contests, this model revealed that individuals should choose persistence times essentially at random (Bishop and Cannings 1978b), hence, generating a negative exponential distribution for contest duration. The first empirical tests revealed two critical points; namely, that contests will rarely be symmetric in nature (Parker and Thompson 1980), and that individuals should modify their behavior whenever relevant asymmetries can be assessed (Sigurjonsdottir and Parker 1981).

Since its inception, evolutionary game theory has developed in ways that more accurately characterize real-world animal behavior. Elaborations have dealt with how asymmetries determine optimal tactics (Leimar and Enquist 1984), and how information is assessed and may accumulate in contest situations (Parker and Rubenstein 1981; Enquist and Leimar 1983b; Arnott and Elwood 2009). The progression of theory has been greatly informed by dedicated empirical tests, and inspired by biological intuition. Insect systems, given their tractability for both natural observation and experimental manipulation, have played a pivotal role in testing and reciprocal development of models. Parker (2013) and Kokko (2013) provide excellent treatises of the history, mathematical basis and subsequent development of evolutionary game theory.

[1] 'Hawk' and 'dove' strategies essentially exist in the war of attrition model via their analogy to individuals that either choose to never give-up or give-up immediately.

---

more than haploid DNA packaged in a mobile instrument (i.e. sperm), whereas females produce larger and more nutritionally rich gametes (i.e. eggs). Among insects, the resource investment in a single egg may outweigh that of a single sperm by up to eight orders of magnitude (Alcock 1993). Anisogamy

is theorized to have evolved from ancestrally **disruptive selection** upon gamete size, thereby both defining each sex and predisposing their roles (Bateman 1948; Fritzsche and Arnqvist 2013). Conventional sex roles are, as colourfully expressed by Parker et al. (1972), such that '*males are dependent* *on females and propagate at their expense, rather as in a parasite–host relationship.'* Trivers (1972) expanded upon anisogamy to consider the manifold additional dimensions of **parental investment**, thereby generalizing the theory of sex roles. This theory, and its recent elaborations (e.g. Kokko et al. 2014) explain situations where conventional sex roles are reversed, that is, female-competitive **mating systems**. Cases of both obligatory and facultative sex-role reversal certainly exist in insects (e.g. Gwynne and Simmons 1990; Funk and Tallamy 2000; see also Chapter 13). This chapter is phrased in male-competitive terms throughout, but purely for textual expediency.

Insects are overwhelmingly characterized by the conventional sex-role scheme, with populations typified by many receptive males, but few receptive females, i.e. male-biased **operational sex ratios** (OSRs; Emlen and Oring 1977). This promotes sexual competition among males (Alcock 1993), played out in the form of scrambles, endurance rivalry, contests, mate choice, and **sperm competition** (Darwin 1874; Parker 1970; Andersson 1994). 'Sexual selection' (Darwin 1874) has consequently favoured elaborate visual and auditory displays, weapons, mating plugs, and other traits that enhance fertilization success (Parker 1970; see also Chapter 13). It shapes the behavior of mate-seeking males, encompassing such decisions as which habitats to occupy, when to occupy them, and how to behave in relation to competitors at such locations (Thornhill and Alcock 1983).

6.3 Mating habitats, site selection, and territoriality

Aside from foraging and dispersal, reproductively active male insects are (as noted) generally compelled to maximize their encounters with receptive female conspecifics. Tactical choice of habitats and behaviors will, therefore, be shaped first-and-foremost by the biology of female receptivity (Bradbury 1985; Rutowski 1991; Wickman and Rutowski 1999). This simple tenet is fundamental to our contemporary understanding of insect mating systems (Thornhill and Alcock 1983). Two principal concerns for males are the habitats where receptive females are most predictably located, and the occurrence and behavior of competitors. Both factors

shape site selection at the individual level, and determine the occurrence/economics of notable tactics such as **mate guarding** (Alcock 1994) and site defence (i.e. **territoriality**).

6.3.1 The spatiotemporal basis of mating habitats

The theatre for mate location is framed first-and-foremost by the spatiotemporal distribution of receptive females. This depends, in turn, upon factors such as juvenile population dispersion, female pre-reproductive (refractory) periods and dispersal behavior, habitat specificity and/or **aggregation**, and lifetime mating schedules (Wickman and Rutowski 1999). Given the immense diversity of insect biology, these multiple axes of variation define an enormous range of target habitats and occupation schedules. Thornhill and Alcock (1983) presented the incisive view that mate encounter sites will generally fall within four categories:

- The zone of adult emergence.
- Adult foraging areas.
- Oviposition sites.
- Non-resource-based landmarks.

The first three categories need not be mutually exclusive, nor may they prove practically distinguishable for species that oviposit, develop, mate, and forage in highly localized habitats (e.g. *Eurema* butterflies; Kemp 2008).

When females are receptive upon reaching adulthood and predictably located, mating activity often centres upon the habitats of juvenile development. Males seek to occupy and/or search these habitats prior to female emergence. The term 'protandry' is used to describe either the early emergence (i.e. faster rate of maturation) of males, or in a broader sense as the earlier arrival of males to mate encounter sites (Morbey and Ydenberg 2001). Protandry is seen particularly in species that possess discrete and/or synchronous generations (Wiklund and Fagerstrom 1977). Males subsequently compete to mate with females as they emerge from pupa, burrows, or moult to adulthood. Biases in the timing of female emergence can define specific diel schedules of male site occupation (e.g. as classically shown for

Figure 6.1 Flexibility in male mate searching tactics is favoured by variation in male competitiveness and/or the habitats where receptive females predictably occur. A clear example is given by pupal-mating *Heliconius* butterflies such as *H. charithonia*. (a) Conspecific female pupae pose a major target of mate location. Individual males seek to locate these in the first instance via a visual search of the hostplant habitat, and may traverse circuitous 'trap-line' routes that encompass multiple hostplant patches. Pupal gender recognition is aided by the release of monoterpene volatiles by females in their latter stage of pupal development (Estrada et al. 2010); (b) Once pupa are located, males maintain perching vigils that usually span 24–48 hours, but may last up to 10 days. Multiple males often jostle for perching positions upon pupae, and then to achieve copulation at the moment of female eclosion (Deinert et al. 1994). This image shows two adult males perching on a nascent female pupa; (c) aside from pupal mating, male *Heliconiines* also seek eclosed females by defending locations such as sunny corridors away from eclosion sites (Benson et al. 1989). In *H. sara*, smaller males are more likely to defend such sites, which may reflect their reduced potential to win out in the physical jostling upon pupae (Hernandez and Benson 1998). The image shows a male *H. charithonia* courting a 4–5-day-old female in an experimental greenhouse environment.

Figure credit: D.J. Kemp.

the solitary bee, *Centris pallida*; Alcock et al. 1977). Likewise, focal habitats are defined by features of juvenile biology that determine the spatial distribution of maturing females. Intriguing cases exist where dedicated searching within these habitats enables the location of sub-adult females themselves (Deinert et al. 1994; Estrada et al. 2010). The males of such

species are notable for their pre-copulatory vigils (Estrada et al. 2010), their efforts to competitively exclude rivals during such vigils (Conner and Itagaki 1984), and/or their attempts to secure copulation at the precise moment of eclosion (Elgar and Pierce 1978; Deinert et al. 1994; Figure 6.1).

Away from emergence sites, theory expects that mate-searching males would target habitats where the occurrence of receptive females becomes next predictable. Thornhill and Alcock (1983) recognized that the distribution of adult resources will often determine profitable search locations. Classic support is given by the oviposition site-based mating systems of odonates (i.e. damselflies and dragonflies). Juveniles of this hemi-metabolous order develop in aquatic habitats and ultimately moult to winged adults, but adulthood includes an initial teneral (sub-imago) period that may last several days (Córdoba-Aguilar and Cordero-Rivera 2005). In well-studied genera, such as *Calopteryx*, teneral adults are exclusively concerned with resource acquisition (Kirkton and Schultz 2001), and females refrain from mating until they arrive to oviposition habitats. As expected, such habitats define the spatial focus of mate-locating males (Waage 1988; Marden and Waage 1990).

The fourth category of mate-encounter site consists of symbolic or landmark-based 'rendezvous' locations such as hilltops (Alcock 1987). Given their clear segregation from juvenile habitats and adult resources, the use of such sites typifies insect lekking behavior. They are thought to be favoured in highly-dispersed species, where the sexes are unlikely to otherwise coincide. Landmark- and/or lek-based systems have been well-studied in solitary wasps (e.g. Alcock 1981), flies (e.g. Kaspi and Yuval 1999; Alcock and Kemp 2006) and butterflies (e.g. Wickman 1988), and known to exist across a greater diversity of groups (Cole and Wiernasz 1997; Hill 1999; Córdoba-Aguilar et al. 2009).

Against the background of putatively 'mainstream' mating tactics, it is important to appreciate the high prevalence of adaptive diversity in male behavior (Brockmann 2008). Tactical diversity may stem from variation in the spatio-temporal occurrence of receptive mates and/or male competitive phenotypes. In polyandrous species, the occurrence of receptive females will often differ depending upon mating status (e.g. virgin as opposed to receptive previously-mated individuals). Virgin females in strictly **monandrous** species may likewise exist outside of primary mating habitats (Alcock et al. 1977; Hernandez and Benson 1998). These scenarios pose consequences for optimal mate-location by diversifying the habitats under which females are profitably sought (Thornhill and Alcock 1983). Pre-mated females may constitute the primary target for males that are less successful in open competition, or for all males when virgin females are unavailable. Overall, any fluidity in the availability of receptive females will favour the diversification of mate-seeking behavior (see Alcock 1994).

6.3.2 The occurrence and economics of site defence

Many insects exhibit sexual behavior akin to 'prolonged mate-searching **polygyny**.' (Thornhill and Alcock 1983). Members of the proactive sex—most often males—search rewarding habitats and do not act aggressively towards competitors. Long-range mate orientation may sometimes be mediated via advertisement, as in the pheromone signals of female moths and vocal calling of male crickets. Broadly, sexual selection across these species is levied according to the individual capacity to prevail in scramble competition and/or mate choice (Andersson 1994). Elaborations involve mate-guarding, aimed either to secure copulation (as mentioned previously; Figure 6.1) or to ensure post-copulatory fertilization (Alcock 1994). The latter is championed by male odonates that routinely 'chaperone' their mates via tandem association before and during oviposition (e.g. Córdoba-Aguilar et al. 2009). A distinctive class of sexual behavior is exhibited by species whose males strive to defend discrete encounter sites. Such locations may or may not contain resources (e.g. Alcock 1981, 1987), and do not fit neatly into broader classifications, such as the 'resource defence polygyny' of Emlen and Oring (1977). This chapter considers the dynamics of how male insects attempt to monopolize sites irrespective of their resource basis.

The conditions favouring territorial behavior are theorized in terms of cost-benefit economics (*sensu* the 'economic defensibility' principle; see Chapter 13).

Site defence is expected when the likely pay-off for securing mating opportunities exceeds the cost of intercepting and expelling intruders (Thornhill and Alcock 1983). Costs rise with the rate at which intruders arrive; hence, the viability of this tactic is contingent upon the intensity of competition at encounter sites. Instrumental support comes from studies of hilltopping butterflies, where all individuals abandon territoriality once male density at the hilltop breaches a threshold (e.g. Alcock and O'Neill 1987; Alcock 1994).

The economics of site defence is nevertheless a personal equation because different phenotypes vary in their ability to impose costs upon competitors and/or their perception of site value (see Section 6.4.1 for the formal theory). In proximate terms, the costs of occupying and defending territories are levied in currencies of time, energy, and the risks of injury or death. There is great diversity among and within insect groups regarding the importance of these currencies. Ultimately, how they shape the economics of site defence is determined by the form of contest behavior itself. As per adaptive prediction, males engage in territoriality according to their personal cost-benefit schedules. This is again reflected in behavioral plasticity (Brockmann 2008), clearly seen for example when less competitive phenotypes adopt non-aggressive sneaker or satellite behaviors (Alcock et al. 1977; Waage 1988).

6.3.3 Encounter site fidelity

Territoriality exists hand-in-hand with the sustained individual occupation of particular encounter sites. This phenomenon is commonly termed site 'fidelity' or 'tenacity'. Maximum tenure may span time frames measured in hours, days, weeks, and even months. Notable site faithfulness is shown by territorial butterflies, such as *Hypolimnas bolina*, where males occupy the same perching location for up to 54 days (Kemp 2001). Residents in this species have been seen back at their territory following capture and release several kilometres away (McCubbin 1971), and even after spending several weeks in an insectary (D. J. Kemp, unpublished data). At the population level, fidelity is also expressed via consistent site preferences across seasons and/or years. Such consistency has been demonstrated for the rank attractiveness of hilltop territories in wasps (Alcock 1983) and flies (Alcock and Kemp 2004), as well as the placement of non-aggressive leks (e.g. Svensson and Petersson 1995).

Present knowledge of insect site fidelity has benefited greatly from Paul Switzer's work in the 1990s (Switzer 1995). This included a dynamic-state model of individual fidelity based upon (a) the instantaneous value of site occupation (i.e., spatial variation in site value), (b) the predictability of future site value (i.e., temporal variation in site value), and (c) the cost of seeking to switch sites (Switzer 1993). This simple parameterization cleverly cast the potential psychosensory basis of prior experience and expected future rewards against the danger of pursuing alternative options (including the mortality risk associated with dispersing among sites). Model outcomes accorded with intuitive expectation in that fidelity should increase with the cost of changing sites, but decrease when sites vary more greatly in quality. The latter follows because males can more quickly and accurately gauge site value, leading them to seek, where possible, higher-quality alternatives.

This model also furnished two life history predictions. Namely, that site fidelity should increase with age irrespective of lifespan, but reduce overall in shorter-lived species. These predictions have since found empirical support via tests in dragonflies (Switzer 1997a,b), and have guided the interpretation of site fidelity across broader groups, notably wasps (Alcock 2000) and butterflies (Kemp 2001; Takeuchi and Imafuku 2005).

6.3.4 Contest form

As in the study of site fidelity, territorial contests have been most intensively examined in flies, solitary wasps, butterflies, and odonates (damselflies and dragonflies). Competition in these systems is characterized by consecutive pairwise disputes between residents, their neighbours, temporary interlopers, and incoming non-residents (see later). Unlike the physical duels of some insects (e.g. horned beetles; Eberhard 1979), individuals routinely contest site ownership in a manner akin to endurance tests. These often consist of singular or iterated constant-

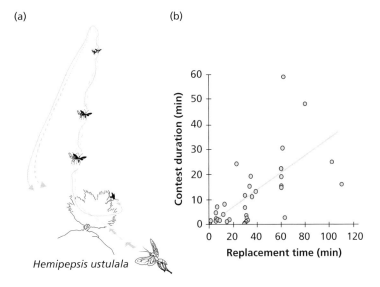

Figure 6.2 Contest persistence has been demonstrated to increase with time-in-residency for a range of territorial species, including this notable example in the tarantula hawk wasp (*Hemipepsis ustulata*). (a) Male *H. ustulata* contest the ownership of prominent tree and shrub locations at hilltops in the Sonoran Desert (USA). Male–male interactions largely consist of chases in which residents quickly vanquish intruders, but occasionally escalate to a series of ascending helical flights. (b) Alcock and Bailey (1997) removed residents and allowed 'replacements' to establish at the site for varying durations. Upon release, initial residents engaged their replacement via greatly escalated contests. Replacement males were ultimately defeated (in sixty of sixty-six cases), but contest duration scaled with the duration of their tenure at the site.

Figure adapted from Alcock and Bailey 1997.

intensity bouts that are terminated when one individual flees the area (i.e. 'gives up'). The ritualized appearance of these interactions (e.g. Figure 6.2a), has made it difficult to interpret how costs accrue (Kemp and Wiklund 2001). In other cases, there are identifiable stages of escalation (Waage 1988) which may include efforts to inflict injury (Ruppell and Hilfert-Ruppell 2013) and/or culminate in outright physical fighting (Carvalho et al. 2016).

Given that territorial male insects typically detect and classify conspecifics via sensory modalities with defined limits, individual recognition among contestants is rarely feasible. With the exception of social insect colonies (Newey et al. 2010), this is true even at close quarters and irrespective of prior encounters. Participants in pairwise contests are therefore considered naïve regarding opponent identity. As explored below, this has important implications for the roles of '**dear enemy**' and '**nasty neighbour**' effects (Ydenberg et al. 1988; Temeles 1994), and for the evolution of conventional rules based upon residency tenure (Alcock and Bailey 1997; Mesterton-Gibbons and Sherratt 2016).

6.4 The logical basis of dyadic contests

6.4.1 Fighting ability, resource value, and motivation

Fundamentally, animal contest theory has emphasized the importance of two asymmetries: **resource-holding power** (or **potential**; hereafter RHP; Parker 1974), and **resource value** (hereafter RV). The former represents the ability to exclude opponents in open competition, whereas the latter influences the motivation to do so. Variation in RHP is clearly seen (or envisaged) via morphological features, such as body size and weaponry that determine the ability to impose costs in physical fighting. Individual-level variation in RV is often more difficult to conceive because of its subjective and contextual basis. At the population level, RV is perhaps best envisaged via clearly-evident differences in the quality of food items, mates, or territories. In game-theoretic modelling, RHP and RV are essentially parameterized in terms of the cost (C) and value (V) of contest participation (Maynard Smith 1982; Mesterton Gibbons et al. 1996).

Importantly, whereas early models assumed fixed values and/or distributions of V and C (Box 6.1), 'self-consistent' modelling has since recognized that population feedbacks will generate fluidity over evolutionary time (Kokko et al. 2006). In other words, these parameters will co-evolve along with the ESS strategies that they favour in ecological time.

Variation in RHP and/or RV are predicted to define the personal cost:benefit schedules of dyadic combatants. By itself, however, empirical knowledge (or estimates) of these parameters can rarely predict the form and outcome of individual contests. There are two main reasons for this. The first is that both RHP and RV are ultimately defined relative to their distribution across the contemporary population (which is expected to change over time; Kokko et al. 2006). One consequence of this is that individuals cannot possess perfect knowledge of either parameter. They do, however, have the potential to gauge information during individual contests, which favours the pairwise assessment of RHP, and can likewise accumulate information regarding population-level RHP and RV via lifetime experience. The second important point is that population 'templates' of contest competition may favour the evolution of behavior based upon conventional rules. Simplistically, the logic behind convention lies in its potential to 'short-circuit' otherwise costly routes to dispute resolution. Such rules may involve restraint, for example, when the fitness rewards of winning a territory are low relative to the costs of ousting incumbents.

6.4.2 The availability and assessment of information

Kokko (2013) usefully categorized dyadic contest models as either 'black box' or 'open box' according to whether behavior is plastic once fighting starts. The former are exemplified by classic hawk–dove and non-assessment war of attrition models (Box 6.1). Here, contestants initially choose a tactic that they adhere to in the manner of an automaton. Tactical choice is assumed to remain sensitive to information gained before each contest; for example, judgements of personal RHP or experiential knowledge of RV (e.g. Sigurjonsdottir and Parker 1981; Bergman et al. 2010). Models such as the non-assessment war of attrition also assume that contestants participate at requisite intensity(ies) prior to reaching their pre-determined 'giving-up' point (Mesterton Gibbons et al. 1996). Open box models, by contrast, consider the potential for behavioral flexibility based upon information gained during contests themselves. Such flexibility makes particular sense when relative RHP is easily perceived and/or signalled (e.g. Fitzstephens and Getty 2000; Moore and Martin 2016).

The potential for pairwise information exchange during contests featured greatly in nascent game theoretic modelling (Parker 1974; Parker and Rubenstein 1981). It was subsequently formalized by the '**sequential assessment**' family of models (Enquist and Leimar 1983a,b, 1987; Leimar and Enquist 1984) and later the '**cumulative assessment**' model (Payne 1998). An excellent historical narrative of the development of assessment-based modelling is provided by Kokko (2013).

Related to intra-contest assessment, males also have the potential to gather information via past contest experience. This possibility has been explored under the banners of 'winner' and 'loser' effects, as synthesized recently by Mesterton-Gibbons et al. (2016a). From an individual perspective, such effects may allow more accurate judgements of the competitive context they find themselves in. Importantly, relevant theory has explicitly dealt with the life-history context in which such information will accumulate (e.g. Fawcett and Johnstone 2010).

6.4.3 Convention

Theory has also dealt extensively with the evolution of conventional rules. A now-famous example emerged from the early game–theoretic treatment of uncorrelated asymmetries (Maynard Smith and Parker 1976). Here, modelling suggested that settlement based upon a reliable cue (akin to 'tossing a coin') could, under certain conditions, prove **evolutionarily stable** (Box 6.1). Two versions of this tactic – termed '**bourgeois**' and '**anti-bourgeois**'—consider residency as such a cue. Mathematically, each version is equally likely to evolve (Kokko et al. 2006). They differ in prescribing whether non-residents always retreat once contested (bourgeois) or residents always retreat (anti-bourgeois). Under a bourgeois-like rule, resident success rates would increase and therefore

generate high site fidelity over discrete time-scales, whereas anti-bourgeois would lead to a situation of fluid site occupation (i.e. fidelity ≈0).

Literal interpretation of the uncorrelated asymmetry idea has fuelled ongoing debate. One of the first empirical tests, conducted in insects, was met with immediate scepticism and then cast in doubt via subsequent experimental work (see Section 6.5.4). The evolutionary stability of such conventions has also been logically questioned. The main criticism of the bourgeois tactic arose from consideration of how it may generate divisive asymmetries that, in turn, favour non-conformant 'desperados' (Grafen 1987). For the anti-bourgeois rule, Maynard Smith himself recognized its potential to drive a complete breakdown of territorial behavior via a process of 'infinite regress' (Maynard Smith 1982).

Although 'anti/bourgeois' rules may rarely (if ever) apply as originally conceived, the notion that contest behavior is open to convention has undisputed value. As explored below, more biologically-realistic models for convention in nature (e.g. Kokko et al. 2006; Mesterton-Gibbons and Sherratt 2014) have particular value for interpreting territorial insect behavior.

6.5 The functional basis of competitive ability

6.5.1 An empirical framework

Territorial insect contests are, by definition, role asymmetric. Unsurprisingly, observation has confirmed residency status as the most overwhelming predictor of contest outcome. Aside from theoretical 'anti-bourgeois' scenarios (Mesterton-Gibbons and Sherratt 2014), this may arise as a logical (if not circular) consequence of how empiricists recognize species as territorial in the first place. In practical terms, most studies are confined to observing sequential, non-independent interactions among cohort members over discrete time-frames. Rates of resident success often approaches 100 per cent. This means that identifying the potential roles and determinants of RHP and RV hinges upon explaining the functional relationship between residency and competitive ability. Kemp and Wiklund (2001) considered four hypotheses of broad relevance to territorial insects:

- Residents possess intrinsically higher RHP;
- Residency itself bestows higher RHP on males in that role;
- Residents place or perceive greater RV of occupied sites, and;
- Residency mediates conventional contest settlement in the manner of the classic bourgeois/anti-burgeois rules (Box 6.1).

For territorial butterflies, Kemp and Wiklund (2001) concluded that knowledge at that time was most consistent with hypothesis (a); that is, the existence of intrinsic differences between resident and non-resident males. This accords with the long-held theory for how intrapopulation RHP variation should—all else being equal—see more competitive phenotypes accumulate in residency roles (Parker 1974; Leimar and Enquist 1984). It implies that RHP exists an intrinsic property independent of role, which is distinguished from hypothesis (b), whereby the resident role itself determines RHP. This is neither to say, however, that the latter situation cannot also apply, nor that perceived RV is not also involved. In other words, at least the first three of these hypotheses are not nearly mutually-exclusive. Hypothesis (d) is perhaps so in its pure form, but not necessarily as a partial rule for reinforcing role-related behavior (Kokko et al. 2006; Mesterton-Gibbons et al. 2016b; Mesterton-Gibbons and Sherratt 2016).

6.5.2 Physical determinants of RHP

The notion that site defence is mediated by intrinsic RHP dates to the very inception of territoriality itself. It follows from observations where discrete items are contested (e.g. females themselves) and in which size, strength, and weaponry are clearly decisive (e.g. Alcock et al. 1977; Eberhard 1979, 1987). For insects that engage in aerial disputes, efforts to define RHP have compared the features of resident with non-resident males, and/or among occupants of sites known to vary in ownership value. One profitable approach involves removing incumbents of fiercely-contested locations, followed by the capture of their immediate replacement (e.g. Kemp and Alcock 2003; Kemp 2005; Takeuchi 2011). This enhancement allows pairwise contrasts and thereby enhances statistical power for detecting contest correlates.

Residency (and/or contest success) across different groups has been correlated with body size or mass, age, energy reserves, parasite load, musculature, muscle power, strength and other morphological features (Vieira and Peixoto 2013; Table 6.1). Body size is the most frequently reported correlate, yet also perhaps the most readily measured (or measurable) trait. In some groups, such as odonates, there is evidence for the general importance of energetic-based features in determining RHP (Córdoba-Aguilar and Cordero-Rivera 2005). Overall, the widespread occurrence of RHP asymmetries in arthropods has been supported by a recent meta-analysis (Vieira and Peixoto 2013). Interestingly, this analysis suggested that putative determinants of RHP may not vary systematically between species that compete via strength- versus persistence-based interactions.

6.5.3 Subjective RV and motivation

As noted before (Section 6.4.1), the individually-subjective basis of RV makes it a far more cryptic parameter to define than RHP (or RHP correlates). Asymmetries may arise when individuals make investments that increase their intrinsic RV (e.g. Field et al. 1998), or when contestants differ verifiably in their personal knowledge of site quality. The latter possibility is best examined via manipulation, a point exemplified by two pivotal studies. First, Switzer (1997b) manipulated the experiences of male dragonflies (*Perithemis tenera*) at standardized field-based territories such that 'experimental' males were prevented from mating, but their paired 'control' counterparts could mate naturally. Along with other potential cues of site quality (i.e. female oviposition resources), the paired nature of this design controlled for male phenotype, site tenure, and prior territorial experience. Contests themselves were not assayed, but experimental males were far more likely to abandon their site the next day than control males.

Secondly, Bergman et al. (2010) conducted a compelling test of subjective RV/motivation in the speckled wood butterfly. Males of this small European woodland inhabitant contest the ownership of

Table 6.1 Selective representation of the diversity in male encounter site defence.

Order/species	Location	Contest form	Correlate of success/residency	References
Diptera				
Cyrtodiopsis dalmanni	Roosting sites	Physical grappling	Eye stalks Resource value	(Panhuis and Wilkinson 1999) (Small et al. 2009)
Cuterebra austeni	Hilltops	Aerial pursuits	Lipid reserves	(Kemp and Alcock 2003)
Hymenoptera				
Hemipepsis ustulata	Hilltops	Aerial (helical flights)	Size	(Alcock 1979)
Amegilla dawsoni	Emergence sites	Grappling	Size	(Alcock 1996)
Centris pallida	Emergence sites	Grappling	Size	(Alcock et al. 1977)
Lepidoptera				
Charis cadytis	Forest openings	Aerial flights wrestling	Unknown	(Chaves et al. 2006)
Pararge aegeria	Forest sunspots	Aerial (circling flights)	Site valuation Loser effects	(Bergman et al. 2007) (Kemp and Wiklund 2004)
Hypolimnas bolina	Flyways	Aerial (circling flights)	Age	(Kemp 2002)
Odonata				
Calopteryx maculata	Oviposition sites	Aerial	Lipid reserves Colouration	(Marden and Waage 1990) (Fitzstephens and Getty 2000)
Libellula pulchella	Oviposition sites	Aerial	Energetics/parasite load	(Convey 1989)
Diastatops obscura	Oviposition sites	Aerial	Body mass	(Lopes and Peixoto 2013)
Orthoptera				
Acheta domesticus	Burrows (+ others)	Grappling	Weaponry Energetics Loser effects	(Judge and Bonanno 2008) (Hack 1997) (Condon and Lailvaux 2016)
Ligurotettix coquilletti	Bushes	Grappling	Physical strength	(Wang and Greenfield 1991)

sunlit patches on the forest floor. Importantly, residency roles can be readily manipulated, either in the wild (Davies 1978) or captivity. Bergman et al. (2010) used a semi-natural insectary environment to conduct a three-step experiment:

- Two naïve males were released, thereby eliciting a role symmetric contest in which winners were clearly identified.
- Winners were removed, allowing losers to establish residency. Over the next 30 minutes, 'experimental' losers were either allowed to interact with conspecific females, whereas 'control' losers inhabited the site without any such encounters.
- Original winners were then reintroduced back into the site to elicit a second, role-asymmetric contest.

Original winners won most (21 of 60) contests once re-introduced, which agrees with prior results gained under similar circumstances (Kemp and Wiklund 2004; see later). However, takeovers were five times more likely for 'experimental' dyads; that is, those involving losers allowed to interact with females. This strongly supports a role for subjective site valuation in determining the intrinsic motivation of males to persist in war-of-attrition contests.

6.6 Residency and role-related phenomena

6.6.1 Residency-based convention

Territorial insects have played an important and colourful role in testing whether residency may

(a)

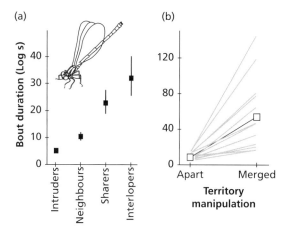

(b)

Figure 6.3 Classic insight into how perceived site ownership influences contest behavior arose from studies of *Calopteryx maculata* (Waage 1988). (a) Contest intensity varies greatly depending upon whether residents interact with newly-arriving males (intruders), residents of adjacent territories (neighbours), or sharers and interlopers. Males in the latter two categories are similar in the sense that they settled at a site without being immediately challenged by the incumbent resident. (b) Contests between the same two males greatly increased from when their territories were perceptually distinct (i.e., 'apart') to merged via the manipulation of landmarks. Halftone lines indicate the change in contest intensity for individual pairs, and box points indicate the mean increase (data from Waage 1988).

actually serve as a decisive convention (Section 6.4.1). The classic example involves a string of studies on the speckled wood butterfly (Davies 1978; Austad et al. 1979; Wickman and Wiklund 1983; Stutt and Willmer 1998; Kemp and Wiklund 2004; Bergman et al. 2010). Figure 6.3 outlines two important experiments in the progression of this work; the first which indicated apparent support for a bourgeois rule, and the second that demonstrated the opposite. The difference between experiments lay in how motivational state was preserved among captured residents (see Kemp 2013 for an expanded discussion).

Despite the attention given to 'pure' rules, there are greater opportunities for convention to shape animal contests. Indeed, prominent features of territorial insect behavior suggest the existence of convention. These features, including 'confusion over residency' and 'time in residency' effects, will be explored further later.

6.6.2 Contestant roles: know thy challenger?

Territory residents routinely encounter conspecific competitors in four main classes:

- *Intruders:* non-owner aspirants detected upon arrival.
- *Interlopers:* non-owner aspirants that settle undetected prior to detection.
- *Neighbours:* established holders of adjacent territories.
- *Sharers:* individuals that share the occupation of territories in subordinate roles.

This classification defines (at least) four main dispute contexts. As studied to famous effect by Jonathan Waage (Waage 1988), these contexts often yield starkly different degrees of contest intensity/escalation. By observing natural contests among male *Calopteryx* damselflies, Waage first demonstrated that far briefer contests ensued when residents faced intruders versus interlopers/sharers (Figure 6.4a). Secondly, whereas the neighbours of adjacent territories routinely settled their disputes rather quickly, bout intensity increased drastically when each individual was made to perceive himself as the resident (Figure 6.4b). Waage was able to demonstrate this using a clever approach to manipulating territory landmarks. Analogous effects have been shown in other taxa by capturing then releasing an incumbent resident once a replacement establishes site residency (e.g. Kemp and Alcock 2003; Bergman et al. 2010). Given that role *per se* is unlikely to directly affect RHP in these landmark-defending systems, the so-called 'confusion over residency' effect is viewed as evidence for a strong role-related component of aggression.

6.6.3 Time-in-residency effects

Whereas Waage's approach generated role uncertainty by manipulating spatial perception, resident-replacement experiments have achieved this by manipulating temporal perception of site ownership. Such experiments have consistently shown that newly-arriving males become more aggressive with increasing tenure. Alcock and Bailey (1997) demonstrated this elegantly for the landmark-defending

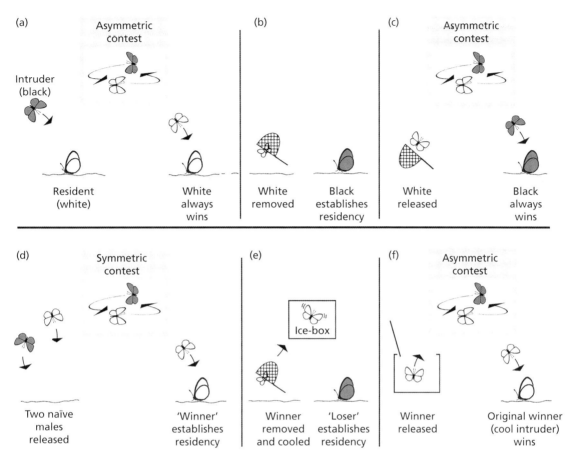

Figure 6.4 Schematics of resident-removal experiments conducted for the speckled wood butterfly (*Pararge aegeria*). Males of this species establish and defend sunspot-based territories in woodland environments. The upper panels illustrate an early experiment (Davies 1978) where (a) an incumbent resident was (b) removed and held in a net to allow a replacement to settle, then (c) released to observe the subsequent interaction. The lower panels represent a subsequent experiment that elaborated upon this approach, primarily via how captured residents were handled and re-introduced at the site (Kemp and Wiklund 2004). Here, (d) two site-naïve individuals were first allowed to compete for ownership of a sunspot, after which (e) the successful resident was removed and stored in a cooler, then (f) carefully re-introduced to re-contest the same site but in the immediate intruder role. Whereas replacement males always chased original residents away in Davies' (1978) experiment, original winners/residents defeated their replacements after escalated contests in Kemp and Wiklund's (2004) experiment.

Figure adapted from Kemp 2013.

wasp, *Hemipepsis ustulata* (Figure 6.2b). They also posed an interesting verbal hypothesis; namely, that the duration of unchallenged tenure at an encounter site may in itself inform ownership status. This has broad relevance for territorial insects unable to gauge site ownership from a distance. In *H. ustulata*, residency is linked with large body size, which may indicate an underlying RHP asymmetry. From the perspective of an incoming male, Alcock and Bailey (1997) therefore hypothe-sized that greater latency until first challenge would be associated with lower probability of that challenge coming from a high RHP incumbent. Subsequent game-theoretic modelling (Mesterton-Gibbons and Sherratt 2016) has confirmed the potential stability of this dynamic. Likewise, time-in-residency effects will be strongly favoured whenever intrinsic RV rises from site ownership itself (Mesterton-Gibbons and Sherratt 2016).

6.6.4 Dear enemies or nasty neighbours?

The above issues regarding role-related effects are relevant to two broadly formulated hypotheses, coined 'dear enemy' and 'nasty neighbour' effects. The first is based on start-up costs involved with the initial negotiation of territory boundaries. This can reduce the cost of site defence for both parties (Fisher 1930), and influence the cost:benefit ratio of leaving to find an alternative site (*sensu* Switzer 1993). Alternatively, the 'nasty neighbour' hypothesis predicts increased aggression, largely due to an ongoing failure among neighbours to negotiate site boundaries (Temeles 1994). Both effects have a strong legacy of study with regard to colony aggression in social Hymenoptera (Newey et al. 2010; reviewed by Benedek and Kobori 2014).

The dear enemy effect is contingent upon whether signalling and/or cognitive systems allow for the recognition of individual identity (Ydenberg et al. 1988). Visual recognition among individuals has been heavily studied in vespid wasps, as demonstrated by the elegant experiments of Tibbetts (2002). Virtually nothing is known for this capacity outside of social insects (for a recent review, see Giurfa 2015). Conventional thought implies that individual recognition—and hence, dear enemy effects—are unlikely to occur in solitary site-defending insects, although this remains to be closely examined.

Analogies to dear enemy/nasty neighbour effects may be informed by how contest intensity differs across known pairwise contexts (as at Section 6.6.2). Interestingly, notable variation exists among/within taxa in their intensity of neighbour–neighbour contests. For example, in highly site-tenacious butterflies, such as *H. bolina*, neighbours engage in repeated high-intensity contests irrespective of how long each male has held his territory (e.g. Kemp 2000). Neighbour–neighbour contests in other butterflies are no more intense than those of other contexts (Kemp 2003). In odonates, such contests also routinely vary from brief (Figure 6.4a; Waage 1988) to relatively prolonged (Munguía-Steyer et al. 2016), and known to be costly (Eason and Switzer 2004). Work in this group, however, suggests a candidate basis for such variation; namely, the reliability with which territorial boundaries are stable or recognizable (Munguía-Steyer et al. 2016). This merges somewhat with the confused-residency principle, wherein Waage's experiment (Figure 6.4b) could be interpreted as having turned 'dear enemies' into 'nasty neighbours' by manipulating perceived site location.

6.7 Future research prospects

The enormity of insect habitat selection has meant that this chapter can merely skim the surface, even for a subtopic such as mating site occupation. This treatment has therefore selectively emphasized key concepts and examples for where theory and empiricism have coincided notably. Additional vibrant areas of enquiry deal with how competitive information is gathered by individuals, both via pairwise assessment strategies during contests (Lopes and Peixoto 2013) and/or via experience accumulated from prior outcomes (Kemp and Wiklund 2004; Benelli et al. 2015; Condon and Lailvaux 2016). Arnott and Elwood (2009) classify the proliferation of assessment-based models and review the avenues for testing them across animals more generally.

An outstanding challenge (hence, opportunity) lies in synthesizing the expansion of knowledge for how male insects select and/or defend mating sites. This requires two things: first, a greater effort to broadly integrate the outcomes of individual studies, and secondly, a relevant framework for doing so. To the first point, it is notable that the study of contest behavior has proceeded largely via a series of treatments of focal species (or pairs of related species in the case of interspecific competition). As emphasized in this chapter, such work has proven highly successful in yielding 'proof of concept' type evidence for specific phenomena (e.g. Figures 6.2–6.4). Understanding the generality of such insights will however demand a measured synthesis across groups, such as Vieira and Peixoto's (2013) meta-analysis of insect RHP. Likewise, as insect phylogenies become increasingly resolved, so too will the opportunity for comparative investigation of contest-related phenomena. Interesting questions concern the evolutionary pathways between RHP, assessment, and convention (e.g. Kokko et al. 2006) and the potential for conservatism in the proximate basis of aggression (e.g. Stevenson and Rillich 2012).

Successfully integrating knowledge across disparate taxa will depend upon an appropriate framework. There are two reasons why traditional contest theory by itself may not prove entirely adequate. The first, as noted by Kokko et al. (2006), is that theoretical modelling deals in slightly different terms (i.e. the underlying motivational basis of tactical decisions) than typically observed by empiricists (i.e. contest outcomes). The second reason is that whereas modelling is borne from the consideration of discrete tactics or strategies, real-world contests involve individuals selected to maximize lifetime fitness. Life history theory predicts lifetime variation in optimal reproductive effort according to a trade-off between contemporary versus future reproductive potential (Williams 1966). This implies that fighting tactics should stand for modification according to features such as age, RHP, and/or experience (e.g. Baxter and Dukas 2017). One strategy that has support in long-lived butterflies, for example, is the apparent reluctance of younger individuals to enter into costly contests (Kemp 2002), perhaps as a consequence of their higher residual reproductive value (i.e. the 'asset protection' principle; Clark 1994). In other species, contest performance or aggressive motivation may decrease with age due to a reduction in RHP and/or greater frequency of losing experiences. Moreover, individuals of different quality are expected to possess 'personal' lifetime schedules for optimising reproductive value (Reznick et al. 2000). For territorial male insects, individual quality may be wholly prescribed by intrinsic RHP (or the ability to acquire high RHP), but it may also be influenced by phenotypic attributes that determine mating success in other contexts (e.g. mate attractiveness). Models that explicitly account for such phenomena are notably rare (Kokko 2013), yet may prove critical for realizing the full collective potential of empirical efforts.

References

Alcock, J. (1979). The behavioural consequences of size variation among males of the territorial wasp *Hemipepsis ustulata* (Hymenoptera: Pompilidae). *Behaviour*, **71**, 322.

Alcock, J. (1981). Lek territoriality in the tarantula hawk wasp *Hemipepsis ustulata* (Hymenoptera, Pompilidae). *Behavioral Ecology and Sociobiology*, **8**(4), 309.

Alcock, J. (1983). Consistency in the relative attractiveness of a set of landmark territorial sites to two generations of male tarantula hawk wasps (Hymenoptera, Pompilidae). *Animal Behaviour*, **31**(2), 74.

Alcock, J. (1987). Leks and hilltopping in insects. *Journal of Natural History*, **21**(2), 319.

Alcock, J. (1993). *Animal Behaviour: An Evolutionary Approach*. Sinauer, Sunderland, MA.

Alcock, J. (1994). Alternative mate-locating tactics in *Chlosyne californica* (Lepidoptera, Nymphalidae). *Ethology*, **97**(2), 103.

Alcock, J. (1994). Postinsemination associations between males and females in insects: the mate-guarding hypothesis. *Annual Review of Entomology*, **39**, 1.

Alcock, J. (1996). The relation between male body size, fighting, and mating success in Dawson's burrowing bee, *Amegilla dawsoni* (Apidae, Apinae, Anthophorini). *Journal of Zoology*, **239**, 663.

Alcock, J. (2000). Possible causes of variation in territory tenure in a lekking pompilid wasp (*Hemipepsis ustulata*) (Hymenoptera). *Journal of Insect Behavior*, **13**(3), 439.

Alcock, J., and Bailey, W.J. (1997). Success in territorial defence by male tarantula hawk wasps, *Hemipepsis ustulata*: the role of residency. *Ecological Entomology*, **22**, 377.

Alcock, J., Jones, C.E., and Buchmann, S.L. (1977). Male mating strategies in the bee *Centris pallida* Fox (Anthophoridae: Hymenoptera). *American Naturalist*, **111**, 145.

Alcock, J., and Kemp, D.J. (2004). Long-term stability in the mating system of the bot fly *Cuterebra austeni* (Cuterebridae). *Journal of Insect Behavior*, **17**(3), 273.

Alcock, J., and Kemp, D.J. (2006). The hilltopping mating system of *Leschenaultia adusta* (Loew) (Diptera: Tachinidae). *Journal of Insect Behavior*, **19**(5), 645.

Alcock, J., and O'Neill, K.M. (1987). Territory preferences and intensity of competition in the grey hairstreak *Strymon melinus* (Lepidoptera, Lycaenidae) and the tarantula hawk wasp *Hemipepsis ustulata* (Hymenoptera, Pompilidae). *American Midland Naturalist*, **118**, 128.

Andersson, M.B. (1994). *Sexual Selection*. Princeton University Press, Princeton, NJ.

Arnott, G., and Elwood, R.W. (2009). Assessment of fighting ability in animal contests. *Animal Behaviour*, **77**(5), 991.

Austad, S.N., Jones, W.T., and Waser, P.M. (1979). Territorial defence in speckled wood butterflies: why does the resident always win? *Animal Behaviour*, **27**, 960.

Bateman, A.J. (1948). Intrasexual selection in *Drosophila*. *Heredity*, **2**, 349.

Baxter, C.M., and Dukas, R. (2017). Life history of aggression: effects of age and sexual experience on male aggression towards males and females. *Animal Behaviour*, **123**, 11.

Benedek, K., and Kobori, O.T. (2014). Nasty neighbour effect in *Formica pratensis* Retz. (Hymenoptera: Formicidae). *North-Western Journal of Zoology*, **10**(2), 245.

Benelli, G., Desneux, N., Romano, D., et al. (2015). Contest experience enhances aggressive behaviour in a fly: when losers learn to win. *Scientific Reports*, **5**, 10.

Benson, W.W., Haddad, C.F.B., and Zikan, M. (1989). Territorial behavior and dominance in some heliconiine butterflies (Nymphalidae). *Journal of the Lepidopterists Society*, **43**(1), 33.

Bergman, M., Gotthard, K., Berger, D., et al. (2007). Mating success of resident versus non-resident males in a territorial butterfly. *Proceedings of the Royal Society of London (B)*, **274**, 1659.

Bergman, M., Olofsson, M., and Wiklund, C. (2010). Contest outcome in a territorial butterfly: the role of motivation. *Proceedings of the Royal Society of London (B)*, **277**, 3027.

Bishop, D.T., and Cannings, C. (1978a). A generalized war of attrition. *Journal of Theoretical Biology*, **70**, 85.

Bishop, D.T., and Cannings, C. (1978b). The war of attrition with random rewards. *Journal of Theoretical Biology*, **74**, 377.

Bradbury, J.W. (1985). Contrasts between insects and vertebrates in the evolution of male display, female choice, and lek mating. In: B. Holldobler, and M. Lindauer (Eds), *Experimental Behavioural Ecology and Sociobiology*, pp. 273–89. Fisher, New York, NY.

Brockmann, H.J. (2008). Alternative reproductive tactics in insects. In: R.F. Oliveira, M. Taborsky, and H.J. Borckmann (Eds), *Alternative Reproductive Tactics: An Integrative Approach*, pp. 177–223. Cambridge University Press, Cambridge.

Carvalho, M.R.M., Peixoto, P.E.C., and Benson, W.W. (2016). Territorial clashes in the Neotropical butterfly *Actinote pellenea* (Acraeinae): do disputes differ when contests get physical? *Behavioral Ecology and Sociobiology*, **70**(1), 199.

Chaves, G.W., Patto, C.E.G., and Benson, W.W. (2006). Complex non-aerial contests in the lekking butterfly *Charis cadytis* (Riodinidae). *Journal of Insect Behavior*, **19**(2), 179.

Clark, C.W. (1994). Antipredator behavior and the asset-protection principle. *Behavioral Ecology*, **5**, 159.

Cole, B.J., and Wiernasz, D.C. (1997). Inbreeding in a lek-mating ant species, *Pogonomyrmex occidentalis. Behavioral Ecology and Sociobiology*, **40**(2), 79.

Condon, C., and Lailvaux, S.P. (2016). Losing reduces maximum bite performance in house cricket contests. *Functional Ecology*, **30**(10), 1660.

Conner, W.E., and Itagaki, H. (1984). Pupal attendance in the crabhole mosquito *deinocerites cancer*: the effects of pupal sex and age. *Physiological Entomology*, **9**(3), 263.

Convey, P. (1989). Influences on the choice between territorial and satellite behaviour in male *Libellula quadrimaculata* Linn. (Odonata: Libellulidae). *Behaviour*, **109**, 125.

Córdoba-Aguilar, A., and Cordero-Rivera, A. (2005). Evolution and ecology of Calopterygidae (Zygoptera: Odonata): status of knowledge and research perspectives. *Neotropical Entomology*, **34**(6), 861.

Córdoba-Aguilar, A., Raihani, G., Serrano-Meneses, M.A., and Contreras-Garduno, J. (2009). The lek mating system of *Hetaerina* damselflies (Insecta: Calopterygidae). *Behaviour*, **146**, 189.

Darwin, C. (1874). *The Descent of Man and Selection in Relation to Sex*. John Murray, London.

Davies, N.B. (1978). Territorial defence in the speckled wood butterfly (*Pararge aegeria*): the resident always wins. *Animal Behaviour*, **26**(1), 138.

Deinert, E.I., Longino, J.T., and Gilbert, L.E. (1994). Mate competition in butterflies. *Nature*, **370**(6484), 23.

Eason, P.K. and Switzer, P.V. (2004). The costs of neighbors for a territorial dragonfly, *Perithemis tenera. Ethology*, **110**(1), 37.

Eberhard, W.G. (1979). The function of horns in *Podischnus agenor* (Dynastinae) and other beetles. In: M. S. Blum, and N. A. Blum (Eds), *Sexual Selection and Reproductive Competition in Insects*, p. 231. Academic Press, New York, NY.

Eberhard, W.G. (1987). The use of horns in fights by the dimorphic males of *Ageopsis nigricollis* (Coleoptera, Scarabeidae, Dynastinae). *Journal of the Kansas Entomological Society*, **60**, 504.

Elgar, M.A., and Pierce, N.E. (1978). Mating success and fecundity in an ant-tended Lycaenid butterfly. In: T.H. Clutton-Brock (Ed.), *Reproductive success: individual variation in complex breeding systems*, p. 59. University of Chicago Press, Chicago, IL.

Emlen, S.T., and Oring, L.W. (1977). Ecology, sexual selection, and the evolution of mating systems. *Science*, **197**, 215.

Enquist, M., and Leimar, O. (1983a). Evolution of fighting behavior: decision rules and assessment of relative strength. *Journal of Theoretical Biology*, **102**(3), 387.

Enquist, M., and Leimar, O. (1983b). Evolution of fighting behaviour: decision rules and assessment of relative strength. *Journal of Theoretical Biology*, **102**, 387.

Enquist, M., and Leimar, O. (1987). Evolution of fighting behavior: the effect of variation in resource value. *Journal of Theoretical Biology*, **127**(2), 187.

Estrada, C., Yildizhan, S., Schulz, S., and Gilbert, L.E. (2010). Sex-specific chemical cues from immatures facilitate the evolution of mate guarding in *Heliconius* butterflies. *Proceedings of the Royal Society B: Biological Sciences*, **277**(1680), 407.

Fawcett, T.W., and Johnstone, R.A. (2010). Learning your own strength: winner and loser effects should change with age and experience. *Proceedings of the Royal Society of London (B)*, **277**, 1427.

Field, S.A., Calbert, G., and Keller, M.A. (1998). Patch defence in the parasitoid wasp *Trissolcus basalis* (Insecta: Scelionidae): the time structure of pairwise contests, and the 'waiting game'. *Ethology*, **104**(10), 821.

Fisher, R.A. (1930). *The Genetical Theory of Natural Selection*, 2nd edn. Clarendon Press, Oxford.

Fitzstephens, D.M., and Getty, T. (2000). Colour, fat and social status in male damselflies, *Calopteryx maculata*. *Animal Behaviour*, **60**, 851.

Fritzsche, K., and Arnqvist, G. (2013). Homage to Bateman: sex roles predict sex differences in sexual selection. *Evolution*, **67**, 1926.

Funk, D.H., and Tallamy, D.W. (2000). Courtship role reversal and deceptive signals in the long-tailed dance fly *Rhamphomyia longicuada*. *Animal Behaviour*, **59**, 411.

Gintis, H. (2009). *Bounds of Reason: Game Theory and the Unification of the Behavioral Sciences*. Princeton University Press, Princeton, NJ.

Giurfa, M. (2015). Learning and cognition in insects. *Wiley Interdisciplinary Reviews–Cognitive Science*, **6**, 383.

Grafen, A. (1987). The logic of divisively asymmetric contests: respect for ownership and the desperado effect. *Animal Behaviour*, **35**, 462.

Gwynne, D.T., and Simmons, L.W. (1990). Experimental reversal of sex roles in an insect. *Nature*, **346**, 172.

Hack, M.A. (1997). The energetic costs of fighting in the house cricket, *Acheta domesticus* L. *Behavioral Ecology*, **8**, 28.

Hernandez, M.I.M., and Benson, W.W. (1998). Small-male advantage in the territorial tropical butterfly *Heliconius sara* (Nymphalidae): a paradoxical strategy? *Animal Behaviour*, **56**, 533.

Hill, P.S.M. (1999). Lekking in *Gryllotalpa major*, the prairie, mole cricket (Insecta: Gryllotalpidae). *Ethology*, **105**(6), 531.

Judge, K.A., and Bonanno, V.L. (2008). Male weaponry in a fighting cricket. *Plos One*, **3**(12), 10.

Kaspi, R., and Yuval, B. (1999). Mediterranean fruit fly leks: factors affecting male location. *Functional Ecology*, **13**(4), 539.

Kemp, D.J. (2000). Contest behavior in male butterflies: does size matter? *Behavioral Ecology*, **11**, 591.

Kemp, D.J. (2001). Age-related site fidelity in the territorial butterfly *Hypolimnas bolina* (L.) (Lepidoptera: Nymphalidae). *Australian Journal of Entomology*, **40**, 65.

Kemp, D.J. (2002). Sexual selection constrained by life history in a butterfly. *Proceedings of the Royal Society of London (B)*, **269**(1498), 1341.

Kemp, D.J. (2003). Twilight fighting in the evening brown butterfly, *Melanitis leda* (L.) (Nymphalidae): age and residency effects. *Behavioral Ecology and Sociobiology*, **54**(1), 7.

Kemp, D.J. (2005). Contrasting lifetime patterns of territorial success in the nymphalid butterflies *Hypolimnas bolina* and *Melanitis leda*: a question of flight physiology? *Australian Journal of Zoology*, **53**, 361.

Kemp, D.J. (2008). Female mating biases for bright ultraviolet iridescence in the butterfly *Eurema hecabe* (Pieridae). *Behavioral Ecology*, **19**(1), 1.

Kemp, D.J. (2013). Contest behaviour in butterflies: fighting without weapons. In: I. C. W. Hardy, and M. Briffa (Eds), *Animal Contests*, p. 134. Cambridge University Press, Cambridge.

Kemp, D.J., and Alcock, J. (2003). Lifetime resource utilization, flight physiology, and the evolution of contest competition in territorial insects. *American Naturalist*, **162**(3), 290.

Kemp, D.J., and Wiklund, C. (2001). Fighting without weaponry: a review of male-male contest competition in butterflies. *Behavioral Ecology and Sociobiology*, **49**(6), 429.

Kemp, D.J., and Wiklund, C. (2004). Residency effects in animal contests. *Proceedings of the Royal Society of London (B)*, **271**, 1707.

Kirkton, S.D., and Schultz, T.D. (2001). Age-specific behavior and habitat selection of adult male damselflies, *Calopteryx maculata* (Odonata: Calopterygidae). *Journal of Insect Behavior*, **14**(4), 545.

Kokko, H. (2013). Dyadic contests: modelling fights between two individuals. In: I. C. W. Hardy, and M. Briffa (Eds), *Animal Contests*, p. 5. Cambridge University Press, Cambridge.

Kokko, H., Lopez-Sepulcre, A., and Morrell, L.J. (2006). From hawks and doves to self-consistent games of territorial behavior. *American Naturalist*, **167**(6), 901.

Kokko, H., Klug, H., and Jennions, M.D. (2014). Mating systems. In: D. M. Shuker and L. W. Simmons (Eds), *The Evolution of Insect Mating Systems*, pp. 42–58. Oxford University Press, Oxford.

Lancaster, J., Downes, B.J., and Arnold, A. (2011). Lasting effects of maternal behaviour on the distribution of a dispersive stream insect. *Journal of Animal Ecology*, **80**(5), 1061.

Leimar, O., and Enquist, M. (1984). Effects of asymmetries in owner–intruder conflicts. *Journal of Theoretical Biology*, **111**(3), 491.

Lopes, R.S., and Peixoto, P.E.C. (2013). Males of the dragonfly *Diastatops obscura* fight according to predictions from game theory models. *Animal Behaviour*, **85**(3), 663.

Marden, J.H., and Waage, J.K. (1990). Escalated damselfly territorial contests are energetic wars of attrition. *Animal Behaviour*, **39**, 954.

Maynard Smith, J. (1972). *Game Theory and the Evolution of Fighting*. Edinburgh University Press, Edinburgh.

Maynard Smith, J. (1974). The theory of games and the evolution of animal conflicts. *Journal of Theoretical Biology*, **47**, 209.

Maynard Smith, J. (1982). *Evolution and the Theory of Games*. Cambridge University Press, Cambridge.

Maynard Smith, J., and Parker, G.A. (1976). The logic of asymmetric contests. *Animal Behaviour* **24**(1), 159.

Maynard Smith, J., and Price, G.R. (1973). The logic of animal conflict. *Nature*, **246**(5427), 15.

McCubbin, C. (1971). *Australian Butterflies*. Nelson, Melbourne.

Mesterton-Gibbons, M., Dai, Y., and Goubault, M. (2016a). Modeling the evolution of winner and loser effects: a survey and prospectus. *Mathematical Biosciences*, **274**, 33.

Mesterton-Gibbons, M., Karabiyik, T., and Sherratt, T.N. (2016b). On the evolution of partial respect for ownership. *Dynamic Games and Applications*, **6**(3), 359.

Mesterton-Gibbons, M., and Sherratt, T.N. (2014). Bourgeois versus anti-bourgeois: a model of infinite regress. *Animal Behaviour*, **89**, 171.

Mesterton-Gibbons, M., and Sherratt, T.N. (2016). How residency duration affects the outcome of a territorial contest: complementary game-theoretic models. *Journal of Theoretical Biology*, **394**, 137.

Mesterton Gibbons, M., Marden, J.H., and Dugatkin, L.A. (1996). On wars of attrition without assessment. *Journal of Theoretical Biology*, **181**(1), 65.

Moore, M.P., and Martin, R.A. (2016). Intrasexual selection favours an immune-correlated colour ornament in a dragonfly. *Journal of Evolutionary Biology*, **29**(11), 2256.

Morbey, Y.E., and Ydenberg, R.C. (2001). Protandrous arrival timing to breeding areas: a review. *Ecology Letters*, **4**(6), 663.

Munguía-Steyer, R., Córdoba-Aguilar, A., and Maya-García, J. (2016). Rubyspot territorial damselflies behave as "nasty neighbors". *Journal of Insect Behavior*, **29**(2), 143.

Newey, P.S., Robson, S.K., and Crozier, R.H. (2010). Weaver ants *Oecophylla smaragdina* encounter nasty neighbors rather than dear enemies. *Etholsogy*, **91**, 2366.

Panhuis, T.M., and Wilkinson, G.S. (1999). Exaggerated male eye span influences contest outcome in stalk-eyed flies (Diopsidae). *Behavioral Ecology and Sociobiology*, **46**(4), 221.

Parker, G.A. (1970). Sperm competition and its evolutionary consequences in insects. *Biological Reviews of the Cambridge Philosophical Society*, **45**(4), 525.

Parker, G.A. (1974). Assessment strategy and the evolution of fighting behavior. *Journal of Theoretical Biology*, **47**, 223.

Parker, G.A. (2013). A personal history of the development of animal contest theory and its role in the 1970s. In: I. C. W. Hardy and M. Briffa (Eds), *Animal Contests*, p. 11. Cambridge University Press, Cambridge.

Parker, G.A., Baker, R.R., and Smith, V.G.F. (1972). The origin and evolution of gamete dimorphism and the male-female phenomenon. *Journal of Theoretical Biology*, **36**, 529.

Parker, G.A. and Rubenstein, D.I. (1981). Role assessment, reserve strategy, and acquisition of information in asymmetric animal conflicts. *Animal Behaviour*, **29**(2), 221.

Parker, G.A., and Thompson, E.A. (1980). Dung fly struggles: a test of the war of attrition. *Behavioral Ecology and Sociobiology*, **7**, 37.

Payne, R.J.H. (1998). Gradually escalating fights and displays: the cumulative assessment model. *Animal Behaviour*, **56**, 651.

Reznick, D.N., Nunney, L., and Tessier, A. (2000). Big houses, big cars, superfleas and the costs of reproduction. *Trends in Ecology & Evolution*, **15**(10), 421.

Ruppell, G., and Hilfert-Ruppell, D. (2013). Biting in dragonfly fights. *International Journal of Odonatology*, **16**(3), 219.

Rutowski, R.L. (1991). The evolution of male mate-locating behavior in butterflies. *American Naturalist*, **138**, 1121.

Sigurjonsdottir, H., and Parker, G.A. (1981). Dung fly struggles: evidence for assessment strategy. *Behavioral Ecology and Sociobiology*, **8**(3), 219.

Small, J., Cotton, S., Fowler, K., and Pomiankowski, A. (2009). Male eyespan and resource ownership affect contest outcome in the stalk-eyed fly, *Teleopsis dalmanni*. *Animal Behaviour*, **78**(5), 1213.

Stevenson, P.A. and Rillich, J. (2012). The decision to fight or flee–insights into underlying mechanism in crickets. *Frontiers in Neuroscience*, **6**, 12.

Stutt, A.D., and Willmer, P. (1998). Territorial defence in speckled wood butterflies: do the hottest males always win? *Animal Behaviour*, **55**(5), 1341.

Svensson, B.G., and Petersson, E. (1995). Diurnal and seasonal variations in swarming and mating behaviour of the dance fly *Empis borealis* (Diptera; Empididae). *Annales Zoologici Fennici*, **32**(4), 403.

Switzer, P.V. (1993). Site fidelity in predictable and unpredictable habitats. *Evolutionary Ecology*, **7**(6), 533.

Switzer, P.V. (1995). *Influences on the Site Fidelity of Territorial Animals: Theoretical and Empirical Studies*. Unpublished thesis (PhD), University of California, Davis.

Switzer, P.V. (1997a). Factors affecting site fidelity in a territorial animal, *Perithemis tenera*. *Animal Behaviour*, **53**, 865.

Switzer, P.V. (1997b). Past reproductive success affects future habitat selection. *Behavioral Ecology and Sociobiology*, **40**(5), 307.

Takeuchi, T. (2011). Body morphologies shape territorial dominance in the satyrine butterfly *Lethe diana*. *Behavioral Ecology and Sociobiology*, **65**(8), 1559.

Takeuchi, T., and Imafuku, M. (2005). Territorial behavior of a green hairstreak *Chrysozephyrus smaragdinus* (Lepidoptera: Lycaenidae): site tenacity and wars of attrition. *Zoological Science*, **22**, 989.

Temeles, E.J. (1994). The role of neighbors in territorial systems: when are they dear enemies? *Animal Behaviour*, **47**(2), 339.

Thornhill, R., and Alcock, J. (1983). *The Evolution of Insect Mating Systems*. Harvard University Press, Cambridge, MA.

Tibbetts, E.A. (2002). Visual signals of individual identity in the wasp, *Polistes fuscatus*. *Proceedings of the Royal Society of London (B)*, **269**, 1423.

Trivers, R. (1972). Parental investment and sexual selection. In: B. Campbell (Ed.), *Sexual Selection and the Descent of Man 1871–1971*, p. 139. Aldine Press, Chicago, IL.

Vieira, M.C., and Peixoto, P.E.C. (2013). Winners and losers: a meta-analysis of functional determinants of fighting ability in arthropod contests. *Functional Ecology*, **27**(2), 305.

Waage, J.K. (1988). Confusion over residency and the escalation of damselfly territorial disputes. *Animal Behaviour*, **36**, 586.

Wang, G.Y., and Greenfield, M.D. (1991). Effects of territory ownership on dominance in the desert clicker (Orthoptera: Acrididae). *Animal Behaviour*, **42**, 579.

Wickman, P.O. (1988). Dynamics of mate-searching behavior in a hilltopping butterfly, *lasiommata megera* (L.): the effects of weather and male density. *Zoological Journal of the Linnean Society*, **93**(4), 357.

Wickman, P.O. and Rutowski, R.L. (1999). The evolution of mating dispersion in insects. *Oikos*, **84**(3), 463.

Wickman, P.O., and Wiklund, C. (1983). Territorial defense and its seasonal decline in the speckled wood butterfly (*Pararge aegeria*). *Animal Behaviour*, **31**(4), 1206.

Wiklund, C., and Fagerstrom, T. (1977). Why do males emerge before females? A hypothesis to explain the incidence of protandry in butterflies. *Oecologia*, **31**, 153.

Williams, G.C. (1966). Natural selection, the costs of reproduction, and a refinement of Lack's principle. *American Naturalist*, **100**, 687.

Ydenberg, R.C., Giraldeau, L.A., and Falls, J.B. (1988). Neighbors, strangers, and the asymmetric war of attrition. *Animal Behaviour*, **36**, 343.

CHAPTER 7

Long-range migration and orientation behavior

Don R. Reynolds[1] and Jason W. Chapman[2]

[1] *Natural Resources Institutem University of Greenwich, Chatham, Kent, UK*

[2] *Centre for Ecology and Conservation, and Environment and Sustainability Institute, University of Exeter, Penryn, Cornwall, UK*

7.1 Introduction

Migrations speak to us, not just as observers of nature but as integral parts of it. The world moves and, deep inside, we long to move with it (Mike Bergin, 2009, of 10,000 birds)

Migratory animals have long been powerful symbols of change and renewal in human cultures.

(Ben Hoare, 2009)

Mass migrations of insects have fascinated mankind throughout recorded history. The sudden appearance of huge numbers of migrants have caused a range of emotions—delight, wonder, and awe at the marvels of nature, annoyance at the irritations of clouds of small insects, and trepidation at the approach of locust swarms. Presumably, it has always been so. To members of human hunter-gatherer societies, highly attuned to the natural world, dramatic seasonal movements of insects may well have been incorporated into folk wisdom, just as the present-day San (Bushman) of Botswana reportedly use the migrations of certain butterflies (*Belenois aurota, Catopsilia florella*) to predict the arrival, from the same direction, of herds of ungulates (Larsen 1992). After the development of agriculture, farming societies would have good reason to fear migratory insect pests, such as locusts and armyworms, which might

well bring famine in their wake. In fact, the locust swarms so entered the communal psyche of the Middle Eastern civilizations that these insects constituted one of the 'Ten Plagues of Egypt', which formed, perhaps, the earliest written record of insect migration (*Exodus* 10, 1–20, perhaps written around 600 BCE). Without modern means of famine relief, large numbers of people could die of starvation following locust invasions [e.g. 800,000 died in North Africa in 125 BC (Williams 1958)]. On a more joyful note, the monsoon rains essential for agriculture in many parts of India are heralded by migrating dragonflies (Corbet 1999, p. 404). In Nigeria, after the migration of *Libythea labdaca* butterflies early in the rains, farmers knew that sowing could safely begin (Farquharson, quoted in Williams 1930, p. 417).

Characteristics that have brought migrations to the attention of people throughout the ages, namely, the movement itself combined with the huge numbers involved, continue to amaze even experienced modern naturalists who must abandon their restrained scientific prose as they struggle to communicate the sheer wonder of what they see. One notes, for example, descriptions of immense butterfly migrations in C. B. Williams' books (1930, 1958); or the accounts of the seasonal passage of hordes of

Reynolds, D. R. and Chapman, J. W., *Long-range migration and orientation behavior.* In: *Insect Behavior: From mechanisms to ecological and evolutionary consequences.* Edited by Alex Córdoba-Aguilar, Daniel González-Tokman, and Isaac González-Santoyo: Oxford University Press (2018). © Oxford University Press.
DOI: 10.1093/oso/9780198797500.003.0007

insects through the Portachuelo Pass in north-central Venezuela (Beebe 1949; and see references in Johnson 1969, p. 574 ff.) or through high mountain passes in the Pyrenees (Lack and Lack 1951).

Butterfly migrations are notable for enormous numbers involved, at least in former years, where in some places great clouds of butterflies 'cast shadows on the ground', 'held up motor cars' or necessitated people 'walking with their heads bent to the storm' (Williams 1930). Nonetheless, for prodigious numbers of large migrant insects, giving rise to a huge biomass of accumulated individuals, little can compare with the swarms of locusts produced during plague periods. The swarms of the desert locust (*Schistocera gregaria*) observed by radar around Delhi in July 1962, for example, comprised ~100 billion individuals (Rainey 1989, p. 153) weighing 200,000 tonnes. These moving biomasses rival those of the largest herds of mammals, such as those comprising the Serengeti wildebeest migration (Holland et al. 2006).

As well as the huge numbers of individuals on the move, our wonder at the dramatic nature of migrations is kindled by the astonishing distances covered by some insect migrants. The migrations of the monarch butterfly (*Danaus plexippus*) in eastern North America are, without doubt, one of the great phenomena of the natural world. Every year butterflies from breeding areas as far north as the Canadian border migrate to very restricted areas in the central Mexican mountains, where they overwinter in dense, quiescent clusters in the oyamel fir forests. In March, the surviving butterflies become reproductively active, mate, and fly back to the southern USA, so that some individuals have probably travelled as far as 5000–6000 km (Brower et al. 2006).

The longest non-stop insect migration recorded seems to have been the crossing of the Atlantic, from the west coast of Africa to various eastern islands in the Caribbean and neighbouring parts of South America, by the desert locust swarms in October and November 1988—a movement of some 5000 km, with an apparent minimum duration of 93 hours (Rosenberg and Burt 1999). It is still unclear how the locusts sustained flight long enough to make this tremendous journey. The 1988 crossing was unusual (and disastrous for the locusts in that all the migrants either died at sea or failed to breed successfully on arrival), but over-land swarm movements covering

distances of 3000–4000 km within one generation are not uncommon (Pedgley 1981). When the longest locust migration distances are expressed in terms of body length (e.g. 5000×10^3 m$/0.06$ m $= 83$ million body lengths), the value exceeds that for virtually all other taxa, with the possible exception of one or two birds (c.f. Table 2 in Alerstam et al. 2003).

Unlike the seasonal 'to-and-fro' migration of most birds between breeding and non-breeding ranges, individual insect migrants (being relatively short-lived as adults) will normally complete only part of each migration circuit, i.e. the circuits are multi-generational. Nonetheless, these circuits can cover stupendous distances. For instance, in the painted lady butterfly, the completion of the Western Palaearctic/West Africa circuit takes a succession of at least six generations shifting populations over 60° of latitude annually (Stefanescu et al. 2013, 2016)—a complete round-trip of up to 15,000 km. There may be an even longer migration circuit in the wandering glider dragonfly, *Pantala flavescens*, which traverses the western Indian Ocean from India to East Africa (Anderson 2009; Hobson et al. 2012); if the putative return leg back to India is confirmed, the annual circuit would then, over four generations, cover a total distance of about 14,000–18,000 km!

We should emphasize, however, that most migrations are not spectacular nor by large species, but occur in insects such as aphids, small flies (Diptera), and beetles (Coleoptera), and parasitic Hymenoptera (Hu et al. 2016a). The enormous diversity of insects and other arthropods high in the air can be seen from aerial trapping studies (e.g. Glick 1939; Chapman et al. 2004). In fact, the surface of the earth lies 'at the bottom of a vault of insect-laden air' from which it receives 'a continuous rain of insects' as C. G. Johnson puts it (Johnson 1969, p. 297), so that few places escape invasion.

The amazing evolutionary success of insects is partly due to their ability to fly and, although flight is important for a range of short-range 'foraging' activities, the ability to relocate over considerable distances in order to exploit seasonal resources must be of great significance. Even within a restricted taxonomic group, the highly migrant species are frequently the most successful, widespread, and cosmopolitan—highly adaptable, colonizing and opportunistic, many of these have become pests (Chapman et al. 2015). We note, however, that migration by means other than flapping flight does occur in wingless insects and in other terrestrial arthropods (e.g. mites, spiders, pseudoscorpions), sometimes over surprising distances. The various modes of this 'non-volant' migration (including aerial migration by wingless species, pedestrian and waterborne migration, and **phoresy**) have been reviewed by Reynolds et al. (2014).

The perspective on migration mentioned above, often emphasizing its mystery and wonder, its 'heroic odysseys' over huge distances, has been termed the 'View from natural history' (V. A. Drake 2008). It has provided a huge amount of accurate, basic information on the phenomena of migration, and has a continuing and highly positive corollary: the effect the 'mysteries of migration' view has in stimulating the public to take an interest in the natural world and conservation, and perhaps to get actively involved in, for example, biological monitoring schemes (Van Swaay et al. 2008). Arousing public interest in maintaining the integrity and diversity of nature is becoming increasingly important at a time when extinctions may be occurring at 1000 times the average 'background' rates over geological time (De Vos et al. 2015).

Nonetheless, the natural history viewpoint tells only part of the story, and we now examine migration as an adaptive biological phenomenon implemented through a correlated suite of traits—behavioral (with which we are particularly concerned), physiological, morphological, and life-history—which together form a '**migration syndrome**' (Dingle and Drake 2007; Dingle 2014; Chapman and Drake 2017; for syndromes, see Chapter 16). As we shall see, the syndrome itself, together with migration components at other organizational levels (the underlying genetic complex; and the **population trajectory** through space and time) plus the environment within which migration occurs, collectively make up the '**migration system**' (Drake et al. 1995; Dingle and Drake 2007; Dingle 2014).

7.2 What is migration?

Migration, like other biological phenomena, involves a hierarchy of organizational or integrative levels (e.g. genetic, physiological, behavioral, ecological, evolutionary)—higher levels have emergent properties not found in the lower levels, and definitions and descriptions made at one level have to be explained by reference to a higher level. Historically, the confounding of the ecological and behavioral levels has led to serious conceptual difficulties when defining animal migration (Rogers 1983; Kennedy 1985; Dingle 1996). There is now widespread agreement that the definition of what constitutes migration is most satisfactorily made in behavioral terms at the level of the individual, as this is where natural selection predominantly operates (Dingle 2014). The consequent redistributive population movements can then be viewed as ecological consequences or outcomes of the individual displacements (Gatehouse 1987; Dingle and Drake 2007, Dingle 2014). The ultimate, evolutionary level feeds back through natural selection on the genome controlling migratory behavior and other traits associated with the syndrome.

Another consideration is that, as with any other important biological process, any definition should encompass all animal taxa—definitions that apply only to certain taxonomic groups (birds, say) are unsatisfactory and, we believe, unnecessary. So the question becomes: can a definition be devised, which distinguishes migrations from other forms of animal movement, which highlights features shared among apparently disparate phenomena, and which applies over as broad a taxonomic range as possible? After a lifetime's consideration of animal movement the eminent behaviorist, John S. Kennedy, formulated a general, objective, testable definition (see Box 7.1) which has withstood scrutiny well.

To put this definition in context, we need to briefly consider types of animal movement. These can be categorized as station-keeping, '**ranging**' and migration (see Dingle 1996, 2014). Station-keeping covers a variety of appetitive or resource-directed behaviors, which localize and retain an animal within its home range—the habitat patch within which it will spend most of its life, and which will provide the resources required for survival, somatic growth, and reproduction. Locomotory activities within the home range will thus be associated with feeding, finding shelter, avoiding predation, mating, egg-laying, or giving birth, etc., and will include various subcategories of movement, e.g. searching, territorial behavior, and commuting (see Chapter 1 in Dingle 2014). In all these cases, the movement will cease when the resource item is discovered or sought-after outcome (e.g. an intruder chased away) is achieved.

Box 7.1 A definition of migration

A succinct and widely accepted definition of migratory behavior, which clearly differentiates it from other types of movement (see main text), is that proposed by Kennedy (1985):

Migratory behavior is persistent and straightened-out movement effected by the animal's own locomotory exertions or by its active embarkation on a vehicle. It depends on some temporary inhibition of station-keeping responses, but promotes their eventual disinhibition and recurrence.

Accordingly, the migration process is defined in terms of *specialized behavior*, which is quite different from that shown during an animal's everyday foraging or exploratory movements. Migration behavior allows an animal to escape from its original habitat patch (in insects, often its natal location) by temporarily depressing responsiveness to the 'vegetative' stimuli (i.e. those associated with growth and reproduction), which give rise to highly sinuous and interrupted movement paths. Freed from such distractions, migrants are able to achieve persistent, rectilinear displacement away from a localized area, by various modes of transport including:

- phoretic attachment to another animal;
- launching into a wind or water current;
- self-propelled and (to some extent) self-steered movement by flight, swimming, or the forms of terrestrial locomotion.

Note that while persistent locomotor activity is characteristic of most migrations, this is not universal—phoretic migrants may be virtually motionless for most of the migratory journey.

As stated in the definition, a period of migratory travel serves to weaken the suppression of appetitive responses, and promotes behavior that will eventually bring the migration to an end. For example, a bout of migratory flight tends to promote landing and settling responses in aphids, and a period of quiescence on a host primes disembarkation behavior in a phoretic animal.

Migrants often leave habitats *before* key resources disappear; in other words, there is an anticipatory component to the migration, elicited by cues such as photoperiod that act as surrogates for habitat change.

We note there is nothing in the definition about the duration or scale of the migratory movement—migrations in different insect taxa might cover a few tens of metres or hundreds of kilometres, and there is also no reference to 'to-and-fro' movements—insect migrations are very largely one-way.

Although the definition above is couched in behavioral terms there are, of course, many other aspects of the migration syndrome (morphological, physiological, life-history, and other behavioral traits) involved in an adequate characterization of migration in an animal. In addition, the complete syndrome has to be *explained* by reference to other integrative levels, e.g. in terms of ecology and evolution (see main text).

At some point in its life, however, an animal may undertake extended movement away from its current habitat patch, and these exodus movements fall into two categories. The animal may explore a wider area—often explicitly to find a new home range. This behavior has been termed ranging by Hugh Dingle (Dingle 2014). As with station-keeping, the resources encountered are being assessed as the individual proceeds, and the ranging movement will stop as soon as a suitable new habitat patch is discovered. The second category of extended movement, migration, is quite different in that responses to local resources are temporarily inhibited, i.e. the individual is undistracted by appetitive stimuli to which it would normally respond. The migrant exhibits specialized behavior that, in various ways, leads

to persistent, straightened-out motion, carrying it away from its former habitat patch. In general, short-lived animals such as insects, this distinctive behavior may be seen only once during an individual's life (see Reynolds et al. 2014, and 'Initiation of migration' see late). Where animals migrate by means of their own exertions, there may a high level of locomotory excitability driving the migratory movement, maybe for hours or even days on end. In contrast, in animals that migrate by phoretic attachment to another (host) animal, the specialized behavior may involve a frenetic bout of movement to board a suitable host followed by a period of quiescence until the time comes for disembarkation (Reynolds et al. 2014). J. S. Kennedy, in his classic studies of migration in the black bean aphid, *Aphis fabae*, found that, while the enhanced movement was dependent on the inhibition of certain station-keeping behaviors, the movement itself lowered thresholds of other station-keeping responses. In other words, it primed the eventual occurrence of settling behavior. In fact, there was a subtle and complex system of reciprocal excitatory and inhibitory responses, promoting migratory flight and settling behavior (see summary in Dingle 1996). It is important to note that, particularly on long migratory journeys, suppression of appetitive activities may well be episodic (butterflies such as the monarch stop to feed and roost during their migrations), and migratory and appetitive activities may be segregated into separate periods of the daily cycle (e.g. moths may migrate on several sequential nights, but will land and hide-away during daylight hours).

7.2.1 The migration syndrome

As mentioned previously, distinctive behavior is just one component of a complex of co-adapted traits, the migration syndrome, which enable and implement migration. An early proponent of this idea was C. G. Johnson who placed migration within a life-history context by pointing out that it often occurs soon after adult eclosion and before sexual maturity, particularly in females (Johnson 1969). This so-called **oogenesis-flight syndrome** allows migrants to maximize their reproductive potential within a newly colonized habitat. In some species, this strict temporal partitioning between flight and oogenesis may be less clear, and females may migrate while partially gravid, e.g. the spruce budworm moth, *Choristoneura fumiferana* (Greenbank et al. 1980).

The migration syndrome concept in insects has been greatly extended by others, particularly H. Dingle working on hemipteran bugs, and D. A. Roff, working on crickets (Dingle 2001, 2014; Roff and Fairbairn 2007). Examples of morphological traits that contribute to the migration syndrome are wing polymorphisms where there is genetic control of the flight apparatus (wings and/or wing musculature) giving rise to flight-capable or flightless morphs. Often, however, genetic expression is subject to substantial environmental influences so that discrete alternative phenotypes are produced from the same genotype, for example, the wing polyphenisms found in parthenogenetic female aphids (Simpson et al. 2011).

There is, of course, a raft of physiological traits facilitating migration. First, there is the basic partitioning of energy resources for growth, movement, and reproduction at different stages of the life cycle, modulated by exogenous factors (particularly day length and sometimes temperature) and endogenous influences (particularly circadian and seasonal 'clocks' and endocrinology; Dingle 2014; Zera 2016). For example, circadian clocks in the antennae provide a timing mechanism for sun compass orientation in migratory monarchs (see later). Then, there are mechanisms to direct an individual's metabolism towards providing fuel for migration (e.g. the accumulation of flight fuels, often in the form of fat reserves) and to mobilize them for the journey. The hormonal, enzymatic, and biomechanical influences on migratory physiology and behavior are outside the scope of this review (but see Dudley 2000; Goldsworthy and Joyce 2001; Van der Horst 2003; see also Chapter 4).

The overarching concept here is that natural selection does not act on movement alone, but rather on the suite of functionally coordinated traits, and that selection can act on the direction and strength of genetic covariance between traits, as well as on single-gene contributions.

7.2.2 A Holistic Model—the Migration System

The holistic 'migration system' model (Drake et al. 1995; Drake and Gatehouse 1996) provides a conceptual framework within which can be integrated all the components and processes associated with a migratory population. The model enables us to identify gaps in our knowledge, to formulate questions about the relationships between elements of the system, and thus better understand the functioning of the system as whole.

Four primary components of a generalized migration system were identified, namely:

- *The 'migration arena'*: the three-dimensional geographical space within which a population's migration takes place, incorporating all the environmental factors (abiotic and biotic) that will impinge on the population.

- *The 'population trajectory'*: the changing population demography that results from migration through space and time (the long-term average trajectory is termed the 'population pathway').
- *The 'migration syndrome'* of co-evolved traits that implement migration and determine the fitness of the migrants.
- *The 'genetic complex'*: the genes that underlie the migration syndrome, their interactions and modes of inheritance.

A modified version of the model is shown in Figure 7.1. The comparatively well-studied migration system of the oriental armyworm, *Mythimna separata*, provides an illustration of how the model might be used (see Drake and Gatehouse 1996). The migration systems of the highly migratory pest moths, *Helicoverpa punctigera* and *H. armigera*, and the plague locust, *Chortoicetes terminifera*, in Australia have also been investigated within the context of the holistic model, examining, for example,

the interactions between aspects of the environmental 'arena' (e.g. wind-flow patterns) and the 'migration syndrome' (e.g. flight behavior) (Rochester 1999, Drake et al. 2001).

7.3 Migration through the atmosphere

The quintessential means of insect migration is by active flight through the atmosphere. The various atmospheric influences acting on airborne insects have been recently reviewed by Drake and Reynolds (2012) and so here we outline a few main points. As mentioned previously, distinctive behavior is an important part of the migration syndrome, and insects usually show highly specific behaviors for the initiation (take-off) and termination (settling) phases of the migratory journey. Between these phases there will be a period, sometimes extensive, of 'transmigration', which equates to the steady horizontal 'cruising' phase in flying migrants.

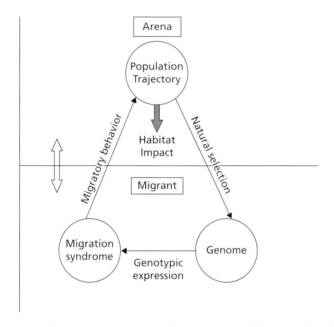

Figure 7.1 A holistic conceptual model of an insect 'migration system'. The model was originally developed by Drake et al. (1995); this version is one incorporating modifications by Dingle (2014). (Upper panel) The 'arena' is the total environment (biotic and abiotic) in which the population trajectory—the progression in space and time of a migratory population—takes place. The migratory life cycle affects habitats through processes such as exhaustion of resources and the arena environment modifies the migrant's genetic complex through natural selection. (Lower panel) The individual migrant whose genome produces the migration syndrome, including appropriate responses to the environment and migratory behavior under the influence of natural selection. The phenotypic syndrome steers individuals and their genes along varying trajectories through the 'arena' with complex interactions between the migrants and the 'arena' environment as indicated by the double-headed arrow.

From Dingle 2014.

A primary distinction occurs between individuals (usually of relatively large, dayflying species) which migrate, at least partly, within their **flight boundary layer** (FBL; Taylor 1974)—the layer of air near the surface where flying insects can control their movement relative to the ground (see Figure 7.2)—as opposed to migrants that ascend out of their FBL so that their movement will inevitably have a large downwind component. Another important distinction is between daytime and nocturnal migrants, where the timing of emigration will result in important differences in the atmospheric conditions experienced, due to diurnal changes in the **atmospheric boundary layer** (ABL). The ABL is the lowest layer of the atmosphere, which is directly influenced (particularly via friction or heating/cooling effects) by its contact with the Earth's surface, over time scales of about 1 day (Garratt 1994). In sunny weather, solar heating of the ground causes thermal plumes to rise and mixing motions typically engulf the lowest

1–2 km of the atmosphere. This daytime ABL becomes full of small migrant insects utilizing the updrafts to carry them to varying heights in the air, during which time the horizontal component of the wind can advect them over considerable distances (see later).

In contrast, under clear fair-weather conditions over land at night, the ABL is represented by a stable *nocturnal boundary layer* which may be only ~100–300 m deep, with temperatures increasing with height, i.e. there is a *nocturnal surface inversion*. At the top of this inversion, there may be a boundary layer wind-speed maximum (= low-level jet). The stratified nature of the nocturnal ABL is often paralleled by stratifications in the vertical distributions of insect migrants which tend to occur in layers of narrow altitudinal (~50–200 m), but broad horizontal, extent (Chapter 10 in Drake and Reynolds 2012).

Wind-fields and weather systems on various scales will obviously be of central importance in the transport of long-range airborne migrants. These

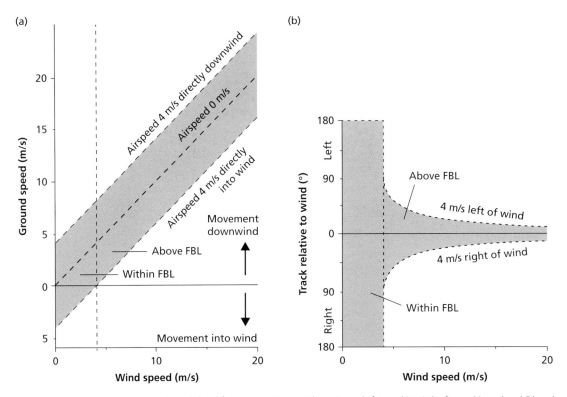

Figure 7.2 Range of (a) possible ground speeds (grey) for an insect migrant with an airspeed of 4 m s^{-1} in winds of up to 20 m s^{-1}, and (b) track directions (relative to the wind direction) (grey) for the same migrant.

From Drake and Reynolds 2012; see also Glossary, this volume.

scales range from 'global' (e.g. the trade winds), through the 'synoptic-scale' (typical of mid-latitude anticyclones and depressions with a horizontal length of ~1000 km), to the various 'meso-scales' (~ a few hundred metres to about 2 km). Reviews of weather influences on insect migratory flight, and examples of the medium- and long-range displacement in various weather systems can be found in Johnson (1969), Pedgley (1982), Drake and Farrow (1988), Rainey (1989), Isard and Gage (2001), and Drake and Reynolds (2012). In disturbed weather, insects may be more affected by smaller-scale weather features embedded in the larger wind system, e.g. thunderstorm outflows may transport insects for ~1 hour in directions quite different from the large-scale airflow. Strong *wind convergence* (horizontal inflows of air at low altitudes) along the leading edge of storm outflows, sea breeze fronts, and other *gravity currents* can 'sweep up' large numbers of insects, which accumulate because they tend to resist being carried up in the rising air at the frontal interface. This process can lead to rapid increases in insect aerial density of 1–2 orders of magnitude (Drake and Reynolds 2012).

7.4 Orientation behavior of insect migrants

Assuming that immediate environmental factors (particularly illumination levels, temperature, and wind speed) are suitable, the migrant is ready to initiate migration. The specialized nature both in preparation (e.g. a pre-flight warm-up of the flight muscles in some insects: Heinrich 1993; Dudley 2000) and in the take-off itself are evident from general accounts (Johnson 1969; Gatehouse 1997; Reynolds et al. 2014). These emphasize repeatedly that the movements concerned are 'intentional' and not the result of inadvertent dislodgement.

7.4.1 The initiation of migration and the orientation at take-off

Some highly distinctive behaviors are used to achieve 'lift-off' in small wingless insects and terrestrial arthropods migrating on air currents. The basic mechanisms are either the use of silken lines to lift the migrant off some exposed surface (an activity known as

'kiting' or 'ballooning') or the adoption of 'rearing-up' postures so that the wind drag on the migrant's body is enough to carry it away (Reynolds et al. 2014). In most of these cases, and in more conventional take-off by small winged insects, there is a positive phototaxis, which causes the potential migrant to climb to an elevated point, such as the periphery of a single leaf in the case of a small insect. Take-off in the migrant aphid is described by Taylor (1986):

It commences as a rapid walk up to a leaf tip or similar platform, preferably in sunshine and clear of immediately adjacent hazards, after a momentary antennal check of wind-direction and speed and hesitant lifting of individual feet, as if to check the mechanism for retraction, lift-off is sudden, upwardly directed and takes the aphid clear of surrounding vegetation. Those that are successful in take-off continue in strong upward flight to the unilateral stimulus of light.

The ascent of aphids is made in response to diffuse short-wavelength light from the sky, which temporarily dominates the response to tactile, chemical, and visual stimuli from the host plant (Kennedy and Fosbrooke 1973).

Small day-flying insects maintaining steep upwardly-directed flight will quickly enter the atmospheric boundary layer where, in sunny weather, they can utilize convective thermals to ascend high into the air; they may then be circulated around in the atmosphere by up- and downdrafts for periods of several hours (Johnson 1969), during which they will undergo considerable horizontal translocation. A recent radar/lidar study (Wainwright et al. 2017) provides an overview of what is happening—small insects are borne upwards within updrafts, but do not keep pace with the air flow; in fact, they are descending with respect to the surrounding air at an average rate of 0.4 m s⁻¹. The most likely possibility is that they continue wing-beating, but with not enough effort to maintain altitude (small insects like aphids do not glide), and this is presumably a widespread energy-saving adaptation for flight under convective conditions. It also explains why these migrants do not ascend if the atmosphere is stable; then they are confined to layers quite close to the ground.

Small weakly-flying insects will make a negligible contribution to their **ground speed** and **ground track** unless they are flying in virtually still air very close to the vegetation canopy. Consequently, these migrants can only exert control over their direction of movement by choosing when to fly. Despite that, wind-borne movements of most small insects, such as aphids, seem to be initiated irrespective of wind direction (apart from, perhaps, in airstreams associated with significantly different air temperatures). This apparently rather indiscriminate scattering is sustainable because of the exceptional rate of reproduction in aphid clonal migrants.

Some small insects are, however, able to fly selectively in response to certain weather factors, e.g. the potato leafhopper (*Empoasca fabae*) in North America, which in autumn shows enhanced emigration in conditions favouring southward movement towards overwintering areas (Taylor and Reling 1986; Shields and Testa 1999).

Nocturnal migrants, taking off at dusk, cannot rely on convection to assist their ascent, but must generally climb to altitude by their own efforts. Radar observations show they are well able to do this (see Chapter 10 in Drake and Reynolds 2012); even a small planthopper like *Nilaparvata lugens* can maintain an ascent rate of 0.2 m s^{-1} up to heights of 1000 m or more. Some migrants have very vigorous initial ascents—the corn earworm moth, *Helicoverpa zea*, ascended in a tight spiral with a climb rates of 3.6 m s^{-1} over the first ~50 m of altitude before it adopted a shallower climb angle. At altitudes when migrants are detectable on radar, large nocturnal insects such as migratory grasshoppers and moths are typically climbing at ~0.5 m s^{-1}.

7.4.2 Orientation in the 'transmigration' phase

A categorization of the orientation behaviors of large insect migrants in the 'transmigration' phase (i.e. after the initial ascent to altitude), with notes on mechanisms and cues, and on advantages and disadvantages of the various orientation strategies, can be found in Table 1 of Chapman et al. (2015) and in Reynolds et al. (2016). Day-flying insects migrating within their FBL, frequently just a few metres from the ground, are suitable for visual observation and for some species there is a huge number of anecdotal reports of migrants flying over open country and maintaining a persistent (seasonally-dependent) orientation direction 'as if guided by a compass' (e.g. Williams 1930, 1958).

It seems likely that there will be a hierarchy of orientation cues, but the dominant one in these cases is very probably a solar compass, based on the sun's azimuth and/or sun-linked patterns of light polarization, intensity or spectral gradient (el Jundi et al. 2014). The compass may or may not be time-compensated for the daily movement of the solar azimuth (Mouritsen and Frost 2002; Srygley and Dudley 2008; Nesbit et al. 2009; Reppert et al. 2010).

Considerable progress has been made in understanding the molecular and cellular basis of the time-compensated solar compass in the monarch butterfly during its autumn migration south to wintering sites in central Mexico and its remigration north in the spring (see Reppert et al. 2016). To maintain the correct bearing, monarchs must compensate for the azimuthal movement of the sun (or the associated polarized light patterns) across the sky during the day, and the timing mechanism was found to be a light-entrained circadian clock situated in the antennae. Processing and integration of the directional and the timing information seems to occur in circuitry in the monarch's midbrain central complex, which is therefore the putative site of the sun compass. These studies have been facilitated by detailed work on the morphological and functional organization of central neural pathways in the brain of the desert locust *Schistocerca gregaria*, which show how sky polarization vision is integrated with spectral gradient cues for the purposes of celestial compass orientation (e.g. Homberg et al. 2011).

Mouritsen et al. (2013) found that monarchs use a simple 'clock-and-compass navigation' (= vector navigation) strategy to orient towards their Mexican overwintering areas, but that they did not compensate directionally when subjected to large longitudinal displacements, and so appear to lack the internal 'map' possessed by 'true navigators'. Large-scale landmarks like the Rocky Mountains and Atlantic Ocean/Gulf of Mexico coastlines might help to 'funnel' migrants into southern Texas. It should be noted that monarchs do not migrate exclusively within their FBL—if winds are favourable they can ascend on thermal updrafts or by using 'slope lift', then glide across country in the appropriate direction. The final approach to the roosting locations may be facilitated by odour plumes produced by the use of the highly-restricted overwintering areas (specific groves of trees) over thousands of years. Similarly, olfactory cues probably guide nocturnally-migrating bogong moths (*Agrotis infusa*) on the last stages of their flights to communal aestivation sites in mountain-top caves and rock crevices in south-eastern Australia (Warrant et al. 2016).

In other FBL migrants, directionality is clearly guided by landmarks and 'leading lines', although these may be abandoned if they diverge too much from a preferred compass direction. In Florida, the great southern white butterfly, *Ascia monuste*, whose migrants fly *en masse* in a narrow stream close to the ground, is strongly guided by linear features such as roads, railways, and shorelines over distances of up to 130 km (Nielsen 1961; summarized in Johnson 1969). These migrations were rectilinear, and more-or-less independent of the wind (the butterflies avoided strong crosswinds by flying in

the lee of dunes), but directions were not necessarily seasonally-fixed as butterflies from the same outbreak sometimes migrated in opposite directions.

Radar observations have led to significant increases in our knowledge of the 'transmigration phase' behavior in larger high-flying migrants in two main respects. First, there is the question of whether migrants are able to migrate selectively on seasonally-appropriate winds; this is particularly important in autumn in the higher-latitude temperate zone because taking to the air indiscriminately could lead to further poleward displacement and no subsequent chance of survival for the migrant and/or its offspring. There is now much evidence, in Britain at least, of tailwind selectivity in many medium-sized and large insects migrating both during the daytime and at night (Chapman et al. 2010; Hu et al. 2016a,b; see Figure 7.3). In some areas of the world,

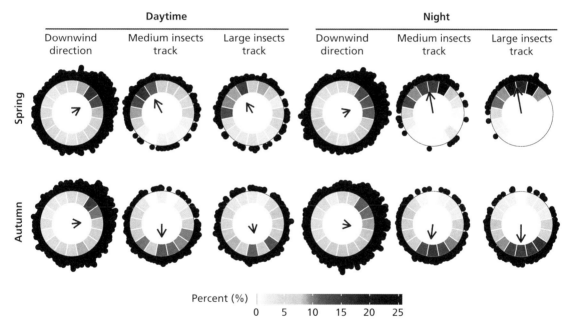

Figure 7.3 Migration tracks (displacement directions with respect to the ground) of medium (10–70 mg) and large (70–500 mg) high-flying insects during 'spring' (May–June), and 'autumn' (August–September) compared with downwind directions at the same altitude. The track directions were measured for insect migration events over a 10-year period with vertical-looking entomological radars in southern England. Small black circles on the periphery of the circular histograms represent mean directions of individual mass migrations, and the grey-scale bar indicates the percentage of migrations in each 22.5° bin. The bearing of the arrow indicates the overall mean direction, and arrow length represents the circular resultant length, *r* (a measure of the 'tightness' of the distribution). As can be seen, the prevailing winds blew towards the northeast or east, but a diverse array of insect migrants attained significantly different—and much more seasonally-beneficial—migration directions. The migrants selected days/nights with seasonally favourable tailwinds, and some also partially compensated for drift from their preferred headings (see text).

Modified from Hu et al. 2016a.

easily detectable characteristics (e.g. differences in temperature and humidity) of opposing air-masses would allow migrants on the ground to decide on the favourability of winds aloft. Meteorological factors of this sort do not seem to be reliably present in Britain, and as a consequence, migrants may ascend to altitude, but then descend again if they find themselves being displaced in the wrong direction. This strongly implies that high-flying insects have a compass sense (Chapman et al. 2008, 2010; Hu et al. 2016a,b). In day-fliers it is highly likely that the compass is solar-based, as it is in migrants that fly within their FBL (see earlier), but the compass mechanism in night-fliers is unknown. The most likely possibilities are the time-compensated use of the light gradient of the Milky Way (Warrant and Dacke 2016), and/or cues from the Earth's magnetic field (Reppert et al. 2016).

In addition to the compass sense, radar observations have revealed the prevalence of wind-related orientations in high-flying migrants (Drake and Reynolds 2012; Hu et al. 2016b; Reynolds et al. 2016), and the fact that these flight headings can have a discernible effect on migrants' trajectories (Chapman et al. 2010), and that they therefore influence migration outcomes. The simplest strategy, found in many medium-sized insects, is to adopt an approximately downwind heading, so that an individual's **airspeed** is added to the wind speed, thus maximizing the distance covered in a given time. A much more sophisticated strategy is shown by *Autographa gamma* (the silver Y moth), where the migrant's heading lies between the downwind direction and a seasonal 'preferred direction of movement' (Chapman et al. 2008, 2010, 2015)—in other words, the moths are partially compensating for crosswind drift from a favoured migration direction (toward the north in spring and the south in autumn). Full compensation is not generally seen, presumably because the migrant would make increasingly little headway in the preferred direction, as this diverged from the downwind direction.

In both the straightforward downwind, and the partially compensating, orientation behaviors it is, of course, necessary for the migrant to be able to determine the ambient wind direction, and to do this when flying at altitudes of several hundreds of metres on very dark nights. Reynolds et al. (2016) have recently reviewed the sensory mechanisms which might be used; the obvious candidate, at least in daylight and in brighter moonlight, is the visual perception of apparent ground movement, but insects also orientate when the combination of low illumination and very slow angular velocities of ground features seem to rule out an optical mechanism. In these cases, and maybe in others, migrants may determine the mean wind direction by mechano-reception of tiny turbulent velocity and acceleration cues within the airflow itself (see Reynolds et al. 2016).

7.4.3 The termination of migration: fallout and settling behavior

Small day-flying insects, like aphids, riding convection currents to various heights in the atmosphere will eventually find themselves in a downdraft; the migrants may then re-enter their FBL and engage in a post-migratory 'alighting flight' prior to landing on potential hosts. These descents are generally not thought to be due to fatigue or depletion of flight fuels because, in aphids, after an investigatory probe with the mouthparts into the host's epidermal cells, the migrant will often take off again even on a preferred host plant. There might be several cycles of reciprocal inhibitory and excitatory interactions between migratory flight and landing, before the aphid finally settles on a suitable host (see chapter 2 in Dingle 1996), assuming one is found.

The termination of migration due to flight fuel exhaustion may occur in some circumstances, for example, when migrants are compelled to continue in flight for long periods because they happen to be over the sea at a time when they would normally land. Prolongation of flight might also occur if migrants were inhibited from descent into the colder air below them when flying at the top of a temperature inversion (Farrow 1986).

In the case of nocturnal long-distance migrants, there is often progressive fallout as the night continues and the aerial density of insects in a layer, for example, gradually thins out. Consequently, there are few radar observations showing a concerted end of a migration. Mass landings can occur, for example, under the influence of heavy downpours of rain (see chapter 11 in Drake and Reynolds 2012), but because precipitation interferes with radar observations, these events are usually deduced, rather than directly observed. The extended fallout of airborne migrants seems to result in landings occurring in rather a haphazard way—newly-arrived immigrants are sometimes found in highly unsuitable habitats, necessitating re-location by further migration or by post-migratory 'appetitive' flights. All this indicates that nocturnal migrants cannot pick out suitable habitats from high altitudes. An intriguing counter-example exists: Greenbank et al. (1980) cite evidence that migrating eastern spruce budworm moths (*Choristoneura fumiferana*) could detect their host trees from altitudes of 100–200 m, and flights ended with the moths descending in a rapid vertical plunge, rather than a gradual descent, into the forest stands.

7.5 Some examples of long-range insect flight trajectories and population trajectories

Documentation of migrations can consist of counts of migrants moving past a point, or over a vertical-profiling radar (the Eulerian approach) or following the actual path of an individual tagged with some telemetric device (the Lagrangian approach). However, as active telemetric devices are presently too heavy for all but the largest insect species (Kissling et al. 2014), a long-distance **flight trajectory** (= flight path) has to be estimated. Estimates are usually made from wind-field information in conjunction with a known migration start or endpoint (derived, for example, from mark-recapture data, departures observed by radar, or evidence of mass arrivals from visual observation or trapping; Pedgley 1982; Drake and Reynolds 2012). Formerly, trajectories would be constructed manually from synoptic wind-field maps (Johnson 1969; Pedgley 1982), but now they are likely to be computed from numerical weather prediction models such as NAME (Chapman et al. 2010) or HYSPLIT (Westbrook et al. 2016). As well as computing air parcel trajectories, models may also estimate the amount of dispersion of the windborne insects due to atmospheric turbulence, and there may be additional submodels to take account of the self-powered flight vector of the migrants; this can make a significant difference to the migration trajectories of the larger species

(Chapman et al. 2010). Backtracking to identify the source of large nocturnal migrants, which are likely to have taken off at dusk, will usually be more accurate than backtracks for small daytime migrants, which may have started flight anytime in the day after the onset of convection (Drake and Reynolds 2012).

Censuses of seasonal redistribution of a species over many years, combined with multiple trajectory analyses, will allow estimates of the variability associated with these spatiotemporal processes (Drake et al. 2001). Eventually, an adequate characterization of the population trajectory for the species concerned (see Section 7.2) should emerge. For most species where there has been some investigation, our knowledge of the population trajectory remains highly fragmentary, but there are exceptions—the monarch butterfly across eastern North America (Miller et al. 2012; Flockhart et al. 2013), and certain locusts, e.g. the desert locust (Pedgley 1981; Magor et al. 2007) and the Australian plague locust (Deveson and Walker 2005). Considerable recent progress has been made in determining the migration cycle of the painted lady butterfly (*V. cardui*) in the Western Palearctic and West Africa (Stefanescu et al. 2013, 2016; Talavera and Vila 2017; see Figure 7.4), the fall armyworm moth (*Spodoptera frugiperda*) in North America (Westbrook et al. 2016), and some migratory planthoppers and noctuid moths in East Asia (Drake and Gatehouse 1996, see references in Chapter 13 in Drake and Reynolds 2012).

7.6 Population consequences of migration

Ultimately, migratory movements are an evolved response to environmental heterogeneity, and particularly to the changing pattern of resource availability, in both space and time; arrivals often coincide with environments becoming more hospitable, e.g. through seasonal vegetation development. Migratory insect species associated with very ephemeral or patchy micro-habitats (dung, carrion, rotten fruit, small bodies of water, e.g. tree-holes) will have to move—staying-put in some form of diapause is not an option here, and movements facilitate the rapid colonization of newly-available environments. Considering more extensive habitats, highly

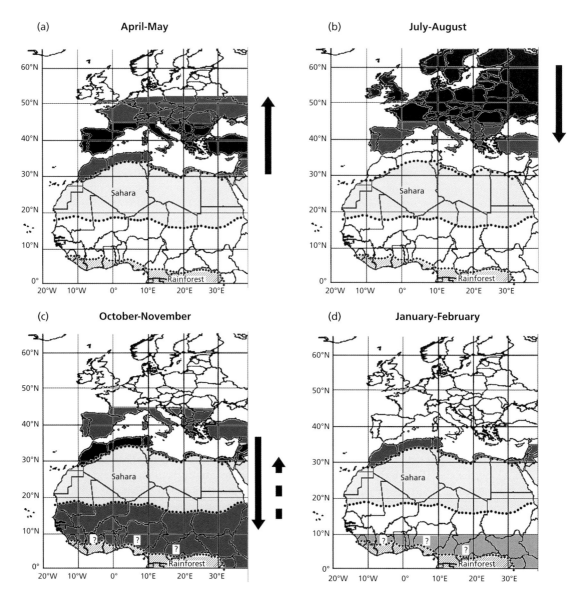

Figure 7.4 A 'population pathway', the pattern of movements through the habitat arena followed by successive generations of migrants. The pathway is a long-term average of a series of population trajectories (see Figure 7.1 above). The figure shows the annual migration circuit of the painted lady butterfly (*Vanessa cardui*) in the Western Palearctic and African Regions. (Data from Stefanescu et al. 2013, 2016; C. Stefanescu personal communication; Talavera and Vila 2017.) The principle latitudinal breeding areas at four selected times of the year are shown with the regions of the heaviest breeding in black, and moderately abundant breeding in dark grey, etc. In addition to the marked areas, heavy, but sporadic rains in the Sahara desert can occasionally lead to large population outbreaks there, e.g. in Western Sahara and Mauritania in the autumn. The rainforest (cross-hatched) areas in tropical Africa are generally unsuitable for *V. cardui*. The black arrows show the main seasonal directions of migration. (a) Progressive seasonal northwards movements from the Maghreb through Europe in spring and early summer is followed (b, c) by southward movements in late summer and autumn back to the Maghreb with some further movement into the Sahelian region south of the Sahara. Some northward movement from the Sahel to the Maghreb also appears to occur in autumn—shown in (c) as a broken arrow. (d) The main overwintering area appears to be in the Maghreb, and the extent of overwintering in the savannah of tropical Africa is uncertain.

migratory species are likely to inhabit 'disturbed-environments', such as ruderal vegetation, intensively-managed grasslands, cultivations, and drier biotopes, which are more likely to be transient (Chapman and Drake 2017). Species which live on long-lived hosts (trees) and stable habitats (woodland, moorland, marshes) are, in contrast, under-represented in the migrant fauna (Reynolds et al. 2013).

Often there will be strong trade-offs between a migratory and a more sedentary mode of life; that is, a trade-off between the energetic and mortality costs of migration, and the potential benefits which may accrue to successful migrants (Chapman et al. 2015). These trade-offs have not been fully elucidated for any one species, although substantial benefits have been identified in a number of species. For instance, seasonal migrations result in increased reproductive output in migrant lineages in the silver Y moth (Chapman et al. 2012), in lower pathogen loads in the monarch butterfly (Altizer et al. 2011) and in reduced mortality from parasitoid wasps in the painted lady butterfly (Stefanescu et al. 2012). The costs associated with long-range migration are less quantified, and the ways by which migrants overcome the physiological challenges and risks of such impressive journeys is a field worthy of further research.

7.7 Environmental change and migration

Behavioral responses by insects to environmental change have been reviewed by Blanckenhorn (see Chapter 19), and the impact of climate and land-use changes on **range shifts**, migration pathways, population trends, and flight **phenology** in insects have been discussed by Parmesan (2001), Sparks et al. (2005, 2007) and Bell et al. (2015).

It seems likely that highly migratory and adaptable species with high rates of reproduction, and with a range of individual flight durations (some relatively short, others very prolonged), will discover and colonize suitable habitats even if the locations of these change. In fact, some migrant Lepidoptera including pest species are showing increased frequency of migration into northern temperate regions, under the influence of climate change (Sparks et al. 2005, 2007; Bebber et al. 2013).

Conservation problems lie, first, with relatively sedentary species where subpopulations may become trapped in increasingly isolated 'islands' of deteriorating habitat and eventually become extinct. Sometimes evolution toward greater dispersal ability (if this is heritable) (e.g. Thomas et al. 2001; Hughes et al. 2007) may help in extending range boundaries, but in other cases, human-assisted 'migration' (i.e. managed relocation) might be necessary for conservation purposes. Secondly, environmental, particularly anthropogenic, changes may subject some migrant species to new wide-scale mortality factors. One example is the decline in breeding habitat of the migratory population of monarch butterflies in the USA due to the loss of larval host plants (milkweeds) arising from the adoption of herbicides used in association with genetically-modified crops (Flockhart et al. 2015; Semmens et al. 2016). Clearly, the prolonged migration and highly specific requirements in the overwintering sites, make this species more vulnerable to environmental changes compared to many migratory insects.

7.8 Some outstanding questions and issues in insect migration behavior

Compared with birds and many other vertebrates, the basic elements of the migration 'pathway' in nearly all insects are virtually unknown, in the sense of ascertaining where the major populations and subpopulations are likely to be at all seasons of the year, and the probabilities of migrants taking various routes between the occupied sites. A few exceptions to this were mentioned earlier. The combination of citizen science surveys and telemetric tracking of individuals (Kissling et al. 2014) will help to elucidate the migratory pathways of some charismatic and large insects, but they do not seem applicable to the vast majority of insect species as these are too small to carry the tagging devices and too 'cryptic' for monitoring by the public. In this case, standardized ground trapping networks (e.g. the Rothamsted Insect Survey, Harrington 2014), the combination of trapping and radar (Drake and Reynolds 2012; Hu et al 2016a), and stable isotopic analyses (Hobson et al. 2012) will have a role to play.

As discussed previously, both day- and night-flying insect migrants have been shown to display sophisticated flight orientations, and disentangling the influence of orientation and wind drift on migrants' destinations is an intriguing issue (Reynolds et al. 2016). What would an omniscient view of the migration pathways and destinations of large insects look like—would it owe more to wind systems or migrant orientations? This is particularly interesting where a migrant (e.g. the painted lady butterfly) shows strong seasonally 'directed' migrations in, for example, Europe, but its range also includes desert areas such as the Sahara, where the optimal strategy would appear to be downwind drift towards wind convergence zones where seasonal rains are likely to fall (Rainey 1989) thus providing vegetation for the larvae.

As the internal neurobiological mechanisms involved in the orientation of nocturnal, high-flying insect migrants are still largely 'black boxes', considerable multi-disciplinary research on these processes is evidently required.

7.9 Concluding remarks

This chapter shows how specialized movement behavior, and associated aspects of physiology and morphology, have a key role in defining animal migration and are, in fact, more indicative than, say, the actual distance moved or the specific mode of locomotion (aerial, terrestrial, aquatic, or phoretic). Migration, so defined, should be distinguishable from other categories of movement or, failing that, the definition should suggest testable hypotheses that allow such distinctions to be made (Dingle 2014). Behavior will have a strong influence on whether an individual migrant is successful in reaching a favourable habitat or not, and so natural selection will act immediately and powerfully on the suite of traits forming the migration syndrome. Environmental variability will act to maintain variation in syndrome traits, possibly allowing different subpopulations to develop divergent behaviors (Drake et al. 1995).

As regards the typical mode of insect migration, by flapping flight, radar-based studies have revealed the ubiquity of flight orientations that tend to maximize distance travelled in seasonally beneficial directions in both medium-sized and large insects migrating at altitude. Uncovering the mechanisms involved in this sophisticated behavior (particularly the compass-sense and wind-finding abilities in nocturnal fliers), and the molecular basis of the migration-enabling traits in general, promise to be exciting areas of research over the next few years.

References

Alerstam, T., Hedenström, A., and Åkesson, S. (2003). Long-distance migration: evolution and determinants. *Oikos*, **103**, 247.

Altizer, S., Bartel, R., and Han, B.A. (2011). Animal migration and infectious disease risk. *Science*, **331**, 296.

Anderson, R.C. (2009). Do dragonflies migrate across the western Indian Ocean? *Journal of Tropical Ecology*, **25**, 347.

Bebber, D.P., Ramotowski, M.A.T., and Gurr, S.J. (2013). Crop pests and pathogens move polewards in a warming world. *Nature Climate Change*, **3**, 985.

Beebe, W. (1949) Insect migration at Rancho Grande in North-central Venezuela. General Account. *Zoologica (New York)*, **34**, 107–10.

Bell, J.R., Alderson, L., Izera, D., et al. (2015). Long-term phenological trends, species accumulation rates, aphid traits and climate: five decades of change in migrating aphids. *Journal of Animal Ecology*, **84**, 21.

Bergin, M. 2009. I and the bird. Available at: http://www.10000birds.com/i-and-the-bird-99.htm (accessed 23 February 2018).

Brower, L.P., Fink, L.S., and Walford, P. (2006). Fueling the fall migration of the monarch butterfly. *Integrative and Comparative Biology*, **46**, 1123.

Chapman, J.W., Bell, J.R., Burgin, L.E., et al. (2012). Seasonal migration to high latitudes results in major reproductive benefits in an insect. *Proceedings of the National Academy of Sciences, USA*, **109**(37), 14924.

Chapman, J.W., and Drake, V.A. (2017). Insect migration. In: *Reference Module in Life Sciences*. Oxford: Elsevier. Available at: http://www.sciencedirect.com/science/article/pii/B9780128096338012486 (accessed 16 February 2018).

Chapman, J.W., Nesbit, R.L., Burgin, L.E., et al. (2010). Flight orientation behaviors promote optimal migration trajectories in high-flying insects. *Science*, **327**, 682.

Chapman, J.W., Reynolds, D.R., Hill, J.K., et al. (2008). A seasonal switch in compass orientation in a high-flying migrant moth. *Current Biology*, **18**, R908.

Chapman, J.W., Reynolds, D.R., Smith A.D., Smith, E.T., and Woiwod, I.P. (2004). An aerial netting study of insects migrating at high-altitude over England. *Bulletin of Entomological Research*, **94**, 123.

Chapman, J.W., Reynolds, D.R., and Wilson, K. (2015). Long-range seasonal migration in insects: mechanisms, evolutionary drivers and ecological consequences. *Ecology Letters*, **18**, 287.

Corbet, P.S. (1999). *Dragonflies: Behaviour and Ecology of Odonata*. Harley Books, Great Horkesley.

De Vos, J.M., Joppa, L.N., Gittleman, J.L., Stephens, P.R., and Pimm, S.L. (2015). Estimating the normal background rate of species extinction. *Conservation Biology*, **29**, 452.

Deveson, E.D., and Walker, P.W. (2005). Not a one-way trip: historical distribution data for Australian plague locusts supports frequent seasonal exchange migrations. *Journal of Orthoptera Research*, **14**, 95.

Dingle, H. (1996). *Migration: The Biology of Life on the Move*. Oxford University Press, Oxford.

Dingle, H. (2001). The evolution of migratory syndromes in insects. In: I. P. Woiwood, D.R. Reynolds, and C.D. Thomas (Eds) *Insect Movement: Mechanisms and Consequences*, pp. 159–81. CABI, Wallingford.

Dingle, H. (2014). *Migration: The Biology of Life on the Move*, 2nd edn. Oxford University Press, Oxford.

Dingle, H., and Drake, V.A. (2007). What is migration? *BioScience*, **57**, 113.

Drake, V.A. (2008). What is migration [presentation, March 2008]. Available at: https://www.researchgate.net/publication/299595880_What_is_Migration_presentation

Drake, V.A., and Farrow, R.A. (1988). The influence of atmospheric structure and motions on insect migration. *Annual Review of Entomology*, **33**, 183.

Drake, V.A., and Gatehouse, A.G. (1996). Population trajectories through space and time: a holistic approach to insect migration. In: R.B. Floyd, A.W. Sheppard, and P.J. De Barro (Eds), *Frontiers of Population Ecology*, pp. 399–408. CSIRO Publishing, Melbourne.

Drake, V.A., Gatehouse, A.G., and Farrow, R.A. (1995). Insect migration: a holistic conceptual model. In: V.A. Drake and A.G. Gatehouse (Eds), *Insect Migration: Tracking Resources through Space and Time*, pp. 427–57. Cambridge University Press, Cambridge.

Drake, V.A., Gregg, P.C., Harman, I.T., et al. (2001). Characterizing insect migration systems in inland Australia with novel and traditional methodologies. In: I.P. Woiwood, D.R. Reynolds, and C.D. Thomas (Eds), *Insect Movement: Mechanisms and Consequences*, pp. 207–33. CABI, Wallingford.

Drake, V.A., and Reynolds, D.R. (2012). *Radar Entomology: Observing Insect Flight and Migration*. CABI, Wallingford.

Dudley, R. (2000). *The Biomechanics of Insect Flight: Form, Function, Evolution*. Princeton University Press, Princeton.

el Jundi, B., Pfeiffer, K., Heinze, S., and Homberg, U. (2014). Integration of polarization and chromatic cues in the insect sky compass. *Journal of Comparative Physiology A*, **200**, 575.

Farrow, R.A. (1986). Interaction between synoptic scale and boundary layer meteorology on micro-insect migration. In: W. Danthanarayana (Ed.), *Insect Flight: Dispersal and Migration*, pp. 185–95. Springer-Verlag, Berlin.

Flockhart, D.T.T., Pichancourt, J-B., Norris, D.R., and Martin, T.G. (2015). Unravelling the annual cycle in a migratory animal: breeding season habitat loss drives population declines of monarch butterflies. *Journal of Animal Ecology*, **84**, 155.

Flockhart, D.T.T., Wassenaar, L.I., Martin, T.G., et al. (2013). Tracking multi-generational colonization of the breeding grounds by monarch butterflies in eastern North America. *Proceedings of the Royal Society B*, **280**, 20131087.

Garratt, J.R. (1994). *The Atmospheric Boundary Layer*. Cambridge University Press, Cambridge.

Gatehouse, A.G. (1987). Migration: a behavioural process with ecological consequences? *Antenna*, **11**, 10.

Gatehouse, A.G. (1997). Behavior and ecological genetics of wind-borne migration by insects. *Annual Review of Entomology*, **42**, 475.

Glick, P.A. (1939). *The Distribution of Insects, Spiders and Mites in the Air*. United States Department of Agriculture, Technical Bulletin no. 673, 1–150. United States Department of Agriculture, Washington, DC.

Goldsworthy, G., and Joyce, M. (2001). Physiology and endocrine control of flight. In: I.P. Woiwood, D.R. Reynolds, and C.D. Thomas (Eds), *Insect Movement: Mechanisms and Consequences*, pp. 65–86. CABI, Wallingford.

Greenbank, D.O., Schaefer, G.W., and Rainey, R.C. (1980). Spruce budworm (Lepidoptera: Tortricidae) moth flight and dispersal: new understanding from canopy observations, radar, and aircraft. *Memoirs of the Entomological Society of Canada*, **110**, 1.

Harrington, R. (2014). The Rothamsted Insect Survey strikes gold. *Antenna*, **38**, 158.

Heinrich, B. (1993). *The Hot-Blooded Insects: Strategies and Mechanisms of Thermoregulation*. Springer-Verlag, Berlin.

Hoare, B. (2009) *Animal Migration: Remarkable Journeys by Air, Land and Sea*. University of California Press, Berkeley, CA.

Hobson, K.A., Anderson, R.C., Soto, D.X., and Wassenaar, L.I. (2012). Isotopic evidence that dragonflies (*Pantala flavescens*) migrating through the Maldives come from the Northern Indian. Subcontinent. *PLoS One*, **7**(12), e52594.

Holland, R.A., Wikelski, M., and Wilcove, D.S. (2006). How and why do insects migrate? *Science*, **313**, 794.

Homberg, U., Heinze, S., Pfeiffer, K., Kinoshita, M., and El Jundi, B. (2011). Central neural coding of sky polarization in insects. *Philosophical Transactions of the Royal Society B: Biological Sciences*, **366** (1565), 680.

Hu, G., Lim, K.S., Horvitz, N., et al. (2016a). Mass seasonal bioflows of high-flying insect migrants. *Science*, **354** (6319), 1584.

Hu, G., Lim, K.S., Reynolds, D.R., Reynolds, A.M., and Chapman, J.W. (2016b). Wind-related orientation patterns in diurnal, crepuscular and nocturnal high-altitude insect migrants. *Frontiers in Behavioral Neuroscience*, **10**, 32.

Hughes, C.L., Dytham, C., and Hill, J.K. (2007). Modelling and analysing evolution of dispersal in populations at expanding range boundaries. *Ecological Entomology*, **32**, 437.

Isard, S.A., and Gage, S.H. (2001). *Flow of Life in the Atmosphere: An Airscape Approach to Understanding Invasive Organisms*. Michigan State University Press, East Lansing, MI.

Isard, S.A., Irwin, M.E., and Hollinger, S.E. (1990). Vertical distribution of aphids (Homoptera: Aphididae) in the planetary boundary layer. *Environmental Entomology*, **19**, 1473.

Johnson, C.G. (1969). *Migration and Dispersal of Insects by Flight*. Methuen, London.

Kennedy, J.S. (1985). Migration: behavioral and ecological. In: M. A. Rankin (Ed.), *Migration: Mechanisms and Adaptive Significance*, pp. 5–26. *Contributions in Marine Science*, vol. 27 (Supplement). Port Aransas, Texas: Marine Science Institute, University of Texas at Austin.

Kennedy, J.S., and Fosbrooke, I.H.M. (1973). The plant in the life of an aphid. In: H. F. van Emden (Ed.), *Insect/Plant Relationships*. Symposia of the Royal Entomological Society of London, No. 6, pp. 129–40. Blackwell Scientific Publications, Oxford.

Kissling, W.D., Pattemore, D.E., and Hagen, M. (2014). Challenges and prospects in the telemetry of insects. *Biological Reviews*, **89**, 511.

Lack, D. and Lack, E. (1951). Migration of insects and birds through a Pyrenean pass. *Journal of Animal Ecology*, **20**, 63.

Larsen, T.B. (1992). Migration of *Catopsilla florella* in Botswana (Lepidoptera: Pieridae). *Tropical Lepidoptera*, **3**, 2.

Magor, J.I., Ceccato, P., Dobson, H.M., Pender, J., and Ritchie. L. (2007). *Preparedness to prevent Desert Locust plagues in the Central Region, an historical review. Part 1: Text and Part 2: Appendices*. Desert Locust Technical Series No. AGP/DL/TS/35. Rome: Food and Agriculture Organization of the United Nations. Available at: http://www.fao.org/ag/locusts/en/publicat/docs/tech/1288/index.html (accessed 16 February 2018).

Miller, N.G., Wassenaar, L.I., Hobson, K.A., and Norris D.R. (2012). Migratory connectivity of the monarch butterfly (*Danaus plexippus*): patterns of spring re-colonization in eastern North America. *PLoS One* **7**(3), e31891.

Mouritsen, H., Derbyshire, R., Stalleicken, J., et al. (2013). An experimental displacement and over 50 years of tag-recoveries show that monarch butterflies are not true navigators. *Proceedings of the National Academy of Sciences, USA*, **110**, 7348.

Mouritsen, H. and Frost, B.J. (2002). Virtual migration in tethered flying monarch butterflies reveals their orienta-tion mechanisms. *Proceedings of the National Academy of Sciences, USA*, **99**, 10162.

Nesbit, R.L., Hill, J.K., Woiwod, I.P., et al. (2009). Seasonally adaptive migratory headings mediated by a sun compass in the painted lady butterfly, *Vanessa cardui*. *Animal Behaviour*, **78**, 1119.

Nielsen, E.T. (1961). On the habits of the migratory butterfly, *Ascia monuste* L. *Biologiske Meddelelser*, **23 (11)**, 1.

Parmesan, C. (2001). Coping with modern times? Insect movement and climate change. In: I. P. Woiwood, D.R. Reynolds and C.D. Thomas (Eds), *Insect Movement: Mechanisms and Consequences*, pp. 387–413. CABI Publishing, Wallingford.

Pedgley, D.E. (1981). *Desert Locust Forecasting Manual, Volumes 1 and 2*. Centre for Overseas Pest Research, London. Available at: http://gala.gre.ac.uk/11862/1/Doc-0563.pdf (accessed 16 February 2018).

Pedgley, D.E. (1982). *Windborne Pests and Diseases. Meteorology of Airborne Organisms*. Ellis Horwood, Chichester.

Rainey, R.C. (1989). *Migration and Meteorology: Flight Behaviour and the Atmospheric Environment of Locusts and other Migrant Pests*. Clarendon Press, Oxford.

Reppert, S.M., Gegear, R.J., and Merlin, C. (2010). Navigational mechanisms of migrating monarch butterflies. *Trends in Neurosciences*, **33**, 399.

Reppert, S.M., Guerra, P.A., and Merlin, C. (2016). Neurobiology of monarch butterfly migration. *Annual Review Entomology*, **61**, 25.

Reynolds, D.R., Nau, B.S., and Chapman, J.W. (2013). High-altitude migration of Heteroptera. *European Journal of Entomology*, **110**, 483.

Reynolds, D.R., Reynolds, A.M., and Chapman, J.W. (2014). Non-volant modes of migration in terrestrial arthropods. *Animal Migration*, **2**, 8.

Reynolds, A.M., Reynolds, D.R., Sane, S.P., Hu, G., and Chapman, J.W. (2016). Orientation in high-flying migrant insects in relation to flows: mechanisms and strategies. *Philosophical Transactions of the Royal Society B*, **371**, 20150392283.

Rochester, W.A. (1999). The migration systems of *Helicoverpa punctigera* (Wallengren) and *Helicoverpa armigera* (Hübner) (Lepidoptera: Noctuidae) in Australia. PhD thesis, University of Queensland, Brisbane, Australia.

Roff, D.A. and Fairbairn, D.J. (2007). The evolution and genetics of migration in insects. *Bioscience*, **57**, 155.

Rogers, D. (1983). Pattern and process in large-scale animal movement. In: I. R. Swingland and P. J. Greenwood (Eds), *The Ecology of Animal Movement*, pp. 160–80. Clarendon Press, Oxford.

Rosenberg, J., and Burt, P.J.A. (1999). Windborne displacements of desert locusts from Africa to the Caribbean and South America. *Aerobiologia*, **15**, 167.

Semmens, B.X., Semmens, D.J., Thogmartin, W.E., et al. (2016). Quasi-extinction risk and population targets for the Eastern, migratory population of monarch butterflies (*Danaus plexippus*). *Scientific Reports*, **6**, article 23265.

Shields, E.J., and Testa, A.M. (1999). Fall migratory flight initiation of the potato leafhopper, *Empoasca fabae* (Homoptera: Cicadellidae): observations in the lower atmosphere using remote piloted vehicles. *Agricultural and Forest Meteorology*, **97**, 317.

Simpson, S.J., Sword, G.A., and Nathan Lo, N. (2011). Polyphenism in insects. *Current Biology*, **21**(18), R738.

Sparks, T.H., Dennis, R.L.H., Croxton, P.J., and Cade, M. (2007). Increased migration of Lepidoptera linked to climate change. *European Journal of Entomology*, **104**, 139.

Sparks, T.H., Roy, D.B., and Dennis, R.L.H. (2005). The influence of temperature on migration of Lepidoptera into Britain. *Global Change Biology*, **11**, 507.

Srygley, R.B. and Dudley, R. (2008). Optimal strategies for insects migrating in the flight boundary layer: mechanisms and consequences. *Integrative and Comparative Biology*, **48**, 119.

Stefanescu, C., Askew, R.R., Corbera, J., and Shaw, M.R. (2012). Parasitism and migration in southern Palaearctic populations of the painted lady butterfly, *Vanessa cardui* (Lepidoptera: Nymphalidae). *European Journal of Entomology*, **109**, 85.

Stefanescu, C., Páramo, F., Åkesson, S., et al. (2013). Multi-generational long-distance migration of insects: studying the painted lady butterfly in the Western Palaearctic. *Ecography*, **36**, 474.

Stefanescu, C., Soto, D.X., Talavera, G., Vila R., and Hobson, K.A. (2016). Long-distance autumn migration across the Sahara by painted lady butterflies: exploiting resource pulses in the tropical savannah. *Biology Letters*, **12**(10), 20160561.

Talavera, G. and Vila, R. (2017). Discovery of mass migration and breeding of the butterfly *Vanessa cardui* in the Sub-Sahara: the Europe–Africa migration revisited. *Biological Journal of the Linnean Society*, **120**, 274.

Taylor, L.R. (1974). Insect migration, flight periodicity and the boundary layer. *Journal of Animal Ecology*, **43**, 225.

Taylor, L.R. (1986). The distribution of virus disease and the migrant vector aphid. In: G.D. McLean, R.G. Garrett, and W.G. Ruesink (Eds), *Plant Virus Epidemics: Monitoring, Modelling and Predicting Outbreaks*, pp. 35–57. Academic Press, Sydney.

Taylor, R.A.J., and Reling, D. (1986). Preferred wind direction of long-distance leafhopper (*Empoasca fabae*) migrants and its relevance to the return migration of small insects. *Journal of Animal Ecology*, **55**, 1103.

Thomas, C.D., Bodsworth, E.J., Wilson, R.J., et al. (2001). Ecological and evolutionary processes at expanding range margins. *Nature*, **411**, 577.

Van der Horst, D.J. (2003). Insect adipokinetic hormones: release and integration of flight energy metabolism. *Comparative Biochemistry & Physiology B*, **136**, 217.

Van Swaay, C.A.M., Nowicki, P., Settele, J., and Van Strien, A.J. (2008). Butterfly monitoring in Europe: methods, applications and perspectives. *Biodiversity and Conservation*, **17**(14), 3455.

Wainwright, C.E., Stepanian, P.M., Reynolds, D.R., and Reynolds, A.M. (2017). The movement of small insects in the convective boundary layer: linking patterns to processes. *Scientific Reports*, **7**(1), 5438.

Warrant, E., and Dacke, M. (2016). Visual navigation in nocturnal insects. *Physiology*, **31**, 182.

Warrant, E., Frost, B., Green, K., et al. (2016). The Australian bogong moth *Agrotis infusa*: a long-distance nocturnal navigator. *Frontiers in Behavioral Neuroscience*, **10**, article 77.

Westbrook, J.K., Nagoshi, R.N., Meagher, R.L., Fleischer, S.J., and Jairam, S. (2016). Modeling seasonal migration of fall armyworm moths. *International Journal of Biometeorology*, **60**, 255.

Williams, C.B. (1930). *The Migration of Butterflies*. Oliver & Boyd, Edinburgh.

Williams, C.B. (1958). *Insect Migration*. Collins, London.

Zera, A.J. (2016). Evolutionary endocrinology of hormonal rhythms: juvenile hormone titer circadian polymorphism in *Gryllus firmus*. *Integrative and Comparative Biology*, **56**, 159.

CHAPTER 8

Feeding behavior

Stephen J. Simpson[1], Carlos Ribeiro[2], and Daniel González-Tokman[3]

[1] *Charles Perkins Centre, The University of Sydney, NSW, Australia*
[2] *Champalimaud Research, Champalimaud Centre for the Unknown, Lisbon, Portugal*
[3] *CONACYT, Red de Ecoetología, Instituto de Ecología, A. C. Xalapa, México*

8.1 Introduction

Feeding behavior is a means to a fundamentally important end—the acquisition of a balanced complement of nutrients to support development, reproduction, and survival. Nutrition directly or indirectly influences all aspects of the life of individuals, their offspring, and the symbionts, commensals, and parasites to which the individual plays host. Nutrition also fashions the interactions between individuals within groups, populations, and societies, and in turn, shapes the structure and dynamics of species assemblages and ecosystems (Simpson and Raubenheimer 2012).

No other group of animals exploits as wide a range of food items to achieve this end as the insects (Chapman 2014). The task of obtaining balanced nutrition is complicated by the fact that nutritional requirements are multidimensional and change over time. Additionally, foods can be time- and energy-consuming to find, are not necessarily nutritionally balanced when found, and are often chemically and physically protected, hence dangerous to subdue and eat. Because one individual forager is another forager's potential meal, locating and eating food can also be hazardous.

There have been several intertwined research strands in the history of the study of insect feeding behavior, which arose from posing a question of considerable practical, as well as academic interest, namely 'What determines the food choices of plant-feeding insects?' Early workers, such as Reginald Painter (1936), argued that nutrients were of greatest significance in herbivorous insect food choices and performance. There was then a shift to the view that plant **secondary metabolites**, rather than nutrients, primarily determine food choices, and have driven the co-evolutionary relationships between insects and their plant foods (e.g. Fraenkel 1959). Nutrients slowly returned to prominence, initially with the advent of insect dietetics and quantitative nutrition (Dadd 1963; Waldbauer 1968), leading to the development of the field of 'classical' insect nutritional ecology (Scriber and Slansky 1981; Slansky and Rodriguez 1987).

Meanwhile, another primary research strand pursued the mechanisms controlling feeding behavior and **appetite** in insects. Early model systems here included blow flies, locusts, caterpillars, aphids, and blood-feeding species (Bernays and Simpson 1982). The towering figure in this area was Vincent Dethier, whose book, *The Hungry Fly* (1976) remains essential reading for any student of insect behavior. Another landmark compendium was the volume edited by Reginald Chapman and

Simpson, S. J., Ribeiro, C., and González-Tokman, D., *Feeding behavior.* In: *Insect Behavior: From mechanisms to ecological and evolutionary consequences.* Edited by Alex Córdoba-Aguilar, Daniel González-Tokman, and Isaac González-Santoyo: Oxford University Press (2018). © Oxford University Press.
DOI: 10.1093/oso/9780198797500.003.0008

Gerrit de Boer, *Regulatory Mechanisms in Insect Feeding* (1995).

The research strands concerning quantitative nutrition and control of feeding came together with the discovery that insects have nutrient-specific appetite systems that regulate food choices to achieve a balanced nutrient intake (Simpson and Simpson 1990; Waldbauer and Friedman 1991). Recent years have built upon these earlier experimental foundations in two principle ways. The first has been by deploying advanced molecular genetic techniques to the understanding of feeding behavior and appetite, in particular taking advantage of *Drosophila melanogaster* as a model, not only for understanding insect feeding behavior, but also as a biomedical tool (Itskov and Ribeiro, 2013). The second major advance has been the development of a unifying framework for nutritional ecology, the **geometric framework**, which brings together earlier strands of research within a single experimental and conceptual scheme (Simpson and Raubenheimer 2012; Raubenheimer et al. 2014) that is currently shaping research in fields far beyond entomology.

This chapter focuses on the control of food and nutrient intake, and the implications of feeding behavior and nutrition at multiple levels, from individuals to ecosystems. General accounts of the diversity of feeding habits among the insects, and of the mechanics of food handling and digestion can be found in Chapman (2014). We will begin by considering insights gained from detailed recording of the patterning of feeding.

8.2 Patterns of feeding: control of meals and inter-meal intervals

Like other animals, insects feed intermittently, with discrete periods of feeding (meals) interspersed with intervals in which they do other things (inter-meal intervals). In some instances, it is self-evident what constitutes the initiation of a meal, as distinct from a shorter food sampling event. Likewise, it may be clear when a meal has ended, rather than simply interrupted or paused. In other cases, however, these are not so clear, and breaking down a record of the pattern of feeding into meals and inter-meal intervals becomes a statistical question (Simpson 1995).

Analysis of the pattern of feeding has helped answer three key questions about the control of feeding:

- What determines the onset of a meal?
- What controls meal size?
- How do insects adjust their total intake of food in response to changing nutritional requirements or shifts in diet composition by adjusting the frequency, duration, and rate of ingestion of meals?

The start of a meal depends on internal and external influences. The former include residual levels of feeding inhibition remaining from the previous meal, nutritional and developmental state, and endogenous activity rhythms, whereas the latter include food stimuli, light, temperature, and other environmental

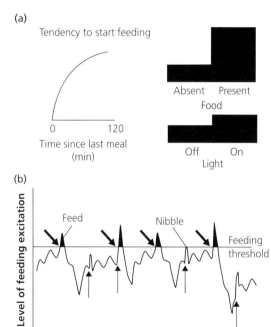

Figure 8.1 (a) The start of feeding depends on the time since the last meal and on environmental factors, such as the presence of food and light. (b) A feeding rhythm starts at the end of a meal and finishes at the beginning of the next meal. Feeding (solid black sections above the threshold) starts when the level of feeding excitation reaches a threshold at which the animal bites the food (oblique arrows). Feeding excitation also increases when the animal defecates (vertical arrows). Modified from Simpson (1995).

factors. Feeding excitation, which Dethier (1976) called 'central excitatory state' (CES), rises with time since the last meal. Other internal and external factors also affect CES and, hence, the probability of commencing feeding. In *Locusta*, a short-term excitatory rhythm was found to multiply baseline feeding excitation, as did the smell of food, and being in the light, rather than in darkness (Simpson and Ludlow 1986; Simpson 1995; Figure 8.1). Feeding commences once a threshold level of excitation is reached and continues if suitable sensory stimuli are provided by the food. Such stimuli include chemical cues (termed **phagostimulants**), such as sugars, amino acids, and in some cases, specific plant secondary metabolites associated with host plants, which are detected by chemosensory receptors located mainly on the tarsi and external mouthparts, and within the pre-oral cavity (Chapman 1995). Other chemical stimuli act as phagodeterrents, either blocking activation of phagostimulant receptors or activating specific inhibitory taste neurons, and therefore inhibiting the onset of feeding (Chapman 1995; Frazier and Chyb, 1995).

Phagostimulation at the beginning of a meal elicits a rapid increase in feeding excitation, providing a positive feedback that stimulates the continuation of feeding. Such positive feedback at the onset of feeding also ensures that if the food runs out before the meal is completed, or contact with the food is lost, the insect exhibits a pattern of movement termed 'local search behavior' (Bell 1990), in which it turns frequently and remains nearby, thereby increasing the likelihood that it will either relocate lost food or find more in the vicinity (Corrales-Carvajal et al. 2016). Dethier (1976) called this behavior in blowflies the 'fly dance'. Such behavior may have been co-opted in the evolution of the bee dance language and provides an effective foraging strategy for insects such as aphid-feeding coccinellid beetles that feed on small (sub-meal) parcels of food that are patchily distributed. Along the same lines, quantitative video tracking-based analysis of 'dances' in *Drosophila* have led to the proposal that these flies use path integration to navigate back to the food spot (Kim and Dickinson 2017) suggesting an evolutionary link between the sophisticated navigational strategies desert ants use to bring food to their nest and Dethier's 'fly dances'.

As feeding continues during a meal there is a build-up in negative feedbacks, which ultimately result in the meal ending once the positive feedback present at the start is negated. Inhibitory feedback comes from stretch receptors variously located on the alimentary tract and body wall, which signal distension. Interrupting these volumetric signals results in overeating (Chapman and de Boer 1995). Other sources of inhibition as a meal continues include falling phagostimulatory input as chemoreceptors adapt, continuing input from slower-adapting chemoreceptors responding to deterrent compounds in the food, rapid changes in blood metabolites during the meal, and feeding-induced release of neurohormones.

The extent by which the feeding threshold is exceeded at the start of a meal sets how much negative feedback is required to terminate the meal. Higher levels of feeding excitation at the outset will therefore result in a larger meal. Higher excitation also results in a faster rate of ingestion. Hence, meal duration tends to vary less than meal size. Because of this, meal duration is often a poor proxy for food intake in feeding studies (Simpson 1995). Overall, the structure of the meal itself is also highly organized and is termed the 'microstructure of feeding'. Given that, in most animals, the motor pattern is highly rhythmic and fixed, animals can regulate the amount of food intake by modulating only two parameters—the length of a feeding bout and the interval between these bouts (Itskov et al. 2014). Similarly to what we described for search strategies, the regulation of meals by changing these two parameters of the microstructure of food intake appears to be conserved across phyla.

More slowly-induced, blood-bore negative feedbacks, some of which arise after the end of a meal, contribute to continuing satiety and contribute to the duration of the next inter-meal interval. As time progresses, these decline, as do more rapidly induced volumetric and other inhibitory signals as the gut empties and metabolite levels return to baseline.

8.3 Automating the recording of feeding behavior

Feeding is one of the most significant of all behaviors, yet among the most difficult to record automatically.

Laborious observational recording of insects feeding in the laboratory (e.g. Simpson 1982) and field (e.g. Raubenheimer and Bernays 1993) has been useful, but has practical limitations—although there is no better way to get a deep understanding of a study species than by spending hours watching it. Lately, this tradition is being revived and further empowered by the advent of automated video tracking approaches and the resulting quantitative analysis and description of behavior (Corrales-Carvajal et al. 2016; Kim and Dickinson, 2017). Although 'computational ethology' often fails to replace the intimate understanding of behavior, which can be obtained through personal observation. Also, video tracking often fails to directly quantify feeding per se. This is most evident in the cases where observation is simply not possible, as is the case for phloem-feeding insects such as aphids, in which the mouthparts are hidden within the plant tissues. Electrical record-

ing of feeding patterns was pioneered in phloem-feeding Homoptera, using both AC- (McLean and Kinsey 1964) and DC-voltage systems based on measuring impedance changes in the circuit made between electrodes placed in the plant and on a tethered insect as its mouthparts penetrate into the plant, travelling between and within different cell types, secreting saliva and ingesting fluid (see Tjallingii 1995). These techniques required the painstaking association of electrical waveforms and key behaviors using detailed histological reconstruction of the pathway of the mouthpart stylets within the plant. Electronic monitoring systems have been developed for feeding in other insect species as well, but most recent efforts have been applied to adult *Drosophila* (Itskov et al. 2014; Ro et al. 2014) allowing the effortless analysis of feeding at the level of the microstructure of feeding (Box 8.1).

8.4 Regulation of multiple nutrient intakes

Insects must ingest multiple nutrients, each at appropriate levels, to attain a balanced diet. Behavioral experiments on caterpillars, locusts, and cockroaches were among the first to demonstrate that insects have a widespread capacity to regulate their separate intakes of multiple nutrients, and to begin to explore the physiological control mechanisms involved (Simpson 1994; Simpson and Raubenheimer 2012). For example, *Locusta migratoria* was shown to regulate intake of protein and carbohydrate in the face of several experimental challenges, by selecting among foods of different nutritional composition, adjusting amounts eaten, and adopting patterns of food selection according to the frequency of foods of different nutritional composition available in the environment (Chamber et al. 1995; Behmer et al. 2001). These remarkable feats of **homeostasis** were found to extend to regulation of salt versus **macronutrient** intake (Trumper and Simpson 1993).

Nutritional choices reflect the changing needs of the insect. They change with recent nutritional experience and level of activity. For instance, locusts and caterpillars select a protein-rich food following a short experience (only one meal in the case of the locust) of a protein-depleted food, and show a similar

---

Box 8.1 Electronic systems can be used to monitor feeding in fruit flies

For example, the flyPAD (fly proboscis and activity detector) or the FLIC (fly liquid food interaction counter) measure the electric signals across two electrodes that contain the food item just between them (Itskov et al. 2014; Ro et al. 2014). As the fly stands on one of the electrodes, any contact with the food either closes the connection to the other electrode or changes its electric properties. These systems are connected to digital converters, which send the processed signals to a computer where feeding parameters, such as timing and duration of feeding, can be computed with high sensitivity, in the case of the flyPAD down to the volume of ingested food during individual sips. These systems have been used to show that, similar to locusts, caterpillars, aphids, rodents, and humans, flies exhibit highly rhythmic feeding patterns of sips and inter-sip intervals.

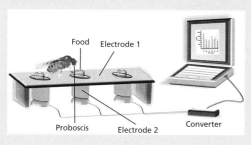

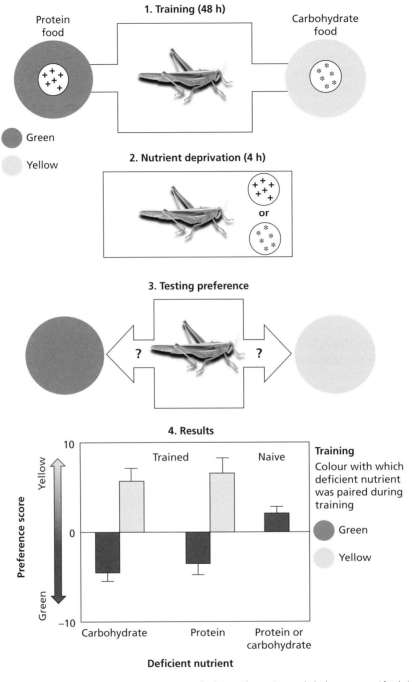

Figure 8.2 *Locusta migratoria* nymphs can be trained to associate meal colours with protein or carbohydrate content. After being deprived of either proteins or carbohydrates for only 4 hours, nymphs make use of coloured cues to choose their next meal, containing the deficient nutrient. Naïve locusts did not receive any training. This experiment demonstrated nutrient-specific associative learning based on visual cues (modified from Raubenheimer & Tucker, 1997).

preference for carbohydrate-rich food after a 4-hour period of carbohydrate deprivation. Nutrient requirements also vary with stage of development, reproduction, and diapause (Barton Browne 1995). Over a longer timescale, nutritional requirements evolve to track changing nutritional environments and life histories (Simpson and Raubenheimer 1993, 2012; Raubenheimer and Simpson 1999).

Regulating nutrient intake requires two sources of information, the first being the composition of the food and the second the nutritional state of the animal. These then must be integrated to elicit regulatory food selection and intake. Assessing the nutritional quality of the food involves directly tasting key nutrients, including sugars, amino acids, salts and water (Chapman 1995). Additionally, learning may be involved. For example, locusts can learn to associate olfactory and visual cues with nutritional consequences of consuming a particular food, then respond selectively to those cues when in a state of specific nutrient deficit for either protein or carbohydrate (Raubenheimer and Tucker 1997; Figure 8.2).

The responsiveness of an insect to a food of given composition should reflect the animal's nutritional and developmental state. Barton Browne (1995) hypothesized that two types of mechanism could be involved, 'demand driven' and 'non-demand driven'. In the first of these, changing nutritional demands elicit signals which increase the salience of cues associated with these nutrients in the environment, leading to selective feeding on foods rich in these nutrients. For example, studies on locusts, caterpillars, blow flies and *Drosophila* have shown that the responsiveness of taste receptors to sugars and amino acids varies with nutritional state, as represented by nutrient concentrations in the haemolymph (e.g. Simpson et al. 1991, 1992; Kim et al. 2017; Steck et al. 2018). Such modulation of sensory responses has also been shown for non-nutrient compounds, such as the plant secondary metabolites, pyrrolizidine alkaloids (PAs), which confer protection to tiger moth caterpillars against parasites and for which there is increased gustatory sensitivity leading to selective foraging for PAs when caterpillars are parasitized (Bernays and Singer 2005; Singer et al., 2009). Learned nutrient-specific appetites can also be involved, as discussed above. Other sources of feedback engage central nutrient signalling

pathways in the brain and other tissues. As we shall detail below, these mechanisms are a subject of current research using advanced molecular genetic techniques in *Drosophila* (see below).

Non-demand driven mechanisms involve nutrient-specific appetites that are activated by neural and hormonal signals at an appropriate stage in development, such as following maturation or mating to anticipate an increased protein requirement for egg production (Barton Browne 1995). We next showcase an example where both demand and non-demand-driven mechanisms are involved in the control of protein and salt feeding in mated female *Drosophila*.

8.5 Physiological and molecular mechanisms of appetite in *Drosophila*

In order to match feeding decisions with current nutritional needs, diet choice involves the action of sensory mechanisms, neuromodulators, metabolites, and receptors that drive the appetite for specific nutrients. *Drosophila* is considered an excellent model to study the genetic and physiological mechanisms regulating diet choice in animals. This is not only because of its importance as a model system in insect physiology and genetics, but also because some neuronal pathways regulating food intake are highly conserved in animals, because *Drosophila* shows specific appetites for key nutrients (protein, sugar, and salt) and because this fly has the ability to adaptively chose food containing the required nutrients, just as other insects and vertebrates, including humans (Cota et al. 2006; Ribeiro and Dickson 2010).

Different methods have been used to determine the physiological and genetic mechanisms responsible for food choice in insects. In *Drosophila*, these include neuron silencing (Cognigni et al. 2011), the use of mutants lacking particular receptors (Ribeiro and Dickson 2010) and gene silencing using RNA interference (Garlapow et al. 2015). Using these methods, it has been shown that food choice in *Drosophila* is driven by a combination of nutrient-specific modulation of gustatory responsiveness, neurons in the intestine and the central nervous system that sense circulating nutrients and track nutrient availability, by insulin-releasing neurons that conform a gut-brain neural circuit, and by enteroendocrine cells expressing gustatory receptors in the fly's intestine that regulate the secretion of hormones in response to current nutritional state (Cognigni et al. 2011; Lemaitre and Miguel Aliaga 2013; Wang et al. 2016; Zhan et al. 2016; Liu et al. 2017; Steck et al. 2018). Two endocrine organs—the fat body and the corpora cardiaca—are also known to secrete systemic signals in response to nutritional state and regulate feeding decisions (Pool and Scott 2014). After sensing current nutritional status, some neuromodulators promote feeding (i.e. small neuropeptide F, sNPF), others inhibit feeding (i.e. drosulfakinin, allatostatin, and insulin-like peptides) and others, such as dopamine, can have both functions (Pool and Scott 2014).

Nutritional requirements change dramatically during the animal's life. In response, appetite also changes. In female *Drosophila*, the mechanisms regulating appetite after mating are 'demand driven' and 'non-demand driven' (see previously) (Simpson and Raubenheimer 2015). For example, during reproduction, females expend both energy and nutrients in egg production and increase their appetite for protein-rich yeast, particularly if they have been fed a low-protein diet (Ribeiro and Dickson 2010; Figure 8.3). Prior to egg-laying, the act of mating also increases the female's appetite for protein-rich yeast, which will be demanded for egg production (Ribeiro and Dickson 2010). Mating not only induces the female's preference for protein, but also for salt, which is also required for egg production and increases egg laying (Walker et al. 2015). The onset of

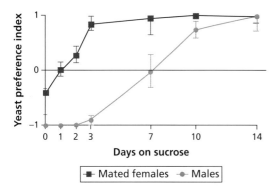

Figure 8.3 The act of mating increases the preference of female *Drosophila* for protein-rich yeast in the demand for resources needed for egg production. This craving for protein increases with the time that flies spent feeding on protein-deficient, sucrose-rich medium (modified from Ribeiro & Dickson 2010).

the appetite for yeast and salt in mated female *Drosophila* is not triggered by nutrient demands, but rather by the sex peptide (SP), a small protein transferred by males in the seminal fluid that binds SP receptors (SPR) in a small set of neurons located in the female reproductive tract (Häsemeyer et al. 2009), which relay the mating information to downstream neurons that project to the brain where they modulate taste information processing (Walker et al. 2015).

The attachment of SP to SPR not only manipulates female appetites, but also prevents females from re-mating (Yapici et al. 2008), potentially leading to sexual conflict (see Chapter 13, by Olzer et al.). Experiments with female flies lacking SPR or with males lacking the SP-producing gene show that females stop eating salt and protein-rich food after being mated, while genetically mimicking SPR activation at the level of the post-mating neuronal circuit induces both salt and protein appetites, confirming the role of SP and SPR and the corresponding neural circuitry in fly 'non-demand' appetite for protein. However, once protein appetite is activated, the actual amount of protein consumed is subject to nutrient feedbacks, and is therefore 'demand-driven' (Ribeiro and Dickson 2010; Vargas et al. 2010). 'Non-demand' appetites can be simply seen as evolutionarily encoded feed-forward predictive systems, which induce a change in appetite upon the occurrence of a stimulus (SP) that almost always leads to a lack of a nutrient (protein and salt upon the concomitant increase in egg production; Walker et al., 2017). They allow the animal to change food intake in anticipation, without having to incur the costs associated with the lack of nutrients that induce 'demand-driven' appetites. Control theory, as well as classical physiology, has always emphasized the power of feed-forward regulation. It is, therefore, not surprising that feed-forward mechanisms are emerging as important features of neuronal systems regulating feeding in insects, as well as vertebrates.

8.6 The geometric framework

The discovery that insects and other organisms possess separate appetites for multiple nutrients initiated the development of an integrative framework for nutrition, known as the geometric framework (GF) (Raubenheimer and Simpson 1993, 1998,

2018; Simpson and Raubenheimer, 1993, 2012). The GF was developed to allow multiple nutrient requirements, the behavior and physiology of the organism, and the nutritional environment to be represented and interrelated within the same models, and then to use simple graphical representations to explore the function, evolution and ecology of nutritional systems across scales from individuals to ecosystems. Central concepts within GF include:

- the nutritional requirements of the animal as multidimensional targets that move over time;
- foods as nutritional vectors that are balanced, imbalanced, or complementary with respect to these targets;
- nutritional compromises on imbalanced diets, in which an animal that is unable to reach its intake target balances under-eating some nutrients against over-eating others, reaching some 'point of compromise';

- topological landscapes that map the association between multidimensional nutrient intakes and physiological, life-history, and ecological responses.

These same models have been used to understand the interactions between nutrients and plant secondary metabolites (Behmer et al. 2002; Raubenheimer and Simpson, 2009). For example, if the dietary protein and carbohydrate ratio and concentrations are near optimal, locusts are not affected adversely by the presence of tannic acid in their food, even up to 10 per cent dry weight. When foods contain less than an optimal protein:carbohydrate ratio, however, tannic acid serves as a powerful feeding deterrent, causing high mortality and extended development, while at higher than optimal protein to carbohydrate ratios, tannic acid does not reduce intake, but instead results in high mortality by disrupting gut function and protein utilization (Simpson and Raubenheimer 2001).

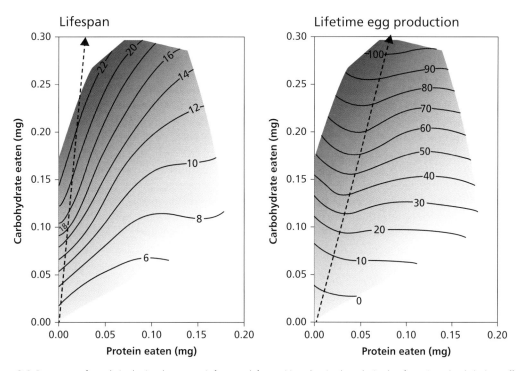

Figure 8.4 Response surfaces derived using the geometric framework for nutrition, showing how the intake of protein and carbohydrate affects lifespan and reproductive output in mated female *Drosophila melanogaster*. Flies were given ad libitum access to one of twenty-eight diets varying in the ratio and total concentration of protein to carbohydrate (P:C). Fitted surfaces for longevity and reproductive output are plotted onto nutrient intake arrays. Dashed lines show the dietary P:C that maximized lifespan (1:16) and egg production (1:4). Flies allowed to select their diet chose a P:C ratio of 1:4 (after Lee et al., 2008).

8.7 Using the geometric framework to map the consequences of feeding behavior for individuals

The consequences for an insect of attaining a particular nutrient intake can effectively be visualized as response surfaces mapped onto arrays of nutrient intakes, typically derived from experiments in which insects (and other organisms) have been restricted to one of a large number of diet formulations (Figure 8.4). Such surface plots have helped define the relationships between nutrient intakes and variables that include lifespan, reproduction, immunity, resistance to infection, and the structure of the populations of microorganisms living in association with the insect (see Simpson and Raubenheimer 2012). Such plots have also been used to define and predict optimal foraging behavior when regulating intake of multiple nutrients in different feeding environments (Simpson et al. 2004; Jensen et al. 2012).

8.8 Microbial associations, parasites, and immunity

Insects are host to an abundance of microorganisms, with which they have evolved complex nutritional associations that allow them to exploit a huge variety of nutritionally unpromising food sources, from leaf litter and wood, to blood, phloem, and crude oil (Douglas 2010; Chapman 2014). Microbial associations have not only shaped physiology and diet niche, they directly and indirectly influence feeding and food choice behavior (Leitão-Gonçalves et al. 2017; Wong et al. 2017; Fischer et al. 2017).

Whereas some microbial associations are beneficial to the host, others are detrimental and there is evidence of an evolutionary battle for control of feeding behavior between host and parasites, both microbial and macro-parasites, such as parasitoids. Feeding behavior offers protection against parasites by allowing insects to self-medicate (see Raubenheimer et al., 2009; see Chapter 18). Examples of **self-medication** include consuming protective plant secondary compounds (Lozano 1998; Singer et al. 2009), selecting macronutrient ratios that enhance immunity and compensate for the nutrient-specific

costs of infection (Lee et al. 2006; Ponton et al. 2011; Povey et al. 2014), or in fruit flies consuming ethanol from rotten fruit in response to infection by some parasitoids (Milan et al. 2012; Lynch et al. 2017).

Just as some insects show compensatory feeding or self-medication when sick, others show a lack of appetite. This does not seem convenient, given the high energetic and nutritional demands of the host's immune responses to combat infection (Schmid-Hempel 2005). However, the immune response requires exactly the same protein, apolipophorin III, needed for lipid transport during digestion. Therefore, when this protein is used for lipid transport, as occurs when lipids are consumed in the diet, it is no longer available for immune function. Illness-induced anorexia is a behavioral mechanism to avoid facing physiological trade-offs between digestion and immunity, that has been observed in many insects, including crickets, lepidopterans, and fruit flies (Ayres and Schneider 2009; Adamo et al. 2010; Povey et al. 2013).

8.9 Beyond the individual: social interactions

Because nutrition impacts the behavior, reproductive success, and survival of individuals, it in turn influences behavioral interactions between individuals within groups, populations and species (Simpson and Raubenheimer 2012; Lihoreau et al. 2015).

The abundance and distribution of nutrients within a habitat can directly influence population dynamics through the nutrient-specific appetites of individuals. One example is mass migration of flightless, cannibalistic Mormon crickets, *Anabrus simplex*, in north-west USA (Simpson et al., 2006). At high local population densities, crickets aggregate, which provides anti-predatory benefits to individuals, but also increases local competition and the risk of cannibalism. Field experiments showed that crickets in large aggregations are deficient in and seek protein and mineral salts, the most abundant and available source of which is other crickets. Local cannibalistic interactions—as crickets chase others to eat them, while also avoiding being cannibalized themselves—provides a local interaction rule that scales up to produce cohesive marching behavior across an entire population of millions of crickets (Buhl et al. 2006). Supplementing the diet of individual crickets with protein reduced locomotion and also crickets' propensity to cannibalize, thereby disrupting mass migration. Cannibalism can be promoted by the mother herself, as conspecifics can be the best source of required nutrients under risky situations. When mother chrysomelid beetles, *Leptinotarsa decemlineata*, are exposed to risk of predation by bugs, females lay unviable eggs for the offspring to cannibalize to satisfy nutritional demands without being exposed to predators (Tigreros et al. 2017).

In other insect species, the nutritional states and behavioral responses of individuals fashion the foraging behavior of an entire group, for example, as seen in the forest tent caterpillar, *Malacosoma disstria*, in which the protein and energy status of individuals determines the movement patterns of the entire group (McClure et al. 2011; Jeanson et al. 2012). Such responses are seen in their most extreme in the colony-level feats of nutrient balancing found in the eusocial insects, in which there is a marked division of labour in respect of feeding behavior. Work on green-headed ants, *Rhytidoponera metallica*, for example, has shown that the colony acts as a nutrient-balancing 'superorganism' at the level of food collection, with larvae providing sources of nutrient feedback to guide the types and amounts of different foods collected and returned to the nest by workers (Dussutour and Simpson 2009). Differences in nutritional requirements among individuals may even have contributed to the evolution of division of labour not only in terms of food collection, but also reproduction (Lihoreau et al. 2015).

8.10 Trophic interactions and ecosystem dynamics

Nutrition and feeding behavior also shape species assemblages and ecosystems. Insects across different trophic levels have been shown to regulate their intake of macronutrients, including predatory species (e.g. Mayntz et al. 2005; Raubenheimer et al. 2007; Jensen et al. 2012). Because of a combination of differences in the nutritional targets of species at different trophic levels and in the composition of their habitual foods (plants, prey), it is likely that

nutrient balancing behavior directly and indirectly impacts food chain length and food web dynamics (Denno and Fagan 2003; Raubenheimer et al. 2009; Wilder et al. 2013). An example of an indirect trophic interaction is where '**fear**' of spider predation (without predation necessarily actually occurring) increases the carbohydrate requirements of grasshoppers, with resulting effects on plant community composition and nutrient cycling (Hawlena and Schmitz 2010, 2011; Hawlena et al., 2012).

Nutrition and feeding behavior can also drive success of invasive species. Fire ant (*Solenopsis*) invasions in the USA have been shown to be linked to greater access to carbohydrate-rich exudates produced by sap feeders, which are less available to fire ants in their native regions of Central and South America due to competition from arboreal ant species (Wilder et al. 2011). Another example is where overgrazing of Eurasian grasslands by livestock has been linked to increased outbreaks of the locust, *Oedaleus asiaticus*, due to grasses in these highly grazed areas being more closely matched to the low protein, high carbohydrate requirements of this locust species (Cease et al. 2012).

8.11 Contribution of insects beyond entomology

As indicated at the outset of this chapter, the study of feeding behavior in insects was initially inspired by the agricultural and environmental imperatives provided by insect herbivory, but latterly it has made a significant contribution more generally. In part, this has been through providing new insights into the molecular genetics and physiology of appetite control and nutritional homeostasis, made possible by advances in molecular genetics for model systems such as *Drosophila* (Leulier et al. 2017). Another contribution of insects has been as a means to a larger end—the unifying framework for nutritional biology provided by the geometric framework was devised and tested using insects, but has since been applied to a wide range of organisms, from slime moulds to humans, and used to address problems from aquaculture and conservation biology to the dietary causes of human **obesity** and **ageing** (Simpson and Raubenheimer, 2012, Raubenheimer and Simpson, 2016).

Current human-induced environmental degradation, habitat loss, contamination, and temperature shifts are causing pronounced changes in species phenology, abundances, distributions, and interactions, and as a consequence, resource availability changes and diets, too. Several ecosystem processes, including pollination, natural pest control, and soil fertilization, result from feeding activity of insects such as bees, butterflies, and dung beetles, and are threatened by human disturbance (Memmott et al. 2007; Nichols et al. 2007). Adaptation and phenotypic plasticity in feeding behavior are fundamental for maintaining ecosystem functioning in the face of global change and deserve further research (see Chapter 19). Notable examples of plasticity are found in the interactions occurring between nutrition and temperature. For example, locusts are able to select microclimates that best allow the optimal ratio of protein and carbohydrate to be assimilated from a meal to best support current nutritional needs (Coggan et al. 2011; Clissold et al. 2013).

8.12 Conclusions

Modern methods in molecular genetics and a unifying theoretical framework for studying feeding behavior have revealed that the mechanisms regulating insect feeding are similar to other animals, including humans in many aspects. Studies in *Drosophila*, locusts, blow flies, and other classic insect model systems have been possible under controlled laboratory conditions, leading to considerable advances in the field of insect nutrition. These advances have been helpful in many areas, from the understanding of evolutionary processes driving dietary niches to pest dynamics, and even human nutrition and disease. It is therefore surprising that, given the importance of nutrients for the fitness of all animals, our understanding of even simple questions, such as what an ideal diet is remains difficult to answer for most animals, notably humans. Likewise, our understanding of the neuronal mechanisms controlling feeding is largely restricted to nutrients providing energy such as sugars. When it comes to proteins, vitamins, and other important nutrients, we lack even a simple understanding of how these nutrients might affect the brain. In this sense, work on the molecular and neuronal circuit mechanisms underlying feeding needs to follow the lead of insect nutrition and start taking into account the diversity of nutrients required for the animal to thrive and survive. Along the same lines, our knowledge of the mechanisms underlying nutrient intake are confined to only a few species. To be able to generalize these findings, studies of non-model organisms are still needed, particularly under natural conditions, where resources are not always available or easily accessed, and where human activity is imposing additional pressure, risking individual nutrition and ecosystem function. Expanding our perspective to take into account the multidimensional aspect of nutrition, as well as the diversity of insect life will not only allow us to gain a more enriching view of the mechanisms underlying feeding, but also will give us the tools to tackle the challenging task of managing our threatened natural resources—and our own health.

Acknowledgements

The authors would like to thank Fernanda Baena-Díaz for helpful comments on the chapter and to Yoshua Esquivel-Soto for help with figure editing.

References

Adamo, S.A., Bartlett, A., Le, J., Spencer, N., and Sullivan, K. (2010). Illness-induced anorexia may reduce trade-offs between digestion and immune function. *Animal Behaviour*, **79**, 3.

Ayres, J.S. and Schneider, D.S. (2009). The role of anorexia in resistance and tolerance to infections in *Drosophila*. *PLoS Biology*, **7**, e1000150.

Barton Browne, L. (1995). Ontogenic changes in feeding behaviour. In: R.F. Chapman and G. de Boer (Eds), *Regulatory Mechanisms in Insect Feeding*, pp. 307–42. Chapman Hall, New York, NY.

Behmer, S.T., Raubenheimer, D., and Simpson, S.J. (2001). Frequency-dependent food selection in locusts: a geometric analysis of the role of nutrient balancing. *Animal Behaviour*, **61**, 995.

Behmer, S.T., Simpson, S.J., and Raubenheimer, D. (2002). Herbivore foraging strategies in chemically heterogeneous environments: allelochemical-nutrient interactions. *Ecology*, **83**, 2489.

Bell, W.J. (1990). Searching behavior patterns in insects. *Annual Review of Entomology*, **35**(1), 447–67.

Bernays, E.A., and Simpson, S.J. (1982). Control of food intake. *Advances in Insect Physiology*, **16**, 59.

Bernays, E.A., and Singer, M.S. (2005). Taste alteration and endoparasites. *Nature*, **436**, 476.

Buhl, J., Sumpter, D.J.T., Couzin, I.D., et al. (2006). From disorder to order in marching locusts. *Science*, **312**, 1402.

Cease, A.J., Elser, J.J., Ford, C.F., et al. (2012). Heavy livestock grazing promotes locust outbreaks by lowering plant nitrogen content. *Science*, **335**, 467.

Chapman, R.F. (1995). Mechanics of food handling by chewing insects. In: R.F. Chapman and G. de Boer (Eds), *Regulatory Mechanisms in Insect Feeding*, pp. 3–31. Chapman Hall, New York, NY.

Chapman, R.F. (2014). *The Insects: Structure and Function*, 5th edn. S. J. Simpson and A.E. Douglas (Eds). Cambridge University Press, Cambridge.

Chapman R.F., and de Boer G. (Eds) (1995). *Regulatory Mechanisms in Insect Feeding*. Chapman and Hall, New York, NY.

Clissold, F.J., Coggan, N., and Simpson, S.J. (2013). Insect herbivores can choose microclimates to achieve nutritional homeostasis. *Journal of Experimental Biology*, **21**, 2089.

Coggan, N., Clissold, F.J., and Simpson, S.J. (2011). Locusts use dynamic thermoregulatory behaviour to optimize nutritional outcomes. *Proceedings of the Royal Society B*, **278**, 2745.

Cognigni, P., Bailey, A.P., and Miguel-Aliaga, I. (2011). Enteric neurons and systemic signals couple nutritional and reproductive status with intestinal homeostasis. *Cell Metabolism*, **13**, 92.

Corrales-Carvajal, V.M., Faisal, A.A., and Ribeiro, C. (2016). Internal states drive nutrient homeostasis by modulating exploration-exploitation trade-off. *eLife*, **5**, e19920.

Cota, D., Proulx, K., Smith, K., and Kozma, S. (2006). Hypothalamic mTOR signaling regulates food intake. *Science*, **312**, 927.

Dadd, R.H. (1963). Feeding behaviour and nutrition in grasshoppers and locusts. *Advances in Insect Physiology*, **1**, 47.

Denno, R.F., and Fagan, W.F. (2003). Might nitrogen limitation promote omnivory among carnivorous arthropods? *Ecology*, **84**, 2522.

Dethier, V.G. (1976). *The Hungry Fly*. Princeton University Press, Princeton.

Douglas, A.E. 2010. *The Symbiotic Habit*. Princeton University Press, Princeton, NJ.

Dussutour, A., and Simpson, S.J. (2009). Communal nutrition in ants. *Current Biology*, **19**, 740.

Fischer, C., Trautman, E.P., Crawford, J.M., et al. (2017). Metabolite exchange between microbiome members produces compounds that influence *Drosophila* behaviour. *eLife*, **6**, e18855.

Fraenkel, G. (1959). The raison d'être of secondary plant substances. *Science*, **129**, 1466.

Frazier, J.L., and Chyb, S. (1995). Use of feeding inhibitors in insect control. In: R.F. Chapman and G. de Boer (Eds), *Regulatory Mechanisms in Insect Feeding*, pp. 364–81. Chapman Hall, New York, NY.

Garlapow, M.E., Huang, W., Yarboro, M.T., Peterson, K.R., and Mackay, T.F.C. (2015). Quantitative genetics of food intake in *Drosophila melanogaster*. *PLoS One*, **10**, e0138129.

Häsemeyer, M., Yapici, N., Heberlein, U., and Dickson, B.J. (2009). Sensory neurons in the *Drosophila* genital tract regulate female reproductive behavior. *Neuron*, **61**(4), 511.

Hawlena, D., and Schmitz, O.J. (2010). Herbivore physiological response to predation risk and implications for ecosystem nutrient dynamics. *Proceedings of the National Academy of Sciences, USA*, **107**, 15503.

Hawlena, D., and Schmitz, O.J. (2011). Herbivore physiological response to predation risk and implications for ecosystem nutrient dynamics. *Proceedings of the National Academy of Sciences, USA*, **107**, 15503.

Hawlena, D., Strickland, M.S., Bradford, M.A., and. Schmitz, O.J. (2012). Fear of predation slows plant-litter decomposition. *Science*, **336**, 1434.

Itskov, P.M., Moreira, J-M., Vinnik, E., et al. (2014). Automated monitoring and quantitative analysis of feeding behaviour in *Drosophila*. *Nature Communications*, **5**, 4560.

Itskov, P.M., and Ribeiro, C. (2013). The dilemmas of the gourmet fly: the molecular and neuronal mechanisms of feeding and nutrient decision making in *Drosophila*. *Frontiers in Neuroscience*, **7**, 12.

Jeanson, R., Dussutour, A., and Fourcassié, V. (2012). Key factors for the emergence of collective decision in invertebrates. *Frontiers in Neuroscience*, **6**, 121.

Jensen, K., Mayntz, D., Toft, S., et al. (2012). Optimal foraging for specific nutrients in predatory beetles. *Proceedings of the Royal Society B*, **279**, 2212.

Kim, I.S., and Dickinson, M.H. (2017). Idiothetic path integration in the fruit fly *Drosophila melanogaster*. *Current Biology*, **27**, 2227.

Kim, S.M., Su, C.Y., and Wang, J.W. (2017). Neuromodulation of innate behaviors in *Drosophila*. *Annual Review of Neuroscience*, **40**, 327.

Lee, K.P., Cory, J.S., Wilson, K., Raubenheimer, D., and Simpson, S.J. (2006). Flexible diet choice offsets protein costs of pathogen resistance in a caterpillar. *Proceedings of the Royal Society B*, **273**, 823.

Leitão-Gonçalves, R., Carvalho-Santos, Z., Francisco, A.P., et al. (2017). Commensal bacteria and essential amino acids control food choice behavior and reproduction. *PLoS Biology*, **15**, e2000862.

Lemaître, B., and Miguel-Aliaga, I. (2013). The digestive tract of *Drosophila melanogaster*. *Annual Review of Genetics*, **47**, 377.

Leulier, F., MacNeil, L.T., Lee, W., et al. (2017). Integrative physiology: at the crossroads of nutrition, microbiota, animal physiology and human health. *Cell Metabolism*, **25**, 522.

Lihoreau, M., Buhl, J., Charleston, M.A., et al. (2015). Nutritional ecology beyond the individual: a conceptual framework for integrating nutrition and social interactions. *Ecology Letters*, **18**, 273.

Liu, Q., Tabuchi, M., Liu, S., et al. (2017). Branch-specific plasticity of a bifunctional dopamine circuit encodes protein hunger. *Science*, **356**, 534.

Lozano, G.A. (1998). Parasitic stress and self-medication. *Advances in the Study of Behavior*, **27**, 291.

Lynch, Z.R., Schlenke, T.A., Morran, L.T., and De Roode, J.C. (2017). Ethanol confers differential protection against generalist and specialist parasitoids of *Drosophila melanogaster*. *PLoS One*, **12**, e0180182.

Mayntz, D., Raubenheimer, D., Salomon, M., Toft, S., and Simpson, S.J. (2005). Nutrient-specific foraging in invertebrate predators. *Science*, **307**, 111.

McClure, M., Ralph, M., and Despland, E. (2011). Group leadership depends on energetic state in a nomadic collective foraging caterpillar. *Behavioral Ecology and Sociobiology*, **65**, 1573.

McLean, D.L., and Kinsey M.G. (1964). A technique for electronically recording aphid feeding and salivation. *Nature*, **202**, 1358.

Memmott, J., Craze, P.G., Waser, N.M., and Price, M.V. (2007). Global warming and the disruption of plant-pollinator interactions. *Ecology Letters*, **10**, 710.

Milan, N.F., Kacsoh, B.Z., and Schlenke, T.A. (2012). Alcohol consumption as self-medication against blood-borne parasites in the fruit fly. *Current Biology*, **22**, 488.

Nichols, L., Larsen, T.H., Spector, S., et al. (2007). Global dung beetle response to tropical forest modification and fragmentation: a quantitative literature review and meta-analysis. *Biological Conservation*, **137**, 1.

Painter, R.H. (1936). The food of insects and its relation to resistance of plants to insect attack. *American Naturalist*, **70**, 547.

Ponton, F., Wilson, K., Cotter, S.C., Raubenheimer, D., and Simpson, S.J. (2011). Nutritional immunology: a multi-dimensional approach. *PLoS Pathogens*, **7**, e1002223.

Pool, A.H., and Scott, K. (2014). Feeding regulation in *Drosophila*. *Current Opinion in Neurobiology*, **29**, 57.

Povey, S., Cotter, S.C., Simpson, S.J., and Wilson, K. (2013). Dynamics of macronutrient self-medication and illness-induced anorexia in virally infected insects. *Journal of Animal Ecology*, **83**, 245.

Povey, S., Cotter, S., Simpson, S.J., and Wilson, K. (2014). Dynamics of macronutrient self-medication and illness-induced anorexia in virally-infected insects. *Journal of Animal Ecology*, **83**, 245.

Raubenheimer, D., and Bernays, E.A. (1993). Patterns of feeding in the polyphagous grasshopper *Taeniopoda eques*: a field study. *Animal Behaviour*, **45**, 153.

Raubenheimer, D., and Simpson, S.J. (1993). The geometry of compensatory feeding in the locust. *Animal Behaviour*, **45**, 953.

Raubenheimer, D., and Simpson, S.J. (1998). Nutrient transfer functions: the site of integration between feeding behaviour and nutritional physiology. *Chemoecology*, **8**, 61.

Raubenheimer, D., and Simpson, S.J. (1999). Integrating nutrition: a geometrical approach. *Entomologia Experimentalis et Applicata*, **91**, 67.

Raubenheimer, D., and Simpson, S.J. (2009). Nutritional PharmEcology: doses, nutrients, toxins, and medicines. *Integrative and Comparative Biology*, **49**, 329.

Raubenheimer, D., and Simpson, S.J. (2016). Nutritional ecology and human health. *Annual Review of Nutrition*, **36**, 603.

Raubenheimer, D., and Simpson, S.J. (2018). Nutritional ecology and foraging theory. *Current Opinion in Insect Science*, **27**, 38.

Raubenheimer, D., and Tucker, D. (1997). Associative learning by locusts: pairing of visual cues with consumption of protein and carbohydrate. *Animal Behaviour*, **54**, 1449.

Raubenheimer, D., Mayntz, D., Simpson, S.J., and Toft, S. (2007). Nutrient specific compensation following overwintering diapause in a generalist predatory invertebrate: implications for intraguild predations. *Ecology*, **88**, 2598.

Raubenheimer, D., Simpson, S.J., and Mayntz, D. (2009). Nutrition, ecology and nutritional ecology: toward an integrated framework. *Functional Ecology*, **23**, 4.

Raubenheimer, D., Rothman, J.M., Pontzer, H., and Simpson, S.J. (2014). Macronutrient contributions of insects to the diets of hunter-gatherers: a geometric analysis. *Journal of Human Evolution*, **71**, 70.

Ribeiro, C., and Dickson, B.J. (2010). Sex peptide receptor and neuronal TOR/S6K signaling modulate nutrient balancing in *Drosophila*. *Current Biology*, **20**, 1000.

Ro, J., Harvanek, Z.M., and Pletcher, S.D. (2014). FLIC: high-throughput, continuous analysis of feeding behaviors in *Drosophila*. *PLoS One*, **9**, e101107.

Schmid-Hempel, P. (2005). Evolutionary ecology of insect immune defenses. *Annual Review of Entomology*, **50**, 529.

Scriber, J.M., and Slansky, F. (1981). The nutritional ecology of immature Insects. *Annual Review of Entomology*, **26**, 183.

Simpson, S.J. (1982). Patterns in feeding: a behavioural analysis using *Locusta migratoria* nymphs. *Physiological Entomology*, **7**, 325.

Simpson, S.J. (1994). Experimental support for a model in which innate taste responses contribute to regulation of salt intake by nymphs of *Locusta migratoria*. *Journal of Insect Physiology*, **40**, 555.

Simpson S.J. (1995). Regulation of a meal: chewing insects. In: R.F. Chapman and G. de Boer (Eds) *Regulatory Mechanisms in Insect Feeding*. Springer, Boston, MA.

Simpson, S.J., and Ludlow, A.R. (1986). Why locusts start to feed: a comparison of causal factors. *Animal Behaviour*, **34**, 480.

Simpson, S.J., and Raubenheimer, D. (1993). A multi-level analysis of feeding behaviour: the geometry of nutritional decisions. *Philosophical Transactions of the Royal Society B*, **342**, 381.

Simpson, S.J., and Raubenheimer, D. (2001). The geometric analysis of nutrient-allelochemical interactions: a case study using locusts. *Ecology*, **82**, 422.

Simpson, S.J., and Raubenheimer, D. (2012). *The Nature of Nutrition: A Unifying Framework from Animal Adaptation to Human Obesity*. Princeton University Press, Princeton, NJ.

Simpson, S.J., and Raubenheimer, D. (2015). Nutritional physiology: sex elicits a taste for salt in *Drosophila*. *Current Biology*, **25**, 980.

Simpson, S.J., and Simpson, C.L. (1990). The mechanisms of nutritional compensation by phytophagous insects. *Journal of Insect-Plant Interactions*, **2**, 111.

Simpson, S.J., James, S., Simmonds, M.S.J., and Blaney, W.M. (1991). Variation in chemosensitivity and the control of dietary selection behaviour in the locust. *Appetite*, **17**, 141.

Simpson, S.J., Sibly, R.M., Lee, K.P., Behmer, S.T., and Raubenheimer, D. (2004). Optimal foraging when regulating intake of multiple nutrients. *Animal Behaviour*, **68**, 1299.

Singer, M.S., Mace, K.C., and Bernays, E.A. (2009). Self-medication as adaptive plasticity: Increased ingestion of plant toxins by parasitized caterpillars. *PLoS One*, **4**, e4796.

Slansky, F., Jr, and Rodriguez, J.G. (Eds) (1987). *Nutritional Ecology of Insects, Mites, Spiders, and Related Invertebrates*. John Wiley, New York, NY.

Steck, K., Walker, S.J., Itskov, P.M., Baltazar, C., Moreira, J.M., & Ribeiro, C. (2018). Internal amino acid state modulates yeast taste neurons to support protein homeostasis in Drosophila. Elife, 7, e31625.

Tjallingii, W.F. (1995). Regulation of phloem sap feeding by aphids. In: R. F. Chapman and G. de Boer (Eds), *Regulatory Mechanisms in Insect Feeding*, pp. 190–209. Chapman Hall, New York, NY.

Tigreros, N., Norris, R.H., Wang, E.H., and Thaler, J.S. (2017). Maternally induced intraclutch cannibalism: an adaptive response to predation risk? *Ecology Letters*, **20**, 487.

Trumper, S., and Simpson, S.J. (1993). Regulation of salt intake by nymphs of *Locusta migratoria*. *Journal of Insect Physiology*, **39**, 857.

Vargas, M., Luo, N., Yamaguchi, A., and Kapahi, P. (2010). A role for S6 kinase and serotonin in postmating dietary switch and balance of nutrients in *D. melanogaster*. *Current Biology*, **20**, 1006.

Waldbauer, G.P. (1968). The consumption and utilization of food by insects. *Advances in Insect Physiology*, **5**, 229.

Waldbauer, G.P., and Friedman, S. (1991). Self-selection of optimal diets by insects. *Annual Review of Entomology*, **36**, 43.

Walker, S.J., Corrales-Carvajal, V.M., and Ribeiro, C. (2015). Postmating circuitry modulates salt taste processing to increase reproductive output in *Drosophila*. *Current Biology*, **25**, 2621.

Walker, S.J., Goldschmidt, D., & Ribeiro, C. (2017). Craving for the future: the brain as a nutritional prediction system. *Current Opinion in Insect Science*, **23**, 96–103.

Wang, Q.P., Lin, Y.Q., Zhang, L., et al. (2016). Sucralose promotes food intake through NPY and a neuronal fasting response. *Cell Metabolism*, **24**, 75.

Wilder, S.M., Holway, D.A., Suarez, A.V., LeBrun, E.G., and Eubanks, M.D. (2011). Intercontinental differences in resource use reveal the importance of mutualisms in fire ant invasions. *Proceedings of the National Academy of Sciences, USA*, **108**, 20639.

Wilder, S.M., Norris, M., Lee, R.W., Raubenheimer, D., and Simpson, S.J. (2013). Arthropod food webs become increasingly lipid-limited at higher trophic levels. *Ecology Letters*, **16**, 895.

Wong, A.C.N., Wang, Q.P., Morimoto, J., et al. (2017). Gut microbiota modifies olfactory-guided microbial preferences and foraging decisions in *Drosophila*. *Current Biology*, **15**, 2397.

Yapici, N., Kim, Y.J., Ribeiro, C., & Dickson, B.J. (2008). A receptor that mediates the post-mating switch in *Drosophila* reproductive behaviour. *Nature*, **451** (7174), 33.

Zhan, Y.P., Liu, L., and Zhu, Y. (2016). Taotie neurons regulate appetite in *Drosophila*. *Nature Communications*, **7**, 13633.

CHAPTER 9

Anti-predator behavior

Thomas N. Sherratt[1] and Changku Kang[1,2]

[1] Department of Biology, Carleton University, Ottawa, Canada, K1S 5B6
[2] Department of Biosciences, Mokpo National University, Muan, Jeollanam-do, 58554, Republic of Korea

9.1 Overview

Go into any backyard and you are likely to see numerous examples of adaptations to avoid being eaten—moths that resemble their backgrounds, warningly-Coloured ladybugs, and harmless hover flies that gain protection by resembling stinging wasps or bees. The preponderance of these adaptations demonstrates that there is intense selection to avoid predation. Many anti-predator defences are morphological, with insects rendered unprofitable to attack by would-be predators through hard cuticle, stings, spines, and toxins, and/or rendered hard to detect or recognize through cryptic and mimetic colour patterns (see Edmunds 1974; Endler 1991; Ruxton et al. 2004). However, in many of these cases, behavior plays an important role in improving the effectiveness of these physical defences. For example, *Catocola* moths are not only cryptically patterned, but orientate themselves on tree trunks to reduce their detectability (Pietrewicz and Kamil 1977). Likewise, some hover flies not only look like wasps, but also behave like them, through waving their legs in front of their heads to create the appearance of long antennae and mock sting (Waldbauer 1970). In other cases, the behavior itself—ranging from fleeing the scene to a retaliatory bite, is the basis of the anti-predatory strategy.

This chapter considers how and why behavior is employed by insects as part of their anti-predator defences. We will not simply list behaviors, but instead seek to classify them while seeking to understand why some insect species have evolved a given anti-predator behavioral trait, but not others. With the generation of reliable phylogenies and accumulation of data, comparative methods are increasingly being used to understand the diversity of anti-predator defences. Recent applications of the comparative method are described, with particular emphasis on understanding why some insect species exhibit given behavioral defences, but other species do not.

Of course, insects exhibit a wide range of anti-predator behavioral defences and most of our examples are drawn from this group. However, it is worth noting that while insects are highly tractable subjects to study, many of the arguments we make can be generalized across taxonomic groups.

9.2 Some simple ways of classifying anti-predator defences

Anti-predator defences may be crudely classified in terms of whether they operate before the predator attacks the prey ('primary defences') or after a predator initiates an attack ('secondary defences'; Table 9.1). Primary defences therefore include **crypsis**,

Sherratt, T. N. and Kang, C., *Anti-predator behavior*. In: *Insect Behavior: From mechanisms to ecological and evolutionary consequences*. Edited by Alex Córdoba-Aguilar, Daniel González-Tokman, and Isaac González-Santoyo: Oxford University Press (2018). © Oxford University Press.
DOI: 10.1093/oso/9780198797500.003.0009

warning signals, and mimicry, while secondary defences include deflection, startle, **death feigning**, and retaliation (Edmunds 1974; Ruxton et al. 2004). Nestled within the primary/secondary dichotomy is a classification based on the exact stage of a predation sequence the defence has evolved to interrupt (Endler 1991; Jeschke 2006). Thus, defences could be thought of as having evolved to hinder:

- encounter (e.g. by remaining areas the predator is unlikely to visit);
- detection (e.g. through **camouflage**);
- attack (e.g. through warning signals or mimicry);

- subjugation (e.g. through fleeing, armor, or fighting back).

By this definition the first two of the aforementioned defences are primary defences, while the second two defences are secondary (see Figure 9.1).

Sometimes behavior is one of the single most important components of the defence. For example, when caterpillars of the moth *Manduca sexta* are physically contacted, then they respond with a mandible strike (van Griethuijsen et al. 2013). Behavior can also serve a secondary role, improving the effectiveness of a given physical defence.

Table 9.1 Summary of the predation sequence and antipredator behaviors that hinder each stage.

Predation stage	Behavioral defence that hinders each stage	Protective mechanism
Encounter	Seek refuge	Being active in a different time of a day than most predators Excavating a burrow and staying there The use of a predator-free habitat
Detection/identification	Microhabitat selection	Staying on a background that confers camouflage Adopting a body orientation/posture that enhances the disruptive or cryptic effect
Attack	Fleeing Warning display Behavioral mimicry Startle display	Running/flying away from predators The display of threatening/deceiving signals to ward off predators. Behaviorally mimicking unprofitable prey (which reinforces the morphological mimicry) to deceive predators Frightening predators by suddenly unleashing unexpected hidden conspicuous colours
Subjugation	Death feigning Fighting back	Adopting a characteristic body posture mimicking being dead Attacking the predator or struggling physically

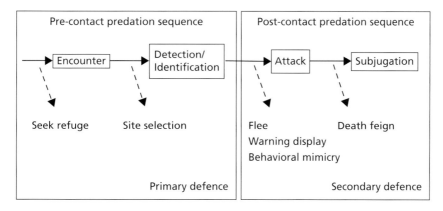

Figure 9.1 The classical predation sequence from search to subjugation (continuous lines) and examples of anti-predator behaviors that have evolved to inhibit specific stages of the sequence (dashed lines). These behaviors can arise pre-contact (as primary defences) or post-contact (as secondary defences) with the predator. Behavioral responses range from inhibiting search (by fighting back for instance) to inhibiting consumption (by death feigning for instance).

For example, rather than responding with a mandible strike as above, many caterpillars with eyespots adopt a defensive posture in which they swell their anterior body segments (Hossie and Sherratt 2014).

An important implication of the fact that predation is sequential is that many prey can be thought of having a 'Swiss army knife' portfolio of defences that they call up selectively (Britton et al. 2007). Given the range of defences at an organism's disposal, behavior thereby allows the organism the flexibility to change its strategy rapidly—for example, switching from crypsis to startle if its first line of defence fails. In instances of life-or-death, clearly it pays to switch defences quickly if one line of defence is not working and behavioral responses allow this precipitous change in strategy.

The portfolio approach to anti-predator defences may also help understand why certain anti-predator traits and associated behaviors have evolved in some species, but not in related species. For example, as Box 9.1 argues, you would only expect to see a costly secondary defence evolving in a prey species if the predation pressure was intense, the primary defence often breaks down and/or the secondary defence was cheap. This simple model also shows that whether (and what) secondary defence is selected will depend on the efficacy of the defences in question, a subject we return to in Section 9.6.

Although there is a range of ways of classifying anti-predator behaviors, we feel that it is natural to introduce them within the context of a predation sequence not only because it has a logical start and end, but also because individual species may use behaviors to escape predation at more than one step along the way (e.g. Edmunds 1974). We therefore begin by surveying the range of anti-predator's behavior involved in primary defence of insects. We then consider the role of behavior in facilitating

---

Box 9.1 Conditions for the evolution of a back-up defence

Consider a palatable prey species with a single reproductive episode at the end of its life (i.e. it is semelparous). Let p_p represent the probability that the primary defence of the species (such as its crypsis) is successful in deterring an attack following a putative encounter with a predator. If the species only has a single line of defence then the probability of it surviving any particular encounter (e.g. not being seen) will be p_p. Let predators encounter prey at random, so that putative encounters with individual prey are Poisson distributed with mean and variance λ. Under these conditions, the expected probability of an individual prey surviving a series of random encounters with predators before reproducing is:

$$\sum_{i=0}^{\infty} \left(\frac{e^{-\lambda}\lambda^i}{i!} \right) p_p^i = e^{\{\lambda(p_p-1)\}} \qquad [B9.1]$$

so its fitness will simply be:

$$w_p = b e^{\{\lambda(p_p-1)\}} \qquad [B9.2]$$

where b is the mean fecundity of those prey that survive to reproduce.

Imagine now the possibility of the prey species evolving a secondary defence with fecundity cost τ. We assume that if the primary defence fails, then the secondary defence will be successful in deterring attack with probability p_s. Under these conditions, the probability s that an individual survives a single putative encounter with a predator is therefore:

$$s = p_p + \left(1 - p_p\right) p_s$$

and its probability of surviving to the end of the season is therefore:

$$\sum_{i=0}^{\infty} \left(\frac{e^{-\lambda}\lambda^i}{i!} \right) s^i = e^{-\{\lambda(p_p-1)(p_s-1)\}}$$

generating a mean fitness (after paying the fecundity cost) of:

$$w_s = e^{-\{\lambda(p_p-1)(p_s-1)\}} \left(b - \tau \right)$$

Clearly, a secondary defence will be selected for whenever $w_s > w_p$, which reduces to:

$$p_p < \frac{\lambda p_s + ln\left(\frac{b-\tau}{b} \right)}{\lambda p_s}$$

Thus, if the secondary defence is cost free ($\tau = 0$), then it will always be selected for whenever the primary defence is imperfect ($p_p < 1$). However, in those cases where the development, maintenance and/or deployment of secondary defence is costly then such a defence will only be selected for if the primary defence is ineffective (p_p is low), the frequency of potential attacks (λ) is high and/or the effectiveness of the secondary defence (p_s) is high. Similar arguments apply to iteoparus species (not shown).

a secondary defence when the primary defence (assuming it has one) fails. We start with inhibiting encounter, rather than inhibiting search because most of the behaviors that render the prey unprofitable to attack (such as fighting back or fleeing) are those behaviors that are employed should any predator mistakenly decide to attack them (Endler 1991; Jeschke 2006).

9.3 Anti-predator behavior as part of a primary defence

9.3.1 Seek (or create) a refuge

Perhaps the simplest form of anti-predator defence is to remain hidden from predators in crevices, beneath bark or in holes in the ground, a phenomenon that has been termed 'anachoresis' (from the Greek word meaning 'one who has withdrawn himself from the world') (Edmunds 1974). Other insects, ranging from tent caterpillars to caddis flies build their own shelters to protect themselves from predators (Hansell 2007).

Clearly, behavior is an essential element of this primary withdrawal strategy. For example, Pierce (1988) found that larvae of some dragonfly species increase their use of cover when in the presence of fish predators. Likewise, in an intriguing paper Yack and Fullard (2000) describe one instance in which behavior may have provided the organism a significant temporal refuge from predation—ultrasound hearing is absent in most butterflies (with the exception of their early nocturnal ancestors), but prevalent in moths, leading the authors to suggest that diurnal activity in butterflies may have evolved largely to avoid predation by bats. While being spatially or temporally separated from predators is an obvious solution to avoiding being eaten, sometimes it is not a viable one if the organism has to pay significant opportunity costs to stay out of sight. As we will see in the next section, behavior is also key component of other primary defences, such as the strategy of remaining hidden, while in plain view.

9.3.2 Micro-habitat selection

Micro-habitat selection is a behavioral process by which insects show a tendency to stay on/near a specific background substrate. An appropriate choice of background substrate is especially crucial for visually cryptic insects living in heterogeneous environments (i.e. when there are many different types of substrates) because their crypsis is largely influenced by the degree of colour pattern-matching between their body and the background substrate. Patterns on the body may also have maximum disruptive effect if the organism orientates itself in such a way that conceals its characteristic outline. Many insects show phenotype-environment matching in that their colour patterns match the substrates on which the insects are commonly found (Endler 1984). Note, however, that phenotype-environment matching can be also achieved by processes other than habitat selection, including background-mediated colour change (Stevens et al. 2014). Here, we adopt a relatively narrow definition of micro-habitat selection and consider the cases where *an active behavioral choice* is involved.

Background choice has been most extensively studied in bark-resting cryptic moths. Classical studies have focused on testing the ability of moths to choose to rest on a background substrate that exhibits the similar colour patterns as their wings, and whether moths adopt a specific resting posture/orientation that reinforces their crypsis. The accumulated evidence suggests that many cryptic moths do not choose their background substrate at random, but instead prefer certain substrates (Kettlewell 1955; Sargent 1968, but also see Sargent 1969 for a conflicting case). In Kettlewell's (1955) pioneering study with polymorphic geometrid species *Biston betularia*, for example, black melanic forms preferred to settle on black backgrounds, while pale forms chose to settle against white backgrounds.

In addition to their behavioral site selection, many moth species adopt specific resting orientations. Sargent and Keiper (1969) describe the species-specific resting orientations of 25 of the most common moth species found on tree trunks in Massachusetts, USA. The adaptive significance of these species-specific orientations was further investigated by Pietrewicz and Kamil (1977) who presented blue jays (*Cyanocitta cristata*) with slides of *Catocala* sp. in different orientations and evaluated how different orientations of moths influence visual detection by birds. More recently, Kang et al. (2012) showed direct evidence of microhabitat choice in moths. Using two geometrid moths, *Hypomecis roboraria* and *Jankowskia fuscaria*, they showed that, after landing on a tree bark, moths re-orientate and re-position to a nearby spot where their detectability is lower.

Lepidopteran caterpillars also show adaptive microhabitat choice that improves their crypsis. For example, the twig-mimicking caterpillar *Selenia dentaria* selectively choose to rest on twigs of comparable size (Skelhorn and Ruxton 2013), and prefer to stay on branches with higher twig densities which reduces predator detection rates (Skelhorn et al. 2011). Other caterpillars that adopt a **masquer-ade** strategy are known to select microhabitats in which their models are abundant (e.g. Herrebout et al. 1963). Naturally, however, microhabitat choice is not exclusive to Lepidoptera. For example, the dragonfly larvae *Pachydiplax longipennis* selectively occupy the leaf axil area of their aquatic plants, which reduces attacks by predatory fish (Wellborn and Robinson 1987). Likewise, different morphs of the pygmy grasshopper *Tetrix undulate* show microhabitat selection that improves their match to their backgrounds (Ahnesjö and Forsman 2005). These findings collectively demonstrate microhabitat selection is prevalent in cryptic and masquerading insects, and they have evolved to improve their camouflage in natural habitats.

9.3.3 Behavioral mimicry

While site-selection and orientation behavior can be used to avoid detection and/or recognition, it can also be used to enhance the effectiveness of protective mimicry. Examples of protective mimicry include:

- the resemblance of the species to an object of no inherent interest to a potential predator such as leaves, thorns, sticks, stones or bird droppings ('masquerade', Skelhorn 2015);
- the resemblance of species to other unpalatable or otherwise defended species ('**Batesian mimicry'**, Bates 1862);
- the evolution of a shared warning signal among defended prey types ('**Müllerian mimicry'**, Müller 1879)—see Ruxton et al. (2004) for review.

Although one might anticipate that resembling a leaf, thorn, or stone would not involve any behavioral

adaptations, this is not always the case. For example, females of the phytophagous stick insect (*Extatosoma tiaratum*) typically hang inverted among the foliage with curled abdomen and have been frequently observed rocking from side-to side. In a recent experiment, Bian et al. (2016) investigated this behavior, finding that wind would initiate (but not indefinitely maintain) a swaying response in the phasmid, and that its movement had quantitatively similar properties to that of the wind-blown plants. Since a motionless object on a moving plant is likely to attract attention, while a moving object on a motionless plant will do the same, it seems highly likely that the behavior has evolved to enhance its camouflage.

Many Batesian mimics also engage in behavioral mimicry. For example, the tomentose burying bee-tle, *Nicrophorus tomentosus* (Silphidae), engages in a twist and flip manoeuvres that exposes the yellow underside during flight, thereby enhancing its resemblance to the flight behavior and appearance of a bumble bee (Heinrich 2012). Likewise, the ant-like jumping spider, *Myrmarachne melanotarsa*, not only resembles ants from the genus *Crematogaster*, but it also forms aggregations that appear to enhance the overall protective effect of the mimicry (Nelson and Jackson 2009). In an intriguing recent paper, Kitamura and Imafuku (2015) quantified the movement behavior of mimetic and non-mimetic females of the polymorphic swallowtail butterfly *Papilio polytes*. They found that the flight paths of mimetic and non-mimetic females were different from each other, but the locomotory behavior of the mimetic females was not significantly different from that of their model (*Pachliopta aristolochiae*) they morphologically resemble. In each of the above cases, it seems clear that behavior has augmented the effectiveness of their disguises.

Above we described the way certain eye-spotted caterpillars adopt a defensive posture when approached that increases their resemblance to snakes (Hossie and Sherratt 2014). Resembling a model in more than one manner may increase the likelihood that predators will be deceived by the mimicry. It may also dupe different predators that use different sensory modalities to detect their prey (Pekár et al. 2011). Despite its potential usefulness, however, it is clear that not all species of a given group engage in behavioral mimicry, so one might wonder why some species have evolved behavioral mimicry and others have not. One possibility is that poor mimics use behavior to compensate for their ease of visual discriminability. For example, Pekár et al. (2011) suggested that because spiders that mimic ants are also selected to run quickly like ants, then there may be little or no selective advantage to them more closely resembling the ants. By contrast, behavioral mimicry might provide little selective benefit if imperfect mimics are readily visually discriminated. Instead, behavioral mimicry might be restricted to cases of good morphological mimicry, reflecting overall stronger selection for mimetic fidelity on all levels. Indeed, Penney et al. (2014) surveyed the incidence of behavioral mimicry (mock stinging, wing wagging, leg waiving) in fifty-seven field-caught species of hoverfly, and found that these highly associated traits occurred in two genera (*Spilomyia* and *Temnostoma*)—see Figure 9.2. Intriguingly, the behavioral mimics were all good morphological mimics of wasp (to human eyes at least), not all good mimics were behavioral mimics. This led the authors to conclude that while the behavioral mimicry may have evolved to augment good morphological mimicry, it does not advantage all good mimics.

Behavioral mimicry is also found in Müllerian mimics, i.e. unpalatable or otherwise defended species that all share common warning signals. For example, Srygley (2007) examined the locomotory behavior of *Heliconius* butterflies, which are well known group of neotropical Müllerian mimics. He suggested that members of one mimicry group (*Heliconius cydno* and *H. sapho*) beat their wings more slowly and their wing strokes were more asymmetrical than their sister species (*Heliconius melpomene* and *H. erato*, respectively) in another mimicry group. Likewise, the slow movement and lack of fleeing response in the chemically defended tiger beetle, *Pseudoxycheila tarsalis*, may enhance its similarity to the stinging mutillid wasp it resembles (Schultz 2001). In each of these instances, if a given behavior is sufficiently characteristic of noxious prey then the more species that adopt it, the greater will be the selection on other defended species to adopt these characteristic behavioral traits.

9.3.4 Warning displays

Many insects that are unprofitable to attack exhibit conspicuous colour patterns (generally interpreted as 'warning signals'; Wallace 1867) to ward off predators. Indeed, the cross-species association between conspicuous colour patterns and unprofitability has been referred to as '**aposematism**' (Poulton 1890). Although warning displays are typically thought of in terms of visual signals, they are often multimodal, such as both visual and auditory, or both visual and chemical (Rowe and Guilford 1999) and in these cases the facultative nature of the display necessarily involves some behavioral control.

Warning signals appear to have evolved in unprofitable prey for several reasons including:

- enhancing predators' ability to learn about and subsequently avoid the same (or similar) prey in subsequent encounters;
- exploiting predators' innate caution with conspicuous signals;
- decreasing the chance that the predator misclassify the prey as palatable (Ruxton et al. 2004).

Of course, while warning displays are often used as an honest signal of unprofitability, as the previous section demonstrates they are also exploited by palatable insects to ward off (and consequently deceive) predators.

Insects have evolved various forms of warning displays. At its simplest, a show of weaponry such as a sting or spines is sufficient to honestly signal to a would-be predator that it would be foolhardy to attack it (Eisner et al. 2005). In other cases, the nonchalant way an unfamiliar species behaves, coupled with its conspicuous appearance (that profitable prey would never have outside mimicry) may deter a wary predator.

Aggregation is a common behavioral defence in many aposematic insects. In his classical book, Fisher (1930) proposed that aggregation evolves in such species because local predators learn to avoid them, thereby conferring survivorship advantages to closely-related individuals within the group that carry the same traits (an early invocation of **kin selection**). Alatalo and Mappes (1996) novel world experiments supported Fisher's kin aggregation hypothesis, but also pointed to more general benefits

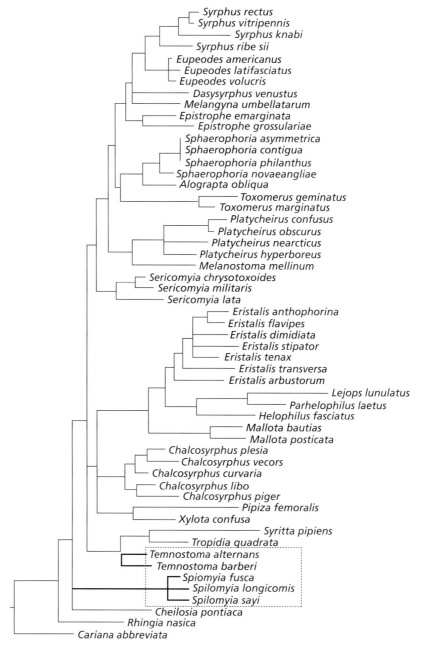

Figure 9.2 Phylogenetic tree of fifty-five field-tested hover fly species with behavioral mimics (those that wing wag, leg wave, or mock sting to improve their similarity to hymenoptera) highlighted in bold within the dotted box. While the tree has been orientated to show behavioral mimics within one box, the genera *Temnostoma* and *Spilomyia* are not sister taxa.

to gregariousness in unprofitable prey. In follow-up experimental studies, Riipi et al. (2001) confirmed that aggregation can benefit aposematic prey despite their higher detectability through several mechanisms, including dilution effects and faster avoidance learning for aggregated aposematic prey. Of course, not all unprofitable species are gregarious, and the widely-observed cross-species association between gregariousness and unprofitability can arise for 'negative' reasons. Thus, aggregation is not seen in palatable prey without some form of defence simply because such gatherings attract attention with little benefit—indeed, they would represent 'gold mines' to predators (Ruxton and Sherratt 2006).

Although they can be classified in a variety of ways, eyespots (and associated behaviors) are one of the most prevalent forms of intimidating warning displays in insects, and they are particularly frequent in the order Lepidoptera (Stevens 2005). Eyespots are generally defined as circular or quasi-circular patterns on the wings, with at least two concentric rings, or with a single colour disc and a central pupil (Monteiro 2008). Intriguingly, there continues to be a debate as to whether eyespots benefit prey by virtue of their conspicuousness (Stevens et al. 2008) or through their resemblance to predator's eyes (Bona et al. 2015), or both. Although eyespots may be effective in deterring predators when constantly displayed (Kodandaramaiah et al. 2009), many insects hide eyespots when resting, but transiently reveal them only when physically disturbed as an active secondary defence, a subject we cover in the next section. Organisms can use behavior to increase the effectiveness of the eyespot by making it more prominent or by changing its shape. Indeed, there are even caterpillar species with eyespots that 'blink' (Hossie et al. 2013). So the 'eyespot' defence could be considered a warning signal in unprofitable prey, a mimetic signal in profitable prey (both examples of primary defences) or as a post-contact startle signal (a secondary defence). These functions are not mutually exclusive.

In rare cases, warning displays take a form of **bioluminescence**. Cock and Matthysen (2003), for example, found that toads learned to avoid bioluminescent glow-warm larvae, *Lampyris noctiluca*, more rapidly than non-luminescent prey. However, it is unclear why these insects use this energy-costly bioluminescence as a warning signal. One possibility is that bioluminescence enables the prey to use a facultative aposematism strategy in that they can have cryptic appearance normally, but show conspicuous warning light signals only when detected and threatened by a predator. By doing so, they can enjoy the benefits of both crypsis and aposematism.

9.4 Anti-predator behavior when the primary defence fails

9.4.1 Flee

Perhaps the most obvious option open to a potential prey individual is to flee. Almost by definition, fleeing occurs after the prey has encountered the potential predator, although it may arise before an actual attack and even before the predator has seen the prey. So the first dilemma that may face a concealed organism is whether it should continue to remain in position, or move away as a potential predator draws near. Despite the advancing threat, there is, of course, a chance that the predator has not discovered the prey, but it will have an increasing likelihood of doing so as it gets closer. Moreover, the closer the predator is to the prey when it discovers it, the more likely the predator will be to successfully capture the prey. A graphical model by Ydenberg and Dill (1986) was the first to explore the optimal time to flee from a trade-off perspective. Broom and Ruxton (2005) formalized and extended this earlier model, albeit with slightly different assumptions (and, hence, predictions). The authors adopted a game theoretical perspective since the prey's pay-off from adopting a given anti-predator strategy depends on the foraging strategy adopted by the predator and vice-versa. In short, they argued that the optimal strategy for the prey is either to flee immediately on seeing the predator or only to flee when the predator initiates an attack, a result that holds true whether the predator attacks immediately on discovering the prey, or whether the predator hides the fact that it has discovered the prey and delays its attack until it has closed the distance to the prey.

Many papers have investigated the fleeing response of insect prey when their primary defence appears compromised, and we do not have the space to review these studies here. Butler (2013), for example, quantified the flight initiation distances (FID), their net displacement before stopping, and angles of escape of nine sympatric species of acridid grasshoppers. Somewhat at odds with the predictions made by Broom and Ruxton (2005), the escape and fleeing distances of individuals tended to be positively correlated within species, suggesting that slow moving individuals had less to gain by fleeing early. In a related paper that further emphasizes behavioral flexibility, as well as inter-specific differences, Bateman and Fleming (2014) evaluated the responses of two acridid grasshoppers on repeated approaches. *Schistocerca alutacea* did not increase its FID upon repeated approach, but tended to flee further over successive escapes. By contrast, *Psinidia fenestralis* increased its FID on the second (but not successive) approach, but the distance it fled showed a downward trend with the number of escapes. Why different species tend to adopt a different 'plan B' is unclear, although escape behavior will clearly be shaped by a number of factors, including physiological condition and ability to hide after fleeing.

While different mathematical models have yielded different predictions, if it is clear to the prey that the predator has detected it and is about to pursue it, then it should flee (or adopt some alternative strategy) immediately, since there is no advantage to delay—the predator will be at the maximal distance from the predator and the cost of the escape will be the same or even higher. Many chemically defended tiger moths click in response to echo locating bats (Corcoran et al. 2009). The reasons they do so has been a matter of debate, with some proposing that it is an acoustic aposematic signal (Ratcliffe and Fullard 2005) and others proposing that it is jamming the radar (Fullard et al. 1994). However, in an intriguing paper, Corcoran et al. (2013) describe a combination of field and laboratory experiments, which indicate that the tiger moth, *Bertholdia trigona*, was capable of distinguishing false-alarms, and initiated their defensive clicks only when it was clear that they have been detected and targeted by a bat. In this way, the moths were maintaining their primary defence until there was clear evidence these defences had failed, at which point they switched to a new form of defence.

9.4.2 Startle defences

Startle display (also known as **deimatic display**) is a behavior in which prey suddenly unleash unexpected defences or conspicuous signals which elicits a reflexive response of a predator (see also Chapter 11). This behavioral defence is mostly deployed only when the prey is under an imminent threat (i.e. when the detection by a predator is certain and is approached/attacked by the predator). Startle displays are seen in a wide range of insect taxa. For example, when some praying mantids sense a predation threat, they turn their bodies towards the threatening source, abduct forelegs, and raise their bodies and wings which makes them look larger and reveal the previously hidden brightly coloured hindwings (Edmunds 1972). Mountain katydid, *Acripeza reticulata*, also uses startle display as a secondary display in that they open their wings and reveal their strikingly coloured abdomens only when they are physically attacked (Umbers and Mappes 2015). The spotted lanternfly, *Lycorma delicatula*, employs this defence as a 'tertiary' defence. They primarily rely on crypsis, but when attacked, they try to flee immediately by jumping away. If the attempt of jumping fails, then they deploy a startle display (Kang et al. 2017).

Startle displays may open a short window of opportunity that may allow the insect to escape (Edmunds 2008), but (if the display is threatening) it may also induce the predator to flee the area entirely. For instance, the peacock butterfly, *Inachis io*, resembles a leaf when resting, but shifts to an active defence when physically disturbed by unfolding their wings and revealing four large eyespots on their wings, a display accompanied by hissing. Vallin et al. (2005) demonstrated that the display of eyespot patterns on wings increases the survival of the peacock butterfly when confronted with an avian predator by effectively forcing the predator to back away. Similar results were obtained by Bona et al. (2015) when they presented artificial prey models with 'owl-mimicking' eyespots to birds. These studies clearly demonstrate the survival value of startle display when combined with eyespots. Many laboratory experiments do not allow captive birds to flee the area, but the fact that predators have been observed to produce alarm calls when insects flash eyespots (Olofsson et al. 2013) suggests that at least some would leave the area entirely. Indeed, if there is a chance the startle is a genuine threat, then any predator that took its time to decide whether to hang around could be lunch.

9.4.3 'Death feigning' (tonic immobility)

An alternative response of prey when the primary defence has failed is to death feign. Death-feigning is an umbrella term for a wide variety of behaviors involving predator initiated immobility so it has become known under a variety of terms including tonic immobility or thanatosis (Edmunds 1974). While death feigning may sometimes involve mimicry of a dead subject (feigning death or playing possum), this is not always the case. Indeed, in a posthumously published essay, Darwin (1883) described the immobility of beetles when touched and noted, 'in several instances the attitude of the feigners and of the really dead were as unlike as they could possibly be'.

Death feigning is extremely widespread in insects and is particularly well known in beetles. There are a variety of reasons why an organism might render itself immobile when touched by a predator or as it draws closer. In some circumstances, it may simply render the prey harder to handle. For example, Honma et al. (2006) noted that when faced with a predator, the pygmy grasshopper, *Criotettix japonicas*, stretched its pronotum, hind legs, and lateral spines in different directions. This increase in functional body size made it difficult for a gape-limited predator, the frog, *Rana nigromaculata*, to swallow such prey. In other cases, if the organism appears dead then predators may refrain from attacking it entirely to avoid the potentially harmful parasites it may contain or to postpone eating it in order to chase moving individuals as part of a foraging strategy. In other cases, if it has dropped away from a predator on first approach, then it may remain motionless to enhance its concealment.

In some instances, tonic immobility may be an adaptation that has evolved to exploit the tendency of motion-orientated predators to focus their attention only on moving prey. Alternatively, the immobility may be a by-product of the prey switching to other forms of defence. Miyatake et al. (2004) con-

ducted artificial selection experiments on red flour beetles, *Tribolium castaneum*, over ten generations and showed that one can select for long or short durations of death-feigning, demonstrating that the behavioral trait is highly heritable. Moreover, the strains that were selected to engage in long durations of death feigning survived exposure to jumping spider, *Hasarius adansoni*, than the short-duration strains that rarely death feigned, either because the predator lost interest when its prey was no longer moving and/or because it was better able to express its defensive chemicals. In some cases, trade-offs may be involved. For instance, in an ambitious experiment, Ohno and Miyatake (2007) imposed selection on the adzuki bean beetle, *Callosobruchus chinensis*, for short or long death-feigning duration, as well as for poor or good flight performance. The strains selected for shorter (longer) duration of death feigning also had higher (lower) flying ability, while the strains selected for lower (higher) flying ability showed longer (shorter) duration of death feigning, indicating a pleiotropic effect with different anti-predator strategies at the end of the continuum. Under these conditions, a mixed strategy of feigners and fliers may be selected for. Indeed, experiments have shown that that the survival rates of flour beetle feigners when exposed to jumping spiders were higher when in the presence of non-feigners or prey of a different species (Miyatake et al. 2009), although evidence that fliers do well when feigner's are common is lacking.

9.4.4 Fighting back

When camouflage, warning, fleeing, or any attempts to ward off the attacking predator fail, many insects resist and fight back against the predator, as a last resort for survival. As might be expected, many predatory insects use their hunting weapons for their defence (such as mandibles of predatory katydid or forelegs of praying mantis), but some others have evolved specialized morphological/behavioral tactics to defend themselves. The fighting back strategy includes kicking, struggling, wing flapping, biting, regurgitation, and stinging, but also chemical defences. When caterpillars of the swallowtail butterfly, *Papillio machaon*, are disturbed, for instance, then they adopt a threatening pose by extending and waving a defensive organ, which secretes noxious chemicals that further protects the caterpillar from predatory (and possibly parasitic) attacks (Eisner and Meinwald 1965).

The chemical secretion strategy will be adaptive when the survivorship benefit of secreting chemicals exceeds the costs of producing them. The predators' mode of attack may have facilitated the evolution of this strategy. Many predators use 'taste-and-rejection' strategy, which involves the non-fatal tasting (attacking) of the prey before making a decision to consume the prey (Skelhorn and Rowe 2006). Exuding a chemical from their body would let predators assess the unprofitability of the prey quickly and accurately, which can protect the prey from a fatal subsequent attack. Caterpillars of the common silkmoth caterpillar produce clicks with their mandibles just before and, while they vomit noxious deterrent chemicals (a case of acoustic aposematism—see Section 9.3.4), but reduce the cost of chemical production by re-imbibing their regurgitant after the threat has passed (Brown et al. 2007). Another example of anti-predatory regurgitation can be found in eastern tent caterpillars, *Malacosoma americanum*. When disturbed, this caterpillar regurgitates defensive fluid. Peterson et al. (1987) found that regurgitation of the caterpillars that fed cyanogenic foliage of black cherry was more repellent to predatory ants than those of the caterpillars that fed less cyanogenic foliage. They also demonstrated that the chemical compound benzaldehyde, derived from the host foliage, was the key element to repel the ant predators.

Bombardier beetles (family Carabidae) are renowned for their ability to eject a hot noxious chemical spray from their abdomen towards the predator (Eisner 1958). These beetles store hydroquinone and hydrogen peroxide separately in two reservoirs, respectively, in their abdomen. When disturbed, these two compounds reach the vestibule and the reactions of these two chemical generates near 100°C of heat and gas, which drives the ejection of the mixture from its abdomen (Aneshansly and Eisner 1969). Some bombardier species, such as *Stenaptinus insignis*, can aim the spray in virtually any direction, possibly an adaptation to defend themselves against ants that can attack the beetles from any bearing (Eisner and Aneshansly 1999).

Some eusocial insects in the orders Hymenoptera, like termites, ants, and wasps, use collective defence strategy against intruders. For example, some species of stingless bees in the genus *Trigona*, despite the lack of a sting, employ various aggressive behaviors including biting, the release of **alarm pheromones**, buzzing, and angular flights to defend the nest against predators (Shackleton et al. 2015). Japanese giant hornets, *Vespa mandarinia japonica*, are important predators of honey bees, but the Japanese honey bees, *Apis cerana japonica*, has evolved a unique behavioral/physiological defence to protect themselves. When hunting for prey, a scout hornet will begin by marking the honey bee colony with pheromones to recruit other wasps. Stings are ineffective against these predators, so when a (scout) hornet approaches their hive and tries to leave the pheromone, the honeybee workers return back to the hive, leaving an opening to allow the hornet to enter. When the hornet enters hundreds of workers surround the hornet, making a ball of bees, and vibrate their flight muscles violently. This raises the temperature and carbon dioxide level inside the ball, which eventually kills the intruder (Ono et al. 1995).

9.5 The comparative approach to understanding variation in anti-predator

Over the past decade considerable research effort has been devoted to addressing specific questions relating to anti-predator defence in insects—such as how **disruptive colouration** reduces detectability of moths and how conspicuous warning signals evolve from rarity. However, in treating each of these defensive adaptations independently, more fundamental questions have been overlooked. So, we can ask whether defensive traits evolved just once or many times. We can also ask why do some species evolve one form of defence, while others evolve another? With increasing ecological data on the anti-predator defences of insects and an increasing number of phylogenetic studies, we are now at a stage where we can begin to test 'big picture' hypotheses relating to the diversity of defensive traits we see in the natural world. Of course, when asking these questions it is essential to control for phylogeny because certain species may share attributes (such as large size and a tendency to engage in behavioral mimicry) not as independent outcomes of selection, but more simply through **shared ancestry**.

Phylogenetic analyses are increasingly revealing the origins of anti-predator traits. For example, Marek et al. (2011) showed that bioluminescence has a single evolutionary origin in the millipede, *Motyxia sp.*, serving as an aposematic signal against nocturnal mammalian predators. In an early phylogenetically controlled study of insect defences, Prudic et al. (2007) noted that most swallowtail (*Papilio*) caterpillar species masquerade as bird dropping in their early instars. However, as they enter their fourth or fifth instar different species adopt alternative phenotypes including an eye-spotted form that putatively resembles a snake and a conspicuous form (likely aposematic, since it is likely that many *Papilio* caterpillar species are unpalatable). Intriguingly, the tendency to become aposematic was not related to the species' extent of diet specialization, but their signal environment—those that develop on narrow-leaved plants were more likely to evolve aposematism since exposure of caterpillars on such plants would tend to render caterpillars conspicuous. In each of these cases, it seems highly likely that there would be concomitant selection for appropriate behavior.

In a particularly ambitious study, Lichter-Marck et al. (2015) compared the effectiveness of anti-predator traits, such as warning signals and camouflage, in an assemblage of thirty-eight caterpillar species monitored over 8 years in multiple ways (including a bird-exclusion experiment). After fitting phylogenetically controlled models to compare predation risk across species, the authors concluded that the caterpillar species possessing warning signals with the least variable (hence stereo-typical) resting substrate experienced the lowest bird predation risk, which once again serves to highlight the role of prey behavior in the visual signalling of predators. Indeed, given the significant synergistic role of behavior it is essential to note that field and laboratory experiments investigating camouflage and signal evolution in relatively artificial prey, such as mealworms, almonds, and pastry baits may only tell part of the story.

In another study comparing defences among caterpillars, Bura et al. (2016) found that twenty of sixty-one species of Bombycoidea moths (Saturnidae

and Sphingidae) investigated produced sounds following simulated attacks (in which subjects were lightly pinched with forceps). Why give this behavioral response and why is there inter-species variation? The authors proposed that the sounds produced either warn predators of their noxiousness (acoustic aposematism), or simply serve to intimidate them (startle displays). If true, one might expect that acoustic aposematism would occur in species that use a chemical defence, and the sounds would precede or accompany chemical release to enhance the predator's association with the chemical. Although no attempt was made to control for phylogeny, after classifying the sound-producers in terms of the presence/absence and timing of chemical release, all 'high chemical' species produced short-duration clicks or chirps. By contrast, sounds produced by 'low chemical' species were significantly longer in duration and higher in energy, which may function in startle or intimidation—effectively an 'acoustic eyespot'.

In a recent study of morphological eye-like markings, Hossie et al. (2015) examined the distribution of eyespots across final instar hawkmoth caterpillars (Macroglossinae). After controlling for phylogeny, they found that eyespots were significantly more likely to occur in larger caterpillars. Why would this be so? Complementary laboratory work confirmed that naïve chicks showed a greater latency to attack large (pastry) caterpillars with eyespots, than those without eyespots. By contrast, chicks more rapidly attacked small caterpillars with eyespots than those without. Eyespots on small prey may not be especially intimidating and yet serve to enhance the prey's conspicuousness. However, since they may pose a legitimate threat, large prey items are generally more intimidating to small birds, and the addition of eyespots further enhances the effect. This cross-species pattern is entirely consistent with the observation that eyespots tend to be substantially reduced or absent altogether during early instars of hawkmoths, only becoming prominent in late instars, when the caterpillars are much larger.

9.6 Conclusions

As this brief survey demonstrates, behavior is an important component of insect prey's anti-predator defences. Behavioral mechanisms that govern the choice of habitat or timing of activity can limit exposure to certain predators altogether. Moreover, even if the predators and prey co-occur then the appropriate choice of microhabitat or resting orientation can reduce the chance of sympatric predators encountering these prey. Once detected, then behavior can be an integral component of any secondary defence that is elicited, ranging from biting back to tonic immobility or startle. It also ensures that prey can move between strategies when certain defensive strategies appear to be failing.

As more data are gathered about the diversity of anti-predator behavior in insects, new questions arise, such as how characteristic they are of any given group and why some species have adopted it, but not others. For example, since *Catocola* moths flash conspicuous hindwings when touched by a predator (Sargent 1978), one might wonder why some species have evolved this form of display, but not others. Comparative analyses are beginning to address these questions and have the potential to place a whole new perspective on the evolution of anti-predator behavior within insects.

References

Ahnesjö, J., and Forsman, A. (2005). Differential habitat selection by pygmy grasshopper color morphs: interactive effects of temperature and predator avoidance. *Evolutionary Ecology*, **20**, 235.

Alatalo, R.V., and Mappes, J. (1996). Tracking the evolution of warning signals. *Nature*, **382**, 708.

Aneshansly, D.J., and Eisner, T. (1969). Biochemistry at 100°C: explosive secretory discharge of bombardier beetles (Brachinus). *Science*, **165**, 61.

Bateman, P.W., and Fleming, P.A. (2014). Switching to Plan B: changes in the escape tactics of two grasshopper species (Acrididae: Orthoptera) in response to repeated predatory approaches. *Behavioral Ecology and Sociobiology*, **68**, 457.

Bates, H.W. (1862). Contributions to an insect fauna of the Amazon valley. Lepidoptera: Heliconidae. *Transactions of the Linnean Society of London*, **23**, 495.

Bian, X., Elgar, M. A., and Peters, R.A. (2016). The swaying behavior of *Extatosoma tiaratum*: motion camouflage in a stick insect? *Behavioral Ecology*, **27**, 83.

Bona, S.D., Valkonen, J.K., López-Sepulcre, A., and Mappes, J. (2015). Predator mimicry, not conspicuousness, explains the efficacy of butterfly eyespots. *Proceedings of the Royal Society of London B: Biological Sciences*, **282**, 20150202.

Britton, N.F., Planqué, R., and Franks, N.R. (2007). Evolution of defence portfolios in exploiter–victim systems. *Bulletin of Mathematical Biology*, **69**, 957.

Broom, M., and Ruxton, G.D. (2005). You can run—or you can hide: optimal strategies for cryptic prey against pursuit predators. *Behavioral Ecology*, **16**, 534.

Brown, S.G., Boettner, G.H., and Yack, J.E. (2007). Clicking caterpillars: acoustic aposematism in *Antheraea polyphemus* and other Bombycoidea. *Journal of Experimental Biology*, **210**, 993.

Bura, V.L., Kawahara, A.Y., and Yack, J.E. (2016). A comparative analysis of sonic defences in Bombycoidea caterpillars. *Scientific Reports*, **6**, 31469.

Butler, E.M. (2013). Species-specific escape behaviour in grasshoppers. *Behaviour*, **150**, 1531.

Cock, D.C., and Matthysen, E. (2003). Glow-worm larvae bioluminescence (Coleoptera: Lampyridae) operates as an aposematic signal upon toads (*Bufo bufo*). *Behavioral Ecology*, **14**, 103.

Corcoran, A.J., Barber, J.R., and Corner, W.E. (2009). Tiger moth jams bat sonar. *Science*, **325**, 325.

Corcoran, A.J., Wagner, R.D., and Conner, W.E. (2013). Optimal predator risk assessment by the sonar-jamming Arctiine moth *Bertholdia trigona*. *Plos One*, **8**, e63609.

Darwin, C.R. (1883). Essay on instinct. In: G.J. Romanes (Ed.), *Mental Evolution in Animals. With a Posthumous Essay on Instinct by Charles Darwin*, pp. 355–84. Kegan Paul Trench & Co., London.

Edmunds, M. (1972). Defensive behaviour in Ghanaian praying mantids. *Zoological Journal of the Linnean Society*, **51**, 1.

Edmunds, M. (1974). *Defence in Animals: A Survey of Anti-Predator Defences*. Longman, Harlow.

Edmunds, M. (2008). Flash colors. In: J.L. Capinera (Ed.). *Encyclopedia of Entomology*, Springer, Dordrecht.

Eisner, T. (1958). The protective role of the spray mechanism of the bombardier beetle *Brachynus ballistarius* Lec. *Journal of Insect Physiology*, **2**, 215.

Eisner, T., and Aneshansly, D.J. (1999). Spray aiming in the bombardier beetle: photographic evidence. *Proceedings of the National Academy of Sciences of the USA*, **96**, 9705.

Eisner, T., Eisner, M., and Siegler, M. (2005). *Secret Weapons: Defenses of Insects, Spiders, Scorpions, and Other Many-legged Creatures*. Harvard University Press, Cambridge, MA.

Eisner, T., and Meinwald, Y.C. (1965). Defensive secretion of a caterpillar (Papillio). *Science*, **150**, 1733.

Endler, J.A. (1984). Progressive background in moths, and a quantitative measure of crypsis. *Biological Journal of the Linnean Society*, **22**, 187.

Endler, J.A. (1991). Interactions between predators and prey. In: J.R. Krebs and N.B. Davies (Eds), *Behavioural*

Ecology: An Evolutionary Approach, pp. 169–96. Oxford University Press, Oxford.

Fisher, R.A. (1930). *The Genetical Theory of Natural Selection*. Clarendon Press, Oxford.

Fullard, J.H., Simmons, J.A., and Saillant, P.A. (1994). Jamming bat echolocation: the dogbane tiger moth *Cycnia tenera* times its clicks to the terminal attack calls of the big brown bat *Eptesicus fuscus*. *Journal of Experimental Biology*, **194**, 285.

Hansell, M. (2007). *Built by Animals*. Oxford University Press, Oxford.

Heinrich, B. (2012). A heretofore unreported instant color change in a beetle, *Nicrophorus tomentosus* Weber (Coleoptera: Silphidae). *Northeastern Naturalist*, **19**, 345.

Herrebout, W.M., Kuyten, P.J., and de Ruiter, L. (1963). Observations on colour patterns and behaviour of caterpillars feeding on Scots pine. *Archives Néerlandaises de Zoologie*, **14**, 315.

Honma, A., Oku, S., and Nishida, T. (2006). Adaptive significance of death feigning posture as a specialized inducible defence against gape-limited predators. *Proceedings of the Royal Society B-Biological Sciences*, **273**, 1631.

Hossie, T.J., and Sherratt, T.N. (2014). Does defensive posture increase mimetic fidelity of caterpillars with eyespots to their putative snake models? *Current Zoology*, **60**, 76.

Hossie, T.J., Sherratt, T.N., Janzen, D.H., and Hallwachs, W. (2013). An eyespot that 'blinks': an open and shut case of eye mimicry in *Eumorpha* caterpillars (Lepidoptera: Sphingidae). *Journal of Natural History*, **47**, 2915.

Hossie, T.J., Skelhorn, J., Breinholt, J.W., Kawahara, A.Y., and Sherratt, T.N. (2015). Body size affects the evolution of eyespots in caterpillars. *Proceedings of the National Academy of Sciences of the United States of America*, **112**, 6664.

Jeschke, J.M. (2006). Density-dependent effects of prey defenses and predator offenses. *Journal of Theoretical Biology*, **242**, 900.

Kang C., Moon H., Sherratt T.N., Lee S., and Jablonski, P.G. (2017). Multiple lines of anti-predator defence in the spotted lanternfly, *Lycorma delicatula*. *Biological Journal of the Linnean Society*, **120**, 115.

Kang, C.K., Moon, J.Y., Lee, S.I., and Jablonski, P.G. (2012). Camouflage through an active choice of a resting spot and body orientation in moths. *Journal of Evolutionary Biology*, **5**, 1695.

Kettlewell, H.B.D. (1955). Recognition of appropriate backgrounds by the pale and black phases of Lepidoptera. *Nature*, **175**, 943.

Kitamura, T., and Imafuku, M. (2015). Behavioural mimicry in flight path of Batesian intraspecific polymorphic butterfly *Papilio polytes*. *Proceedings of the Royal Society B*, **282**, 20150483.

Kodandaramaiah, U., Vallin, A., and Wiklund, C. (2009). Fixed eyespot display in a butterfly thwarts attacking birds. *Animal Behaviour*, **77**, 1415.

Lichter-Marck, I.H., Wylde, M., Aaron, E., Oliver, J.C., and Singer, M.S. (2015). The struggle for safety: effectiveness of caterpillar defenses against bird predation. *Oikos*, **124**, 525.

Marek, P., Papaj, D., Yeager, J., Molina, S., and Moore, W. (2011). Bioluminescent aposematism in millipedes. *Current Biology*, **21**, 680.

Miyatake, T., Katayama, K., Takeda, Y., et al. (2004). Is death-feigning adaptive? Heritable variation in fitness difference of death-feigning behaviour. *Proceedings of the Royal Society B-Biological Sciences*, **271**, 2293.

Miyatake, T., Nakayama, S., Nishi, Y., and Nakajima, S. (2009). Tonically immobilized selfish prey can survive by sacrificing others. *Proceedings of the Royal Society B: Biological Sciences*, **276**, 2763.

Monteiro, A. (2008). Alternative models for the evolution of eyespots and of serial homology on lepidopteran wings. *BioEssays*, **30**, 358.

Müller, F. (1879). Ituna and Thyridia: a remarkable case of mimicry in butterflies. *Transactions of the Entomological Society*, **1879**, 10.

Nelson, X.J., and Jackson, R.R. (2009). Collective Batesian mimicry of ant groups by aggregating spiders. *Animal Behaviour*, **78**, 123.

Ohno, T., and Miyatake, T. (2007). Drop or fly? Negative genetic correlation between death-feigning intensity and flying ability as alternative anti-predator strategies. *Proceedings of the Royal Society B: Biological Sciences*, **274**, 555.

Olofsson, M., Løvlie, H., Tibblin, J., Jakobsson, S., and Wiklund, C. (2013). Eyespot display in the peacock butterfly triggers antipredator behaviors in naïve adult fowl. *Behavioral Ecology*, **24**, 305.

Ono, M., Igarashi, T., Ohno, E., and Sasaki, M. (1995). Unusual thermal defence by a honeybee against mass attack by hornets. *Nature*, **377**, 334.

Pekár, S., Jarab, M., Fromhage, L., and Herberstein, M.E. (2011). Is the evolution of inaccurate mimicry a result of selection by a suite of predators? A case study using myrmecomorphic spiders. *American Naturalist*, **178**, 124.

Penney, H.D., Hassall, C., Skevington, J.H., Lamborn, B., and Sherratt, T.N. (2014). The relationship between morphological and behavioral mimicry in hover flies (Diptera: Syrphidae). *American Naturalist*, **183**, 281.

Peterson, S.C., Johnson, N.D., and LeGuyader, J.L. (1987). Defensive regurgitation of allelochemicals derived from host cyanogenesis by eastern tent caterpillars. *Ecology*, **68**, 1268.

Pierce, C.L. (1988). Predator avoidance, microhabitat shift, and risk-sensitive foraging in larval dragonflies. *Oecologia*, **77**, 81.

Pietrewicz, A.T., and Kamil, A.C. (1977). Visual detection of cryptic prey by blue jays (*Cyanocitta cristata*). *Science*, **195**, 580.

Poulton, E.B. (1890). *The Colours of Animals: their meaning and use especially considered in the case of insects*, The International Science Series. Kegan Paul, Trench, Trubner & Co. Ltd, London.

Prudic, K.L., Oliver, J.C., and Sperling, F.A.H. (2007). The signal environment is more important than diet or chemical specialization in the evolution of warning coloration. *Proceedings of the National Academy of Sciences of the* USA, **104**, 19381.

Ratcliffe, J.M., and Fullard, J.H. (2005). The adaptive function of tiger moth clicks against echolocating bats: an experimental and synthetic approach. *Journal of Experimental Biology*, **208**, 4689.

Riipi, M., Alatalo, R.V., Lindström, L., and Mappes, J. (2001). Multiple benefits of gregariousness cover detectability costs in aposematic aggregations. *Nature*, **413**, 512.

Rowe, C., and Guilford, T. (1999). The evolution of multimodal warning displays. *Evolutionary Ecology*, **13**, 655.

Ruxton, G.D., and Sherratt, T.N. (2006). Aggregation defence and warning signals: the evolutionary relationship. *Proceedings of the Royal Society B: Biological Sciences*, **273**, 2417.

Ruxton, G.D., Sherratt, T.N., and Speed, M.P. (2004). *Avoiding Attack: The Evolutionary Ecology of Crypsis, Warning Signals and Mimicry*. Oxford University Press, Oxford.

Sargent, T.D. (1968). Cryptic moths: effects on background selections of painting the circumocular scales. *Science*, **159**, 100.

Sargent, T.D. (1969). Background selections of the pale and melanic forms of the cryptic moths, *Phigalia titea* (Cramer). *Nature*, **222**, 585.

Sargent, T.D. (1978). On the maintenance of stability in hindwing diversity among moths of the genus *Catocala* (Lepidoptera: Noctuidae). *Evolution*, **32**, 424.

Sargent, T.D., and Keiper, R.R. (1969). Behavioral adaptations of cryptic moths. I. preliminary studies on barklike species. *Journal of the Lepidopterists Society*, **23**, 1.

Schultz, T.D. (2001). Tiger beetle defenses revisited: alternative defense strategies and colorations of two neotropical tiger beetles, *Odontocheila nicaraguensis* Bates and *Pseudoxycheila tarsalis* Bates (Carabidae: Cicindelinae). *Coleopterists Bulletin*, **55**, 153.

Shackleton, K., Toufailia, H.A., Balfour, N.J., et al. (2015). Appetite for self-destruction: suicidal biting as a nest defense strategy in *Trigona* stingless bees. *Behavioral Ecology and Sociobiology*, **69**, 273.

Skelhorn, J. (2015). Masquerade. *Current Biology*, **25**, 643.

Skelhorn, J., and Rowe, C. (2006). Avian predators taste-reject aposematic prey on the basis of their chemical defence. *Biology Letters*, **2**, 348.

Skelhorn, J., Rowland, H., Delf, J., Speed, M.P., and Ruxton, G.D. (2011). Density-dependent predation influences the evolution and behavior of masquerading prey. *Proceedings of the National Academy of Sciences of the USA*, **108**, 6532.

Skelhorn, J., and Ruxton, G.D. (2013). Size-dependent microhabitat selection by masquerading prey. *Behavioral Ecology*, **24**, 89.

Srygley, R.B. (2007). Evolution of the wave: aerodynamic and aposematic functions of butterfly wing motion. *Proceedings of the Royal Society B*, **274**, 913.

Stevens, M. (2005). The role of eyespots as anti-predator mechanisms, principally demonstrated in the Lepidoptera. *Biological Reviews*, **80**, 573.

Stevens, M., Hardman, C.J., and Stubbins, C.L. (2008). Conspicuousness, not eye mimicry, makes "eyespots" effective antipredator signals. *Behavioral Ecology*, **19**, 525.

Stevens, M., Lown, A.E., and Wood, L.E. (2014). Colour change and camouflage in juvenile shore crabs *Carcinus maenas*. *Frontiers in Ecology and Evolution*, **2**, 14.

Umbers, K.D.L., and Mappes, J. (2015). Post attack deimatic display in the mountain katydid, *Acripeza reticulata*. *Animal Behaviour*, **100**, 68.

Vallin, A., Jakobsson, S., Lind, J., and Wiklund, C. (2005). Prey survival by predator intimidation: an experimental study of peacock butterfly defence against blue tits. *Proceedings of the Royal Society of London B: Biological Sciences*, **272**, 1203.

van Griethuijsen, L.I., Banks, K.M., and Trimmer, B.A. (2013). Spatial accuracy of a rapid defense behavior in caterpillars. *Journal of Experimental Biology*, **216**, 379.

Waldbauer, G.P. (1970). Mimicry of hymenopteran antennae by Syrphidae. *Psyche*, **77**, 45.

Wallace, A.R. (1867). *Proceedings of the Entomological Society of London* March 4th, lxxx–lxxxi.

Wellborn, G.A., and Robinson, J.V. (1987). Microhabitat selection as an antipredator strategy in the aquatic insect *Pachydiplax longipennis* Burmeister (Odonata: Libellulidae). *Oecologia*, **71**, 185.

Yack, J.E., and Fullard, J.H. (2000). Ultrasonic hearing in nocturnal butterflies. *Nature*, **403**, 265.

Ydenberg, R., and Dill, L. (1986). The economics of fleeing from predators. *Advances in the Study of Behaviour*, **16**, 229.

CHAPTER 10

Chemical communication

Bernard D. Roitberg

Biology, Simon Fraser University, Burnaby, Canada

10.1 What is communication?

Communicating is something that nearly all of us engage in, nearly all the time, and that would be the case for nearly all living beings, from arthropods to plants to birds, and so on. While it is easy to describe communication, it is not so easy to define it to the satisfaction of behavioral biologists. In fact, after decades of discussion, no consensus has been reached regarding a formal definition of communication (see Maynard Smith and Harper 2003; Scott-Phillips 2008; Font and Carazo 2010). Part of this disagreement is semantic and technical, but it also illustrates the subtle nature of communication, something that will be seen again and again in this chapter.

In the previous paragraph, the word 'engage' is used to describe communication and this is something that nearly everyone agrees upon, i.e. communication is an activity that involves two or more individuals. Individuals send signals that are received and interpreted by a recipient. Signals (which may be visual, acoustic, chemo, or tactile) have been defined, in an adaptationist manner by Maynard Smith and Harper (2003) as acts or structures that have evolved to cause (evolved) effects on the recipient. Thus, a signal is very different from a cue as illustrated by the following example. If you or I were to inadvertently cast a shadow (the cue) on a resting butterfly, causing it to fly away, we would not say

that we signalled it to flee. There was no intent on our part; we accidentally provided a (danger) cue by our presence. If instead, we waved our hands back and forth causing that same event to ensue, we can easily see that we were signalling danger to that butterfly; we were communicating with it. Of course, it is not clear why we would spend our time and energy to signal danger to some non-threatening, non-competitive organism. Recall, such signals have evolved to generate a particular response; they should be adaptive and it is not at all clear how this particular example is adaptive from the sender's perspective though it likely is so for the recipient. As students of behavior, it is our job to elucidate the function(s) of such signals as obscure as they may be in this example.

The lack of consensus noted above revolves around the information content of a signal. Adaptationists argue that signals must contain useful information, otherwise they would not have evolved, thus, they argue, there is no reason to include information content as part of a definition of communication. Another approach would be to consider the influence that such signals would confer and, again, it would be redundant to specifically include information content as part of the definition of signal (Owren et al. 2010). Informationalists, by contrast, argue that signals are essentially messages encoded by senders and it is that distinction (from information

Roitberg, B. D., *Chemical communication*. In: *Insect Behavior: From mechanisms to ecological and evolutionary consequences.* Edited by Alex Córdoba-Aguilar, Daniel González-Tokman, and Isaac González-Santoyo: Oxford University Press (2018).
© Oxford University Press.
DOI: 10.1093/oso/9780198797500.003.0010

per se) that should be included in the definition. I will not take an explicit stand on this disagreement. However, you might detect the author's preference in what follows.

In addition to the above, is the notion of **eavesdropping**. This topic will be discussed later, but here it is posited that, should an individual listen in on a conversation of others (e.g., a parasitoid responds to the sex pheromone of its host), this would not be considered communication, by human definition. This would be equivalent to the shadow casting discussed in the first paragraph in that the conversation acts as a cue, not a signal, at least not between the parasitoid and its host.

This chapter discusses various forms of communication and, in particular, chemical communication, which is manifested in smell or taste. As the concepts and examples are worked through, it is important to keep in mind, the previous discussion. If a moth releases a sex pheromone, it is only communicating if it engages in conversation with a conspecific recipient who responds; here, the old tree-falling-in-a-forest analogy applies.

10.2 What makes chemical communication special?

All of the senses that organisms employ to communicate fall under the same general rubric. However, when arthropods are discussed, the primary mode of communication is via chemicals, the evidence for multi-modal communication notwithstanding (e.g., Ravi et al. 2016). Arthropods may well be predisposed to employing chemical signals due to their small size. As such, their ability to transmit sound or visual signals should be limited (Greenfield 2002) and thus favour chemical communication as discussed in the following paragraphs.

A way to think about chemical signals is to compare them with other modalities, for example, sound (see Table 10.1 for a comparison of different signal modalities). Next follows some comparisons.

10.2.1 Specificity

Chemical signals can be highly specific. A chemical structure may be exact and confer an exact meaning. It is also possible to employ different combinations of chemical signals to confer different meanings; the number of permutations from such combinations could be large. It can be seen that many insects release pheromones that are blends of several components, which should increase their signal-to-noise ratio (Linn and Roelofs 1989). Sound signals may also be highly specific in relation to frequency, tempo, etc. (see Chapter 12). Specificity should be important with regard to communication regarding reproduction (e.g., sex pheromones), although specificity might not be a high-priority aspect for all signals. For example, for alarm pheromones, speed might trump specificity (i.e. one must act

Table 10.1 Characteristics of different modes of communication signals.

	Chemo	Photo	Sound	Tactile
Speed	+	+++	++	+++
Specificity	+++	+++	+++	+++
Production cost	+	+	++	+
Range	+++	++	+++	+
Directionality	+	++	+++	+++
Persistence	+++	+	+	+
Susceptibility to eavesdropping	+++	++	++	+
Detectability	+++	+	++	+
Environmental constraints	Air turbulence, barriers	Most are only effective by day or night	Barriers, transmission medium	Can be prevented by barriers
Some representative orders	All orders	Diurnal—Lepidoptera Nocturnal—Coleoptera	Orthoptera, Diptera, Isoptera	Araneae, Hymenoptera, Isoptera

quickly to avoid predator attacks regardless of the threat) and here there are obvious advantages to sound (see later).

10.2.2 Cost

The cost to producing chemical signals is thought to be relatively low, especially if precursors of such odours are produced for other purposes, e.g., when signallers employ secondary metabolites. On the other hand, the allocation of resources to odour-producing structures may be significant. The production of sound is likely more energetically costly than for chemicals and then, again, allocation of resources to sound-producing organs could be very high.

10.2.3 Directionality

The ability to control direction of propagated signals is much greater for sound (or light) than for odours. Also, for the odour plumes that emanate from an emitter, their shape, and directionality are greatly affected by environmental conditional, including perch size, height wind, and objects in the environment (e.g., trees that make up a forest—Murlis et al. 1992). By their very nature, sound being propagated by waves, carries information in a directional manner.

Odour disperses from its source, but is often greatly impacted by air turbulence (Murlis et al. 1992). As such, an insect flying through the air is likely to encounter bursts of odour packets. Once an odour plume containing such odour bursts has been located, a flying or walking insect may locate the odour source by moving upwind in a zigzag pattern. By employing these search strategies, insects can communicate readily over long distance via this modality (Carde and Willis 2008). Similarly, by use of binaural information, insects can readily locate emitters of sounds. Pollack (2000) cites the case of male grasshoppers wherein the small, binaural intensity difference of only 1.5 db is sufficient to orientate them to calling females with very high accuracy. We can conclude that the location of both odour and sound sources does not pose a severe constraint on communication.

10.2.4 Speed

Depending upon distance, it can take considerable time for an odour to reach its recipient and this can be problematic. Consider an alarm pheromone versus a vocal alarm call, where time may be of the essence. The movement of the pheromone (gas) occurs at approximately 1 cm/sec, whereas sound, in air, moves at approximately 300 m/sec, a difference of several orders of magnitude. Of course, in a very small space, absolute speed might not matter. Take, for example, a small cluster of aphids on a leaf, all of whom might be spaced within a centimetre or less from one another. By the time a predator has elicited release of the pheromone and consumed its prey, the other members of the group would probably have received the signal and abandoned the site.

10.2.5 Persistence

Chemical signals are likely to endure in the environment much longer than sound or light signals. This can be particularly advantageous when such signals confer an advantage over time. Think about a trail-marking pheromone that is used by foraging ants to recruit others to a foodstuff. Of course, longest need not be best. There should be an optimal persistence time such that the trail mark will no longer recruit others once a food resource has been depleted. This problem will be discussed later in reference to oviposition-marking pheromones, which can vary greatly in half-life.

10.2.6 Susceptibility to eavesdropping

It can be expected that the longer that a signal persists in the environment, the more likely that it is exploited by others for nefarious purposes relative to the intent of the signaller. Many of the signals discussed in this chapter could be described as public information, i.e. they are released into the environment with no specific individual in mind. Visual, auditory, and chemical signals could all be released as public information and thus would be susceptible to eavesdropping (and they are); however, as noted above, chemicals often persist longest and would thus be most susceptible for exploitation by enemies.

10.2.7 Physical and energetic limits

It is physically and energetically difficult for organisms to produce sound waves much larger than their body (Reinhard 2004; see Chapter 12). More specifically, the lower the frequency, the larger size of the sound-producing organ required for efficient sound dissemination (Cocroft and de Luca 2005). Such organs will scale to body size, although not necessarily in a linear fashion (i.e. an **allometric** relationship); however, there will a limit to organ size as a function of body size and competition for space from other organs in the insect's body. A further complication is that low frequency sounds travel more readily through the environment because they are less readily absorbed and deflected. Unfortunately, insect body size is limited by the ability to move oxygen through the trachea. Thus, most arthropods are constrained from producing low-frequency sounds due to small body size (but not crickets see Kostarakos and Romer 2010). To produce high-frequency sounds, however, requires muscles to twitch at extremely high rates, which again may be physically constraining for arthropods. Examples of evolutionary work-arounds for this problem involve modulation via stridulation. In other words, a single moderate-frequency twitch that causes a run along a leg comb can generate a high-frequency sound. Again, however, resources must be allocated to such structures and may compromise morphology.

By contrast with sound, any constraint on production of pheromones is obviated by the fact that such odours can be perceived at tiny quantities. For example, the cabbage looper, *Trichoplusiani*, can detect sex pheromone at concentrations of 10 molecules in a cubic millimetre of air. Nevertheless, there is some evidence for a positive correlation between body size and primary sex pheromone components in moths (Jaffe et al. 2007). Thus, larger moths release more pheromone, but it is not clear if this relationship is linear or decelerating.

10.2.8 Chemical diversity

Arthropod pheromones rarely if ever comprise a single chemical structure, but rather, they are made of blends of various components. These components are drawn from a broad range of structural classes, including hydrocarbons, ketones, aldehydes, and amines to name just some (see www. pherobase. com). Most components are small and have low molecular weights, well under 500 g/mol. By varying blend ratios instead of producing components *de novo*, new species-specific pheromones can arise relatively easily. Thus, the few-thousand pheromone components that have been identified to date may be combined to produce exponentially more blends and, as such, should not constrain pheromone evolution (see Symonds and Elgar 2008, for an excellent review on this topic). The diversity of arthropod pheromones is impressive, but as noted previously, not surprising.

10.3 The detection versus reliability problem

The discussion that immediately follows, borrows from **kairomone**-based work on parasitoids. Recall, that it was stated earlier that when a parasitoid or predator locates a host or prey by identity cues (e.g., odour) that this is not communication. Nevertheless, many of the same physical and physiological principles hold, whether there is overt signalling or exploitation of host odours. This is something that Peter Price et al. (1997) referred to as the body odour problem.

Vet and Dicke (1992) made the important point that insect parasitoids face the problem of detection versus reliability. In their scenario, plants that herbivores feed on are relatively easy to detect, but such plants are not necessarily reliable indicators that the appropriate insect herbivores (the hosts for the parasitoid) are present. On the other hand, based upon their chemical signatures, direct information on the identity of an insect host is quite reliable. However, herbivores have been favoured by natural selection to make themselves difficult to detect. We can extend the argument of reliability-detection to communication in that such engagement between or among tiny, hidden individuals will be difficult to achieve unless their signals are strong and obvious, e.g., broadcast release of a sex pheromone with a high signal-to-noise ratio. Of course, the solution to this problem only makes the releaser more detectable to their predators, parasites or competitors. In fact, there are examples in the literature of parasitoids using their host's sex pheromones to locate them (Fatouros et al. 2008). As such, a further solution is required to ameliorate this new problem, for example signallers might call in groups to dilute the threat to any one individual (Hamilton 1971). The point here is that there are significant trade-offs when signals are public.

Insights into the detection versus reliability problem can be gained by considering trail or recruitment pheromones. Ants, termites and some caterpillars apply pheromones to substrates that can act as a recruiting mechanism. Ant **trail pheromones** are just that, species-specific compounds, and thus reliable indicators of their propagators. Trail pheromones can endure for considerable lengths of time and thus have the potential to also be highly detectable. Of course, species-specific in propagation need not mean species-specific in recognition. In fact, a recent study by Binz et al. (2014) demonstrated interspecific recognition of trail pheromones and odour footprints, i.e. body odours inadvertently left behind by foraging ants. Interestingly, subordinate species of ants use the odours to avoid confrontation with their superiors whereas dominant ant species are attracted to subordinates presumably to defend their territories from intruders. This is an excellent example of contextual **chemical ecology**, wherein the response to an odour depends upon context under which it was received.

10.4 Chemical compounds as mediators of conflict and resolution

A conflict occurs when two individuals have opposing needs; for example, optima for frequency of mating are often in conflict between males and females. How conflicts get resolved in biological systems has intrigued behavioral ecologists for some time. This chapter will concentrate on chemically mediated conflicts and chemically mediated resolutions.

A classic conflict occurs in many hymenopteran colonies (e.g., ants) between workers and their queen. Hymenopterans are **haplo-diploid**, wherein unfertilized eggs develop into haploid males and fertilized eggs develop into diploid females. As a result, within colonies where the queen has mated just once, worker sisters will be related by 75 per cent to one another (100 per cent genetic contribution from their haploid father and 50 per cent from their diploid mother), but only 25 per cent to their brothers (no contribution from the father). By contrast, the queen is equally related by 50 per cent to each of her children. As such, she should favour equal investment in offspring by gender, whereas her female offspring should favour 3:1 investment in daughters versus sons. These ideas were first formulated by Hamilton (1964) in his **inclusive fitness** theory and expanded by Trivers and Hare (1976). Further analysis of this theory leads to the conclusion that workers should favour a much higher investment into sexuals versus maintenance of the colony, compared with the queen (Pamilo 1991). Social insects are well known for their ability to cooperate, yet as pointed out by Ratnieks et al. (2006), the same theory that explains **cooperation** also predicts **conflict**, mostly in a quantitative manner, i.e. optimal investments differ between the cooperating parties.

The previous paragraph deals with the functional or the 'why' question, i.e. why there might be a conflict between the queen and workers? How the conflict is resolved is more a 'how' or a causal question. In the case of the ant, *Aphaenogaster senilis*, the answer is chemical (pheromonal) mediation. Boulay et al. (2007) showed that queens may suppress production of sexuals through an as-yet unidentified contact pheromone, possibly released from the Dufour's or post-pharyngeal glands. The authors used a clever single- and double-mesh queen-separation design to eliminate the possibility that the pheromone is volatile. When all contact between the queen and her workers is prevented by the double mesh, suppression is lost, but only partly, so when a single mesh is used that would allow limited contact. Furthermore, adding context to the problem, when colonies attain large size, queen suppression also declines, probably because of dilution of the suppression pheromone. Boulay et al. (2007) speculated that queens tend to overproduce the suppression pheromone because their sexual production optimum is much lower and in conflict with that of the workers.

Another form of conflict concerns mating (see Chapter 13). It has been hypothesized and demonstrated that the optimal mating rate often differs between males and females; males are most often selected to maximize mating quantity, whereas females are most often selected to maximize quality. Put simply, most females are limited in the number of offspring that they can produce (via eggs) and males are generally much less limited in the number of offspring that they might sire (via sperm; but see Boivin 2013, for exceptions in some hymenoptera). If there is a cost to mating (e.g., increased rates of death or injury from predators while mating), we might expect that females would gain much less overall from multiple matings than would males because females are egg limited. As such, one can calculate sex-specific mating frequency optima, which should differ based upon the arguments mentioned previously. This difference in optima generates a mating antagonism between the sexes (see Parker 1979). This is particularly true when mating has physical consequences. Nowhere is this issue more obvious than in the bed bug, *Cimex lectularius*, where mating occurs via **traumatic insemination**, wherein males pierce their partner's abdomen and ejaculate into the haemocoel. To mitigate possible serious injury from traumatic insemination, females have evolved a morphology that lessens the impact (Reinhardt and Siva-Jothy 2007). However, as noted previously, males in general are often selected to mate as frequently (read: indiscriminately) as possible and, here, males attempt to mate (traumatically) with any large individual, which could include other males—gender recognition does not occur until after abdominal piercing in bed bugs. Of course, the mountee gains no benefits, only costs from these piercings and they have apparently evolved a mechanism to reduce such occurrences. Males, but never females, release alarm pheromones when mounted by another male. Would this chemical communication reduce traumatic interactions? Ryne (2009) employed a clever experimental design wherein nail polish was applied to the alarm pheromone glands, effectively turning them off and rendering males female-like, and then compared them with sham-operated males. Ryne also created male-like females by releasing alarm pheromone during a mounting attempt. The results confirmed that alarm pheromone released by in male–male interactions reduce traumatic piercing.

Sexual antagonism can be expressed in many different manners. For example, it may be in the interest of males to prevent their female mates from acquiring sperm from competitors (recall that females may benefit from multiple matings, but usually at a lower rate than males). In some *Drosophila*, males transfer pheromones to females that signals them non-virgins and thus less attractive to mate-seeking males (Kurtovic et al. 2007).

10.5 Honest signals

Signals may be used to inform recipients of various aspects (e.g., quality) of the sender. When such signals are true indicators of the sender, they are referred to as honest signals. For example, in the sagebrush cricket, *Cyphoderris strepitans*, males sing to attract females. However, whether a female accepts a male for mating depends upon the honest chemical signal of male size, cuticular hydrocarbons (Steiger et al. 2013).

How and why honest signals evolve has been a matter of discussion in behavioral ecology for some time. However, it is best to provide a functional context. If both parties had a common interest, it would be easy to imagine that they would communicate honestly. If, however, their interests conflict (e.g., male mating frequency versus female mate quality), one might question the conditions under which honest signals would evolve. The key criterion, it seems is that there must be some cost associated with deceptive signalling that leads to negative net benefits, i.e. it pays to be honest (see Higham 2014 for a cogent discussion on this topic). For example, small individuals cannot pay the cost of advertising themselves as large ones, but large individuals can. Thus, high cost alone should not be enough to maintain honest signals, but rather relative high cost and net benefits (see discussions on Zahavi's (1975) handicap principle).

To better understand honest signals in an ecological context, examples from the author's own work on marking pheromones will be used. Many insects, including herbivores and parasitoids, place a chemical mark on their hosts either during or following oviposition. Frequently, but not always, conspecifics or even the same individual that re-encounters such marked hosts will reject them and search for others. The 'why' explanation for this rejection or host discrimination is that the eggs already present in infected hosts will eclose and act as competitors for the focal forager's offspring thus lowering the quality of such hosts. Hence, the insect mother is usually better off searching for uninfected/unparasitized hosts, where competition for her offspring is absent. This would particularly be the case when older hosts are superior competitors and competition is intense, which is often the case in parasitoid larvae. There will, however, be some conditions under which it is adaptive to ignore the oviposition signal and oviposit into/onto already parasitized hosts, for example, when hosts are exceedingly rare (van Alphen and Visser 1990).

One question arises from the description above. Why would females publically, honestly mark their hosts with a message that says: My (older, competitively superior) baby is already present, please look elsewhere? How could it be adaptive for mothers to expend time and energy to aid others in their foraging decisions? The answers lies in applying ecological context. Marc Mangel and the author showed that, for insects that engage in success-motivated local search, there is a high net pay-off if foragers are likely to re-encounter hosts they have been previously parasitized and thus avoid sib–sib offspring competition, even if such public information benefits others (Roitberg and Mangel 1988). Secondly, although older individuals are normally competitively superior, superiority is rarely 100 per cent, and so it might pay to warn others in order to avoid offspring competition. What if re-encounter was low and superiority approached unity, would females then act so apparently altruistically? Work with former student Judy Nelson on the leaf miner parasitoid, *Opius dimidiatus*, showed that the putative host mark decayed very quickly, in less than 3 hours (Nelson and Roitberg 1993). Why provide public information that lasts such a short time? Nelson showed that after 3 hours (the better part of a foraging day) re-encounters were unlikely, thus negating the net benefit of avoiding sib–sib competition. Even more intriguing is the aspect of competitive advantage. In *O. dimidiatus*, within 24 hours, the first instar larva ecloses as a fully mandibulate individual; any subsequent egg that is laid in an already

parasitized host is almost certain to die before hatching. The answer to the question raised previously is 'Do not release honest public signals unless you benefit and clearly, after eclosion, there is little or no benefit from warning others'. Mangel and Roitberg used this help-yourself first principle to calculate optimal decay rates for marking pheromones.

A further complication to the marking pheromone story regards an added cost from production of these private/public marking pheromones. In some communities, parasitoids or hyperparasitoids have evolved the ability to read the marks of their hosts. Imagine the advantage that accrues to a forager whose host's eggs are buried within hosts (recall the detectability–reliability problem) if she can determine that an ovipositing host has recently visited a site. Colleagues Bob Lalonde and Thomas Hoffmeister and the author have shown that in the wasp, *Halticoptera rosae* (Figure 10.1a), the egg/first instar parasitoid of the rose hip fly, *Rhagoletis basiola* (Figure 10.1b), uses the fly's oviposition marks to locate eggs hidden beneath the skin of rose hips (Hoffmeister et al. 2000).

Figure 10.2 shows that the longer that the wasp searches on a fruit the more likely it is that an *R. basiola* egg will be parasitized, but this occurs at a higher rate on a fly-marked fruit than on an unmarked one.

This evolutionary cost from eavesdropping must add to the physical cost of employing oviposition marks, and should also select for faster decay rates. The story becomes even more intriguing and complicated in that Hoffmeister and Roitberg showed that *H. rosae* also employs its own marking pheromone to avoid sib–sib offspring competition within *R. basiola* hosts, not surprising given its own success-motivated search tactics. An even more complicated layer to this co-evolutionary cake is that the fly, *R. basiola*, seems to have evolved the ability to read *H. rosae's* marking pheromone and adjust its oviposition patterns (Hoffmeister and Roitberg 1997). To summarize, *R. basiola* host marks to reduce competition within family and within conspecifics, but its honest signal is exploited by its enemy, *H. rosae. H. rosae* also employs its own oviposition marks and its public signal is, in turn, exploited by its prey.

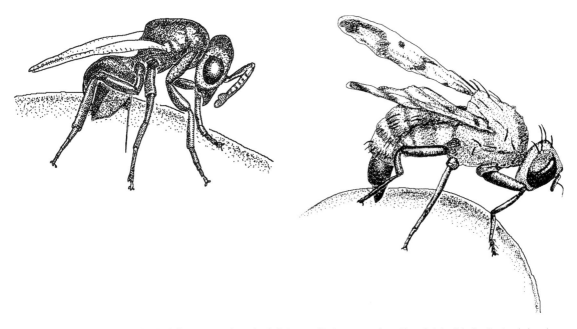

Figure 10.1 *Halticoptera rosae* (on the left) ovipositing through a hole in a rosehip into an egg from *Rhagoletis basiola* that lies just below the surface of the fruit. This tiny hole was created by the ovipositing fly (on the right) who then pheromone marks the rosehip fruit.

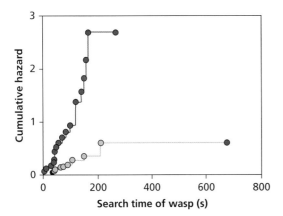

Figure 10.2 The search time-dependent hazard that a host egg will be detected by a searching parasitoid wasp, expressed as cumulative hazard functions of the Cox proportional hazard model for hosts on host-marking pheromone-marked rosehips (open circles) and unmarked fruits (filled circles). Note that the hazard function is not a probability, but gives the death rate of host eggs per unit of parasitoid search time, and thus need not be less than unity. The hazard function thus gives information on how likely it is that a host will be detected at time *t*, given it was not detected until time *t* (Norusis 1994). The circles represent the cumulative hazard at the search times found in the first experiment for the respective treatments.

Source: Hoffmeister et al. (2000).

10.6 Deceptive signals

If honest signals are true indicators of signaller quality or ability then deceptive signals must be the inverse. Deception can take many forms, from false eyespots (e.g., owl butterflies—Nymphalidae), camouflage by covering oneself with debris (e.g., some assassin bug nymphs), and sound mimicry [e.g., some (innocuous) syrphids mimic the sounds of (dangerous) vespids].

One of most intriguing forms of deception that sits in the chemical realm can be coined **sexual deception**. In a large number of orchids found through Europe, Australia, South America, and Asia, non-rewarding orchids signal availability to pollinating bees without providing reproductive pay-offs to the recipient. This deception is remarkable for at least two reasons:

- *Frequency*: currently, at least eleven orchid genera are known to sexually deceive and an even greater number (by far) practice food deception (Gaskett 2011).
- *Specificity* of this deceit.

In the latter case, the orchid exquisitely mimics the size, shape, and odour of a female pollinator thereby attracting a male to the flower. Subsequent attempts to mate with the pseudo-female causes a transfer of pollen to the orchid with no reward to the deceived male. The odour deceit in this case is the mimicking of the pollinator's sex pheromone.

Two fascinating questions arise from the sexual deception problem:

- How does a plant synthesize an animal's odours for its benefit?
- How can deceit persist over evolutionary time?

The answer to the first question comes from Ayasse et al. (2011), who argued that most of the raw materials required to synthesize the sex pheromone already exist within the plant's chemical milieu (e.g., its acetate pool) and it is just a matter of fine-tuning these pre-adapted enzymes for conversion, rather than creating enzymes *de novo*.

Answering the second question is a bit more complicated. First, in order for this question to be non-trivial, it is necessary to show that there is a cost to the recipient from being deceived. Gaskett et al. (2008) showed that males spend time, energy, and copious amounts of sperm during mating attempts at false mates. One might question why males have not evolved discrimination behaviors. First, recall a point made earlier that in many species, male reproductive success is based upon mating frequency. As such, we should expect selection against innate mate discrimination, i.e. it pays not to be too choosy. If, however, males could learn to discriminate against non-rewarding orchids that might feed into a co-evolutionary arms race, wherein orchids would be selected to provide better and better mate mimics over time. In fact, some pollinators have experimentally been shown to learn to avoid orchid patches where they were sexually deceived via the use of dummy wasps (plastic beads) coated with synthetic sex pheromone (chiloglottone—Schiestl et al. 2003; Whitehead and Peakall 2013). The learned response was, however, short-lived (< 24 hours) and wasps could readily be recaptured in the vicinity for several days, post-deception. From the plant's perspective this short-term avoidance can be thought of as a useful tool for promoting outcrossing for these self-incompatible plants.

Returning to the pollinator's perspective, however, it still appears that there should be costs to being deceived so it is not clear why innate or learned discrimination is not more evident. Aside from the sexual selection context discussed previously is the notion of how to measure fitness in these deception systems. One can imagine that female wasps in competition with deceitful orchids will suffer reduced fertilization rates and that lineages without discrimination would disappear over time. This notion, however, neglects the fact that nearly all pollinators in sexually deceptive orchid systems are haplo-diploid (Gaskett et al. 2008) thus, virgin females will still reproduce, but will be restricted to producing sons only. As such, sex ratios will be skewed toward males, thereby providing sufficient males to fertilize females and pollinate orchids.

10.7 Chemical communication and higher-order processes

In Section 10.6, the fact that chemical communication could impact dynamics at levels of organization above that of individual interactions is alluded to, e.g., at the population or the community level. This final section focuses on concepts and examples from where this might be most evident—the chemical ecology of fear.

The population dynamics of prey can be impacted both directly and indirectly by their natural enemies. Removal of prey via predation, referred to as consumptive effects (CE's) has an obvious and direct impact on population numbers. Less obvious are non-consumptive effects (NCE's). Here, the very presence of natural enemies can cause prey to alter their phenotype (e.g., reduce rates of grazing by hiding more frequently), and thus impact birth and/or death rates. It has been argued that NCE's have great potential to impact dynamics, but are often overlooked because of the obvious parallel culling of prey by predators. Peckarsky et al. (2008) suggest that poor reproductive performance of snowshoe hare during the low predator phase of the snowshoe hare–lynx cycle could be due to lag effects (via stress hormones) from earlier periods of high predator density. This phenomenon of lagged response is essential to generate population cycles (Myers and Cory 2013). Furthermore, in keeping with the theme in this section, NCE's have the potential to generate trophic cascades, wherein the presence of a top predator impacts the dynamics two or more trophic levels below by altering traits of an intermediate organism. For example, in a classic study, Oz Schmitz et al. (1997) demonstrated that the presence of spiders (even with their jaws glued) caused a significant change in grasshopper feeding time budgets, as well as in food choice and, as such, spiders had a positive impact on grass biomass.

How might chemical communication factor into NCE's and trophic cascades? There are several means by which this might occur. First, predator odour might be perceived by prey, who then alter their trait expression accordingly. This phenomenon would not fit into the earlier definition of communication-mediated effects if the sender (the predator) did not intentionally signal its presence. In fact, predators should be selected to be stealthy, all else being equal. There is, however, another scenario in which intent is pervasive, alarm calls and, in particular,

alarm pheromones. Alarm pheromones can be found in a variety of arthropods including aphids, ants, honey bees, stink bugs, and thrips among others. In a number of aphid species, (E)–β–farnesene is released from the cornicles of individuals who have been attacked by predators or parasitoids (Edwards et al. 1973). The benefit that the signaller receives is via inclusive fitness, wherein nearby, identical clone-mates are warned of imminent danger. Upon receipt of the alarm signal, aphids respond in variety of manners from running, to dropping from the host plant, to waggling their abdomens, all of which are thought to increase probability of survival. Of course, alarm pheromones are public information and thus susceptible to eavesdropping, especially by aphid predators, although the degree to which that happens is up for debate (see Vosteen et al. 2016).

Might aphid alarm pheromones act as a causal agent in NCE's? Given that responders move to new feeding sites in response to alarm, that seems

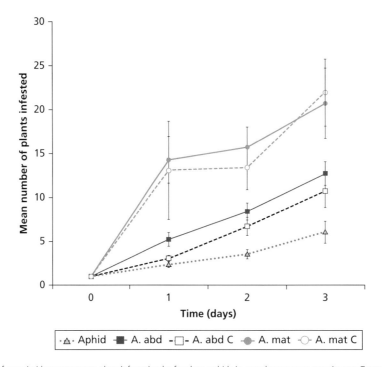

Figure 10.3 Effect of parasitoid treatments on plant infestation by foxglove aphids in caged mesocosm experiments. Treatments were (parasitoid 1) *Aphelinus abdominalis* (A. abd), (parasitoid 2) *Aphidius matricariae* (A. mat), aphids only (A), or cue treatments that simulate either *A. abdominalis* (A. abd C) or *A. matricariae* (A. mat C).

Source: Henry et al. (2010).

like a reasonable conjecture. With former students Lee Henry and Jordan Bannerman, and colleague Dave Gillespie, this possibility has been investigated using the foxglove aphid, *Aulacorthum solani*, and two parasitoid species, each with different foraging tactics (Henry et al. 2010). The first parasitoid (*Aphidius matricariae*) is aggressive and elicits a strong response from aphids, while the second (*Aphelinus abdominalis*) is furtive and elicits a weak response from aphid colonies. This contrast provided an ecological context for the possible impact of NCE's. Henry and Bannerman simulated parasitoid attacks using insect pins and in the case of *A. matricariae*, release of alarm pheromone. A further NCE control was the use of male parasitoids that would forage on plants, but never attack aphids. Results from the experiments showed the simulated attacks caused aphids to disperse in a very similar manner to that caused by the female parasitoids and that the more aggressive parasitoid caused greater dispersal (Figure 10.3).

Why is this important? Because aphids and, in particular, foxglove aphids can generate qualitative damage to plants as opposed to simple quantitative effects, such as those generated by grazers (e.g., grasshoppers). Thus, the treatment with parasitoids or parasitoid-disturbance cues led to more plants infested and damaged than the treatment with aphids alone a somewhat counterintuitive result. Thus, we can conclude that trophic cascades can be generated by chemical communication among prey, but that the end result (positive or negative) might depend upon the identity of the top predator, as well as the form of damage that the meso-predator, in this case the aphid herbivore, inflicts on its host plant.

The notion of NCE's and trophic cascades may be important in the field of biological control, where one employs natural enemies to maintain pest insects below some economically-determined threshold. If one simply prescribes release rates of natural enemies from measuring consumption in the laboratory, they may err greatly in the number of enemies they need to release in nature.

10.8 Conclusions

This chapter has attempted to blend a number of concepts and examples to illustrate how insects communicate via **semiochemicals**. The important take-home message is that the impacts from chemical communication are often contextual and not always intuitive. However, a functional approach to studying arthropod chemical communication can be powerful and insightful.

References

Ayasse, M., Stokl, J., and Francke, W. (2011). Chemical ecology and pollinator-driven speciation in sexually deceptive orchids. *Phytochemistry*, **72**, 1667.

Binz, H., Foitzik, S., Staab, F., and Menzel, F. (2014). The chemistry of competition: exploitation of heterospecific cues depends on the dominance rank in the community. *Animal Behaviour*, **94**, 45.

Boivin, G. (2013). Sperm as a limiting factor in mating success in Hymenoptera parasitoids. *Entomologia Experimentalis et Applicata*, **146**, 149–55.

Boulay, R., Hefetz, A., Cerdá, X., et al. (2007). Production of sexuals in a fission-performing ant: dual effects of queen pheromones and colony size. *Behavioural Ecology and Sociobiology*, **61**, 1531.

Carde R.T., and Willis M.A. (2008). Navigational strategies used by insects to find distant, windborne sources of odor. *Journal of Chemical Ecology*, **34**, 854.

Cocroft, R.B., and de Luca, P. (2005). Size-frequency relationships in insect vibatory signals. In: S. Drosopoulos and M.F. Claridge (Eds), *Insect Sounds and Communication: Physiology, Behaviour, Ecology, and Evolution*, pp. 99–110. Taylor and Francis, Abingdon.

Edwards, L., Siddall, J., Dunham, L., Uden, P., and Kislow, C. (1973). Trans-β-farnesene, alarm pheromone of the green peach aphid, *Myzus persicae* Sulzer. *Nature*, **241**, 126.gy.

Fatouros, N.E. Dicke, M., Mumm, R., Meiners, T., and Hilker, M. (2008). Foraging behavior of egg parasitoids exploiting chemical information. Behavioural. *Ecology*, **19**, 677.

Font, E., and Carazo, P. (2010). Animals in translation: why there is meaning (but probably no message) in animal communication. *Animal Behaviour*, **80**, e1.

Gaskett, A.C. (2011). Orchid pollination by sexual deception: pollinator perspectives. *Biology Reviews*, **86**, 33.

Gaskett, A.C., Winnick, C.G., and Herberstein, M.E. (2008). Orchid sexual deceit provokes ejaculation. *American Naturalist*, **171**, E206.

Greenfield, M.D. (2002). *Signalers and Receivers: Mechanisms and Evolution of Arthropod Communication*. Oxford University Press, Oxford.

Hamilton, W.D. (1964). The genetical evolution of social behaviour, parts I and II. *Journal of Theoretical Biology*, **7**, 1.

Hamilton, W.D. (1971). Geometry for the selfish herd. *Journal of Theoretical Biology*, **31**, 295.

Henry, L.M., Bannerman, J.A., Gillespie, D.R., and Roitberg, B. (2010). Predator identity and the nature and strength of food web interactions. *Journal of Animal Ecology*, **79**, 1164.

Higham, J.P. (2014). How does honest costly signaling work? *Behavioral Ecology*, **25**, 8.

Hoffmeister, T.S., and Roitberg, B.D. (1997). Counterespionage in an insect herbivore-parasitoid system. *Naturwissenschaften*, **84**, 117.

Hoffmeister, T.S., Roitberg, B.D., and Lalonde, R.G. (2000). Catching Ariadne by her thread: how a parasitoid exploits the herbivore's marking trails to locate its host. *Entomologia Experimentalis et Applicata*, **95**, 77.

Jaffe, K., Mirás, B., and Cabrera, A. (2007). Mate selection in the moth *Neoleucinodes elegantalis*: evidence for a supernormal chemical stimulus in sexual attraction. *Animal Behaviour*, **73**(4), 727–34.

Kostarakos, K., & Römer, H. (2010). Sound transmission and directional hearing in field crickets: neurophysiological studies outdoors. *Journal of Comparative Physiology A*, **196**(9), 669–81.

Kurtovic A., Widmer A., and Dickson, B.J. (2007). A single class of olfactory neurons mediates behavioural responses to a *Drosophila* sex pheromone. *Nature*, **446**, 542.

Linn, C.E., and Roelofs, W.L. (1989). Response specificity of male moths to multicomponent pheromones. *Chemical Senses*, **14**(3), 421–37.

Maynard Smith, J., and Harper, D.G.C. (2003). *Animal Signals*. Oxford University Press, Oxford.

Murlis, J., Elkinton, J.S., and Carde, R.T. (1992). Odor plumes and how insects use them. *Annual Review of Entomology*, **37**, 505.

Myers, J.H., and Cory, J.S. (2013). Population cycles in forest lepidoptera revisited. *Annual Review of Ecology, Evolution, and Systematics*, **44**, 565.lo.

Nelson, J.M., and Roitberg, B. (1993). Factors governing host discrimination by *Opius dimidiatus* (Ashmead) (Hymenoptera: Braconidae). *Journal of Insect Behavior*, **6**, 13.

Norusis, M.J. (1994). *SPSS Professional Statistics 6.1*. SPSS Incorporated, Chicago, IL.

Owren, M.J, Rendall, D., and Ryan, M.J. (2010). Redefining animal signaling: influence versus information in communication. *Biology and Philosophy*, **25**, 755.

Pamilo, P. (1991). Evolution of colony characteristics in social insects. 1. Sex allocation. *American Naturalist*, **137**, 83.

Parker, G.A. (1979). Sexual selection and sexual conflict. In: M.S. Blum and N. A. Blum (Eds) *Sexual Selection and Reproductive Competition in Insects*, pp. 123–66. Academic Press, New York, NY.

Peckarsky, B.I., Abrams, P.A., Bolnick, D.I., et al. (2008). Revisiting the classics: considering nonconsumptive effects in textbook examples of predator–prey interactions. *Ecology*, **89**, 2416.

Pollack, G. (2000). Who, what, where? Recognition and localization of acoustic signals by insects. *Current Opinion in Neurobiology*, **10**, 763.

Price, P.W., Roininen, H., and Carr, T. (1997). Landscape dynamics, plant architecture and demography, and the response of herbivores. In: K. Dettner, G. Bauer, and W. Völkj (Eds), *Vertical Food Web Interactions*, pp. 319–33. Springer, Berlin.

Ratnieks, F.L.W., Foster, K.R., and Wenseleers, T. (2006). Conflict resolution in insect societies. *Annual Review of Entomology*, **51**, 581.

Ravi. S., Garcia, J.E., Wang, C., and Dyer, A.G. (2016). The answer is blowing in the wind: free-flying honeybees can integrate visual and mechano-sensory inputs for making complex foraging decisions. *Journal of Experimental Biology*, **219**, 3465.

Reinhard, J. (2004). Insect chemical communication. *ChemoSense*, **6**, 1.

Reinhardt, K., and Siva-Jothy, M.T. (2007). Biology of the bed bugs (Cimicidae). *Annual Review of Entomology*, **52**, 351.

Roitberg, B., and Mangel, M. (1988). On the evolutionary ecology of marking pheromones. *Evolutionary Ecology*, **2**, 289.eco

Ryne, C. (2009). Homosexual interactions in bed bugs: alarm pheromones as male recognition signals. *Animal Behaviour*, **78**, 1471.

Schiestl, F.P., Peakall, R., Mant, J. G., et al. (2003). The chemistry of sexual deception in an orchid-wasp pollination system. *Science*, **302**, 437.

Schmitz, O.J., Beckerman, A.P., and O'Brien, K.M. (1997). Behaviorally mediated trophic cascades: effects of predation risk on food web interactions. *Ecology*, **78**, 1388.

Scott-Phillips, T.C. (2008). Defining biological communication. *Journal of Evolutionary Biology*, **21**, 387.

Steiger, S., Ower, G.D., Stökl, J., et al. (2013). Sexual selection on cuticular hydrocarbons of male sagebrush crickets in the wild. *Proceedings of the Royal Society B*, **280**, 1773.

Symonds, R.E. and Elgar, M. (2008). The evolution of pheromone diversity. *Trends in Ecology and Evolution*, **23**, 220.

Trivers, R.L., and Hare, H. (1976). Haplodiploidy and the evolution of the social insects. *Science*, **191**, 249.

van Alphen, J.J.M., and Visser, M.E. (1990). Superparasitism as an adaptive strategy for insect parasitoids. *Annual Review of Entomology*, **35**, 59.

Vet, L.E.M., and Dicke, M. (1992). Ecology of infochemical use by natural enemies in a tritrophic context. *Annual Review of Entomology*, **37**, 141.

Vosteen, I., Weisser, W.W., and Kunert, G. (2016). Is there any evidence that aphid alarm pheromones work as prey and host finding kairomones for natural enemies? *Ecological Entomology*, **41**(1), 1–12.

Whitehead, M.R., and Peakall, R. (2013). Short-term but not long-term patch avoidance in an orchid-pollinating solitary wasp. *Behavioural Ecology*, **24**, 162.

Zahavi A. (1975). Mate selection-a selection for a handicap. *Journal of Theoretical Biology*, **53**, 205.

CHAPTER 11

Visual communication

James C. O'Hanlon[1], Thomas E. White[2], and Kate D.L. Umbers[3,4]

[1] School of Environmental and Rural Science, University of New England, Armidale, Australia
[2] School of Life and Environmental Sciences, The University of Sydney, Sydney, New South Wales, 2006, Australia
[3] School of Science and Health Western Sydney University, Hawkesbury, Richmond, NSW, Australia
[4] Hawkesbury Institute for the Environment, Western Sydney University, Hawkesbury, Richmond, NSW, Australia

11.1 Introduction

In a rainforest in Southeast Asia, a worker bee leaves her hive and flies in search of flowers. Dotted throughout the forest landscape, flowers shine in bright **hues**, signalling their presence to flying insects that will collect and transfer pollen. With each flower visit, the bee is rewarded with rich nectar.

As she buzzes through the understory, she perceives the world around her as a complex array of reflected light in varying intensities, hues, and shades. Her eyes are honed to see a wide array of colours, perfect for detecting flowers. She senses the movement of leaves blowing in the wind and the terrain passing her by. She navigates around obstacles and over the undulating terrain, while maintaining her bearings for the critical flight home. As the sun passes across the sky, as clouds pass, and as the bee flies from shade into open sun, the light environment changes together with the information the light contains. As she forages she remembers which flowers she has visited and avoids them in favour of new unexplored flowers. She begins to associate particular colours, shapes, and sizes with nectar rewards, learning to distinguish which flower

types are more profitable than others. The colours, shapes, and patterns of every object, the shadows they cast and the form of the landscape itself presents visual information that the bees sensory system must collect and process accurately so that she can fulfil her role of collecting food for the hive.

As the bee meanders along her route she detects a bright spot amongst the green vegetation. As she flies closer her eyes are able to take in more detail, the bright spot is revealed to have the symmetrical shape of a bright white flower. Flying in a series of arcs the bee carefully inspects the flower, gradually getting closer with each pass. She distinguishes white petals radiating outwards from the centre, but the bee has been deceived, for these are not flower petals, but the broad hind-legs of a patient orchid mantis (Figure 11.1).

The visual system of the orchid mantis has evolved for a very different function, to catch fast-moving prey. The large eyes stationed high up on her head give the mantis a wide panoramic view of the world. The mantis does not need to distinguish a varied palette of colours, instead her eyes are finely attuned to movement. Beneath her head the mantis'

O'Hanlon, J. C., White, T. E., and Umbers, K. D. L., *Visual communication.* In: *Insect Behavior: From mechanisms to ecological and evolutionary consequences.* Edited by Alex Córdoba-Aguilar, Daniel González-Tokman, and Isaac González-Santoyo: Oxford University Press (2018). © Oxford University Press.
DOI: 10.1093/oso/9780198797500.003.0011

Figure 11.1 The orchid mantis (*Hymenopus coronatus*). Photo: James O'Hanlon

raptorial forelimbs are poised ready to strike at any unsuspecting prey that may pass. This particular praying mantis will not have to wait very long. It has evolved a unique signalling strategy that lures in other animals as prey. It is not a dull green colour like so many of her close relatives; instead her exoskeleton is bright white—like a flower. She does not hide in ambush; instead her bright colours stand out like a beacon against the dark green vegetation on which she sits—an irresistible signal to her prey.

The mantis' eyes detect the contrast of a moving target against its background. Her body remains motionless while her head gently turns to follow the arcing flight path of the approaching bee. In a blur of motion the mantis lunges forward and ensnares the bee with the pointed spines of its forelimbs. With no chance of escape the bee is devoured and the mantis' deception is complete (O'Hanlon et al. 2014c).

How insects see the world around them has intrigued scientists for centuries. With the possible exception of primates, the eyes and optical neurons of model species, such as honey bees and fruit flies have been studied more than any other animal. The ability of insects to see things that we cannot, such as light in the **ultraviolet** spectrum or the polarization plane of light, has been the subject of much research and has highlighted the limitations of our own perception. Ongoing research continues to provide inspiration for the development of bioinspired robotics and remote sensing systems. While insect visual systems process information on colour, luminance, shape, pattern, polarization, and movement, insects can also emit their own visual signals and

---

Box 11.1 The exoskeleton: nature's most dynamic canvas

Probably due to their immense diversity, no group of organisms can change their body colour in as many different ways as the arthropods and the insects are, as usual, major contributors (Umbers et al. 2014). Although rarely as fast as famous colour changers such as cephalopods and chameleons, insects undergo all the major kinds of colour change: mechanistic, morphological, and physiological (Umbers et al. 2014), which can each be either reversible or irreversible. Mechanistic colour change is often very fast as it is achieved by the simple movement of body parts or reposturing. For example, mountain katydids lift their wings to reveal bright red, blue, and black abdomens (Umbers and Mappes 2015) and praying mantises rearing up on hind legs lift their wings to reveal conspicuous colour patches on their thoraxes, raptorial forelegs, and wings. Counting this as colour change is perhaps cheating, in that the colours are static and are simply hidden, then revealed, but then colour change *is* in the eye of the beholder. Morphological colour change, on the other hand, occurs when insects slowly build-up or break-down molecules that cause them to change colour over days to weeks. A famous example is in locusts that, when they transition from solitary to gregarious, change colour from generally cryptic green-brown, to black, orange, and yellow; colours classically associated with warning signals (Pener and Simpson 2009). Finally, physiological colour changes are rapid changes that involve the rapid rearrangement of pigments or structures and are usually reversible. There are many wonderful examples, including the Panamanian tortoise beetle, which has a switchable reflector that changes its colour from gold to red when touched (Vigneron et al. 2007), and the many damselflies and grasshoppers that turn from black to turquoise when their body temperatures increase (Veron 1974, Umbers 2011, Umbers et al. 2013). Colour change is not rare, but rather a fascinating, widespread adaptation found in many insects. The outstanding challenges in this field are to elucidate the mechanisms and functions of colour change to understand how colour change influences fitness.

cues, combining behavior with the structures and **pigments** of their exoskeletons (Box 11.1).

Communicating and interpreting information is essential for survival, but as the tale above shows it is not always perfect or honest. There are many cases of visual communication in insects where deceptive signallers exploit the sensory biases of other animals. Being the most diverse group of animals, it is not surprising that there are seemingly endless ways in which insects receive and transmit visual information. Rather than attempt an exhaustive review in this chapter we will provide an overview of the major adaptations involved in insect visual communication and the functions for which they have evolved. We first consider the physiology of visual sensory systems in insects, then examine how signals and cues are generated, propagated, and received amidst ubiquitous noise and varying environmental conditions. Finally, we discuss the evolution of visual communication in insects and its critical role in foraging, defence, mate choice, competition, and sociality, citing modern and classic examples.

11.2 Physiology: structure and optics of the compound eye

All insects with sight have compound eyes that, while functionally similar, bear little resemblance to our own lens eye. The basic modular structure of the compound eye has been specialized and modified across insect species, depending on their ecological niche. This adaptability has probably contributed to the success of insects as a group. The surface of the compound eye appears as a honeycomb pattern of repeated 'facets'. Each individual facet is the entry to a single light-sensing structure termed an **ommatidium** (plural: ommatidia). A compound eye is a cluster of these long, tube-like structures, radiating outwards so that each ommatidium points in a slightly different direction (Figure 11.2). The number of ommatidia making up a compound eye varies immensely between species; *Drosophila* fruit flies have around 750 individual facets per eye (Paulk et al. 2013), whereas honey bees, *Apis mellifera*, have around 4500 facets per eye (Srinivasan 2010). In the diminutive moth, *Stigmella microtheriella* each compound eye only contains 123 facets (Fischer et al. 2012), whereas certain dragonflies can have a whopping 30,000 facets per eye (Sherk 1978).

Figure 11.2 shows the general structure of ommatidia within a compound eye. On the outer end sits the transparent cornea and crystalline cone, which focuses incoming light onto the sensitive structures beneath. At the base of the ommatidia, the 'retinula' is made up of a ring of photoreceptor cells. Extending inwards from the photoreceptor

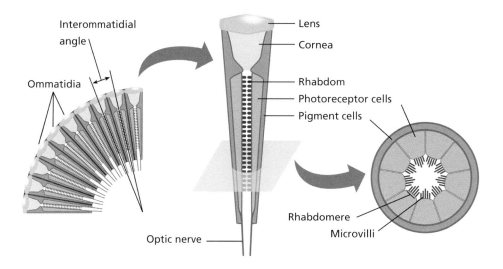

Figure 11.2 Structure of the compound eye.

cells towards the centre of the ommatidium, are numerous finger-like projections called 'microvilli'. Each cells' cluster of microvilli forms a structure called a 'rhabdomere'; the combined rhabdomeres form a tube-like structure called the '**rhabdom**'. The orientation of microvilli within the rhabdom contributes to the sensitivity of insect eyes to polarization (Marshall and Cronin 2011, Horváth and Horváth 2014). Ommatidia specialized for detecting polarized light appear to have groups of microvilli aligned perpendicular to each other, whereas eyes that are not sensitive to polarity can have curved or less regularly aligned microvilli.

The rhabdom is the light sensing structure within the ommatidium. Distributed throughout the rhabdom are pigment molecules (rhodopsin) consisting of a light sensitive **chromophore**, bound to an **opsin** protein. Opsin proteins can come in many forms and determine the wavelength sensitivity of the visual pigment. For example, honey bees have three different opsin genes that code for three different opsin proteins (Hempel de Ibarra et al. 2014). This results in bee eyes having three different photoreceptor cell types that are sensitive to either UV, blue, or green light. Dragonflies can exhibit extreme variation in opsins, with some species (e.g. *Anax parthanopes*) having over thirty-three different opsin genes (Futahashi et al. 2015). How this influences their ability to perceive and respond to colourful stimuli is still unclear.

Light is focused onto the rhabdom by the lens and crystalline cone. When a photon of light is intercepted by an opsin molecule it sets into action a rapid chemical cascade (termed phototransduction), which results in the transmission of an electrochemical signal via the nerves at the base of the ommatidium (Hardie 2012). The frequency of nerve impulses contains information on the intensity of light. Information on colour can be calculated, based on the relative excitation of wavelength-specific photoreceptors.

Each ommatidium transmits a single unit of visual information to the brain. An insect's neural networks then piece together the individual elements from every ommatidium to form a coherent image of the world around them, analogous to the way in which a digital image is pieced together from smaller individual pixels. The resolution of the image formed is then dependent on the number of ommatidia within the compound eye, as well as their size and the angles at which ommatidia are spaced (Campbell and Green 1965; Land 1997). Additionally, due to the radially orientated structure of ommatidia in the compound eye, the resolving power of the insect eye is dependent on viewing distance. The field of view of each ommatidium is determined by its acceptance angle. As ommatidia are wider at the top than they are at the base, this field of view expands outwards as you move further away from the compound eye. At further distances the amount of visual space subtended by the acceptance angle is larger and any visual detail within that space is then generalized into the single signal transmitted by the ommatidium. Only by moving closer to an object can the insect increase its visual resolution. Experiments with bees have demonstrated that bees can learn to identify simple visual patterns, such as black and white horizontal stripes (Srinivasan and Lehrer 1988). When these stripes are large the bees can detect the pattern, although when the lines are small and condensed the bee's eyes do not have the resolution to differentiate the black and white stripes.

The multifaceted nature of compound eyes also dictates how insects detect movement. An object moving left to right across the visual field of an insect will initially be detected by the ommatidia on the left of the compound eye. As it moves across the insect's view it will then be detected by other ommatidia. Where neighbouring ommatidia experience changes in photoreceptor excitation that are temporally offset, this 'optic-flow' can be interpreted as a moving stimulus (Wardill et al. 2012). This process can be used to detect the motion of a stimulus of interest, such as the movement of a nearby prey item, or it can be used to detect the apparent motion of stationary objects passing by an insect moving through its environment (Srinivasan 1992). Unfortunately, there has been a tendency to consider compound eyes as 'simple' compared with lens eyes. However, compound eyes are capable of facilitating incredible feats, such as the ability of paper wasps to recognize nest mates based on subtle facial features (Sheehan and Tibbetts 2011), or the ability of a hunting dragonfly to catch flying insects mid-air (Olberg 2011; Box 11.2).

Box 11.2 The mysterious pseudo-pupil

Should you ever take a close look at a large insect like a dragonfly or grasshopper you may notice in the centre of their eyes a small black 'pupil-like' dot. As you move around the insect this small dot will follow you around the room like the eerie gaze of the Mona Lisa. This dot is not a true pupil; it is a phenomenon called a 'pseudo-pupil' and is an artefact of the structure of the 'compound eye' that is common to all insects. At a given viewing angle there are a small number of ommatidia pointed directly towards your line of sight. When looking directly into the tubular structure of these ommatidia, they appear as a dark spot. On other areas of the eye, the pigmented walls of separate ommatidia, positioned obliquely to the line of sight, are seen. As the viewer shifts from side to side, different ommatidia will be positioned directly towards them, thus giving the illusion of a black spot moving across the surface of the compound eye.

Ongoing research continues to unveil insect visual capacities previously thought impossible. In a series of ingenious experiments, researchers discovered that praying mantises are capable of 3D stereoscopic vision similar to vertebrates (Nityananda et al. 2016). Our brain is able to compare the disparity between the images captured by our left and right eyes, and use this to estimate the distance of objects. It is this phenomenon that we use to trick our brains into seeing depth when watching 3D movies, by presenting the left and right eyes with slightly different images. When researchers equipped mantises with miniature 3D glasses and presented them with footage of a moving object they were able to elicit prey capture strikes from mantises, tricking them into thinking that the object was closer than it actually was. The ability to perceive depth using binocular vision was something that was previously unheard of in insects.

Understanding the physiological function of insect eyes gives us critical information on how they perceive the visual information available to them. How their neurological systems process this information is a further level of enquiry altogether (see also Chapter 3). As explored in Section 11.3, the ability of insects to detect and interpret visual information plays an integral role in all aspects of their ecology, behavior, and evolution.

11.3 Ecology: Senders, receivers, and signalling environments

Transmitting a salient signal to its intended destination is a goal shared across domains, from the cells within individuals, through rivals competing for mates, to eusocial colonies exploring untapped resources (Bradbury and Vehrencamp 1998; Searcy and Nowicki 2005). Irrespective of context, natural selection favours the evolution of efficacy in communication (Endler 1992; Pierce 2012), which entails the reliable generation, transmission, reception, and processing of visual information (Figure 11.3). The suites of traits associated with these processes—such as the structure of signals, or receivers' visual systems—are not evolutionarily independent, and will coevolve to define the form of a given communication system (Endler 1992, 1993a). Of course, selection will also affect the information or 'content' being sent and received (Guilford and Dawkins 1993; Smith and Harper 2003). Form and content are often shaped by different processes, however, we explore the latter in finer detail in the following section.

Although communication systems are subject to strong selection for efficacy, visual signals are not

Signals

Spectral reflectance
Pigment absorbance
Micro/nano-surface
architecture
Signalling behavior

Environments

Ambient intensity
Habitat light
Transmission media
Viewing backgrounds

Vision

Optical and
retinal anatomy
Screening pigments
Receptor sensitivity

Perception

Attention
Signal recognition
Colour versus
luminance channels
Colour opponency
Colour categorization

Figure 11.3 The ecology of insect visual communication. The viewing environment provides the setting in which a given signal is viewed, and the structure of the visual system determines what is detected and encoded for subsequent processing in the viewer's brain. The myriad interactions between signal production, transmission, and reception drive coevolutionary feedback between each component, to shape the ultimate structure of communication systems.

always received precisely as they are sent (see also Chapter 10). This is due to the ubiquitous presence of noise, which broadly describes any unpredictable modifications of the 'intended' signal (Pierce 2012). Noise accumulates across all stages of communication as signals are generated and propagated through variable environments, and then received and processed via limited sensory systems (Cronin et al.

2014). The random firing of photoreceptors, for example, sets a fundamental threshold for the visual information that can be reliably perceived (Vorobyev and Osorio 1998; Vorobyev et al. 2001). At a broader scale, temporal variation in light environments (e.g. as a product of changing habitat structure or weather conditions) will alter the subjective appearance of a signal, often in unpredictable ways (Endler 1993b). The ubiquity of noise across all stages of visual communication has driven the evolution of exceptionally diverse biological solutions, and insects showcase some particularly striking innovations.

11.3.1 Signal generation

Visual communication is a dynamic process, and there are few areas in which this may be more readily appreciated than in the structure of insect visual signals. Visual signals are invariably composed of multiple components, including colour (i.e. hue and chroma), luminance, polarization, pattern, shape, and motion (Dyer 2002; Osorio and Vorobyev 2008). Some of these features will directly provide information—such as potential mate quality—to receivers (e.g. Kemp and Rutowski 2011; Barry et al. 2015), while other components may enhance the detectability of the signal (Sweeney et al. 2003), attract the attention of receivers (Whitney et al. 2016), or otherwise modify a signal's information content (Stavenga et al. 2004). The entrancing displays of fireflies, for example, combine bioluminescence with ritualized flights to attract potential mates (Seliger and McElroy 1964; Lewis and Cratsley 2008). Each firefly species has a characteristic display that informs conspecifics of their identity and quality, and the sexes engage in a protracted call-and-response at twilight (Seliger and McElroy 1964; Lewis et al. 2004).

Colour is an intensively studied feature of insect signal structure, with a history reaching back to the very inception of modern biology (Poulton 1890; Wallace 1877). This is not surprising given the conspicuous diversity on show among insects (Seago et al. 2009; Osorio and Vorobyev 2008; Lloyd 1983) and their tractability as study organisms. As in most animals, insects chiefly produce colour signals via three mechanisms. One, touched on above, is bioluminescence, which describes the chemical generation of light through an oxygenation reaction between a 'luciferin' substrate and a 'luciferase' enzyme (Day et al. 2004; Kahlke and Umbers 2016). Although fireflies present one of the most charismatic examples of bioluminescent signalling in nature, the phenomenon is actually rare in insects.

The other two mechanisms, pigmentary and **structural colouration**, typically coincide, and form the basis of almost all insect colouration. As outlined in our discussion of vision, pigments are macromolecules that selectively absorb particular wavelengths of light, and colours primarily arising from pigments are typically diffuse and matte (Johnsen 2012). They may be produced by the insects themselves by combining simpler chemical compounds, like the melanic blacks and browns seen throughout the group (Wittkopp and Beldade 2009), or environmentally acquired, as in the carotenoid-based red and orange warning signals of ladybird beetles (Bezzerides et al. 2007). Structural colours, in contrast, arise from the reflectance of light by micro- or nanoscale structures, and can generate some of the most brilliant signals in nature (Kinoshita 2008; Seago et al. 2009). The shimmering, metallic blues of *Morpho* butterflies, for example, are a product of alternating layers of air and chitin that constructively reflect mid-wavelength light (Kinoshita et al. 2002). Compared with pigment-based colours, structural colours present several features that make them particularly interesting (and useful; discussed in the next paragraph) as visual signals. They are capable of unrivalled brightness and spectral purity, may produce hues that are otherwise unachievable through the use of pigments, and may exhibit auxiliary features including polarization and iridescence (Kinoshita 2008). While the rudiments of structural colouration are well understood, exemplary groups such as butterflies and beetles continue to serve as sources of physical and evolutionary insight (Kinoshita 2008; Seago et al. 2009).

11.3.2 Signal transmission

Signals must be displayed, transmitted, and received in environments filled with noise. For insects, these environments are primarily terrestrial, but span a breadth of changeable conditions that present challenges for the effective transmission of information. Spatial and temporal variation in light environments (i.e. time of day, cloud cover, fog, habitat complexity) can vastly alter the subjective appearance of a signal to receivers. Forest-dwelling insects, for example, must contend with sunlight filtered through a canopy. The intensity of light can vary several orders of magnitude across small spatial scales between open and closed canopies, and its colour may shift from the rich 'blue' of direct, open skylight, to a deep 'green' beneath heavy vegetation (Endler 1993b, Cronin et al. 2014). Since all, but a few insect signals rely on interactions with ambient light, its spectral quality and intensity will fundamentally affect how signals are subjectively perceived (Figure 11.3), which is the ultimate canvas for visual signal evolution.

As with their varied signal designs, insects have evolved numerous solutions to the inherent challenges of effectively transmitting signals. These include the use of precise presentation behaviors (e.g. White et al. 2015), the utilization of displays in favourable conditions (e.g. O'Hanlon et al. 2014c), or the active modification of viewing environments (e.g. Heiling et al. 2003). Particular extremes are showcased by the Lepidoptera, among whom the use of iridescent signals (colours that change with orientation and viewing angle) is particularly widespread (Vukusic et al. 2001; Douglas et al. 2007; Wilts et al. 2011). For example, males of the common eggfly, *Hypolimnas bolina*, have iridescent violet-on-white spots that are visible only from a restricted range of viewing angles, and females prefer males bearing brighter ornamentation (Kemp and Macedonia 2006; Kemp 2007). During courtship, a male will engage in a highly ritualized fluttering flight immediately below a female, which effectively 'beams' a maximally conspicuous signal directly to their potential mate (White et al. 2015). The restricted-view nature of these signals, combined with such precise behavioral delivery, also provides some degree of protection from unintended eavesdroppers—a perennial hazard during signal transmission (Peake 2005). More broadly, recent work suggests that iridescence in visual signals may confer some general protection from predators, as their shifting appearance makes them difficult to track when in motion (Pike 2015).

11.3.3 Signal reception and processing

As insects navigate their environments—often on the wing—they are bombarded with visual information. This includes abundant useless noise interspersed with vital information about potential mates, rivals, food sources, and predators (Döring and Chittka 2007; De Ibarra et al. 2014; Lunau 2014). In response, selection has favoured a suite of sensory filters tasked with completing the final stage of visual communication; the perceptual separation of signal from noise. As we previously explored, the architecture of the insect eyes is itself the first of such 'filters', and typically contains adaptations for extracting information about space, motion, colour, and luminance (Land and Nilsson 2012).

More specialized filters are a useful way to further extract visual information, and the need to reproduce has driven the evolution of some remarkable examples (Warrant 2016). As noted previously, the photoreceptor arrays of many insects are sensitive to natural polarization patterns (i.e. the orientation of light waves; Johnsen 2012). Since the polarization of light is largely independent of colour and intensity, it offers a particularly reliable channel of information in trying environmental conditions (Sweeney et al. 2003; Johnsen 2012). Although seemingly rare among animals more broadly, some butterflies have incorporated polarization into their signals that, when coupled with dedicated sensory structures, offers an effective means of detecting mates in their visually noisy forest environments (Kelber 1999; Sweeney et al. 2003). Indeed, the effectiveness of matching polarization filters and signals has been forwarded as an explanation for the fact that polarized colour patches occur most commonly in forest-dwelling insects (Douglas et al. 2007). In more extreme examples of specialization, some male flies possess entirely separate eyes dedicated to the detection of the visual signature of females. Male mayflies, for instance, have a bi-lobed eye with a flattened dorsal region (Zeil 1983a). This dramatically increases retinal sampling within an upward field of view, which allows males to readily detect the distinctive visual signature of darkened females against the bright sky (Zeil 1983a,b).

As we explore below, the use of such filters for mining visual information may also leave receivers open to deception. Innate biases for particular colour and luminance cues (Peng et al. 2013; White and Kemp 2016; White et al. 2017), motion signatures (Lloyd 1965; Woods Jr et al. 2007), and patterns (Herberstein et al. 2000; Heiling et al. 2003; White 2017) all offer evolutionary opportunities for deception in visual signalling systems. Few examples rival the aforementioned orchid mantis, *Hymenopus coronatus* (O'Hanlon 2016). By mimicking the colour and morphology of floral resources, juvenile mantises are able to tap into millions of years of coevolution between flower signals and the visual systems of honeybees to attract their prey (O'Hanlon et al. 2013, 2014a, 2014b).

11.4 Evolution: forms and function of visual communication

Over millennia, evolutionary drivers such as natural and sexual selection have given rise to a stupendous diversity of form and function in insect visual communication. As discussed earlier, visual communication involves the sending and receiving of visual information through a medium, and the reception of this information by the eyes of a receiver that may perceive colour, luminance, polarization, pattern, shape, and motion. Insect signals have evolved in response to many competing and complementary forces, and are shaped by the constraints of their bodies, the penetrability of the environmental medium through which they must send their signal, and the perception of their enemies and friends. By understanding the evolution of insect visual signalling we can unveil how and why insects have come to exhibit certain traits and what functions those traits perform.

To complicate things, traits used in signalling may have more than one function and these functions are not necessarily mutually exclusive (Seago et al. 2009; Umbers et al. 2014). Apparent signalling traits may also serve non-signalling functions including **thermoregulation** (May 1979; Clusella Trullas et al. 2007; Kearney et al. 2009), physical reinforcement, and immune responses (Kanost and Gorman 2008; Nappi and Christensen 2005). Also, signals can be enhanced by behaviors, such as resting in sun or shade, or movement, to maximize or minimize conspicuousness (Kemp 2003; Rutowski et al. 2010; White et al. 2015). Of course, convergent evolution of the same signalling traits by species from different phylogenetic groups is ubiquitous (Jiggins et al. 2005; Seago et al. 2009; Umbers et al. 2014; Viviani 2002). Thus, the selective advantages of signals are difficult to elucidate as they can be varied, complex, context dependant, and often perform multiple functions. There are even cases where apparent visual signals seem to have no clear function at all (Umbers et al. 2013a,b).

Generally speaking, signalling functions occur in two main contexts, within species (intraspecific) and between species (interspecific). Within species, individuals might send signals about their need for a mate or their quality as a mate, they may advertise purported invincibility to rivals, or they may offer or ask for mutually beneficial services. Between species, insects might send warning signals to predators or emit visual cues that enable them to hide from predators in plain sight. It is in these broad contexts that we explore insect visual communication in Section 11.4.1.

11.4.1 Visual communication between mates

Visual communication between mates may convey direct or indirect information about quality (see also Chapter 13). Whether it be cues regarding overall body size (condition dependence; e.g. Cotton et al. 2010), ability to display high chroma colours (precise development hypothesis: Kemp and Rutowski, 2007), ability to sport ridiculously over-sized appendages and still survive to reproduce (handicap hypothesis; e.g. Sauer et al. 1998), or a pre-existing quirk of the sensory system that makes a mate irresistible (supernormal stimulus/sensory exploitation; e.g. Gwynne and Rentz 1983; Endler and Basolo 1998). A wonderful variety of mating signals exist in the insects including colour patterns, elaboration, and elongation of body parts, and the production of light (bioluminescence; Kahlke and Umbers 2016; Viviani 2002).

Differences in the shape and size of body parts can provide receivers with insight into the quality of their potential mate. One classic example in insects is stalk-eyed flies in which the length of the eye-stalks of males is seemingly only constrained by the physical properties of the eye-stalks themselves (see Figure 11.4a). Female stalk-eyed flies prefer to mate with males with greater eye spans (Cotton et al. 2010). Yet, while wide-eyed females prefer wide eyed-males, is it also true that narrow-eyed females prefer narrow eyed males (Wilkinson and Reillo 1994). In addition to eye-stalk width, eye shape varies among stalk-eyed flies, but whether the shape itself is a cue or just a trade-off of ever-larger eye stalk construction is yet to be shown (Worthington et al. 2012).

Some insects communicate through the production of light, such as the remarkable bioluminescent displays of firefly beetles, in which the luciferin-luciferase reaction takes place in their abdomens (Lewis and Cratsley 2008; Lloyd 1983; Viviani 2002). These beetles famously emit soft pulses of light of around 500–580 nm in wavelength through the forest on summer evenings (Figure 11.4b). In some beetle species (Lampyridae) light is used in place of sex pheromones to attract mates and seems to work in two distinct ways, one immobile sex (usually the female) produces a luminescent signal and the other (usually male) is attracted to it. Alternatively, both sexes can produce a luminescent signal with a species-specific flashing pattern (Branham and Wenzel 2003).

11.4.2 Visual communication between rival conspecifics

Conflict among individuals within species can be triggered by limited resources, including food, shelters, territories, heat sources, and mates. Many studies on insects focus on rivalries played out between males for maintaining territories or access to females (see Chapter 6). Visual identification and assessment of conspecifics can lead to individuals either avoiding particular habitat patches (e.g. O'Hanlon 2011), or deciding to engage with those conspecifics. Some insects fight without weapons, for example, many male butterflies fight to maintain territories (Kemp, 2013). Others have elaborated or exaggerated traits that increase apparent body size and/or act as weapons that signal fighting ability (Emlen, 2008). Males of the territorial American rubyspot damselflies, *Hetaerina americana* signal their quality to other males in the area with a red spot on their wings. Males with larger spots are better quality in that they have higher

Figure 11.4 Examples of visual communication in insects. (a) Stalk-eyed fly, *Cyrtodiopsis dalmanni* (Photo: Wikicommons); (b) firefly, *Lampyridae* (Photo: Ben Pfeiffer); (c) New Zealand giraffe weevil, *Lasiorhynchus barbicornis* (Photo: Christina Painting); (d) honey bee, *Apis mellifera* (Photo: Wikicommons); (e) paper wasp, *Polistes fuscatus* (Photo: Wikicommons); (f) peppered moth, *Biston betularia* (Wikicommons); (g) dead leaf mantis, *Deroplatys dessicata* (Photo: James O'Hanlon); (h) ladybird beetle, *Coccinella semtempunctata* (Photo: Wikicommons); (i) mountain katydid, *Acripeza reticulata* (Photo: Kate Umbers).

immune responses and greater fat reserves that allow them to more successfully defend larger territories (Córdoba-Aguilar and González-Tokman 2014). If we revisit stalk-eyed flies, we discover that not only do females prefer males with ever-wider eyes (as mentioned previously), but male stalk-eyed flies 'size each other up' by lining up their eye stalks to see which has the longest stalks (Panhuis and Wilkinson 1999; Figure 11.4a). Elaborate body parts can be used for both conspecific assessment and as weapons. In the New Zealand giraffe weevil, *Lasiorhynchus barbicornis*, some males have massively elongated rostra with which they fight other large males (Figure 11.4c). While these weapons have a mechanical function, they also send visual signals regarding a male's ability to defend their females. Other males in the species have tiny rostra, the appearance and size of which are thought to allow them to gain sneaky copulations (Painting and Holwell 2014). So, bigger is not always better.

11.4.3 Visual communication in cooperation among conspecifics

Conveying information accurately is critical for functional groups. In the fertile meadows of the Austrian alps, von Frisch et al. (1967) famously discovered that, unlike our unfortunate friend in the introductory paragraph, worker bees that return safely from foraging trips perform ritualized behaviors that he described as the 'waggle dance' and the 'round dance' (von Frisch et al. 1967; Figure 11.4d). Famous Nobel prize-winning experiments showed that these displays convey precise information about the distance, direction, and quality of resources to nest mates, which subsequently visit the described patch.

Communication between individuals can also be utilized in distinguishing kin from non-kin. In many cases, social insects are able to distinguish nest-mates based on chemical cues, although recent research into social wasps has shown that they can recognize each other simply by looking at their facial features (Sheehan and Tibbetts, 2011; Figure 11.4e). Paper wasps can learn and remember individual faces by distinguishing variation in shape and colour patterns. Remarkably, wasps can learn to recognize images of faces faster than they can learn generic

patterns, such as an image of contrasting stripes or shapes, suggesting that they have independently evolved specialized face learning similar to humans and other mammals. Individual facial recognition is present in wasp species where colonies are formed by cooperating queens, whereas other species that have colonies founded by solitary queens are not able to recognize individual faces. This research is providing our first insight into specialized face learning in insects, and it is possible that this phenomenon is more common than we currently realize.

11.4.4 Protective signalling to avoid predation

Insects use myriad visual strategies to avoid predation (see also Chapter 9). These fall into two main categories: signals that prevent the insect from being detected, and signals that advertise their real or purported unprofitability to predators (Edmunds 1974; Ruxton et al. 2004). Avoiding detection through camouflage can be accomplished by crypsis or masquerade, and insects provide some spectacular examples of both (Skelhorn et al. 2010). Cryptic animals tend to 'blend in' to their backgrounds through special patterning that breaks up their body outline, sometimes rendering them near imperceptible, like a moth on a tree trunk (Figure 11.4f). This strategy is highly context-dependant as when they are removed from their natural background the illusion fails. In its natural environment, a grasshopper with brown and grey mottling that sits on the dry ground it resembles (e.g. Tsurui et al. 2010) may be very difficult to see, but if placed on a white background it is conspicuous and is shaped just as one expects a grasshopper to be shaped. On the other hand, while species that masquerade also blend in with their surroundings, when they are viewed in isolation they can still resemble their 'model' such as a dead leaf or stick (Wedmann 2010). Insects may achieve such trickery by taking the three-dimensional shape of their model (dead leaf mantis, *Deroplatys*; Figure 11.4g) or they may create the illusion of a three-dimensional shape in just two dimensions (Kelley and Kelley 2014).

Instead of avoiding detection, some species are aposematic, they actively advertise their real or purported unprofitability through conspicuous and/or memorable colour patterns that predators learn to associate with unprofitability (Mappes et al. 2005). Ladybird beetles are classic examples of aposematic insects, their red and black pattern adversities that they are an unprofitable meal and predators should not eat them (Arenas et al. 2015). Interestingly, it seems that the shininess of an insect integument can have the same effect as a colour. When faced with harlequin bugs, *Tectocoris diopthalmus*, that are either iridescent or non-iridescent, birds avoid the bugs with iridescent patches (Fabricant et al. 2014). The protection afforded by aposematic signals in one species can be reinforced or exploited by others through mimicry. As indicated in Chapter 9, Müllerian mimics are two (or more) species that share the same or a very similar signal and are both unprofitable; they are sometimes called 'honest' mimics. Batesian mimics, on the other hand, are 'dishonest'; they have similar colour patterns to unprofitable species although they themselves are profitable prey. These two forms of mimicry are not always easy to distinguish. For a long time it was assumed that the viceroy butterfly (*Limenitis archippus*) was mimicking the unprofitable monarch butterfly, *Danaus plexippus*, and taking advantage of the monarch's aposematic colouration. However, Ritland and Brower (1991) showed that viceroys are also unprofitable, and are therefore better described as Müllerian, not Batesian mimics as was long thought (Ritland 1991; Ritland and Brower 1991). It is mysterious then how, in some species, several different patterns can each effectively deter predators, depending on fluctuating ecological factors (Mallet and Joron, 1999; Nokelainen et al. 2014). Thus, the evolution of colour-polymorphic aposematism presents an exciting problem for evolutionary biologists to solve.

Some species exhibit deimatic signals, defending themselves against predators by incorporating a sudden visual transformation transition into their defence (Umbers et al., 2017; Figure 11.4i; see also Chapter 9). In many species this transition involves avoiding detection in the first instance by a kind of camouflage, and performing a display only when they realize they have been discovered. The performance may then reveal a further defence, such as predator-mimicking eyespots (De Bona et al. 2015). These secret defences usually, but not exclusively, invoke the visual mode and in this mode can be comprised of movement, bright colour patterns, and apparent change in body shape and size (Umbers et al., 2015, 2017). Across the insects, **deimatism** is most widely represented in the Orthoptera, Phasmatodea, Dictyoptera, and Lepidoptera, but can be found in the Hemiptera, and Coleoptera as well. Perhaps the most spectacular examples of deimatism in insects come from the *Deroplatys* mantises where we see elaborate dead leaf mimicry at rest, and genuinely striking posturing and colour patterns during display (Crane 1952; Edmunds 1972; Figure 11.4g). Precisely when and how deimatism provides protection by inciting a startle reflex or by overwhelming the predator's senses is poorly understood. More work in this area is required to uncover all the secrets of this defence strategy (Kang et al. 2016; Skelhorn et al. 2016; Umbers et al. 2015, 2017).

Who exactly are these 'predators' we keep mentioning? Insects are preyed upon by many types of animals, with birds often accused of being their main predators, but there is little empirical data on the predators of most species. There is no doubt that countless species of vertebrates, spiders, centipedes, and velvet worms, among others, rely on insects as prey, but what is often underappreciated is that there are many formidable insect predators that prey on other insects (like the praying mantis mentioned earlier). Important predatory groups include praying mantises, ants and wasps, assassin bugs, some katydids, dragonflies and damselflies, some beetles, antlions, scorpionflies, and so on. Despite this list, the roles that insect predators have played in shaping insect defences remains largely unexplored (Fabricant et al. 2014).

11.5 Conclusion

The initial evolution of the eye—from light-detecting ocelli to image-forming lens eyes—has had a great impact on the physical form of living things. The organ itself demands neural networks and specialized organs, precise development, and significant processing power, while sight, the primary function of the eye, provides animals with a valuable sense with which to interact with the world around them. The mind-boggling diversity of form and function in visual signals in insects reflects millions of years of evolution driven by exploitation of the limits of visual communication. This chapter and the references herein, hopefully have provided a broad overview of the main themes within insect visual communication as a jumping-off point for diving deeper into this kaleidoscopic world.

References

Arenas, L.M., Walter, D., and Stevens, M. (2015). Signal honesty and predation risk among a closely related group of aposematic species. *Scientific Reports*, **5**, 11021.

Barry, K.L., White, T.E., Rathnayake, D.N., Fabricant, S.A., and Herberstein, M.E. (2015). Sexual signals for the colour-blind: cryptic female mantids signal quality through brightness. *Functional Ecology*, **29**, 531.

Bezzerides, A.L., McGraw, K.J., Parker, R.S., and Husseini, J. (2007). Elytra color as a signal of chemical defense in the Asian ladybird beetle *Harmonia axyridis*. *Behavioral Ecology and Sociobiology*, **61**, 1401.

Bradbury, J.W., and Vehrencamp, S.L. (1998). *Principles of Animal Communication*. Oxford University Press, Oxford.

Branham, M.A., and Wenzel, J.W. (2003). The origin of photic behavior and the evolution of sexual communication in fireflies (Coleoptera: Lampyridae). *Cladistics*, **19**, 1.

Campbell, F.W., and Green, D.G. (1965). Optical and retinal factors affecting visual resolution. *Journal of Physiology*, **181**(3), 576–93.

Clusella Trullas, S., van Wyk, J.H., and Spotila, J.R. (2007). Thermal melanism in ectotherms. *Journal of Thermal Biology*, **32**, 235.

Córdoba-Aguilar, A., and González-Tokman, D.M. (2014). The behavioral and physiological ecology of adult rubyspot damselflies (*Hetaerina*, Calopterygidae, Odonata). *Advances in the Study of Behavior*, **46**, 311.

Cotton, S., Small, J., Hashim, R., and Pomiankowski, A. (2010). Eyespan reflects reproductive quality in wild stalk-eyed flies. *Evolutionary Ecology*, **24**, 83.

Crane, J. (1952). A comparative study of innate defensive behavior in Trinidad mantids (Orthoptera, Mantoidea). *Zoologica*, **37**, 259.

Cronin, T.W., Johnsen, S., Marshall, N.J., and Warrant. E.J. (2014). *Visual ecology*. Princeton University Press, Princeton, NJ.

Day, J.C., Tisi, L.C., and Bailey M.J. (2004). Evolution of beetle bioluminescence: the origin of beetle luciferin. *Luminescence*, **19**, 8.

De Bona, S., Valkonen, J.K., López-Sepulcre, A., and Mappes, J. (2015). Predator mimicry, not conspicuousness, explains the efficacy of butterfly eyespots. *Proceedings of the Royal Society B*, **282**, 20150202.

De Ibarra, N.H., Vorobyev, M., and Menzel, R. (2014). Mechanisms, functions and ecology of colour vision in the honeybee. *Journal of Comparative Physiology A*, **200**, 411.

Döring, T., and Chittka, L. (2007). Visual ecology of aphids—a critical review on the role of colours in host finding. *Arthropod–Plant Interactions*, **1**, 3.

Douglas, J.M., Cronin, T.W., Chiou, T-H., and Dominy, N.J. (2007). Light habitats and the role of polarized iridescence in the sensory ecology of neotropical nymphalid butterflies (Lepidoptera: Nymphalidae). *Journal of Experimental Biology*, **210**, 788.

Dyer, F.C. (2002). The biology of the dance language. *Annual Review of Entomology*, **47**, 917.

Edmunds, M. (1972). Defensive behavior in Ghanaian praying mantids. *Zoological Journal of the Linnean Society*, **51**, 1.

Edmunds, M. (1974). *Defence in Animals: a Survey of Anti-predator Defences*. Longman, London.

Emlen, D.J. (2008). The evolution of animal weapons. *Annual Review of Ecology, Evolution, and Systematics*, **39**, 387.

Endler, J.A. (1992). Signals, signal conditions, and the direction of evolution. *American Naturalist*, **139**, S125.

Endler, J.A. (1993a). Some general comments on the evolution and design of animal communication systems. *Philosophical Transactions of the Royal Society B: Biological Sciences*, **340**, 215.

Endler, J.A. (1993b). The color of light in forests and its implications. *Ecological Monographs*, **63**, 1.

Endler, J.A., and Basolo, A.L. (1998). Sensory ecology, receiver biases and sexual selection. *Trends in Ecology & Evolution*, **13**(10), 415–20.

Fabricant, S.A., Exnerova, A., Jezova, D., and Stys, P. (2014). Scared by shiny? The value of iridescence in aposematic signalling of the hibiscus harlequin bug. *Animal Behavior*, **90**, 315.

Fischer, S., Meyer-Rochow, V.B., and Müller, H.G. (2012). Challenging limits: ultrastructure and size-related functional constraints of the compound eye of *Stigmella microtheriella* (Lepidoptera: Nepticulidae). *Journal of Morphology*, **273**, 1064.

Futahashi, R., Kawahara-Miki, R., Kinoshita, M., et al. (2015). Extraordinary diversity of visual opsin genes in dragonflies. *Proceedings of the National Academy of Sciences*, **112**, 1247.

Guilford, T., and Dawkins, M.S. (1993). Receiver psychology and the design of animal signals. *Trends in Neurosciences*, **16**, 430.

Gwynne, D.T., and Rentz, D.A. (1983). Beetles on the bottle: male Buprestids mistake stubbies for females. *Austral Entomology*, **22**, 79.

Hardie, R.C. (2012). Phototransduction mechanisms in *Drosophila* microvillar photoreceptors. *WIREs Membrane Transport and Signaling*, **1**, 162.

Heiling, A.M., Herberstein, M.E., and Chittka, L. (2003). Pollinator attraction: crab-spiders manipulate flower signals. *Nature*, **421**, 334.

Hempel de Ibarra N., Vorobyev, M., and Menzel, R. (2014). Mechanisms, functions and ecology of colour vision in the honeybee. *Journal of Comparative Physiology A*, **200**, 411.

Herberstein, M., Craig C., Coddington J., and Elgar, M. (2000). The functional significance of silk decorations of orb-web spiders: a critical review of the empirical evidence. *Biological Reviews of the Cambridge Philosophical Society*, **75**, 649.

Horváth, G. and Horváth, G. (Eds) (2014). *Polarized Light and Polarization Vision in Animal Sciences*. Springer, Berlin.

Jiggins, C.D., Mavarez, J., Beltrán, M., et al. (2005). A genetic linkage map of the mimetic butterfly *Heliconius melpomene*. *Genetics*, **171**, 557.

Johnsen, S. (2012). *The Optics of Life: A Biologist's Guide to Light in Nature*. Princeton University Press, Princeton, NJ.

Kahlke, T., and Umbers, K.D.L. (2016). Bioluminescence. *Current Biology*, **26**, R313.

Kang, C-K., Cho, H-J., Lee, S-I., and Jablonski, P.G. (2016). Post-attack aposematic display in prey facilitates predator avoidance learning. *Frontiers in Ecology and Evolution*, **4**, 1.

Kanost, M.R., and Gorman, M.J. (2008). *Phenoloxidases in Insect Immunity*. Elsevier Academic Press Inc., San Diego, CA.

Kearney, M., Shine, R., and Porter, W.P. (2009). The potential for behavioral thermoregulation to buffer 'cold-blooded' animals against climate warming. *Proceedings of the National Academy of Sciences of the United States of America*, **106**, 3835.

Kelber, A. (1999). Why 'false' colours are seen by butterflies. *Nature*, **402**, 251.

Kelley, L.A., and Kelley, J.L. (2014). Animal visual illusion and confusion: the importance of a perceptual perspective. *Behavioral Ecology*, **25**, 450.

Kemp, D.J. (2013).Contest behaviour in butterflies: fighting without weapons. In: I. C. W. Hardy and M. Briffa (Eds), *Animal Contests*, pp. 134–46. Cambridge University Press, Cambridge.

Kemp, D., and Macedonia, J. (2006). Structural ultraviolet ornamentation in the butterfly *Hypolimnas bolina* l. (Nymphalidae): visual, morphological and ecological properties. *Australian Journal of Zoology*, **54**, 235.

Kemp, D.J. (2003). Twilight fighting in the evening brown butterfly, *Melanitis leda* (L.) (Nymphalidae): age and residency effects. *Behavioral Ecology and Sociobiology*, **54**, 7.

Kemp, D.J. (2007). Female butterflies prefer males bearing bright iridescent ornamentation. *Proceedings of the Royal Society of London B*, **274**, 1043.

Kemp, D.J., and Rutowski, R.L. (2007). Condition dependence, quantitative genetics, and the potential signal content of iridescent ultraviolet butterfly coloration. *Evolution*, 61, 168.

Kemp, D.J., and Rutowski, R.L. (2011). The role of coloration in mate choice and sexual interactions in butterflies. *Advances in the Study of Behavior*, **43**, 55.

Kinoshita, S. (2008). *Structural Colors in the Realm of Nature*. World Scientific, Singapore.

Kinoshita, S., Yoshioka, S., and Kawagoe, K. (2002). Mechanisms of structural colour in the *Morpho* butterfly: cooperation of regularity and irregularity in an iridescent scale. *Proceedings of the Royal Society of London B: Biological Sciences*, **269**, 1417.

Land, M.F. (1997). Visual acuity in insects. *Annual Review of Entomology*, **42**, 147.

Land, M.F., and Nilsson, D.E. (2012). *Animal Eyes*. Oxford University Press, Oxford.

Lewis, S.M., and Cratsley, C.K. (2008). Flash signal evolution, mate choice, and predation in fireflies. *Annual Review of Entomology*, **53**, 293.

Lewis, S.M., Cratsley, C.K., and Demary, K. (2004). Mate recognition and choice in *Photinus* fireflies. *Annales zoologici fennici*, **41**, 809.

Lloyd, J.E. (1965). Aggressive mimicry in *Photuris*: firefly femmes fatales. *Science*, **149**, 653.

Lloyd, J.E. (1983). Bioluminescence and communication in insects. *Annual Review of Entomology*, **28**, 131.

Lunau, K. (2014). Visual ecology of flies with particular reference to colour vision and colour preferences. *Journal of Comparative Physiology A*, **200**, 497.

Mallet, J., and Joron, M. (1999). Evolution of diversity in warning color and mimicry: polymorphisms, shifting balance, and speciation. *Annual Review of Ecology and Systematics*, **30**, 201.

Mappes, J., Marples, N., and Endler, J.A. (2005). The complex business of survival by aposematism. *Trends in Ecology and Evolution*, **20**, 598.

Marshall, J., and Cronin, T.W. (2011). Polarization vision. *Current Biology*, **21**, R101.

May, M.L. (1979). Insect thermoregulation. *Annual Review of Entomology*, **24**, 313.

Nappi, A.J., and Christensen, B.M. (2005). Melanogenesis and associated cytotoxic reactions: applications to insect innate immunity. *Insect Biochemistry and Molecular Biology*, **35**, 443.

Nityananda, V., Tarawneh, G., Rosner, R., Nicolas, J., Chrichton, S., and Read, J. (2016). Insect stereopsis demonstrated using a 3D insect cinema. *Scientific Reports*, **6**, 18718.

Nokelainen, O., Valkonen, J., Lindstedt, C., and Mappes, J. (2014). Changes in predator community structure shifts the efficacy of two warning signals in Arctiid moths. *Journal of Animal Ecology*, **83**, 598.

O'Hanlon, J.C. (2011). Intraspecific interactions and their effect on habitat utilisation by the praying mantid *Ciulfina biseriata* (Mantodea: Liturgusidae). *Journal of Ethology*, **29**, 47.

O'Hanlon, J.C. (2016). Orchid mantis. *Current Biology*, **26**, R145.

O'Hanlon, J., Herberstein, M., and Holwell, G. (2014a). Habitat selection in a deceptive predator: maximizing resource availability and signal efficacy. *Behavioral Ecology*, **26**, 194.

O'Hanlon, J.C., Holwell, G.I., and Herberstein, M.E. (2014b). Predatory pollinator deception: Does the orchid mantis resemble a model species? *Current Zoology*, **60**, 90.

O'Hanlon, J.C., Holwell, G.I., and Herberstein, M.E. (2014c). Pollinator deception in the orchid mantis. *American Naturalist*, **183**, 126.

O'Hanlon, J., Li, D., and Norma-Rashid, Y. (2013). Coloration and morphology of the orchid mantis *Hymenopus coronatus* (Mantodea: Hymenopodidae). *Journal of Orthoptera Research*, **22**, 35.

Olberg, R.M. (2011). Visual control of prey–capture flight in dragonflies. *Current Opinion in Neurobiology*, **22**, 1.

Osorio, D., and Vorobyev, M. (2008). A review of the evolution of animal colour vision and visual communication signals. *Vision Research*, **48**, 2042.

Painting, C.J., and Holwell, G.I. (2014). Exaggerated rostra as weapons and the competitive assessment strategy of male giraffe weevils. *Behavioral Ecology*, **25**, 1223.

Panhuis, T.M., and Wilkinson, G.S. (1999). Exaggerated male eye span influences contest outcome in stalk-eyed flies (Diopsidae). *Behavioral Ecology and Sociobiology*, **46**, 221.

Paulk A., Millard, S.S., and van Swinderen, B. (2013). Vision in *Drosophila*: seeing the world through a model's eyes. *Annual Review of Entomology*, **58**, 313.

Peake, T.M. (2005). Eavesdropping in animal networks. In: P.K. McGregor (Ed.) *Animal Communication Networks*, pp. 13–38. Cambridge University Press, London.

Pener, M.P., and Simpson, S.J. (2009). Locust phase polyphenism: an update. *Advances in Insect Physiology*, **36**, 1.

Peng, P., Blamires, S.J., Agnarsson, I., Lin, H-C., and Tso, I.M. (2013). A color-mediated mutualism between two arthropod predators. *Current Biology*, **23**, 172.

Pierce, J.R. (2012). A*n Introduction to Information Theory: Symbols, Signals and Noise*. Courier Corporation, Mineola, NY.

Pike, T.W. (2015). Interference coloration as an anti-predator defence. *Biology Letters*, **11**, 20150159.

Poulton, E.B. (1890). *The Colors of Animals*. Trübner and Co Ltd., London.

Ritland, D.B. (1991). Unpalatability of viceroy butterflies (*Limenitis archippus*) and their purported mimicry models, Florida queens (*Danaus gilippus*). *Oecologica*, **88**, 102.

Ritland, D.B., and Brower, L.P. (1991). The viceroy butterfly is not a batesian mimic. *Nature*, **350**, 497.

Rutowski, R.L., Nahm, A.C., and Macedonia, J.M. (2010). Iridescent hindwing patches in the pipevine swallowtail: differences in dorsal and ventral surfaces relate to signal function and context. *Functional Ecology*, **24**, 767.

Ruxton, G.D., Sherratt, T.N., and Speed, M.P. (2004). *Avoiding Attack: the Evolutionary Ecology of Crypsis, Warning Signals, and Mimicry*. Oxford University Press, Oxford.

Sauer, K.P., Lubjuhn, T., Sindern, J., et al. (1998). Mating system and sexual selection in the scorpionfly *Panorpa vulgaris* (Mecoptera: Panorpidae). *Naturwissenschaften*, **85**, 219.

Seago, A.E., Brady, P., Vigneron J-P., and Schultz, T.D. (2009). Gold bugs and beyond: a review of iridescence and structural colour mechanisms in beetles (Coleoptera). *Journal of the Royal Society Interface*, **6**, S165.

Searcy, W.A., and Nowicki, S. (2005). *The Evolution of Animal Communication: Reliability and Deception in Signaling Systems*. Princeton University Press, Princeton, NJ.

Seliger, H., and McElroy, W. (1964). The colors of firefly bioluminescence: enzyme configuration and species specificity. *Proceedings of the National Academy of Sciences*, **52**, 75.

Sheehan, M.J., and Tibbetts, E.A. (2011). Specialized face learning is associated with individual recognition in paper wasps. *Science*, **334**, 1272.

Sherk, T.E. (1978). Development of the compound eyes of dragonflies (Odonata). III. Adult compound eyes. *Journal of Experimental Zoology*, **203**, 61.

Skelhorn, J., Holmes, G.G., and Rowe, C. (2016). Deimatic or aposematic? *Animal Behavior*, **113**, E1.

Skelhorn, J., Rowland, H.M., Speed, M.P., and Ruxton, G.D. (2010). Masquerade: camouflage without crypsis. *Science*, **327**, 51.

Smith, J.M., and Harper, D. (2003). *Animal Signals*. Oxford University Press, Oxford.

Srinivasan, M. (2010). Honey bees as a model for vision, perception and cognition. *Annual Review of Entomology*, **55**, 267.

Srinivasan, M., and Lehrer, M. (1988). Spatial acuity of honeybee vision and its spectral properties. *Journal of Comparative Physiology*, **162**, 159.

Srinivasan, M.V. (1992). How bees exploit optic flow: behavioral experiments and neural models. *Philosophical Transactions: Biological Sciences*, **337**, 253.

Stavenga, D.G., Stowe, S., Siebke, K., Zeil, J., and Arikawa, K. (2004). Butterfly wing colours: Scale beads make white pierid wings brighter. *Proceedings of the Royal Society B Biological Sciences*, **271**, 1577.

Sweeney, A., Jiggins, C., and Johnsen, S. (2003). Insect communication: polarized light as a butterfly mating signal. *Nature*, **423**, 31.

Tsurui, K., Honma, A., and Nishida, T. (2010). Camouflage effects of various colour-marking morphs against different microhabitat backgrounds in a polymorphic pygmy grasshopper *Tetrix japonica*. *PLoS One*, **5**, e11446.

Umbers, K.D.L. (2011). Turn the temperature to turquoise: cues for colour change in the male chameleon grasshopper (*Kosciuscola tristis*) (Orthoptera: Acrididae). *Journal of Insect Physiology*, **57**, 1198.

Umbers, K.D.L., De Bona, S., White, T.E., et al. (2017). Deimatism: a neglected form of antipredator defence. *Biology Letters*, **13**, 20160936.

Umbers, K.D.L., Fabricant, S.A., Gawryszewski, F.M., Seago, A.E., and Herberstein, M.E. (2014). Reversible colour change in Arthropoda. *Biological Reviews*, **89**, 820.

Umbers, K.D.L., Herberstein, M.E., and Madin, J.S. (2013). Colour in insect thermoregulation: empirical and theoretical tests in the colour-changing grasshopper, *Kosciuscola tristis*. *Journal of Insect Physiology*, **59**, 81.

Umbers, K.D.L., Lehtonen, J., and Mappes, J. (2015). Deimatic displays. *Current Biology*, **25**, R58.

Umbers, K.D.L., and Mappes, J. (2015). Post-attack deimatic display in the mountain katydid (*Acripeza reticulata*). *Animal Behavior*, **100**, 68.

Umbers, K.D.L., Tatarnic, N.J., Holwell, G.I., and Herberstein, M.E. (2013). Bright turquoise as an intraspecific signal in the chameleon grasshopper (*Kosciuscola tristis*). *Behavioral Ecology and Sociobiology*, **67**, 439.

Veron, J.E.N. (1974). The role of physiological colour change in the thermoregulation of *Austrolestes annulosus* (Selys) (Odonata). *Australian Journal of Zoology*, **22**, 457.

Vigneron, J.P., Pasteels, J.M., Windsor, D.M., et al. (2007). Switchable reflector in the Panamanian tortoise beetle *Charidotella egregia* (Chrysomelidae: Cassidinae). *Physical Review E*, **76**, 31907.

Viviani, V.R. (2002). 'The origin, diversity, and structure function relationships of insect luciferases. *Cellular and Molecular Life Sciences*, **59**, 1833.

von Frisch, K., Wenner, A.M., and Johnson, D.L. (1967). Honeybees: do they use direction and distance information provided by their dancers? *Science*, **158**, 1072.

Vorobyev, M., Brandt, R., Peitsch, D., Laughlin, S.B., and Menzel, R. (2001). Colour thresholds and receptor noise: behavior and physiology compared. *Vision Research*, **41**, 639.

Vorobyev, M., and Osorio, D. (1998). Receptor noise as a determinant of colour thresholds. *Proceedings of the Royal Society of London. Series B: Biological Sciences*, **265**, 351.

Vukusic, P., Sambles, R., Lawrence, C., and Wakely, G. (2001). Sculpted-multilayer optical effects in two species of *Papilio* butterfly. *Applied Optics*, **40**, 1116.

Wallace, A.R. (1877). The colors of animals and plants. *American Naturalist*, **11**, 641.

Wardill T.J., List, O., Li, X., et al. (2012). Multiple spectral inputs improve motion discrimination in the *Drosophila* visual system. *Science*, **336**, 925.

Warrant, E.J. (2016). Matched filtering and the ecology of vision in insects. In: G. von der Emde and E. Warrant (Eds) *The Ecology of Animal Senses*, pp. 143–67, Springer, Berlin.

Wedmann, S. (2010). A brief review of the fossil history of plant masquerade by insects. *Palaeontographica Abteilung B*, **283**, 175.

White, T.E. (2017). Jewelled spiders manipulate colour lure geometry to deceive prey. *Biology Letters*, **13**, 20170027.

White, T.E., Dalrymple, R.L., Herberstein, M.E., and Kemp, D.J. (2017). The perceptual similarity of orb-spider prey lures and flower colours. *Evolutionary Ecology*, **31**, 1–20.

White, T.E., and Kemp, D.J. (2016). Colour polymorphic lures target different visual channels in prey. *Evolution*, **70**, 1398.

White, T.E., Zeil, J., and Kemp, D.J. (2015). Signal design and courtship presentation coincide for highly biased delivery of an iridescent butterfly mating signal. *Evolution*, **69**, 14.

Whitney, H.M., Reed, A., Rands, S.A., Chittka, L., and Glover, B.J. (2016). Flower iridescence increases object detection in the insect visual system without compromising object identity. *Current Biology*, **26**, 802.

Wilkinson, G., and Reillo, P. (1994). Female choice response to artificial selection on an exaggerated male trait in a stalk-eyed fly. *Proceedings of the Royal Society B-Biological Sciences*, **255**, 1.

Wilts, B.D., Michielsen, K., De Raedt, H., and Stavenga, D.G. (2011). Iridescence and spectral filtering of the gyroid-type photonic crystals in *Parides sesostris* wing scales. *Interface Focus*, **2**, 681.

Wittkopp, P.J., and Beldade, P. (2009). Development and evolution of insect pigmentation: genetic mechanisms and the potential consequences of pleiotropy. *Seminars in Cell and Developmental Biology*, **20**, 65.

Woods Jr, W.A., Hendrickson, H., Mason, J., and Lewis, S.M. (2007). Energy and predation costs of firefly courtship signals. *American Naturalist*, **170**, 702.

Worthington, A.M., Berns, C.M., and Swallow, J.G. (2012). Size matters, but so does shape: quantifying complex shape changes in a sexually selected trait in stalk-eyed flies (Diptera: Diopsidae). *Biological Journal of the Linnean Society*, **106**, 104.

Zeil, J. (1983a). Sexual dimorphism in the visual system of flies: the free flight behavior of male bibionidae (Diptera). *Journal of Comparative Physiology A: Neuroethology, Sensory, Neural, and Behavioral Physiology*, **150**, 395.

Zeil, J. (1983b). Sexual dimorphism in the visual system of flies: the compound eyes and neural superposition in bibionidae (Diptera). *Journal of Comparative Physiology A: Neuroethology, Sensory, Neural, and Behavioral Physiology*, **150**, 379.

CHAPTER 12

Acoustic communication

Heiner Römer

Department of Zoology, University of Graz, Graz, Austria

12.1 Introduction

This chapter is about acoustic communication, but a definition of such is not as straightforward as one might imagine. While we can define communication as the production and transmission of a signal by the sender, which elicits a response in a receiver that is beneficial to both the signaller and receiver (Bradbury and Vehrencamp 2011; see also Chapters 10 and 11), we still need to precisely define the signal modality. Consider, for example, a male cricket that is sitting on a leaf of a plant and producing a calling song to attract females. This sound is a mechanical vibration that travels as alternating waves of high and low pressure through the air. The pressure variation is accompanied by air particle movements. Close to a sound source, the energy due to the particle displacements may be much greater than the sound pressure. The amplitude of the particle displacement decreases by $1/r^2$, where r = distance from the source; this differs from the propagated pressure variation, which decreases by only $1/r$. Therefore, we can define near-field sound as the region within which the particle displacement is greater than the pressure component, and far-field sound as the region within which the pressure becomes the dominating component.

Therefore, the same motor act of the male cricket may cause three different mechanical disturbances that propagate over different distances:

- a pressure variation that propagates as a far-field sound, which can be detected by tympanal hearing organs over tens or even hundreds of metres;
- a near-field medium motion, which travels over distances of only a few centimetres and can be detected by wind-sensitive filiform hairs located on the cerci or by the antennae of a receiver. In addition, the movement of the wings that occurs during singing also causes the insect to vibrate and, because it is standing on the leaf, this produces
- a substrate vibration, which results in the propagation of torsional and/or bending waves perpendicular to the surface of the substrate.

This vibration can be detected by different kinds of 'ears', including vibration receptors, such as **subgenual organs** in the legs, if the receiver has contact with the same substrate. Although most authors of textbooks and reviews would place these different mechanical disturbances under different headings or in separate chapters, we address all kinds of sounds and vibrations here and, thus, with different types of communication signals and receptor organs.

12.2 The behavioral context for signalling

12.2.1 Mate attraction

If individuals of a species live at relatively low densities, the probability of finding a mate just by chance is rather low. Without some kind of

Römer, H., *Acoustic communication*. In: *Insect Behavior: From mechanisms to ecological and evolutionary consequences.* Edited by Alex Córdoba-Aguilar, Daniel González-Tokman, and Isaac González-Santoyo: Oxford University Press (2018).
© Oxford University Press.
DOI: 10.1093/oso/9780198797500.003.0012

signalling, this would only occur due to random encounters. The evolution of a long-range signal that propagates well in an environment and conveys information about the sex, species identity, and location of the signaller would greatly increase the chance that the signaller finds a mate. Far-field sound is ideal for this purpose: it can be used during day or night; it is not blocked by vegetation; it fades out quickly so that fast temporal modulation may contain information about species identity (Figure 12.1); and receivers equipped with ears providing some degree of directionality (Section 12.5.) can locate the sender. Therefore, it is not surprising that some insect taxa, such as grasshoppers, crickets, katydids, cicadas, and even some groups of moths, have evolved airborne sound signals for acoustic communication (von Helversen and von Helversen 1994; Greenfield 2014).

Several signalling and mate-finding strategies have evolved (Bradbury and Vehrencamp 2011; see Chapter 13). A typical example is that of a stationary singing male and a female that performs sensory-guided movement (i.e. **phonotaxis**) towards the male. Note that, in this case, males are 'speculative signallers' since they do not receive sensory feedback that indicates the presence of a female within hearing distance of the signal. Such males would be expected to produce songs that have far-reaching properties in order to attract as many potential females as possible. Particularly in situations in which females occur in low densities, a male could

increase his chances of finding a mate via calling by moving from one calling site to another.

In duetting communication systems, both males and females produce sound. Duetting has been thoroughly described in some phaneropterine and ephippigerine katydids, grasshoppers, and cicadas (review in Bailey 2003). Duetting can also involve the precise temporal interaction of air-borne sound and vibrational signals (Rajaraman et al. 2015). In almost all cases reported, the male starts with a call, and the female then replies, often with one or a few extremely short sound pulses. The female reply is typically produced with a predictable, species-specific latency, and males respond by phonotaxis toward the female only when her reply is received within a narrow time window (Heller and von Helversen 1986). However, in other duetting systems, the female performs phonotaxis, or even members of both sexes move to achieve pair formation (Bailey 2003). Note that the signalling and mate-finding strategies, and in particular which sex performs the signalling and movement, have important consequences with regard to the costs involved in communication (Section 12.7.).

12.2.2 Agonistic interactions between males

Prior to displaying overt aggression, males may compete with other males for favourable calling sites, increasing the active space of their signals and, thus, their chances of attracting females (Arak and Eiriksson 1992). Males of many species form aggregations where they are within earshot of other males, driving acoustic rivalry to attract females. In these chorus situations, males may change their calls in response to those produced by other males in a way favoured by females in a choice of several call alternatives. Changes in acoustic signals may be made in terms of their duration, rate, or timing with respect to competing calls. The latter may result in spectacular patterns of call synchrony or alternation in choruses of males (Greenfield 2015).

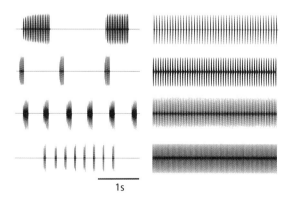

Figure 12.1 The temporal pattern of calling songs of eight species of crickets recorded in the nocturnal Panamanian rainforest. Note that all recordings are presented at the same time scale, demonstrating that each temporal pattern is characteristic for each species.
From Schmidt et al. (2012).

12.2.3 Spacing

Signalling males in aggregations often maintain a certain inter-male distance, either as a result of direct physical aggression or by responding to acoustic cues produced by competing males (Römer and Bailey

1986). A regular distribution of males within such aggregations may be the result of the attraction toward distant males and repulsion away from males calling nearby. One simple interpretation of such spacing behavior is that it allows each male in a chorus to broadcast his signal within a zone that is free of competing conspecifics.

12.2.4 Courtship

When attraction to a potential mate via acoustic signalling is successful, a multimodal courtship display usually precedes mating. Acoustic and vibratory signals, apart from visual and/or olfactory cues, can be components of these displays. Because male and female are in close proximity of one another, both near-field and far-field sound can be involved. For example, field crickets produce a close-range courtship song, which differs from the calling song in its temporal structure, frequency, and intensity (Alexander 1961). The courtship song is important for the male's mating success, because it induces the female to mount the male (Balakrishnan and Pollack 1996). Most *Drosophila* species produce two types of songs during courtship—a continuous 'sine song' and a train of pulses, the so-called 'pulse song' (review Tauber and Eberl 2003). The inter-pulse intervals in the pulse song are species-specific and, thus, can be used for species identification, whereas the function of the sine song appears to be to acoustically prime the females prior to courtship. Since the vibrating wings of the male are only 2.5–5 mm away from the female and produce a directional sound output that is aimed towards the female, this communication will remain mostly 'private' to a pair of flies, without stimulating other nearby males. One important aspect of the courtship is that the female signals her readiness for copulation by becoming immobile. When the male shakes his abdomen, substrate-borne vibrations are generated that induce the female to stop walking (Fabre et al. 2012).

12.3 Signal production

12.3.1 Far-field sound

Once again, we need to differentiate between the production of near-field and far-field sound, as well as substrate vibration. Stridulation is the most common method used by insects to produce sound signals in the acoustic far field. This is not surprising, given that the surface of an insect's cuticle may form into ridges and spines, and if one roughened surface is moved across another, sound is produced. More elaborate mechanisms involve the movement of a plectrum (scraper) across a file. Even genitalia may be modified into a stridulatory apparatus, as in some pyrallid and sphingid moths. More familiar examples of stridulation are the calling and courtship songs of grasshoppers, crickets, and katydids. The former use both the hind legs and wings to produce sound, whereby a file located on the inner surface of the femur is moved across the edge of the wing (the plectrum). In another subfamily of grasshoppers (Oedipodinae), the legs function as the plectrum, and the wings as files. The impact of each of the cuticular pegs on the wing produces a rapidly damped sound pulse, with the entire movement of the leg producing a sound syllable. The species-specific pattern of amplitude modulation of the song is produced by the neural network, which activates the up- und down movement of the hind legs. This temporal pattern is complicated by the fact that both legs stridulate at the same time and are rarely synchronous. Instead, they are in a constant phase relationship (for various mechanisms of sound production see Elsner 1983 and Gerhardt and Huber 2002 for review).

The stridulatory apparatus in crickets and katydids consists of the primary sound generator (the file and scraper on the forewings) and the secondary sound radiator (specific forewing membranes). In almost all species, sound is produced only through the closing stroke of the forewings. During this closure, the scraper hits the file teeth, resulting in the production of a wing vibration. Because muscles cannot produce vibrations in the kilohertz range, stridulation is a process of frequency multiplication (Ewing 1989). By measuring the wing vibrations during singing in *Gryllus bimaculatus* the two wings were shown to resonate at one identical frequency—the song carrier frequency—although their resonant frequencies were significantly different when studied in isolation (Montealegre-Z et al. 2011). Thus, cricket wings work as coupled oscillators that control the mechanical oscillations together to generate the pure tone, species-specific song.

Unlike crickets, katydids have strongly asymmetrical forewings. Their left forewing has a reduced role in sound radiation (Montealegre-Z and Mason 2005). Katydids can either produce pure-tone sounds with narrow frequency spectra (with carrier frequencies below 2 kHz and up to extremely high ultrasonics above 100 kHz) or broadband sounds (noisy signals with broad frequency spectra). Pure tone signals result from resonant stridulation, while broadband signals result from a non-resonant mechanism (Montealegre-Z 2009).

Unlike these frequency-multiplier devices, tymbal organs produce sound in a different way. In the well-studied cicadas, the tymbal is located in the first abdominal segment, the lateral side of which is modified into a pair of ribbed cuticular membranes (i.e. tymbals). Females lack tymbals. A large tymbal muscle is attached to the V-shaped apodeme of the sternum. Sound is produced when the powerful tymbal muscle contracts. The convex tymbal is then loaded with mechanical energy and, after a few milliseconds, develops enough tension to overcome its stiffness so that it buckles in and produces a short sound pulse, the frequency of which is determined by the damped free vibration of the tymbal (Fonseca 2014). In the case of the Australian cicada *Cystosoma australasiae*, this low frequency vibration is strongly amplified by resonant air sacs surrounding the tymbal, so that incredible sound pressure levels of 120 dB can be produced (Young and Bennet-Clark 1995).

12.3.2 Near-field sound

In taxa such as Diptera and Hymenoptera, communication in the acoustic near-field plays an important role in the contexts of courtship and other social interactions. Consider, for example, an insect as small as the fruit fly. When in contact with a female, the male *Drosophila melanogaster* vibrates his wings to sing the 'sine' and 'pulse' song to females. During courtship, the male's vibrating wing produce a particle velocity well above the thresholds for antennal receivers that are sensitive to the particle motion (Section 12.8.2.1.). In a similar way, near-field sound appears to play a major role in the recruitment of follower bees (*Apis mellifera*) during the wagging dances they perform on vertical combs in the bee

hive. The dancer vibrates her wings at a frequency of 200–300 Hz with peak velocities of about 1 m/s, producing oscillating air currents close to the abdomen of the dancing bee. The air current amplitudes decrease rapidly with distance. Most of the follower bees hold their antennae (the particle velocity receivers) near the dancer in the zone of maximum velocity of the oscillating air currents (Michelsen et al. 1987).

The example of the honeybee also shows that the seeming disadvantage of near-field sound, namely the fact that it rapidly declines with distance, can become an advantageous property when considering the social situations in which communication takes place. Several dancers may perform their wagging dances in the hive simultaneously, separated by only a few centimetres. In this case, the followers need to listen to one dancer bee at a time in order to avoid confusing the information with that of a neighbouring dancer. The strong decrease in the particle velocity with the third power of distance guarantees that 'private' communication can be maintained with a single dancer (Michelsen et al. 1987). In a similar way, the particle velocity of the sound radiated by the *Drosophila* wing drops steeply by 18 dB per doubling distance (Bennet-Clark, 1971). Thus, a *Drosophila* male that communicates with a female using his 'love song' avoids alerting another competing male located some distance away to the presence of a courting female.

12.3.3 Substrate vibration

Substrate vibration has received much less attention in the past as a means to communicate, simply because special devices and methods are needed to record these vibrations (e.g. laser Doppler vibrometry or accelerometers based on piezoelectric principles). However, as compared to air-borne sound, the ability to produce and detect substrate vibrations is far more widespread in insects, with an estimated 195,000 described insect species across eighteen orders (Cocroft and Rodríguez 2005). For more details about taxa using substrate vibrations for communication, see Virant-Doberlet and Čokl (2004), Cocroft and Rodríguez (2005) and Hill (2008). Unlike what we have learned in the context of sound production in far-field sound, insects can produce substrate vibrations without specialized structures. In principle, any body part can be involved, and four different mechanisms for the generation of vibrational signals are known: stridulation, **tremulation**, drumming, and tymbal buckling.

In the case of the cricket example, stridulation involves rubbing the forewings, which are specialized for the production of air-borne sound, against each other. However, rubbing of non-specialized body parts, such as mandibles against a substrate (body-substrate stridulation), can also produce signals that are transmitted as bending waves through the substrate. Tremulation does not involve the impact of a body part against the substrate—as is the case during drumming—but rather the process of grasping a structure and shaking or trembling using the whole body or abdomen. The vibrations induced are transferred to the substrate via the insects legs if the body mass can generate sufficient energy to overcome the inertia of the substrate. Surprisingly, however, even insects with a body mass well below 0.5 mg may produce such vibrational signals for communication (Fabre et al. 2012). While drumming (also called percussion), the insect hits the substrate with one of its body parts to induce vibrations (see Yack 2016 for review). Tymbals were described earlier as elaborate structures found in cicadas that can be used to produce high-amplitude, airborne sounds. However, tymbals are also common in many small Homoptera, but their design has been modified so that they produce negligible air-borne sound, but instead transmit their mechanical energy via the substrate. A very special type of communication with vibrational signals has been observed in water striders and whirligig beetles. Their domain is the elastic interface between air and water on the water surface. They extract information from the ripples on the water surface, which can be produced by prey that are trapped on the surface, but also by conspecific mates. For example, males of *Gerris remigis* produce surface ripples at 80–90 Hz in competition with a competing male, and low-frequency ripples of 3–10 Hz are produced to attract females (see Bleckmann 1985, for review).

Beyond these four categories of vibrational signal production, insects may produce substrate vibrations while singing, when the wing or leg movements may simultaneously transmit a vibration to the substrate through direct contact. Although these vibrations are incidental by-products of singing, nevertheless they are important for receivers and allow them, for example, to localize mates in a complex environment.

12.4 The transmission channel for the signal

The signal produced by the sender is rarely the one perceived by receivers because it may undergo substantial changes in terms of its amplitude, temporal properties, and frequency content on the transmission channel. Except for near-field sound communication, where sender and receiver are almost in direct contact with one another, we need to consider the physical properties of the transmission channel for far-field sound and substrate vibration, which are strikingly different. Such differences also result in large differences in the active space of signals or in the way signals are localized (Section 12.5.). In addition, the acoustic background by abiotic noise and/or sound signals from other species may render the signal less detectable due to masking interference.

12.4.1 Transmission of air-borne sound: geometric spreading, excess attenuation, and degradation of temporal cues

When a sound wave propagates away from a source, its pressure inevitably decreases with increasing distance, as a result of the geometry of the air space occupied by the sound energy. As a rule of thumb, geometric spreading causes a 6 dB decrease for each doubling of distance (6 dB/dd), which is the minimum attenuation for a signal under ideal conditions (Figure 12.2a). This is a rather counterintuitive relationship because it means that a sound wave that travels from 1 to 2 m suffers the same amount of 6 dB attenuation as when the same sound, after arriving at a point 250 m from its origin travels another 250 m.

Attenuation in excess to geometric spreading can result from absorption, scattering by vegetation and turbulence, or sound interference in stratified environments. Absorption is frequency dependent,

(a)

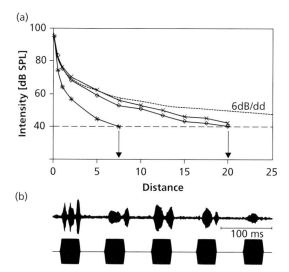

(b)

Figure 12.2 (a) Attenuation of a sound signal over distance. Compared with the geometric spreading of sound under ideal conditions (6 dB/per doubling of distance) sound signals suffer from excess attenuation in real insect habitats, depending on the density of vegetation and/or the location of the receiver. Thus, the detection distance for the signal for a receiver with a hearing threshold of 40 dB SPL (stippled line) would be either 7.5 or 20 m (arrows). (b) Comparison of the broadcast and perceived temporal pattern of a sound signal (lower and upper panel, respectively) after transmission over 10 meters.

and absorption coefficients have been calculated at 0.5 dB/100 m for a frequency of 1 kHz and 10 dB/100 m, for 10 kHz. Thus, this phenomenon may be of little relevance for many insects that call at these lower frequencies and have hearing distances that are much less than 50 m. However, for frequencies as high as 100 kHz, attenuation due to absorption may amount to 10 dB/m, which is of particular interest in terms of the sensory arms race between insectivorous bats that hunt for insects using ultrasonic echolocation, and the insect prey listening to these sounds (Römer 1998, for a review of these effects for insect sound communication).

Insects often live within vegetation, with the consequence that sound signals have a strong advantage over visual signals because they can circumvent obstacles on the transmission channel. However, the vegetation produces scattering (or diffraction), reducing the intensity along the broadcast beam direction by redirecting the sound energy into other directions. This causes attenuation in excess of geometric spreading (Figure 12.2a). Again, there is a strong positive correlation between sound frequency and the degree of excess attenuation due to scattering. Due to the frequency-dependency nature of absorption and scattering, the transmission channel essentially acts as a low-pass filter, which can have negative consequences for communication since the information in the filtered frequency components is lost over greater distances. In general, any kind of attenuation will reduce the active space of a signal, which is defined as the area within which the signal amplitude is above the detection threshold of potential receivers. The size of this space is determined by:

- the amplitude of the signal at the source;
- the attenuation during transmission through the medium;
- the amplitude of the background noise;
- the threshold of the sensory system of receivers (Brenowitz 1982).

Since information about species identity is primarily encoded in the temporal structures of the songs of insects (Figure 12.1), any degradation of temporal parameters would impose limitations on long-range communication. Reverberations and amplitude fluctuations induced by wind or atmospheric turbulence produce distortions in the time domain (Figure 12.2b) and vary in magnitude according to the weather conditions, time of day and carrier frequency (Wiley and Richards 1978).

12.4.2 Noise in the air-borne sound channel

An important characteristic of the transmission channel for air-borne sound and vibration signals is the amount of masking noise, which stems from abiotic sources, such as wind, running water, or rain, but also biotic sources and, in particular, the signalling of other species in the same modality at the same time and location (Brumm and Slabbekoorn 2005; Cocroft and Rodriguez 2005; Römer 2013). Masking occurs when the detection or discrimination threshold for a signal is increased in the presence of background noise. An increasing amount of evidence is accumulating from different taxa including insects that anthropogenic noise (noise generated by human activities such as transportation) presents animals with a novel challenge with respect to communication (Morley et al. 2014). Particularly strong masking occurs in habitats in which many species are signalling at the same time, such as in the case of the nocturnal choruses heard in tropical rainforests, which consist primarily of crickets, katydids, and cicadas calling at the same time and place (see Figure 12.3). Depending on other ecological variables such as moonlight intensity or rain, the background noise level increases strongly after sunset and can reach values of 65 dB SPL.

The fact that different species are able to communicate successfully under these noisy conditions suggests that they have developed strategies to deal with the detection and recognition of relevant signals in the presence of high levels of masking noise (Römer 2013; Schmidt and Balakrishnan 2015). Adaptive changes in signal structures over evolutionary time may be one solution, while the immediate signalling behavior of the sender, such as signalling at times when, or locations where there is less noise, is another one. Finally, the selection pressure imposed by the noisy environment may result in adaptations in the sensory system of receivers that allow them to cope with the challenge (Römer 2013). With respect to solutions based on the signaller, the ecological niche concept has been used as a framework, and researchers have argued that the transmission channel represents an ecological resource for acoustically-communicating species.

Sharing this resource with an increasing number of sound sources will increase the probability of interference and masking (Schmidt et al. 2012; Jain et al. 2014). Although several studies have been conducted, mainly examining calling in crickets, katydids, and cicadas, and some evidence has been found for the partitioning of the **acoustic niche** into time, space, and frequency spectra, the general theme that emerges is that a high degree of overlap exists among these niche dimensions (Schmidt and Balakrishnan 2015; Balakrishnan 2016). One should keep in mind, however, that selection pressures other than masking interference (e.g. gradients in predation pressure) exist that may force insects to live and communicate at given times and locations.

12.4.3 Transmission of substrate vibrations

As is the case in air-borne sound transmission, where we have seen that insects' habitats may be highly variable in terms of their transmission properties, many insects select their host plants for purposes other than vibrational signalling, such as for feeding, and these plants can differ greatly in their mechanical properties. The waves responsible for the transmission of vibrational signals are usually bending waves. The speed of wave propagation in the substrate depends on the thickness of the structure and the square root of the wave frequency. The transmission is slow as compared to the speed of sound in air or water. In their attempt to compare vibrational transmission in different plants, Michelsen et al. (1982) found similar propagation velocities in six plant species. Because the propagation velocity changes with the square root of frequency, high frequencies travel faster than low frequencies, with the consequence that the time-frequency structure of a vibrational signal with a complex spectrum gradually changes (a phenomenon called **dispersive propagation**). The authors found no clear relationship between the signal amplitude and the distance from the source. The recorded songs of bugs or plant hoppers often had higher amplitudes at the top of the plant or on a far-away leaf than on the stem close to the signaller. Thus, the signal amplitude does not decrease monotonically with distance (Miklas et al. 2001). The recorded absolute acceleration amplitudes

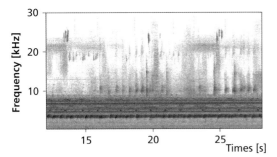

Figure 12.3 Sonogram of the multi-species chorus of insects recorded after sunset in the Panamanian rainforest. Note the different frequency bands between 3 and 7.5 kHz produced predominantly by crickets, and those at higher frequencies produced by katydids.

of the signals produced by the insects also varied strongly between 0.06–8 ms⁻², but were well above the known perception threshold of the vibration receptors. When stimulating the plant with short sine waves, the plant responded with much longer vibrations, which travelled up and down the plant several times due to reflections at the ends and low attenuation rates.

12.4.4 Noise in the vibratory channel

The strong masking effects as in the air-borne sound channel are not expected in the vibratory transmission channel, since this would only include different species and/or individuals signalling on the same plant. This channel, nevertheless, represents a complex vibrational environment, and animals living in this environment experience interference from abiotic sources, such as wind and rain, as well as biotic sources, such as the signals of competing individuals. Wind is the major form of abiotic noise experienced by insects that communicate using substrate vibrations. The wind's strength varies over time and is stronger in open habitats such as grasslands or in the forest canopy. Most of the energy in wind-generated noise has low frequencies, which renders communication at these frequencies more difficult (Cocroft and Rodriguez 2005).

12.5 Localization of the signal

Identifying a signal as species-specific is only one task performed during communication. In almost all situations in which long-range signals are involved, they also need be correctly localized. This is a problem for small insects for two reasons: Being equipped with bilateral pairs of ears (with the exception of the mantid ear, which exists as a single, 'cyclopean' ear; Yager 1999) allows the insect to principally make use of **binaural cues** for sound localization. However, because the distances between the ears are usually so small, interaural time differences are also minute. In addition, since the ratio of body size to the wavelength of the sound signal is rather unfavourable in terms of diffractive effects, the interaural intensity differences are also quite small. These difficulties have favoured the evolution of amazingly complex biophysical solutions for directional hearing, such

as pressure and pressure difference receivers, or mechanically coupled ears, that have an acuity with a spatial resolution of 1–2° under ideal laboratory conditions (parasitoid flies: Mason et al. 2001; field crickets: Schöneich and Hedwig 2010). In addition to providing such excellent acuity in the azimuth, these ears also allow the localization of elevated (or depressed) sound sources (Rheinlaender et al. 2007).

The small size of most insects also presents a problem in terms of localizing a vibratory signal. Although directional searching for substrate vibrations has been reported in various taxa (Virant-Doberlet et al. 2006), the actual sensory cues involved are still a matter of debate. An intriguing possibility would be to evaluate the arrival times of signals at the receptors located on spatially-separated legs. However, although the propagation speed of vibrations in plants is less than that in air, the speed is still between 40 and 80 m/s. For a stink bug with a maximum leg span of 1 cm, this speed would create time-of-arrival differences between 125 and 250 μs, and for many smaller insects even smaller ones. These small time differences in the microsecond range are the lowest behaviorally determined thresholds found in scorpions, but given the exceptional time resolution found, for example, in the directional hearing of the parasitoid fly *Ormia* (Mason et al. 2001), future studies on the sensory basis of time processing of vibrational signals may reveal surprising results.

12.6 The costs of acoustic communication: 'unintended receivers' (see 'eavesdropping' in Chapter 10)

Whereas the fitness benefits of attracting mates using an acoustic signal are obvious, we also have to consider the costs of signalling. Apart from the substantial amount of energy invested in muscle activity that is required to produce sound (Prestwich 1994), and the metabolic energy required to maintain the sensory and central nervous systems to process the signal (Laughlin et al. 1998), the same signal may also be detected by so-called 'unintended receivers' such as predators, parasitoids, or male rivals, with dramatic consequences with regard to the survival and fitness of the signaller (Zuk and Kolluru 1998). For example, calls of neotropical katydids not only attract females, but also bats, which passively listen for such calls (Belwood and Morris 1987). It has been suggested that the short duration, low-duty- cycle calls produced by forest katydids, which are under strong selection from these bats species, represent an evolutionary adaptation to this selection pressure because they are more difficult to detect or locate. Although this may be true, katydid calls recorded in a paleotropical rain forest assemblage were neither of short duration nor had low-duty cycles (Raghuram et al. 2015). If acoustic signalling alone causes predation by bats and other listening predators, non-calling females would be free of such costs. However, when females move towards calling males, their risk of predation can be as high as or even higher than that of males (Raghuram et al. 2015).

Acoustic signals can also be conspicuous to parasitoids such as the fly *Ormia ochracea*, which deposits its larvae on male field crickets. These larvae then develop in the male and eventually kill it at the end of their developmental cycle (Cade 1975). To find their host, female flies localize the male cricket by his call and have evolved a very sensitive ear only for this purpose (Robert et al. 1992; Lakes-Harlan and Lehmann 2015). Acoustically-signalling katydids and cicadas also suffer from parasitoid flies. In Europe, the fly *Therobia leonidei* parasitizes species of the genus *Poecilimon* (Phaneropterinae) (Lakes-Harlan and Lehmann 2015). When a comparison between two host species, which differed in the number of pulses per chirp was made, the species with the higher number of pulses/chirp was parasitized at higher levels (Lehmann and Heller 1998). As in the case of predation by passively listening bats, this indicates that insects that produce more conspicuous calls are easier to detect and/or localize.

12.7 The receiver: insect ears

For successful communication to occur, receivers must have sensory organs and the associated nervous system to detect and localize a signal, discriminate the signal from noise and discriminate among different signal variants. We can examine insect audition from different viewpoints: phylogeny, the behavioral context in which ears are used, the basic *bauplan*, or the nature of the physical stimulus for their evolved function. Given that a broader view of acoustic communication is presented in this chapter, including the topics of substrate vibration and communication by near-field sound, we also must differentiate between the sensory organs used to detect these different stimuli. Excellent in-depth reviews exist on various aspects of hearing organs, which are highly recommended for interested readers who wish to explore this topic in detail (Hoy and Robert 1996; Yager 1999; Gerhardt and Huber 2002; Yack 2004; Robert 2005; Strauß and Lakes-Harlan 2014; Göpfert and Hennig 2016).

12.7.1 Evolution of ears in two behavioral contexts: intraspecific communication and predator avoidance

A proposed scenario for the evolution of hearing organs is as follows. After insects took to land about 400 million years ago, receptors rapidly evolved that allowed the detection of substrate vibrations. These insects were already equipped with **mechanoreceptors** that functioned as proprioceptors, for example, to monitor the movements of body parts. Since vibrations in the substrate induce movements of the insect body and appendages such as legs and antennae, which are in direct contact with the substrate, these mechanoreceptors were pre-adapted for the transition from a proprioceptive to an exteroceptive function monitoring vibrations of the substrate. By contrast, the transition from a **proprioreceptor** to a receptor monitoring far-field airborne sound was more complicated and involved the evolution of tympana and associated structures (see later). This may explain that the use of substrate vibrations for communication is far more widespread in insects as compared with the use of airborne sound (Cocroft and Rodríguez 2005).

The ability to hear airborne sound evolved much later and independently more than twenty times in several insect orders (Orthoptera, Mantodea, Blattodea, Hemiptera, Hymenoptera, Coleoptera, Neuroptera, Lepidoptera, and Diptera; for reviews, see Fullard and Yack 1993; Hoy and Robert 1996; Yager 1999; Greenfield 2016). Two major types of selection pressure drove the development of structures that allowed insects to hear far-field sound with tympanal ears—intraspecific communication and the detection of predators.

Since species in some of these groups use both elaborate acoustic communication and auditory-evoked predator avoidance behavior (mainly in response to ultrasound), the question is whether hearing evolved first and communication later, or whether the primary function of auditory systems was to communicate, and these systems later became adapted to function in acoustically-mediated predator avoidance. The latter appears to be the case in katydids, where fossil evidence could be used to trace the evolution of hearing and communication. Katydids could already produce sounds and may have communicated acoustically about 165 million years ago, long before the appearance of bats (sixty-five million years ago), which then may have driven the evolution of ultrasonic hearing and predator avoidance behavior (Conner and Corcoran 2012; Pollack 2016). The situation is even more complicated because hearing has secondarily evolved as a function of intraspecific communication in many groups in which hearing seems to have initially been used for predator detection (such as the Lepidoptera, Greenfield 2014).

The example of hearing in mantids also demonstrates the difficulties faced when trying to reconstruct the selection pressure(s) driving the evolution of ears (Yager and Svenson 2008). While all other insects have bilateral pairs of ears, the ear of mantids is a single, 'cyclopean' ear located in the ventral metathorax. It functions as an effective 'bat-detector' and is highly sensitive to the ultrasound frequencies in the echolocation calls of bats (Yager 1999). By collecting data on functional anatomy and physiology, and superimposing these onto a phylogenetic tree constructed using molecular data, researchers could demonstrate that the cyclopean mantis ear evolved only once, approximately 120 million years ago. The neurophysiological response to ultrasound and the avoidance behavior were remarkably consistent across all taxa tested. Thus, mantids have an ancient, highly-conserved auditory system, which suggests that it evolved due the high selection pressure imposed by echolocating bats. This is, however, not the case: mantid

hearing predates the appearance of bats (by nearly sixty-five million years), so it may have originally functioned in communication, prey detection, or avoidance of non-bat predators (Yager and Svenson 2008).

12.7.2 Receptor organs for near-field and far-field sound and substrate vibrations

Given the differences in the nature of the physical stimuli that result in near-field and far-field sound and substrate vibrations, an overview of receptor organs needs to include those that are used to detect sound waves in air, water, or solids, as well as at the water/air interface. Some of the mechanoreceptors involved in this task may not even be specialized for detecting a particular kind of mechanical disturbance. An insect sensillum that usually monitors stress or strain in the cuticle as a proprioceptor may also respond to substrate vibrations or even higher-amplitude sound (Shaw 1994; Strauß and Lakes-Harlan 2014). The sensory organ located in the second segment of insect antennae (i.e. Johnston's organ) has several functions and can be used to detect acoustic near-field sound in mosquitoes and fruit flies (Albert and Kozlov 2016). Water surface waves function in auto-communicative echolocation in gyrinid beetles (Tucker 1969); to regulate insect flight or swimming (Gewecke et al. 1974); or in detection of the direction of substrate vibrations. One would expect that the anatomical features of receptor organs that detect these stimuli and their location on the insect body would be completely different. Indeed, we find such sensory organs distributed all over the insect's body, from the head and mouthparts to the thorax and abdomen, as well as on the wings or legs. Despite this variability, however, the basic sensory units that make up all these organs are similar and consist of so-called scolopidia. A single scolopidium includes one or more sensory neurons with a distal dendrite which extends through a tubular space formed by the scolopale cell and inserts into the extracellular scolopale cap. An attachment cell surrounds the distal portion of the scolopale cell and the scolopale cap, and it connects the scolopidium directly to the vibrating membrane, or indirectly via a ligament (reviewed in Field and Matheson 1998; Yager 1999;

Yack 2004). Deformation of the dendritic cilium depolarizes the sensory cell. Scolopidia are usually organized into groups called chordotonal organs and respond to substrate vibrations (such as subgenual organs in the legs), near-field sound (Johnston's organs in the antennae), or far-field sound (tympanal ears distributed at various locations of the insect body) depending on the accessory structures and where they are located.

Antennal ears for near-field sound

The antennal ears of fruit flies and mosquitoes detect the sound-induced particle motion of the surrounding air. They operate at frequencies below 1 kHz and only at short distances (a few centimetres). In *Drosophila melanogaster*, the antennal ear consists of two components: the scolopidia of the Johnston's organ (JO) in the pedicellus (second segment) and the third antennal segment (funiculus) together with an appendage called arista, both of which act as the sound receiver proper. The sensory neurons of JO are activated when the funiculus and arista begin to rotate about their longitudinal axis in response to sound stimulation. Unlike the *Drosophila* antenna, the mosquito antenna acts as an inverted pendulum that can swing in all directions within its plane of suspension. The other difference is that the organ contains about 200 scolopidia (corresponding to 500 sensory neurons) in *Drosophila* as compared to the 7500 scolopidia (15,000 sensory neurons) found in mosquitos (for review, see Albert and Kozlov 2016).

Single sensory hairs, located on the body surface of insects, represent another type of near-field receptor. An example of such is the filiform hairs on the cercus, which also respond to the air particle displacement that accompanies the pressure variation of a sound wave. Although these hairs have been primarily associated with a number of escape reactions in various insects (e.g. Casas and Dangles 2010), the same type of receptor could also be involved in intraspecific communication.

Tympanal ears for far-field sound

Despite the variation of body positions where tympanal ears are found in different insect orders, the basic *bauplan* of these ears is surprisingly similar and consists of three elements:

- A thin cuticular membrane (tympanum) that vibrates in response to sound.
- An air-filled cavity behind the tympanum.
- A chordotonal organ that is either directly or indirectly mechanically coupled to the tympanum (for reviews on tympanal hearing, see Robert and Hoy 1998; Yager 1999; Strauß and Lakes-Harlan 2014).

A classic example is the tympanal ear of the locust, which is located on each side of the first abdominal segment (Figure 12.4a). The tympanum is encircled

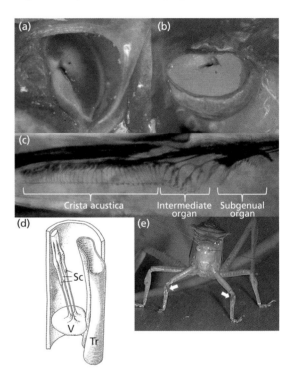

Figure 12.4 (a) The tympanal ear of locusts is located at the lateral body wall of the first abdominal segment. (b) The view from inside shows the receptor organ (Müller's organ) directly attached to the tympanum. (c) Histological staining of receptors of the complex tibial organ in the forelegs of a katydid (arrows in (e)), comprising those in the so-called *crista acustica* for air-borne sound perception, receptors in the intermediate organ responding to both sound and vibration, and the subgenual organ. (d) Subgenual organ in the green lacewing Chrysoperla carnea. (e) Forelegs of a Katydid *Chrysoperla carnea* (arrows) where is located the complex tibial organ. SC, three scolopidia; Tr, trachea; V, velum.

(c) Image courtesy of Johannes Strauß; unpublished. (d) Image modified from Devetak and Pabst (1994); *Tissue and Cell*, 26, 249–57.

by a rigid, sclerotized ring. The receptor organ (the so-called Müller's organ) contains 60–80 receptor cells and is directly attached to the inner side of the tympanum (Figure 12.4b).

In katydids, the ears are located on the prothoracic tibia. Each ear has a pair of tympanal membranes backed by an air-filled tracheal tube that is connected through the leg to the first spiracle on each side of the body. The mechanosensory organ is the *crista acustica* (Figure 12.4c), a linear array of receptors on the dorsal wall of the anterior tracheal branch. In the katydid forelegs, the *crista acustica* is associated with two other mechanosensory organs: the intermediate organ, which responds to both low-frequency sound and substrate vibration, and the subgenual organ. The linear array of sensilla in the *crista acustica* is tonotopically organized, whereby those on the proximal end are tuned to lower frequencies, and those at the distal end, to high-sonic, and ultrasonic frequencies. The total range of frequencies covered by these three organs is thus extremely wide, from less than 100 Hz for receptors in the subgenual organ, to more than 100 kHz for receptors in the *crista acoustica*. The number of receptors in this organ varies from about twenty to fifty among different katydid species, with a notable exception seen in a duetting katydid species, which displays sexual dimorphism: 115 receptors in males and eighty-six in females (Scherberich et al. 2016). Katydid mechanoreceptors are not in direct contact with the tympanal membranes, unlike those in the ears of locusts and other insects.

Striking differences exist in the number of auditory sensilla in the hearing organs, which varies from only one in some moths to more than 1500 and 2000 in cicadas and bladder grasshoppers, respectively. The highest numbers have been found in the antennal hearing organs of mosquitoes, the Johnston's organ, which has about 15,000 sensilla. The high number of sensilla in the primitive bladder grasshopper genus *Bullacris* and the much lower number in modern grasshoppers raises the question of whether high sensilla numbers represent the ancestral state and, if so, why reduction occurred (review Strauß and Stumpner 2015).

Vibration receptors

In order to be responsive to substrate vibrations, receptors and/or their accessory structures need to be coupled to the substrate in such a way that they can respond to vertical displacements or acceleration caused by substrate movement. Therefore, unlike tympanal ears, which have appeared at almost any position of the insect body over the course of evolution, an ideal position for vibration receptors is the distal leg region. Subgenual organs are the main vibration receptors and are located in the proximal tibia of all six legs in most insect taxa. In species where the organ is particularly sensitive to substrate vibrations, the scolopidia are perpendicularly attached to a septum that spans the leg cavity. In green lacewings (Neuroptera), substrate vibrations are involved in the context of mating and species recognition, and their reception is mediated by the subgenual organ (Yack 2016). The organ is comprised of only three sensilla, and their attachment cells form a septum that is loosely attached to the integument and trachea of the leg (Devetak and Pabst 1994; Figure 12.4d). Substrate vibrations transmitted to the leg cause an acceleration of the haemolymph against the septum, which stimulates the attached sensilla. The subgenual organ is the most sensitive vibration receptor that has been identified in insects, with thresholds (i.e. the lowest displacement or acceleration that elicits a response in sensory cells) of 0.22–5 nm in the cockroach, *Periplaneta americana* (Shaw 1994), or acceleration thresholds of 0.02 m/s^2 in green lacewings (Devetak and Amon 1997). For a more in-depth review of the diversity of the subgenual organs in different insect taxa, and other mechanosensory sensilla that are most commonly employed as vibration receptors, see Cokl and Virant-Doberlet 2003, Lakes-Harlan and Strauß 2014; Yack 2016).

12.8 Conclusion

Insects have evolved in an amazing variety of ways to communicate with mechanical disturbances, either as sound waves or bending waves in the substrate. Depending on the social and ecological context the broadcast range of communication signals may be limited to a few millimetres, such as in fruit flies or mosquitoes using near-field sound. Some taxa can hear far-field sound, so that the range over which communication can take place can reach several hundreds of meters. The diversity of insects as a taxonomic group of animals is reflected in the variety of ears, which evolved in the context of predator detection or intraspecific communication. As in all hearing animals, insects must accomplish the basic tasks of detecting, discriminating, and localizing sound sources. For signals propagated over some distance, this is challenging due to signal attenuation and degradation, and due to masking interference from biotic or abiotic sound sources. Despite the relative simplicity of their ears and nervous systems when compared with vertebrates, insects evolved solutions for all these tasks, so that insects may provide important models for studies of all aspects of hearing, from mechanisms of auditory transduction to niche partitioning.

Acknowledgements

For the author's own research cited in this review he would like to acknowledge the Smithsonian Tropical Research Institute (STRI) and the National Authority for Environment for logistical support on Barro Colorado Island, Panama, and for providing research permits. Funding was provided by the Austrian Science Fund (FWF) through grants P17986-B06, P20882-B09, P23896-B24, and P26072-B25. Thank also go to S. Crockett for proofreading the manuscript and Erik Schneider for support with the figures.

References

Albert, J.T., and Kozlov, A.S. (2016). Comparative aspects of hearing in vertebrates and insects with antennal ears. *Current Biology*, **26**, R1050.

Alexander, R.D. (1961). Aggressiveness, territoriality, and sexual behavior in field crickets (Orthoptera: Gryllidae). *Behaviour*, **17**, 130.

Arak, A., and Eiriksson, T. (1992). Choice of singing sites by male bushcrickets *(Tettigonia viridissima)* in relation to signal propagation. *Behavioral Ecology and Sociobiology*, **30**, 365.

Bailey, W.J. (2003). Insect duets: Underlying mechanisms and their evolution. *Physiological Entomology*, **28**, 157.

Balakrishnan, R., and Pollack, G.S. (1996). Recognition of courtship song in the field cricket, *Teleogryllus oceanicus. Animal Behaviour*, **51**, (2) 353.

Belwood, J., and Morris, G.K. (1987). Bat predation and its influence on calling behavior in neotropical katydids. *Science*, **238**, 64.

Bennet-Clark, H.C. (1971). Acoustics of insect song. *Nature*, **234**, 255.

Bleckmann, H. (1985). Perception of water surface waves: how surface waves are used for prey identification, prey localization, and intraspecific communication. In: H. Autrum, D. Ottoson, E.R. Perl, et al. (Eds) *Progress in Sensory Physiology*, pp. 147–66. Springer-Verlag, Berlin.

Bradbury, J.W., and Vehrencamp, S.L. (2011). *Principles of Animal Communication*. Sinauer Associates, Sunderland, MA.

Brenowitz, E.A. (1982). The active space of red-winged blackbird song. *Journal of Comparative Physiology*, **147**, 511.

Brumm, H. and Slabbekoorn, H. 2005. Acoustic communication in noise. *Advances in the Study of Behavior*, **35**, 151.

Cade, W.H. (1975). Acoustically orienting parasitoids: fly phonotaxis to cricket song. *Science*, **190**, 1312.

Casas, J., and Dangles, O. (2010). Physical ecology of fluid flow sensing in Arthropods. *Annual Review of Entomology*, **55**, 505.

Cocroft, R.B., and Rodríguez, R.L. (2005). The behavioral ecology of insect vibrational communication. *Bioscience*, **55**, 323.

Cokl, A., and Virant-Doberlet, M. (2003). Communication with substrate-borne signals in small plant-dwelling insects. *Annual Review of Entomology*, **48**, 29.

Conner, W.E., and Corcoran, A.J. (2012). Sound strategies: the 65–million-year-old battle between bats and insects. *Annual Review of Entomology*, **57**, 21.

Devetak, D., and Amon, T. (1997). Substrate vibration sensitivity of the leg scolopidial organs in the green lacewing, *Chrysoperla carnea. Journal of Insect Physiology*, **43**, 433.

Devetak, D., and Pabst, M. (1994). Structure of the subgenual organ in the green lacewing, *Chrysoperna carnea. Tissue and Cell*, **26**, 249.

Elsner, N. (1983). A neuroethological approach to the phylogeny of leg stridulation in gomphocerine grasshoppers. In: F. Huber & H. Markl (Eds), *Neuroethology and Behavioral Physiology*, pp. 54–68. Springer, Berlin.

Ewing, A.W. (1989). *Arthropod Bioacoustics: Neurobiology and Behavior*. Cornell University Press, Ithaca, NY.

Fabre, C.C.G., Hedwig, B., Conduit, G., et al. (2012). Substrate-borne vibratory communication during courtship in *Drosophila melanogaster. Current Biology*, **22**, 2180.

Field, L.H., and Matheson, T. (1998). Chordotonal organs of insects. *Advances in Insect Physiology*, **27**, 1.

Fonseca, P.J. (2014). Cicada acoustic communication. In: B. Hedwig (Ed.), *Insect hearing and acoustic communication*, pp. 81–100. Springer-Verlag, Berlin.

Fullard, J.H., and Yack, J.E. (1993). The evolutionary biology of insect hearing. *Trends in Ecology and Evolution*, **8**, 248.

Gerhardt, H.C., and Huber, F. (2002). *Acoustic Communication in Insects and Anurans: Common Problems and Diverse Solutions.* University of Chicago Press, Chicago, IL.

Gewecke, M., Heinzel, H.G., and Philippen, J. (1974). Role of antennae of the dragonfly *Orethrum cancellarum* in flight control. *Nature (London)*, **249**, 584.

Göpfert, M.C., and Hennig, R.M. (2016). Hearing in insects. *Annual Review of Entomology*, **61**, 257.

Greenfield, M.D. (2014). Acoustic communication in the nocturnal Lepidoptera. In: B. Hedwig (Ed.) *Insect Hearing and Acoustic Communication,* pp. 81–100. Springer-Verlag, Berlin.

Greenfield, M.D. (2015). Signal interactions and interference in insect choruses: singing and listening in the social environment. *Journal of Comparative Physiology A*, **201**, 143.

Greenfield, M.D. (2016). Evolution of acoustic communication in insects. In: G.S. Pollack, A. C. Mason, A.N. Popper, and R. R. Fay (Eds), *Insect Hearing*, Springer Handbook of Auditory Research, **55**, pp. 17–47. Springer International Publishing, Basel.

Heller, K.G., and Helversen, D. von (1986). Acoustic communication in phaneropterid bushcrickets: species-specific delay of female stridulatory response and matching male sensory time window. *Behavioral Ecology Sociobiology*, **18**, 189.

Helversen, O. von, and Helversen, D. von (1994). Forces driving coevolution of song and song recognition in grasshoppers. In: K. Schildberger and N. Elsner (Eds) *Neural Basis of Behavioral Adaptations,* Fortschritte der Zoologie **39**, pp. 253. Gustav Fischer Verlag, Stuttgart.

Hill, P.S.M. (2008). *Vibration Communication in Animals.* Harvard University Press, Cambridge, MA.

Hoy, R.R., and Robert, D. (1996). Tympanal hearing in insects. *Annual Review of Entomology*, **41**, 433.

Jain, M., Diwakar, S., Bahuleyan, J., Deb, R., and Balakrishnan, R. (2014). A rain forest dusk chorus: cacophony or sounds of silence?' *Evolutionary Ecology*, **28**, 1.

Lakes-Harlan, R., and Lehmann, G.U. (2015). Parasitoid flies exploiting acoustic communication of insects—comparative aspects of independent functional adaptations. *Journal of Comparative Physiology A: Neuroethology, Sensory, Neural, and Behavioral Physiology*, **201**(1), 123.

Laughlin, S.B., de Ruyter van Steveninck, R.R., and Anderson, J.C. (1998). The metabolic cost of neural information. *Nature Neuroscience*, **1**, 36.

Lehmann, G.U., and Heller, K.G. (1998). Bushcricket song structure and predation by the acoustically orienting parasitoid fly *Therobia leonidei* (Diptera: Tachinidae: Ormiini). *Behavioral Ecology and Sociobiology*, **43**(4–5) 239.

Mason, A.C., Oshinsky, M.L., and Hoy, R.R. (2001). Hyperacute directional hearing in a microscale auditory system. *Nature*, **410**, 686.

Michelsen, A., Fink, F., Gogala, M., and Traue, D. (1982). Plants as transmission channels for insect vibrational songs. *Behavioral Ecology Sociobiology*, **11**, 269.

Michelsen, A., Towne, W.F., Kirchner, W.H., and Kryger, P. (1987). Acoustic near field of a dancing honey bee. *Journal of Comparative Physiology A*, **161**, 633.

Miklas, N., Stritih, N., Cokl, A., Virant-Doberlet, M., and Renou, M. (2001). The influence of substrate on male responsiveness to the female calling song in *Nezara viridula*. *Journal of Insect Behavior*, **14**, 313.

Montealegre-Z, F. (2009). Scale effects and constraints for sound production in katydids (Orthoptera: Tettigoniidae): correlated evolution between morphology and signal parameters. *Journal of Evolutionary Biology*, **22**, 355.

Montealegre-Z, F., Jonsson, T., and Robert, D. (2011). Sound radiation and wing mechanics in stridulating field crickets (*Orthoptera: Gryllidae*). *Journal of Experimental Biology*, **214**, 2105.

Montealegre-Z, F., and Mason, A.C. (2005). The mechanics of sound production in *Panacanthus pallicornis* (Orthoptera: Tettigoniidae: Conocephalinae): the stridulatory motor patterns. *Journal of Experimental Biology*, **208**, 1219.

Morley, E.L., Jones, G., and Radford, A.N. (2014). The importance of invertebrates when considering the impacts of anthropogenic noise. *Proceedings of the Royal Society B: Biological Sciences*, **281**, 20132683. http://dx.doi.org/10.1098/rspb.2013.2683 (accessed 20 February 2018).

Pollack G.S. (2016). Hearing for defense. In: : G.S. Pollack, A.C. Mason, A.N. Popper, and R.R. Fay (Eds), *Insect Hearing*, Springer Handbook of Auditory Research, **55**. Springer International Publishing, Basel.

Prestwich, K.N. (1994). The energetics of acoustic signaling in anurans and insects. *American Zoologist*, **34**, 625.

Raghuram, H., Deb, R., Nandi, D., and Balakrishnan, R. (2015). Silent katydid females are at higher risk of bat predation than acoustically signalling katydid males. *Proceedings of the Royal Society of London B: Biological Sciences*, **282**, 20142319.

Rajaraman, K., Godthi, V., Pratap, R., and Balakrishnan, R. (2015). A novel acoustic-vibratory multimodal duet. *Journal of Experimental Biology*, **218**, 3042.

Rheinlaender J., Hartbauer M., and Römer H. (2007). Spatial orientation in the bushcricket *Leptophyes punctatissima* (Phaneropterinae; Orthoptera): I. Phonotaxis to elevated and depressed sound sources. *Journal of Comparative Physiology*, **193**, 313.

Robert, D. (2005). *Directional hearing in insects.* In: A.N. Popper and R. R. Fay (Eds) *Sound Source Localization*, pp.6–35. Springer, New York, NY.

Robert, D., Amoroso, J., and Hoy, R.R. (1992). The evolutionary convergence of hearing in a parasitoid fly and its cricket host. *Science*, **258**, 1135.

Robert, D., and Hoy, R.R. (1998). The evolutionary innovation of tympanal hearing in Diptera. In R.R. Hoy, A.N. Popper, and R. R. Fay (Eds) *Comparative Hearing: Insects*, p. 197. Springer, New York, NY.

Römer, H. (1998). The sensory ecology of acoustic communication in insects. In: R.R. Hoy, A.N. Popper, and R. R. Fay (Eds), *Comparative Hearing: Insects*, Springer Handbook of Auditory Research, pp. 63–96. Springer, New York, NY.

Römer, H. (2013). Masking by noise in acoustic insects: Problems and solutions. In: H. Brumm (Ed.), *Animal Communication and Noise, Animal Signals and Communication*, **2**, p. 33. Springer, New York, NY.

Römer, H., and Bailey, W.J. (1986). Insect hearing in the field. II. Male spacing behavior and correlated acoustic cues in the bushcricket *Mygalopsis marki*. *Journal of Comparative Physiology*, **159**, 627.

Scherberich, J., Hummel, J., Schöneich, S., and Nowotny, M. (2016). Auditory fovea in the ear of a duetting katydid shows male-specific adaptation to the female call. *Current Biology*, **26**, R1205.

Schmidt, A.K.D., and Balakrishnan, R. (2015). Ecology of acoustic signalling and the problem of masking interference in insects. *Journal of Comparative Physiology A*, **201**, 133.

Schmidt, A.K.D., Römer, H., and Riede, K. (2012). Spectral niche segregation and community organization in a tropical cricket assemblage. *Behavioural Ecology*, **24**. 470.

Schöneich, S., and Hedwig, B. (2010). Hyperacute directional hearing and phonotactic steering in the cricket (*Gryllus bimaculatus* deGeer). *PLoS One* **5**, e15141.

Shaw, S.R. (1994). Re-evaluation of the absolute threshold and response mode of the most sensitive known 'vibration' detector, the cockroach's subgenual organ: a cochlea-like displacement threshold and a direct response to sound. *Journal of Neurobiology*, **25**, 1167.

Strauß, J., and Lakes-Harlan, R. (2014). Evolutionary and phylogenetic origins of tympanal hearing organs in insects. In: B. Hedwig (Ed.), *Insect Hearing and Acoustic Communication, Animal Signals and Communication* **1**, p. 5. Springer-Verlag, Berlin.

Strauß, J., and Stumpner, A. (2015). Selective forces on origin, adaptation and reduction of tympanal ears in insects. *Journal of Comparative Physiology A*, **201**, 155.

Tauber, E., and Eberl, D.F. (2003). Acoustic communication in Drosophila. *Behavioural Processes*, **64**, 197.

Tucker, V.A. (1969). Wave-making by whirligig beetles (*Gyrinidae*). *Science*, **166**, 897.

Virant-Doberlet, M., and Čokl, A. (2004). Vibrational communication in insects. *Neotropical Entomology*, **33**, 121.

Virant-Doberlet, M., Cokl, A., and Zorovic, M. (2006). Substrate vibrations for orientation: from behavior to physiology. In: S. Drosopoulos and M.F. Claridge (Eds) *Insect Sound and Communication: Physiology, Behaviour, Ecology and Evolution*, pp. 81–97. Taylor and Francis, Boca Raton, FL.

Wiley, R.H., and Richards, D.G. (1978). Physical constraints on acoustic communication in the atmosphere: implications for the evolution of animal vocalizations. *Behavioral Ecology and Sociobiology*, **3**, 69–94.

Yack, J. (2016). Vibrational signaling. In: G. S. Pollack, A.C. Mason, A.N. Popper, and R.R. Fay (Eds), *Insect Hearing*, Springer Handbook of Auditory Research **55**, pp. 99–123. Springer International Publishing, Basel.

Yack, J.E. (2004). The structure and function of auditory chordotonal organs in insects. *Microscopy Research and Technique*, **63**, 315.

Yager, D.D. (1999). Structure, development, and evolution of insect auditory systems. *Microscopy Research and Technique*, **47**, 380.

Yager, D.D., and Svenson, G.J. (2008). Patterns of praying mantis auditory system evolution based on morphological, molecular, neurophysiological, and behavioural data. *Biological Journal of the Linnean Society*, **94**, 541.

Young, D., and Bennet-Clark, H.C. (1995). The role of the tymbal in cicada sound production. *Journal of Experimental Biology*, **198**, 1001.

Zuk, M., and Kolluru, G.R. (1998). Exploitation of sexual signals by predators and parasitoids. *Quarterly Review of Biology*, **73**, 415.

CHAPTER 13

Reproductive behavior

Rachel Olzer, Rebecca L. Ehrlich, Justa L. Heinen-Kay, Jessie Tanner, and Marlene Zuk

Department of Ecology, Evolution, and Behavior, University of Minnesota, Twin Cities, St Paul, Minnesota 55108, USA

13.1 Introduction to reproductive behavior

As is true in many animals, male and female insects markedly differ in their morphology and behavior. Unlike many animals, however, insects have a high reproductive rate and short generation times that give them the potential to quickly evolve. Additionally, the ease of rearing and manipulation of insects in the laboratory has provided key experimental insights into the operation of sexual selection. Because insects are so strikingly diverse, their adaptations are often beyond what other animals exhibit. This chapter shows how the diversity of insects has provided key insights into the complexity of reproductive behavior (Box 13.1).

To understand the complex reproductive behaviors of insects, we invoke the **economic defendability principle**, which states that resource defencedefence has costs (e.g. energy, time, risk) and benefits (e.g. priority access to mates; Brown 1964; see also Chapter 6). In reproduction, resources can be thought of as anything that is relatively limited, difficult to produce, renders an individual vulnerable to predation or is otherwise costly. For instance, males often use sexual signals to gain access to females for mating. These signals are often difficult to produce, energetically expensive, and may put the signalling

sex at a greater risk of predation (Zuk and Kolluru 1998). However, these elaborate signals also provide males with priority access to choosy females (Andersson 1994). Thus, through the lens of economic defendability, females can be thought of as a resource that males defend by signalling. We use economic defendability throughout this chapter to illustrate the ways that male and female behaviors increase the benefits of mating, while decreasing the costs. We begin by providing an overview of insect anatomy and physiology, followed by an examination of how males and females invest in reproduction. We then discuss the evolutionary forces acting on reproductive behavior and morphology, and close by describing the ways that insect systems are structured in relation to their sexual behavior. Throughout this we examine the behaviors exhibited before, during, and after copulation.

13.1.1 Basic anatomy and physiology

Insects employ two basic reproductive methods—sexual and asexual. Like much of the animal kingdom, most insects reproduce sexually—one sperm from a male fuses with one egg from a female to produce a zygote.

Female insects have a pair of ovaries containing dozens of ovarioles, within which egg cells are pro-

Olzer, R., Ehrlich, R. L., Heinen-Kay, J. L., Tanner, J., and Zuk, M., *Reproductive behavior.* In: *Insect Behavior: From mechanisms to ecological and evolutionary consequences.* Edited by Alex Córdoba-Aguilar, Daniel González-Tokman, and Isaac González-Santoyo: Oxford University Press (2018). © Oxford University Press.
DOI: 10.1093/oso/9780198797500.003.0013

Box 13.1 'Extreme' insect reproductive behaviors and morphologies

Parental care

Japanese redbug mothers work tirelessly to provision their offspring. In addition to providing unfertilized eggs as a food source, they trek 5–10 m at a time, carrying relatively massive fruits of their only host tree back to their nests (Filippi et al. 2002). Their voracious nymphs not only eat the fruits, but sometimes their younger siblings. Finally, when the mother has worked herself to death caring for her offspring, they devour her before leaving the nest for good (Hironaka et al. 2005).

Nuptial gifts

Katydid males produce large spermatophores that function as nuptial gifts. These nutrient-rich spermatophores are costly to produce and can represent huge portions of an individual male's body mass. For example, in *Poecilimon thessalicus*, the spermatophore makes up 36.7% of a male's body weight (McCartney et al. 2008).

Female genitalia

Although it is rare, some female insects have an erectile organ, which they use to collect and burst the spermatophore. Female cave insects of the genus *Neotrogla* have a genital organ, referred to as a gynosome (Yoshizawa et al. 2014). In addition to collecting sperm, the gynosome has spines that anchor the female to the male during copulation.

Female sperm storage

Some ant queens can store viable sperm for up to 30 years, producing viable offspring throughout that time (Ingram et al. 2013).

Sperm gigantism

Sperm size is likely the most extreme sexual ornament, as allometric slopes of sperm far exceeded slopes of other sexual selected traits (Lupold et al. 2016). *Drosophlia bifurca*, for example, produce single sperm that are approximately 6 cm long (Bjork and Pitnick 2006).

Mating plugs

Male honeybee drones have an 'explosive penis' that detaches during mating and functions as a mating plug (Oldroyd 2006).

---

duced (see Nation 2015 for details). As the egg cells mature, they pass through the reproductive tract and are stored in a genital chamber, known as the bursa copulatrix, where accessory glands create a protein-rich shell that surrounds each mature egg. Eggs are the functional reproductive unit for females. Male insects have a pair of testes containing hundreds of follicles that serve as the functional unit of the testes, much like the ovarioles in females (see Nation 2015 for details). Within the follicles, germ cells divide and pass through the reproductive tract into the accessory glands where a gelatinous shell encases each spermatozoon. This sperm package, called the spermatophore, can have one or many sperm, and serves as the functional reproductive unit of males.

During copulation, sperm are either directly or indirectly transferred into a spermatheca, where it can be stored for weeks, months, or even years at a time (Kaulenas 1992). Upon ovulation, an egg passes across the spermatheca and stimulates the release of sperm (Kaulenas 1992). Fertilization occurs as one sperm fuses with one egg. In addition to the spermatophore, some male insects also produce a gelatinous bolus called the spermatophylax, which functions as a nutritive supplement for females during copulation (Boggs and Gilbert 1979; Vahed 1998).

Many insects asexually reproduce via several mechanisms. Hermaphroditism, in which eggs and sperm are produced, and contained within one individual, can be seen in several scale insects and polyembryony, in which multiple clonal offspring are produced from a single egg cell, occurs in Hymenopteran parasitoids and twisted-wing parasites. Perhaps the best-understood mechanism for asexual reproduction is **parthenogenesis**, in which an egg grows and develops into an embryo in the absence of fertilization by sperm (Bell 1982). Offspring that are produced via parthenogenesis may be only female, as in aphids; only male, as in most Hymenoptera; or both. In most bees, ants, and wasps, females simultaneously employ sexual and asexual mechanisms of reproduction. Here, males are produced from unfertilized eggs laid by a single queen and females are produced via fertilization (Bell 1982).

13.1.2 Parental investment

Any expenditure, such as time, energy, or physical resources that an individual makes in an effort to increase the success of their offspring at the cost of investing in other components of an individual's fitness is termed parental investment (Trivers 1972). Parental investment is initially evident in the different contributions males and females make toward gamete production. The functional units of reproduction are eggs and sperm. In many species, females produce few, large and costly eggs, while males produce many, small and less expensive sperm. Because of the difference in gamete production, known as anisogamy, males and females often use different strategies to maximize reproductive success. In accordance with economic defendability, females are expected to be choosy about whom they mate with because eggs are a limiting resource in reproduction, and the benefits of mating with a particular male must outweigh the costs of producing an egg. Typically, females will mate less than males, while males will mate with many females. Additionally, females will often spend more time caring for offspring, as compared with males. However, because male insects sometimes provide nutrient-rich nuptial gifts that can be expensive to produce, eggs are not always the most expensive reproductive resource. Rather, female insects tend to produce small and inexpensive eggs, while sperm packages are typically very large and costly. Additionally, insects are unique in that males and females rarely provide any parental care to their offspring (Clutton-Brock 1991; see Chapter 14).

13.1.3 Sexual selection

Sexual selection was first proposed by Darwin (1871) in response to a major biological puzzle: why do individuals possess traits that seem to reduce their survival? In natural selection, ecological competition arises when individuals compete for access to resources. In sexual selection, competition arises when individuals compete for priority access to mates (Andersson 1994).

There are two primary forms of sexual selection:

- Intersexual selection via mate choice.
- Intrasexual selection via mate competition.

Sexual selection via mate choice and mate competition can lead to the elaboration of secondary sexual characteristics that are not directly involved in reproduction, but may help individuals monopolize the opposite sex. Mate choice involves one sex choosing among members of the opposite sex based on the attractiveness of certain phenotypic traits. This can lead to the evolution of sexual signals, referred to as ornaments, that may or may not be honest signals, or accurate indicators, of individual fitness (Arnold 1983). Mate competition involves one sex competing with members of the same sex for priority access to mates (Andersson 1994). Here, individuals may employ the use of weapons, such as the horns on many beetles, that can help them succeed in fights (e.g. Gwynne and Jamieson 1998). As in much of the animal kingdom, insects are subject to intense sexual selection in which ornaments and weapons can become so exaggerated that they affect an individual's ability to escape predators, find food, or otherwise perform behaviors essential to their survival.

An important question in sexual selection is why so many animals exhibit mate choice (Lande 1981; Kirkpatrick 1982). This is especially perplexing in insects because they are relatively small, and by virtue of their wings, often capable of dispersing far from each other. This means that initially finding a mate can be difficult because males and females may be dissociated in space, and refusing a mate once one is found puts individuals at a risk of remaining unmated (Andersson 1994). There are several, non-exclusive models of sexual selection that explain how and why mating preferences evolve. Broadly, two types of fitness benefits are thought to drive the evolution of mate choice— **direct** and **indirect benefits**. Direct benefits increase the fitness of choosy individuals through material resources (Kirkpatrick 1996; Møller and Jennions 2001). In insects, direct benefits most often take the form of nuptial gifts that are typically edible, nutrient packages presented to females before or during copulation (Boggs 1995; see box 13.1). These gifts may be food items such as the salivary secretions that many scorpionflies present to females prior to copulation (Sauer et al. 1998) or the spermatophylax, described previously, that males transfer during copulation. In ground crickets, males offer their mates blood meals by allowing them to chew on a specialized hind leg structure during copulation (Lewis et al. 2011). In exchange for these nuptial gifts, males are allowed to copulate for longer periods of time, during which they transfer more sperm leading to a potential increase in reproductive success (Lewis et al. 2011). For example, male *Photinus* fireflies transfer gelatinous spiral-shaped spermatophores to females during copulation. Radiolabelling studies have shown

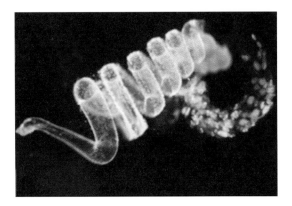

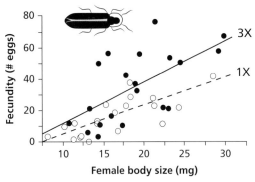

Figure 13.1 Male *Photinus* fireflies manufacture spiral-shaped spermatophores, which are transferred to the female reproductive tract (top panel). Researchers showed that these spermatophores function as nuptial gifts that increase female fecundity: females who mated three times laid more eggs than females that were mated only once (bottom panel).

From Lewis et al. (2004).

that male-derived proteins in the spermatophore are used to help provision the female's developing offspring (Figure 13.1; Lewis et al. 2004). Furthermore, multiply-mated female *Photinus* fireflies receive several spermatophores that can increase her fecundity (Figure 13.1; Lewis et al. 2004).

Mate choice can also evolve when choosy individuals receive indirect benefits, such as genetically superior for their offspring (Fisher 1958). Under 'good genes' models of sexual selection, individuals produce offspring with greater fitness by preferentially mating with individuals that display some signal of viability, such as body condition, fighting ability, or foraging efficiency (Iwasa et al. 1991). Because lower quality individuals may benefit by posing as high quality individuals, the sexual signal should be an indicator of superiority, such that only high quality individuals are capable of producing it (Fisher 1958). For example, female *Drosophila montana* preferentially mate with males that produce high frequency songs, because these males produce offspring with higher survival rates. This, in turn, provides females with the indirect benefit of increased offspring fitness, without receiving any direct, material benefit (Hoikkala et al. 1998).

Frequently, males and females have conflicting optimal reproductive fitness strategies and sexual conflict can arise (Chapman et al. 2003; see also Chapter 10). Because of anisogamy, sexual conflict commonly occurs in relation to the mode and frequency by which each sex mates. As such, sexual conflict can arise in various stages of reproduction in which males and females may bias paternity in favour of their own interests. Early work by Bateman (1948) on *Drosophila* showed that male, but not female, reproductive success increased with the number of mates acquired, suggesting that males and females have different reproductive fitness optima. This chapter provides examples of antagonistic sexual conflict, in which males and females are prevented from achieving the optimal outcome of an interaction such as copulation or parental care (Parker 2006).

13.1.4 Mating systems

The type of association between males and females during the breeding period is the mating system

(Thornhill and Alcock 1983). We often divide mating systems into two broad categories:

- Monogamy, in which females and males form exclusive pairs
- **Polygamy**, in which females and/or males mate multiply within a breeding season.

Because not all individuals of a population mate successfully, information is usually needed about the variance in male and female mating success to determine a population's mating system.

Monogamy occurs when individuals of both sexes have only one mate during a breeding season. This arrangement is fairly uncommon in insects, but does occur in species such as *Nicrophorus* burying beetles (Fetherston et al. 1990). Burying beetle reproduction requires a small animal carcass, on which the young will subsist during development. When a suitable carcass becomes available, intrasexual fighting occurs among burying beetles until only one male and one female prevail (Figure 13.2). The new couple then cooperates to move the carcass to an acceptable burial site, remove its fur or feathers, and bury it. The female lays her eggs in the soil nearby, and the young, once hatched, feed on the carcass until they complete development. Both parents remain in the brood chamber and care for the young. Monogamy in burying beetles was first attributed to the need for mate assistance in rearing young (Thornhill and Alcock 1983), but more recent evidence suggests that it may have evolved because males defend the brood chamber from invading burying beetles that may commit infanticide and re-inseminate the female, or may steal the carcass

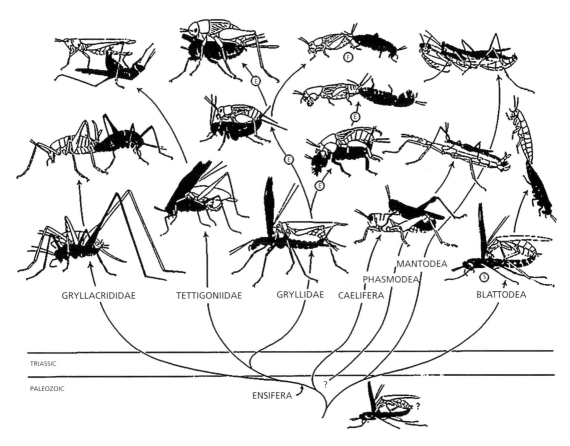

Figure 13.2 Insects with indirect sperm transfer display a diverse range of mating positions. Evolutionary biologists can study the copulatory behavior and phylogenetic relationships between species to infer ancestral mating positions.

From Alexander (1964, pp. 88).

to rear their own offspring (Scott 1989, 1990). Monogamy may also occur in the absence of parental care by both partners (Fincke 1987), suggesting that this type of mating system can occur in a variety of social contexts.

Polygamy, in which individuals multiply-mated, can be further divided based on observed patterns of mating behavior particular to each sex. Polygynandry occurs when both males and females have multiple mates during a breeding season. Across insects, this is the most common arrangement (Snook 2014). Polygyny occurs when some males mate with multiple females, as in the alpine weta, *Hemideina maori*. In this species, males defend harems containing as many as seven females by controlling territories (Gwynne and Jamieson 1998). Males with large mandibles are more likely to have larger harems and higher mating success (Gwynne and Jamieson 1998). Polyandry occurs when some females mate with multiple males. Scientists long assumed that if a single mating provided enough sperm to fertilize all of her eggs, a female would have little to gain with multiple matings. We now have widespread evidence in insects that females can gain indirect benefits when they mate multiply (Arnqvist and Nilsson 2000). For example, in honeybees, *Apis mellifera*, females that mate with multiple males have more genetically diverse colonies with measurably higher rates of foraging, food storage, and population growth that lead to better production of drones and increased winter survival (Mattila and Seeley 2007).

To understand how mating systems evolve, we can think about available mates as resources. Ecological factors can affect the ability of individuals to monopolize access to potential mates, and therefore affect the strength of sexual selection (Emlen and Oring 1977). Polygamy is expected to evolve when environmental conditions promote the defense of multiple mates or if sufficient resources are available to control access to multiple mates. The operational sex ratio (OSR) is the ratio of fertilizable females to sexually active males within a population at a given time; it provides a measure of the degree to which one sex can monopolize the other (Kvarnemo and Ahnesjo 1996). When the OSR is skewed toward females, polygyny is expected to evolve; when the OSR is skewed toward males, polyandry is favoured by selection. When the OSR is highly skewed, intrasexual competition for mates is expected to be very strong.

13.2 Pre-copulatory behavior

Before copulation can occur, a suitable mate must be located. Insects display an incredible diversity of strategies to solve this problem; most involve signalling (see Chapters 10, 11, and 12). An **animal signal** is any stimulus produced by a sender that modifies the behavior of one or more receivers (Bradbury and Vehrencamp 2011). An important feature of a signal is that it has evolved specifically in the context of communication. Sexual signals can occur in one or more sensory modalities, including chemical (see Chapter 10), visual (see Chapter 11), and acoustic (see Chapter 12), as well as vibratory and tactile.

Sexual signals are often categorized as advertisement signals, in which one sex broadcasts information about its species identity, location, and physical condition to potential mates within the area where the signal can elicit responses from a receiver. Typically, the sender of a signal is the limited sex, often males, and the receiver is the limiting or choosy sex, often females. In some cases, sexual signals involve duetting between male and female pairs, as in treehoppers, *Enchenopa binotata*. Male treehoppers produce vibratory signals that travel along the stems of plants and female treehoppers produce their own, somewhat less elaborate, signal that helps males localize them.

Because insects can disperse quite far, males and females can be dissociated in space, making mate location difficult. As such, many insects have developed effective long distance communication to facilitate mate identification and localization. Orthopterans provide good examples of long-range signallers—male eastern sword-bearing katydids, *Neoconocephalus ensiger*, can call at amplitudes as high as 100 dB SPL (Faure and Hoy 2000)—about as loud as a motorcycle or jackhammer sounds to a human observer. This helps receivers detect signals from far away.

In addition to advertisement signals, many insects use short-range courtship signals, which convey information to nearby receivers. Short-range signals can be advantageous because they are harder for unintended receivers, such as predators and parasites, to detect. Courtship signals are often produced in the presence of a potential mate to elicit mating behavior, and often prevent rivals from intercepting a prospective mate. Asian corn-borer moths, *Ostrinia furnacalis*, are a good example. After females attract males using long-range chemical signals, males produce an ultrasonic song that functions as a courtship signal. The female moths are more likely to mate with courting males than non-courting males, but this tendency is absent in deafened females. This demonstrates the importance of male courtship behavior to male mating success (Nakano et al. 2006).

In the majority of species that practice mate choice, females are the choosy sex, while males indiscriminately mate when given the opportunity. However, when females exhibit variance in their quality as potential mates, male mate choice is expected to evolve. Males are especially likely to exercise mate choice when the number of eggs produced by individual females is highly variable, which is often the case in insects (Bonduriansky 2001). Male fruit flies, *Drosophila melanogaster*, have adaptive mating preferences for larger, and more fecund females (Long et al. 2009). The study of male mate choice is an exciting addition to our growing understanding of reproductive behavior in nature. Further work can shed light on the ways that male mate choice is shaping female behavior, including research on female alternative reproductive tactics.

Sexual signalling is not always a cooperative behavior, and mate choice does not necessarily lead to direct or indirect benefits for the choosy sex (see Chapters 10 and 11). Indeed, in some species, males exploit existing features of female sensory systems in order to gain more matings (Ryan and Rand 1993). The underlying idea is that novel male traits that stimulate pre-existing sensory biases in females are more likely to become established as sexual traits (Endler and Basolo 1998). These pre-existing sensory biases are often adaptive in non-sexual contexts, and may relate to an individual's ability to find food or avoid predators. For example, neotropical, Euglossini orchid bees, feed on nectar and pollen from flowers (Dressler 1982). Male orchid bees exploit female sensory systems by collecting, storing, and processing fragrant substances, then vibrating their wings to spray aerosol clouds of perfumes (Peruquetti 2000). Females are attracted to these perfumes, which they naturally seek while foraging. Males can also exploit sensory biases

Figure 13.3 Male genitalia in the cowpea weevil, *Callosobruchus maculatus*, contains spines that may injure females during copulation. This is known as traumatic insemination.

Adapted from Lange et al. 2013.

during copulation to influence how a female's eggs are fertilized. In damselflies, *Calopteryx haemorrhoidalis asturica*, males rub their genitalia against sensory receptors in the female reproductive tract that mimic egg movements, thereby eliciting sperm ejection during fertilization (Córdoba-Aguilar 2002).

13.3 Copulatory behavior

Copulation in insects is arguably the most critical component of reproduction. For many, the terms copulation, insemination, and fertilization are synonymous. However, to illustrate the variation in copulatory behaviors between insect systems, we distinguish between these terms. Whereas insemination is any transfer of sperm to the female reproductive tract, copulation refers to the physical coupling, or direct genital contact, between males and females. This distinction is important because it underscores how copulation can occur without insemination. For instance, females may reject males before sperm are transferred, but after the two have physically coupled. As discussed earlier, fertilization occurs when sperm and egg fuse. Thus, insemination can occur without fertilization, as is the case when females 'cryptically' reject sperm after it has been transferred by a male (see Section 13.4).

13.3.1 Sperm transfer

Male insects employ two modes of insemination or sperm transfer—direct and indirect. These modes are important in the context of economic defendability because they determine the likelihood of a given male's sperm being taken up by a female. Direct sperm transfer occurs in the majority of insects and involves internal transfer of sperm from male genitalia to females (Alexander and Otte 1967). This primarily occurs during copulation when male and female genitalia are already in direct contact. For males, direct sperm transfer provides greater insurance that his sperm resources are going into a female. Males of most holometabolous insects, or insects undergoing complete metamorphosis with a pupal stage, use direct sperm transfer (Alexander and Otte 1967).

Indirect sperm transfer is employed by many hemimatabolous insects, or insects with incomplete metamorphosis. Here, a male will deposit a spermatophore outside of his body, after which the spermatophore is taken up by a female (Alexander and Otte 1967). This process is often quite dissociated from copulation, allowing females to have greater control over insemination. There are many examples of indirect sperm transfer in insects. For instance, male springtails will deposit spermatophores on substrates without any indication that a female is nearby (Proctor 1998). In silverfish, males will spin silk threads and deposit a spermatophore on top (Proctor 1998). Female silverfish then search for the spermatophore either by following the threads or attending to signals from the male (Proctor 1998). Some bristletail males also use silk threads to attract females to spermatophores; however, these males directly guide the female's ovipositor to the spermatophore (Proctor 1998).

13.3.2 Mating positions

Although sperm transfer can be accomplished without copulation, physical coupling between males and females is common. Insects adopt a variety of mating positions, or orientations of males relative to females during mating. These include female-above, male-below, side-to-side, and end-to-end positions (Figure 13.3; Alexander 1964). The evolution of mating positions among insects is thought to be the result of intrasexual selection (Parker 1970). Recall that males typically compete for access to mates before, during, and after copulation. Thus, males sometimes develop alternative mating positions, or mating positions other than male-above-female, that allow them to have more control over insemination by protecting females from encroaching males (Parker 1970; Thornhill and Alcock 1983).

13.3.3 Copulatory sexual conflict

Because females are typically choosy with regard to whom they mate with, they often reject mating attempts from courting males. As such, males occasionally go to great lengths to ensure that they successfully inseminate a female. This can lead to sexual conflict where males evolve traits to force copulate with females, and females evolve traits that help them resist these attempts. Here, three examples of antagonistic sexual conflict are highlighted—forced copulation, traumatic insemination, and sexual cannibalism.

Like other animals (Clutton-Brock and Parker 1995), male insects maximize their reproductive fitness sometimes by coercing females to copulate. Often, this is accomplished with specialized genital structures that allow males to hold onto resisting females and forcefully inseminate them. In many water striders, for example, males use a grasping apparatus to hold females and force them to mate, despite female resistance (Rowe et al. 1994). These claspers, which are present on males and females, have made water striders an ideal system for studying the evolution of sexual coercion (Arnqvist and Rowe 2002). Forced copulation can occur with both mature and immature females. Teneral mating occurs when males force copulate with sexually immature females. In *Diploptera punctata* cockroaches, males force females to copulate days before they become sexually receptive (Roth and Willis 1955). Teneral mating has also been documented in *Drosophila melanogaster,* in which males force copulate with females within 1 hour of their final eclosion (Seeley and Dukas 2011). To achieve teneral matings, males sometimes engage in pre-copulatory mate-guarding, finding immature individuals and waiting for them to eclose. Several species of *Heliconius* butterflies practice an extreme version of pre-copulatory mate-guarding in which males will perch on pupae and wait for females to emerge (Deinert 2003). Some males go further and insert their genitalia through the pupal cuticle and mate with females the moment they begin to emerge (Deinert 2003). This mating strategy is referred to as pupal mating (Deinert 2003).

Males sometimes force their sperm inside females, causing harm in the process. The key difference between forced copulation and traumatic insemination is that, rather than merely holding females and forcing copulations, traumatic insemination involves the insertion of male genitalia into the non-genital tissue of females. As with forced copulation, males have evolved specialized genital structures that aid in traumatic insemination. For example, male bed bugs have specialized genital structures that can pierce through and inject sperm into female abdomens (Carayon 1966). Following traumatic insemination, sperm migrate through the female's body to her ovaries, eventually resulting in fertilization.

Traumatic insemination imposes fitness costs on females, such as increased risk of infection through the puncture wound and immune defense against sperm fluids that are introduced directly into their circulatory system (Arnqvist and Rowe 2002). For these reasons, traumatic insemination can lead to reduced lifespan and decreased reproductive output (Arnqvist and Rowe 2002). In the cowpea weevil, *Callosobruchus maculatus*, male genitalia are covered in small spines that often damage the female reproductive tract during copulation and can reduce female lifespan (Figure 13.4; Lange et al. 2013). As a result, females of some species have evolved specialized structures that mitigate the costs of traumatic insemination. In bed bugs, females have evolved a pair of specialized reproductive organs at the site of penetration that serve as sperm-receptacles from which sperm directly migrate to the ovaries (Reinhardt et al. 2003).

While not common, sexual cannibalism—where females eat their mate during or after copulation—occurs in a few insects, including nearly all mantids (Elgar and Schneider 2004). A large conundrum in evolutionary biology is how such a bizarre behavior evolves. For female mantids, the fitness benefits of sexual cannibalism are fairly straightforward—prey consumption provides resources that are especially

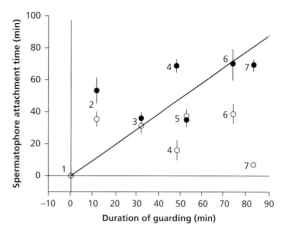

Figure 13.4 Studies of crickets show that females retain spermatophores for longer periods of time and thus receive more sperm, when males exercised mate-guarding behaviors (filled circles) compared with males that were experimentally removed after sperm transfer (open circles). Numbers indicate different gryllid species: (1) *Cycloptiloides canariensis* (this species does not practice mate guarding); (2) *Balamara gidya*; (3) *Acheta domesticus*;, (4) *Gryllus bimaculatus*; (5) *Gryllodes sigillatus*; (6) *Teleogryllus natalensis*; (7) *Teleogryllus commodus*.

From Simmons (2001, pp. 173).

beneficial for provisioning offspring. However, the fitness benefits for males are not as clear. By succumbing to female cannibalism, males eliminate any chance of re-mating. Although males often evaluate the risks of mating with a female in an effort to avoid being eaten (Lelito and Brown 2006; Jayaweera et al. 2015; Scardamaglia et al. 2015), some evidence suggests that male mantids continue to copulate with and inseminate females despite the initiation of cannibalism. This is because the benefits of boosting female fecundity outweigh the risks of not subsequently re-mating (Buskirk et al. 1984). Some hypothesize that nuptial feeding, including sexual cannibalism, evolves via a sensory bias on female foraging behavior or gustatory responses (Sakaluk 2000).

13.4 Post-copulatory behavior

If females mate with multiple males, so that the sperm from several individuals is present in the female's body, sexual selection can continue even after mating. Here, fertilization rather than access to mates is the critical resource. This phenomenon is known as post-copulatory sexual selection, and occurs in two forms—**sperm competition** and **cryptic female choice**. Sperm competition, an extension of mate competition, is defined as competition between sperm of two or more males for fertilization of the ova (Parker 1970). Cryptic female choice describes the process by which females bias fertilization during or after copulation (Eberhard 1996). Here, females defend their own egg resources by influencing paternity in favour of preferred males.

13.4.1 Sperm competition

As in pre-copulatory mate competition, two main strategies exist in sperm competition:

- reducing the opportunity for direct competition;
- actively engaging in competition.

Males can attempt to avoid sperm competition through a suite of behavioral, physiological, and morphological adaptations, most of which reduce the likelihood that a female will re-mate.

Many males engage in mate-guarding behaviors that actively prevent females from re-mating, and ensure a successful sperm transfer. These behaviors may include associating with a single female for some time after copulation, actively defending a female from other males, or copulating for periods that extend beyond the time necessary for a successful sperm transfer. In the aptly named lovebug, *Plecia nearctica*, copulation can last for 56 hours, even though sperm transfer requires only about 12 hours of physical contact (Thornhill 1976). Although mate guarding can be costly, it can also benefit males by reducing the total amount of sperm needed, as well as the number of competitors. Additionally, mate guarding allows males to influence the order of mating, which can affect male reproductive success as the last male to inseminate a female is often capable of fertilizing the majority of her eggs (Smith 1979). Mate guarding can also ensure a successful transfer of sperm. For example, when male crickets engage in mate-guarding behavior, females retain spermatophores for a longer period of time, thus increasing the amount of sperm that is transferred (Figure 13.5; Simmons 2001).

In many insects, males may block the female reproductive tract to reduce re-mating, bias sperm

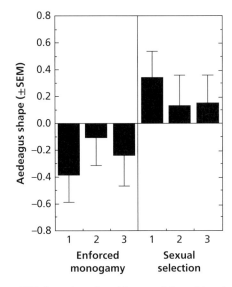

Figure 13.5 Researchers allowed three populations of dung beetles, *Onthophagus taurus*, to breed for nineteen generations under either enforced monogamy or polygamy. Researchers discovered that sexual selection was responsible for the rapid increase in aedeagus (i.e. male genital) shape in polygamous beetles.
From Simmons et al. (2009).

use, and influence female behavior and physiology. One way that males do this is by depositing a mating plug in the female genital opening during copulation. This plug is often a physiological product that coagulates or hardens inside the female, blocking her reproductive tract (Simmons 2001), and may be incorporated into the spermatophore that males transfer during copulation. Mating plugs can affect females in a number of ways. In butterflies, females have two genital openings so mating plugs do not hinder oviposition, but they have been shown to reduce copulation attempts from rival males (Dickinson and Rutowski 1989). Females can sometimes remove the plug by pulling it out of the reproductive tract, and may subsequently consume the plug. Interestingly, traumatic insemination may have evolved in response to mating plugs. By inserting their genitals into non-genital tissue of females, males can bypass the mating plugs left by their competitors (Arnqvist and Nilsson 2000).

Other physiological products function in sperm competition. In some insects, seminal fluid proteins transferred in the ejaculate during copulation can modify female behavior and physiology in a way that favours the male that transferred them (Avila et al. 2011). For example, in most *Drosophila* species, seminal fluid proteins can reduce a female's propensity to re-mate, reduce her attractiveness toward other males, and can increase her egg production and oviposition behavior (Kalb et al. 1993; Tram and Wolfner 1998). Recent work on some *Drosophila* and neriid flies, *Telostylinus angusticollis*, shows that the influence of seminal fluids can be independent of fertilization and can impact the traits of offspring sired by a different male (Crean et al. 2016).

In species with internal fertilization, males often evolve complex genital organs that are often adapted to reduce sperm competition (Simmons 2001). For example, dung beetles engaging in polygamy over multiple generations evolved more complex genitals than those experiencing enforced monogamy (Figure 13.5; Simmons et al. 2009). Morphological adaptations can also allow males to remove or reposition sperm from other males in order to place their own sperm in a more favourable location within the female reproductive tract. For instance, morphological adaptations on damselfly genitalia remove sperm deposited by previous males in the female reproductive tract during copulation (Waage 1979).

Of course, sperm competition cannot always be avoided. Engagement in sperm competition has selected for a number of adaptations in sperm traits, namely sperm number, size, and flexibility to strategically invest in the ejaculate depending on current conditions (Simmons 2001). Theoretical and empirical evidence has shown that many, tiny sperm are favoured in sperm competition, as only a single sperm is required for fertilization (Parker 1982; García-González and Simmons 2007). Other insects, notably *Drosophila* species, invest more heavily in sperm size and, as such, have evolved giant sperm that can be up to twenty times the body length of the male that produced it (Pitnick 1996). Giant sperm appear to have evolved in response to a combination of female choice and sperm competition, and probably represent the most extreme case of an exaggerated sexual ornament (Lüpold et al. 2016). Some males can also adjust ejaculate investment depending on perceived sperm competition risk. For instance, under male-biased conditions, the golden egg bug, *Phyllomorpha laciniata*, copulates for a longer duration of time and transfers larger

ejaculates than in the absence of rival males (García-González and Gomendio 2003).

13.4.2 Cryptic female choice

Cryptic female choice represents an additional mechanism of post-copulatory sexual selection, where females can bias paternity after copulation. This can include female influences over sperm storage and utilization that may depend on male relatedness, phenotype, or nuptial gift quality. For instance, Australian field cricket females bias paternity against relatives by storing less sperm when guarded by a related male (Tuni et al. 2013). In red flour beetles, *Tribolium castaneum*, more copulatory courtship by males results in greater fertilization success, but only when females are able to perceive this behavior (Edvardsson and Arnqvist 2005). Thus, the differences in fertilization success can be attributed to female-driven, rather than male-driven processes. Post-copulatory female choice is considered 'cryptic' because it occurs within the female reproductive tract and cannot be directly observed. Because of this, cryptic female choice is difficult to test and has yet to be widely studied.

Relatively non-cryptic female behaviors, such as terminating copulation, removing spermatophores and mating plugs, and re-mating, can also influence paternity (Eberhard 1996, 2011). Female black field crickets, *Teleogryllus commodus*, prematurely dislodge sperm products, more often from less attractive males (Bussiere et al. 2006). Despite a growing number of examples, cryptic female choice largely remains a black box. More work is needed to understand its driving mechanisms and evolutionary significance, including questions regarding how cryptic female choice interacts with sperm competition. Further work can also help us to understand whether post-copulatory sexual selection amplifies precopulatory sexual selection, or whether these episodes of selection act in different directions.

13.5 Conclusions

Insects are often considered simple, but as we have seen, insect reproductive behavior is incredibly complex and diverse (see also Chapter 2). From pre- to post-copulatory behavior, insects provide ideal systems for studying the evolutionary processes shaping male and female morphologies and behaviors. Through the lens of economic defendability, we can understand what the limiting resource is at each step, and how this affects male and female reproductive behavior. Differences in reproductive physiology between males and females constrain initial investment in offspring. Furthermore, disparities in parental investment have consequences for insect reproductive behavior, resulting in the evolution of female choosiness, the elaboration of male secondary sexual characteristics, and courtship behaviors.

The numerous and remarkable reproductive behaviors of insects continue to provide a useful lens for biologists interested in evolution by sexual selection. Many questions remain unanswered, however. Recently developed detailed genetic analyses are just beginning to provide further insight into the mechanisms of sexual selection, particularly under good genes models. Furthermore, while we now recognize that cryptic female choice occurs in a number of animal systems, more research is needed to understand the prevalence and adaptive significance of this type of sexual selection.

References

Alexander, R.D. (1964). The evolution of mating behavior in arthropods. *Symposium of the Royal Entomological Society London*, **2**, 78.

Alexander, R.D., and Otte, D. (1967). The evolution of genitalia and mating behavior in crickets (Gryllidae) and other Orthoptera. *Miscellaneous Publications of the Museum of Zoology University of Michigan*, **133**, 1.

Andersson, M.B. (1994). *Sexual Selection*. Princeton University Press, Princeton.

Arnold, S.J. (1983). Sexual selection: the interface of theory and empiricism. In: P. Bateson (Ed.) *Mate Choice*, pp. 67–107. Cambridge University Press, Cambridge.

Arnqvist, G. and Nilsson, T. (2000). The evolution of polyandry: multiple mating and female fitness in insects. *Animal Behaviour*, **60**, 145.

Arnqvist, G. and Rowe, L. (2002). Antagonistic coevolution between the sexes in a group of insects. *Nature*, **415**, 787.

Avila, F.W., Sirot, L.K., LaFlamme, B.A., Rubenstein, C.D., and Wolfner, M.F. (2011). Insect seminal fluid proteins: identification and function. *Annual Reviews in Entomology*, **56**, 21.

Bateman, A.J. (1948). Intra-sexual selection in *Drosophila*. *Heredity*, **2**, 349.

Bell, G. (1982). *The Masterpiece of Nature: The Evolution and Genetics of Sexuality*. University of California Press, Berkeley, CA.

Bjork, A., and Pitnick, S. (2006). Intensity of sexual selection along the anisogamy–isogamy continuum. *Nature*, **441**(7094), 742.

Boggs, C.L. (1995). Male nuptial gifts: phenotypic consequences and evolutionary implications. In: S.R. Leather and J. Hardie (Eds), *Insect Reproduction*, pp. 215–42. CRC Press, New York, NY.

Boggs, C.L. and Gilbert, L.E. (1979). Male contribution to egg production in butterflies: evidence for transfer of nutrients at mating. *Science*, **206**, 83.

Bonduriansky, R. (2001). The evolution of male mate choice in insects: a synthesis of ideas and evidence. *Biological Reviews*, **76**, 305.

Bradbury, J.W. and Vehrencamp, S.L. (2011). *Principles of Animal Communication*. Sinauer Associates, Sunderland, MA.

Brown, J. (1964). The evolution of diversity in avian territorial systems. *Wilson Bulletin*, **76**, 160.

Buskirk, R.E., Frohlich, C., and Ross, K.G. (1984). The natural selection of sexual cannibalism. *American Naturalist*, **123**, 612.

Bussiere, L.F., Hunt, J., Jennions, M.D., and Brooks, R. (2006). Sexual conflict and cryptic female choice in the black field cricket, *Teleogryllus commodus*. *Evolution*, **60**, 792.

Carayon, J. (1966). Paragenital system. In: R.L. Usinger (Ed.) *Monograph of the Cimicidae*, pp. 81–167. Entomological Society of America, Philadelphia, PA.

Chapman, T., Arnqvist, G., Bangham, J., and Rowe, L. (2003). Sexual conflict. *Trends in Ecology & Evolution*, **18**, 41.

Clutton-Brock, T.H. (1991). *The Evolution of Parental Care*. Princeton University Press, Princeton, NJ.

Clutton-Brock, T.H., and Parker, G.A. (1995). Sexual coercion in animal societies. *Animal Behaviour*, **49**, 1345.

Córdoba-Aguilar, A. (2002). Sensory trap as the mechanism of sexual selection in a damselfly genitalic trait. *American Naturalist*, **160**, 594.

Crean, A.J., Adler, M.I., and Bonduriansky, R. (2016). Seminal fluid and mate choice: new predictions. *Trends in Ecology and Evolution*, **31**, 253.

Darwin, C. (1871). *The Descent of Man and Selection in Relation to Sex*. John Murray, London.

Deinert, E.I. (2003). Mate location and competition for mates in a pupal mating butterfly. In: C.L. Boggs, W.B. Watt, and P.R. Ehrlich (Eds) *Butterflies: Ecology and Evolution Taking Flight*, pp. 91–108. University of Chicago Press, Chicago, IL.

Dickinson, J.L., and Rutowski, R.L. (1989). The function of the mating plug in the chalcedon checkerspot butterfly. *Animal Behaviour*, **38**, 154.

Dressler, R.L. (1982). Biology of the orchid bees (Euglossini). *Annual Review of Ecology and Systematics*, **13**, 373.

Eberhard, W.G. (1996). *Female Control: Sexual Selection by Cryptic Female Choice.*: Princeton University Press, Princeton, NJ.

Eberhard, W.G. (2011). Experiments with genitalia: a commentary. *Trends in Ecology and Evolution*, **26**, 17.

Edvardsson, M., and Arnqvist, G. (2005). The effects of copulatory courtship on differential allocation in the red flour beetle *Tribolium castaneum. Journal of Insect Behavior*, **18**, 313.

Elgar, M.A., and Schneider, J.M. (2004). The evolutionary significance of sexual cannibalism. *Advances in the Study of Behavior*, **34**, 135.

Emlen, S.T., and Oring, L.W. (1977). Ecology, sexual selection, and the evolution of mating systems. *Science*, **197**, 215.

Endler, J.A., and Basolo, A.L. (1998). Sensory ecology, receiver biases, and sexual selection. *Trends in Ecology & Evolution*, **13**, 415.

Faure, P.A., and Hoy, R.R. (2000). The sounds of silence: cessation of singing and song pausing are ultrasound-induced acoustic startle behaviors in the katydid *Neoconocephalus ensiger* (Orthoptera: Tettigoniidae). *Journal of Comparative Physiology A*, **186**, 129.

Fetherston, I.A., Scott, M.P., and Traniello, J.F.A. (1990). Parental care in burying beetles: the organization of male and female brood-care behavior. *Ethology*, **85**, 177.

Filippi, L., Hironaka, M., and Nomakuchi, S. (2002). Risk-sensitive decisions during nesting may increase maternal provisioning capacity in the subsocial shield bug *Parastrachia japonensis*. *Ecological Entomology*, **21**, 152.

Fincke, O.M. (1987). Female monogamy in the damselfly *Ischnura verticalis* Say (Zygoptera: Coenagrionidae). *Odonatologica*, **16**, 129.

Fisher, R.A. (1958). *The Genetical Theory of Natural Selection,* 2nd edn. Dover, New York, NY.

García-González, F., and Gomendio, M. (2003). Adjustment of copula duration and ejaculate size according to risk of sperm competition in the golden egg bug (*Phyllomorpha laciniata*). *Behavioral Ecology*, **15**, 23.

García-González, F., and Simmons, L.W. (2007). Shorter sperm confer higher competitive fertilization success. *Evolution*, **61**, 816.

Gwynne, D.T., and Jamieson, I. (1998). Sexual selection and sexual dimorphism in a harem-polygynous insect, the alpine weta (*Hemideina maori*, Orthoptera Stenopelmatidae). *Ethology Ecology & Evolution*, **10**, 393.

Hironaka, M., Nomakuchi, S., Iwakuma, S., and Filippi, L. (2005). Trophic egg production in a subsocial shield bug, *Parastrachia japonensis* Scott (Heteroptera: Parastrachiidae), and its functional value. *Ethology*, **111**, 1089.

Hoikkala, A., Aspi, J., and Suvanto, L. (1998). Male courtship song frequency as an indicator of male genetic quality in an insect species, *Drosophila montana*. *Proceedings of the Royal Society B*, **265**, 503.

Ingram, K.K., Pilko, A., Heer, J., and Gordon, D.M. (2013). Colony life history and lifetime reproductive success of red harvester ant colonies. *Journal of Animal Ecology*, **82**, 540.

Iwasa, Y., Pomiankowski, A., and Nee, S. (1991). The evolution of costly mate preferences II. The 'handicap' principle. *Evolution*, **45**, 1431.

Jayaweera, A., Rathnayake, D.N., Davis, K.S., and Barry, K.L. (2015). The risk of sexual cannibalism and its effect on male approach and mating behaviour in a praying mantid. *Animal Behaviour*, **110**, 113.

Kalb, J.H., DiBenedetto, A.J., and Wolfner, M.F. (1993). Probing the function of *Drosophila melanogaster* accessory glands by directed cell ablation. *Proceedings of the National Academy of Sciences USA*, **90**, 8093.

Kaulenas, M.S. (1992). Insect accessory reproductive structures: function, structure, and development. *Zoophysiology*, **31**, 80.

Kirkpatrick, M. (1982). Sexual selection and the evolution of female choice. *Evolution*, **36**, 1.

Kirkpatrick, M. (1996). Good genes and direct selection in the evolution of mating preferences. *Evolution*, **50**, 2125.

Kvarnemo, C., and Ahnesjo, I. (1996). The dynamics of operational sex ratios and competition for mates. *Trends in Ecology & Evolution*, **11**, 404.

Lande, R. (1981). Models of speciation by sexual selection on polygenic traits. *Proceedings of the National Academy of Sciences USA*, **78**, 3721.

Lange, R., Reinhardt, K., Michiels, N.K., and Anthes, N. (2013). Functions, diversity, and evolution of traumatic mating. *Biological Reviews*, **88**, 585.

Lelito, J.P. and Brown, W.D. (2006). Complicity or conflict over sexual cannibalism? Male risk taking in the praying mantis *Tenodera aridifolia sinensis*. *American Naturalist*, **168**, 263.

Lewis, S.M., Cratsley, C.K., and Rooney, J.A. (2004). Nuptial gifts and sexual selection in *Photinus* fireflies. *Integrative and Comparative Biology*, **44**, 234.

Lewis, S., South, A., Burns, R., and Al-Wathiqui, N. (2011). Nuptial gifts. *Current Biology*, **21**, R644.

Long, T.A.F., Pischedda, A., Stewart, A.D., and Rice, W.R. (2009). A cost of sexual attractiveness to high-fitness females. *PLoS Biology*, **7**, 1.

Lüpold, S., Manier, M.K., Puniamoorthy, N., et al. (2016). How sexual selection can drive the evolution of costly sperm ornamentation. *Nature*, **533**, 535.

Mattila, H.R., and Seeley, T.D. (2007). Genetic diversity in honey bee colonies enhances productivity and fitness. *Science*, **317**, 362.

McCartney, J., Potter, M.A., Robertson, A.W., et al. (2008). Understanding nuptial gift size in bush-crickets: an analysis of the genus Poecilimon (Tettigoniidae: Orthoptera). *Journal of Orthoptera Research*, **17**, 231.

Møller, A., and Jennions, M. (2001). How important are direct benefits of sexual selection? *Naturwissenschaften*, **88**, 401.

Nakano, R., Ishikawa, Y., Tatsuki, S., et al. (2006). Ultrasonic courtship song in the Asian corn borer moth, *Ostrinia furnacalis*. *Naturwissenschaften*, **93**, 292.

Nation, J.L. (2015). *Insect Physiology and Biochemistry*. CRC Press, Boca Raton, FL.

Oldroyd, B.P. (2006). *Asian Honey Bees: Biology, Conservation, and Human Interactions*. Harvard University Press, Cambridge.

Parker, G.A. (1970). Sperm competition and its evolutionary consequences in the insects. *Biological Reviews*, **45**, 525.

Parker, G.A. (1982). Why are there so many tiny sperm? Sperm competition and the maintenance of two sexes. *Journal of Theoretical Biology*, **96**, 281.

Parker, G.A. (2006). Sexual conflict over mating and fertilization: an overview. *Philosophical Transactions of the Royal Society B*, **361**, 235.

Peruquetti, R.C. (2000). Function of fragrances collected by Euglossini males (Hymenoptera: Apidae). *Entomologia generalis*, **25**, 33.

Pitnick, S. (1996). Investment in testes and the cost of making long sperm in *Drosophila*. *American Naturalist*, **148**, 57.

Proctor, H.C. (1998). Indirect sperm transfer in arthropods: behavioral and evolutionary trends. *Annual Review of Entomology*, **43**,153.

Reinhardt, K., Naylor, R., and Siva-Jothy, M.T. (2003). Reducing a cost of traumatic insemination: female bedbugs evolve a unique organ. *Proceedings of the Royal Society B*, **270**, 2371.

Roth, L.M., and Willis, E.R. (1955). Intra-uterine nutrition of the 'beetle-roach' *Diploptera dytiscoides* (Serv.) during embryogenesis, with notes on its biology in the laboratory (Blattaria: Diplopteridae). *Psyche*, **62**, 55.

Rowe, L., Arnqvist, G., Sih, A., and Krupa, J.J. (1994). Sexual conflict and the evolutionary ecology of mating patterns: water striders as a model system. *Trends in Ecology & Evolution*, **9**, 289.

Ryan, M.J. and Rand, A.S. (1993). Sexual selection and signal evolution: the ghost of biases past. *Philosophical Transactions: Biological Sciences*, **340**, 187.

Sauer, K.P., Lubjuhn, T., Sindern, J., et al. (1998). Mating system and sexual selection in the scorpionfly *Panorpa vulgaris* (Mecoptera: Panorpidae). *Naturwissenschaften*, **85**, 219.

Scardamaglia, R.C., Fosacheca, S., and Pompilio, L. (2015). Sexual conflict in a sexually cannibalistic praying mantid: males prefer low-risk over high-risk females. *Animal Behaviour*, **99**, 9.

Scott, M.P. (1989). Male parental care and reproductive success in the burying beetle, *Nicrophorus orbicollis*. *Journal of Insect Behavior*, **2**, 133.

Scott, M.P. (1990). Brood guarding and the evolution of male parental care in burying beetles. *Behavioral Ecology and Sociobiology*, **26**, 31.

Seeley, C., and Dukas, R. (2011). Teneral matings in fruit flies: male coercion and female response. *Animal Behaviour*, **81**, 595.

Simmons, L.W. (2001). *Sperm Competition and its Evolutionary Consequences in the Insects*. Princeton University Press, Princeton.

Simmons, L.W., House, C.M., Hunt, J., and García-González, F. (2009). Evolutionary response to sexual selection in male genital morphology. *Current Biology*, **19** (17), 1442.

Smith, R.L. (1979). Repeated copulation and sperm precedence: paternity assurance for a male brooding water bug. *Science*, **205**, 1029.

Snook, R.R. (2014). The evolution of polyandry. In: D.M. Shuker and L.W. Simmons (Eds), *The Evolution of Insect Mating Systems*, pp. 159–80. Oxford University Press, Oxford.

Thornhill, R. (1976). Reproductive behavior of the lovebug *Plecia neartica* (Diptera: Bibionidae). *Annals of the Entomological Society of America*, **69** (5), 843.

Thornhill, R., and Alcock, J. (1983). *The Evolution of Insect Mating Systems*. Harvard University Press, Cambridge, MA.

Tram, U., and Wolfner, M.F. (1998). Seminal fluid regulation of female sexual attractiveness in *Drosophila melanogaster*. *Proceedings of the National Academy of Sciences USA*, **95**, 4051.

Trivers, R.L. (1972). Parental investment and sexual selection. In: B. Campbell (Ed.), *Sexual Selection and the Descent of Man, 1871–1971*, pp. 139–79. Aldine Publishing Company, Chicago, IL.

Tuni, C., Beveridge, M., and Simmons, L.W. (2013). Females crickets assess relatedness during mate guarding and bias storage of sperm towards unrelated males. *Journal of Evolutionary Biology*, **26**, 1261.

Vahed, K. (1998). The function of nuptial feeding in insects: review of empirical studies. *Biological Reviews*, **73**, 43.

Waage, J.K. (1979). Dual function of the damselfly penis: sperm removal and transfer. *Science* **203**, 916.

Yoshizawa, K., Ferreira, R.L., Kamimura, Y., and Lienhard, C. (2014). Female penis, male vagina, and their correlated evolution in a cave insect. *Current Biology*, **24**, 1006.

Zuk, M., and Kolluru, G. (1998). Exploitation of sexual signals by predators and parasitoids. *Quarterly Review of Biology*, **73**, 415.

CHAPTER 14

Parental care

Glauco Machado[1] and Stephen T. Trumbo[2]

[1] LAGE do Departamento de Ecologia, Instituto de Biociências, Universidade de São Paulo, Rua do Matão, trav. 14, no 101, Cidade Universitária, 05508–090, São Paulo, SP, Brazil

[2] Department of Ecology and Evolutionary Biology, University of Connecticut, Waterbury, CT 06702, USA

14.1 Forms of parental care

Parental care can be divided into two main episodes, pre- and post-oviposition. Pre-ovipositional care occurs before egg laying, including provisioning of nutrients and defensive structures or chemicals into eggs, selection of oviposition sites, nest building, mass food provisioning, and modification of environmental conditions (Danks 2002; Smiseth et al. 2012). These forms of care probably represent the **plesiomorphic state** in most insect orders (Hinton 1981; but see Tallamy and Schaeffer 1997). Post-ovipositional care occurs after egg laying, including egg and offspring attendance, egg and offspring carrying, viviparity, progressive provisioning, and care after nutritional independence (Smiseth et al. 2012). These forms of care are restricted to a relatively smaller number of insect species. An exhaustive review of parental care in insects would demand an entire book (e.g. Costa 2006); thus, here, a more synthetic approach is adopted. The following two sections define each form of pre- and post-ovipositional parental care, and illustrate them with selected examples.

14.1.1 Pre–ovipositional parental care

Although it is difficult to determine whether the nutrients or other resources provisioned into eggs exceed the minimum required for successful fertilization, studies with arthropods show that egg size has a positive effect on juvenile growth and survival when food is limited or when predation risk is high (Fox and Czesak 2000). Perhaps because this form of pre–ovipositional care is ubiquitous, yet difficult to quantify, it has attracted less study. More easily accessible and tractable in experimental studies are defensive structures or chemicals associated with the eggs. Lacewings (Chrysopidae), for instance, oviposit at the tip of stalks, keeping the eggs protected from ants and other predators (Figure 14.1a). Among resin bugs (Reduviidae), females store plant resins and transfer them to the eggs as a way to increase protection from predation and desiccation (Forero et al. 2011). Finally, females of several lepidopterans transfer to the eggs chemical deterrents sequestered from the host plants or synthesized by the female (Bernays and Chapman 2007).

Machado, G. and Trumbo, S. T., Parental care. In: Insect Behavior: From mechanisms to ecological and evolutionary consequences. Edited by Alex Córdoba-Aguilar, Daniel González-Tokman, and Isaac González-Santoyo: Oxford University Press (2018).
© Oxford University Press.
DOI: 10.1093/oso/9780198797500.003.0014

Selection of oviposition sites is the rule among herbivorous insects because larvae and nymphs are usually less vagile than adults, and are easily singled out by predators due to their smaller body size and soft exoskeleton. Thus, females lay their eggs on host plants where the offspring will find appropriate food, amenable abiotic conditions, and low density of competitors and natural enemies (Bernays and Chapman 2007). Females of the butterfly *Eunica bechina* (Nymphalidae), for instance, avoid plants with high density of ants, an important predator of their larvae. When there is no branch free of ants, females are able to visually distinguish ant species and oviposit in branches visited by species that feed mostly on liquids and which will not harm their larvae (Sendoya et al. 2009). Oviposition site selection is also the rule among parasitoids (Godfray 1994), and even strictly predatory insects may exhibit this behavior. Female ladybugs (Coccinellidae), for instance, lay eggs on plants containing high density of prey (aphids) for their larvae (Seagraves 2009).

The simplest form of a nest occurs when females bury their eggs in the substrate, a common behavior among hemipterans (Tallamy and Schaeffer 1997). In web spinners (Embiidina) and some bark lice (Psocoptera), parents build their own nests using silk (Edgerly 1997). Among wasps and bees, parents use mud, processed plant matter, wax, and other materials to build complex structures where the eggs are laid. In some thrips (Phlaeothripidae), active feeding by adults induces the formation of galls in the host plant. The gall can be considered as a special type of nest that offers food and shelter for both the adults and their offspring (Costa 2006). Regardless of the group, nests may protect the offspring from harsh abiotic conditions and natural enemies. Some vespid wasps, for instance, build nests that are attached to the substrate by a narrow pedicel that prevents ants from accessing the brood cells (Figure 14b). This nest pedicel may be coated with a glandular substance that works as an ant repellent (London and Jeanne 2000).

Mass food provisioning occurs when one or both parents stock all food necessary for offspring development in a small chamber or cell before oviposition, which is abandoned after oviposition (Eickwort

Figure 14.1 (a) Stalked eggs of a lacewing (Chrysopidae) on a branch in southeastern United States (photo: Eric Hunt). (b) Nest of the Brazilian wasp *Mischocyttarus* sp. (Vespidae) hanging from a boulder by a nest pedicel nearly 4 cm in length (photo: Fernando B. Noll). (c) Female of the dung beetle *Gymnopleurus virens* (Scarabaeidae) rolling a dung ball in a South African savannah (photo: Bernard Dupont). (d) Female of the sawfly *Neodiprion* sp. (Diprionidae) laying an egg in a ponderosa pine needle in southwestern United States (photo: Michael R. Wagner). The inset (bottom left) shows an egg laid in a slit cut in the edge of a living pine needle (photo: Donald Owen).

1981). In dung beetles (Scarabaeidae), the main food source for the offspring is the excrement of herbivorous vertebrates. Excrement on the soil is exposed to intense competition from conspecifics, flies, nematodes, and fungi, and also to rapid desiccation, making manipulation by the dung beetles difficult. Thus, parents relocate excrement to their nests either by packing pieces into the end of a tunnel dug beneath the food source or by forming a ball at the food source and rolling it to be buried inside a nearby cavity (Halffter 1997; Figure 14.1c). Mass provisioning also occurs in wasps and bees of several families (Eickwort 1981). Given that nest building and mass food provisioning represent a great investment of time and energy, females of some species avoid the costs of nest building and/or food provisioning by exploiting the investment of conspecifics or heterospecifics, a behavior known as **parental–care parasitism**.

Finally, parents can modify environmental conditions to increase offspring survival. In addition to attenuating stressful abiotic conditions by egg covering, nest building, and nest provisioning described above, Danks (2002) includes other specialized behaviors. Females of some sawflies (Hymenoptera) use their ovipositors to cut resin canals of selected needles above the future ovipositional slits that will hold the eggs (Figure 14.1d). The cuts drain resin and reduce the host plant's response to injury from the egg slits themselves. A similar behavior is also found among membracids (Hemiptera), in which plant wounding by the female ovipositor blocks host plant defences and also gives access to feeding sites by the hatchlings (Danks 2002).

14.1.2 Post-ovipositional parental care

Table 14.1 provides updated numbers on the taxonomic distribution of the different forms of post-ovipositional care in insects. Although eusocial species exhibit post-ovipositional care, they were not included in Table 14.1, which is mostly focused on **subsocial** species. Subsociality occurs when one or both parents care for the eggs until hatching, and then care for nymphs or larvae for some period after hatching (Eickwort 1981). Table 14.1 also includes many viviparous and ovoviviparous species, as well as species with post-ovipositional care

restricted to the eggs, which cannot be regarded as subsocial.

According to this search, post-ovipositional care in insects occurs in nineteen orders, 164 families, and 2041 genera. The most common form of post-ovipositional care is 'internal egg carrying' (including viviparity and ovoviviparity), characterized by the retention of fertilized eggs within the female reproductive tract. The defining distinction between viviparity and ovoviviparity is the occurrence of nutrient exchange between the female and the embryo, which is found in the former (Hagan 1951). In turn, ovoviviparity has been used in the literature to designate many reproductive strategies. These range from true ovoviviparity, when a chorionated egg develops in the maternal oviduct, having the yolk as its only food supply and hatching before deposition, to eggs that hatch a few minutes or hours after deposition (Hagan 1951). We lumped all such cases as internal egg carrying because, despite the diversity of developmental routes, eggs that are carried internally in the female's body are well protected against natural enemies and harsh environmental conditions (Smiseth et al. 2012).

Nearly 72 per cent of cases of internal egg carrying occur in seven taxa in which viviparity is a **synapomorphic trait**—blaberid cockroaches (158 genera), aphids (209 genera), strepsipterans (forty-five genera), and the dipteran families Hippoboscidae (twenty-one genera), Nycteribiidae (eleven genera), Sarcophagidae (130 genera), and Tachinidae (388 genera). Because of the taxonomic value of viviparity, information on this form of post-ovipositional care has accumulated faster than for other forms. Moreover, viviparity is a non-behavioral trait that can be inferred, based on preserved individuals and morphological studies of the female reproductive tract—the same is not true for behavioral forms of post-ovipositional care. Thus, the high frequency of internal egg carrying in insects should be attributed, in part, to a possible bias in the way information on viviparity is collected and reported in the literature.

The most common behavioral form of post-ovipositional care is egg attendance, which occurs when parents remain close to the eggs at a fixed location after egg laying (Smiseth et al. 2012; Figure 14.2a). Egg attendance is particularly frequent in

Table 14.1 Occurrence of the different forms of post–ovipositional parental care in insects. This compilation is based on an extensive search in the literature including comprehensive reviews on the subject, and backward and forward searches of original papers found in Google Scholar. Although eusocial species exhibit post-ovipositional parental care, they were not included here.

Order	Occurrence		Caring sex			Form of parental care (number of genera)						
	Families	Genera	Female	Male	Both	Egg attendance	Internal egg carrying	External egg carrying	Food provisioning	Offspring attendance	Offspring carrying	Nutritional independence
Blattodea	4	168	166	0	2	0	159	5	14	20	6	1
Coleoptera	17	331	130	0	206	300	25	11	127	290	1	30
Dermaptera	9	28	28	0	0	22	6	0	6	23	0	0
Diplura	1	1	1	0	0	1	0	0	0	1	0	0
Diptera	21	676	676	0	0	1	675	0	0	0	0	0
Embidiina	4	6	6	0	0	6	0	0	3	6	0	0
Ephemeroptera	1	3	3	0	0	0	3	0	0	0	0	0
Hemiptera	51	599	587	13	0	118	367	159	9	73	28	1
Hymenoptera	14	85	79	0	6	85	0	0	76	79	0	6
Lepidoptera	11	23	23	0	0	1	22	0	0	0	0	0
Mantodea	4	12	12	0	0	12	0	0	0	4	0	0
Odonata	1	1	1	0	0	0	1	0	0	0	0	0
Orthoptera	4	6	6	0	0	6	0	0	0	6	0	0
Phthiraptera	1	1	1	0	0	0	1	0	0	0	0	0
Plecoptera	6	10	10	0	0	0	10	0	0	0	0	0
Psocoptera	3	3	3	0	0	3	2	0	0	1	0	0
Strepsiptera	10	45	45	0	0	0	45	0	0	0	0	0
Thysanoptera	1	42	37	6	2	30	16	0	0	28	0	1
Trichoptera	1	1	1	0	0	0	1	0	0	0	0	0
Totals	164	2041	1815	19	216	585	1333	175	235	531	35	39

Figure 14.2 (a) Female of the Senegalese praying mantis *Tarachodes* sp. (Tarachodidae) attending eggs laid on a branch (photo: Joke van den Heuvel). (b) Female of the Japanese bug *Parastrachia japonensis* (Parastrachiidae) carrying a drupe of the plant *Schoepfia jasminodora* (Olacaceae) to provision her nymphs (photo: Takahiro Hosokawa). (c) Matriphagy in the earwig *Anechura harmandi* (Forficulidae) from Japan (photo: Seizi Suzuki). The nymphs are scattered around parts of the mother's body, and the arrows indicate remains of her forceps (left) and thorax (right). (d) Encumbered male of the water bug *Abedus* sp. (Belostomatidae) aerating eggs close to the water surface in central United States (photo: Greg Mayberry). Note that some nymphs are hatching from the eggs attached to the dorsum of the male. (e) Female of the leaf beetle *Omaspides brunneosignata* (Chrysomelidae) guarding her offspring that pupated on a branch in southeastern Brazil (photo: Paula Akeho). (f) Female of the Australian shield bug *Peltocopta crassiventris* (Tessaratomidae) carrying first instar nymphs on her venter (photo: Jeff Wright).

Coleoptera and Hemiptera (Table 14.1). Offspring attendance, which occurs when parents remain with their offspring after hatching (Smiseth et al. 2012), is also frequent in Coleoptera, Hemiptera, and Hymenoptera (Table 14.1). Both forms of care protect young against natural enemies, including predators, parasitoids, and fungi. In fact, Wilson (1975) proposed that intense predation on eggs and recently hatched young imposed by conspecifics and ants,

as well as the high risk of fungal infection in tropical forests, may have been the major forces selecting for the evolution of post-ovipositional care in arthropods. Although this hypothesis cannot explain why egg/offspring attendance evolved in some species and not others living in the same place, it may be used to understand how the benefits of these two forms of care vary geographically in response to biotic conditions. Assuming that biotic interactions

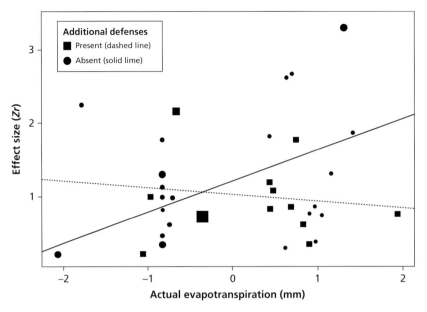

Figure 14.3 Relationship between actual evapotranspiration (centred and scaled) and the effect size *Zr* of the intensity of egg/offspring mortality in thirty-six terrestrial arthropod species including arachnids, crustaceans, and insects. Actual evapotranspiration is a proxy of the intensity of biotic interactions (e.g., predation and parasitism), so that the higher the value of actual evapotranspiration, the higher the mortality of unprotected eggs imposed by natural enemies. The effect size *Zr* represents a measure of the intensity of egg/offspring mortality in experiments of parental removal, so that the higher the value of *Zr*, the higher the benefits of egg/offspring protection. Additional lines of egg/offspring defence include silk, mud or mucus surrounding the eggs, faecal shields surrounding larvae, nests, and biotic interaction with ants. When additional defences are absent, actual evapotranspiration has a significant positive effect on the magnitude of egg/offspring mortality (slope = 0.422). In turn, when additional defences are present, actual evapotranspiration has no effect on the magnitude of egg/offspring mortality (slope = −0.087). Modified from Santos et al. (2017).

(e.g., predation and parasitism) are stronger in tropical climates, the benefits of parental protection should be higher in tropical than in temperate climates. Santos et al. (2017) tested this prediction using a phylogenetic meta-regression based on experimental studies of parent removal under natural conditions. They found that mortality of unguarded broods was higher in tropical climates, where egg/offspring attendance in the control group was more beneficial than in temperate climates (Figure 14.3). Moreover, additional lines of defence (e.g., egg coatings, nests, and association with mutualistic ants) significantly reduced offspring mortality in tropical climates (Figure 14.3), implying that egg/offspring attendance alone may not be enough to cope with stronger predation/parasitism in tropical climates. These results may explain the high frequency of egg/offspring attendance in tropical arthropods.

Food provisioning occurs in five orders and it is the fourth most common form of post–ovipositional

care in insects (Table 14.1). To be considered as a form of post–ovipositional care, food provisioning must be progressive, which occurs when adults repeatedly bring food to the nest as required by the offspring (Eickwort 1981). In some insects, females may also provision the offspring with trophic eggs, which are usually unfertilized eggs laid by the mother to feed her young (Trumbo 2012). Progressive provisioning is the most common type of food provisioning in Table 14.1, accounting for nearly 87 per cent of the genera in which parents feed the offspring. However, we also included some species exhibiting mass provisioning because they also exhibit other forms of post-ovipositional care, such as:

- parasitoid wasps in which females defend the food source against other individuals (Godfray 1994);
- dung beetles in which females and sometimes males remain with their offspring to prevent fungal infection, parasitism, and desiccation (Halffter 1997);

- burying beetles (Silphidae) in which males and females alter microbial communities on the carcass, and regurgitate food to the larvae (Trumbo 2012);
- several *Trypoxylon* wasps (Sphecidae) in which males guard the nest entrance, preventing parasitic flies from entering the nest and laying their own eggs on the provisioned food (Eickwort 1981).

Because the presence of the parents close to the food may improve offspring fitness in these mass provisioners, we included them as a form of post-ovipositional care, accounting for nearly 13 per cent of the provisioning genera in Table 14.1.

There is a great diversity of ways that parents provide food to their offspring among taxa that exhibit progressive provisioning. Several cockroach species feed their nymphs with body fluids, including hemolymph, tergal or external exudates, and hindgut fluids (Nalepa and Bell 1997). Females of the bug *Parastrachia japonensis* (Parastrachidae) provision their nests with drupes of the host plant, which is the sole source of food for the nymphs (Figure 14.2b). Nymphs die if the mother is removed, but a small number of unattended nymphs may reach adulthood when drupes are experimentally provided, suggesting that food provisioning is crucial for offspring survival (Tsukamoto and Tojo 1992). Larvae and adults of ambrosia beetles (Curculionidae) feed on cultivated fungi cultivated on woody tissues by the parents (Kirkendall et al. 2015). The most extreme type of food provisioning is matriphagy, which occurs when the hatched offspring consume their mother. According to our search, the only unequivocal case of matriphagy in insects occurs in the forficulid earwig *Anechura harmandi* (Suzuki et al. 2005; Figure 14.2c).

External egg carrying occurs when parents carry the eggs after laying. Although this form of care occurs in 175 genera, it is restricted to only three orders (Table 14.1). In Blattodea, females of some species belonging to the families Blattellidae and Blattidae carry an ootheca attached externally to the body. External egg carrying in cockroaches is considered an intermediate evolutionary step between species that attach their ootheca to the substrate and ovoviviparous species that internalized the egg–case (Nalepa and Bell 1997). In Coleoptera, females of at

least five genera of Hydrophilidae carry an egg-case attached to the venter (Eickwort 1981). There are also at least eighteen families of Hemiptera in which females carry egg–cases externally (Table 14.1). In several families of the suborder Coccoidea, females attach eggs to their bodies, either inside a waxy egg–case or under a modified scale that covers the mother's body (Spodek et al. 2014). In Belostomatinae water bugs (Belostomatidae), eggs are attached to the males' dorsum by the females immediately after copulation (Figure 14.2d). Because eggs are large, with a low surface:volume ratio, gas exchange between the embryo and the water is difficult. To improve oxygenation, males expose the eggs to the air close to the water surface (Figure 14.2d), or pump their body to promote oxygen flow (Munguía–Steyer et al. 2008).

Care for offspring after they reach the age of nutritional independence is rare among insects, with most cases occurring in Coleoptera (Table 14.1). Among chrysomelids, females of many species protect the offspring throughout egg and larval development, and also remain close to the pupae, a behavior known as pupal guarding (Chaboo et al. 2014; Figure 14.2e). All passalids live in family groups inside a tunnel system built in rotten wood. Last instar larvae are unable to build a pupal case without adult help, a very rare form of post-ovipositional care (Schuster and Schuster 1997). In the cockroach *Cryptocercus punctulatus* (Cryptocercidae), brood care may last more than 3 years, during which the mating pair defends the family, enlarge and sanitize the nest built in rotten wood, and feed the nymphs. Although nymphs are nutritionally independent after the third instar, the familial groups are maintained until the death of the parents (Nalepa and Bell 1997).

Finally, offspring carrying, which occurs when parents carry the young after hatching, is the rarest form of post-ovipositional care in insects, being recorded for only thirty-five genera in three orders (Table 14.1). Most of the cases of offspring carrying occur in Hemiptera, particularly in the families Phloeidae and Tessaratomidae (Monteith 2006). Phloeids are flattened dorso-ventrally and well-camouflaged bugs, looking like patches of lichens on the bark. Females of *Phloea quadrata* stand over their eggs laid on the bark, and after the eggs hatch, the first instar nymphs climb onto the ventral surface of their mothers. Apparently, the nymphs do not feed until they moult to the second instar, when they leave the female permanently and feed independently on sap sucked from the bark. A similar behavior has convergently evolved in some Australian tessaratomids, in which the female's body is also flattened dorso–ventrally (Figure 14.2f).

14.2 Evolution of parental care

14.2.1 Origin of care

The vast majority of insects do not provide post-ovipositional care. This rarity has been related to the evolution of two traits:

- an amniotic egg, which resists desiccation, but permits adequate transfer of oxygen;
- the ovipositor, which allows insects to lay eggs in safe places (Hinton 1981, Zeh et al. 1989).

The high frequency of post-ovipositional care in two insect groups without ovipositors (Dermaptera and Thysanoptera, Table 14.1), as well as in other terrestrial arthropods that lack ovipositors (e.g. centipedes, isopods, spiders), points to the correlation of these traits (Costa 2006). Moreover, all models of the evolution of parental care require care to increase offspring survival (Klug et al. 2013). If egg survival without care is relatively high, then there is a ceiling on the added value of post-ovipositional care (from now on referred to only as 'parental care').

Why are the extraordinary adaptations of the insect egg and ovipositor sometimes not enough? Wilson (1975) identified four environmental prime movers favouring the evolution of parental care in insects—stable and structured habitats, harsh environments, use of rich ephemeral resources, and intense predation of eggs. Tallamy and Wood (1986) associated parental care with food resources, with a focus on their persistence, dispersion, and nutritional value. In fact, parental care is over-represented among insects that exploit wood and among those using ephemeral resources, such as carrion and dung. However, there is a limit to this approach as the vast majority of insects with these nutritional associations cope with environmental challenges with non–parental adaptations rather than with parental care.

A second approach is to examine insects with relatively simple, **uni-dimensional care** to identify the benefits that may have been important at the origin of care. Our compilation suggests that insects with the simplest forms of care often protect eggs against mobile, biotic agents, such as predators, parasitoids, and competitors (including conspecifics). Mechanistically, protecting eggs by retaining them, by covering them with the body or by active defense would be straightforward to evolve. When a non–provisioning female insect chooses where to oviposit, there may be a trade-off between safety on the one hand and proximity to food on the other hand, to maximize offspring feeding (Gardner and Smiseth 2011). An alternative to abandoning eggs further from the resource in a safe location is for the parent to stay in place with eggs closer to where offspring feed. The costs of care as measured by heightened parent mortality and loss of future breeding opportunities have received less attention than the benefits of care (see Santos et al. 2017). A prime difficulty in experimental work with insects is that they are usually difficult to follow in the field, a task necessary to estimate mortality rates and the probability of survivors breeding again, data essential to examine the decision to care.

14.2.2 Transitions in care

The most common transition in systems of care among insects has been from no care to maternal care (Gilbert and Manica 2015; Figure 14.4). Internal egg carrying and egg attendance are the most common forms of maternal care, found in 1333 and 585 genera, respectively (Table 14.1). Internal egg carrying is a relatively stable endpoint among animals, with few reversals (e.g., Mank et al. 2005). The taxonomic overlap between species with biparental care and uniparental female care (Table 14.1) suggests that some groups may have evolved female care from biparental care. A similar route to maternal care has been proposed for birds (Cockburn 2006) and cichlid fish (Gonzalez–Voyer et al. 2008), and may be favoured by male abandonment when re-mating opportunities are high. This route to maternal care has been taken by some holometabolous insects (Gilbert and Manica 2015; Figure 14.4), but explains fewer cases of uniparental female care than the no care to maternal care route. The transition from uniparental male care directly to female care, as occurs in some anurans (Lehtinen et al. 2003),

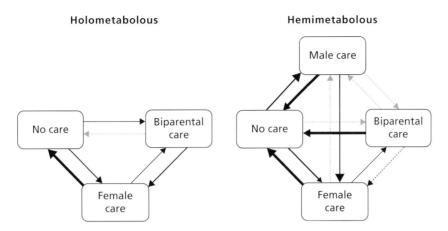

Figure 14.4 Evolutionary transitions in the systems of care in 2013 insect species. Data are presented for holometabolous (in which there is no case of exclusive paternal care) and hemimetabolous separately (in which all systems of care are found). Arrow width is directly proportional to the evolutionary transition rate, and grey dashed lines indicate rates in which standard deviation of the rate distribution overlaps zero.
Modified from Gilbert and Manica (2015).

appears to be quite rare in insects (Gilbert and Manica 2015; Figure 14.4).

Among insects, species with uniparental male care and biparental care differ in taxonomic distribution, types of resources exploited, the complexity of care, and nesting behavior, suggesting that neither system of care evolved from the other (Trumbo 2012; Table 14.2). This was confirmed by the comparative analysis of Gilbert and Manica (2015), who concluded that male care among hemimetabolous insects evolved only from no care (but see section on uniparental male care below) and biparental care evolved largely from maternal care (Figure 14.4). Wesolowski (1994) suggested that uniparental male care in birds evolved in two ways: ancestrally from species with no care via the opportunity to obtain multiple mates (similar to modern ratites and megapodes), but secondarily from biparental care via female desertion, as hypothesized for waders (Cockburn 2006). Both the data compiled here and the comparative analysis by Gilbert and Manica (2015) suggest that insects have taken the former but not the latter route to uniparental male care (Figure 14.4).

14.2.3 Which sex cares?

The preponderance of female care over male care in insects is related to opportunity. While it is not unprecedented for males to provide care internally in other taxa (e.g., anurans and fish), egg retention and viviparity are largely maternal (and exclusively so in insects) because of the ready availability of a female reproductive system that is adapted to accommodate eggs (Gross and Sargent 1985). Internal fertilization is also an important asymmetry to explain the preponderance of female over male care. The female is present at egg-laying, while the male is often not (Williams 1975). Multiple mating and sperm storage, as occurs in many insects (see Chapter 13), can also select against male care when it results in reduced paternity (Queller 1997). The group with the greatest number of transitions from no care to male care is fish with external fertilization and indeterminate growth, facilitating factors for paternal care that are not present in insects. Indeterminate growth can result in a larger cost of care for females compared to males because parental effort diverts resources that could be put into growth in body size and females have the steeper fecundity to size relationship (Gross and Sargent 1985).

Additional factors alter the cost of care between the sexes. Among *Rhinocoris* bugs (Reduviidae), male care is found in *R. tristis*, which has a high population density and female preference for caregiving males, minimizing lost mating opportunities for caregiving males. Female care is found in the closely related *R. carmelita*, which has a low population density and a higher promiscuity cost if males were to stay in place with eggs (Gilbert et al. 2010). The evolutionary dynamics between the multiple factors that influence the decision to stay with young suggest that the contingencies that promoted the origin and maintenance of parental care will be difficult to work out, perhaps even more so than for other aspects of sex roles (Royle et al. 2016).

14.2.4 Uniparental male care

Perhaps no aspect of the application of theory to parental care has changed in the past years as much as the perspective on paternal care. Among insects, uniparental male care occurs among species that are all **iteroparous**, hemimetabolous, and in which care terminates at the egg stage (Tables 14.1 and 14.2). Females tend to lay eggs at common oviposition sites, whether that is a space guarded by a male or on the male himself. These correlates alone suggest difficulties for the hypotheses that male care is favoured by limited mating opportunities for males (Maynard Smith 1977) and low population density (Owens 2002). This reasoning followed from an expected association between paternal care and a reversal of sexual selection (Trivers 1972). While sex role reversal in parenting may occur, other aspects of mating systems in insects with uniparental male care are not reversed, including female choice, male courtship behavior, and strategies to pursue polygyny (Requena et al. 2013). Uniparental male care, rather than limiting mating, may increase opportunities for polygyny when females prefer caregiving

males (sexy dad hypothesis), a pattern that has been observed empirically in some insects (Gilbert et al. 2010; Ohba et al. 2016), and supported by both comparative analyses (Gilbert and Manica 2015) and theoretical models (Manica and Johnstone 2004). A similar route to uniparental male care has been hypothesized for fish (Blumer 1979). Uniparental male care and polygyny may necessitate less complex care: no elaborate nest, minimal provisioning, and not extending to post–hatching young (Tables 14.1 and 14.2). This is in contrast to uniparental female care, in which nesting, food provisioning, and other forms of care vary from simple to highly complex, and biparental care, which is always associated with complex, **multi–dimensional care**.

Parental care is widespread in the haplodiploid Phlaeothripidae, the tube-tailed thrips lacking an ovipositor, in which at least five genera have uniparental male care, thirty-seven genera have uniparental female care, and some uncertainty about whether any species are truly biparental (Table 14.1). The reproductive biology of thrips is quite varied, with some genera (e.g., *Elaphrothrips*) exhibiting oviparity, ovoviviparity, and viviparity. In *Tiarothrips*, females guard their clutch and males guard the females or oviposition sites (Costa 2006). In *Idolothrips*, *Sporothrips*, and *Haplothrips*, females also oviposit communally, but leave the eggs with a male who defends both eggs and oviposition sites (Tallamy 2001). This suggests that one evolutionary route to uniparental male care is when females abandon clutches in the territory of a dominant male (Alissa et al. 2017).

Table 14.2 Correlates of uniparental male care and biparental care among insects.

Features	Uniparental male care	Biparental care
Taxonomic distribution	19 of 19 genera are hemimetabolus	212 of 216 genera are holometabolous
Stage at which care is terminated	Egg or at eclosion[1]	Larva
Nesting	Minimal or none	Well developed
Offspring development post-eclosion	Precocial	Altricial
Forms of care	Egg attendance or external egg carrying	Multi–dimensional
Food resource	Live prey (11 of 18), fungi (5 of 18), folivore (2 of 18)	Wood and associated fungi (197 of 216), carrion, dung, detritus (9 of 216), live prey (6 of 216), fungi (4 of 216)
Parity	All iteroparous	Both semelparous and iteroparous

[1]The example of a uniparental male staying with and feeding his brood in the genus *Zelus* (Hemiptera: Reduviidae) turned out to be a case of sex misidentification. The actual genus is *Atopozelus* and the caring sex is female (see Tallamy 2001).

Male care would be selected by the ability to attract additional females to the favoured oviposition site, a high population density of females (Manica and Johnstone 2004), paternity assurance, and a benefit to his young from attendance. In arthropods, fathers are known to protect eggs from desiccation, predators, parasitoids, and fungal attack (Requena et al. 2013). In inbred populations, haplodiploidy may also favour paternal care (Davies and Gardner 2014).

In some taxa, uniparental male care via territory–based polygyny is thought to evolve through an intermediary stage of female care (Alissa et al. 2017). Gilbert and Manica (2015) did not find support for this hypothesis among insects, associating a female to male care transition with selection for enhanced female fecundity (Figure 14.4). We agree that enhanced female fecundity was probably not the primary driving force for male care in insects, but feel that sexual selection is a separate route for a female to male care transition. The rarity of uniparental male care among insects, as well as gaps in natural history and evolutionary relationships, increase the uncertainty around phylogenetic methods. The occurrence of both female care and male care among thrips, *Rhinocoris*, and pentatomid bugs, and the similar forms of care within each of these groups, suggest that further work is needed. Moreover, evidence for **amphisexual care** among *Rhinocoris* and in other arthropods (see Alissa et al. 2017) suggests that the genetic architecture for caregiving in one sex may allow an easier selective pathway for caregiving to evolve in the other sex.

Although high paternity favours male care generally, it may not be necessary for the evolution of uniparental male care if the potential for polygyny is sufficiently high (Alonzo and Klug 2012). In fact, males without eggs may steal or adopt unrelated eggs in an apparent attempt to increase their attractiveness as mates. These behaviors, reported for some arthropods with exclusive male care, suggest an important role of sexual selection in the evolution or maintenance of paternal care in the group (Requena et al. 2013). Males that carry eggs on their body, such as the water bugs, present more of a challenge for the evolution of male care by sexual selection. Body size constrains the number of eggs that can be cared for and thus the number of mates. Although the potential for polygyny is more limited than in egg attenders, recent work demonstrates that males carry eggs of multiple females simultaneously and are preferred as mates (Ohba et al. 2016). The more limited potential for polygyny may necessitate high paternity if carrying eggs is to be evolutionarily stable. Paternity assurance mechanisms can be elaborate in egg carriers, including frequent interruption of oviposition for repeated matings (Smith 1997). Intriguingly, egg-carrying by male water bugs may have evolved from egg attendance. Two genera of belostomatids (*Kirkaldyia* and *Lethocerus*) exhibit the plesiomorphic behavior of attending multiple clutches of eggs laid on emergent vegetation, while also engaging in repeated matings with each female. Smith (1997) hypothesized that when females dispersed to bodies of water without emergent vegetation, there was selection to oviposit on males as an alternative. Thus, the potential for polygyny may have been established before the origin of egg carrying.

14.2.5 Biparental care

The correlates of uniparental male care and biparental care are strikingly different among insects (Table 14.2), suggesting that neither system of care evolved from the other. The comparison also points toward the different selective pressures that led to these parenting systems. In contrast to uniparental male care, biparental care is associated with complex, multi-dimensional care that extends beyond the egg stage, and includes direct provisioning or facilitation of feeding within a nest or burrow (Tables 14.1 and 14.2). Most cases of biparental care among insects are found in holometabolous groups (212 of 216 genera), that have **altricial** young. Among the few hemimetabolous genera with biparental care, there is a tendency toward altriciality (Nalepa and Bell 1997). There is also an association of biparental care and altriciality in birds (Cockburn 2006), with some question about the direction of causation between the two. While Burley and Johnson (2002) suggest that they coevolve in birds (and probably in hemimetabolous insects), the strong holometabolous bias suggests that altriciality facilitated the subsequent evolution of biparental care. Once biparental care evolves, a reduction in juvenile mortality would subsequently select for longer developmental times and increased altriciality, an association that could then coevolve (Royle et al. 2016).

Our compilation provides support for the hypothesis that biparental care is favoured by multi-dimensional care (Barta et al. 2014; Table 14.2). Cooperation between the male and female partner can be favoured when their efforts are not additive, which includes, but is not limited to, task specialization (Trumbo 2012). Although task specialization has been hypothesized to promote the stability of biparental care (Grafen and Sibly 1978), most models begin with the assumption that the parental effort of the two parents are interchangeable, i.e., uni-dimensional and additive (see Houston et al. 2013). Models assuming uni-dimensional care reveal that biparental systems can be unstable. Stability requires restrictive values of **parental compensation** or **parental negotiation** over partner effort (Johnstone and Hinde 2006) because the partner that is less efficient in care or has superior future breeding opportunities is expected to desert early (Grafen and Sibly 1978). Even more problematic is that over evolutionary time, an initial sex difference in providing uni-dimensional parental care will be subject to disruptive selection. The sex with a lower cost or higher benefit from care will be under stronger selection to increase in competency (McNamara and Wolf 2015) and tendency to care (Fromhage and Jennions 2016), which will reinforce initial sex differences in care. Furthermore, a strong tendency for males to abandon early may select for females to produce smaller clutches that they could rear on their own without overwhelming costs, further diminishing the value of male care. This latter scenario may be expected to play out in provisioners with depreciable care (Gilbert and Manica 2010) if both sexes were providing the same form of care.

Models of parental negotiation that assume interchangeable, additive care may help us to understand how the sexes allocate effort for a particular task, but may not explain the evolutionary stability of biparental care. While biparental care is hypothesized to be selected by high offspring need (Maynard Smith 1977; Thomas and Székely 2005), a high value of uni-dimensional care alone may do little to promote its evolution. Large offspring to care for or large litters are not thought to promote biparental care among mammals, but once evolved, biparental care

may facilitate the evolution of larger litters in systems in which males feed young (Royle et al. 2016). Thus, an assumption of uni-dimensional care/additive effort appears to miss the *raison d'être* of biparental care.

When care is multi-dimensional, disruptive selection on the sexes should stabilize, rather than destabilize biparental care. It is expected that males and females will have initial differences in their ability to perform specific components of care for reasons unrelated to selection for parental tasks (Lessels 2012; Barta et al. 2014). For example, it has been hypothesized that a greater tendency for the male parent to guard the nest emerged out of prior selection to guard females or favourable oviposition sites (Alcock 1975). A sex difference in guarding probably occurs in wood-boring beetles (Curculionidae), which constitute the majority of insect genera with biparental care (Kirkendall et al. 2015; Table 14.1). The sex with a greater tendency to perform a specific task is under stronger selection to improve the ability to perform that task, and may experience trade-offs for the ability to perform other parental tasks, reinforcing initial differences (McNamara and Wolf 2015). While multi-dimensional care may be a prerequisite of biparental care, it is not sufficient for its evolution nor does it prevent reversals to uniparental female care, as has likely occurred among Scolytinae beetles (Gilbert and Manica 2015).

In diverse social contexts, within-population variation in the tendency to perform a task may lead to the emergence of pronounced division of labour without a fixed difference in roles (Pruitt et al. 2012). Operational division of labour is compatible with overlapping abilities to perform tasks and flexibility of work (Trumbo 2012). The dynamics of biparental interactions may, in fact, promote the emergence of flexibility for both task allocation and task competency. Burying beetles cover a small vertebrate carcass with anal exudates to manipulate the microbial community in their nest. When a male or female *Nicrophorus vespilloides* provides care on their own, anal exudates have similar levels of antimicrobial activity. When they provide care together, however, the female has greater antimicrobial activity, while the male has less (Cotter and Kilner 2010). Such dynamics in paired and unpaired caregivers need more study to understand the evolution of division of labour. It is thought that division of labour in eusocial insects can emerge early in the evolutionary process by differences in response thresholds to care-prompting stimuli (Robinson and Page 1989). A slight difference in the response threshold to a task-related stimulus can result in one co-foundress ant to complete most bouts for that task (Fewell and Page 1999). In biparental insects, evolutionary feedbacks would be expected to promote associations between response thresholds for tasks and task ability, which would select for further sexual divergence in response thresholds. With task specialization, it is axiomatic that one parent cannot fully compensate for its partner without incurring higher costs, a situation that should stabilize cooperative care of offspring (Trumbo 2006). Species with biparental care may be models for examining the emergence of cooperation in social systems in which individuals have both competing and overlapping interests.

Partner cooperation in biparental systems is not limited to task specialization. Important synergies can also occur when parents need to be at two places at the same time (many burrowers) or when simultaneous action is needed to defend a resource or to complete a task under a severe time constraint, i.e., a **superadditive effect** (Motro 1994). Desert arthropods that must complete a burrow system in a brief window of time are prominent examples of the latter, in which the biparental system is also favoured by a lack of breeding opportunities for deserters (Linsenmair 1987; Rasa 1999). When parents care for active young within an extensive nest or burrow system, parental care and social interactions are necessarily complex even if task specialization is not pronounced (Nalepa and Bell, 1997). Passalid beetles, for example, use many distinct acoustical signals in familial interactions within rotting logs (Schuster and Schuster 1997). Each sex will defend the burrow system against same-sex infanticidal intruders more aggressively than against opposite-sex intruders, probably because a successful intrusion has greater costs for the same-sex resident. An extended biparental association is then promoted because each parent must stay to protect its reproductive interests (Kirkendall et al. 2015). Defence of the nest against a male versus a female conspecific can be considered separate parental tasks for which each partner has differentially been selected (Trumbo

2006; Hopwood et al. 2015). Studies of other taxa provide support for an association of biparental care with multi-dimensional care. Uniparental male care is far more common than biparental care in fish, many of which attend eggs (Gross and Sargent 1985). Although uniparental care is rare in birds, it is associated with **precocial** young (Cockburn 2006). The group with the greatest number of transitions from female care to biparental care is the primates (Reynolds et al. 2002), whose offspring have perhaps the most complex needs of any taxa.

14.3 Concluding remarks

While pre-ovipositional care is widespread among insects, post-ovipositional care is rare, occurring in a little more than 2000 genera, distributed in nineteen orders (Table 14.1). This is only a fraction of the enormous insect diversity, estimated to be over 1 million species. Although rare, post-ovipositional care has evolved many times independently, thus insects are an ideal group to employ the comparative approach to test hypotheses on parental care (e.g., Gilbert and Manica 2010, 2015). Two ideas proposed many years ago remain untested, and deserve attention—the role of the four environmental prime movers proposed by Wilson (1975) for the evolution of parental care, and the effect of resource distribution and predictability on the forms of parental care in insects (Tallamy and Wood 1986).

This chapter shows that uniparental female care is by far the most common system of parental care in insects, being found both in holo- and hemimetabolous. In most species, female care is restricted to eggs and newly hatched young, but in some groups females care for offspring after their nutritional independence. Uniparental male care is found only in hemimetabolous species with precocial young, which usually disperse after hatching. As occurs with fish, male care in insects seems to increase opportunities for polygyny when females prefer caregiving males, but studies are restricted to only a few hemipteran species (Gilbert et al. 2010; Ohba et al. 2016). Biparental care is found primarily in holometabolous species with altricial young, and parental activities always extend beyond the egg stage. The differences between uniparental male care and biparental care suggest that neither system of care evolved from the other. Uniparental male care likely evolved either from no care or uniparental female care on the ability to attract additional mates. The correlates of biparental care support the hypothesis that multi-dimensional care is a prerequisite for its evolution.

Despite the great potential of insects as model systems for studies on the evolution of parental care, detailed information on the behavioral forms of post-ovipositional parental care in the group is still fragmentary. If we want to understand how biotic and abiotic conditions influence the costs and benefits of parental care, we need a broad geographical coverage, especially in extreme habitats such as deserts, caves, and very cold places. Biological information is also not evenly distributed taxonomically, and some orders have received much more attention than others. Finally, although the benefits of post-ovipositional care have been investigated in many species, studies on the costs are scarce, which limits conclusions on the adaptive function of parental behaviors. We hope the information presented here stimulates more studies with insects, especially in poorly explored places and with poorly explored groups.

References

Alcock, J. (1975). Territorial behavior by males of *Philanthus multimaculatus* (Hymenoptera: Sphecidae) with a review of male territoriality in male sphecids. *Animal Behaviour*, **23**, 889.

Alissa, L.M., Muniz, D.G., and Machado, G. (2017). Devoted fathers or selfish lovers? Conflict between mating effort and parental care in a harem–defending arachnid. *Journal of Evolutionary Biology*, **30**, 191.

Alonzo, S.H. (2012). Sexual selection favours male parental care, when females can choose. *Proceedings of the Royal Society of London. Biological Sciences*, **279**, 1784.

Alonzo, S.H., and Klug, H. (2012). Paternity, maternity, and parental care. In: N.J. Royle, P.T. Smiseth, and M. Kölliker (Eds), *The Evolution of Parental Care*, pp. 189–205. Oxford University Press, Oxford.

Barta, Z., Székely, T., Liker, A., and Harrison, F. (2014). Social role specialization promotes cooperation between parents. *American Naturalist*, **183**, 747.

Bernays, E.A., and Chapman, R.F. (2007). *Host–plant Selection by Phytophagous Insects*. Chapman & Hall, New York, NY.

Blumer, L.S. (1979). Male parental care in the bony fishes. *Quarterly Review of Biology*, **54**, 149.

Burley, N.T., and Johnson, K. (2002). The evolution of avian parental care. *Philosophical Transactions of the Royal Society of London, Biological Sciences*, **357**, 241.

Chaboo, C.S., Frieiro–Costa, F.A., Gómez–Zurita, J., and Westerduijn, R. (2014). Origins and diversification of subsociality in leaf beetles (Coleoptera: Chrysomelidae: Cassidinae: Chrysomelinae). *Journal of Natural History*, **48**, 2325.

Cockburn, A. (2006). Prevalence of different modes of parental care in birds. *Proceedings of the Royal Society of London, Biological Sciences*, **273**, 1375.

Costa, J.T. (2006). *The Other Insect Societies*. Harvard University Press, Cambridge, MA.

Cotter, S.C., and Kilner, R.M. (2010). Sexual division of antibacterial resource defence in breeding burying beetles, *Nicrophorus vespilloides*. *Journal of Animal Ecology*, **79**, 35.

Danks, H.H. (2002). Modification of adverse conditions by insects. *Oikos*, **99**, 10.

Davies, N., and Gardner, A. (2014). Evolution of paternal care in diploid and haplodiploid populations. *Journal of Evolutionary Biology*, **27**, 1012.

Edgerly, J.S. (1997). Life beneath silk walls: a review of the primitively social Embiidina. In: J. Choe and B. Crespi (Eds), *The Evolution of Social Behavior in Insects and Arachnids*, pp. 14–25. Cambridge University Press, Cambridge.

Eickwort, G.C. (1981). Presocial insects. In: H.R. Hermann (Ed.) *Social Insects*, vol 2, pp. 199–280. Academic Press, New York, NY.

Fewell, J.H., and Page, R.E. (1999). The emergence of division of labour in forced associations of normally solitary ant queens. *Evolutionary Ecology Research*, **1**, 537.

Forero, D., Choe, D.H., and Weirauch, C. (2011). Resin gathering in neotropical resin bugs (Insecta: Hemiptera: Reduviidae): functional and comparative morphology. *Journal of Morphology*, **272**, 204.

Fox, C.W., and Czesak, M.E. (2000). Evolutionary ecology of progeny size in arthropods. *Annual Review of Entomology*, **45**, 341.

Fromhage, L., and Jennions, M.D. (2016). Coevolution of parental investment and sexually selected traits drives sex–role divergence. *Nature Communications*, **7**, 12517.

Gardner, A., and Smiseth, P.T. (2011). Evolution of parental care driven by mutual reinforcement of parental food provisioning and sibling competition. *Proceedings of the Royal Society of London, Biological Sciences*, **278**, 196.

Gilbert, J.D.J., and Manica, A. (2010). Parental care trade–offs and life–history relationships in insects. *American Naturalist*, **176**, 212.

Gilbert, J.D.J., and Manica, A. (2015). The evolution of parental care in insects: a test of current hypotheses. *Evolution*, **69**, 1255.

Gilbert, J.D.J., Thomas, L.K., and Manica, A. (2010). Quantifying the benefits and costs of parental care in assassin bugs. *Ecological Entomology*, **35**, 639.

Godfray, H.C.J. (1994). *Parasitoids: Behavioral and Evolutionary Ecology*. Princeton University Press, Princeton, NJ.

Gonzalez–Voyer, A., Fitzpatrick, J.L., and Kolm, N. (2008). Sexual selection determines parental care patterns in cichlid fishes. *Evolution*, **62**, 2015.

Grafen, A., and Sibly, R. (1978). A model of mate desertion. *Animal Behaviour*, **26**, 645.

Gross, M.R., and Sargent, R.C. (1985). The evolution of male and female parental care in fishes. *American Zoologist*, **25**, 807.

Hagan, H.R. (1951). *Embryology of the Viviparous Insects*. Ronald Press Company, New York, NY.

Halffter, G. (1997). Subsocial behavior in Scarabaeinae beetles. In: J. Choe and B.J. Crespi (Eds) *The Evolution of Social Behavior in Insects and Arachnids*, pp. 237–59. Cambridge University Press, Cambridge.

Hinton, H.E. (1981). *Biology of Insect Eggs*. Pergamon Press, Oxford.

Hopwood, P.E., Moore, A.J., Tregenza, T., and Royle, N.J. (2015). Male burying beetles extend, not reduce, parental care duration when reproductive competition is high. *Journal of Evolutionary Biology*, **28**, 1394.

Houston, A.I., Székely, T., and McNamara, J.M. (2013). The parental investment models of Maynard Smith: a retrospective and prospective view. *Animal Behaviour*, **86**, 667.

Johnstone, R.A., and Hinde, C.A. (2006). Negotiation over offspring care—how should parents respond to each other's efforts? *Behavioral Ecology*, **17**, 818.

Kirkendall, L.R., Biedermann, P.H.W., and Jordal, B.H. (2015). Evolution and diversity of bark and ambrosia beetles. In: F.E. Vega and R.W. Hofstetter (Eds), *Bark Beetles: Biology and Ecology of Native and Invasive Species*, pp. 85–156. Academic Press, New York, NY.

Klug, H., Bonsall, M.B., and Alonzo, S.H. (2013). The origin of parental care in relation to male and female life history. *Ecology and Evolution*, **3**, 779.

Kokko, H., and Jennions, M.D. (2008). Parental investment, sexual selection and sex ratios. *Journal of Evolutionary Biology*, **21**, 919.

Kokko, H., and Jennions, M.D. (2012). Sex differences in parental care. In: N. J. Royle, P. T. Smiseth, and M. Kölliker (Eds), *The Evolution of Parental Care*, pp. 101–16. Oxford University Press, Oxford.

Lehtinen, R.M., Nussbaum, R.A., and Jamieson, B. (2003). Parental care: a phylogenetic perspective. *Reproductive Biology and Phylogeny of Anura*, **2**, 343.

Lessels, C.M. (2012). Sexual conflict. In: N.J. Royle, P.T. Smiseth, and M. Kölliker (Eds), *The Evolution of Parental Care*, pp. 150–70. Oxford University Press, Oxford.

Linsenmair, K.E. (1987). Kin recognition in subsocial arthropods, in particular in the desert isopod *Hemilepistus reamuri*. In: D.J.C. Fletcher and C.D. Michener (Eds), *Kin Recognition in Animals*, pp. 121–208. Wiley, New York, NY.

London, K.B., and R.L. Jeanne. (2000). The interaction between mode of colony founding, nest architecture and ant defence in polistine wasps. *Ethology, Ecology & Evolution*, **12**, 13.

Manica, A., and Johnstone, R.A. (2004). The evolution of paternal care with overlapping broods. *American Naturalist*, **164**, 517.

Mank, J.E., Promislow, D.E.L., and Avise, J.C. (2005). Phylogenetic perspectives in the evolution of parental care in ray–finned fishes. *Evolution*, **59**, 1570.

Maynard Smith, J. (1977). Parental care: a prospective analysis. *Animal Behaviour*, **25**, 1.

McNamara, J.M., and Wolf, M. (2015). Sexual conflict over parental care promotes the evolution of sex differences in care and the ability to care. *Proceedings of the Royal Society of London, Biological Sciences*, **282**, 20142752.

Monteith, G.B. (2006). Maternal care in Australian oncomerine shield bugs (Insecta, Heteroptera, Tessaratomidae). In: W. Rabitsch (Ed.) *Hug the Bug—For Love of True Bugs*, Festschrift zum 70 Geburtstag von Ernst Heiss, Denisia **19**, 1135–52. Biologiezentrum, Linz.

Motro, U. (1994). Evolutionary and continuous stability in asymmetric games with continuous strategy sets—the parental investment conflict as an example. *American Naturalist*, **144**, 229.

Munguía–Steyer, R.,Favila, M.E., and Macías–Ordóñez, R. (2008). Brood pumping modulation and the benefits of paternal care in *Abedus breviceps* (Hemiptera: Belostomatidae). *Ethology*, **114**, 693.

Nalepa, C.A., and Bell, W.J. (1997). Postovulation parental investment and parental care in cockroaches. In: J. Choe and B.J. Crespi (Eds), *The Evolution of Social Behavior in Insects and Arachnids*, pp. 26–51. Cambridge University Press, , UK.

Ohba, S., Okuda, N., and Kudo, S. (2016). Sexual selection of male parental care in giant water bugs. *Royal Society Open Science*, **3**, 150720.

Owens, I.P.F. (2002). Male-only care and classical polyandry in birds: phylogeny, ecology and sex differences in remating opportunities. *Philosophical Transactions of the Royal Society of London, Biological Series*, 357(1419), 283–93.

Pruitt, J.N., Oufiero, C.E., Avilés, L., and Riechert, S.E. (2012). Iterative evolution of increased behavioral variation characterizes the transition to sociality in spiders and proves advantageous. *American Naturalist*, **180**, 496.

Queller, D.C. (1997). Why do females care more than males? *Proceedings of the Royal Society of London, Biological Sciences*, **264**, 1555.

Rasa, O.A.E. (1999). Division of labour and extended parenting in a desert tenebrionid beetle. *Ethology*, **105**, 37.

Requena, G.S., Munguía–Steyer, R., and Machado, G. (2013). Paternal care and sexual selection in arthropods. In: R.H. Macedo and G. Machado (Eds), *Sexual Selection Perspectives and Models from the Neotropics*, pp. 201–34. Elsevier, Amsterdam.

Reynolds, J.D., Goodwin, N.B., and Freckleton, R.P. (2002). Evolutionary transitions in parental care and live bearing in vertebrates. *Philosophical Transactions of the Royal Society B, Biological Sciences*, **357**, 269.

Robinson, G.E., and Page, R.E. (1989). Genetic determination of nectar foraging, pollen foraging, and nest–site scouting in honey bee colonies. *Behavioral Ecology and Sociobiology*, **24**, 317.

Royle, N.J., Alonzo, S.H., and Moore, A.J. (2016). Co–evolution, conflict and complexity: what have we learned about the evolution of parental care behaviours? *Current Opinion in Behavioral Sciences*, **12**, 30.

Santos, E.S.A., Bueno, P.P., Gilbert, J.D.J., and Machado, G. (2017). Macroecology of parental care in arthropods: higher mortality risk leads to higher benefits of offspring protection in tropical climates. *Biological Reviews*, **92**, 1688.

Schuster, J.C., and Schuster, L.B. (1997). The evolution of social behavior in Passalidae (Coleoptera). In: J. Choe and B.J. Crespi (Eds), *The Evolution of Social Behavior in Insects and Arachnids*, pp. 260–9. Cambridge University Press, Cambridge.

Seagraves, M. (2009). Lady beetle oviposition behavior in response to the trophic environment. *Biological Control*, **51**, 313.

Sendoya, S.F., Freitas, A.V.L., and Oliveira, P.S. (2009). Egg–laying butterflies distinguish predaceous ants by sight. *American Naturalist*, **174**, 134.

Smiseth, P.T., Kölliker, M., and Royle, N.J. (2012). What is parental care? In: N.J. Royle, P.T. Smiseth, and M. Kölliker (Eds), *The Evolution of Parental Care*, pp. 1–20. Oxford University Press, Oxford.

Smith, R.L. (1997). Evolution of paternal care in the giant water bugs (Heteroptera: Belostomatidae). In: J. Choe and B.J. Crespi (Eds), *The Evolution of Social Behavior in Insects and Arachnids*, pp. 116–49. Cambridge University Press, Cambridge.

Spodek, M., Ben-Dov, Y., and Mendel, Z. (2014). The scale insects (Hemiptera: Coccoidea) of oak trees (Fagaceae: *Quercus* spp.) in Israel. *Israel Journal of Entomology*, **43**, 95–124.

Suzuki, S., Kitamura, M., and Matsubayashi, K. (2005). Matriphagy in the hump earwig, *Anechura harmandi* (Dermaptera: Forficulidae), increases the survival rates of the offspring. *Journal of Ethology*, **23**, 211.

Tallamy, D.W. (2001). Evolution of exclusive paternal care in arthropods. *Annual Review of Entomology*, **46**, 139.

Tallamy, D.W., and Schaefer, C. (1997). Maternal care in the Hemiptera: ancestry, alternatives, and current adaptive value. In: J.C. Choe and B.J. Crespi (Eds), *The Evolution of Social Behavior in Insects and Arachnids*, pp. 94–115. Cambridge University Press, Cambridge.

Tallamy, D.W., and Wood, T.K. (1986). Convergence patterns in subsocial insects. *Annual Review of Entomology*, **31**, 369.

Thomas, G.H., and Székely, T. (2005). Evolutionary pathways in shorebird breeding systems: sexual conflict, parental care, and chick development. *Evolution*, **59**, 2222.

Trivers, R.L. (1972). Parental investment and sexual selection. In: B. Campbell (Ed.) *Sexual Selection and the Descent of Man*, pp. 136–79. Aldine, Chicago, IL.

Trumbo, S.T. (2006). Infanticide, sexual selection and task specialization in a biparental burying beetle. *Animal Behaviour*, **72**, 1159.

Trumbo, S.T. (2012). Patterns of parental care in invertebrates. In: N.J. Royle, P.T. Smiseth, and M. Kölliker (Eds), *The Evolution of Parental Care*, pp. 81–100. Oxford University Press, Oxford.

Tsukamoto, L., and Tojo, S. (1992). A report of progressive provisioning in a stink bug, *Parastrachia japonensis* (Hemiptera: Cydnidae). *Journal of Ethology*, **10**, 21–9.

Wesolowski, T. (1994). On the origin of parental care and the early evolution of male and female parental roles in birds. *American Naturalist*, **143**, 39–58.

Williams, G.C. (1975). *Sex and Evolution*. Princeton University Press, Princeton, NJ.

Wilson, E.O. (1975). *Sociobiology*. Harvard University Press, Cambridge, MA.

Zeh, D.W., Zeh, J.A., and Smith, R.L. (1989). Ovipositors, amnions and eggshell architecture in the diversification of terrestrial arthropods. *Quarterly Review of Biology*, **64**, 147–68.

CHAPTER 15

Sociality

Jennifer Fewell[1] and Patrick Abbot[2]

[1] *School of Life Sciences, Arizona State University, Tempe, AZ, USA*
[2] *Department of Biological Sciences, Vanderbilt University, Nashville, TN, USA*

15.1 Introduction

Like animal social taxa more broadly, insect societies and their distributions reflect an interplay between phylogenetic history, and the physiological, life history, and ecological parameters that shape sociality. Thus, exploring sociality in insects allows us to touch upon all the diverse beauties and challenges of the social domain. They also offer an intriguing set of challenges that draw us into the most fundamental question of social living: 'why be social?' The class Insecta includes some of the most numerous and diverse social taxa on earth. Indeed, insects beat vertebrates by orders of magnitude in the number of social species. The highly eusocial insects have been particularly successful; they dominate the natural ecology of almost every habitat on earth (Wilson 1971; Hölldobler and Wilson 1990). Despite this, the vast majority of insect species are actually solitary, even within most of the families containing eusocial species. The exceptions are the ants and termites, in which all extant species are eusocial (Hölldobler and Wilson 1990; Thorne 1997; Korb and Hartfelder 2008).

The social insects inform us about how different forms of sociality evolve, particularly about the ecological and social contexts favouring cooperation. They provide important testbeds for the roles of mutual cooperation and **altruism** in social evolution

(Hamilton 1964), as well as for exploring levels of selection (Wade et al. 2010). Social evolution involving strong altruism is clearly facilitated by kin-based relatedness (kin selection), as evidenced by the clustering of highly social taxa within the haplo-diploid Hymenoptera (Hamilton 1964; Eberhard 1975). The importance of kinship also receives strong support from the consideration that eusociality (defined by altruism and reproductive division of labour) evolves almost exclusively from family groups headed by singly mated females or monogamous pairs (Hughes et al. 2008; Boomsma 2009). However, kinship alone does not explain the diversity of social types within and across insect taxa (Michener 1974; Eberhard 1975). Nor does it completely explain the transformation of some eusocial groups into the mega-social colonies of the ants and termites—cities comprising millions of workers that are often highly genetically diverse and sometimes only minimally related to each other (Oldroyd and Fewell 2007; Helanterä et al. 2009).

It is impossible to capture the diversity that is insect sociality in a single chapter. To provide an initial framework to build upon, a simplified categorization of the different types of insect social living most commonly seen will be presented here. Some of the core issues of insect social evolution will also be considered in the contexts of:

Fewell, J. and Abbot, P., *Sociality*. In: *Insect Behavior: From mechanisms to ecological and evolutionary consequences*. Edited by Alex Córdoba-Aguilar, Daniel González-Tokman, and Isaac González-Santoyo: Oxford University Press (2018). © Oxford University Press.
DOI: 10.1093/oso/9780198797500.003.0015

- fitness frameworks for cooperation and altruism;
- the ecological and behavioral contexts under which cooperation may be more beneficial;
- the unique physiological and life history trade-offs delimiting insect versus vertebrate social groups.

Although this is a chapter about insects, many of the types of sociality discussed here can extend more broadly to vertebrate societies as well.

15.2 What is social living?

Insect social groups are extremely diverse (Costa 2006), to the degree that assigning them to 'types' becomes a lesson in categorizing the uncategorizable (Wcislo 1997a). With this in mind, it is still useful to consider the qualitative differences between different forms of social life. To do so, first the difference between social *behavior* and social *living* must be defined. Arguably, all sexually reproducing animals have social behaviors, because it is necessary to interact with others of the same species in order to mate or compete for mates. The transition from social behavior to social living, however, requires a more complex set of interactions built primarily around cooperation (Bergmüller et al. 2007).

It is also possible to distinguish broadly between group living and the subset of groups considered as cooperative sociality. A large number of species form aggregations, temporary to longer-term associations in which individuals cluster in space. Aggregations form for a number of behavioral and ecological reasons, including ephemeral aggregations during mating, feeding or migration, individuals clustered together for predator protection, or nests packed into patchy, but valuable sites. In contrast, cooperative sociality is characterized by social behaviors that benefit the group, whether performed together as coordinated behaviors, or as individual acts contributing to group benefits [including mutually cooperative and altruistic behaviors (Clutton-Brock 2002; Bergmüller et al. 2007)]. One useful way to conceptualize sociality is that it occurs when:

- individuals remain with others of the same species for an extended part of their lifespans;
- when the interactions among group members are generally mutually cooperative and/or altruistic;

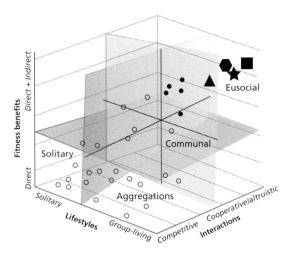

Figure 15.1 Insect species vary along multiple axes related to solitary and social lifestyles, including (a) the degree to which conspecific individuals live in groups, (b) how their interactions vary along a spectrum of competition or cooperation, and (c) how relatedness and ecological factors interact to translate into inclusive fitness advantages provided by the group in which they live (unfilled circles). Species not recognized as social insects, including those that do or do not aggregate (filled circles). Insects in which grouping and social interactions constitute major lifestyle features, including those that are primitively or facultatively eusocial (square, star, triangle, and hexagon). The eusocial insects that form multi-generational family groups with strong division of labour, castes of workers that may be sterile and that help to defend the colony and/or to rear offspring, including the ants, honey bees and other corbiculate bees, vespine wasps, and termites.

individuals benefit via inclusive fitness advantages provided by the group in which they live (Figure 15.1). This is messy, and there are many species spanning the boundaries between sociality and other forms of grouping; however, it gives us a starting point for considering the large diversity of insect social groups, and animal social groups more broadly.

15.2.1 Fitness contexts for sociality

Paper wasps of the genus *Polistes* occur worldwide, and their nests may be familiar sights on the eaves of houses or in backyards (Figure 15. 2). Their nests are constructed out of a mixture of saliva and plant material, and resemble paper (Downing and Jeanne 1988; Hunt 2007). In one species (*Polistes metricus*), the nests are constructed in the spring by a single

Figure 15.2 A *Polistes metricus* foundress at her nest. She has deposited a single pale egg within some of the nest cells (photo credit: J. Hunt).

reproductive female (the foundress); her first brood become workers who help to rear subsequent broods over the course of summer. Her later brood will mature into reproductives. They leave the nest, mate, and overwinter as adults, emerging the following spring to start the cycle again (Hunt 2007). To appreciate why the life history of a species like *P. metricus* needs special explanation, it is useful to keep in mind that providing a benefit to another in a resource-limited world is costly. Why would the first brood of *P. metricus* help their mother rear their siblings, rather than reproduce themselves?

There are different ways to approach the question(s) of why *P. metricus* daughters stay and help. When considering the hormonal and behavioral changes associated with delaying dispersal, or the specific ways that helpers coordinate tasks, we are examining this social strategy from a proximate level. If we consider it in terms of the consequences to individual reproductive output or to group competitive success, we are answering it at an

ultimate level. The term 'cooperation' is often mistakenly applied across both levels. To avoid confusion, it is used only at the ultimate level, defining it in terms of the fitness costs and benefits of social behaviors.

As per Hamilton (1964), the fitness outcomes of a behavioral strategy accrue in two generalized ways; first by the direct fitness benefits of a given strategy on the reproductive success of the individual performing it, and secondly, via indirect fitness benefits, which derive from the survival and/or reproductive consequences of helping or being helped (depending on the directional perspective of that relationship), multiplied by the relatedness (or degree of covariance) between those individuals. The inclusive fitness of an individual that interacts socially with neighbours is the relatedness-weighted sum of **direct and indirect fitness** (Queller 1992; Grafen 2009). The social group itself also has an important role in shaping the expressed phenotypes of group members and, consequently, their fitness outcomes (Mcglothlin et al. 2010; Wade et al. 2010). Indirect genetic effects capture the idea that individual social phenotypes and their resulting fitness outcomes are highly dependent on the genotypes of other group members. By considering the entire range of potential costs and benefits, the concept of inclusive fitness guides our understanding of why individuals cooperate with others.

Cooperation and fitness outcomes

Most categorizations of the relationships between behavioral interactions and fitness outcomes focus on the direct fitness impacts of that behavior on the individual performing it (the actor), and on the recipient(s). Competitive interactions, in which one individual benefits at the expense of another (+/−), are at the core of solitary living. Interactions within a cooperative social group, however, generally include a mix of competitive and cooperative behaviors. A large proportion of interactions within cooperative societies generate mutually shared, direct benefits (+/+), as when tent caterpillars work collectively to construct a silk nest that shades them from solar influx and protects them from predators. Their group activity allows them to construct a more functional structure that improves survival

(Fitzgerald and Costa 1999; Costa and Ross 2003). By extension, relationships where contributions are reciprocated, but delayed in time are also considered cooperative (+/+).

Altruism occurs when an actor performs behaviors directed at benefiting others, with direct fitness cost to themselves (−/+). The boundary between cooperation and weaker forms of altruism can be difficult to delineate for more cohesive and/or complex social groups (Figure 15.1). However, strongly altruistic behaviors (in which one individual clearly decreases reproductive output to help another), occur almost exclusively in family groups, and in the context of indirect fitness benefits. This becomes starkly clear in the case of most eusocial insects, because a majority of the group do not reproduce, and rather assist the queen in brood care (Wilson 1971). As a caution against categorical labelling, though, some eusocial relationships that seem altruistic actually generate delayed direct fitness gains (Thorne 1997). For example, the workers of primitively eusocial termites are juveniles (nymphs) that retain the ability to mature into reproductive adults. When resources become depleted, workers accelerate development and disperse. Inclusive fitness estimates suggest that in some contexts, delaying maturation and dispersal may provide a higher direct than indirect fitness pay-off (Korb and Hartfelder 2008; Korb and Thorne 2017).

Individuals exhibiting altruistic behavior do not always do so voluntarily. Sister foundresses of the socially polymorphic carpenter bee, *Ceratina australensis*, establish nests in which one forages and reproduces, and the other helps by guarding the nest. The inclusive fitness outcomes for the reproductive (via direct fitness gain) are higher than for helpers (via indirect fitness gain) (Rehan et al. 2014). In cases of sibling-based groups such as this, the helpers are often directly reproductively suppressed (West 1967; Premnath et al. 1996; Ratnieks and Wenseleers 2008). Reproductives may also manipulate their own offspring into becoming helpers. In small eusocial wasp or bee societies, offspring that develop into helpers are often underfed by the mother, producing individuals with lower capacities for reproduction should they try to disperse and reproduce independently (Kapheim et al. 2015).

15.3 Insect social diversity

With an estimated two million identified species, insects are the largest and most diverse group of terrestrial animals in the world, and arguably also the most diverse in terms of life history and ecological strategies (Costa 2006; Capinera 2008). Only a small subset of this wealth of species, however, is social (Table 15.1).

The Hymenoptera or 'social insects' are well known for their complex sociality (Figure 15.3). However, they contain only one family that is essentially completely eusocial, the ants (Formicidae; Hölldobler and Wilson 1990). The majority of bees (in superfamily Apoidea) and wasps (groups in suborder Aculeata, other than ants and bees) are actually solitary. Eusocial bees are limited primarily to the sweat bees (Halictidae), and the corbiculate (pollen basket) bees, which include the honey bees, bumblebees, and the stingless bees (Michener 1974). The eusocial wasps are clustered in one family, the Vespidae (Hunt and Toth 2017).

The roaches are a phylogenetically diverse group (Jeon and Park 2015); essentially all roaches can be considered as **group-living** in that almost all cluster as aggregations and almost all show parental care (subsociality) (Nalepa and Bell 1997). However, they rarely form more complex societies (Tallamy and Wood 1986). The termites (Isoptera) represent a monophyletic lineage, ancestrally related to wood roaches (Engel et al. 2009). All termites are eusocial (Thorne 1997; Korb and Thorne 2017). The poor quality and indigestibility of their primary food source is thought to have been a primary ecological driver of the elaboration of the sociality in termites, from a basic subsocial ground plan in roaches built around extended parental care (Nalepa 2015; Korb and Thorne 2017).

Four other orders are notable for containing a small number of species with distinctive cooperatively social types: Hemiptera, Thysanoptera, Coleoptera, and Lepidoptera (Table 15.1). A subset of gall-forming aphids in the Hemiptera (hemimetabolous insects with sucking mouthparts and incomplete metamorphosis) and Thysanoptera (thrips) consist

Table 15.1 Major insect taxonomic groups and their sizes that have species expressing various forms of sociality, from simple aggregations to advanced eusociality.

Order	Common name(s)	Estimated # families[a]	Estimated # species[b]	Social forms
Blattodea (excluding Isoptera)	Roaches	7 All families have some form of group living	3700–4000	Aggregations: all roaches live gregariously Extended parental care: most roaches exhibit some parental care Communal: cohesive mixed family groups found rarely
Isoptera	Termites	9 All families eusocial	2600–2800	Eusocial (all species)
Coleoptera	Beetles	172 Cooperative sociality primarily in Silphidae	360,000–	Primitively eusocial: primarily wood-dwelling taxa Advanced eusociality: wood and soil dwelling taxa
			>1,000,000	Solitary (most species)
				Aggregations: Ephemeral adult mating aggregations Biparental family groups: burying beetles (Silphidae) and some passalids
				Cooperatively communal groups:
		Scolytidae, Platypotidae and in superfamily Scaraboidea (includes Pallisidae)		some burying beetles, passalids and ambrosia beetles (Scolytidae and Platypotidae)
Diptera	Flies	209	153,000	Solitary (most species)
Ephemeroptera	Mayflies		2500–3000	Aggregations: ephemeral feeding and mating aggregations Aggregations: ephemeral adult mating aggregations
Hemiptera	True bugs (aphids)	168 Eusociality only in Aphididae	80,000–88,000	Solitary (most species)
				Eusocial: some gall-forming aphids with soldier castes
Hymenoptera: Suborder Symphyta	Sawflies	14	6200[c]	Solitary (most species and all adults) Aggregations: larval clusters for huddling Communally cooperative: larval associations with cooperative foraging
Hymenoptera: Suborder Apocrita (excluding ants and bees)	Wasps	11 Cooperative sociality only in Vespidae	110,000[c]	Solitary or parasitoid (most species)
				Aggregations: nesting aggregations Semisocial groups: sister groups found in Vespidae
				Eusocial:
				primitive eusociality (more common form)
				advanced eusociality (rare: found in Polistes)

(Continued)

Table 15.1 Continued

Order	Common name(s)	Estimated # families[a]	Estimated # species[b]	Social forms
Hymenoptera: Superfamily Apoidea (excluding wasps)	Bees	7[d] Sociality rare or absent in Melittidae, Andrenidae, Colletidae, and Stenotritidae; Limited cooperative sociality in Megachilidae; All forms of sociality in Apidae (including corbiculates) and Halictidae	20,000–30,000	Solitary (most species) Aggregations: nesting aggregations in all families except Apidae[f] Communal: small social groups of non-relatives (occasionally found in mining bees, Andrenidae; leafcutter bees, Megachilidae, Halictidae) Semisocial: sister groups (found in Halictidae and some Apidae) Eusociality: primitively eusociality (more common) Advanced eusociality: limited to corbiculate bees, and some sweat bees, Halictidae
Hymenoptera: Family Formicidae	Ants	1 20 subfamilies, all eusocial[e]	12,000–15,000	Eusociality (all species) Primitively eusociality: primarily in Ponerines Advanced eusociality (common)
Lepidoptera	Moths and butterflies	132 Some form of group living estimated in 27 families[f]	112,000	Aggregations: ephemeral adult mating aggregations Aggregations: larval clusters Communally cooperative: larval associations for foraging and tent construction (in family Lasiocampidae)
Orthoptera	Grasshoppers and crickets	39	24,400	Solitary (most species) Aggregations: foraging and migratory aggregations
Thysanoptera	Thrips	9 Eusociality only in Phlaeothripidae		Solitary (most species) Eusocial: primitive eusociality with soldier caste in some *Acacia* gall-forming species

[a] Family number estimates primarily from, except Hymenoptera. [B] Species number estimates primarily from (Chapman 2009.)

[c] Wasp and sawfly family and species estimates calculated from Encyclopedia of Entomology (Capinera 2008). [D] Bee families from (Hedtke et al. 2013).

[e] Ant subfamily and species estimates from AntBase (www.AntBase.org), and Chapman (2009). [F] Social caterpillar estimates from Fitzgerald (web.cortland.edu/fitzgerald), and from (Costa 2006).

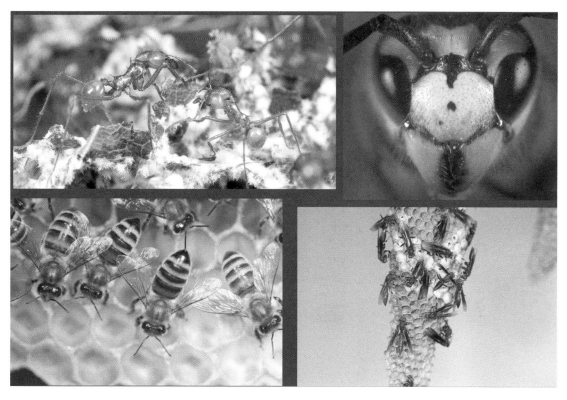

Figure 15.3 Examples of eusocial Hymenoptera (clockwise from upper left): workers of the leaf-cutter ant *Atta sextans* tending their fungal gardens (photo credit: C. Currie); the vespine wasp *Vespula germanica* (photo credit: B. Klein); a colony of *Polistes canadensis* paper wasps, including a dominant reproductive and subordinates, tending brood; honey bee *Apis mellifera* workers on a comb (photo credit: B. Klein).

of cohesive family groups with reproductive division of labour; they are eusocial (Abbot and Chapman 2017). Although the majority of Coleoptera (beetles) are solitary, there are distinct clusters of social species, including the Passalidae (Schuster and Schuster 1997), which are tropical beetles with extensive parental care (Schuster and Schuster 1985), and the fungus-tending ambrosia beetles (Curculionidae; Kirkendall et al. 2015). Burying beetles, *Nicrophorus* (in the carrion beetle family Silphidae) rely on a rare bonanza of a dead vertebrate to rear their offspring, often requiring extensive parental care (Trumbo 1992; Scott 1998). Finally, some sawflies (Hymenoptera) and moths (Lepidoptera) have the unusual case of juveniles being cooperatively social, while adults are generally solitary (Fitzgerald 1995; Costa 1997). These examples illustrate that there are many forms of insect sociality beyond the 'classic' eusocial colonies typified by

the ants and honey bees. Costa (2006) provides in-depth descriptions of many of the social and group-living insects outside of the ants, bees, wasps, and termites.

15.3.1 A taxonomy of insect social groups

The primary sources categorizing insect sociality can be found in Michener (1974) and Wilson (1971). In a manner similar to that illustrated in Figure 15.1, social insect groups can be further subdivided by:

- whether groups cooperate versus simply aggregate;
- whether they are formed as family groups from delayed or limited offspring dispersal, or are formed by individuals (usually adults) that cluster and cooperatively nest together;

Figure 15.4 A colony of the wood-dwelling termite *Cryptotermes secundus* exposed in an experimental arena (left) and within a split log (right), revealing a mix of individuals at different developmental stages (photo credit: J. Korb).

• whether group members are close relatives, or non-relatives and/or a mix of both.

Layered onto these categorizations, we can also ask the question of whether most individuals in the society reproduce, or whether reproduction is dominated by one or a few individuals (Ross 2001; Dew et al. 2016). Unsurprisingly, the second category generally aligns with the category of family-based groups.

The largest number of insect taxa fall immediately into the category of aggregation, including most of the group-living Lepidoptera, Orthoptera, Siphonoptera, Diptera, and Hemiptera (Table 15. 1; Costa 2006). The cooperatively social species, however, have much more interesting social relationships. Cooperative groups formed outside of societies

Figure 15.5 Examples of 'other' social insects, including from top (a) a social aphid colony (*Pemphigus obesinymphae*) within its poplar petiole gall (left), with a mix of smaller aggressive nymphs and their larger, maturing clone mates (photo credit: P. Abbot); (b) a gallery of the ambrosia beetle, *Xyleborinus saxesenii*, with adult and larval individuals (photo credit: P. Biedermann); and (c) a female burying beetle, *Nicrophorus* spp., feeding larval offspring (photo credit: S. Trumbo).

based on parent–offspring relationships have traditionally been clumped together as '**parasocial**' (Michener 1974), but they actually form two fairly distinct groups based on whether the social group's core is built around a family group (usually sisters) or a society of non- or mixed relatives. Parent–offspring groups also show two major types—families in which parents provide the care (subsocial), and those with extensive **alloparental care**, including foraging contributions, feeding, and/or defence by siblings (eusocial). In some species, the two categories facultatively merge together (Schwarz et al. 2010; Wcislo and Fewell 2017).

15.3.1.1 Communal groups: cooperation without strong altruism

With the interesting exception of the social caterpillars and sawflies, which cooperate only as larvae (Costa 2006), communally cooperative social groups are generally associations of adults (Michener 1974). These groups were originally divided by Michener (1974) into groups with (**quasisocial**) or without (communal) cooperative brood care. This distinction is likely a consequence of life history variation, rather than representing levels of sociality (Dew et al. 2016), and we will discuss them together as being communally cooperative. These groups also should not be assumed to be an evolutionary intermediary between solitary and eusocial. Phylogenetic analyses of social evolution in bees (Wcislo and Danforth 1997; Danforth 2002; Schwarz et al. 2007) provide strong evidence that family groups (subsocial or **semisocial**), may act as precursors in the evolution of eusociality, but that communally social groups generally do not.

Communally social groups often contain non-relatives, with most group members reproducing (Ross 2001; Moore and Kukuk 2002). Group members collectively share in some subset of those tasks that would normally be performed by an individual when alone. They often work together in nest construction and group defence; less often they may cooperatively forage or food share. In general, communally cooperative societies seem most similar to joint nesting societies of birds (Vehrencamp 2000), or to social mammal groups such as prides and fission-fusion societies (Lehmann et al. 2007; Smith et al. 2008; Silk et al. 2014).

In communal sweat bees (*Lasioglossum zephyrum*), for example, small groups (5–10) of unrelated adult females form social groups upon emergence each

spring. They collectively excavate a ground nest, and alternate in the task of guarding the nest entrance (Kukuk and Schwarz 1987; Kukuk and May 1991; Kukuk et al. 1998). The sharing of excavation and guarding provides mutual individual benefits, and can be performed socially because the bees occupy these parts of the nest as shared space. The bees do not cooperatively tend brood, however, probably because eggs are laid in separate side tunnels and individually provisioned there (Kukuk and Schwarz 1987).

The ambrosia beetles represent a different kind of communal lifestyle. They live in damp wood environments, and cultivate fungus as their primary food source (Kirkendall et al. 2015). Females lay clusters of eggs around fungus in galleries throughout the nest structure. Reproducing females generally excavate their own galleries and care for their own brood piles (French and Roeper 1972). In some taxa, males stay with the female to provide parental care (Kirkendall 1983). The nests themselves are rare, but valuable sites, and female offspring often delay dispersal to help tend brood or to establish their own galleries. This generates a structured system with mixed family groups sharing the nest as a whole, and individuals within family groups sharing of the tasks of fungus gardening, gallery maintenance, and brood care (Biedermann and Taborsky 2011; Kirkendall et al. 2015). The relationships between food source, nesting choice, and sociality in ambrosia beetles have distinct analogies to some wood-dwelling roaches (Nalepa 1994), and one species has been found to be primitively eusocial (Kent and Simpson 1992).

15.3.1.2 Subsocial or family groups

A large number of social systems are built around parent–offspring relationships, beginning with extended parental care of brood, with offspring delaying dispersal until adulthood; groups with extensive parental care are categorized as subsocial (Tallamy and Wood 1986). Parental care in insects is different from most terrestrial vertebrates (see Chapter 14). Vertebrate taxa with extensive sociality, almost without exception, have some level of sustained parental care; in birds both parents commonly provide care (Black and Hulme 1996). Most insects express essentially no parental care, beyond the oviposition of eggs, and perhaps provisioning with food (see Chapter 14). However, virtually all cooperative insect societies involve some level of continued parental care, either in the form of directly feeding and grooming developing brood, or by defending them (Trumbo 1996, 2012).

Subsocial groups represent the primary intermediate step between solitary and eusocial living. When adult offspring forgo dispersal to help with care of siblings, subsociality transitions into eusociality (Schwarz et al. 2010). The co-occurrence of subsociality and eusociality has been reported within multiple bee species, with both strategies sometimes occurring within the same population (Schwarz et al. 2007; Rehan et al. 2011; Wcislo and Fewell 2017). Indirect fitness effects are key in eusocial species, as the helper relinquishes reproduction (Figure 15. 1). However, in subsocial species, adult siblings remaining at the nest may have opportunities to achieve reproduction later (Schwarz et al. 2010). The facultative nature of sociality, as illustrated by the case of subsocial bees, highlights the difficulty of establishing clear divisions between social types (Figure 15.1).

15.3.1.3 Semisocial groups, groups of adult siblings

In semisocial groups, adult close relatives (usually sisters) nest together. Semisocial groups have been identified most often in the wasps and bees (Hunt and Toth 2017;Wcislo and Fewell 2017). They generally have dominance hierarchies and reproductive division of labour, in which one individual becomes the primary reproductive. In other cases, colonies begin as semisocial groups, then become eusocial when offspring delay dispersal to help. The wasp, *Ropalidia marginata*, serves as an exemplar of this mixed strategy. Queens may initiate nests alone or as part of sibling groups. Within-group dominance generates a queen, and her subordinates help rear worker brood (Gadagkar 1980; Premnath et al. 1996).

15.3.1.4 Eusociality

Ants, termites, honey bees and other corbiculate bees, and some vespine wasps form multi-generational family groups composed of one or a few reproductives, and multiple workers that help in various ways to defend the colony and/or to rear offspring (Wilson 1971; Crespi and Yanega 1995). They have

'true' worker castes that are either morphologically or functionally sterile (unable to produce queens). In more derived eusocial species, workers can exhibit anatomical, physiological, or behavioral specializations that reflect their complex patterns of division of labour (Oster and Wilson 1978; Hölldobler and Wilson 2009). As originally defined, eusocial colonies were characterized by the overlapping of generations as offspring delay or fail to disperse, reproductive division of labour, and alloparental (worker) brood care. This pattern describes the eusocial Hymenoptera; however, closer study of such species as the wood-dwelling termites, thrips, and aphids have expanded the concept of eusociality to include other behaviors, contributing to colony success, in particular the presence of a sterile soldier caste (Crespi 1994; Abbot and Chapman 2017; Korb and Thorne 2017; Figure 15.4).

Eusocially living species can range from a queen and a few functionally sterile helpers, to thousands or millions of morphologically sterile workers in the highly eusocial 'superorganism' colonies (Hölldobler and Wilson 2009). Although the term 'eusociality' generally brings to mind large and complex colonies, most eusocial species are primitively eusocial, with non-reproducing, but not morphologically sterile workers. Most polistine wasps (O'Donnell 1998) and halictine bees (Wcislo 1997b, Schwarz et al. 2010) for example, have colonies with helpers that retain the ability to mate and independently found new nests, even though in practice they rarely do (Crespi and Yanega, 1995). In contrast, colonies of the New World tropical leafcutter ant *Atta sextans*, contain several million sterile workers, with specialized morphological and age-based worker subcastes (Hölldobler and Wilson 2009, 2010). Similarly, colonies of the mound-building termite *Macrotermes bellicosus* in Africa reach millions of individuals, including major, minor, and soldier castes. These live in underground nests where, like leaf cutter ants, they farm fungi on dead and decaying plant material (Collins 1981).

Altogether, eusociality has evolved independently at least 24 times in insects (Bourke 2011), with four phylogenetically deep origins of obligate, complex eusociality (ants, bees, wasps, and termites; Crespi 2007). The number of eusocial origins will surely grow as new groups are characterized, but eusociality of the sort expressed by species like ants, bees, wasps and termites is clearly taxonomically rare. Yet these species tend to be unusually important in the ecological communities in which they occur (Wilson 1971, Hölldobler and Wilson 1990).

15.3.1.5 Eusociality: direct and indirect fitness benefits

Although lifetime monogamy is rare in insects, the ancestral condition for advanced eusocial insects was single mating by females, emphasizing the role of relatedness in the evolution of eusociality (Hughes et al. 2008, Boomsma 2009). In the highly derived social insects, the colony functions as the primary unit of selection. Often these colonies are paradoxically polyandrous, which reduces relatedness between workers, as when honey bees, *Apis mellifera,* mate with 10-20 males in a single mating flight (Laidlaw and Page 1984). In other species, multiple queens may form cooperative associations that persist through the colony lifespan (polygyny), as do colonies of the termite *Nasutitermes corniger,* which may contain anywhere from one to over thirty queens (Thorne 1984). The resulting genetic diversity among workers has multiple functional benefits, from enhancing division of labour (Oldroyd and Fewell 2007), to promoting disease resistance (Shykoff and Schmid-Hempel 1991, Seeley and Tarpy 2007). However, polyandry in eusocial systems is essentially limited to groups containing sterile workers, with selection acting primarily at the colony level (Hölldobler and Wilson 2009).

In primitively eusocial colonies, colony behavior often reflects a tension between cooperation and conflict (Bourke 1999). The primitively eusocial ant, *Harpegnathus saltator*, serves as an example. Colonies are initiated by a queen, but several workers in any colony retain the ability to mate (gamergates). In the event that the founding queen dies, the gamergates fight for reproductive dominance. Even in advanced eusocial colonies of the Hymenoptera, workers often retain the ability to produce males via unfertilized eggs, because of haplo-diploidy, generating a potential pathway to direct fitness gain. These eggs are generally eaten by other workers, a form of policing that mitigates reproductive conflict within the colony (Ratnieks and Wenseleers, 2005).

15.4 Behavioral and ecological contexts for sociality

New England buck moth caterpillars (*Hemileuca lucina*) huddle in tight aggregations to bask. The thermal benefits of huddling mean that eggs can hatch earlier in the spring, when predators are less common (Stamp and Bowers 1990). The tent caterpillars move from aggregation to communal sociality because they additionally coordinate in nest construction, such that individual efforts contribute to the group (Fitzgerald 1995). Cooperative behaviors like those exhibited by *H. lucina* can be generalized as fitting into the categories of: (a) parental and alloparental care; (b) foraging and food sharing; and (c) group, offspring and nest defence.

Parental care: By definition, extended parental care is central to subsociality and eusociality (see also chapter 14); however, it is also a general component of insect social organization more widely. Insects form social alliances when they need direct help in offspring care, as in the social Hymenoptera, or when they need to stay in place for extended periods to rear offspring. Beetles illustrate the latter case nicely. Burying beetles rear offspring on the carcass of a mouse or similar carrion, a rare and precious find (Figure 15.5). They meticulously process the carcass and defend it fiercely; offspring are fed with the partially predigested remains. The requirements for this work are extensive. The parent must excavate a site to bury the carcass, and maintain it through larval development as they process the carcass to feed the larvae (Trumbo 1992, Scott 1998). The costs of offspring care and associated food defence are a major factor determining burying beetle social strategy. This can range from a single female exhibiting extended parental care, to biparental care (unusual in insects), with males particularly aiding in corpse guarding and offspring defence, to rarer cases of communally cooperative societies, in which multiple pairs cooperatively utilize and protect the carcass (Trumbo 1992, Scott 1998, Trumbo 2012).

Wood-consuming taxa illustrate the theme that offspring care, type of food, and the need for defence, intertwine as drivers of social evolution. Developing offspring of *Cryptocerus* wood roaches (sister group to the termites) benefit from the protection and food provided by the rotten wood nests they occupy (Figure 15.3). However, as hemimetabolous insects consuming a high-cellulose diet, they require periodic inoculation of microbes for digestion. Their developmental rates are also constrained by the poor food quality, and they may take up to five years to develop into adults. Parents must provide extensive care for this life history strategy to work, from creating a chamber for the developing brood, processing food for offspring, and performing **trophallaxis** to provide symbiotic microbes (Bell et al. 2007). As an interesting parallel, the passalid beetles also live in rotten wood; although they are holometabolous, they still require extended parental care including help with food pre-digestion. They live in cooperative associations, in which parents construct individual galleries and care for developing larva, and the group defends the nest (Schuster and Schuster 1985).

Foraging and food sharing: Groups can dominate a food resource, but the benefits of social foraging depend on the distribution and temporal availability of food. Thus, taxa specializing on a temporally patchy or seasonally limited food source, whether a specific plant or prey, are generally solitary (Wcislo and Fewell 2017). Conversely, highly social central place foragers tend to be generalists. There is, for example, a negative relationship between sociality and level of pollen specialization in bees (Wcislo and Fewell 2017). More broadly, although the social Hymenoptera are usually categorized by food types—for example, granivore, predator; nectivore (Chapter 8)—they tend to be generalist foragers within those categories.

Food can also create a barrier to social evolution when limitation intensifies potential competition among group members (Korb and Hartfelder 2008). Resource availability is particularly a problem for central place foragers, which live in a fixed location and forage from it. The army ants have broken through this barrier in a unique way; they carry their 'nest' with them as they chase down prey, by building temporary bivouacs constructed of workers (Hölldobler and Wilson 1990). Other social species solve the resource problem by growing their own food. Leafcutting ants (*Atta* and *Acromyrmex*) (Hölldobler and Wilson 2010), some social beetles (Curculionidae), and some higher termites (Macroter-

mitinae) all cultivate fungus gardens, which offer a more nitrogen-rich food source (Breznak 1973) and break down the cellulose of the media on which it grows (Breznak 1994), allowing exploitation of a resource not otherwise available. Sociality enhances the ability to exploit this resource. This can be seen dramatically in the leafcutter ants, some of which mobilize huge foraging forces to retrieve plant material for the fungus (Hölldobler and Wilson 2010).

Group defence: Defence is perhaps the most universally performed cooperative behavior (see Chapter 9). Sociality involving more than simple group defence (as seen in aggregating caterpillars) often revolves around a centralized and defensible nest. Social cooperation clearly provides a defence benefit even to small social groups. Communal groups of ground nesting sweat bees (*Lasioglossum spp.*) can continuously guard the entrance to their nests against parasites and predators (Lin 1964), while a lone sweat bee must alternate between defence, nest construction and foraging (Kukuk and Schwarz 1987).

Defence presents as a central theme for the evolution of eusociality in gall-forming aphids and thrips, as well as in termites. Caste specialization in the aphids, thrips and other 'fortress defence' societies comes in the form of a soldier caste, and worker castes that provide other forms of brood care are often entirely absent (Tian and Zhou 2014). The need for defence is characteristic of all eusocial insects, and many highly eusocial species have evolved extreme collective defence strategies including the highly effective defensive swarms of Africanized honey bees (Collins et al. 1982), and the 'death-grip' attacks by stingless bees (Shackleton et al. 2015). Perhaps most impressive are the morphologically specialized defenders of some advanced ants and termites. The *Pheidole* super-majors use their disproportionately massive heads to fight and to plug the nest entrance in response to invaders such as army ants (Huang 2010). The *Nasutitermes* are named for the heads of their soldiers, which function as nozzles that shoot a sticky irritating liquid at invaders (again often ants). These extreme morphological specializations are limited to highly derived eusocial species, which have colonies large enough to afford them (Oster and Wilson 1978), and also to those in which workers are not morphologically constrained by the need for winged flight. As far as is known, there seem to be no wasps or bees sporting allometrically enlarged heads with nozzles.

15.4.1 Insect life history and constraints on social evolution

Although they share many of the same ecological drivers of sociality, insects distinctly differ from social vertebrates in a few key ways that influence how they are social. One of these is development. In holometabolous insects, such as the Hymenoptera, Coleoptera and the Lepidoptera, the juvenile or larval forms are typically reduced and soft-bodied, often with limited mobility and a diet that differs from adults. The transition from larval stages to adult requires a pupal stage and complete metamorphosis (Chapman 1998). Some holometabolous insects have completely different social forms between juvenile and adult, such as has been noted for cooperative tent caterpillars and sawflies (Costa 2006). However, most social forms of holometabolous insects involve extended parental care of the juvenile phase. In the eusocial Hymenoptera particularly, social altruism revolves around the care and protection of brood (Wilson 1971).

In the hemimetabolous insects, including the Hemiptera, Thysanoptera, and Blattodea, juveniles or nymphs resemble adults, and growth occurs via incomplete metamorphosis; i.e. without a pupal stage (Chapman 1998). Juveniles and adults do not typically differ in diet, and juveniles are typically more mobile. In the hemimetabolous eusocial insects, juveniles are normally the helpers. Termite workers, for example, are nymphs that delay or terminate their development into adult reproductives (Korb and Thorne 2017). Although the particular shapes of sociality are affected by developmental type, the range of social types, from aggregations to communally cooperative groups, to eusociality, are found in both hemimetabolous and holometabolous taxa.

Most social insect taxa are found in the tropics, suggesting a constraint of temperature on insect sociality; even within the tropics, most social taxa consist of small groups limited by seasonal patterns of temperature, moisture and food resource availability. The perennial colonies of advanced eusocial species have escaped these constraints by finding ways to maintain stable micro-environments for brood across temporal and resource fluxes. Almost all eusocial species (and most social species generally) construct nests in protective spaces and use their workers to further thermoregulate the queen and brood. Almost all perennial colonies also store food, or live in it, providing a buffer against resource fluctuations. Although most bee species have little food storage capacity, the stingless bees (Meliponini) and honey bees (*Apis*) have evolved the capacity to convert nectar into storable honey (Wille 1983, Winston 1991, Roubik 2006). This adaptation was likely a key factor in the transformation of ancestrally small eusocial colonies into large units that can survive resource shortfalls and persist in more temperate regions.

Finally, the social insect societies offer interesting answers to the question of what it means to be socially complex (Bourke 1999, Anderson and McShea 2001). Insect societies, particularly eusocial groups, accomplish extremely complex group level feats when compared to non-human vertebrates, from the construction of elaborate nest architectures to their large-scale collective decision making (Camazine et al. 2002). Social insects do this, however, with much smaller brains and using much simpler individual behavioral rules than do most birds or mammals. They accomplish this because of their extremely high reproductive potential, allowing some taxa to achieve colony sizes of thousands to millions. Queens in ants, honeybees, and termites also have unusually long lives for insects, up to a decade, and in rarer cases over 30 years (Keller 1998, Remolina and Hughes 2008).

This strength in numbers allows large eusocial colonies to organize in ways that smaller social groups simply cannot (Anderson and McShea 2001). Colony behaviors are organized using a combination of sophisticated communication networks, coupled with self-organizational mechanisms built around simple behavioral rules for each individual worker (Seeley 1995, Camazine et al. 2002). We could generalize that highly complex group patterns in insect societies are generated from the collective actions of more simple individuals (Gronenberg and Riveros 2009, O'Donnell et al. 2015), while in the vertebrate world, somewhat less complex group level patterns are achieved by smaller sets of larger, longer-lived, and possible more complex individuals.

15.5 Conclusion

The breadth and complexity of insect social forms make the study of their sociality both rewarding and challenging. Insect societies range from simple aggregations, to cooperative groups in which non-relatives collaborate for behaviors that bring individual benefit, to eusocial colonies with extensive altruism. This social diversity is framed around the general paradox that many social taxa are extremely successful, but sociality itself is rare. As with all social taxa, any exploration of insect social evolution needs to consider the selective and ecological contexts around which social groups are formed. Communal societies are built from mixes of relatives and/or non-relatives that individually benefit from cooperation, whether in defence, nest construction, brood care, or other tasks. Other groups form when offspring and/or siblings delay dispersal to help relatives. When they do so, selection can enhance the division of labour between those who reproduce, and those who forage, defend, and perform other tasks that benefit the colony. From these groups, we see the elaboration of astoundingly complex societies, such as the eusocial colonies of ants and termites.

A hallmark of social behavior in insects is that common ecological and behavioral themes reappear despite diverse taxonomic origins. All complex social groups require some form of parental care and alloparental support, whether it involves direct brood care, or defence against predators. Extended parental care is the major driver of sociality in the ants, bees, and wasps, and is primary to many beetle social groups as well. Virtually all insect societies cooperate in some form of group defence; in many cases, for example, in the termites, thrips and aphids, defence forms the primary basis for reproductive division of labour. Food is also a central organizing principle; if an insect group does live directly in or on their food, they forage for it, and the relative benefits or costs of collective foraging can enhance or constrain social evolution. For researchers new to social insects, this diversity offers opportunities. Despite the rich history of social insect biology, practically nothing is known about most social insects in any detail; one can pick a favourite, knowing it is sure to lead to somewhere good. No matter what the theme in social biology, or the preferred technique, the world of insect social behavior tent is wide open for exploration.

References

Abbot, P., and Chapman, A.D. (2017). Sociality in aphids and thrips. In: D. R. Rubenstein and P. Abbot (Eds), *Social Evolution*, pp. 154–87. Cambridge University Press, Cambridge.

Anderson, C., and McShea, D.W. (2001). Individual versus social complexity, with particular reference to ant colonies. *Biological Reviews*, **76**, 211.

Bell, W.J., Roth, L.M., and Nalepa, C.A. (2007). *Cockroaches: Ecology, Behavior, and Natural History*. Johns Hopkins University Press, Baltimore, MD.

Bergmüller, R., Johnstone, R.A., Russell, A.F., and Bshary, R. (2007). Integrating cooperative breeding into theoretical concepts of cooperation. *Behavioural Processes*, **76**, 61.

Biedermann, P.H., and Taborsky, M. (2011). Larval helpers and age polyethism in ambrosia beetles. *Proceedings of the National Academy of Sciences*, **108**, 17064.

Black, J.M., and Hulme, M. (1996). *Partnerships in Birds: The Study of Monogamy*. Oxford University Press, Oxford.

Boomsma, J.J. (2009). Lifetime monogamy and the evolution of eusociality. *Philosophical Transactions of the Royal Society B: Biological Sciences*, **364**, 3191.

Bourke, A. (1999). Colony size, social complexity and reproductive conflict in social insects. *Journal of Evolutionary Biology*, **12**, 245.

Bourke, A.F. (2011). *Principles of Social Evolution*, Oxford University Press, Oxford.

Breznak, J.A. (1973). Nitrogen fixation in termites. *Nature*, **244**, 577.

Breznak, J.A. (1994). Role of microorganisms in the digestion of lignocellulose by termites. *Annual Review of Entomology*, **39**, 453.

Camazine, S., Deneubourg, J.L., Franks, N., et al. (2002). *Self-organization in Biological Systems*. Princeton University Press, Princeton, NJ.

Capinera, J.L. (2008). *Encyclopedia of Entomology*. Springer Science & Business Media, Berlin.

Chapman, A.D. (2009). *Numbers of living species in Australia and the World*, 2nd edn. Australian Biodiversity Information Services, Canberra.

Chapman, R.F. (1998). *The Insects: Structure and Function*, Cambridge University Press, Cambridge.

Clutton-Brock, T. (2002). Breeding together: kin selection and mutualism in cooperative vertebrates. *Science*, **296**, 69.

Collins, A.M., Rinderer, T.E., Harbo, J.R., and Bolten, A.B. (1982). Colony defense by Africanized and European honey bees. *Science (Washington)*, **218**, 72.

Collins, N. (1981). Populations, age structure and survivorship of colonies of *Macrotermes bellicosus* (Isoptera: Macrotermitinae). *Journal of Animal Ecology*, **50**(1), 293.

Costa, J.T. (1997). Caterpillars as social insects: largely unrecognized, the gregarious behavior of caterpillars is changing the way entomologists think about social insects. *American scientist*, **85**, 150.

Costa, J.T. (2006). *The Other Insect Societies*. Harvard University Press, Cambridge, MA.

Costa, J.T., and Ross, K.G. (2003). Fitness effects of group merging in a social insect. *Proceedings of the Royal Society of London B: Biological Sciences*, **270**, 1697.

Crespi, B. (1994). Three conditions for the evolution of eusociality: are they sufficient? *Insectes Sociaux*, **41**, 395.

Crespi, B.J., and Yanega, D. (1995). The definition of eusociality. *Behavioral Ecology*, **6**, 109.

Danforth, B.N. (2002). Evolution of sociality in a primitively eusocial lineage of bees. *Proceedings of the National Academy of Sciences*, **99**, 286.

Dew, R., Tierney, S., and Schwarz, M. (2016). Social evolution and casteless societies: needs for new terminology and a new evolutionary focus. *Insectes Sociaux*, **63**, 5.

Downing, H.A., and Jeanne, R.L. (1988). Nest construction by the paper wasp, Polistes: a test of stigmergy theory. *Animal Behaviour*, **36**, 1729.

Eberhard, M.J.W. (1975). The evolution of social behavior by kin selection. *Quarterly Review of Biology*, **50**, 1.

Engel, M.S., Grimaldi, D.A., and Krishna, K. (2009). Termites (Isoptera): their phylogeny, classification, and rise to ecological dominance. *American Museum Novitates*, 1.

Fitzgerald, T.D. (1995). *The Tent Caterpillars*. Cornell University Press, Ithaca, NY.

Fitzgerald, T.D., and Costa, J.T. (1999). Collective behavior in social caterpillars. In: C. Detrain, J.L. Deneubourg, and J.M. Pasteels (Eds), *Information Processing in Social Insects*, pp. 379–400. Springer, Birkhauser.

French, J.R., and Roeper, R.A. (1972). Interactions of the ambrosia beetle, *Xyleborus dispar* (Coleoptera: Scolytidae), with its symbiotic fungus *Ambrosiella hartigii* (Fungi Imperfecti). *Canadian Entomologist*, **104**, 1635.

Gadagkar, R. (1980). Dominance hierarchy and division of labour in the social wasp, *Ropalidia marginata* (Lep.) (Hymenoptera: Vespidae). *Current Science*, **49**, 772.

Grafen, A. (2009). Formalizing Darwinism and inclusive fitness theory. *Philosophical Transactions of the Royal Society of London B: Biological Sciences*, **364**, 3135.

Gronenberg, W., and Riveros, A.J. (2009). Social brains and behavior: past and present. In: J. Gadau, and J.H. Fewell (Eds) *Organization of Insect Societies: from Genome to Sociocomplexity*, pp. 377–401. Harvard University Press, Cambridge, MA.

Hamilton, W.D. (1964). The genetical theory of social behavior. I and II. *Journal of Theoretical Biology*, **7**, 1–52.

Helanterä, H., Strassmann, J.E., Carrillo, J., and Queller, D.C. (2009). Unicolonial ants: where do they come from, what are they and where are they going? *Trends in Ecology & Evolution*, **24**, 341.

Hedtke S.M., Patiny S., and Danforth B.N. (2013). The bee tree of life: a supermatrix approach to apoid phylogeny and biogeography. *BMC Evolutionary Biology*, **13**, 138.

Hölldobler, B., and Wilson, E.O. (1990). *The Ants*. Springer Verlag, Berlin.

Hölldobler, B., and Wilson, E.O. (2009). *The Superorganism: the Beauty, Elegance, and Strangeness of Insect Societies*. WW Norton & Company, New York, NY.

Hölldobler, B., and Wilson, E.O. (2010). *The Leafcutter Ants: Civilization by Instinct*, WW Norton & Company, New York, NY.

Huang, M.H. (2010) Multi-phase defence by the big-headed ant, Pheidole obtusospinosa, against raiding army ants. *Journal of Insect Science*, **10**, 1.

Hughes, W.O., Oldroyd, B.P., Beekman, M., and Ratnieks, F.L. (2008). Ancestral monogamy shows kin selection is key to the evolution of eusociality. *Science*, **320**, 1213.

Hunt, J.H. (2007). *The Evolution of Social Wasps*. Oxford University Press, Oxford.

Hunt, J.H., and Toth, A.L. (2017). Sociality in wasps. In: D. R. Rubenstein and P. Abbot (Eds), *Comparative Social Evolution*, pp. 84–123. Cambridge University Press, Cambridge.

Jeon, M.G., and Park, Y.C. (2015). The complete mitogenome of the wood-feeding cockroach *Cryptocercus kyebangensis* (Blattodea: Cryptocercidae) and phylogenetic relations among cockroach families. *Animal Cells and Systems*, **19**, 432.

Kapheim, K.M., Nonacs, P., Smith, A.R., Wayne, R.K., and Wcislo, W.T. (2015). Kinship, parental manipulation and evolutionary origins of eusociality. *Proceedings of the Royal Society of London B: Biological Sciences*, **282**, 20142886.

Keller, L. (1998). Queen lifespan and colony characteristics in ants and termites. *Insectes Sociaux*, **45**, 235.

Kent, D., and Simpson, J. (1992). Eusociality in the beetle *Austroplatypus incompertus* (Coleoptera: Curculionidae). *Naturwissenschaften*, **79**, 86.

Kirkendall, L. (1983). The evolution of mating systems in bark and ambrosia beetles (Coleoptera: Scolytidae and Platypodidae). *Zoological Journal of the Linnean Society*, **77**, 293.

Kirkendall, L.R., Biedermann, P.H., and Jordal, B.H. (2015). Evolution and diversity of bark and ambrosia beetles. In: F.E. Vega and R.W. Hofstetter (Eds), *Bark Beetles: Biology and Ecology of Native and Invasive Species*, pp. 85–156. Academic Press, San Diego, CA.

Korb, J., and Hartfelder, K. (2008). Life history and development-a framework for understanding developmental plasticity in lower termites. *Biological Reviews*, **83**, 295.

Korb, J., and Thorne, B.L. (2017). Sociality in termites. In: D.R. Rubenstein and P. Abbot (Eds), *Comparative Social Evolution*, pp. 124–53. Cambridge University Press, Cambridge.

Kukuk, P.F., and May, B. (1991). Colony dynamics in a primitively eusocial Halictine bee, *Lasioglossum (Dialictus) Zephyrum* (Hymenoptera, Halictidae). *Insectes Sociaux*, **38**, 171.

Kukuk, P.F., and Schwarz, M. (1987). Intranest behavior of the communal sweat bee, *Lasioglossum (Chilalictus) erythrurum* (Hymenoptera, Halictidae). *Journal of the Kansas Entomological Society*, **60**, 58.

Kukuk, P.F., Ward, S.A., and Jozwiak, A. (1998). Mutualistic benefits generate an unequal distribution of risky activities among unrelated group members. *Naturwissenschaften*, **85**, 445.

Laidlaw, H.H., and Page, R.E. (1984). Polyandry in honey bees (*Apis mellifera* L.): sperm utilization and intracolony genetic relationships. *Genetics*, **108**, 985.

Lehmann, J., Korstjens, A.H., and Dunbar, R. (2007). Fission–fusion social systems as a strategy for coping with ecological constraints: a primate case. *Evolutionary Ecology*, **21**, 613.

Lin, N. (1964). Increased parasitic pressure as a major factor in the evolution of social behavior in halictine bees. *Insectes Sociaux*, **11**, 187.

Mcglothlin, J.W., Moore, A.J., Wolf, J.B., and Brodie Iii, E.D. (2010). Interacting phenotypes and the evolutionary process. III. Social evolution. *Evolution*, **64**, 2558.

Michener, C.D. (1974). *The Social Behavior of the Bees: A Comparative Study*. Harvard University Press, Cambridge, MA.

Moore, A.J., and Kukuk, P.F. (2002). Quantitative genetic analysis of natural populations. *Nature Reviews Genetics*, **3**, 971.

Nalepa, C.A. (1994). Nourishment and the origin of termite eusociality. In: J.H. Hunt and C.A. Nalepa (Eds), *Nourishment and Evolution in Insect Societies*, pp. 26–51. Westview Press, Boulder, CO.

Nalepa, C.A. (2015). Origin of termite eusociality: trophallaxis integrates the social, nutritional, and microbial environments. *Ecological Entomology*, **40**, 323.

Nalepa, C.A., and Bell, W.J. (1997). Postovulation parental investment and parental care in cockroaches. In: J.C.

Choe and B.J. Crespi (Eds), *The Evolution of Social Behavior in Insects and Arachnids*, pp. 26–51.. Cambridge University Press, Cambridge.

O'Donnell, S. (1998). Reproductive caste determination in eusocial wasps (Hymenoptera: Vespidae). *Annual Review of Entomology*, **43**, 323.

O'Donnell, S., Bulova, S.J., Deleon, S., et al. (2015). Distributed cognition and social brains: reductions in mushroom body investment accompanied the origins of sociality in wasps (Hymenoptera: Vespidae). *Proceedings of the Royal Society of London B: Biological Sciences*, **282**, 20150791.

Oldroyd, B.P., and Fewell, J.H. (2007). Genetic diversity promotes homeostasis in insect colonies. *Trends in Ecology & Evolution*, **22**, 408.

Oster, G.F., and Wilson, E.O. (1978). *Caste and Ecology in the Social Insects*. Princeton University Press, Princeton, NJ.

Premnath, S., Sinha, A., and Gadagkar, R. (1996). Dominance relationship in the establishment of reproductive division of labour in a primitively eusocial wasp (*Ropalidia marginata*). *Behavioral Ecology and Sociobiology*, **39**, 125.

Queller, D.C. (1992). A general model for kin selection. *Evolution*, **46**, 376.

Ratnieks, F.L., and Wenseleers, T. (2005). Policing insect societies. *Science*, **307**(5706), 54–6.

Ratnieks, F.L., and Wenseleers, T. (2008). Altruism in insect societies and beyond: voluntary or enforced? *Trends in Ecology and Evolution*, **23**, 45.

Rehan, S.M., Richards, M.H., Adams, M., and Schwarz, M.P. (2014). The costs and benefits of sociality in a facultatively social bee. *Animal Behaviour*, **97**, 77.

Rehan, S.M., Schwarz, M.P., and Richards, M.H. (2011). Fitness consequences of ecological constraints and implications for the evolution of sociality in an incipiently social bee. *Biological Journal of the Linnean Society*, **103**, 57.

Remolina, S.C., and Hughes, K.A. (2008). Evolution and mechanisms of long life and high fertility in queen honey bees. *Age*, **30**, 177.

Ross, K.G. (2001). Molecular ecology of social behaviour: analyses of breeding systems and genetic structure. *Molecular Ecology*, **10**, 265.

Roubik, D.W. (2006). Stingless bee nesting biology. *Apidologie*, **37**, 124.

Schuster, J.C., and Schuster, L.B. (1985). Social behavior in Passalid beetles (Coleoptera: Passalidae): cooperative brood care. *Florida Entomologist*, 68(2), 266.

Schuster, J.C., and Schuster, L.B. (1997). The evolution of social behavior in Passalidae (Coleoptera). In: J.C. Choe and B.J. Crespi (Eds), *The evolution of social behavior in insects and arachnids*, pp. 260–9. Cambridge, UK: Cambridge University Press.

Schwarz, M.P., Richards, M.H., and Danforth, B.N. (2007). Changing paradigms in insect social evolution: insights from halictine and allodapine bees. *Annual Review of Entomology*, **52**, 127.

Schwarz, M.P., Tierney, S.M., Rehan, S.M., Chenoweth, L.B., and Cooper, S.J. (2010). The evolution of eusociality in allodapine bees: workers began by waiting. *Biology Letters*, rsbl20100757.

Scott, M.P. (1998). The ecology and behavior of burying beetles. *Annual Review of Entomology*, **43**, 595.

Seeley, T.D. (1995). *The Wisdom of the Hive*. Harvard University Press, Cambridge, MA.

Seeley, T.D., and Tarpy, D.R. (2007). Queen promiscuity lowers disease within honeybee colonies. *Proceedings of the Royal Society of London B: Biological Sciences*, **274**, 67.

Shackleton, K., Al Toufailia, H., Balfour, N.J., Nascimento, F.S., Alves, D.A., and Ratnieks, F.L. (2015). Appetite for self-destruction: suicidal biting as a nest defence strategy in Trigona stingless bees. *Behavioral Ecology and Sociobiology*, **69**, 273.

Shykoff, J.A., and Schmid-Hempel, P. (1991). Parasites and the advantage of genetic variability within social insect colonies. *Proceedings of the Royal Society of London B: Biological Sciences*, **243**, 55.

Silk, M.J., Croft, D.P., Tregenza, T., and Bearhop, S. (2014). The importance of fission–fusion social group dynamics in birds. *Ibis*, **156**, 701.

Smith, J.E., Kolowski, J.M., Graham, K E., Dawes, S.E., and Holekamp, K.E. (2008). Social and ecological determinants of fission–fusion dynamics in the spotted hyaena. *Animal Behaviour*, **76**, 619.

Stamp, N.E., and Bowers, M.D. (1990). Variation in food quality and temperature constrain foraging of gregarious caterpillars. *Ecology*, **71**, 1031.

Tallamy, D.W., and Wood, T.K. (1986). Convergence patterns in subsocial insects. *Annual Review of Entomology*, **31**, 369.

Thorne, B.L. (1984). Polygyny in the Neotropical termite *Nasutitermes corniger*: life history consequences of queen mutualism. *Behavioral Ecology and Sociobiology*, **14**, 117.

Thorne, B.L. (1997). Evolution of eusociality in termites. *Annual Review of Ecology and Systematics*, **28**, 27.

Tian, L. and Zhou, X. (2014). The soldiers in societies: defence, regulation, and evolution. *International Journal of Biological Sciences*, **10**, 296.

Trumbo, S.T. (1992). Monogamy to communal breeding: exploitation of a broad resource base by burying beetles (*Nicrophorus*). *Ecological Entomology*, **17**, 289.

Trumbo, S.T. (1996). Parental care in invertebrates. *Advances in the Study of Behavior*, **25**, 3.

Trumbo, S.T. (2012). Patterns of parental care in invertebrates. In: N.J. Royle, P.T. Smiseth, and M. Kolliker (Eds), *The Evolution of Parental Care*, pp. 81–100. Oxford University Press, Oxford.

Vehrencamp, S.L. (2000). Evolutionary routes to joint-female nesting in birds. *Behavioral Ecology*, **11**, 334.

Wade, M.J., Wilson, D.S., Goodnight, C., et al. (2010). Multilevel and kin selection in a connected world. *Nature*, **463**, E8.

Wcislo, W.T. (1997a). Are behavioral classifications blinders to studying natural variation. In: J.C. Choe and B.J. Crespi (Eds), *Social Behavior in Insects and Arachnids*, pp. 8–13. Cambridge University Press, Cambridge.

Wcislo, W.T. (1997b). Behavioral environments of sweat bees (Halictinae) in relation to variability in social organization. J.C. Choe and B.J. Crespi (Eds), *Social Behavior in Insects and Arachnids*, pp. 316–33. Cambridge University Press, Cambridge.

Wcislo, W.T., and Danforth, B.N. (1997). Secondarily solitary: the evolutionary loss of social behavior. *Trends in Ecology & Evolution*, **12**, 468.

Wcislo, W.T., and Fewell, J.H. (2017). Sociality in bees. In: D.R. Rubenstein and P. Abbot (Eds), *Comparative Social Evolution*, pp. 50–83. Cambridge University Press, Cambridge.

West, M.J. (1967). Foundress associations in polistine wasps: dominance hierarchies and the evolution of social behavior. *Science*, **157**, 1584.

Wille, A. (1983). Biology of the stingless bees. *Annual Review of Entomology*, **28**, 41.

Wilson, E.O. (1971). *The Insect Societies*. Harvard University Press, Cambridge, MA.

Winston, M.L. (1991). *The Biology of the Honey Bee*. Harvard University Press, Cambridge, MA.

CHAPTER 16

Personality and behavioral syndromes in insects and spiders

Carl N. Keiser, James L.L. Lichtenstein, Colin M. Wright,
Gregory T. Chism, and Jonathan N. Pruitt

*Department of Ecology, Evolution and Marine Biology, University of California at Santa Barbara,
Santa Barbara, CA, USA*

16.1 History and novelty

Consistent individual differences in behavior have far-reaching implications for ecology and evolutionary biology (Sih et al. 2012; Wolf and Weissing 2012). It is upon such variation that natural selection has acted to produce the behavioral diversity observed today across the animal kingdom. The last 16 years have seen a substantial rise in the number of papers devoted to the study of individual variation in behavior. This emphasis stems from two key events. First, within general ecology, there has been growing interest in how individual variation of any sort can impact individual niches and how the milieu of traits presents within populations impact community level outcomes (Bolnick et al. 2002, 2011; Hughes and Stachowicz 2004; Crutsinger et al. 2006). The second key event comes from within behavioral ecology itself, from the simple observation that individual variation in multiple behavioral traits are often intercorrelated. Thus, how an individual behaves while foraging often predicts how it behaves in other ecological contexts, like predator escape or mating (Dall et al. 2004; Sih et al. 2004). This latter point inspired interest because behavioral

correlations have the potential to generate cross-contextual performance trade-offs. Such correlations may therefore conceivably constrain individuals' ability to deploy their ideal phenotype in any one context. Constellations of correlated traits are called **behavioral syndromes**. We will refer to temporally consistent individual in behavior here as **animal personality**.

Animal personality and behavioral syndromes have had an influence on behavioral ecology because they raise our consciousness to a few key points. First, they draw attention to individual variation. Despite a few notable exceptions (e.g., alternative mating tactics within species; Dominey 1984), individual differences have tended to be understudied by ecologists, who often suggested that such variation might merely represent statistical noise (Martin and Bateson 1993). The popularization of the animal personality framework and new statistical tools used to analyse individual variation have changed this view. The advancement of statistical tools helped to not only account for this inter-individual variance, but also analyse it specifically, thus playing a crucial role in the field of animal personalities. Secondly, the topic of behavioral syndromes raises our conscious-

Keiser, C. N., Lichtenstein, J. L. L., Wright, C. M., Chism, G. T., and Pruitt, J. N., Personality and behavioral syndromes in insects and spiders. In: *Insect Behavior: from mechanisms to ecological and evolutionary consequences*. Edited by Alex Córdoba-Aguilar, Daniel González-Tokman, and Isaac González-Santoyo: Oxford University Press (2018). © Oxford University Press.
DOI: 10.1093/oso/9780198797500.003.0016

Figure 16.1   (a) The fruit fly, *Drosophila melanogaster* (Photo: Thomas Wydra). (b) Water striders, *Aquaris remigis* (Photo: Cory Chiappone). (c) Grass spider, *Agelenopsis aperta* (Photo: R.J. Adams). (d) Social spiders, *Stegodyphus dumicola* (Photo: Graham Montgomery). (e) Field cricket, *Gryllus integer* (Photo: Nicholas DiRienzo). (f) Pea aphids, *Acyrthosiphon pisum* (Photo: Shipher Wu). (g) Flour beetle, *Tribolium* sp. (Photo Credit CSIRO).

ness to the fact that behaviors deployed by an individual in any one context may appear suboptimal because an individual must balance the costs and benefits of having a particular behavioral tendency *across* contexts. In addition to a few important vertebrate test systems, insects and spiders have played a critical role in developing the methodological and

conceptual frameworks of the animal personality field (some important groups represented in Figure 16.1).

For example, female grass spiders exhibit characteristic differences in their aggressiveness that manifest in a variety of contexts. More aggressive females are more likely to secure high quality territories and to defend those territories from intrusion by conspecifics (Riechert and Hedrick 1993b). More aggressive females are also more likely to capture a higher proportion of prey that are intercepted by their webs, because aggressive females attack prey more quickly. However, aggressive females often exhibit a maladaptively high incidence of **precopulatory sexual cannibalism**, where extremely aggressive females, although superior in many other respects,

attack all of the males that ever court them (Riechert et al. 2001). This results in some females remaining unmated. Thus, cross-contextual correlations in aggressiveness generate performance trade-offs and ensure that no one behavioral tendency consistently enjoys high performance across all contexts. Though rarely so well documented, such trade-offs draw our attention to the point that attempts to account for

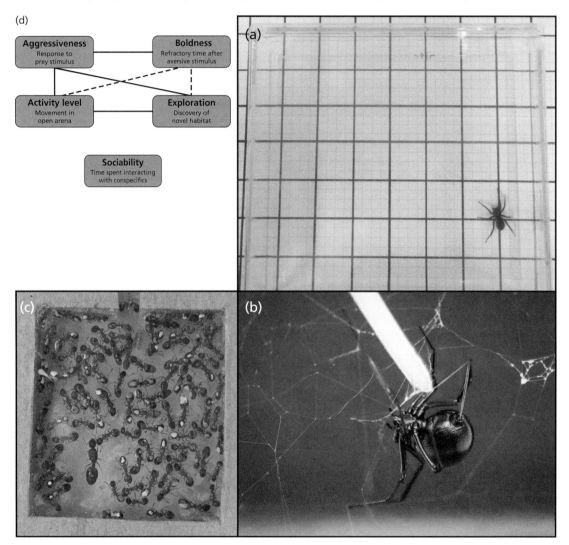

Figure 16.2 (a) Wolf spider (*Tigrosa* sp.) in an activity assay grid where researchers record the number of lines crossed over a period of time (Photo credit: Carl Nick Keiser); (b) *Temnothorax rugatulis* ants individually painted to track social interactions and task participation (Photo credit: Jennifer Jandt). (c) Black widow (*Latrodectus hesperus*) in a foraging aggressiveness assay attacking a vibratory stimulus used to minimize variation in responses that may arise from differences in prey behavior (Photo credit: Nicholas DiRienzo); (d) Hypothetical structure of a behavioral syndrome describing the pairwise relationships between different personality axes. In this example, solid lines represent positive relationships between two axes and dotted lines represent negative relationships. In this example, sociability is not significantly related to any other trait. (presentation inspired by Bengston and Dornhaus 2014).

the performance consequences of individual variation in behavior in only a single context is likely to produce an incomplete and potentially misleading understanding of the system. Cross-contextual performance trade-offs result in situations where the optimal phenotype often varies situationally or across time. When one integrates observations made across time and context, in effect, these trade-offs flatten out the fitness surface experienced by animals, thereby maintaining behavioral trait variation.

16.2 Behavioral traits and their assays

The animal personality literature has descended upon five key behavioral axes that are examined again and again across diverse taxonomic groups. The 'Big Five' of animal personality are boldness, aggressiveness, sociability, exploration, and activity (Sih and Bell 2008). Virtually every study designed to test for the presence of animal personality examines at least one of these axes and, without fail, recovers evidence for consistent individual differences in at least one of them. We briefly define each behavioral axis below, describe some of the most common ways each metric is evaluated in terrestrial arthropods (Figure 16.2), and a brief rationale for why each test is used.

16.2.1 Behavioral axes

Boldness

An individuals' willingness to place itself in risky situations.

Common assays

Latency to emerge from a safe retreat, latency to resume movement following an aversive stimulus, willingness to traverse an exposed (presumed to be high risk) environment, willingness to forage or remain active in the presence of predator cues, or latency/willingness to accept dangerous prey.

Rationale

The key to all boldness assays is to subject individuals to conditions perceived by the animal to be risky or merely unknown, then to measure individuals' willingness to enter or explore that environment. Individuals that enter those environments earlier or

move about more within them are deemed bolder. For example, the field cricket *Gryllus integer* exhibits stable differences in individuals' latency to become active inside safe refugia, their latency to emerge from a refuge after disturbance, and their time spent inactive after an alarm cue (Kortet et al. 2007).

Aggressiveness

The degree to which individuals exhibit agonistic or combative behavior towards rival conspecifics or heterospecifics.

Common assays

Individuals' response towards a mirror image, a caged stimulus animal, or behavior under natural agonistic interactions, for instance, during territory defence or intrusion. Individuals' latency of approach, number of chases, bites, or defensive gestures are often used as indicators of their aggressiveness (Figure 16.2).

Rationale

Aggressiveness as a term is used more promiscuously in the literature, although the overall desire is to capture the voracity with which individuals approach and contend with conflict. The spider literature occasionally uses the term 'sexual aggressiveness' to refer to females' probability of attacking mates (Johnson and Sih 2005; Kralj-Fišer et al. 2013), whereas sexual aggressiveness in male water striders is used to describe aggression directed towards males and females in regards to mating tactics (Sih and Watters 2005; Chang and Sih 2013). Studies on predators occasionally use the phrase 'foraging aggressiveness' to refer to speed of attack or the number of bites delivered before subduing prey (Pruitt et al. 2010). Number of consecutive prey killed or the proportion of prey killed and left uneaten (superfluous or wasteful killing) are occasionally used as metrics of aggressiveness as well (Maupin and Riechert 2001).

Sociability

Individuals' tendency to engage in social interactions or social settings.

Common assays

Time spent near versus away from conspecifics in staged choice trials, interindividual distances in open field tests, proximity to conspecifics *in situ* using remote tracking, or highly connected positions in social networks (Figure 16.2).

Rationale

To some degree, sociability can be considered the polar opposite to aggressiveness. However, this depends on the context of the behavioral traits studied. For example, individuals' willingness to engage with conspecifics can be unrelated to their degree of aggressiveness towards prey, as has been shown in the lady beetle *Eriopis connexa* (Rodrigues et al. 2016).

Exploration

Individuals' tendency to explore their environment, often in the context of a novel environment.

Common assays

Animals are subjected to unfamiliar settings where the distance travelled is measured, the number of new chambers entered is counted, or the time spent investigating novel objects is recorded. Novel object tests are occasionally used to evaluate boldness as well.

Rationale

Exploration is thought to be linked with the information quality that individuals obtain about their environment and the speed with which that information is obtained. Individuals that move around in novel environments are thought to acquire more information than sedentary individuals. In other cases, exploration may be a product of individuals' state or experiences, for instance, 'searching behavior' for food or other resources may play into measurements of exploration. For example, in the mustard leaf beetle, *Phaedon cochleariae*, individuals' measurements of non-targeted exploration in an open-field were repeatable but not associated with diet quality (Tremmel and Müller 2013).

Activity level

The proportion of time individuals appear active.

Common assays

Time spent moving in an open field environment, the distance travelled over a discrete time period, or the inverse of the amount of time individuals spend in apparent quiescence (Figure 16.2).

Rationale

For prey species, activity level can dictate individuals' interaction rates with parasites, predators, and resources, and has a physiological influence on the amount of resources individuals need merely to survive. For predators, activity level have a similar effect, but has the added influence that activity level underlies differences in hunting mode and is, therefore, the defining difference between active foraging versus ambush predators (Huey and Pianka 1981; Miller et al. 2014). For example, the wolf spider, *Pardosa milvina*, exhibits consistent individual variation in activity level, which determines the type of insect prey individuals consume (Royauté and Pruitt 2015).

16.2.2 Methodological considerations

When attempting to test for behavioral syndromes, one must take care to measure at least semi-independent traits and to design tests that make some ecological sense for the species in question. When behavioral traits are measured in similar experimental settings with similar metrics, while one is likely to observe strong correlations, the degree to which these traits represent independent measures is often questionable. An example of a potentially problematic study might include three open-field assays where investigators measure the distance individuals travel in an open field, the time they spend near a novel object placed in the center of an open field, and the time spent near a potential mate, trapped in a transparent chamber, also placed in the middle of an open field. These measurements will likely be correlated and not represent an actual behavioral syndrome between distinct behavioral traits. In an ideal situation, one should take care to design tests that do not appear too similar. However, a final counterpoint to this legitimate concern is that the very idea of a behavioral syndrome is that what may appear to be very different behavioral traits might not be. Although a behaviorist may deem aggressiveness towards conspecifics, tendency to explore novel environments, and tendency to engage in affiliative social interactions different ecological contexts, a behavioral syndromes mindset reminds us that behavior displayed in all three settings could be driven by some central underlying trait.

Another concern is the desire for investigators to design behavioral tests that map to the focal species' ecology. For instance, time spent investigating a novel object, say, a rubber duck, in the middle of an open field might be very highly repeatable in a good number of organisms. However, what this variation could possibly mean and why one would even think to measure it is not always obvious. Researchers of terrestrial arthropods have the special challenge of devising assays for animals that are quite unlike many of the established models for animal personality research (e.g., great tits, sticklebacks, chipmunks, mice, primates). Thankfully, new assays can also provide an avenue for communicating interesting aspects of the focal species' biology to interested readers (e.g., precopulatory cannibalism). Perhaps most importantly, critically revisiting assay design may help us to break free from an ever-present question in animal personality research: What are we really measuring about these animals?

16.3 Intraindividual variability and behavioral reaction norms

There is an occasional tendency for researchers to view personality and behavioral plasticity as mutually incompatible phenomena. This is not the case. Animal personality research freely admits that individuals' behavior is likely to shift to some degree across contexts and situations, and individuals' personalities are very likely to shift over the course of development or as a consequence of experience. A strong signature of animal personality merely requires that the rank order differences in individuals' behavior are consistent across situations. For example, individuals may reduce their activity level in response to the approach of a predator. However, if there is a strong signature of personality, one would predict that the most active individual in the absence of predator cues will also be the most active individual in the presence of predator cues (Sih et al. 2004; Sih and Bell 2008). Generalized linear mixed effect models provide use with a nice tool set for partitioning the variation explained by individual identity (or personality) versus other environmental

factors in a shared statistical framework (Dingemanse and Dochtermann 2013). Notably, within the past 5 years, two new non-mutually exclusive frameworks have been developed to help unite personality research with research on behavioral plasticity in a more synthetic framework. We briefly discuss these two frameworks below.

16.3.1 Intraindividual variability

The intraindividual variability literature argues that individual variation in behavioral consistency may itself represent a noteworthy axis of individual variation (Stamps et al. 2012). That is, some individuals do consistently display more erratic responses than others upon repeated measuring. Individuals with highly erratic behavioral tendencies would be characterized by high intraindividual variability. A small number of studies have documented stable differences in intraindividual variability in vertebrates and invertebrates alike (Biro et al. 2010; Biro and Adriaenssens 2013). However, the proximate mechanisms governing differences in intraindividual variability and its ecological impacts on the animals themselves remain unclear. Yet, one can imagine how a high degree of intraindividual variability might be beneficial in some instances. For instance, individuals with high intraindividual behavioral variability may more quickly stumble across behavioral solutions to novel tasks, or such erratic individuals may be more difficult for predators to track and subdue. Individuals with more varied behavioral responses may also enjoy mating advantages. For instance, individuals with more varied behavioral responses may be better at flexibly engaging in mating dialogues (Peretti et al. 2006). Conversely, behavioral consistency may have benefits, for instance, it may facilitate more rapid decision-making under high-pressure situations or the maintenance of group conformity during predator evasion. The broader ecology of intraindividual variability is still little explored, and thus, virtually any investigation into this topic is likely to produce important and unanticipated results.

16.3.2 Behavioral reaction norms

The **behavioral reaction norms** approach borrows concepts from quantitative genetics to capture the

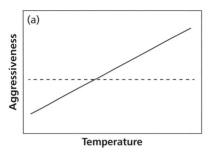 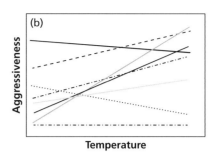

Figure 16.3 A hypothetical relationship between temperature and aggressiveness in a poikilothermic animal. (a) Overall, we expect that aggressiveness would increase with temperature, since insect and spider behavior is directly influenced by temperature (solid line). (b) However, individuals (represented by different lines) will vary in the direction and magnitude by which their aggressiveness changes as temperature increases. Thus, taking the average of each individual's response to increasing temperatures may not actually generate a trend between aggressiveness and temperature (dotted line in [a]).

idea that, in addition to differences in mean behavioral type (personality) and variance (intraindividual variability), individuals may differ in how they respond to changes in their environment (e.g., Figure 16.3; Dingemanse et al. 2010). Consider again the example of aggressive mating tactics in male water striders. Striders' aggressiveness will depend on their physical and social environment (e.g., the size of the pool in which they aggregate and the types of individuals in the group) and a behavioral reaction norm approach raises the point that although the mean aggressiveness of males may increase in smaller pools because of increased competition for mates, not all males might do so at an equivalent rate. Some males might not increase their aggressiveness across environments at all while others might exhibit a pronounced response. This means that the rank-order differences in individuals' aggressiveness (or any other trait) can change based on the conditions under which individuals' behavior is assessed. This individual variation in the environment-behavior relationship can produce rapid shifts in mating system dynamics across different environments (Sih et al. 2017). As with intraindividual variability above, the vital step now for the behavioral reaction norms literature is to document what this variability means in terms of individual function, performance, ecology, and fitness. The major problem with the behavioral reaction norm approach is a practical one: merely characterizing an individuals' behavioral reaction norm requires a large number of observations across variable environments. For many systems, this alone could take an entire field season

or possibly the entire life expectancy of one's beloved research subject (e.g., for short-lived insects or spiders). One must therefore think carefully before delving into the behavioral reaction norms approach.

16.4 Mechanisms of behavioral syndrome emergence

16.4.1 Genetic and neuroendocrine mechanisms of behavioral syndromes

Much research regarding the emergence of consistent individual differences in behavior focuses on the genetic and mechanistic underpinnings of a single axis of personality. Consequently, the mechanisms by which two or more behavioral traits become correlated across contexts are not well understood. The genetic tools available in the fruit fly, *Drosophila melanogaster*, has allowed for the identification of some genes (e.g., the *for* gene) that underlie correlations between axes of behavioral variation like larval activity, adult foraging, and antiparasite behaviors (Sokolowski 2001). Behavioral correlations among suites of traits that emerge via a shared underlying genetic mechanism (pleiotropy) may not only be a constraint on adaptive behavioral plasticity, but may also be more difficult to decouple than syndromes governed by other mechanisms, such as linkage disequilibrium, physical linkage, or correlated selection. Hormones (the functional outcome of gene × environment interaction) often operate on multiple target tissues throughout the body, which has led researchers to investigate the neuroendocrine basis of behavioral syndromes. In arthropods, a set of neurotransmitters called biogenic amines (e.g., dopamine, octopamine, serotonin) mediate many behaviors that appear to be functionally distinct (see Chapter 3). In the division of labour of honey bees, octopamine levels in the antennal lobes and mushroom bodies are associated with shifts in behavioral state (Schulz and Robinson 1999). Octopamine has also been associated with the agonistic contests of male crickets (pictured Figure 16.1), whereas nitric oxide acts to suppress aggressiveness (Stevenson and Rillich 2016). Research on web-building spiders has found that topical applications of biogenic amines can alter some personality traits, but not others (DiRienzo et al. 2015), suggesting that different behaviors may have independent mechanistic origins and that syndromes can be decoupled via changes to the neuroendocrine system.

16.4.2 Development, ontogeny, and experience

Although a hallmark of behavioral syndromes is behavioral consistency, behavioral tendencies may also shift over an individual's lifetime (Sih and Bell 2008). It is often the case that individuals' behavioral tendencies change from one developmental stage to another, often because the selective pressures to which an individual is exposed can change wildly across life stages. If there is behavioral consistency across such developmental changes, it may be that the behavioral tendencies that beget success at developmental stage could result in low performance in another. For example, larval damselflies that are less active are less likely to encounter predators in high-predation risk environments. However, these less-active individuals will emerge as less-active adults and may, therefore, be less likely to disperse and produce offspring elsewhere in sites with reduced predation risk (Brodin 2009).

Although most studies on behavioral syndromes in insects and spiders focus on adult individuals, an understanding of how syndromes develop over time is central to our understanding of why they develop in the first place. A foundational paper by Stamps and Groothius (2010) highlighted the importance of using model systems where information about individuals' genotypes can be used to generate 'replicate individuals' to measure the effects experience on generating behavioral syndromes Again, *D. melanogaster* has emerged as a model system for finely prodding the influence of environmental effects on syndrome development, and also how social interactions can carry over to behaviors expressed in non-social contexts (i.e. social carry-over effects; Niemelä and Santostefano 2015). In the field cricket, *Gryllus integer*, boldness is repeatable across juvenile and adult life stages in females, but not in males, who grow less bold across metamorphosis (Hedrick and Kortet 2012). This may be due to increased predation risk associated with a common adult male trait in crickets: calling for mates. In fact, juvenile *G. integer*, that are exposed to acoustic sexual signals from conspecific adults are less aggressive than juveniles reared in the absence of these signals (DiRienzo et al. 2012), suggesting that boldness and aggressiveness are possibly linked together in a behavioral syndrome dictated by individuals' rearing environment. In the funnel-weaving spider *Agelenopsis pennsylvanica*, a positive boldness–aggressiveness syndrome is observed as animals develop, but only in field-caught animals (Sweeney et al. 2013b). That is, this syndrome never emerges in laboratory-reared animals, which may be due to reduced enrichment/stressors in the laboratory or via the constant stress of captivity, as evidenced from studies on captive tarantulas (Bengston et al. 2014). Researchers have even gone so far as to say that experiences over development, be they affiliative or antagonistic, may be *necessary* for the development of personality differences and behavioral syndrome emergence (Urszán et al. 2015), perhaps because they provide the animal with reliable cues of environmental stressors. Understanding how different axes of behavioral variation co-vary and how they are influenced by developmental experiences is important when designing experiments for the detection of 'pseudo-syndromes' generated by non-permanent social/environmental factors.

As explained in more detail later in this chapter, social insects represent an interesting case for behavioral syndrome studies, because syndromes can emerge at two levels of biological organization: the individual- and colony-level. Individual workers may differ in terms of their boldness, aggressiveness, and activity level, and these tendencies are often associated with individuals' tendency to participate in specific tasks (Pinter-Wollman 2012; Jandt et al. 2014). The ratios of different types of individuals, colony size, and nest architecture then interact to shape differences in colony-level behavior, and have the potential to generate colony-level syndromes between traits such as group foraging, boldness, and exploration (Scharf et al. 2012; Modlmeier et al. 2014a; Wright, Keiser and Pruitt 2015). Because social insect societies are often founded by only a single individual, we propose that the experiences of these females and their initial actions during colony development have the potential to have long-term legacy effects on the collective behavior of their societies (argued in Bengston and Jandt 2014). Furthermore, termites present an interesting case where colonies are typically founded by a queen and king, and their early behavioral type × behavioral type interactions (see 16.8.2) may alter future colony trajectories in unexpected ways. Data on paper wasps confirm that queen personality can be a major driver of colony behavior, further indicating a link between queen personality, colony development, and downstream collective behavior (Wright et al. 2017).

16.5 Repeatability and behavioral syndromes across metamorphosis

Insects represent an interesting and valuable group to evaluate behavioral consistency across development because their growth is punctuated by a series of moulting events (ecdysis). Holometabolous life cycles are characterized by a period of drastic phenotypic reorganization during the pupal stage. That said, most of our understanding of behavioral syndromes in insects is based on hemimetabolous insects with gradual metamorphosis (e.g., crickets). What has been discovered in holometabolous insects is that repeatability within the larval stage and adult stages are interrupted by a lack of repeatability across pupation, which has been observed in leaf beetles (Müller and Müller 2015) and flour beetles (Wexler et al. 2016). This makes sense, because lifelong fixed behavioral correlations may produce performance trade-offs as correlated selection in the juvenile stage could produce behavioral tendencies that are

disadvantageous during the adult stage. However, despite the transformative period between two very different natural histories, behavioral repeatability across pupation has been found in other groups like ladybeetles (Rodrigues et al. 2016). Even some hemimetabolous insects experience drastic changes in life history across the transition from adolescence to adulthood. The Odonata represent a particularly fascinating example, because selective pressures in the aquatic juvenile stage are undoubtedly different from those in the flighted terrestrial adult stage. However, Brodin (2009) found that the activity level and the direction of behavioral syndromes in aquatic damselfly larvae are carried over to adulthood, even when it appears to be disadvantageous.

16.6 Personality and sociality

The ants, social bees and wasps, and termites often exhibit highly specialized reproductive division of labour (see Chapter 15). The organized social structure of many of these insect societies often awards them the term 'superorganisms' (Hölldobler and Wilson 2009). Unlike other organisms, where adaptations arise from natural selection acts on the individual, superorganisms differ in that natural selection can simultaneously act on the colony as a whole as well as the individual, to produce group-level adaptation (Folse and Roughgarden 2010). This complexity has resulted in a large and ever-increasing body of literature devoted to the interplay of animal personality and social arthropods.

Eusociality is characterized by a species having the following traits: cooperation in brood care, overlapping generations within a colony of adults, and reproductive division of labor (i.e. queens and workers; Nowak et al. 2010; see also Chapter 15). What makes the eusocial insects amenable to a personality research approach is the fact that these societies can exhibit behavioral variation at multiple levels of organization (Pinter-Wollman 2012; Jandt et al 2013). Different species, for instance, differ in their behavior (Davidson 1998; Rowles and O'Dowd 2007; Yip 2014; Bertelsmeier et al. 2015). It is also true that colonies within a species behave quite differently in colony tasks (Raine et al. 2006; Raine and Chittka 2008; Gordon et al. 2011; Wray et al. 2011; Wright et al. 2017). Further still, the workers in many eusocial systems show marked behavioral variation, both between and within colonies (Wilson 1976; Holldobler and Wilson 1990; Beshers and Fewell 2001). This behavioral variation can be consistent across time, or can vary predictably as the worker ages—a phenomenon known as **temporal polyethism** (Seeley 1982). In some ant species, workers vary drastically in size, which helps facilitate division of labor within a colony (Oster and Wilson 1979). The size distribution of works is also often linked with their behavior (Jeanne 1986; Beshers and Fewell 2001). In army ants, for example, younger, smaller workers are more docile and often tend to brood, whereas larger 'soldier' workers are highly aggressive and are primarily involved in colony defence (Wilson 1976). For a comprehensive understanding of insect societies, we must not only understand how certain levels of organization operate, but also how each level influences and feeds back to other levels both above and below it (Pinter-Wollman 2012; Jandt et al. 2013; Bengston and Jandt 2014).

16.6.1 Colony composition

Behavioral differences between workers can give rise to division of labour, which means that different individuals within a colony show different propensities to perform certain tasks (Beshers and Fewell 2001). For instance, some workers spend more time foraging while others consistently care for the colony brood (Wilson 1976). By allocating workers to different tasks, work becomes more streamlined and colony productivity increases (Beshers and Fewell 2001; Jeanson et al. 2007). This increased productivity arises because colonies incur fewer costs associated with workers switching tasks. This model helps us to understand why a colony may benefit from harbouring a variety of worker personality types, but what is the origin of these behavioral differences and how are they maintained?

Differences in colony composition may arise from various sources. For instance, genetic diversity among workers may determine the personality types present within a colony (Robinson 1989; Page and Robinson 1991; Robinson 1992; Fewell and Page 1993). Queens mating with multiple males may increase the genetic diversity among workers

within a colony, thereby generating greater behavioral diversity (Page 1980; Page and Robinson 1991; see Chapter 15). In some cases, the workers may have fixed personality types and perform the same task(s) throughout their lives, while other workers are innately drawn to other tasks (Stamps and Groothuis 2010; Wright et al. 2014). In other cases, the workers may act differently according to more flexible **behavioral response thresholds** that allow workers to be allocated to certain tasks based on colony needs (Fewell and Page 1993; Jeanson et al. 2008; Holbrook et al. 2011). For instance, ant workers may differ in their propensity to forage. When foraging ants increasingly return with food at faster rates, other worker that are not currently foraging may be induced to forage (Gordon et al. 2011). Since many workers differ in the threshold required to induce action for various tasks, a self-regulating system emerges that is able to allocate workers to tasks proportional to colony needs. Division of labour may also emerge from differences in how genes are regulated in the worker's brain as they age (e.g., the temporal polyethism described above). For instance, newly eclosed honeybee workers remain inside the colony tending to brood. As the workers age, they take up increasingly risky jobs such as nest guarding and foraging (Seeley 1982; Robinson et al. 1994; Beshers et al. 2001).

In addition to genetic mechanisms governing intracolony behavioral diversity, it has been shown that repeated social interactions with group mates can help generate personality variation, a phenomenon known as **social niche specialization** (Bergmuller and Taborsky 2010; Montiglio et al. 2013; Laskowski and Pruitt 2014; Modlmeier et al. 2014c). Originally developed in the context of foraging theory (Svanback and Bolnick 2007; Araujo et al. 2011), evidence now suggests that members of a group or colony benefit from specializing on different 'social niches' within a colony. Evidence from social groups of *Drosophila melanogaster* suggests that individuals actively shape their social environments (i.e., social niche construction; Saltz and Nuzhdin 2014), and social environment preferences have a genetic basis (Saltz and Foley 2011). In addition to the individual benefits associated with developing social niches, the group as a whole often benefits from the emergence of division of labour for all the reasons mentioned above. An interesting hypothesis stemming from these observations is that personality may serve as a stepping-stone to more advanced forms of division of labour. That is, social niche construction may produce a rudimentary division of labour in non-eusocial societies, thereby increasing the productivity of the group and allowing selection on groups to further facilitate this process (a process called **social heterosis**; Nonacs and Kapheim 2007; Nonacs and Kapheim 2008; Nowak et al. 2010). However, over time, the division of labour may be intensified from selection on genes favouring intracolony personality variation to selection favouring linkages between personality and morphological traits, including sterility.

16.6.2 Social spiders

While social insects are excellent systems for studying the intersection of personality and sociality, they are by no means the only model. Our understanding of how personality interacts with sociality has made notable strides forward from the study of social spiders. Social spiders lack morphological castes but have been shown to be behaviorally diverse (Grinsted et al. 2013; Holbrook et al. 2014; Keiser and Pruitt 2014b; Wright et al. 2015), and the ratio of different personality types within colonies can have large effects on colony behavior and success (Pruitt and Riechert 2011; Wright et al. 2016). For instance, colonies of the African spider, *Stegodyhus dumicola*, containing a high proportion of bold individuals tend to attack prey items faster and with more individuals than shyer colonies (Grinsted et al. 2013). Variation in boldness is also predictive of collective defence against predatory ants (Wright et al. 2016), web repair (Keiser et al. 2016), task differentiation among individuals (Grinsted et al. 2013), and the degree to which microbes are transmitted within a colony (Keiser et al. 2016a,b). Experiments have further shown that colony personality composition can be more important than colony size in predicting collective foraging in *S. dumciola* in situ (Keiser and Pruitt 2014b).

Studies on *S. dumicola* have also shown that particularly bold spiders appear to act as '**keystone individuals**', having an inordinately large impact on group processes compared to other individuals (Modlmeier et al. 2014b). Adding a single, highly bold spider into a group of extremely shy spiders increases collective foraging by catalysing variation in boldness within the colony (Pruitt et al. 2013; Pruitt and Keiser 2014). Interestingly, this catalysing influence is directly proportional to the keystone's boldness, as well as how long the keystones remains within these groups, and these effects persist long after the removal of the very bold individual (Pruitt and Pinter-Wollman 2015). Such keystone individuals also increase colonies' mass gain and survival in laboratory conditions (Pruitt et al. 2016b).

In another social spider, *Anelosimus studiosus*, individuals within colonies fall into a bimodal distribution representing both docile and aggressive spiders. The behavioral composition of incipient colonies in nature has been shown to predict colony survival and extinction in contrasting environments (Pruitt 2012; Pruitt 2013). Docile groups produce many more egg sacs and offspring colonies, but are unable to successfully drive off colony predators and parasites. Aggressive groups have the opposite problem: they can drive away predators and parasites, but spend less time tending to brood. Such colonies also engage in more intracolony aggression (Pruitt 2013). Thus, it is colonies containing a mixture of docile and aggressive spiders that are able to flourish. More recent work has shown that the ideal mixture of personality types for a society differs based on a habitat's resource levels, and that colonies have evolved mechanisms of maintaining their optimal personality mixtures through the selective cessation of reproduction within the society (Pruitt and Goodnight 2014; Pruitt et al. 2017). This cessation of reproduction may, in some ways, bear resemblance to incipient eusociality seen in some social insects.

16.7 Behavioral syndromes and reproduction

16.7.1 Mating assortativity

A central factor in understanding patterns of mate choice is '**assortativity**', the degree to which individuals prefer to mate with individuals that share similar traits (assortativity) or express different traits (disassortativity). Both assortativity and disassortativity based on personality have been documented in spider models. For instance, in the bridge spider, *Larinioides sclopetarius* (Araneidae), aggressive males preferentially mate with aggressive females, and non-aggressive males with non-aggressive females (Kralj-Fišer et al. 2013). Since aggressiveness is a heritable trait in this and other species, assortative mating by aggressiveness likely aids in the maintenance of variation in aggressiveness across generations Alternatively, in extreme cases of sexual conflict where female spiders engage in precopulatory sexual cannibalism (i.e. kill their potential mates before reproduction occurs), interindividual variation in aggressiveness can generate disassortative mating patterns. In *A. studiosus*, aggressive females are more likely to kill aggressive males and mate with docile males. This is because more aggressive males tend to win access to females of all behavioral types, but aggressive males approach females too quickly and this behavior frequently results in an attack by females. In contrast, docile males approach females slowly and are less likely to be killed by aggressive females. Docile females only rarely engage in cannibalism and thus aggressive males easily win access to them and only rarely suffer attacks (Pruitt et al. 2011).

16.7.2 Cross-contextual trade-offs: the aggressive spillover hypothesis

As previously stated, a central tenant of the behavioral syndromes concept is that behavioral traits are correlated across different contexts. This does not mean, however, that behavioral tendencies that are adaptive in one context will be adaptive in another context. A demonstration of this phenomenon is the **aggressive spillover hypothesis**, which posits that precopulatory sexual cannibalism may evolve as a consequence of directional selection on aggressiveness in the context of foraging. This hypothesis emerged as a means to explain the incidence of precopulatory sexual cannibalism in *Dolomedes* fishing spiders by Arnqvist and Henriksson (1997). Simply put, more aggressive females experience greater success during foraging, but this aggressiveness spills over to other contexts, such as interactions with potential mates. Kralj-Fišer et al. (2013) further argued that precopulatory sexual cannibalism may result from a spillover of aggressiveness from juvenile females, and also highlight the important

of incorporating mate-choice theory into the framework of the aggressive spillover hypothesis. The viability of the aggressive spillover hypothesis is still debated within the literature (Johnson 2013; Kralj-Fišer et al. 2013), although it remains one of the flagship examples of how behavioral consistency can potentially generate performance trade-offs.

16.7.3 Individual variation and population voltinism

In some insects, differences in reproductive strategies can determine regional patterns of voltinism, the number of generations produced per year. In environmentally variable locales, local populations can transition from univoltine to multivoltine across years. These are often referred to as **voltinism transition zones** (Blanckenhorn 1994). For example, water striders reside in mixed-sex flotillas on stream surfaces across North America and either reproduce the same summer they were born (multivoltine strategy), or wait until the following spring to mate (univoltine strategy). The univoltine strategy has traditionally been linked to environmental characteristics like colder temperatures or reduced food availability (Blanckenhorn 1991). However, a non-mutually exclusive explanation may be that the behavioral type mixture residing in voltinism transition zones is unfavourable for the immediate reproduction executed in multivoltine populations. Another instance in which behavioral syndromes may influence voltinism patterns at the population level are differences in dispersal syndromes, where disperser and resident phenotypes both exist within populations. For example, correlations among dispersal-related traits [e.g., flight patterns, wing morphology, etc. (Legrand et al. 2016) in the butterfly *Pieris brassicae* change across a latitudinal gradient (Ducatez et al. 2013) and may be related to voltinism (Stevens et al. 2013)].

16.8 Ecological consequences of insect behavioral syndromes

Studies on insects and other arthropods have contributed more to our understanding of the ecological consequences of personality perhaps more than any other taxa (Gosling 2001; Kralj-Fišer and Schuett 2014; Modlmeier et al. 2015). From work on insects, we draw evidence that personality is closely involved with **local adaptation** (Hedrick and Kortet 2006; Jongepier et al. 2014; Bengston and Dornhaus 2015) and dispersal (Gyuris et al. 2011), both of which contribute to the distribution and abundance of species. More recently, however, ecologically-focused studies of animal personality have examined the degree to which personality differences mediate the nature and magnitude of species interactions. This research has generated three key questions: How might personality act differently across different kinds of interactions (predation, mutualism, etc.)? How do the personality types of two different species synergize to predict interaction outcomes? How can personality shape species interactions on the level of populations or large groups within populations? We will detail here how researchers have begun to address these problems, and suggest some future avenues of study.

16.8.1 Species interactions

Personality variation predicts the outcome of a wide variety of species interactions. This is perhaps because many personality tests are designed to estimate how individuals interact with other species. For instance, boldness tests often measure individuals' fear of predators (Riechert and Hedrick 1993a; Ward et al. 2004), and exploration or activity-level often seek to simulate foraging behavior (Dingemanse et al. 2002; Lichtenstein et al. 2015). For many systems, personality variation can predict the outcome of contests over food resources (Webster et al. 2009; Lichtenstein et al. 2015), predator–prey interactions (Griffen et al. 2012; Pruitt et al. 2012; Toscano and Griffen 2014), and even whether an interaction between two species will be mutualistic or parasitic (Pruitt and Ferrari 2011; Keiser and Pruitt 2014a). This last example draws attention to the power of personality to predict the nature of interaction outcomes and is worth a deeper look. Social spider colonies are often home to solitary **inquiline** spider species, that consume the prey of social spiders and sometimes the social spiders themselves (Perkins et al. 2007). The social spider personalities present within the colony dictate whether different inquiline species will either reduce or increase colony success (Keiser and Pruitt 2014a). This is because

whether the inquilines are more aggressive than the social spiders or vice versa determines whether the social spiders scrounge prey from their inquilines or whether the inquilines scrounge prey from the social spiders.

Beyond these dyadic species interactions, personality variation can also predict the outcome of more complex species interaction modules. For instance, predator personality can predict the intensity of trophic cascades (Keiser et al. 2015), because they determine the degree to which top predators effect the behavior of mesopredators and, in turn, how their effects combine to determine the survival of lower trophic levels (Griffen et al. 2012; Toscano and Griffen 2014). These experimental modules are still obviously less complex than many natural food webs, although they hint that the effects of personality are likely to be large even under more complex conditions. Outstanding research questions include: How might the personality of pollinators affect which flowers they visit (Cane and Schiffhauer 2001; Muller and Chittka 2008; Chittka et al. 2009)? How might the behavioral tendencies of parasites (e.g., ectoparasites) affect their own propagation and the fitness of their hosts (Barber and Dingemanse 2010)? The relationships between personality and various species interaction outcomes are potentially innumerable, and insects and other arthropods have already proven to be nimble systems in which to evaluate such effects.

16.8.2 Behavioral type × behavioral type interactions

Situations where the combination of two species' personality types predicts the outcome of a species interaction are called behavioral type by behavioral type interactions (**BT × BT interactions**). They are characterized statistically by an interaction term between two species' personality traits when predicting the outcome of their interaction (Pruitt et al. 2012). One model that explains the emergence of these interaction terms is the **locomotor cross-over hypothesis**. This model predicts that active predators will be more likely to encounter slow-moving or sedentary prey and that sedentary predators will be more likely to encounter active prey (Huey and Pianka 1981). The locomotor cross-over hypothesis appears to also apply to intraspecific variation in predator and prey activity level in spider-cricket (DiRienzo et al. 2013; Sweeney et al. 2013a) and sea star-snail interactions (Pruitt et al. 2012) where individual variation in predator foraging mode predicts the behavioral phenotypes of the prey individuals they encounter. The literature is in need of more models like this one to predict the effects of personality in multiple interacting species and the circumstances under which we expect these patterns to emerge.

Links between personality, nutritional ecology, and metabolism seem likely (Biro and Stamps 2008; Careau et al. 2008). This is in part because individuals with different personality types may differ in their energetic or nutritional requirements (Lichtenstein and Pruitt 2015; Lichtenstein et al. 2016; Toscano et al. 2016). It would be exciting to investigate whether behavioral types differ in their tendency to accept various kinds of food resources or in their willingness to endure risks in order to secure those resources (Sih et al. 2015; Belgrad et al. 2017) due to differences in metabolic or nutritional requirements. Individual variation in nutritional requirements may also help to explain the maintenance of behavioral variation, because variation in foraging needs may reduce intraspecific competition.

16.8.3 Group and population level ecology

Interactions between species are composed of innumerable pairwise interactions between individuals. We have already discussed how personality can predict individual-level interactions, but scaling these interactions up to the group or population level requires additional considerations and analyses. The first step is to describe the personalities of a large group of individuals. So far, there are two approaches for characterizing personality at the group- or population-level: single-axis personality distributions/metrics and **behavioral hypervolumes**. A single-axis personality distribution merely refers to the frequency distribution by which individuals in a group fall along a single axis of personality. The mean/median/variance of these values are often also used as metrics of group-level behavioral tendencies. Already, there is evidence to suggest that average personality types can predict

the top-down effects of predator populations on prey community composition (Werner 1991, 1992; DiRienzo et al. 2013; Royauté and Pruitt 2015; Toscano et al. 2016). However, whether other distribution characteristics, such as variance or skew, can determine the outcome of species interactions remains little explored. Furthermore, while it is clear that populations vary geographically in their personality distributions, the underlying reasons for these differences are often unclear (Pruitt et al. 2010; Bengston and Dornhaus 2015).

Behavioral hypervolumes represent a multi-trait approach for assessing personality diversity at the group, population, or community level. The concept of a behavioral hypervolume is borrowed from niche hypervolumes in ecology. Behavioral hypervolumes are the multidimensional volume captured by a group, population, or community of animals in behavioral trait space. Each axis in this trait space represents a single behavioral trait (transformed so it can be compared with other traits with different dimensions). Thus, each axis represents a different personality assay, and each individual will represent a point along each axis. When lines are drawn between each point, a convex polyhedron is created in that trait space, which can be described by its volume. Groups that are more diverse in their behavioral tendencies will exhibit larger behavioral hypervolumes than more homogenous groups. These descriptions and comparisons can then be used to address ecological phenomena. For instance, the overlap between the behavioral hypervolumes of two interacting species may help us to predict the intensity of their interaction (e.g., greater overlap might result in more competition), or tracking change in the location of a population's behavioral hypervolume in trait space over time may reveal directional or cyclical patterns of behavioral change for populations or entire communities. Because of the recent genesis of this approach, only two studies have probed the relationship between behavioral hypervolume and species interaction outcomes. Multispecies colonies of spiders occupying larger hypervolumes have been shown to gain more mass collectively and are less likely to disband (Pruitt et al. 2016a), and groups of sea stars with larger hypervolumes are more lethal to their snail prey in mesocosm studies (Pruitt et al. 2017). Behavioral hypervolume studies consider a wide swathe of individuals' behavioral repertoires relative to past conceptual approaches, and may begin to shed light on how personality can predict the outcomes of higher order ecological processes, like the quality of ecosystem services, the rate and path of community succession, or how the link between animal personalities and ecological processes may change as a result of climate change.

16.9 Applications for applied insect behavior

16.9.1 Agricultural pests and invasive species

Functional variation in predator foraging mode is important in determining their efficacy as biocontrol agents, though management protocols nearly always ignore intraspecific variation in predators' foraging tactics. Increased intraspecific variation in parasitoid wasp assemblages fosters greater mortality in aphid communities compared to increased parasitoid species diversity (Finke and Snyder 2008). Likewise, in generalist cursorial spiders, the presence of more active predators or a combination of active and inactive behavioral phenotypes among predators can increase pest mortality by over 50 per cent (Keiser et al. 2015; Royauté and Pruitt 2015). Therefore, a modern understanding of the role that behavioral syndromes play in the dynamics of agricultural pest outbreaks requires a multifactorial approach that includes behavioral ecology, community ecology, and potentially ecotoxicology (Montiglio and Royauté 2014).

Insecticide and herbicide regiments are a pervasive stressor in agroecosystems that can alter selection regimes, shift fitness optima, and modify species interactions. Field experiments in apple orchards with different histories of insecticidal applications have shown that some populations of the jumping spider *Eris militaris* exhibit robust behavioral syndromes across five personality traits, although populations in insecticide-treated orchards show a decoupling of some of these correlations (Royauté et al. 2014). This could be because insecticide application selects for trait correlations, or that insecticides shift prey community composition in such a way that behavioral syndromes are no longer adaptive. This, in turn,

could have strong consequences in these predators' ability to suppress pest populations, and **integrated pest management** strategies should therefore attempt to account for how pesticide treatments alter the foraging capabilities of biological control agents.

In regards to invasive species, theoretical models have suggested that an invasion front will spread most rapidly when a population contains a mixture of dispersal-related behavioral phenotypes (Fogarty, Cote and Sih 2011). In one of the most successful invasive ants in North America, the red imported fire ant *Solenopsis invicta*, colonies can vary from acting as primary consumers to secondary predators, a diet range similarly achieved by entire ant communities (Roeder and Kaspari 2017). This variation in trophic breadth between colonies is influenced by intraspecific variation among workers within a colony, and thus the link between individual and colony traits may serve as a mechanism for invasive spread (Roeder and Kaspari 2017) (see previous sections on behavioral syndromes in eusocial insects). In the native hornet, *Vespa crabro*, and its invasive competitor, *V. velutina*, both species exhibit similar behavioral syndromes where activity, boldness, and exploration are correlated in the same directions. However, the invasive wasp is consistently bolder, more active, and more exploratory than its native counterpart (Monceau et al. 2015). It is, therefore, important to note that factors beyond average personality type or syndrome structure may be important in determining differential performance of natives versus invasive species or among exotics themselves.

16.10 Conclusions

Here we traced the thread of animal personality from its mechanistic underpinnings to its implications for mating systems, social evolution, species interactions, and applied topics in insect and spider behavior. The reoccurring inference throughout this discussion is that the effects of personality seem ever-present, and appear to influence many aspects of the behavioral and evolutionary ecology of insects and spiders. This is simultaneously intriguing, from the perspective of a basic scientist, and alarming, from an applied angle. From a basic science perspective, this new framework allows us to enhance the predictability of a wide variety of ecological and evolutionary processes. From an applied angle, animal personality is a cryptic kind of biodiversity invisible to us, unless we go through the laborious task of observing and recording the behavioral tendencies of large numbers of individuals. Insect and other arthropod models, for their part, have been and will continue to be front-running systems for studying animal personality from multiple perspectives (Kralj-Fišer and Schuett 2014). Yet, the literature is poised for new investigators to make fundamental steps towards enhancing the field. We hope that some of the ideas expressed herein, or ideas like them, will prove too tempting for some readers to ignore.

References

Araujo, M.S., Bolnick, D.I., and Layman, C.A. (2011). The ecological causes of individual specialisation. *Ecology Letters*, **14**, 948.

Arnqvist, G.R., and Henriksson, S. (1997). Sexual cannibalism in the fishing spider and a model for the evolution of sexual cannibalism based on genetic constraints. *Evolutionary Ecology*, **11**, 255.

Barber, I., and Dingemanse, N.J. (2010). Parasitism and the evolutionary ecology of animal personality. *Philosophical Transactions of the Royal Society of London B: Biological Sciences*, **365**, 4077.

Belgrad, B.A., Karan, J., and Griffen, B.D. (2017). Individual personality associated with interactions between physiological condition and the environment. *Animal Behaviour*, **123**, 277.

Bengston, S., and Dornhaus, A. (2015). Latitudinal variation in behaviors linked to risk tolerance is driven by nest-site competition and spatial distribution in the ant *Temnothorax rugatulus*. *Behavioral Ecology and Sociobiology*, **69**, 1265.

Bengston, S.E., and Jandt, J.M. (2014). The development of collective personality: the ontogenetic drivers of behavioral variation across groups. *Frontiers in Ecology and Evolution*, **2**, 35.

Bengston, S.E., Pruitt, J.N., and Riechert, S.E. (2014). Differences in environmental enrichment generate contrasting behavioural syndromes in a basal spider lineage. *Animal Behaviour*, **93**, 105.

Bergmuller, R., and Taborsky, M. (2010). Animal personality due to social niche specialisation. *Trends in Ecology and Evolution*, **25**, 504.

Bertelsmeier, C., Avril, A., Blight, O., et al. (2015). Different behavioural strategies among seven highly invasive ant species. *Biological Invasions*, **17**, 2491.

Beshers, S.N., and Fewell J.N. (2001). Models of division of labor in social insects. *Annual Review in Entomology*, **46**, 413.

Beshers, S.N., Huang, Z.Y., Oono, Y., and Robinson, G.E. (2001). Social inhibition and the regulation of temporal polyethism in honey bees. *Journal of Theoretical Biology*, **213**, 461.

Biro, P.A., and Adriaenssens, B. (2013). Predictability as a personality trait: consistent differences in intraindividual behavioral variation. *The American Naturalist*, **182**, 621.

Biro, P.A., Beckmann, C., and Stamps, J.A. (2010). Small within-day increases in temperature affects boldness and alters personality in coral reef fish. *Proceedings of the Royal Society of London B: Biological Sciences*, **277**, 71.

Biro, P.A., and Stamps, J.A. (2008). Are animal personality traits linked to life-history productivity? *Trends in Ecology and Evolution*, **23**, 361.

Blanckenhorn, W.U. (1991). Life-history difference in adjacent water strider populations: phenotypic plasticity or heritable responses to stream temperature? *Evolution*, **45**, 1520.

Blanckenhorn, W.U. (1994). Fitness consequences of alternative life histories in water striders, *Aquarius remigis* (Heteroptera: Gerridae). *Oecologia*, **97**, 354.

Bolnick, D.I., Amarasekare, P., Araújo, M.S., et al. (2011). Why intraspecific trait variation matters in community ecology. *Trends in Ecology and Evolution*, **26**, 183.

Bolnick, D.I., Svanbäck, R., Fordyce, J.A., et al. (2002). The ecology of individuals: incidence and implications of individual specialization. *American Naturalist*, **161**, 1.

Brodin, T. (2009). Behavioral syndrome over the boundaries of life—carryovers from larvae to adult damselfly. *Behavioral Ecology*, **20**, 30.

Cane, J.H., and Schiffhauer, D. (2001). Pollinator genetics and pollination: do honey bee colonies selected for pollen-hoarding field better pollinators of cranberry *Vaccinium macrocarpon*? *Ecological Entomology*, **26**, 117.

Careau, V., Thomas, D., Humphries, M., and Réale, D. (2008). Energy metabolism and animal personality. *Oikos*, **117**, 641.

Chang, A.T., and Sih, A. (2013). Multilevel selection and effects of keystone hyperaggressive males on mating success and behavior in stream water striders. *Behavioral Ecology*, **24**, 1166.

Chittka, L., Skorupski, P., and Raine, N.E. (2009). Speed–accuracy tradeoffs in animal decision making. *Trends in Ecology and Evolution*, **24**, 400.

Crutsinger, G.M., Collins, M.D., Fordyce, J.A., Gompert, Z., Nice, C.C., and Sanders, N.J. (2006). Plant genotypic diversity predicts community structure and governs an ecosystem process. *Science*, **313**, 966.

Dall, S.R., Houston, A.I., and McNamara, J.M. (2004). The behavioural ecology of personality: consistent individual differences from an adaptive perspective. *Ecology Letters*, **7**, 734.

Davidson, D.W. (1998). Resource discovery versus resource domination in ants: a functional mechanism for breaking the trade-off. *Ecological Entomology*, **23**, 484.

Dingemanse, N.J., Both, C., Drent, P.J., Van Oers, K., and Van Noordwijk, A.J. (2002). Repeatability and heritability of exploratory behaviour in great tits from the wild. *Animal Behaviour*, **64**, 929.

Dingemanse, N.J., and Dochtermann, N.A. (2013). Quantifying individual variation in behaviour: mixed-effect modelling approaches. *Journal of Animal Ecology*, **82**, 39.

Dingemanse, N.J., Kazem, A.J., Réale, D., and Wright, J. (2010). Behavioural reaction norms: animal personality meets individual plasticity. *Trends in Ecology and Evolution*, **25**, 81.

DiRienzo, N., McDermott, D.R., and Pruitt, J.N. (2015). Testing the effects of biogenic amines and alternative topical solvent types on the behavioral repertoire of two web-building spiders. *Ethology*, **121**, 801.

DiRienzo, N., Pruitt, J.N., and Hedrick, A.V. (2012). Juvenile exposure to acoustic sexual signals from conspecifics alters growth trajectory and an adult personality trait. *Animal Behaviour*, **84**, 861.

DiRienzo, N., Pruitt, J.N., and Hedrick, A.V. (2013). The combined behavioural tendencies of predator and prey mediate the outcome of their interaction. *Animal Behaviour*, **86**, 317.

Dominey, W.J. (1984). Alternative mating tactics and evolutionarily stable strategies. *American Zoologist*, **24**, 385–96.

Ducatez, S., Baguette, M., Trochet, A., et al. (2013). Flight endurance and heating rate vary with both latitude and habitat connectivity in a butterfly species. *Oikos*, **122**, 601.

Fewell, J.H., and Page, R.E. (1993). Genotypic variation in foraging responses to environmental stimuli by honey-bees, *Apis mellifera*. *Experientia*, **49**, 1106.

Finke, D.L., and Snyder, W.E. (2008). Niche partitioning increases resource exploitation by diverse communities. *Science*, **321**, 1488.

Fogarty, S., Cote, J., and Sih, A. (2011). Social personality polymorphism and the spread of invasive species: a model. *American Naturalist*, **177**, 273.

Folse, H.J., and Roughgarden, J. (2010). What is an individual organism? A multilevel selection perspective. *Quarterly Review of Biology*, **85**, 447.

Gordon, D.M., Guetz, A., Greene, M.J., and Holmes, S. (2011). Colony variation in the collective regulation of foraging by harvester ants. *Behavioral Ecology*, **22**, 429.

Gosling, S.D. (2001). From mice to men: what can we learn about personality from animal research? *Psychological bulletin*, **127**, 45.

Griffen, B.D., Toscano, B.J., and Gatto, J. (2012). The role of individual behavior type in mediating indirect interactions. *Ecology*, **93**, 1935.

Grinsted, L., Pruitt, J.N., Settepani, V., and Bilde, T. (2013). Individual personalities shape task differentiation in a social spider. *Proceedings of the Royal Society B-Biological Sciences*, **280**, 20131407.

Gyuris, E., Feró, O., Tartally, A., and Barta, Z. (2011). Individual behaviour in firebugs (*Pyrrhocoris apterus*). *Proceedings of the Royal Society of London B: Biological Sciences*, **278**, 628.

Hedrick, A.V., and Kortet, R. (2006). Hiding behaviour in two cricket populations that differ in predation pressure. *Animal Behaviour*, **72**, 1111.

Hedrick, A.V., and Kortet, R. (2012). Sex differences in the repeatability of boldness over metamorphosis. *Behavioral Ecology and Sociobiology*, **66**, 407.

Holbrook, C.T., Barden, P.M., and Fewell, J.H. (2011). Division of labor increases with colony size in the harvester ant *Pogonomyrmex californicus*. *Behavioral Ecology*, **22**, 960.

Holbrook, C.T., Wright, C.M., and Pruitt, J.N. (2014). Individual differences in personality and behavioural plasticity facilitate division of labour in social spider colonies. *Animal Behaviour*, **97**, 177.

Hölldobler, B., and Wilson, E.O. (1990). *The Ants*. Belknap Press, Cambridge, MA.

Hölldobler, B., and Wilson, E.O. (2009). *The Superorganism: the Beauty, Elegance, and Strangeness of Insect Societies*. WW Norton and Company, New York, NY.

Huey, R.B., and Pianka, E.R. (1981). Ecological consequences of foraging mode. *Ecology*, **62**, 991.

Hughes, A.R., and Stachowicz, J.J. (2004). Genetic diversity enhances the resistance of a seagrass ecosystem to disturbance. *Proceedings of the National Academy of Sciences of the United States of America*, **101**, 8998.

Jandt, J.M., Bengston, S., Pinter-Wollman, N., et al. (2014). Behavioural syndromes and social insects: personality at multiple levels. *Biological Reviews*, **89**, 48.

Jandt, J.M., Bengston, S., Pinter-Wollman, N., et al. (2013). Behavioural syndromes and social insects: personality at multiple levels. *Biological Reviews*, **89**, 48.

Jeanne, R.L. (1986). The organization of work in *Polybia occidentalis*—costs and benefits of specialization in a social wasp. *Behavioral Ecology and Sociobiology*, **19**, 333.

Jeanson, R., Clark, R.M., Holbrook, C.T., et al. (2008). Division of labour and socially induced changes in response thresholds in associations of solitary halictine bees. *Animal Behaviour*, **76**, 593.

Jeanson, R., Fewell, J.H., Gorelick, R., and Bertram, S.M. (2007). Emergence of increased division of labor as a function of group size. *Behavioral Ecology and Sociobiology*, **62**, 289.

Johnson, J.C. (2013). Debates: challenging a recent challenge to the aggressive spillover hypothesis. *Ethology*, **119**, 811.

Johnson, J.C., and Sih, A. (2005). Precopulatory sexual cannibalism in fishing spiders (*Dolomedes triton*): a role for behavioral syndromes. *Behavioral Ecology and Sociobiology*, **58**, 390.

Jongepier, E., Kleeberg, I., Job, S., and Foitzik, S. (2014). Collective defence portfolios of ant hosts shift with social parasite pressure. *Proceedings of the Royal Society of London B: Biological Sciences*, **281**, 20140225.

Keiser, C.N., Howell, K.A., Pinter-Wollman, N., and Pruitt, J.N. (2016a). Personality composition alters the transmission of cuticular bacteria in social groups. *Biology Letters*, **12**, 20160297.

Keiser, C.N., Pinter-Wollman, N., Augustine, D.A., et al. (2016b). Individual differences in boldness influence patterns of social interactions and the transmission of cuticular bacteria among group-mates. *Proceedings of the Royal Society B-Biological Sciences*, **283**, 20160457.

Keiser, C.N., and Pruitt, J.N. (2014a). Spider aggressiveness determines the bidirectional consequences of host–inquiline interactions. *Behavioral Ecology*, **25**, 142.

Keiser, C.N., and Pruitt, J.N. (2014b). Personality composition is more important than group size in determining collective foraging behaviour in the wild. *Proceedings of the Royal Society B*, **281**, 20141424.

Keiser, C.N., Slyder, J.B., Carson, W.P., and Pruitt, J.N. (2015). Individual differences in predators but not producers mediate the magnitude of a trophic cascade. *Arthropod-Plant Interactions*, **9**, 225.

Keiser, C.N., Wright, C.M., and Pruitt, J.N. (2016). Increased bacterial load can reduce or negate the effects of keystone individuals on group collective behaviour. *Animal Behaviour*, **114**, 211.

Kortet, R., Rantala, M.J., and Hedrick, A. (2007). Boldness in anti-predator behaviour and immune defence in field crickets. *Evolutionary Ecology Research*, **9**, 185.

Kralj-Fišer, S., Sanguino Mostajo, G.A., Preik, O., Pekár, S., and Schneider, J.M. (2013). Assortative mating by aggressiveness type in orb weaving spiders. *Behavioral Ecology*, **24**, 824.

Kralj-Fišer, S., Schneider, J.M., and Kuntner, M. (2013). Challenging the aggressive spillover hypothesis: is precopulatory sexual cannibalism a part of a behavioural syndrome? *Ethology*, **119**, 615.

Kralj-Fišer, S., and Schuett, W. (2014). Studying personality variation in invertebrates: why bother? *Animal Behaviour*, **91**, 41.

Laskowski, K.L., and Pruitt, J.N. (2014). Evidence of social niche construction: persistent and repeated social interactions generate stronger personalities in a social spider. *Proceedings of the Royal Society B-Biological Sciences*, **281**, 20133166.

Legrand, D., Larranaga, N., Bertrand, R., et al. (2016). Evolution of a butterfly dispersal syndrome. *Proceedings of the Royal Society B: Biological Sciences*, **283**, 20161533.

Lichtenstein, J.L., Pruitt, J.N., and Modlmeier, A.P. (2015). Intraspecific variation in collective behaviors drives interspecific contests in acorn ants. *Behavioral Ecology*, **27**, 553.

Lichtenstein, J.L.L., and Pruitt, J.N. (2015). Similar patterns of frequency-dependent selection on animal personalities emerge in three species of social spiders. *Journal of Evolutionary Biology*, **28**, 1248.

Lichtenstein, J.L.L., Wright, C.M., Luscuskie, L.P., Montgomery, G.A., Pinter-Wollman, N., and Pruitt, J.N. (2016). Participation in cooperative prey capture and the benefits gained from it are associated with individual personality. *Current Zoology*, **189**(3), 254–66.

Martin, P., and Bateson, P. (1993). *Measuring Behaviour: an Introductory Guide*. Cambridge University Press, New York, NY.

Maupin, J.L., and Riechert, S.E. (2001). Superfluous killing in spiders: a consequence of adaptation to food-limited environments? *Behavioral Ecology*, **12**, 569.

Miller, J.R., Ament, J.M., and Schmitz, O.J. (2014). Fear on the move: predator hunting mode predicts variation in prey mortality and plasticity in prey spatial response. *Journal of Animal Ecology*, **83**, 214.

Modlmeier, A.P., Keiser, C.N., Shearer, T.A., and Pruitt, J.N. (2014a). Species-specific influence of group composition on collective behaviors in ants. *Behavioral Ecology and Sociobiology*, **68**, 1929.

Modlmeier, A.P., Keiser, C.N., Watters, J.V., Sih, A., and Pruitt, J.N. (2014b). The keystone individual concept: an ecological and evolutionary overview. *Animal Behaviour*, **89**, 53.

Modlmeier, A.P., Keiser, C.N., Wright, C.M., Lichtenstein, J.L., and Pruitt, J.N. (2015). Integrating animal personality into insect population and community ecology. *Current Opinion in Insect Science*, **9**, 77.

Modlmeier, A.P., Laskowski, K.L., DeMarco, A.E., et al. (2014c) Persistent social interactions beget more pronounced personalities in a desert-dwelling social spider. *Biology Letters*, **10**, 20140419.

Monceau, K., Moreau, J., Poidatz, J., Bonnard, O., and Thiéry, D. (2015). Behavioral syndrome in a native and an invasive hymenoptera species. *Insect Science*, **22**, 541.

Montiglio, P.O., Ferrari, C., and Reale, D. (2013). Social niche specialization under constraints: personality, social interactions and environmental heterogeneity. *Philosophical Transactions of the Royal Society B-Biological Sciences*, **368**, 20120343.

Montiglio, P-O., and Royauté, R. (2014). Contaminants as a neglected source of behavioural variation. *Animal Behaviour*, **88**, 29.

Muller, H., and Chittka, L. (2008). Animal personalities: the advantage of diversity. *Current Biology*, **18**, 961.

Müller, T., and Müller, C. (2015). Behavioural phenotypes over the lifetime of a holometabolous insect. *Frontiers in Zoology*, **12**, S8.

Niemelä, P.T., and Santostefano, F. 2015. Social carry-over effects on non-social behavioral variation: mechanisms and consequences. *Frontiers in Ecology and Evolution*, **3**, 24.

Nonacs, P., and Kapheim, K.M. (2007). Social heterosis and the maintenance of genetic diversity. *Journal of Evolutionary Biology*, **20**, 2253.

Nonacs, P., and Kapheim, K.M. (2008). Social heterosis and the maintenance of genetic diversity at the genome level. *Journal of Evolutionary Biology*, **21**, 631.

Nowak, M.A., Tarnita, C.E., and Wilson, E.O. (2010). The evolution of eusociality. *Nature*, **466**, 1057.

Oster, G.F., and Wilson, E.O. (1979). *Caste and Ecology in the Social Insects*. Princeton University Press, Princeton, NJ.

Page, R.E. (1980). The evolution of multiple mating-behavior by honey bee queens *Apis mellifera*. *Genetics*, **96**, 263.

Page, R.E., and Robinson, G.E. (1991). The genetics of division-of-labor in honey-bee colonies. *Advances in Insect Physiology*, **23**, 117.

Peretti, A., Eberhard, W.G., and Briceño, R.D. (2006). Copulatory dialogue: female spiders sing during copulation to influence male genitalic movements. *Animal Behaviour*, **72**, 413.

Perkins, T.A., Riechert, S.E., and Jones, T.C. (2007). Interactions between the social spider *Anelosimus studiosus* (Araneae, Theridiidae) and foreign spiders that frequent its nests. *Journal of Arachnology*, **35**, 143.

Pinter-Wollman, N. (2012). Personality in social insects: how does worker personality determine colony personality? *Current Zoology*, **58**, 579.

Pruitt, J.N. (2012). Behavioural traits of colony founders affect the life history of their colonies. *Ecology Letters*, 15, 1026.

Pruitt, J.N. (2013). A real-time eco-evolutionary dead-end strategy is mediated by the traits of lineage progenitors and interactions with colony invaders. *Ecology Letters*, **16**, 879.

Pruitt, J.N., Bolnick, D.I., Sih, A., DiRienzo, N., and Pinter-Wollman, N. (2016a). Behavioural hypervolumes of spider communities predict community performance and disbandment. *Proceedings of the Royal Society B: Biological Sciences*, **283**, 20161409.

Pruitt, J.N., and Ferrari, M.C. (2011). Intraspecific trait variants determine the nature of interspecific interactions in a habitat-forming species. *Ecology*, **92**, 1902.

Pruitt, J.N., and Goodnight, C.J. (2014). Site-specific group selection drives locally adapted colony compositions. *Nature*, **514**, 359.

Pruitt, J.N., Goodnight, C.J., & Riechert, S. E. (2017). Intense group selection selects for ideal group compositions, but selection within groups maintains them. *Animal behaviour*, **124**, 15–24.

Pruitt, J.N., Grinsted, L., and Settepani, V. (2013). Linking levels of personality: personalities of the 'average' and 'most extreme' group members predict colony-level personality. *Animal Behaviour*, **86**, 391.

Pruitt, J.N., Howell, K.A., Gladney, S.J., et al. 2017. Behavioral hypervolumes of predator groups and predator-predator interactions shape prey survival rates and selection on prey behavior. *American Naturalist*, **189**, 254.

Pruitt, J.N., and Keiser, C.N. (2014). The personality types of key catalytic individuals shape colonies' collective behaviour and success. *Animal Behaviour*, **93**, 87.

Pruitt, J.N., and Pinter-Wollman, N. (2015). The legacy effects of keystone individuals on collective behaviour scale to how long they remain within a group. *Proceedings of the Royal Society B: Biological Sciences*, **282**, 89.

Pruitt, J.N., and Riechert, S.E. (2011). How within-group behavioural variation and task efficiency enhance fitness in a social group. *Proceedings of the Royal Society B: Biological Sciences*, **278**, 1209.

Pruitt, J.N., Riechert, S.E., and Harris, D.J. (2011). Reproductive consequences of male body mass and aggressiveness depend on females' behavioral types. *Behavioral Ecology and Sociobiology*, **65**, 1957.

Pruitt, J.N., Riechert, S.E., Iturralde, G., et al. (2010). Population differences in behaviour are explained by shared within-population trait correlations. *Journal of Evolutionary Biology*, **23**, 748.

Pruitt, J.N., Stachowicz, J.J., and Sih, A. (2012). Behavioral types of predator and prey jointly determine prey survival: potential implications for the maintenance of within-species behavioral variation. *American Naturalist*, **179**, 217.

Pruitt, J.N., Wright, C.M., Keiser, C.N., DeMarco, A.E., Grobis, M.M., and Pinter-Wollman, N. (2016b). The Achilles' heel hypothesis: misinformed keystone individuals impair collective learning and reduce group success. *Proceedings of the Royal Society B-Biological Sciences*, **283**, 20152888.

Raine, N.E., and Chittka, L. 2008. The correlation of learning speed and natural foraging success in bumble-bees. *Proceedings of the Royal Society B: Biological Sciences*, **275**, 803.

Raine, N.E., Ings, T.C., Ramos-Rodriguez, O., and Chittka, L. (2006). Intercolony variation in learning performance of a wild British bumblebee population (Hymenoptera: Apidae: *Bombus terrestris audax*). *Entomologia Generalis*, **28**, 241.

Riechert, S.E., and Hedrick, A.V. (1993a). A test for correlations among fitness-linked behavioural traits in the spider *Agelenopsis aperta* (Araneae, Agelenidae). *Animal Behaviour*, **46**, 669.

Riechert, S.E., and Hedrick, A.V. (1993b). A test for correlations among fitness-linked behavioural traits in the spider *Agelenopsis aperta* (Araneae, Agelenidae). *Animal Behaviour*, **46**, 669.

Riechert, S.E., Singer, F.D., and Jones, T.C. (2001). High gene flow levels lead to gamete wastage in a desert spider system. *Genetica*, **112**, 297.

Robinson, G.E. (1992). Regulation of division-of-labor in insect societies. *Annual Review of Entomology*, **37**, 637.

Robinson, G.E., and Page, R.E.J. (1989). Genetic basis for division of labour in an insect society. In: M. D. Breed and R. E. Page (Eds), *The Genetics of Social Evolution*, pp. 61–80. Westview Press, Boulder, CO.

Robinson, G.E., Page, R.E., and Huang, Z.Y. (1994). Temporal polyethis in social insects is a developmental process. *Animal Behaviour*, **48**, 467.

Rodrigues, A.S., Botina, L., Nascimento, C.P., et al. (2016). Ontogenic behavioral consistency, individual variation and fitness consequences among lady beetles. *Behavioural Processes*, **131**, 32.

Roeder, K.A., and Kaspari, M. (2017). From cryptic herbivore to predator: stable isotopes reveal consistent variability in trophic levels in an ant population. *Ecology*, **98**, 297.

Rowles, A.D., and O'Dowd, D.J. (2007). Interference competition by Argentine ants displaces native ants: implications for biotic resistance to invasion. *Biological Invasions*, **9**, 73.

Royauté, R., Buddle, C.M., and Vincent, C. (2014). Interpopulation variations in behavioral syndromes of a jumping spider from insecticide-treated and insecticide-free orchards. *Ethology*, **120**, 127.

Royauté, R., and Pruitt, J.N. (2015). Varying predator personalities generates contrasting prey communities in an agroecosystem. *Ecology*, **96**, 2902.

Saltz, J.B., and Foley, B.R. (2011). Natural genetic variation in social niche construction: social effects of aggression drive disruptive sexual selection in *Drosophila melanogaster*. *American Naturalist*, **177**, 645.

Saltz, J.B., and Nuzhdin, S.V. (2014). Genetic variation in niche construction: implications for development and evolutionary genetics. *Trends in Ecology and Evolution*, **29**, 8.

Scharf, I., Modlmeier, A.P., Fries, S., Tirard, C., and Foitzik, S. (2012). Characterizing the collective personality of ant societies: aggressive colonies do not abandon their home. *PloS One* **7**, e33314.

Schulz, D.J., and Robinson, G.E. (1999). Biogenic amines and division of labor in honey bee colonies: behaviorally related changes in the antennal lobes and age-related changes in the mushroom bodies. *Journal of Comparative Physiology A*, **184**, 481.

Seeley, T.D. (1982). Adaptive significance of the age polyethism schedule in honeybee colonies. *Behavioral Ecology and Sociobiology*, **11**, 287.

Sih, A., and Bell, A.M. (2008). Insights for behavioral ecology from behavioral syndromes. *Advances in the Study of Behavior*, **38**, 227.

Sih, A., Bell, A.M., and Johnson, J.C. (2004). Behavioral syndromes: an ecological and evolutionary overview. *Trends in Ecology and Evolution*, **19**, 372.

Sih, A., Cote, J., Evans, M., Fogarty, S., and Pruitt, J. (2012). Ecological implications of behavioural syndromes. *Ecology Letters*, **15**, 278.

Sih, A., Mathot, K.J., Moirón, M., Montiglio, P-O., Wolf, M., and Dingemanse, N.J. (2015). Animal personality and state–behaviour feedbacks: a review and guide for empiricists. *Trends in Ecology and Evolution*, **30**, 50.

Sih, A., Montiglio, P-O., Wey, T.W., and Fogarty, S. (2017). Altered physical and social conditions produce rapidly reversible mating systems in water striders. *Behavioral Ecology*, **28**, 632.

Sih, A., and Watters, J.V. (2005). The mix matters: behavioural types and group dynamics in water striders. *Behaviour*, **142**, 1417.

Sokolowski, M.B. (2001). *Drosophila*: genetics meets behaviour. *Nature Reviews Genetics*, **2**, 879.

Stamps, J., and Groothuis, T.G. (2010). The development of animal personality: relevance, concepts and perspectives. *Biological Reviews*, **85**, 301.

Stamps, J.A., Briffa, M., and Biro, P.A. (2012). Unpredictable animals: individual differences in intraindividual variability (IIV). *Animal Behaviour*, **83**, 1325.

Stevens, V.M., Trochet, A., Blanchet, S., et al. (2013). Dispersal syndromes and the use of life histories to predict dispersal. *Evolutionary Applications*, **6**, 630.

Stevenson, P.A., and Rillich, J. (2016). Controlling the decision to fight or flee: the roles of biogenic amines and nitric oxide in the cricket. *Current Zoology*, **62**(3), 265–75.

Svanback, R., and Bolnick, D.I. (2007). Intraspecific competition drives increased resource use diversity within a natural population. *Proceedings of the Royal Society B: Biological Sciences*, **274**, 839.

Sweeney, K., Cusack, B., Armagost, F., et al. (2013a). Predator and prey activity levels jointly influence the outcome of long-term foraging bouts. *Behavioral Ecology*, **24**, 1205.

Sweeney, K., Gadd, R.D.H., Hess, Z.L., et al. (2013b). Assessing the effects of rearing environment, natural selection, and developmental stage on the emergence of a behavioral syndrome. *Ethology*, **119**, 436.

Toscano, B.J., Gownaris, N.J., Heerhartz, S.M., and Monaco, C.J. (2016). Personality, foraging behavior and specialization: integrating behavioral and food web ecology at the individual level. *Oecologia*, **182**, 55.

Toscano, B.J., and Griffen, B.D. (2014). Trait-mediated functional responses: predator behavioural type mediates prey consumption. *Journal of Animal Ecology*, **83**, 1469.

Tremmel, M., and Müller, C. (2013). Insect personality depends on environmental conditions. *Behavioral Ecology*, **24**, 386.

Urszán, T.J., Garamszegi, L.Z., Nagy, G., et al. (2015). No personality without experience? A test on *Rana dalmatina* tadpoles. *Ecology and Evolution*, **5**, 5847.

Ward, A.J., Thomas, P., Hart, P.J., and Krause, J. (2004). Correlates of boldness in three-spined sticklebacks (*Gasterosteus aculeatus*). *Behavioral Ecology and Sociobiology*, **55**, 561.

Webster, M., Ward, A., and Hart, P. (2009). Individual boldness affects interspecific interactions in sticklebacks. *Behavioral Ecology and Sociobiology*, **63**, 511.

Werner, E.E. (1991). Nonlethal effects of a predator on competitive interactions between two anuran larvae. *Ecology*, **72**, 1709.

Werner, E.E. (1992). Individual behavior and higher-order species interactions. *American Naturalist*, **140**, S5–S32.

Wexler, Y., Subach, A., Pruitt, J.N., and Scharf, I. (2016). Behavioral repeatability of flour beetles before and after metamorphosis and throughout aging. *Behavioral Ecology and Sociobiology*, **70**, 745.

Wilson, E.O. (1976). Behavioral discretization and number of castes in an ant species. *Behavioral Ecology and Sociobiology*, **1**, 141.

Wolf, M., and Weissing, F.J. (2012). Animal personalities: consequences for ecology and evolution. *Trends in ecology and evolution*, **27**, 452.

Wray, M.K., Mattila, H.R., and Seeley, T.D. (2011). Collective personalities in honeybee colonies are linked to colony fitness. *Animal Behaviour*, **81**, 559.

Wright, C.M., Holbrook, C.T., and Pruitt, J.N. (2014). Animal personality aligns task specialization and task proficiency in a spider society. *Proceedings of the National Academy of Sciences of the United States of America*, **111**, 9533.

Wright, C.M., Keiser, C.N., and Pruitt, J.N. (2015). Personality and morphology shape task participation, collective foraging and escape behaviour in the social spider *Stegodyphus dumicola*. *Animal Behaviour*, **105**, 47.

Wright, C.M., Keiser, C.N., and Pruitt, J.N. (2016). Colony personality composition alters colony-level plasticity and magnitude of defensive behaviour in a social spider. *Animal Behaviour*, **115**, 175.

Wright, C.M., Skinker, V.E., Izzo, A.S., Tibbetts, E.A., and Pruitt, J.N. (2017). Queen personality type predicts nest-guarding behaviour, colony size and the subsequent collective aggressiveness of the colony. *Animal Behaviour*, **124**, 7.

Yip, E.C. (2014). Ants versus spiders: interference competition between two social predators. *Insectes Sociaux*, **61**, 403.

CHAPTER 17

Cognition and learning

Reuven Dukas

Animal Behaviour Group, Department of Psychology, Neuroscience and Behaviour, McMaster University, 1280 Main Street West, Hamilton, Ontario, L8S 4K1, Canada

17.1 Introduction

It is certain that there may be extraordinary mental activity with an extremely small absolute mass of nervous matter: thus the wonderfully diversified instincts, mental powers, and affections of ants are notorious, yet their cerebral ganglia are not so large as the quarter of a small pin's head. Under this point of view, the brain of an ant is one of the most marvellous atoms of matter in the world, perhaps more so than the brain of a man.

(Darwin 1874, p. 54.)

In spite of Darwin's early insight, much of the research on insect behavior through over half of the twentieth century focused on instincts, rather than on 'mental powers' (Fabre et al. 1918; Tinbergen 1951). Extensive investigations in the past few decades, however, have firmly established that insects possess sophisticated cognition that is important in shaping insect behavior (reviewed in Alloway 1972; Papaj and Prokopy 1989; Menzel 1985; Dukas 2008b; Giurfa 2013). Consequently, insects are now widely recognized as useful model systems for cognitive research that is relevant to all animals including humans (O'Kane 2011; Bier 2005; Inlow and Restifo 2004).

This chapter will focus on answering several central questions. First, what is insect cognition and what are its components? Second, what is the adaptive significance of cognitive traits? Third, what is learning and what is special about it? Fourth, how do insects rely on learning to maximize fitness? Fifth, what is special about **social learning** and how important is it in insects? Finally, what issues of insect cognition deserve further study?

17.2 Insect cognition

Cognition is defined as the processes concerned with the acquisition, retention, and use of information. In animals, such information processing is carried out by the nervous system and this discussion focuses on cognition mediated by neurons. Cognition is typically divided into a few interrelated components. Perception involves the translation of sensory information into neuronal representations. Learning is the acquisition and retention of neuronal representations of new information. These include new associations between stimuli and environmental states, spatial configurations of resources, and new motor patterns. Long-term memory comprises representations of information already learned. Working memory refers to the small set of neuronal representations active over some short period of time. Attention is the neuronal representations activated at any given time. Finally, decision consists of the determination of action given the known states of

Dukas, R., *Cognition and learning*. In: *Insect Behavior: From mechanisms to ecological and evolutionary consequences*. Edited by Alex Córdoba-Aguilar, Daniel González-Tokman, and Isaac González-Santoyo: Oxford University Press (2018). © Oxford University Press.
DOI: 10.1093/oso/9780198797500.003.0017

relevant environmental features and experience (Platt 2002; Dukas 2004; Rolls 2014).

Is insect cognition different from cognition in other animals? The ingrained bias that is still prevalent in some disciplines, as well as popular culture is that cognition is unique to humans and, perhaps, some closely related species. Although this opinion has strong historical and emotional origins, it is not based on empirical data (Chittka and Niven 2009). Certainly, there is large between-species variation in cognitive traits, meaning that a close examination of any species will probably reveal unique features. However, cognition, defined above as the processes concerned with the acquisition, retention, and use of information is universal to all animals as well as other organisms (Dukas 1998a). Insects have small brains and short lifespans relative to most vertebrates, and it is tempting to invoke these features as dictating inferior cognition (e.g., Mayr 1974; Staddon 1983). As just noted, however, no critical data have documented fundamental differences between insect and vertebrate cognition.

From first principles, we can deduce one difference between short-lived insects and long-lived vertebrates. Short-lived insects cannot express complex learned skills that are acquired over durations longer than their expected lifespans. The development of expertise has been well studied in humans (Ericsson et al. 2006). While research on honey bees, *Apis mellifera*, indicates similar cognitive processes of expertise development (Dukas and Visscher 1994; Dukas 2008d), the magnitude of such expertise is limited by the short duration of a forager life—only about 1 week (Dukas 2008a). Even with expertise, however, the difference between insects and vertebrates probably reflects magnitude, rather than essential quality. This means that insects can be superb models for research on the development of expertise and its mechanistic bases (see Section 17.7).

17.3 Adaptive significance of insects' cognitive traits

A thorough understanding of insect cognition necessitates analysing cognitive traits within an evolutionary biological framework, which integrates mechanisms and function. It is sensible to assume that most insect cognitive traits are adaptive, meaning that they have evolved via natural selection due to their beneficial effects on fitness. A variety of constraints, however, cannot be ruled out (Dukas 1998b). Ideally, one would be able to show that insect cognitive traits show heritable variation that is associated with fitness, and that they evolve (Dukas 2004). Currently, however, the available data for insects are rather limited and restricted primarily to learning, which I will discuss separately in the next two sections. Here, I will focus on examples of apparent adaptations in the domains of perception, memory and decision making.

17.3.1 Night vision

As noted in Section 17.2, the first stage of cognition is perception, which involves capturing information from the external environment and translating it into neuronal representations. Perception is carried out by specialized receptors tuned to distinct cue attributes including touch, smell, taste, vision, and hearing (see Chapter 3). Receptors are often concentrated in dedicated organs that enhance information capture. Vision in nocturnal insects represents a remarkable example of apparent perceptual adaptation (see Chapter 11). While most insects are diurnal, nocturnal activity has evolved multiple times and has probably provided protection from visual predators and access to unexploited resources. The most dramatic visual adaptation in nocturnal insects is the refracting superposition compound eye, best studied in nocturnal beetles and moths. The major feature distinguishing the refracting superposition compound eye from the ancestral apposition compound eye is the number of lens units providing light to each group of photoreceptors (rhabdom). While that number is always one in the **apposition eye** design, it is usually several hundreds in **superposition eyes**. This allows for up to 1000 times higher optical sensitivity (Warrant and Dacke 2011; Land and Nilsson 2012; Cronin et al. 2014; for a detailed explanation of eye structure and evolution, see Chapter 11).

In addition to the improved optical sensitivity, nocturnal insects also employ temporal and spatial summation. Temporal summation means that both photoreceptors and higher levels of visual neurons

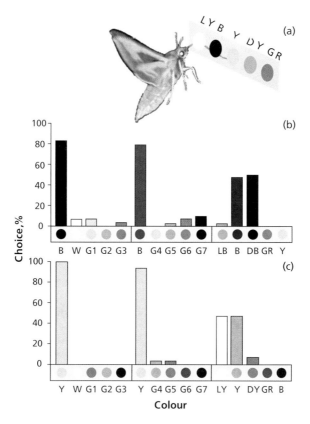

Figure 17.1 Superb colour night vision under starlight in nocturnal hawkmoths, *Deilephila elpenor*. Individual hawkmoths learned to discriminate a training colour from white and seven different shades of grey, as well as from two other colours, but not from brighter or darker shades of the training colour. (a) A hawkmoth choosing one of the stimuli in the set-up. The hawkmoths showed significant colour discrimination when the training colour was blue (b) and yellow (c). Letters indicate white (W), seven shades of grey (G1–G7), three shades each of yellow (LY, Y, and DY) and blue (LB, B, and DB) and green (GR).

Figure modified from Kelber and Roth (2006).

integrate information over longer time at lower light intensities (see Chapter 3). Similarly, spatial summation involves laterally spreading neurons that integrate information from larger groups of photoreceptors under lower light intensities (Warrant and Dacke 2011; Stöckl et al. 2016). The combined outcome of such optical and neuronal adaptations is an extraordinary ability to navigate and perceive colours at night solely under starlight illumination (see Chapter 7). For example, Kelber et al. (2002) trained nocturnal hawkmoths, *Deilephila elpenor*, to feed on either blue or yellow artificial flowers. The hawkmoths could then distinguish these flower colours from both other colours and shades of grey with similar contrasts under light intensity comparable to starlight (Figure 17.1).

17.3.2 Memory

The information encountered at any given time by the sensory receptors of an individual far exceeds the brain's information processing capacity. Consequently, animals must selectively attend to and commit to memory only a small proportion of the incoming information (Broadbent 1965; Dukas 2002). Choosing a particular perceptual focus, while ignoring other information depends on the expected effects on fitness. Two major means used by animals to select information are evolved innate biases and the reoccurrence of relevant information. For example, a naturally foraging bee is more likely to attend to, learn and later recall flower odour and colour than leaf shape owing to innate biases, but, in the laboratory, she can learn to associate reward with novel shapes if they consistently predict the reward (see Garcia and Koelling 1966; Anderson and Schooler 1991; Dunlap and Stephens 2014). Storing information in memory also entails a variety of costs including the necessary metabolically demanding brain tissue and perhaps a negative association between memory load, and the speed and efficiency of recall (Dukas 1999; Bernays 2001; Plaçais and Preat 2013). Animals should thus possess mechanisms that modulate memory consolidation and retention based on evolved biases and experience.

The mechanisms underlying memory consolidation and retention are complex, and still subject to intense research. Nevertheless, the successful identification of genetically distinct memory phases (Davis 2005; Dubnau and Chiang 2013) opens up unique opportunities to link such memory types to animal evolutionary biology. The primary memory phases are short-term memory (STM), middle-term memory (MTM), anaesthesia-resistant memory (ARM), and long-term memory (LTM). These phases vary in longevity, from 1 hour with STM to many days or even a whole lifespan with LTM (Dubnau and Tully 1998). We would expect that animals will optimally allocate information to a specific memory phase based on its expected value and usefulness over time. Lesser information should be retained for shorter periods of time than more important information that will be useful in the long term. Data from parasitoid wasps of the cabbage white butterflies agree with this prediction.

The large cabbage white butterfly (*Pieris brassicae*) lays clusters of over 100 eggs on a single plant, whereas the small cabbage white butterfly (*P. rapae*) lays single eggs on different plants. While the parasitoid wasp, *Cotesia glomerata*, prefers to lay eggs in caterpillars of the large than of the small cabbage white butterfly, it accepts both host species. The wasps are proficient at learning odours of the larval host plant and will subsequently rely on these conspicuous odours for locating additional caterpillars (Vet and Dicke 1992; for chemical communication see Chapter 10). When *C. glomerata* wasps were given a single oviposition experience with either large or small cabbage white butterfly larvae, they

showed significantly better memory of the host plant associated with the large than with the small cabbage white butterfly caterpillars 24 hours later. Further analyses indicated that ovipositing on the small cabbage white butterfly caterpillar produced only ARM, whereas ovipositing on the large cabbage white butterfly caterpillar generated LTM (Kruidhof et al. 2012). This difference is consistent with the prediction that more valuable information is more likely to be committed into a longer-term memory phase. In general, while the mechanisms that allocate information into distinct memory phases and to forgetting are partially known (Placais et al. 2012; Guven-Ozkan and Davis 2014), data linking such mechanisms to animal ecology and evolution are mostly lacking.

17.3.3 Decisions

Decisions have two components. First, an individual has to choose which action to execute given its current information. Second, it has to perform the sequence of behavior constituting a given action. Variation in antipredatory behavior among larvae of closely related *Enallagma* damselfly species is a good example for adaptive variation in both components of decision. In eastern North America, the ancestral aquatic habitat of larval *Enallagma* damselflies contained fish as the top predators. Over the past 100,000 years, there have been three independent habitat shifts of *Enallagma* to lakes lacking fish, in which the top predators are larval dragonflies. Behavioral analyses of larvae indicated that:

- in the absence of predators, dragonfly-lake species walk more than fish-lake species;
- dragonfly-lake species have completely lost their ancestral antipredatory response to fish;
- while fish-lake species freeze in response to an approaching predatory fish, dragonfly-lake species attempt to escape in response to an approaching predatory dragonfly.

This difference is adaptive because the larvae are sufficiently fast for escaping dragonflies, but not fish (McPeek 1990; Brown et al. 2000; Stoks et al. 2003). Larvae from the two lake types that were reared from eggs in the laboratory maintained their behavioral differences, indicating that there is a genetic basis to their distinct antipredatory behavior (M. McPeek, personal communication).

17.4 Insect learning

17.4.1 Variation, change, and phenotypic modification

To understand learning, it is helpful to consider, first, behavior with no learning. All organisms encounter variation in the quantity and quality of resources that affect fitness. In terrestrial systems, physical factors including topography, rock and soil types, wind directions and velocities, solar radiation, and precipitation shape the spatial structure of temperatures, and the availability of minerals and water. These, in turn, generate a complex spatial distribution of species, which may vary continuously or be broken into distinct patches differing in the quality and quantity of a given resource, and surrounded by areas lacking that resource. Further daily and seasonal variation in abiotic factors adds temporal variation in organismal activity, productivity, and interactions. This combination of spatial and temporal variation in the availability of abiotic and biotic factors means that an individual's location in time and space can strongly affect its fitness. Hence, individuals in mobile species attempt to optimize their spatial position at any given time in order to secure the ideal abiotic settings, including temperature, humidity, and sunlight, and the best sites containing essential nutrients, providing shelter, or having social and sexual partners (Hills and Dukas 2012). These abiotic and biotic settings may vary in their priorities throughout individuals' different life stages.

The best studied non-learned behavior for nutrient search is chemotaxis for which the molecular mechanisms from nutrient sensing through information processing to moving are thoroughly understood at the molecular level (Figure 17.2a). Because learning is ubiquitous in insects, this chapter focuses first on simpler organisms that cannot learn. Briefly, in bacteria such as *Escherichia coli*, receptors at the cell surface count relevant molecules, including sugar and amino acids, and control the direction of move-

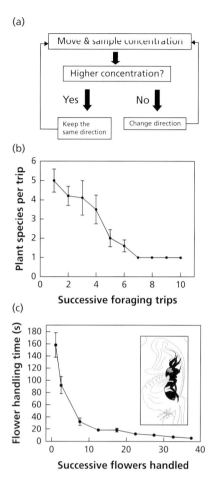

Figure 17.2 (a) All organisms are plastic. Shown here is a schematic of chemotaxis, a ubiquitous plastic mechanism for seeking food (illustrated) and avoiding noxious compounds (not shown). Compared with other types of plasticity, learning involves internal representations of new information, which open a variety of new opportunities. For example, (b) a new forager bumble bee, *Bombus vagans*, encountering a variety of plants in bloom and no competition begins by sampling many of these plants, then gradually restricts her visits to the most rewarding one (data from Heinrich 1979). (c) Bumble bees, *Bombus fervidus*, can also learn new motor skills for handling complex flowers, such as those of monkshood, *Aconitum napellus*, shown in the inset (data and inset from Laverty and Plowright 1988), and the spatial location of their preferred flower patches in relation to their nest (not shown).

From Dukas (2013).

ment of the flagella. If the current direction leads to higher nutrient concentration, the flagellar motors keep spinning in counter-clockwise direction to maintain the same direction. If the present direction is not favourable, the motors switch to clockwise spinning, leading to a change in direction (Koshland 1980; Berg 2003; Wadhams and Armitage 2004; Baker et al. 2006; Eisenbach and Lengeler 2004). Similar plastic mechanisms involving adaptive modulation of directed movement allow a large variety of organisms to locate, then stay in patches containing high levels of desired resources (Hills and Dukas 2012). Bacteria such as *E. coli*, however, can rely on plasticity to do much more than merely follow nutrient gradients. When glucose, which is *E. coli*'s preferred sugar, is unavailable, the bacteria can modulate alternative genetic networks and change their diet to metabolizing lactose. The classic lac operon system (Jacob and Monod 1961; Beckwith and Zipser 1970) is now just one of the many well-studied complex genetic networks that modulate phenotypes.

The basic genetic architecture of all organisms is based on modulation, which is necessary for key life functions, including development, growth, reproduction, achieving optimal internal environment, seeking external resources, and responding to external abiotic and biotic factors. Through the modulation of gene action, physiology, and behavior, all organisms continuously modify their phenotypes. That is, phenotypic modification, or phenotypic plasticity, is an integral feature of life shared by all organisms. The prevalent notion that phenotypic plasticity is an adaptation to some form of environmental change, misrepresents our fundamental biological understanding of organismal life, where plastic phenotypic handling of variation and change is the rule. While phenotypic plasticity occurs in all living things, learning, which is a special type of phenotypic plasticity, is more restricted in its distribution.

17.4.2 Learning as a special type of phenotypic plasticity

Learning involves modulation of neurons resulting in the neuronal representation of new information. That is, while other types of plasticity allow individuals to execute behavior in response to given information based on evolved innate mechanisms, learning is unique because it allows individuals to acquire both newly perceived information and new motor patterns via neuronal modulation. Compared

to other types of plasticity, the added ability to internally represent novel information allows animals to better exploit environmental features unique to certain times and places, respond to a larger variety of features, and to increase their behavioral repertoire. For example, a bee can acquire neuronal representations of the spatial features unique to her nest location, record the spatial location, odour, and colour of her preferred flowers, and learn new motor patterns for optimizing the handling of these flowers (Figure 17.2b, c). In many animal species, including some insects, individuals also gain from learning to identify their parents, neighbours, competitors, potential mates, and offspring (Dukas 2013).

17.4.3 Measuring learning

Subsections 17.4.1 and 17.4.2 make it clear that a common definition for learning as 'change in behavior with experience' is insufficient. Thus, one cannot infer learning based on an individual's change in behavior with experience. On the other hand, measuring learning directly via quantifying a new neuronal representation of information is impractical at this time. The best current compromise is to rely on carefully designed experiments that measure changes in behavior, while including strict controls to exclude non-learning alternatives. For example, experiments should be conducted with observers blind to treatments in order to avoid observer bias. Depending on the type of learning, controls must be included to verify, for instance, that either the presentation of stimuli or environmental states alone do not generate behavioral changes interpreted as learning, or that the passage of time alone does not generate learning-like changes (see later).

Claims for absence of learning in either a certain species or a particular task may also be problematic because they could just reflect either low motivation of the subjects or behavioral deficiency caused by the experimental settings, rather than a genuine inability to learn. Hence, the protocol must also include a control, which tests for some relevant behavioral response in which we expect a significant response by the same subjects. For example, learning in the context of mate choice can influence sexual selection and speciation. In our long-term work on fruit flies, *Drosophila* spp., we repeatedly

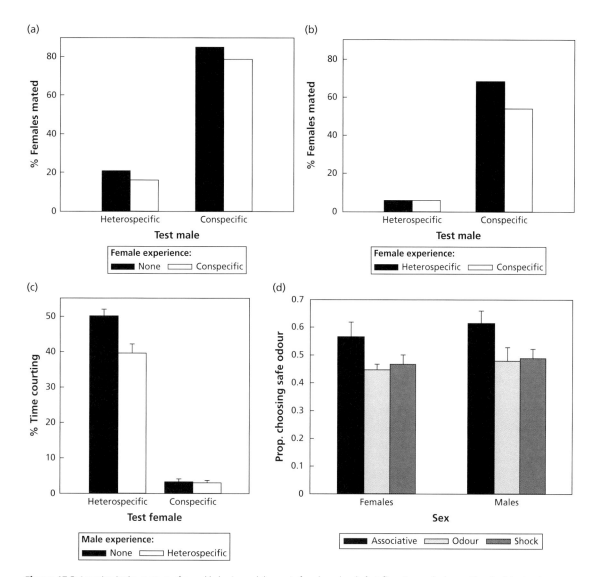

Figure 17.3 Learning in the contexts of sexual behavior and danger in female and male fruit flies, *D. pseudoobscura*. Female did not reduce their receptivity to heterospecific males after experiencing courtship by heterospecific (a) and conspecific (b) males, but males reduced courtship of heterospecific females after experiencing rejection by heterospecific females (c). (d) Unlike the sex difference in learning in the context of sexual behavior, both sexes showed similar, significant associative avoidance learning of odours associated with shock. Both sexes also showed similar random choice of odours after experiencing only odours or only shock.

From Dukas et al. (2012).

documented robust learning in the context of mate choice in males, which could, indeed, influence population divergence (Dukas 2008c). Female learning in the context of mate choice, however, can have stronger effects on incipient speciation than does male learning (Servedio and Dukas 2013). Our tests of female *D. pseudoobscura* revealed no evidence

that early experience with either only conspecific males, or both conspecific and heterospecific males affects females' subsequent receptivity to heterospecific males (Figure 17.3a, b). At the same time, previous findings that male *D. pseudoobscura* that are rejected by heterospecific females later selectively reduce courtship of heterospecific females could readily be replicated (Figure 17.3c). To substantiate these unexpected sex differences in learning, we tested whether females can learn as well as males in a nonsexual context. To this end, we conducted an associative learning test in which blind observers tested male and female learning to avoid odours associated with a mild electric shock. That test also included two controls, one using odours only to assess male and female possible differential responses to odours with experience, and the other using electric shock only to test for male and female possible differential responses to odours after experiencing shock. In that test, we documented similar robust associative learning about shock in both males and females (Figure 17.3d; Dukas et al. 2012).

17.5 Adaptive significance of insect learning

As noted in Section 4.2, one can readily imagine the adaptive advantage of an ability to internally represent novel information (i.e. learning) over merely perceiving and responding to environmental change (i.e. innate behavior). Learning can open up a multitude of new opportunities, including spatial orientation, the acquisition of new motor skills, individual recognition, and associating a variety of cues with resources that either increase or decrease fitness. Such resources include limiting nutrients, shelter, prospective mates, harmful substances, aggressive opponents, and predators. Nevertheless, the adaptive advantage of learning over realistic null models should also be critically tested. There are currently few such tests and one relevant study with rather surprising conclusions will be discussed.

Like all animals, insects face the constant challenge of acquiring a proper balance of essential nutrients from food sources that vary widely in nutritional composition, while minimizing the consumption of harmful ingredients. Deviations from optimal nutritional balance typically decrease growth and reproduction (see Chapter 8). The role of learning in acquiring an optimally balanced diet has been extensively studied in locusts and grasshoppers (Acrididae; Simpson and Raubenheimer 2000). In a controlled laboratory experiment, Dukas and Bernays (2000) assigned grasshoppers, *Schistocerca americana*, into one of two treatments, learning and random (here-

after learning and random grasshoppers), and placed them individually inside cages each containing two synthetic foods and a perch situated next to a heat lamp. One food consisted of a balanced diet, whereas the other was carbohydrate deficient. Because grasshoppers, like most animals, cannot taste complex carbohydrates, there was no perceived difference in the flavour of the two basic foods. However, three types of cues to the food dishes of the learning grasshoppers were provided—taste, colour, and spatial location. These cues remained constant for the duration of the experiment. While providing similar cues to the food dishes of the random grasshoppers, the diet-cue associations were randomly altered twice a day (Figure 17.4a). Consequently, the learning grasshoppers could learn to associate the constant cues with each diet, but the random grasshoppers could not owing to the frequent change in cue-diet association.

As expected, the learning grasshoppers quickly learned to restrict their visits to the nutritionally balanced food, whereas the random grasshoppers kept visiting each food type at similar frequencies (Figure 17.4b). Surprisingly, however, the random grasshoppers gradually increased the proportion of time spent feeding on the balanced diet (Figure 17.4c). While they spent a lower proportion of time on the balanced diet, 87 per cent compared with 99 per cent by the learning grasshoppers, that high proportion suggested that they relied on a non-learning mechanism for food selection. These direct behavioral observations indicated that the random grasshoppers consumed meals of regular duration at the balanced dish, but were quick to abort feeding at the deficient diet. This resulted in longer intervals between meals on the balanced diet. Two probable mechanisms guiding such non-learned behavior are changes in taste-receptor sensitivities based on nutritional needs (Abisgold and Simpson 1988) and changes in behavior based on nutrient sensing in the brain (Domingos et al. 2011). The behavioral differences between the treatments translated into a 20 per cent higher growth rate in the learning grasshoppers than in the random grasshoppers (Figure 17.4d). Throughout the experiment, the learning grasshoppers could restrict their travel to visiting the balanced diet, whereas the random grasshoppers had to keep sampling the two diets

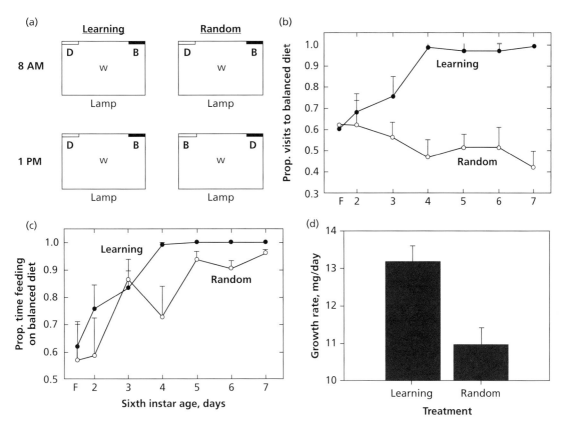

Figure 17.4 (a) The set up used for testing effects of learning on fitness in grasshoppers. Each cage contained a water dish (w) and two dishes, one containing a nutritionally balanced (B) and the other nutritionally deficient (D) food. (b) The average proportion of visits to the dish containing nutritionally balanced food during the first 6 days of feeding. (c) The average proportion of time spent feeding on the nutritionally balanced food during the first 6 days of feeding. Note that 'F' in the X axis legend refers to the first meal observed. (d) The average growth rate of grasshoppers from the learning and random treatments. From Dukas and Bernays (2000). Copyright National Academy of Sciences.

(Figure 17.4b). In nature, the extra travel associated with sampling will likely translate into even higher costs due to predation (Dukas and Bernays 2000).

17.6 Social learning in insects

17.6.1 Social learning: uniqueness and prevalence

Social learning means that a focal individual learns from other individuals, rather than through its own trials and errors. While this seems a miniscule difference, the outcomes of social learning can be dramatically different from those of individual learning. Most notably, in species with no social learning, knowledge acquired via individual learning is lost when individuals die. On the other hand, in species with social learning, information can be transmitted among individuals and across generations. Hence, social learning allows for useful knowledge to accumulate within a population indefinitely. Another distinguishing feature of social learning is that it is faster and perhaps safer than individual learning because most individuals can bypass the trial and error stage through learning from experienced individuals.

How prevalent is social learning among insects? Traditionally, social learning has been studied primarily in mammals and birds, while relevant cases in social insects were classified under distinct categories such as communication. Research on social learning in non-social insects is probably limited to two taxa (see below). Given the limited research effort, it is premature to state how common social learning is in insects. Nevertheless, to help direct future research, we can predict under what conditions social learning can be most beneficial over individual learning and, hence, is most likely to evolve.

Social learning can be most advantageous when there is large variation in knowledge and frequent interactions among individuals. Features associated with such conditions are overlapping generations and gregarious living (Dukas 2010). A special case combining both features is parental care. Although much of the research on social learning has been conducted on vertebrate species with parental care, no insect with parental care has been examined. Finally, another feature that can promote the evolution of social learning is kinship. This is because experienced individuals are more likely to share knowledge with kin when such sharing is costly. Information sharing can be costly if it either takes extra time or reduces the sharer's gain from a resource. Frequent kin interactions occur in species with parental care discussed above as well as in social insects, which indeed provide the most celebrated cases of social learning in insects.

17.6.2 Social learning in social insects

As discussed in Section 17.4.1, the large spatial and temporal variation in the availability of resources means that individuals have to optimally balance the time they spend searching for and exploiting the best locally available resources. For example, the area within a short flight distance from a bee's nest contains numerous plant patches that vary widely in blooming time as well as the quality and quantity of floral rewards. When naïve bumblebees (*Bombus vagans*) initiate foraging, they primarily rely on individual learning for choosing flowers. The bumblebees sample a variety of flowering plants before specializing on the most profitable flowers (Figure 17.2b; Heinrich 1979). That is, each bee individually learns via sampling which flowers she should visit.

Unlike bumblebees, honeybees, *Apis mellifera*, rely on sophisticated social learning to guide their flower choice. When naïve honeybees start foraging, they typically observe the waggle dances of experienced foragers (models), which are conducted on the vertical comb inside the dark nest cavity. The waggle portion of the dance involves the bee moving in a constant direction, while waggling her body sideways and vibrating her wings. At the end of each waggle run, the bee circles back to her starting point, alternating between clockwise and counter clockwise turns, such that each two successive rounds create a figure eight. The angle of the waggle run relative to the upward direction indicates the angle of the flower patch relative to the current sun's position in the sky, and the waggle duration indicates the distance. Finally, the overall number of waggle runs is positively correlated with relative food quality (von Frisch 1967; Seeley, 1996). A variety of experiments, as well as observations using harmonic radar (Riley et al. 2005) indicate that observer bees indeed learn the direction and distance information encoded in the waggle dance and rely on that knowledge to arrive in the general vicinity of the flowers. The bees are further assisted by olfactory and visual cues from the flowers, and perhaps also by directly following model bees and pheromones. Honeybees also rely on waggle dances to inform hive mates about other resources, as well as new potential nest sites when the colony swarms (Seeley 2010). While the waggle dance is unique to honeybees, social insects have a variety of means for conveying information about relevant resources to nestmates. These include odour trails, pheromonal marking of resources, direct following of successful foragers, and alerting nestmates to profitable resource which they can then find based on odour cues (Wilson 1971; Holldobler and Wilson 1990).

In addition to social learning via communication, individuals can learn by observation. Worden and Papaj (2005) placed observer bumblebees, *Bombus impatiens*, one at a time in an observation box adjacent to a foraging arena containing three green and three orange artificial flowers. The experiment consisted of three treatments, model bees foraging on green flowers, model bees foraging on orange flowers or no model bees. After removing the model bees and replacing the flowers with new, non-rewarding flowers of identical colours, each observer bee was allowed to make six visits to the flowers. The observer bees made more visits to the green flowers after observing model bees visiting green flowers

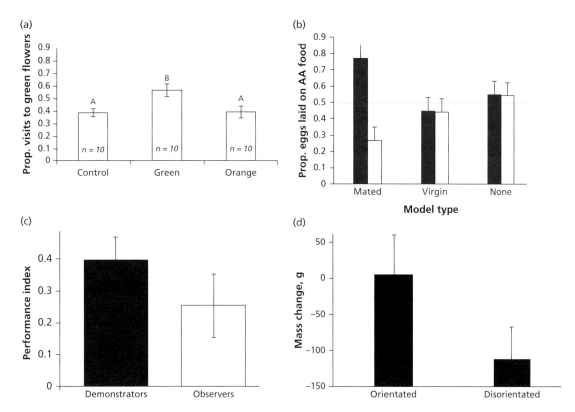

Figure 17.5 Social learning. (a) The proportion of landings on green flowers by focal bees that observed live foraging bee models visiting either green or orange flowers. Control focal bees observed no model bees. From Worden and Papaj (2005). (b) The average proportion of eggs laid on AA-flavoured food by focal females trained with mated models, virgin models or alone on food flavoured with either AA (amyl acetate, black bars) or BA (benzaldehyde, white bars). Only the focals trained with mated models showed significant preference for the flavoured food they had experienced. From Sarin and Dukas (2009). (c) The performance index of demonstrators and observers tested separately in a two-food choice. A value of 1 indicates maximal performance and 0 indicates random choice. Prior to the test, the observers had been housed with demonstrators previously conditioned to prefer sweet over bitter food. Data from Battesti et al. (2012). (d) The average mass change in small bee hives during periods in which bees could perform oriented waggle dances and during periods of disoriented waggle dances. Data from Sherman and Visscher (2002).

compared with the other two treatments. Observing model bees visiting orange flowers, however, did not lead to the observer bees visiting more orange flowers than the control observer bees (Figure 17.5a). That is, watching model bees preferring to visit one flower type over another subsequently led observer bees to show a weak preference for that flower type. While it may not be surprising that social insects attend to social information, the prevalence of social learning in non-social insects may be limited.

17.6.3 Social learning in non-social insects

In 2008, we initiated a research project aiming to elucidate the evolution of social learning in insects. Based on the considerations in Section 6.1, we chose two non-social insect taxa, locusts, *Locusta migratoria* and *Schistocerca gregaria*, and fruit flies, *D. melanogaster*, because they have overlapping generations and frequent interactions owing to their tendencies to aggregate. In both locust species, we found robust individual learning. However, even though we tried a large variety of protocols, we consistently found no evidence for social learning (Dukas and Simpson 2009; Lancet and Dukas 2012). Hence, locusts might represent a basal insect taxon that is gregarious, but lacking the requisite cognitive traits necessary for social information use.

In our work with fruit flies (Sarin and Dukas 2009), we found that focal females that experienced novel food together with mated females, which had laid eggs on that food, subsequently showed a stronger preference for laying eggs on that food over another novel food compared with focal females that experienced the food alone. We observed no social learning, however, when focal females experienced food with potentially more ambiguous social information provided by the presence of either virgin models or the **aggregation pheromone**, cis vaccenyl acetate (Figure 17.5b).

Work on fruit flies in another laboratory (Battesti et al. 2012) has shown that odour cues on demonstrator flies are sufficient for social transmission of information. The experiment involved three phases. In the conditioning phase, eight demonstrator flies experienced regular food with one flavour and a bitter food with another. Flies under such settings prefer the flavour associated with the regular food. In the social transmission phase, the eight demonstrators shared an arena containing no food with four observers not experienced with the flavoured foods. Finally, in the test phase, the four observers were transferred into a new arena containing two regular foods each with a flavour used in the conditioning phase. The observers showed significant preference for laying eggs on the food flavoured with the flavour associated with the standard food in the conditioning phase (Figure 17.5c). That is, even though the observers had never encountered the flavoured standard and bitter foods before, they biased their egg laying choice based on cues on the demonstrators.

17.6.4 Adaptive significance of insect social learning

I am aware of only a single study that has linked social learning in an insect to a fitness surrogate. In honey bees, the waggle dances are done on the vertical comb inside the dark nest cavity and upward represents the sun direction (Section 6.2). When on horizontal combs provided with diffused light, bees still perform waggle dances, but they are disorientated, meaning that they do not provide correct direction and distance information. Observers, however, can still learn from the dances about the presence of an attractive floral source and its odour. Bees can perform orientated waggle dances on horizontal combs if they are provided with a directional light that simulates the sun direction. Sherman and Visscher (2002) relied on this knowledge to test for

the effect of orientated dances on mass gain by small honey bees colonies. They compared mass gain over 8 months in four colonies. At any given time, two colonies could perform orientated dances and two colonies performed disorientated dances, with treatments alternating every 11 days. Mass gain during periods with orientated dances was significantly higher than that during disorientated dances (Figure 17.5d). That is, the social information about direction and distance to food was associated with significantly higher mass gain, which is positively associated with colony fitness.

As with individual learning (Section 17.5), the current experimental data indicating adaptive significance of social learning in insects is limited. Nonetheless, it is likely that social learning contributes significantly to fitness in many social insects. Given the scarcity of data about social learning in non-social insects (Section 17.6.3), it is premature to assess its probable significance.

17.7 Conclusions and prospects

For three reasons, we have much more to learn about insect cognition. First, for much of the history of science, biologists have grossly underestimated insect cognitive abilities. Second, although insect taxonomic diversity is astounding, cognition has been examined in only very few taxa. Third, it is likely that the well-documented diversity in a variety of readily observable traits, such as shape and colour, is associated with comparable myriad cognitive specializations. Nevertheless, our current knowledge allows us to make a few fundamental conclusions. First, insects possess sophisticated cognitive abilities, which sometimes surpass comparable human abilities. Examples include night vision in nocturnal insects (Section 17.3.1), superb navigational abilities in central place foragers, such as ants and bees (e.g., Collett and Collett 2002), and exceptional motor skills (e.g., Muijres et al. 2014). Second, most insects have excellent learning and memory abilities. These fundamental cognitive traits allow them to learn about the spatial locations of and cues associated with preferred resources, such as food, shelter and prospective mates, and to distinguish among individuals they have interacted with. Third, insects are good at integrating experience and newly acquired information to reach optimal decisions in contexts, such as nest selection and mate choice (e.g., Dukas and Dukas 2012; Sasaki and Pratt 2013).

One can readily identify gaps in our current knowledge of insect cognition and there are a few promising areas for future research. First, as noted in Section 17.3, it is important that we investigate cognitive traits within an evolutionary framework. Given the paucity of data, we will gain from acquiring more information on heritable variation in insect cognitive traits and its association with fitness. Furthermore, data on the evolution of insect cognitive traits in the wild will be highly illuminating.

Second, the vast majority of cognitive experiments in insects are brief even relative to insects' short lifespan. This inadvertently reinforces the traditional notion that insects' cognition is limited. The little data available, however, suggest that insects show a gradual increase in performance with long-term experience (Dukas and Visscher 1994; Dukas 2008d), which is similar to expertise development in long-lived vertebrates (Wooler et al. 1990; Ericsson et al. 2006). Hence, to truly appreciate insect cognition, we should develop further experimental protocols that allow us to assess the performance of expert individuals. Furthermore, doing so in a rich model system such as fruit flies will help us understand mechanisms of expertise development relevant to all species including humans.

Third, while there has recently been increased interest in studying social learning in insects, we still know little about the prevalence and importance of social learning in non-social insects. Note that, although I use the term 'social insects' to refer to the classical social insects, which include social hymenoptera and termites (Wilson 1971), one can adopt a broader view accommodating many more insect taxa (Prokopy and Roitberg 2001; Costa 2006; see also Chapter 15). Ultimately, it is sensible to define social behavior as interactions among conspecifics, and sociability as the tendency to engage in friendly activities with conspecifics. These inclusive definitions open up exciting possibilities for research on social behavior and social information use in a large variety of insects. Because parental care is a feature predicted to promote the evolution of social learning (Section 17.6.1), well studied insect species with parental care such as the German cockroach, *Blattella germanica*, and European earwig, *Forficula*

auricularia (Costa 2006) can be especially promising for such future work.

Finally, insects have recently been adopted for research on **pain** and **emotion**, and there are convincing reasons for pursuing and expanding the scope of such exciting work. Emotion is an internal state that guides behavioral decisions and all the available evidence indicates that it is a universal cognitive feature. Examples for research on insect emotion include pessimistic and optimistic **cognitive biases** in bees (Bateson et al. 2011; Perry et al. 2016) and fear and **anxiety** in fruit flies (Gibson et al. 2015; Mohammad et al. 2016). Because pain and emotion guide decisions, we must study them if we wish to advance our understanding of insect cognition. Furthermore, work on model systems, such as fruit flies, can further our general knowledge of pain, emotion, and their disorders in all species including humans.

Acknowledgements

The author would like to thank L. Dukas for comments on the manuscript, and the Natural Sciences and Engineering Research Council of Canada, Canada Foundation for Innovation and Ontario Ministry of Research and Innovation for funding.

References

Abisgold, J.D., and Simpson, S.J. (1988). The effect of dietary-protein levels and hemolymph composition on the sensitivity of the maxillary palp chemoreceptors of locusts. *Journal of Experimental Biology*, **135**, 215.

Alloway, T.M. (1972). Learning in insects except Apoidea. In: W.C. Corning, J. A. Dyal, and A. O. D. Willow, (Eds) *Invertebrate Learning*, pp. 131–71. New York: Plenum.

Anderson, J.R., and Schooler, L.J. (1991). Reflections of the environment in memory. *Psychological Sciences*, **2**, 396.

Baker, M.D., Wolanin, P.M., and Stock, J.B. (2006). Signal transduction in bacterial chemotaxis.' *Bioessays* 28: p. 9.

Bateson, M., Desire, S., Gartside, S.E., and Wright, G.A. (2011). Agitated honeybees exhibit pessimistic cognitive biases. *Current Biology*, **21**, 1070.

Battesti, M., Moreno, C., Joly, D., and Mery, F. (2012). Spread of social information and dynamics of social transmission within *Drosophila* groups. *Current Biology*, **22**, 309–13.

Beckwith, J.R. and Zipser, D. (1970). *The Lactose Operon*. Cold Spring Harbor, Cold Spring Harbor Laboratory.

Berg, H.C. (2003). The rotary motor of bacterial flagella. *Annual Review of Biochemistry*, **72**, 19.

Bernays, E.A. (2001). Neural limitations in phytophagous insects: implications for diet breadth and evolution of host affiliation. *Annual Review of Entomology*, **46**, 703.

Bier, E. (2005). *Drosophila*, the golden bug, emerges as a tool for human genetics. *Nature Reviews Genetics*, **6**, 9.

Broadbent, D.E. (1965). Information processing in the nervous system. *Science*, **150**, 457.

Brown, J.M., McPeek, M.A., and May, M.L. (2000). A phylogenetic perspective on habitat shifts and diversity in the North American *Enallagma* damselflies. *Systematic Biology*, **49**, 697.

Chittka, L., and Niven, J. (2009). Are bigger brains better? *Current Biology*, **19**, R995.

Collett, T.S., and Collett, M. (2002). Memory use in insect visual navigation. *Nature Reviews Neuroscience*, **3**, 542.

Costa, J.T. (2006). *The other insect societies*. Harvard University Press, Cambridge, MA.

Cronin, T.W., Johnsen, S., Marshall, N.J., and Warrant, E.J. (2014). *Visual ecology* Princeton University Press, Princeton, NJ.

Darwin, C. (1874). *The Descent of Man and Selection in Relation to Sex*. Murray, London.

Davis, R.L. (2005). Olfactory memory formation in *Drosophila*: from molecular to systems neuroscience. *Annual Review of Neuroscience*, **28**, 275.

Domingos, A.I., Vaynshteyn, J., Voss, H.U., et al. (2011). Leptin regulates the reward value of nutrient. *Nature Neuroscience*, **14**, 1562.

Dubnau, J., and Chiang, A-S. (2013). Systems memory consolidation in *Drosophila*. *Current Opinion in Neurobiology*, **23**, 84.

Dubnau, J., and Tully, T. (1998). Gene discovery in *Drosophila*: new insights for learning and memory. *Annual Review of Neuroscience*, **21**, 407.

Dukas, R. (Ed.) (1998a). *Cognitive Ecology: The Evolutionary Ecology of Information Processing and Decision Making*. University of Chicago Press, Chicago, IL.

Dukas, R. (1998b). Constraints on information processing and their effects on behavior. In: R. Dukas (Ed.) *Cognitive Ecology*, pp. 89–127. University of Chicago Press, Chicago, IL.

Dukas, R. (1999). Costs of memory: ideas and predictions. *Journal of Theoretical Biology*, **197**, 41.

Dukas, R. (2002). Behavioural and ecological consequences of limited attention. *Philosophical Transactions of the Royal Society of London B*, **357**, 1539.

Dukas, R. (2004). Evolutionary biology of animal cognition. *Annual Review of Ecology Evolution and Systematics*, **35**, 347.

Dukas, R. (2008a). Bee senescence in the wild. *Insectes Sociaux*, **55**, 252.

Dukas, R. (2008b). Evolutionary biology of insect learning. *Annual Review of Entomology*, **53**, 145.

Dukas, R. (2008c). Learning decreases heterospecific courtship and mating in fruit flies. *Biology Letters*, **4**, 645.

Dukas, R. (2008d). Life history of learning—short and long term performance curves of honeybees in settings that minimize the role of learning. *Animal Behaviour*, **75**, 1125.

Dukas, R. (2010). Insect social learning. In: M. Breed and J. Moore (Eds), *Encyclopedia of Animal Behavior*, pp. 176–9. Academic Press, Oxford.

Dukas, R. (2013). Effects of learning on evolution: robustness, innovation and speciation. *Animal Behaviour*, **85**, 1023.

Dukas, R., and Bernays, E.A. (2000). Learning improves growth rate in grasshoppers. *Proceedings of the National Academy of Sciences USA*, **97**, 2637.

Dukas, R., and Dukas, L. (2012). Learning about prospective mates in male fruit flies: effects of acceptance and rejection. *Animal Behaviour* 84: p. 1427.

Dukas, R., Durisko, Z., and Dukas, L. (2012). Learning in the context of sexual behavior and danger in female and male *Drosophila pseudoobscura*. *Animal Behaviour*, **83**, 95.

Dukas, R., and Simpson, S.J. (2009). Locust show rapid individual learning but no social learning about food. *Animal Behaviour*, **78**, 307.

Dukas, R., and Visscher, P.K. (1994). Lifetime learning by foraging honey bees. *Animal Behaviour*, **48**, 1007.

Dunlap, A.S., and Stephens, D.W. (2014). Experimental evolution of prepared learning. *Proceedings of the National Academy of Sciences*, **111**, 11750.

Eisenbach, M., and Lengeler, J.W. (2004). *Chemotaxis*. Imperial College Press, London.

Ericsson, K.A., Charness, N., Feltovich, P.J., and Hoffman, R.R. (2006). *The Cambridge Handbook of Expertise and Expert Performance*. Cambridge University Press, Cambridge.

Fabre, J-H., Teixeira De Mattos, A., and Miall, B. (1918). *The Wonders of Instinct*. The Century Co., New York, NY.

Garcia, J., and Koelling, R.A. (1966). Relation of cue to consequence in avoidance learning. *Psychonomic Science*, **4**, 123.

Gibson, W.T., Gonzalez, C.R., Fernandez, C., et al. (2015). Behavioral responses to a repetitive visual threat stimulus express a persistent state of defensive arousal in *Drosophila*. *Current Biology*, **25**, 1401.

Giurfa, M. (2013). Cognition with few neurons: higher-order learning in insects. *Trends in Neurosciences*, **36**, 285.

Guven-Ozkan, T., and Davis, R.L. (2014). Functional neuroanatomy of *Drosophila* olfactory memory formation. *Learning and Memory*, **21**, 519.

Heinrich, B. (1979). Majoring and minoring by foraging bumblebees, *Bombus vagance*: an experimental analysis. *Ecology*, **60**, 245.

Hills, T.T., and Dukas, R. (2012). The evolution of cognitive search. In: P. M. Todd, T.T. Hills, and T.W. Robbins. (Eds) *Cognitive Search: Evolution Algorithms and the Brain*, pp. 11–24. MIT Press, Cambridge, MA.

Holldobler, B., and Wilson, E. O. (1990). *The Ants*. Harvard University Press, Cambridge, MA.

Inlow, J.K., and Restifo, L.L. (2004). Molecular and comparative genetics of mental retardation. *Genetics*, **166**, 835.

Jacob, F., and Monod, J. (1961). Genetic regulatory mechanisms in the synthesis of proteins. *Journal of Molecular Biology*, **3**, 318–56.

Kelber, A., Balkenius, A., and Warrant, E.J. (2002). Scotopic colour vision in nocturnal hawkmoths. *Nature*, **419**, 922–5.

Kelber, A., & Roth, L.S. (2006). Nocturnal colour vision—not as rare as we might think. *Journal of Experimental Biology*, **209**(5), 781–8.

Koshland, D. (1980). *Bacterial Chemotaxis as a Model Behavioral System*. Raven Press, New York, NY.

Kruidhof, H.M., Pashalidou, F.G., Fatouros, N.E., et al. (2012). Reward value determines memory consolidation in parasitic wasps. *PLoS One*, **7**, e39615.

Lancet, Y., and Dukas, R. (2012). Socially influenced behaviour and learning in locusts. *Ethology*, **118**, 302.

Land, M.F., and Nilsson, D-E. (2012). *Animal Eyes*. Oxford University Press, Oxford.

Laverty, T.M., and Plowright, R.C. (1988). Flower handling by bumblebees: a comparison of specialists and generalists. *Animal Behaviour*, **36**(3), 733–40.

Mayr, E. (1974). Behavior programs and evolutionary strategies. *American Scientist*, **62**, 650.

McPeek, M.A. (1990). Behavioral differences between *Enallagma* species (Odonata) influencing differential vulnerability to predators. *Ecology*, **71**, 1714.

Menzel, R. (1985). Learning in honey bees in an ecological and behavioral context. In: B. Holldobler and M. Lindauer (Eds), *Experimental Behavioral Ecology*, pp. 55–74. Stuttgart: Verlag.

Mohammad, F., Aryal, S., Ho, J., et al. (2016). Ancient anxiety pathways influence *Drosophila* defense behaviors. *Current Biology*, **26**, 981.

Muijres, F.T., Elzinga, M.J., Melis, J.M., and Dickinson, M.H. (2014). Flies evade looming targets by executing rapid visually directed banked turns. *Science*, **344**, 172.

O'Kane, C.J. (2011). *Drosophila* as a model organism for the study of neuropsychiatric disorders. In: M.A. Geyer, B.A. Ellenbroek, C.A. Marsden, T.R.E. Barnes, and S.L. Andersen (Eds), *Current Topics in Behavioral Neurosciences*, pp. 37–60. Springer, Berlin.

Papaj, D.R., and Prokopy, R.J. (1989). Ecological and evolutionary aspects of learning in phytophagous insects. *Annual Review of Entomology*, **34**, 315.

Perry, C.J., Baciadonna, L., and Chittka, L. (2016). Unexpected rewards induce dopamine-dependent positive emotion–like state changes in bumblebees. *Science*, **353**, 1529.

Plaçais, P-Y. and Preat, T. (2013). To favor survival under food shortage, the brain disables costly memory. *Science*, **339**, 440.

Placais, P-Y., Trannoy, S., Isabel, G., et al. (2012). Slow oscillations in two pairs of dopaminergic neurons gate long-term memory formation in *Drosophila*. *Nature Neuroscience*, **15**, 592.

Platt, M.L. (2002). Neural correlates of decisions. *Current Opinions in Neurobiology*, **12**, 141.

Prokopy, R.J., and Roitberg, B.D. (2001). Joining and avoidance behavior in nonsocial insects. *Annual Review of Entomology*, **46**, 631.

Riley, J.R., Greggers, U., Smith, A.D., Reynolds, D.R., and Menzel, R. (2005). The flight paths of honeybees recruited by the waggle dance. *Nature*, **435**, 205.

Rolls, E.T. (2014). *Emotion and Decision-making Explained*. Oxford University Press, Oxford.

Sarin, S., and Dukas, R. (2009). Social learning about egg laying substrates in fruit flies. *Proceedings of the Royal Society of London B: Biological Sciences*, **276**, 4323.

Sasaki, T., and Pratt, S.C. (2013). Ants learn to rely on more informative attributes during decision-making. *Biology Letters*, **9**, 20130667.

Seeley, T.D. (1996). *The Wisdom of the Hive*. Harvard University Press, Cambridge, MA.

Seeley, T.D. (2010). *Honeybee Democracy*. Princeton University Press, Princeton, NJ.

Servedio, M., and Dukas, R. (2013). Effects on population divergence of within-generational learning about prospective mates. *Evolution*, **67**, 2363.

Sherman, G., and Visscher, P.K. (2002). Honeybee colonies achieve fitness through dancing. *Nature* 419: p. 920.

Simpson, S.J., and Raubenheimer, D. (2000). The hungry locust. *Advances in the Study of Behavior*, **29**, 1.

Staddon, J.E.R. (1983). *Adaptive Behavior and Learning*. Cambridge University Press, Cambridge, MA.

Stöckl, A.L., O'Carroll, D.C., and Warrant, E.J. (2016). Neural summation in the hawkmoth visual system extends the limits of vision in dim light. *Current Biology*, **26**, 821.

Stoks, R., McPeek, M.A., and Mitchell, J.L. (2003). Evolution of prey behavior in response to changes in predation regime: damselflies in fish and dragonfly lakes. *Evolution*, **57**, 574.

Tinbergen, N. (1951). *The Study of Instinct*. Oxford University Press, Oxford.

Vet, L.E.M., and Dicke, M. (1992). Ecology of infochemical use by natural enemies in a tritrophic context. *Annual Review of Entomology*, **37**, 141.

Von Frisch, K. (1967). *The Dance Language and Orientation of Bees*. Harvard University Press, Cambridge, MA.

Wadhams, G.H., and Armitage, J.P. (2004). Making sense of it all: bacterial chemotaxis. *Nature Review Molecular and Cellular Biology*, **5**, 1024.

Warrant, E., and Dacke, M. (2011). Vision and visual navigation in nocturnal insects. *Annual Review of Entomology*, **56**, 239.

Wilson, E.O. (1971). *The Insect Societies*. Harvard University Press, Cambridge, MA.

Wooler, R.D., Bradley, J.S., Skira, I.J., and Serventy, D.L. (1990). Reproductive success of short-tailed shearwater *Puffinus tenuirostris* in relation to their age and breeding experience. *Journal of Animal Ecology*, **59**, 161.

Worden, B.D., and Papaj, D.R. (2005). Flower choice copying in bumblebees. *Biology Letters*, **1**, 504.

CHAPTER 18

The influence of parasites

Pedro F. Vale[1], Jonathon A. Siva-Jothy[1], André Morrill[2], and Mark R. Forbes[2]

[1] *Institute of Evolutionary Biology, School of Biological Sciences, University of Edinburgh, Scotland, UK*
[2] *Department of Biology, Carleton University, Ottawa, Canada*

18.1 Background

It is estimated that well over 50 per cent of extant species have a parasitic lifestyle, for at least some stage of their life cycle (Schmid-Hempel 2011). For example, parasitism is thought to have evolved at least once independently in ten of the twenty-one extant insect orders (Weinstein and Kuris 2016). It is perhaps not surprising that infection is widespread among host species from all taxa of life, and hosts have evolved a number of physiological, immunological, and behavioral responses to pathogens and parasites (Schmid-Hempel 2011). This chapter is concerned with behavioral responses to actual or potential infection by pathogens and parasites using insects as model hosts, although instances are highlighted where the players also include insect species that are parasites or parasitoids.

Insects and their pathogens and parasites serve as ideal systems to investigate questions at the interface of infection and behavior. Questions such as how does behavior affect likelihood of infection? Or, is a novel or stereotypical behavior expressed following infection interpretable as a host adaptation, parasite adaptation, or by-product of infection not particularly beneficial to either the parasite or host individual? Insects are ideal

hosts because we know much about disciplines related to their infection and behavioral ecology, including insect natural history, genetics, physiology, immunology, and developmental biology. We generally know a great deal about insect biology because insects are important pests of agricultural crops (see Chapter 20), important vectors of some of the most deadly human diseases (see Chapter 21) and/or particularly amenable to laboratory rearing and protocols. Often our interpretations of insect behavioral response to infection rely on insight gleaned from those other disciplines. For example, consider that there are genes for behaviors and immune expression, and genetic variation in either could explain phenotypic variation in behavior. This is especially important if that phenotypic variation in behavior influences variation in infection risk or costs of infection.

Reviews are defined by the phenomena they include as well as those they ignore. In writing a chapter like this, there are literally hundreds of general observational and specific experimental studies from which to illustrate key concepts in the study of parasitism and host behavior, made much less daunting by the taxonomic focus on insects. We have focused further on some of the more iconic

Vale, P. F., Siva-Jothy, J. A., Morrill, A., and Forbes, M. R., *The influence of parasites*. In: *Insect Behavior: From mechanisms to ecological and evolutionary consequences.* Edited by Alex Córdoba-Aguilar, Daniel González-Tokman, and Isaac González-Santoyo: Oxford University Press (2018). © Oxford University Press.
DOI: 10.1093/oso/9780198797500.003.0018

and recent examples, where interpretations of insect behavioral responses to pathogens and parasites are fairly well understood because the systems are particularly amenable to investigation and interpretation. Although these tractable species associations for study may be seen as a non-random subset of nature, they provide examples of what evolved or plastic host responses are possible. One aspect that is largely ignored herein is what effect co-infecting pathogens and parasites have on insect behavioral responses to focal species. The fact that insect pathogens and parasites often have several possible host species for a given life history (infective) stage is also largely ignored (Rigaud et al. 2010). We fully recognize that these aspects might prove important in determining the types or magnitudes of behavior expressed, but there is still much to discuss in the absence of treatment of these phenomena.

This chapter divides behavior conceptually into two broad categories—those that occur before infection (possibly following encounter, or contact with pathogens and parasites or their cues), and those behaviors that occur following infection (see Box 18.1a). Avoiding infection is the first line of defence and is known to occur in a broad range of taxa, being reasonably well-studied for insect hosts. The evidence that host insects detect and discriminate between clean and potentially infectious environments and whether avoidance behavior is context-dependent is reviewed herein, as well as the importance of this avoidance for the consequences of infection. A second broad line of inquiry concerns behavioral responses to infection. Behavioral 'responses' following infection are widely reported among animals, and can be classified more specifically into the following:

- **sickness behaviors** that benefit the host by conserving energetic resources during infection;
- **host manipulation** by parasites that enhances parasite survival or transmission;
- a by-product of pathogenicity that does not necessarily benefit the host or the parasite.

The second section of this chapter reviews the evidence for adaptive 'sickness' behavior of hosts, whereas the third part of this chapter illustrates key concepts involved in the demonstration and consequences of parasitic manipulation of insect host behavior.

Some of the latest techniques used to investigate insect behavior in relation to infection are also discussed (Box 18.2). In addition to underscoring useful recording and computational techniques in the study and analysis of insect behavior, another common theme shared by the types of studies highlighted is the level of behavioral explanation being sought. Here, we rely on Tinbergen's (1963) four levels of explanation in behavior (Box 18.1b). Most of the work highlighted concerns the expected fitness value to the host or to the parasite of the host's behavioral response being investigated, and only some work concerns the mechanism or causation of behavioral change of infected insects. Box 18.3 is devoted to what is known about the neuroendocrine-immune axis in insects and its mechanistic relation to behavioral expression and immunity. Considerably less time will be spent on what is known about the evolutionary history and developmental trajectory of insect behavior in the context of infection, although both concepts are introduced briefly. The Chapter finishes by highlighting, for each section, tractable yet unresolved questions and issues whose consideration will help inform research for students of insect behavior in relation to infection by pathogens and parasites.

18.2 Infection avoidance behavior in insects

The first line of host defence is **infection avoidance** behavior that prevent parasites from infecting hosts. This **behavioral immunity** can be hugely effective because it results in hosts avoiding the costs of parasite **virulence**, but also because hosts avoid potential negative consequences of immune deployment, such as immunopathology or costly investment in immune mechanisms (Curtis 2014). Avoidance behaviors rely on sensory systems detecting parasites or their cues, mechanisms that integrate this sensory information, and effector systems that, when activated, reduce host–parasite contact. This section discusses insect avoidance mechanisms with respect to the source of infection they avoid and/or the aspect of host ecology they affect.

Box 18.1 Guiding principles for the study of insect behavior in relation to parasitism

Conceptual framework

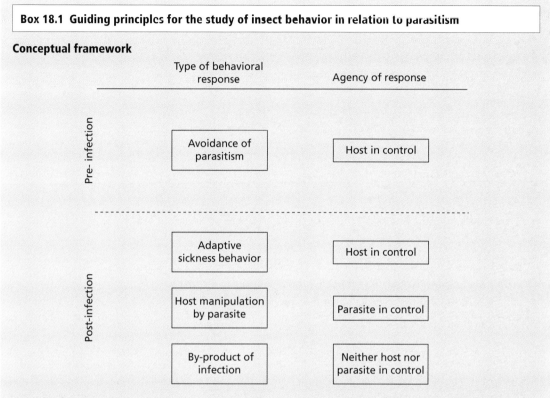

Box 18.1a Categories of behavioral responses to parasitism with their timing in relation to infection and an indication of whether the host or the parasite is in control of the response.

Box 18.1b Tinbergen's four levels of explanation

Nikolaas Tinbergen outlined four modes of questioning animal behaviors as a means of categorizing approaches to their biological explanation (Tinbergen 1963). 'Tinbergen's four questions' summarize the 'lenses' through which behaviors can be viewed, providing four comprehensive and non-exclusive perspectives applicable to any behavioral phenomenon. These levels of explanation are:

- *Biological mechanism (causation)*: the immediate physiological cause(s) of the observed behavior. Separate from any consideration of the behavior's function or its adaptive significance, one might ask what mechanism brings about its expression [e.g., what sorts of molecules are produced to influence the functions of a host's central nervous system during manipulation by a parasite? (Biron et al. 2005)].
- *Function (survival value)*: what advantage(s) is (are) derived from a particular behavior. For example, an insect might increase its grooming as a response to encounter-

ing a parasite infective stage because this reduces the probability of the parasite successfully establishing an infection (Gaugler et al. 1994). In the study of behavioral changes resulting from parasitism, the value of the host behavior may be studied from the host's or the parasite's perspective. Indeed, identifying whether the parasite significantly benefits from the behavior is the crux of differentiating true host manipulation from infection-induced behavioral changes more generally.

- *Evolution (phylogeny)*: the explanation of a behavioral phenomenon from consideration of the organism's evolutionary history. Variation in a behavior expressed by different host species resulting from infection by the same manipulating parasite species may be explained by the degree of phylogenetic relatedness of those hosts (e.g., Malfi et al. 2014). Why individuals of a given species express some particular behavior can also be answered at the level of evolution through addressing how that behavior arose over evolutionary time, and due to what pressures. Selective pressures that favoured the

(continued)

Box 18.1 *Continued*

behavior originally may no longer be relevant, and current survival benefits resulting from the behavior may be relatively new from an evolutionary perspective; therefore, levels of explanation based on function and evolution must be properly distinguished.

- *Ontogeny (development)*: how the observed behavior develops during an individual's lifetime, or is otherwise influenced by the organism's development over the course of its life. For example, one could imagine a scenario wherein an insect's increased exposure to ectoparasites at an earlier life stage increases that individual's propensity to groom at a much later life stage.

Tinbergen's four questions can be divided into two categories depending on whether they deal with 'ultimate' explanations, reasons relating to the purpose of the trait ('why' questions; levels of evolution and function); or whether they address 'proximate' explanations, reasons relating to mechanisms underlying the trait's expression acting within the lifetime of the organism in question ('how' questions; levels of biological mechanism and ontogeny; Klopfer and Hailman 1972). Explanations can alternatively be viewed as either relating to changes occurring over time and to developmental processes (evolution and ontogeny) or, in contrast, relating to more immediate, 'synchronic' perspectives (biological mechanism and function; Bateson and Laland 2013).

Box 18.2 Techniques to study insect behavioral responses to infection

The ability to observe, record, and quantify animal behavior is a considerable challenge in many biological fields. The first challenge arises from the very definition of behavior, which will vary according to the focus of the study (locomotion, courtship, aggression, feeding, etc.) and will therefore require distinct methods and techniques for accurate quantification. The second challenge arises from the inherent complexity of behavioral traits. Historically, measuring behavior has required either laborious and time-consuming manual descriptions of behaviors on a limited number of individuals, or has relied on assays that only allow quantification of a simplified aspect of a more complex behavior. While these approaches have advanced the study of behavior, they are prone to inevitable logistic limitations and inherent biases. Here, we briefly describe recent advances in the measurement of insect behavioral responses to infection, with a focus on methods that allow automated, high throughput quantification of individual behaviors. While many of these techniques have been developed using the fruit fly (*D. melanogaster*), they are applicable to many insect species.

Measuring locomotor activity in insects— Trikinetics® Activity Monitor

One of the most popular instruments employed to study activity levels in insects is the Trikinetics® Drosophila Activity Monitor (DAM). The system works by placing insects inside tubes (individually) or vials (in groups) within an activity monitor, and activity is recorded each time active insects break an infrared (IR) beam within the midpoint of each tube (Pfeiffenberger et al. 2010). Its ease of use, relative

affordability and the ability to automate the recording of activity on a large number of individuals has made the DAM a popular choice for recording lethargy and somnolence in infected insects such as fruit flies (Shirasu-Hiza et al. 2007; Vale and Jardine 2015), mosquitos (Rund et al. 2016), bees, wasps, and other insects of similar size (Giannoni-Guzmán et al. 2014).

Automated image-based behavioral tracking of locomotor activity in groups of insects

While IR-based tracking systems have a number of advantages, they are limited in their spatial and movement resolution (activity is measured as a binary trait) and their scalability, which requires additional activity monitors. These limitations, allied to the desire of quantifying behaviors in more natural settings, has spurred the development of image and video-based tracking of individual insects while interacting in large groups (Gilestro 2012; Mersch et al. 2013). Some of these approaches are relatively simple. For example, the level of social aggregation in groups of insects (which is relevant for the likelihood of disease spread), is easily measured by using still images of insect groups to measure nearest-neighbour distances (Simon et al. 2012). Actual insect movement, however, tends to occur in three dimensions, which increases the challenge of real-time tracking and quantification (Ardekani et al. 2012). Automated video tracking technology, allied with powerful computational analysis of insect activity data (Egnor and Branson 2016) has revolutionized the field of insect behavior by allowing individual-level behaviors to be linked with higher level ecological patterns (Dell et al. 2014). For example, using continuous tracking of individually-tagged ants within a colony, Mersch and colleagues were able to describe temporal and spatial distribution of all individual ants, and identified unique behavioral units within the ant colony that changed over time (Mersch et al. 2013).

Computational analysis of behavior

Automated image-based tracking is complex and extremely data-intensive. Its success as a method to measure behavior will therefore rely heavily on advances in computational

analysis (Egnor and Branson 2016). The main challenge is reconstructing the trajectories of individual flies across video frames while accounting for considerable noise both in the measurement and in the assay environment, compounded by the need to maintain the identities of individual insects over time (Ardekani et al. 2012; Reiser 2009). These technical challenges have benefitted greatly from independent advances in the fields of machine learning, which with a solid mathematical grounding, are now being applied to important biological and behavioral processes (Ardekani et al. 2012; Egnor and Branson 2016).

18.2.1 Spatial avoidance

Spatial avoidance occurs when hosts avoid areas where parasites are detected or likely to be found. Small-scale spatial avoidance of parasites, as seen for many ovipositing insect hosts, is just one type of avoidance behavior (Table 18.1). For example, the spreadwing damselfly, *Lestes sponsa*, can lay eggs in aquatic plants above or below the water's surface. Below-surface oviposition incurs additional energetic costs to the mother, but offers greater protection of eggs from parasitoids. In the presence of high egg parasitism, females preferentially lay eggs below the water's surface (Harabis et al. 2015).

Some larger-scale movements of insects have been explained using infection avoidance. For example, the migration of the monarch butterfly, *Danaus plexippus*, across North America has been explained partly as a mechanism of avoiding infection (Satterfield et al. 2015; see also Chapter 7). However, this example is somewhat controversial because while migrating populations experience less parasitism (Altizer et al. 2011), it is equally true that parasitized individuals have reduced flight capabilities (Bradley and Altizer 2005) making it unclear whether butterflies migrate to avoid infection, or if infected butterflies migrate less or not at all.

18.2.2 Temporal avoidance

In addition to spatial avoidance, insects also can avoid times where infection risk is highest. Almost all examples of temporal avoidance come from interactions between ants and parasitoid Phorid flies.

Table 18.1 Examples of infection avoidance behaviors (classified according to the subsections of Section 18.2), across a range of insect orders

Insect host order	Parasite	Avoidance behavior	Reference
Coleoptera	Bacteria	Medication	Arce et al. (2012)
Coleoptera	Nematode	Grooming	Gaugler et al. (1994)
Coleoptera	Bacteria and Fungus	Grooming	Lusebrink et al. (2008)
Coleoptera	–	Grooming	Valentine (2007)
Diptera	Parasitoid wasp (Hymenoptera)	Medication	Kacsoh et al. (2013)
Hymenoptera	–	Decreased social contact	Bigio et al. (2014)
Hymenoptera	Fungus	Niche construction	Chapuisat et al. (2007)
Hymenoptera	Protozoan/Bacteria	Trophic avoidance	Fouks and Lattorff (2011)
Hymenoptera	Fungus	Medication	Konrad et al. (2012)
Hymenoptera	Parasitoid fly	Trophic avoidance	Orr (1992)
Hymenoptera	–	Niche construction	Pie et al. (2004)
Hymenoptera	Fungus	Grooming	Nielsen et al. (2010)
Isoptera	–	Decreased social contact	Crosland et al. (1997)
Lepidoptera	Multiple (natural populations)	Spatial avoidance	Sadek et al. (2010)
Odonata	Parasitoid wasp	Spatial avoidance	Harabis et al. (2015)

Members of the worker caste of the tropical fire ant, *Solenopsis geminata*, recruit nest mates using pheromone trails to forage *en masse* when large food items are found. Columns of worker ants carrying food can be seen on forest floors, and these high host densities attract parasitoid Phorid flies. In response to the presence of female Phorid flies, worker activity and nest mate recruitment decrease significantly (Feener and Brown 1992). This avoidance behavior also demonstrates the significance of parasite pressure for host ecology, forcing fire ants to balance the cost of reduced foraging efficiency with the benefit of reduced parasitism. A more extreme form of temporal avoidance of Phorid flies is observed in the leaf-cutter ant, *Atta cephalotes*. In the presence of the diurnal Phorid fly, *Neodohrniphora curvinervis*, whole colonies of ants shift their activity from daytime to night-time (Orr 1992). This response to parasitism is co-ordinated by many individuals over a short time period, making the interaction between ants and Phorid flies of fundamental interest to our understanding of adaptation and behavioral plasticity.

18.2.3 Trophic avoidance

Several activities that are essential to survival also increase the risk of acquiring infection. Foraging and consuming food is vital to host health, but offer parasites an ideal route to the internal environment of the host. Many parasites infect the insect digestive system (see also Chapter 21). The cost of this infection is thought to have driven the evolution of trophic avoidance, where individuals avoid eating infectious food items (Alma et al. 2010; Fouks and Lattoroff 2011). Even insects typically thought of as having poor sensory systems, such as the larvae of holometabolous insects, can exhibit trophic avoidance to a range of parasites. For example, larvae of the grapevine moth, *Lobesia botrana*, avoid eating fungus-infected grapes (Tasin et al. 2012), while larvae of the gypsy moth, *Lymantria dispar*, avoid leaves where viruses are detected (Parker et al. 2010).

18.2.4 Altered mate preference

Courtship and mating are activities during which infections can be transmitted between individuals.

As a result, many hosts avoid mating with infected conspecifics. This mate avoidance is commonly seen in vertebrates, but is rarely seen in insects (Abbot and Dill 2001; Rosengaus et al. 2011), despite that it has been tested across several insect host systems (Arbuthnott et al. 2016). The rarity of altered mate preference in insects in relation to parasitism is puzzling as traits that are linked to parasitism have been shown to be targets of mate choice, e.g., a number of sexually selected traits have demonstrable ties to immunocompetence (Siva-Jothy 1999; Tregenza et al. 2006). Therefore, the overall rarity of altered mate preferences must have another explanation, such as parasites evading host detection (Lambardi et al. 2007), parasites generally manipulating host behavior (but see Section 18.3) or the cost of abstinence outweighing the cost of infectious mating.

Microbiota generally have a range of effects on host–parasite dynamics and have been shown to influence mate choice (Arbuthnott et al. 2016; Damodaram et al. 2016). That insects can detect these microorganisms and incorporate this information into mate choice, makes general parasite evasion of host detection less likely (de Roode and Lefèvre 2012). Particularly for microorganisms such as *Wolbachia* causing host cytoplasmic incompatability, there is expected to be intense selection on detection as uninfected female hosts mating with infected males have zero fitness. The cost-of-abstinence argument, in comparison, makes sense in light of insects being relatively short-lived organisms with a premium on mating and given that this form of infection avoidance is so commonly seen in longer-lived vertebrates.

18.2.5 Decreased social contact

Insects also exhibit a range of non-sexual interactions (see Chapter 15). These interactions can be exploited by parasites to infect hosts. Literature detailing the influence of parasites on these social interactions focuses on eusocial insects, due to the significance these interactions have on colony fitness (Cremer et al. 2007). Workers increase their fitness through kin selection; protecting and providing for reproductive nest mates (see Chapter 15). Selection favours workers preventing transmission to the rest of the colony even if that individual increases its own short-term infection risk (Cremer et al. 2007). The termite, *Reticulitermes fukienensis*, for example, undertakes a range of behaviors in order to limit contact with infected workers or workers that have died from infection. One marked example is the burying of nest mates that died from fungal infection (Crosland et al. 1997). In so doing, workers reduce infection risk to other nest mates.

Social organization of eusocial insect colonies also serves to reduce infection risk. For example, worker castes of the carpenter ant, *Camponotus fellah*, take on a range of tasks that have variable infection risk (Mersch et al. 2013). Although the effect of infection risk was not directly tested, worker castes central to colony reproduction, i.e. brood nurses, seldom interacted with castes that performed jobs with a high risk of infection (Mersch et al. 2013). This aids brood nurses in safely rearing the colony's next generation.

18.2.6 Niche construction and maintenance

As many parasites can persist in the environment without hosts, insects can avoid infection by making and maintaining their environment less hospitable to infective stages of parasites. Nest-making insects frequently dedicate time or, in the case of eusocial insects, castes, to the removal of parasites or infectious material from the nest (Cremer et al. 2007; Neoh et al. 2012; Bigio et al. 2014). Other factors that reduce parasitism risk include antimicrobial secretions (Turillazzi et al. 2006; Chapuisat et al. 2007) and nest architecture (Pie et al. 2004).

The burying beetle, *Nicrophorus vespilloides*, raises its offspring on small vertebrate carcasses and employs antimicrobial secretions. The carcasses attract microparasites that can affect offspring survival and development adversely. This cost has driven the evolution of antimicrobials secretion by parents, while preparing their offspring's food source (Arce et al. 2012).

Nest architecture is diverse in many social insects. Nests comprised of multiple chambers limit interactions between castes to focal individuals or specialized castes (Mersch et al. 2013). Additionally, chambers can be dedicated to specific functions, for example, the nests of the social cricket, *Anurogryllus muticus*, have dedicated latrine chambers (Curtis 2014).

18.2.7 Grooming

Grooming is a mechanism of infection avoidance that shares activities such as those described previously but it primarily concerns maintenance of the insect's cuticle. Grooming reduces infection risk by removing parasites from, or preventing their establishment on, the cuticle (Gaugler et al. 1994; Turillazzi et al. 2006). Individuals of many insect species groom themselves (autogrooming) and one another (allogrooming; Valentine 2007). Grooming is even observed between species, as is the case in the aphid-farming ant, *Formica podzolica*, which grooms its aphid-livestock (Nielsen et al. 2010). Insects have a number of adaptations that increase grooming efficiency including specialized bristles (Zhukovskaya et al. 2013) and cuticular secretions (Lusebrink et al. 2008). For example, the rove beetle, *Stenus comma*, secretes the alkaloid, stenusine, from specialized abdominal glands to reduce the growth of fungi, Gram-positive and Gram-negative bacteria on the cuticle (Lusebrink et al. 2008).

Interestingly, secretions also are used by insect parasites frequently to evade hosts. The cuckoo bee, *Bombus bohemicus*, infiltrates host hives by mimicking that colony's cuticular secretions, where it goes on to monopolize resources for reproduction and suppress host reproduction (Kreuter et al. 2012). In comparison, the social parasite, *Acromyrmex insinuator*, infiltrates host nests by appearing 'chemically insignificant', by not secreting cuticular hydrocarbons or any other pheromones (Lambardi et al. 2007). It appears that hosts do not use cuticular secretions to evade parasites for as yet unknown reasons.

18.2.8 Medication

Just as insects produce compounds that reduce infection, they can also apply or consume compounds found in their environment for a similar purpose. When done prior to infection, this is known as **prophylactic** medication and, in insects, typically occurs between siblings or parents and offspring. If insects are already infected, using compounds that help clear infection is known as therapeutic medication (see Section 18.3.4).

Parent–offspring prophylaxis occurs during *D. melanogaster* oviposition in the presence of parasitoid wasps (Kacsoh et al. 2013). In the absence of wasps, mothers avoid laying eggs in sites with a high ethanol content due to the detrimental effect of alcohol on offspring development. However, ethanol confers increased larval avoidance of parasitoid infection. As a result, if mothers see parasitoid wasps, they preferentially oviposit in high-ethanol sites (Kacsoh et al. 2013).

Sibling–sibling prophylactic medication occurs in the ant, *Lasius neglectus*, where individuals infected with the fungal parasite, *Metarhizium anisopliae*, transfer small quantities of the fungus to susceptible nest mates. The transfer of fungus acts as an elicitor for **immune priming** rarely killing workers, but regularly decreasing susceptibility to subsequent fungal infection by upregulating fungal immune genes (Konrad et al. 2012). Socially inoculated colonies are far less susceptible to outbreaks of *M. anisopliae* infection.

18.2.9 Integrated studies of infection avoidance

Studies of infection avoidance typically focus on the broader ecological consequences of the behavior due to its consequences for host–parasite interactions. Studying the mechanisms of these avoidance behaviors enables the discovery of their physiological mediators and, by extension, the constraints on their evolution. Recent advances in neurobiology and endocrinology have significantly increased the detail with which we can study such mechanisms (Kohmura et al. 2015) (see Box 18.3).

Box 18.3 The mechanistic basis of behavioral changes in infected insects: the neuroendocrine-immune axis

Behavioral changes in response to infection appear to be taxonomically conserved within animals and especially within insects. However, two major questions remain (Lopes 2017):

- What are the mechanisms by which behavior and immunity are linked (how)?
- Why would behavior and immunity be linked at all (why has this link evolved)?

Both proximate and ultimate questions can be addressed by considering that infection is essentially an extreme form of physiological stress, and that the immune and central nervous systems are interconnected by the endocrine system, forming the major neuroendocrine-immune axis which regulates both behavior and immunity in response to pathogens (Adamo 2014).

The physiological response of organisms to stress, such as fight or flight behaviors, is mainly coordinated by the release of neuroendocrine factors, which appear to be largely conserved across vertebrates and invertebrates, and similar stress hormones mediate both endocrine and immune systems of many arthropods, including crustaceans, molluscs, and insects (Adamo 2006). In insects, the major neurohormone released during stress and fight-or-flight behaviors is octopamine (OA). Similar to the effect of its mammalian homologue norepinephrine (NO), the release of OA during stress in invertebrates results in the release of energetic resources, and increases responsiveness to external threats. Approximately 30 minutes after the release of OA from the dorsal unpaired medial (DUM) cells (essentially the equivalent to insect neurons), a peptide called adipokinetic hormone (AKH) is released by the corpora cardiacum, an insect endocrine organ. AKH plays an important role in mobilizing lipids and releasing energy for costly activities, such as flight (Adamo 2006; Adamo 2017).

The role of OA during infection in insects has received considerable attention (Adamo 2006, 2014). Haemocytes, invertebrate immune cells, have OA cell surface receptors (Huang et al. 2012), which allows direct neuroendocrine modulation of immune function. Overall, the effects of OA on insect immunity appear to be immune enhancing, although some immune suppressive effects also have been reported. For example, injection of high physiological doses of OA tends to increase haemocytes and phagocytocis, but injections of low physiological doses of OA make the cricket (*Gryllus texensis*) more susceptible to bacterial infection. AKH has also been directly linked to changes in immune function, but again its effects appear to be highly context dependent. For example, in the African migratory locust, *Locusta migratoria*, AKH increased phenoloxidase activity after an immune challenge enhancing immune function but also reduced antimicrobial activity, increased bacterial growth in the hemolymph and increased susceptibility to the fungus *Metarhizium anisopliae* (Goldsworthy et al. 2002).

The OA-AKH cascade therefore provides a possible evolutionary explanation for the link between behavior and immunity—if immune responses are energetically costly, the neuroendocrine mechanisms that have evolved to allocate resources to other costly behaviors (like fight-or-flight) may also have been co-opted to do the same during an immune challenge (Adamo 2017).

Drosophila melanogaster is a model organism particularly well suited to this research due to the suite of molecular tools available in the study of its neurobiology, genetics and behavior. As previously mentioned, prophylactic medication is elicited in *D. melanogaster* by the visual cues of the parasitoid wasp, *Leptopilina heterotoma* (Kacsoh et al. 2013). Using genetically modified flies, with fluorescently-labelled neural circuits, researchers detailed the specific neural pathways and neuropeptide mediating this oviposition choice behavior (Kacsoh et al. 2013). This study went on to also examine the evolution of this avoidance, measuring it in six other *Drosophila* spp., revealing multiple independent evolutions within the genus. *D. melanogaster* has also been used to detail dedicated olfactory circuits for harmful microbes (Stensmyr et al. 2012) alongside the signatures of such sensory systems in the genome (McBride and Arguello 2007).

18.3 Behavioral changes in infected insects: sickness behaviors as host adaptations

Until now, the variation in insect behavior in response to cues of infection from infectious conspecifics or environments that may increase the risk of infection has been described. If avoiding these sources of infection is unsuccessful and individuals become infected, a suite of behavioral changes may also be observed in infected animals. The most common behavioral changes in infected animals are manifested by alterations in activity levels (increased lethargy) and sleep patterns (increased somnolence)—decreased foraging behavior and reduced food intake (anorexia), and a lower libido and reduced investment in sexual reproduction (Hart 1988, 2011; Moore 2013).

In some cases, the simplest explanation for these behaviors is that they arise as a direct consequence of the pathology either due the direct damage caused by pathogen growth, or because of the energetic expenditure arising from mounting an effective immune response (Moore 2013). In other cases, behavioral changes in infected hosts are the result of intricate pathogen strategies to manipulate the way an infected host behaves, which enhance a pathogen's evolutionary fitness by increasing the chances of successful transmission to new hosts

(Poulin 2010). We will describe these **host manipulations** in greater detail in the following section. A third potential explanation for changes in host behaviors during infection, however, is that they reflect adaptive sickness behaviors that allow animals to conserve energetic resources during infection.

This adaptive hypothesis for sickness behaviors was originally proposed by Benjamin Hart (Hart 1988), and posits that energy not expended by infected animals in finding food, or finding a mate could, instead, be reallocated to immunity, thereby increasing the chances of clearing the infection and recovering health. One of the arguments put forward for the adaptive nature of sickness behaviors is that lethargy, somnolence, and reduced reproduction appear to be evolutionarily conserved responses to infection across a wide range of taxa (Moore 2013; Sullivan et al. 2016).

In addition to quantifying the possible fitness benefit of sickness behaviors, it is also important to consider potential indirect fitness costs. For example, reducing activity during infection may conserve energy that is allocated to fighting infection, but also means that animals are forced to reduce other fitness-enhancing activities such as foraging, courtship and mating, parental care, and territorial defence, and may even leave individuals more susceptible to predation (Adelman and Martin 2009; Lopes 2014; Vale and Jardine 2016). As discussed next, insects have become central to addressing both the benefits and potential costs of sickness behaviors (de Roode and Lefèvre 2012; Sullivan et al. 2016).

18.3.1 Infection-induced lethargy

One of the most commonly observed behaviors in sick animals is a decrease in activity following infection (lethargy). For example, fruit flies, *Drosophila melanogaster*, show reduced daily locomotor activity when infected with Drosophila C Virus (DCV; Vale and Jardine 2015). While lethargy may simply be a consequence of being sick, reduced activity may also bring benefits if energetic resources are instead allocated to immune defence. For example, honey bees (*Apis mellifera*) challenged with **lipopolysaccharide (LPS)** mount a strong immune response and subsequently exhibit a reduction in locomotor activity, suggesting that reduced activity results from a reallocation of resources to immunity (Kazlauskas et al. 2016).

18.3.2 Decreased social contact and isolation

Active individuals are more likely to be more gregarious and to partake in activities with other conspecifics, such as courtship, mating, and fighting (Lopes 2014). Reduced activity will therefore lead to decreased social interactions, thereby decreasing the likelihood of encountering sources of infection. We may therefore expect that risk of disease (and the resulting fitness costs of acquiring infection) should select for behaviors that lead to less gregarious individuals. The effects of infection on social aggregation have been found to be especially clear in eusocial insect systems that are strongly influenced by kin selection (Cremer et al. 2007). In the same honey bee example described previously, bees pricked with bacterial-derived LPS spent more time alone, standing still, or self-grooming, and showed reduced social contact with other bees (Kazlauskas et al. 2016). Another popular example of changes in social aggregation following infection occurs in worker ants, *Temnothorax unifasciatus*, which when infected with a pathogenic fungus change their social behavior by leaving the nest permanently (Heinze and Walter 2010). While this could initially appear to be a form of host manipulation by the fungus, by leaving the nest long before death, this behavior is likely to reduce the spread of fungus, possibly curtailing pathogen transmission.

18.3.3 Infection-induced anorexia

The degree to which infection is detrimental to hosts is often affected by their physiological status (Adamo 2009), which in turn is greatly impacted by the quantity and quality of host diet (Ponton et al. 2013; Singer et al. 2014). Behavioral changes affecting the quantity of food intake are particular common and animals commonly exhibit a loss of appetite (anorexia) following infection (Hart 2011; see also Chapter 8). Given the heavy energetic burden of immune responses, this behavioral change effectively reducing caloric intake may seem counterintuitive. However, reduced food intake has been shown to increase recovery from infection, and is a

conserved behavior across many vertebrates and invertebrates (Murray and Murray 1979; Ayres and Schneider 2009; de Roode and Lefèvre 2012).

Work in insect host–pathogen systems is starting to address the complex link between nutritional intake and invertebrate immunity. For example, fruit flies, *D. melanogaster*, infected with either *Salmonella typhimurium* or *Listeria monocytogenes* have been shown to become anorexic (Ayres and Schneider 2009). However, this change in feeding behavior appears to have different effects on fly immunity according to the type of infection. When infected with *L. monocytogenes*, anorexia reduced the ability of the fly to clear infection, while anorexia during *S. typhimurium* infection increased **infection tolerance** because flies did not improve their ability to clear infection, but still survived longer. These contrasting effects of infection-induced anorexia appear to occur because restriction of nutrition affect cellular and humoral immune responses differently, suggesting that in the wild we should expect this sickness behavior to vary with the prevalence of specific pathogen types (Ayres and Schneider 2009).

In addition to the quantity of food, infected animals may also alter the type and quality of food they choose to ingest (see also Chapter 8). For example, diet-induced changes in susceptibility may be determined by the precise ratio of specific macronutrients. The survival of the Egyptian leafworm, *Spodoptera littoralis*, when infected with a nucleopolyhedrovirus (NPV) depends strongly on the protein:carbohydrate ratio in its diet, and insects fed on a high-protein, low-carbohydrate diet showed the highest survival following NPV challenge (Lee et al. 2008). Furthermore, when insects were given a free choice of diets containing different ratios of protein:carbohydrate, larvae that selected a diet containing higher protein content lived longer than those choosing alternative diets. Behavioral changes that affect what and how much an insect chooses to eat when infected can therefore have important effects on how sick it gets and how likely it is to recover (Singer et al. 2014).

18.3.4 Dietary self-medication

Dietary choice following infection suggests that infected insects are capable of dietary self-medication (see Chapter 8), by choosing food sources that enhance their chances of survival (de Roode et al. 2013). Self-medication can occur through nutritional effects that enhance the host's ability to fight infection, or by the direct anti-parasitic properties of the ingested compounds. Insects have emerged as powerful systems to conduct manipulative experimental tests of the occurrence and potential benefits of dietary self-medication (de Roode and Lefèvre 2012).

For instance, when *Grammia incorrupta* caterpillars are infected by a parasitoid, they show an increased preference for alkaloid toxins found in larval food plants, and by choosing to eat these toxic plants when infected, are able to increase survival. Using self-medication to ensure the survival of offspring also has been shown in Monarch butterflies. Several studies have shown that milkweeds—the plant species that Monarch butterflies use as their larval food plants—increase the survival of butterflies infected with the protozoan *Ophryocystis elektroscirrha*. While infected larvae did not preferentially consume medicinal milkweed, preferential oviposition on medicinal milkweeds by infected females resulted in reduced parasite growth in the offspring (Lefèvre et al. 2010).

18.3.5 Behavioral thermoregulation

Mammals and other endotherms can regulate their own body temperature and fever is a commonly observed response to infection. Ectotherms, such as insects and other invertebrates, can elevate temperature when infected by seeking out warm locations that allow their body temperature to rise to levels that may be detrimental to pathogens, called **behavioral fever** (Thomas and Blanford 2003; de Roode and Lefèvre 2012). Behavioral fever is a widespread behavioral response to infection, and has been especially well documented in insects (Stahlschmidt and Adamo 2013). For example, heat-seeking behavior has been documented in house flies, grasshoppers, and in the desert locust infected with entomopathogenic fungi (Blanford et al. 1998; Kalsbeek et al. 2001; Elliot et al. 2002), and in honeybees, who communally raise the temperature of their hive in response to an infection with the heat-sensitive pathogen that causes chalkbrood (Starks et al. 2000).

In addition to increasing body temperature, behavioral thermoregulation can also be used to lower temperature, which can delay parasite growth. Examples of cold-seeking behavior have been reported in fruit flies, *D. melanogaster*, infected with the fungal pathogen, *Metarhizium robertsii*; cold-seeking comes at an initial cost of lower reproduction, but slowing fungal growth ultimately results in higher lifetime reproduction for flies (Hunt et al. 2016). Similar growth-retarding effects of cold-seeking behavior occur in acanthocephalan-infected cockroaches (Moore and Freehling 2002), and in bumblebees infected with thick-headed flies (Conopidae). Instead of spending the night in the warmth of the hive, observational and experimental work has shown that infected bumblebees achieve increased survival by preferentially spending time in cold areas, which reduces parasite growth rate (Müller and Schmid-Hempel 1993).

18.4 Behavioral changes in infected insects: host manipulation as a parasite adaptation

Parasite-induced changes in host behavior, often interpreted as parasite adaptations to aid in their own transmission and/or survival, are widespread amongst parasite–host associations, but not universal (Heil 2016). Some of the most celebrated examples of host manipulation come from associations where insects are the host species, or both the host species and parasitoid. These widespread and recurring phenomena provide considerable scope to examine the significance of host manipulation with respect to Tinbergen's four levels of behavioral explanation (Box 18.1b).

Poulin (1995) initially outlined four criteria indicative of adaptive parasite-induced behavioral changes in hosts. They include the following:

- the induced behavior should be complex;
- it should align with clear a priori expectations for its potential benefit to the parasite, i.e. demonstrate obvious purpose;
- it may arise independently in multiple host and parasite lineages by convergence;
- the induced behavior should increase the fitness of the parasite.

However, it is really only the last criterion, that there be a fitness benefit to the parasite, which decisively designates a behavioral change as adaptive manipulation (Poulin 2010).

In such studies, therefore, we tacitly focus on how the parasite's genes and gene products might enhance its fitness through the extended behavioral phenotype of its host species. By extension, we are examining the degree to which parasites are locally and temporally adapted to their insect hosts. It is important to remember that for any observational or experimental study of behavioral responses of host individuals to parasite or pathogen infection, the intensity of the host's response could be due to parasite-related factors, host-related factors and environmental factors acting alone or in combination. The parasite- and host-related factors can be further subdivided into genetic versus non-genetic characteristics of the parasite or host (Thomas et al. 2012). Disentangling the various causes of the altered behavior can help researchers interpret whether the altered behavior is mainly a host adaptation to curb impacts of infection (see Box 18.1a, host in control), a true parasite manipulation of host behavior benefitting the parasite, or a by-product of infection (as mentioned, neither host nor parasite is in control, and the behavioral change might be costly or beneficial to either or both).

For students of insect behavior, the altered behaviors of parasitized insect hosts that fall under the rubric of host manipulation are sometimes subtle and sometimes strange and curious, as the following examples will illustrate. Each of the examples with insect hosts was chosen to further illustrate a key concept(s) concerning the study host manipulation by parasites.

18.4.1 Manipulation of concealment behaviors of parasitized hosts

The life cycle of the liver fluke, *Dicrocoelium dendriticum*, makes for an enjoyable read. The case for adaptive manipulation of ants by larvae of this worm was made over 55 years ago [by Hohorst and Graefe (1961), cited in Hölldobler 2012]. Importantly, this liver fluke is a cosmopolitan parasite of grazing mammals: adults of this trematode reside in the liver of sheep, cattle, pigs, goats, and cervids

(Goater et al. 2014). The eggs are passed with the faeces (via the bile duct) and ingested by terrestrial snails, which later egest cercaria (larval flatworms) in a mucous mass. Ants of various species eat the mucous and ingest the larval worms, some of which encyst in the ant's abdomen and at least one of those migrates to the suboesophagial ganglion of the ant, but does not encyst and therefore is not transmissible (Goater et al. 2014). This brain worm 'turns on' a stereotypical behavior of the infected ant by its climbing vegetation and clamping its mandibles down on a leaf, flower, or blade of grass. These solitary anchored ants are susceptible to incidental ingestion by grazing mammals, wherein the parasites later excyst in the small intestine and travel to the bile duct to develop and reproduce.

This iconic example of host manipulation illustrates two important points. The first is that apparent maladaptive host behavior can make sense. Such height- or open-seeking (non-concealment) behavior has been described for other diverse associations, e.g., ants parasitized by fungi (*Ophioccordyceps* spp.) and flour beetles parasitized by nematodes. However, researchers still have to test whether the altered behavior actually leads to increased transmission success. For example, Schutgens' et al. (2015) excellent work uncovered behavioral changes in the flour beetle, *Tribolium confusum*, infected with the spirurid nematode, *Protospirura muricola*. Infected beetles took longer to conceal themselves in experimental trials and spent less time concealed. Infected beetles (particularly those with >1 cyst) also were more likely to be in an illuminated part of the trial arena. These differences were observed for hosts with older and not younger cysts. All these findings support the hypothesis that ontogenetic changes in infection-adjusted behavior are adaptive by increasing successful transmission to the definitive host. However, these experiments were not conclusive in this respect—host behavioral changes might simply be related to infection pathology, and no predation experiments with actual definitive hosts were performed, wherein the success of parasite establishment following infection was assessed.

The second point is that there are an astounding number of natural experiments of 'evolution in action', even with just this one parasitic worm species. Given its worldwide distribution, there are probably many different species of snails and ants that act as intermediate hosts. Add to this the fact that wild and domestic grazing mammals also vary from place to place. There is probably a wide array of efficacies in the ant stereotypical behavior depending on how well adapted larval trematodes are to their insect host and the extent to which the infected insects secure a place and a time where/when it is likely to be ingested by suitable definitive host.

18.4.2 Increased defensiveness of moribund hosts protecting parasitoids

It is a hallmark of many systems involving host manipulation by parasites that the manipulation results in the host death, either through ingestion of infected prey by final hosts as described above for the liver fluke, or by other means such as drowning in the iconic case of cricket infected with horsehair (nematomorph) worms (Thomas et al. 2012). A unique form of **behavioral manipulation** is described by Harvey et al. (2011) who show that host armyworm that are moribund (near death) following emergence of parasitoid wasps protect the wasps against hyperparasitism by repelling other parasitoid wasps (see Figure 18.1 for a related example). The initial parasitoids presumably benefit from a moderate level of virulence leaving moribund, but not dead caterpillars to protect them. Another example of this phenomenon concerns a solitary parasitoid wasp, *Dinocampus coccinellae*, parasitizing a ladybird beetle, *Coleomegila maculata*. The larval parasitoid emerges as a late instar from the ladybird beetle and spins a cocoon for pupation between the ladybird's legs. The ladybird beetle is moribund, but defends the cocoon against predatory lacewings, *Chrysoperla carnea*, which are common (Thomas et al. 2011). More recently, there is genetic and microscopic evidence that time-related changes in behavior of the ladybird beetle is due to replication/clearance of a virus acquired from the parasitoid (Dheilly et al. 2015).

18.4.3 Manipulation of host sexual or social behavior

Work by Burand et al. (2004, 2005) on the corn earworm moth infected by Hz-2V virus shows morphological and physiological changes in females, coupled

Figure 18.1 Two cases of behavioral manipulation by parasitoids in which an insect host defends an emerged parasitoid during the latter's pupation, i.e. examples of hosts expressing modified behavior after infection has ceased. (a) A *Thyrinteina* leucocerae caterpillar stands guard below a cluster of braconid (*Glyptapanteles* spp.) pupae, which previously emerged from it following an approximately 2-week period of infection. The affected caterpillar displays defensive behaviors when disturbed (vigorous thrashing of its head) for the 6 or 7 days before it dies following egression of the parasitoid larvae, and otherwise remains practically motionless (Grosman et al. 2008; adapted from a photo by José Lino-Neto, CC BY 2.5). (b) A single braconid (*Dinocampus coccinelae*) pupa is protected underneath the coccinellid (ladybeetle) host from which it emerged. While guarding the parasitoid, the ladybeetle displays tremors, particularly when disturbed, but otherwise remains paralysed. Approximately 25% of these beetles survive the entire parasitoid pupation period (~7 days) and then regain expression of normal behavior (Thomas et al. 2011; adapted from a photo by coniferconifer on Flickr, CC BY 2.0).

with behavioral changes of mate searching males. Infected females continue to attract males following first contact, i.e., they do not become refractory. What is interesting is that these females are agonadal, but still attract males, probably due to the heightened production of pheromones. The virus-infected females are highly attractive to mate searching males, which pick up the highly contagious plugs and transfer them to uninfected females. The virus thus induces a change in the behavior of the males towards infected females—a form of indirect manipulation.

In his provocative review, Heil (2016) raised the issue that there are few examples of host manipulation by sexually transmitted pathogens and parasites (STDs). This might be because the costs of abstinence (see Section 18.2.4) ensure that host sexual activity is high enough to ensure transmission or it might be that the STDs have evolved not to invoke host sickness or asocial behaviors.

18.4.4 Manipulation of behavior of insect vectors

There are several studies in the primary literature showing effects of microparasites on the feeding behavior of their insect vectors of human diseases, including sand flies and Leishmanias, Tsetse flies and African sleeping sickness, and mosquito vectors and malaria (reviewed by Hurd 2003). One of the points made is that the parasite may influence the vector in ways that increases vectoring ability (parasite's interest) even if it poses a risk to the insect vector's survivorship. For example, the study by Botto-Mahan et al. (2006) showed kissing bugs infected with Chagas disease demonstrated more rapid host detection, increased biting rate and short-ened time to defecation (defecation facilitates transmission of the protozoan; see Chapter 21). Chagas disease infects an estimated 10–12 million people in South and Central America (Goater et al. 2014). Some understanding of the subtleties of apparent manipulation of vector risky behavior can help inform epidemiological models. Other studies on potential vector-manipulated behaviors with an applied thrust include studies such as aphids and cucumber plants harbouring cucumber mosaic virus. Here, researchers are interested in uninfected aphids being attracted to virus laden plants and then recording the later behavior of the infected aphids in terms of whether they are attracted to already infected plants versus uninfected plants, as a means to promote viral transmission (Carmo-Sousa et al. 2014). The reader is referred to McMenemy et al. (2012) for similar results with raspberry viruses.

18.4.5 Manipulation in an ecological context

Lafferty and Kuris (2012) outline some examples of where host manipulation by parasites has far-reaching effects on community organization. An insect example of this phenomenon concerns the horse-hair worm manipulation of orthopteran insects. Manipulated crickets and grasshoppers drown and become food for trout. The manipulated insects can account for up to 60 per cent of the trout's energetic needs, according to work by Sato et al. (2012, and references therein). These easy prey result in less predation pressure on benthic invertebrates, which means streams frequented by parasitized insects have more diverse and abundant communities of benthic invertebrates (Sato et al. 2012). Although the host drowning is beneficial for the horsehair worm, ingestion by trout would not be for the free-living adult nematomorph worms. Perhaps this is why the behavioral manipulation is timed to occur at night and the worms exit the drowning hosts, rather quickly lest they also become prey for foraging fish.

18.4.6 Manipulation in an evolutionary context

Ever since Poulin's (1995) seminal paper, researchers have emphasized that the evolution of host manipulation by parasites needs to be understood in a phylogenetic context. Here, the manipulative behavior of a parasite could have been inherited from an ancestor, and might only work partially in the current context and provide examples of 'evolution in action'. Acanthocephalans, for example, are comprised entirely of manipulative species: that is, they have been shown to alter the behavior of their intermediate hosts to make predation by definitive hosts more likely. However, whether predation by appropriate definitive hosts is increased as a result of the manipulation is a tall order often missing from studies, as mentioned already. Another related question is the extent to which different host species differ with respect to behavioral modification following parasitism. Malfi et al. (2014) reported that bumblebee self-burying behavior is observed across several host species infected with the same parasitoid canopid fly (burying enables fly pupation), although there is variability in this response—a higher likelihood of self-burying in response to infection was observed in two closely related species (same subgenus), while lower probability of self-burying was observed in a less closely related bumblebee species.

18.5 Concluding remarks: future directions in the study of insect behavior in relation to parasites

It is clear that insects have provided a wealth of examples of behavioral responses to (risk of) infection that can be classified as avoidance behavior pre-infection, adaptive host sickness behavior following infection, and adaptive parasite manipulation following infection. Each of these areas of investigation is ripe for future research that focuses on integration of mechanism with investigations of proposed function. Only then can serious claims that responses are not a by-product of infection be made. We suspect that there are many cases in nature of 'evolution in action' where partial solutions to the problems of risks or costs of infection or parasite transmission/survival will be observed, and longitudinal studies largely absent from this review are welcomed. It is also important to ask in this context why a predicted behavior is not present when it is expected to be, such as is the case with mate avoidance in insects in relation to parasitism risk.

What follows is a brief overview of some of the main questions still outstanding. In addition to the need for integrated studies such as those cited at the end of the parasite avoidance section, researchers can explore the extent to which behavioral responses are plastic or context dependent, e.g., how an infection avoidance behavior might vary, depending on virulence or infection risk. A plastic infection avoidance response was experimentally demonstrated in the pollination behavior of the bumblebee, *Bombus terrestris* (Fouks and Lattorff 2011). This was done using two parasites of differing virulence, the more virulent bumblebee specialist, *Crithidia bombi* and the less virulent generalist parasite, *Escherichia coli*. Congruous with a plastic infection avoidance response, *B. terrestris* avoided flowers contaminated with *C. bombi*, significantly more than *E. coli* (Fouks and Lattorff 2011). Considering how the plasticity of other infection avoidance behaviors affects their evolution and influence on host–parasite dynamics goes hand-in-hand with furthering our understanding of their mechanistic basis. Plasticity's significance to evolutionary biology has dramatically increased in recent years, with some arguing it warrants a drastic shift in how we conceptualize evolutionary processes (Laland et al. 2014). Not only are these dynamics probably central to infection-avoidance behaviors, but they are also an ideal context with which to test their significance.

Secondly, it is extremely difficult to measure sickness behaviors in the wild on individuals who have contracted infections naturally. However, the aforementioned longitudinal studies are needed to assess variation in prevalence and intensity of parasitism in order to design realistic experiments, but also to assess the costs of retaining particular 'sickness' behavioral responses if the threat of parasitism is at best intermittent. A combination of field assessments and common garden experiments might prove fruitful for understanding the evolutionary dynamics of apparent sickness behaviors. Such studies will also have to consider that the indirect costs of sickness behaviors (reduced time spent foraging or searching for mates) might vary from place to place and time to time.

Thirdly, researchers might wish to explore or at least discuss the evolutionary trajectories that have led to particular responses. For example, trematodes alter the behavior of their hosts by becoming encysted in the host brains. This location might have been favoured early on by providing the parasite with a host immunity barrier. Another question of priority event concerns the hairworm. Adults mate in water and larvae manipulate their insect hosts toward suicidal drowning behavior. Thomas et al. (2012) question whether the behavioral manipulation preceded mating in water or whether mating in water provided the context by which behavioral manipulation was subsequently selected for. It is our hope that these and other issues in the study of insect behavior in relation to parasitism will continue to generate much interest.

References

Abbot, P., and Dill, L.M. (2001). Sexually transmitted parasites and sexual selection in the milkweed leaf beetle, *Labidomera clivicollis*. *Oikos*, **92**, 91.

Adamo, S.A. (2006). Comparative psychoneuroimmunology: evidence from the insects. *Behavioral and Cognitive Neuroscience Reviews*, **5**, 128.

Adamo, S.A. (2009). The impact of physiological state on immune function in insects. In: S. Reynolds and J. Rolff (Eds), *Insect Infection and Immunity*, pp. 173–86. Oxford University Press, Oxford.

Adamo, S.A. (2014). The effects of stress hormones on immune function may be vital for the adaptive reconfiguration of the immune system during fight-or-flight behavior. *Integrative and Comparative Biology*, **54**, 419.

Adamo, S.A. (2017). The stress response and immune system share, borrow, and reconfigure their physiological network elements: evidence from the insects. *Hormones and Behavior*, **88**, 25.

Adelman, J.S. and Martin, L.B. (2009). Vertebrate sickness behaviors: adaptive and integrated neuroendocrine immune responses. *Integrative and Comparative Biology*, **49**, 202.

Alma, C.R., Gillespie, D.R., Roitberg, B.D., and Goettel, M.S. (2010). Threat of infection and threat-avoidance behavior in the predator *Dicyphus hesperus* feeding on whitefly nymphs infected with an entomopathogen. *Journal of Insect Behavior*, **23**, 90.

Altizer, S., Bartel, R., and Han, B.A. (2011). Animal migration and infectious disease risk. *Science*, **331**, 296.

Arbuthnott, D., Levin, T.C., and Promislow, D.E.L. (2016). The impacts of *Wolbachia* and the microbiome on mate choice in *Drosophila melanogaster*. *Journal of Evolutionary Biology*, **29**(2), 461–8.

Arce, A.N., Johnston, P.R., Smiseth, P.T., and Rozen, D.E. (2012). Mechanisms and fitness effects of antibacterial defences in a carrion beetle. *Journal of Evolutionary Biology*, **25**, 930.

Ardekani, R., Biyani, A., Dalton, J.E., et al. (2012). Three-dimensional tracking and behaviour monitoring of multiple fruit flies. *Journal of the Royal Society Interface*, 20120547.

Ayres, J.S., and Schneider, D.S. (2009). The role of anorexia in resistance and tolerance to infections in Drosophila. *PLoS Biology* **7**, p.e1000150.

Bateson, P., and Laland, K.N. (2013). Tinbergen's four questions: an appreciation and an update. *Trends in Ecology and Evolution*, **28**, 712.

Bigio, G., Al Toufailia, H., and Ratnieks, F.L.W. (2014). Honey bee hygienic behaviour does not incur a cost via removal of healthy brood. *Journal of Evolutionary Biology*, **27**, 226.

Biron, D. G., Marché, L., Ponton, F., et al. (2005). Behavioural manipulation in a grasshopper harbouring hairworm: a proteomics approach. *Proceedings of the Royal Society B*, **272**, 2117.

Blanford, S., Thomas, M.B., and Langewald, J. (1998). Behavioural fever in the Senegalese grasshopper, *Oedaleus senegalensis*, and its implications for biological control using pathogens. *Ecological Entomology*, **23**, 9.

Botto-Mahan, C., Cattan, P.E., and Medel, R. (2006). Chagas disease parasite induces behavioural changes in the kissing bug *Mepraia spinolai*. *Acta Tropica*, **98**, 219.

Bradley, C.A., and Altizer, S. (2005). Parasites hinder monarch butterfly flight: implications for disease spread in migratory hosts. *Ecology Letters*, **8**, 290.

Burand, J.P., Rallis, C.P., and Tan, W. (2004). Horizontal transmission of Hz-2V by virus infected *Helicoverpa zea* moths. *Journal of Invertebrate Pathology*, **85**, 128.

Burand, J.P., Tan, W., Kim, W., Nojima, S., and Roelofs, W. (2005). Infection with the insect virus Hz-2v alters mating behavior and pheromone production in female *Helicoverpa zea* moths. *Journal of Insect Science*, **5**, 6.

Carmo-Sousa, M., Moreno, A., Garzo, E., and Fereres, A. (2014). A non-persistently transmitted-virus induces a pull-push strategy in its aphid vector to optimize transmission and spread. *Virus Research*, **186**, 38.

Chapuisat, M., Oppliger, A., Magliano, P., and Christe, P. (2007). Wood ants use resin to protect themselves against pathogens. *Proceedings of the Royal Society B*, **274**, 2013.

Cremer, S., Armitage, S.A.O., and Schmid-Hempel, P. (2007). Social immunity. *Current Biology*, **17**, R693.

Crosland, M.W.J., Lok, C.M., Wong, T.C., Shakarad, M., and Traniello, J.F. (1997). Division of labour in a lower termite: the majority of tasks are performed by older workers. *Animal Behavior*, **54**, 999–1012.

Curtis, V.A. (2014). Infection-avoidance behaviour in humans and other animals. *Trends in Immunology*, **35**, 457.

Damodaram, K.J.P., Ayyasamy, A., and Kempraj, V. (2016). Commensal bacteria aid mate-selection in the fruit fly, *Bactrocera dorsalis*. *Microbial Ecology*, **72**, 725.

de Roode, J.C., and Lefèvre, T. (2012). Behavioral immunity in insects. *Insects*, **3**, 789–820.

de Roode, J.C., Lefèvre, T., and Hunter, M.D. (2013). Self-medication in animals. *Science*, **340**, 150.

Dell, A.I., Bender, J.A., Branson, K., et al. (2014). Automated image-based tracking and its application in ecology. *Trends in Ecology and Evolution*, **29**, 417–28.

Dheilly, N.M., Maure, F., Ravallec, M., et al. (2015). Who is the puppet master? Replication of a parasitic wasp-associated virus correlates with host behaviour manipulation. *Proceedings of the Royal Society B*, **282**, 20142773.

Egnor, S.E.R., and Branson, K. (2016). Computational analysis of behavior. *Annual Review of Neuroscience*, **39**, 217–36.

Elliot, S.L., Blanford, S., and Thomas, M.B. (2002). Host–pathogen interactions in a varying environment: temperature, behavioural fever and fitness. *Proceedings of the Royal Society B*, **269**, 1599–607.

Feener, D.H., and Brown, B. V. (1992). Reduced foraging of *Solenopsis geminata* (Hymenoptera: Formicidae) in the presence of parasitic *Pseudacteon* spp. (Diptera' Phoridae). *Annals of the Entomological Society of America*, **85**, 80.

Fouks, B., and Lattorff, H.M.G. (2011). Recognition and avoidance of contaminated flowers by foraging bumblebees (*Bombus terrestris*). *PLoS One*, **6**(10), e26328.

Gaugler, R., Wang, Y.I., and Campbell, J.F. (1994). Aggressive and evasive behaviors in Popillia japonica (Coleoptera: Scarabaeidae) larvae: defenses against entomopathogenic nematode attack. *Journal of Invertebrate Pathology*, **64**, 193.

Giannoni-Guzmán, M.A., Avalos, A., Perez, J.M., et al. (2014). Measuring individual locomotor rhythms in honey bees, paper wasps and other similar-sized insects. *Journal of Experimental Biology*, **217**, 1307.

Gilestro, G.F. (2012). Video tracking and analysis of sleep in *Drosophila melanogaster*. *Nature Protocols*, **7**, 995.

Goater, T.P., Goater, C.P., and Esch G.W. (2014). *Parasitism: the diversity and ecology of animal parasites*, 2nd edn. Cambridge University Press, Cambridge.

Goldsworthy, G., Opoku-Ware, K., and Mullen, L. (2002). Adipokinetic hormone enhances laminarin and bacterial lipopolysaccharide-induced activation of the prophenoloxidase cascade in the African migratory locust, *Locusta migratoria*. *Journal of Insect Physiology*, **48**, 601.

Grosman, A.H., Janssen, A., De Brito, E.F., et al. (2008). Parasitoid increases survival of its pupae by inducing hosts to fight predators. *PLoS One*, **3**, e2276.

Harabis, F., Dolný, A., Helebrandova, J., and Rusková, T. (2015). Do egg parasitoids increase the tendency of *Lestes sponsa* (Odonata: Lestidae) to oviposit underwater? *European Journal of Entomology*, **112**, 63.

Hart, B.L. (1988). Biological basis of the behavior of sick animals. *Neuroscience & Biobehavioral Reviews*, **12**, 123.

Hart, B.L. (2011). Behavioural defences in animals against pathogens and parasites: parallels with the pillars of medicine in humans. *Philosophical Transactions of the Royal Society B*, **366**, 3406.

Harvey, J.A., Tanaka, T., Kruidhof, M., Vet, L.E.M., and Gols, R. (2011). The 'usurpation hypothesis' revisited: dying caterpillar repels attack from a parasitoid wasp. *Animal Behavior*, **81**, 1281.

Heil, M. (2016). Host manipulation by parasites: cases, patterns, and remaining doubts. *Frontiers in Ecology and Evolution*, **4**, 80.

Heinze, J., and Walter, B. (2010). Moribund ants leave their nests to die in social isolation. *Current Biology*, **20**, 249.

Hohorst, W., and Graefe, G. (1961). Ameisen—obligatorische Zwischenwirte des Lanzettegels (*Dicrocoelium dendriticum*). *Naturwissenschaften*, **48**(7), 229.

Hölldobler, B. (2012). Afterward. In: D. P. Hughes, J. Brodeur, and F. Thomas (Eds), *Host Manipulation by Parasites*, pp. 155–7. Oxford University Press, Oxford.

Huang, J., Wu, S-F., Li, X-H., Adamo, S.A., and Ye, G-Y. (2012). The characterization of a concentration-sensitive α-adrenergic-like octopamine receptor found on insect immune cells and its possible role in mediating stress hormone effects on immune function. *Brain, Behavior, and Immunity*, **26**, 942.

Hunt, V.L., Zhong, W., McClure, C.D., et al. (2016). Cold-seeking behaviour mitigates reproductive losses from fungal infection in *Drosophila*. *Journal of Animal Ecology*, **85**, 178.

Hurd, H. (2003). Manipulation of medically important insect vectors by their parasites. *Annual Review of Entomology*, **48**, 141.

Kacsoh, B.Z., Lynch, Z.R., Mortimer, N.T., and Schlenke, T.A. (2013). Fruit flies medicate offspring after seeing parasites. *Science*, **339**, 947.

Kalsbeek, V., Mullens, B.A., and Jespersen, J.B. (2001). Field studies of entomophthora (Zygomycetes: Entomophthorales)—induced behavioral fever in *Musca domestica* (Diptera: Muscidae) in Denmark. *Biological Control*, **21**, 264.

Kazlauskas, N., Klappenbach, M., Depino, A.M., and Locatelli, F.F. (2016). Sickness Behavior in Honey Bees. *Frontiers in Physiology* 7.

Klopfer, P.H., and Hailman, J.P. (1972). *Function and Evolution of Behavior*. Addison-Wesley Pub. Co., Boston, MA.

Kohmura, H., Hirayama, H., and Ueno, T. (2015). Diving into the water: cues related to the decision-making by an egg parasitoid attacking underwater hosts. *Ethology*, **121**, 168.

Konrad, M., Vyleta, M.L., Theis, F.J., et al. (2012). Social transfer of pathogenic fungus promotes active immunisation in ant colonies. *PLoS Biology*, **10**, e1001300.

Kreuter, K., Bunk, E., Lückemeyer, A., et al. (2012). How the social parasitic bumblebee Bombus bohemicus sneaks into power of reproduction. *Behavioral Ecology and Sociobiology*, **66**, 475.

Lafferty, K.D., and Kuris, A.M. (2012). Ecological consequences of manipulative parasites. In: D.P. Hughes, J. Brodeur, and F. Thomas (Eds), *Host Manipulation by Parasites*, pp. 158–79. Oxford University Press, Oxford.

Laland, K., Wray G.A., and Hoekstra, H.E. (2014). Does evolutionary theory need a rethink? *Nature*, **514**, 161.

Lambardi, D., Dani, F.R., Turillazzi, S., and Boomsma, J.J. (2007). Chemical mimicry in an incipient leaf-cutting ant social parasite. *Behavioral Ecology and Sociobiology*, **61**, 843.

Lee, K.P., Simpson, S.J., and Wilson, K. (2008). Dietary protein-quality influences melanization and immune function in an insect. *Functional Ecology*, **22**, 1052.

Lefèvre, T., Oliver, L., Hunter, M.D., and de Roode, J.C. (2010). Evidence for trans-generational medication in nature. *Ecology Letters*, **13**, 1485–93.

Lopes, P.C. (2014). When is it socially acceptable to feel sick? *Proceedings of the Royal Society B*, **281**, 20140218.

Lopes, P.C. (2017). Why are behavioral and immune traits linked? *Hormones and Behavior*, **88**, 52.

Lusebrink, I., Dettner, K., and Seifert, K. (2008). Stenusine, an antimicrobial agent in the rove beetle genus *Stenus* (Coleoptera, Staphylinidae). *Naturwissenschaften*, **95**, 751.

Malfi, R.L., Davis, S.E., and Roulston, T.H. (2014). Parasitoid fly induces manipulative grave-digging behaviour differentially across its bumblebee hosts. *Animal Behavior*, **92**, 213.

McBride, C.S., and Arguello, J.R. (2007). Five *Drosophila* genomes reveal nonneutral evolution and the signature of host specialization in the chemoreceptor superfamily. *Genetics*, **177**, 1395.

McMenemy, L.S., Hartley, S.E., MacFarlane, S.A, et al. (2012). Raspberry viruses manipulate the behaviour of their insect vectors. *Entomologia Experimentalis et Applicata*, **144**, 56.

Mersch, D.P., Crespi, A., and Keller, L. (2013). Tracking individuals shows spatial fidelity is a key regulator of ant social organization. *Science*, **340**, 1090.

Moore, J. (2013). An overview of parasite-induced behavioral alterations—and some lessons from bats. *Journal of Experimental Biology*, **216**, 11.

Moore, J., and Freehling, M. (2002). Cockroach hosts in thermal gradients suppress parasite development. *Oecologia*, **133**, 261.

Müller, C.B., and Schmid-Hempel, P. (1993). Exploitation of cold temperature as defence against parasitoids in bumblebees. *Nature*, **363**, 65.

Murray, M.J., and Murray, A.B. (1979). Anorexia of infection as a mechanism of host defense. *American Journal of Clinical Nutrition*, **32**, 593.

Neoh, K-B., Yeap, B-K., Tsunoda, K., Yoshimura, T., and Lee, C-Y. (2012). Do termites avoid carcasses? Behavioral responses depend on the nature of the carcasses. *PLoS One*, **7**, e36375.

Nielsen, C., Agrawal, A.A., and Hajek, A.E. (2010). Ants defend aphids against lethal disease. *Biology Letters*, **6**(2), 205–8.

Orr, M. (1992). Parasitic flies (Diptera: Phoridae) influence foraging rhythms and caste division of labor in the leafcutter ant, *Atta cephalotes* (Hymenoptera: Formicidae). *Behavioral Ecology and Sociobiology*, **30**, 395.

Parker, B.J., Elderd, B.D., and Dwyer, G. (2010). Host behaviour and exposure risk in an insect-pathogen interaction. *Journal of Animal* Ecology, **79**, 863.

Pfeiffenberger, C., Lear, B.C., Keegan, K.P., and Allada, R. (2010). Locomotor activity level monitoring using the *Drosophila* activity monitoring (DAM) system. *Cold Spring Harbor Protocols*, **11**, pdb.prot5518.

Pie, M.R., Rosengaus, R.B., and Traniello, J.F.A. (2004). Nest architecture, activity pattern, worker density and the dynamics of disease transmission in social insects. *Journal of Theoretical Biology*, **226**, 45.

Ponton, F., Wilson, K., Holmes, A.J., et al. (2013). Integrating nutrition and immunology: a new frontier. *Journal of Insect Physiology*, **59**, 130.

Poulin, R. (1995). 'Adaptive' changes in the behaviour of parasitized animals: a critical review. *International Journal for Parasitology*, **25**, 1371.

Poulin, R. (2010). Parasite manipulation of host behavior: an update and frequently asked questions. In: T.J.R.H. Jane Brockmann, M. Naguib, K. E. Wynne-Edwards, J. C. Mitani, and L. W. Simmons (Eds), *Advances in the Study of Behavior*, pp. 151–86. Academic Press, Cambridge, MA.

Reiser, M. (2009). The ethomics era? *Nature Methods*, **6**, 413.

Rigaud, T., Perrot-Minnot, M.J., and Brown, M.J. (2010). Parasite and host assemblages: embracing the reality will improve our knowledge of parasite transmission and virulence. *Proceedings of the Royal Society B*, **277**, 3693.

Rosengaus, R.B., James, L-T., Hartke, T.R., and Brent, C.S. (2011). Mate preference and disease risk in *Zootermopsis angusticollis* (Isoptera: Termopsidae). *Environmental Entomology*, **40**, 1554.

Rund, S.S.C., O'Donnell, A.J., Gentile, J.E., and Reece, S.E. (2016). Daily rhythms in mosquitoes and their consequences for malaria transmission. *Insects*, **7**, 14.

Sadek, M.M., Hansson, B.S., and Anderson, P. (2010). Does risk of egg parasitism affect choice of oviposition sites by a moth? A field and laboratory study. *Basic and Applied Ecology*, **11**, 135.

Sato, T., Egusa, T., Fukushima, K., et al. (2012). Nematomorph parasites indirectly alter the food web and ecosystem function of streams through behavioural manipulation of their cricket hosts. *Ecology Letters*, **15**, 786.

Satterfield, D.A., Maerz, J.C., and Altizer, S. (2015). Loss of migratory behaviour increases infection risk for a butterfly host. *Proceedings of the Royal Society B*, **282**, 20141734.

Schmid-Hempel, P. (2011). *Evolutionary Parasitology: the Integrated Study of Infections, Immunology, Ecology, and Genetics*. Oxford University Press, Oxford.

Schutgens, M., Cook, B., Gilbert, F., and Behnke, J.M. (2015). Behavioural changes in the flour beetle *Tribolium confusum* infected with the spirurid nematode *Protospirura muricola*. *Journal of Helminthology*, **89**, 68.

Shirasu-Hiza, M.M., Dionne, M.S., Pham, L.N., Ayres, J.S., and Schneider, D.S. (2007). Interactions between circadian rhythm and immunity in *Drosophila melanogaster*. *Current Biology*, **17**, R353.

Simon, A.F., Chou, M-T., Salazar, E.D., et al. (2012). A simple assay to study social behavior in *Drosophila*: measurement of social space within a group. *Genes, Brain and Behavior*, **11**, 243.

Singer, M.S., Mason, P.A., and Smilanich, A.M. (2014). Ecological immunology mediated by diet in herbivorous insects. *Integrative and Comparative Biology*, **54**, 913.

Siva-Jothy, M.T. (1999). Male wing pigmentation may affect reproductive success via female choice in a calopterygid damselfly Zygoptera). *Behaviour*, **136**, 1365.

Stahlschmidt, Z.R., and Adamo, S.A. (2013). Context dependency and generality of fever in insects. *Naturwissenschaften*, **100**, 691.

Starks, P.T., Blackie, C.A., and Seeley, T.D. (2000). Fever in honeybee colonies. *Naturwissenschaften*, **87**, 229.

Stensmyr, M.C., Dweck, H.K.M., Farhan, A., et al. (2012). A conserved dedicated olfactory circuit for detecting harmful microbes in *Drosophila*. *Cell*, **151**, 1345.

Sullivan, K., Fairn, E., and Adamo, S.A. (2016). Sickness behaviour in the cricket *Gryllus texensis*: comparison with animals across phyla. *Behavioral Processes*, **128**, 134.

Tasin, M., Knudsen, G.K., and Pertot, I. (2012). Smelling a diseased host: grapevine moth responses to healthy and fungus-infected grapes. *Animal Behavior*, **83**, 555.

Thomas, F., Maure, F., Brodeur, J., et al. (2011). The cost of a bodyguard. *Biology Letters*, **7**, 843.

Thomas, F., Rigaud, T., and Brodeur, J. (2012). Evolutionary routes leading to host manipulation by parasites. In: D.P. Hughes, J. Brodeur, and F. Thomas, (Eds), *Host*

Manipulation by Parasites, pp. 16–35. Oxford University Press, Oxford.

Thomas, M.B., and Blanford, S. (2003). Thermal biology in insect–parasite interactions. *Trends in Ecology and Evolution*, **18**, 344.

Tinbergen, N. (1963). On aims and methods of ethology. *Zeitschrift für Tierpsychologie*, **20**, 410.

Tregenza, T., Simmons, L.W., Wedell, N., and Zuk, M. (2006). Female preference for male courtship song and its role as a signal of immune function and condition. *Animal Behavior*, **72**, 809.

Turillazzi, S., Mastrobuoni, G., Dani, F.R., et al. (2006). Dominulin A and B: two new antibacterial peptides identified on the cuticle and in the venom of the social paper wasp *Polistes dominulus* using MALDI-TOF, MALDI-TOF/TOF, and ESI-Ion Trap. *Journal of the American Society for Mass Spectrometry*, **17**, 376.

Vale, P.F., and Jardine, M.D. (2015). Sex-specific behavioural symptoms of viral gut infection and *Wolbachia* in *Drosophila melanogaster*. *Journal of Insect Physiology*, **82**, 28.

Vale, P.F., and Jardine, M.D. (2017). Infection avoidance behavior: viral exposure reduces the motivation to forage in female Drosophila melanogaster. *Fly*, **11**(1), 3–9.

Valentine, B.D. (2007). Mutual grooming in cucujoid beetles (Coleoptera: Silvanidae). *Insecta Mundi* **2007** (0001–0008), 1–3.

Weinstein, S.B., and Kuris, A.M. (2016). Independent origins of parasitism in Animalia. *Biology Letters*, **12**, 20160324.

Zhukovskaya, M., Yanagawa, A., and Forschler, B. (2013). Grooming behavior as a mechanism of insect disease defense. *Insects*, **4**, 609.

CHAPTER 19

Behavioral, plastic, and evolutionary responses to a changing world

Wolf U. Blanckenhorn

Evolutionary Biology & Environmental Studies, University of Zürich-Irchel, Winterthurerstrasse 190, CH-8057 Zürich, Switzerland

19.1 Introduction: types of responses to environmental change

Environmental change is a fact of life, a characteristic of nature, and always has been. It occurs more or less all the time, everywhere. Whether and how much humans are involved in producing environmental change is certainly interesting, but from the viewpoint of the affected organisms, it is merely a secondary question. Many organisms alter their environment such that it becomes inhospitable for others—buffalo graze the prairie such that no trees can grow, beavers flood an area by building a dam, plants produce poisons so they are not eaten by insect larvae, bacteria on a petri dish outgrow other microorganisms by excreting toxins. What matters is whether, and how fast, organisms can respond to a changing environment. In general, there are only few options:

- organisms can move away, a behavioral response;
- they can acclimate their phenotype, a plastic, often physiological response;
- they can adapt, an evolutionary, implying a genetic response;
- they will die and become (locally) extinct (Williams et al. 2008; Table 19.1).

These responses are, of course, not mutually exclusive, so combinations are very likely.

The type and rate of responses of organisms to environmental change will mainly depend on the nature of the trait and the time-scale of the environmental change (Holt 1990; Kokko and López-Sepulcre 2006; Table 19.1). A catastrophic event, such as a meteorite impact, leaves far fewer options to react than a slow and gradual temperature change—the latter permits *in situ* adaptation, the former probably not. Different traits will also respond differently. Traits that are fixed for a given individual, body size in adult insects, for instance, can at the earliest respond in the next generation, whereas behavior, such as dispersal, (micro)habitat or mate choice can be adjusted immediately within individuals; the response of physiological traits such as thermoregulation or diapause induction (Tauber et al. 1986; Huey et al. 2012; Kleckova et al. 2014) will be intermediate (Table 19.1). A behavioral response will always be fastest because it is inherently plastic and flexible (Fagen 1982; Clark and Ehlinger 1987; Wong and Candolin 2012; Sih 2013; Table 19.1), which does not mean, however, that behavior has no heritable component that can evolve in the long term (Mousseau and Roff 1987). The main difficulty in assessing and categorizing these responses lies in distinguishing evolutionary

Blanckenhorn, W. U., *Behavioral, plastic, and evolutionary responses to a changing world*. In: *Insect Behavior: From mechanisms to ecological and evolutionary consequences*. Edited by Alex Córdoba-Aguilar, Daniel González-Tokman, and Isaac González-Santoyo: Oxford University Press (2018). © Oxford University Press.
DOI: 10.1093/oso/9780198797500.003.0019

Table 19.1 A classification of the possible responses of organisms to environmental change according to the trait involved, proximate mechanism, time scale, frequency of occurrence, and the variance component where variation may be detected (examples in the text).

Response	Trait type	Proximate change	Flexibility	Frequency	Time scale of response	Variance component (cf. Figure 19.1)
Evolution	Life history	Gene frequency	Low	Low	Across generations (slow)	V_G or single alleles
Acclimation (plasticity)	Physiology	Physiological/developmental	Intermediate	Intermediate	Minutes to months, within or between generations	V_{GxE} (or V_E)
Dispersal/range- or habitat-shift	Behavior	Behavioral	High	High	Seconds to days (fast), within generations	V_{GxE} or V_E

from plastic responses, and in evaluating the relative importance of immediate behavioral reactions (Gienapp et al. 2008; Sánchez-Guillen et al. 2016; Table 19.1).

19.2 Evaluating the frequency and importance of various responses to environmental change

In an era of human-induced environmental change, recently termed the 'anthropocene' (Finnay and Edwards 2016), biodiversity research should focus on investigating, and ultimately predicting, the relative importance of the three types of fundamental responses listed in Table 19.1 for any given organism. This is no trivial task, such that a formal meta-analysis would be premature at this point in time (but see Stoks et al. 2014; Schilthuizen and Kellermann 2014, for recent reviews of the insect literature, and also other papers in a special issue of *Evolutionary Applications*). Ultimately, this is an empirical, after-the-fact question, though we can formulate some a priori expectations. For this, it is best to take a population genetic view, because dispersal propensity will exert the greatest impact on all types of responses, and with it any characteristic that strongly affects dispersal, most notably the taxon and body size (Gienapp et al. 2008; Travis et al. 2013). The reason is obvious—if I no longer find the environment I am in suitable, I should move away. Mobile animals, such as most vertebrates and insects, can do this more or less effectively and immediately. For immobile organisms, such as plants, it will be more difficult, as

they can typically relocate only cross-generationally by way of seed dispersal.

Insects have the principal ability to directly respond to harsh environmental conditions by behaviors such as dispersal or (micro)habitat choice (cf. Table 19.1), thereby shifting their habitat and, ultimately, their range. That this happens has been documented (e.g., for grasshoppers: Hochkirch and Damerau 2009). Such immediate dispersal will be highly species-specific, but generally more efficient and faster for winged than for unwinged insects (Roff and Fairbairn 1991; Hochkirch and Damerau 2009; Feder et al. 2010), and also easier to perform for larger insects with better flight ability (e.g., dragonflies or grasshoppers versus aphids); it will also be more likely to happen if the environmental challenge is drastic and fast (i.e. extreme) than when it is gradual and slow.

Nevertheless, whether and how often range shifts actually occur through directed dispersal of individuals, and whether it is even part of an insect's repertoire to perceive, anticipate, and respond to slow and gradual environmental changes, is surprisingly under-researched (Travis et al. 2013; Lindström and Lehmann 2015). To distinguish whether range shifts occur by way of behavioral, plastic, or evolutionary responses it is necessary to document whether dispersal happens within as opposed to across generations. This can be estimated by classic mark-recapture studies involving direct and repeated observations of single individuals and/or, at a more anonymous level, by indirect population-genetic methods (Rannala and Mountain 1997; Paetkau et al. 2004; Excoffier et al. 2009; e.g., for grasshoppers: Hochkirch and Damerau 2009; butterflies: Öckinger and van Dyck

2012). Such studies are surprisingly rare in the climate change context. Rather, researchers increasingly rely on cross-sectional sightings, often using data from repositories generated by specialists as well as amateur naturalists (e.g., Altermatt 2010; Phillimore et al. 2012; Karlsson 2014). Such data, however, are typically not individualized and hence do not allow discriminating within- from cross-generational responses, i.e. pure behavior from plasticity or evolution. Note that this is primarily a methodological issue concerning the nature of the data and not a fundamental constraint.

Distinguishing plastic from evolutionary responses is more difficult, as both occur over longer time scales, typically involving multiple generations (Table 19.1; as previously also argued for vertebrates by Gienapp

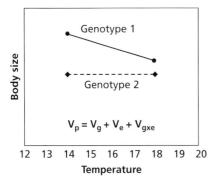

Figure 19.1 A quantitative genetic framework to distinguish genetic (evolutionary) from plastic responses. In the simplest case, the phenotypic variance of a quantitative trait (V_p), e.g., a behavioral or morphological trait, such as foraging, mating, or body size, is partitioned into a genetic (V_g), an environmental (V_e), and a genotype-by-environment component (V_{gxe}): $V_p = V_g + V_e + V_{gxe}$ (Roff 1997). By way of common-garden laboratory or reciprocal-transplant field approaches, these variance components can be separated and estimated using analysis of variance. In this example, genotype 1 is larger than genotype 2 at any temperature, so there is genetic variation in body size in the population ($= V_g$). As temperature increases, genotype 1 decreases in size ($= V_e$), so this genotype is phenotypically plastic. As this is not the case for genotype 2, there is genetic variation in the response of genotypes to temperature in the population ($= V_{gxe}$). Phenotypic plasticity consequently has two components, a purely environmental (V_e) and a heritable (V_{gxe}) component (Via and Lande 1985). The latter evolves, the former not. Such responses of phenotypic traits to any variable environmental factor are called **"reaction norms"** which need not be linear (as drawn here) but can be complex (van Noordwijk and de Jong 1986). Also, here plasticity is viewed as a property of individual genotypes. Alternative interpretations allow viewing such reaction norms as a property of entire populations.

et al. 2008 or Merilä and Hendry 2014). Of course, at least for model organisms and traits suited for laboratory experimentation, plastic and evolutionary (i.e. adaptive genetic) responses can be separated using common-garden laboratory or reciprocal-transplant field approaches as outlined in Figure 19.1.

For example, empirical investigations of temperature effects on various life history and thermal traits in *Drosophila* (e.g., body size, growth rate or development time; **critical thermal maxima**, heat shock responses, **chill coma recovery**, etc.), dung flies or butterflies often specifically incorporate **acclimation** and reveal that such plastic physiological effects are substantial (e.g., Fischer and Karl 2010; Santos et al. 2011; Blanckenhorn et al. 2014; Esperk et al. 2016). In combination with studies of latitudinal or altitudinal populations of the same species (e.g., Huey et al. 2000; Blanckenhorn and Demont 2004; Balanyà et al. 2006; Scharf et al. 2010; Klepsatel et al. 2014; Kapun et al. 2016), this work suggests that plastic trait responses to temperature and other climatic variables are much more substantial than adaptive genetic differentiation among populations. The same holds true for diapause induction (Tauber et al. 1986; Bradshaw and Holzapfel. 2001; Scharf et al. 2010).

How about species that cannot easily be investigated in the laboratory in such a manner, which is the vast majority even for insects, and for which we therefore might only have field data, e.g., from citizen science data banks (e.g., Altermatt 2010; Karlsson 2014)?

Phillimore et al. (2012) recently presented an interesting approach for such cases (summarized in Figure 19.2a), allowing at least qualitative differentiation between plastic and genetic evolutionary responses to climate change. To investigate whether climate warming has extended the season, expected to cause earlier emergence from winter diapause and, consequently, a phenological shift, these authors analysed first spring sightings for many geographic populations of the orange tip butterfly over multiple years in England and Sweden. They argued that geographic (i.e. spatial) variation among populations principally can be caused either by plasticity or local adaptation to the environment due to the long-term action of natural selection. In contrast, short-term (temporal) phenological shifts over few years should clearly reflect plasticity, if only because evolution by natural

(a)

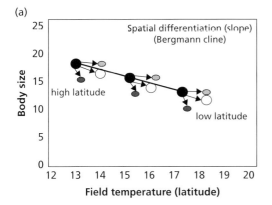

(b)

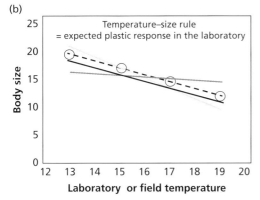

Figure 19.2 Comparison of short-term temporal changes in a trait (here body size) in response to environmental (e.g., climate) change, assumed to reflect merely plasticity, to the existing *spatial* differentiation among populations, assumed to reflect long-term local adaptation to the environment (e.g., climate), as assessed using 'big' data from monitoring repositories. In (a) the black line reflects spatial (i.e. genetic) differentiation among populations, here along a temperature or latitudinal gradient (*x*-axis), suggesting a Bergmann cline (larger size in colder climes or at higher latitudes). If temporal shifts in the body size of local populations over several years (denoted by white circles) occur along the black spatial axis (slope), i.e. spatial = temporal slope, then the null hypothesis that populations are differentiated due to plasticity alone, with no (additional) local adaptation, cannot be rejected. If, however, the temporal slope is shallower (light grey) or steeper (dark grey) than the black spatial slope, then co-gradient or counter-gradient (i.e. compensatory) local adaptation of populations can be inferred, because populations (black circles) are spatially differentiated more and less (respectively) than expected by plasticity alone (adapted from Phillimore et al. 2012). In (b) a similar pattern is (re-)interpreted in light of the a priori plastic temperature-size expectation for ectothermic organisms (smaller when warmer, Atkinson 1994), which can be estimated in the laboratory for many insects (hatched slope, white circles). If spatial populations (black slope) fall along the temperature-size (hatched) slope, plasticity can be inferred as the main (only) cause of differentiation. If the spatial slope is shallower (dark grey) or steeper (light grey) than the black temperature–size slope, counter- and co-gradient local adaptation, respectively, is revealed.

selection is unlikely to result in change so quickly, a reasonable assumption. Thus, by comparing the spatial slope, here representing an expected latitudinal cline in spring phenology of the butterfly from south to north, with the temporal slope in a so-called space-for-time approach, one can evaluate the degree of local adaptation of populations—if the two slopes are similar, all visible geographic differentiation is likely merely due to plasticity (the null hypothesis of no local adaptation); if they differ, there must be some local adaptation beyond the effects of plasticity. Moreover, a steeper spatial than temporal slope indicates a co-gradient, whereas a shallower slope indicates a counter-gradient adaptational pattern (Figure 19.2a; Conover and Present 1990; Blanckenhorn and Demont 2004). In nature, such phenological shifts in seasonal timing should reflect a combination of climatic factors, including season length, temperature, plus possibly many others. In principle, therefore, this conceptual approach is applicable to other traits and environmental variables.

An approach combining the previous two (cf. Figure 19.1, 2a) would also be fruitful. For species that can be bred in the laboratory, the relationship between an environmental variable and any trait of choice can be empirically determined. For example, relationships between traits and temperature are called **thermal performance curves (TPC)** and have a long tradition in scientific research (Angilletta 2009). Most often life history traits are investigated (e.g., growth rate, body size, etc.: Angilletta 2009), but behavioral traits are also interesting in this context (e.g., insect locomotor activity: Kjærsgaard et al. 2010, Figure 19.3d). While TPC are usually expected to be hump-shaped, a quasi-linear relationship often shows in the middle of the temperature range (e.g., for development rate of three fly species in Blanckenhorn 1999, Figure 19.3c), often giving rise to the well-known temperature-size-rule (TSR) describing a negative relationship between body size and temperature in a great majority of ectothermic organisms (Atkinson and Sibly 1997; Figure 19.3 a,b). Although its ultimate cause remains unclear to this day, this relationship is entirely plastic and thus, in fact, defines the null relationship between body size and temperature due to plasticity, against which any field data of the sort investigated by Phillimore et al. (2012) can be evaluated (Figure 19.2b).

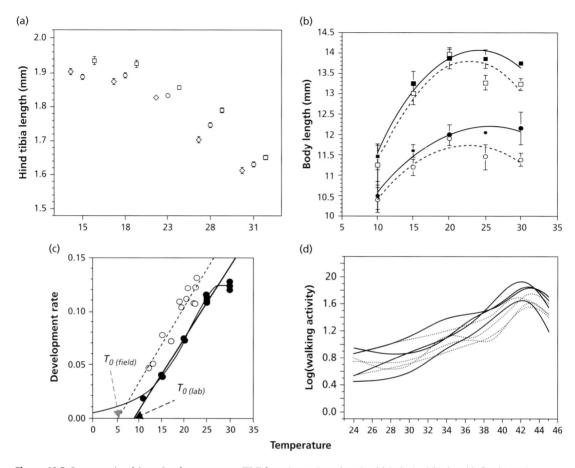

Figure 19.3 Four examples of thermal performance curves (TPC) for various traits and species. (a) Body size (tibia length) of male *Sepsis punctum* flies from southern (diamonds), central (circles) and northern European (squares) populations at 5 constant rearing temperatures in the laboratory (unpublished data from Berger et al. 2013). The pattern represents the right section only of a typical TPC and largely agrees with the temperature-size-rule (smaller at warmer temperatures: Atkinson and Sibly 1997). (b) Body length as a function of five constant laboratory temperatures for female (squares) and male (circles) water striders (*Aquarius remigis*) from a warm stream (black symbols and solid lines) and an adjacent cold stream (white symbols and hatched lines; data from Blanckenhorn 1991). The hump-shaped pattern represents an entire TPC and thus does not agree with the temperature-size-rule. (c) Development rate (= 1/ egg-to-adult development time) at five constant laboratory temperatures (black dots and solid lines), as well as at naturally varying mean temperatures in the field (white dots and hatched line), for the black scavenger fly *Sepsis cynipsea* (data from Blanckenhorn 1999). The increasing, non-linear solid line depicts the classic TPC shape left of the maximum. The x-intercepts of the corresponding linear regressions estimate the different lower thermal thresholds T_0 for the field and the lab at which development ceases. (d) Walking activity (a behavioral trait) at a range of laboratory temperatures of *Sepsis punctum* flies from four replicate laboratory lines evolved at either hot (31° C, solid lines) or cold (15° C, dotted lines) temperatures for twenty to forty generations (Blanckenhorn et al., unpublished data). Full TPCs are displayed.

In principle, temperature–performance relationships (or curves) can and should be systematically estimated in the laboratory for all important environmental parameters in as many species as possible. This is relatively easy to do for the major important environmental variables, such as food availability (cf. Blanckenhorn 1998, Figure 19.4b), while admittedly more difficult for others, for instance season length, responses to which vary strongly in nature and depend greatly on the concrete environmental factor manipulated in the laboratory (photoperiod, temperature, latitude, etc.: Blanckenhorn and Demont 2004). Nevertheless, though tedious and rather unrewarding for the impact factor race, and therefore seldom performed specifically in this context, performance relationships as in Figures

19.3 and 19.4 are easily generated, at least for insects that can be held in the laboratory, and stored in repositories for the benefit of future meta-analyses (e.g., Teder and Tammaru 2005), but also to test null expectations with citizen science and other 'big data' in the environmental change context (cf. Altermatt 2010; Phillimore et al. 2012). Such a concerted effort of the entire scientific community would help implement Holt's (1990) plea for a thorough understanding of a species' behavior, physiology, ecology, and genetics being necessary to best predict what will happen in situations of environmental change.

Finally, as most recently pointed out by Gienapp et al. (2008), Schilthuizen and Kellermann (2014), and Merilä and Hendry (2014), but also multiply before, a successful approach distinguishing evolution from plasticity ultimately must involve assessing genetic as opposed to merely phenotypic evidence. Following what was said above, there are several approaches at hand (summarized in Merilä and Hendry 2014). In the field, so-called animal model studies, which permit investigation of phenotypic and genotypic changes over extended time periods, often in combination with estimation of selection in the wild (Kruuk 2004; Merilä and

Hendry 2014; e.g., Bonnet et al. 2016), clearly are the state-of-the-art method of choice. However, for insects, which cannot be easily tracked individually in the field, and for which there is generally no genetic pedigree information, this approach is next to impossible (Schilthuizen and Kellermann 2014; Stoks et al. 2014; cf. Blanckenhorn 2015). Instead, phenotypic field studies of insects are best combined with quantitative genetic, common-garden, reciprocal transplant, and/or experimental evolution studies in the laboratory or under semi-natural field conditions (as exemplified for dung flies by Blanckenhorn 2009; Scharf et al. 2010; Esperk et al. 2016). Unfortunately, detection of genetic patterns must remain limited and indirect in such approaches lacking pedigree information (Blanckenhorn 2015). Alternatively, studies can focus on particular candidate genes, associated genetic markers, or chromosome inversions that mediate climate adaptation (e.g., Balanyà et al. 2006; Hoffmann and Daborn 2007; Kapun et al. 2016). Finally, distributional (range) changes associated with particular allele frequency changes affected by climate shifts can be studied with modern population genetic methods in combination with the above candidate gene or genomic approaches, in addition to or as alternatives

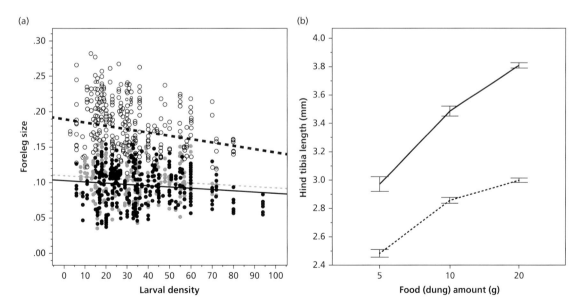

Figure 19.4 Two examples of performance curves for body size in response to other changing environmental variables. (a) Foreleg size of female (grey) and dimorphic male (white circles and hatched line, or black circles and solid line) *Sepsis thoracica* flies decreases with larval density in the laboratory (Busso et al., 2017). (b) Hind tibia length of female (grey hatched) and male (black solid) yellow dung flies increases with the amount of larval food (data from Blanckenhorn 1998).

to mark-recapture studies (Rannala and Mountain 1997; Paetkau et al. 2004; Hoffmann and Daborn 2007; Schilthuizen and Kellermann 2014; Stoks et al. 2014). Modern genomic methods have facilitated advances in this realm of environmental change research. However, identification of direct mechanistic links between particular genes evolving by natural selection in response to environmental changes and the corresponding (behavioral) phenotypes at the whole-organism level are still very rare. The promise, and the hope is that genomics will foster many such breakthroughs in the near future.

Assessing whether insect responses to environmental changes in nature are behavioral, phenotypically plastic or genetic (i.e. evolutionary) thus remains a challenge. This should not keep us from trying to gather such information, which is becoming easier in the age of genomics. Some classic methods to particularly target behavior, such as mark–recapture studies, are currently underused in this context. There is no way around integrative, long-term studies of especially suitable model insects to ultimately quantitatively assess the relative importance of the three fundamental response types listed in Table 19.1. Intuitively, following what was argued previously, and primarily based on the reaction time, purely behavioral responses should be most common, followed by plastic responses, with evolutionary responses being rarest. This requires verification. Combinations might turn out to be most effective over evolutionary time. After all, we must not forget that phenotypic plasticity has its limits, too (DeWitt et al. 1998), and sometimes complex species interactions may constrain the evolution of adaptive responses even further, as nicely exemplified by the de-synchronization and lack of co-evolution of the spring phenology of great tits (top predators), their caterpillar food items, and the plant food of the latter (Nussey et al. 2005).

19.3 Exemplary behavioral responses of insects to environmental change

Having laid out a research approach to study responses of insects to environmental change, I now turn to some notable exemplary studies. After focusing in my examples above largely on climate change and life history, rather than behavioral traits, for which

the literature is richest, the examples discussed below span a broader range of environmental change types and traits, with special focus on behavior. These include **urbanization**, artificial light environments, (de)forestation, habitat fragmentation, river or stream (de)regulation, salinity changes, and pollution. While most of these environmental changes clearly occur naturally, many of them are nowadays closely linked to human activities. I emphasize that this is not a comprehensive review, but a sketch of approaches that can and have been used to study such questions in insects. Similar conceptual frameworks about behavioral responses of animals in a changing world have been recently published (Dukas 2004; Dingemanse et al. 2010; Wong and Candolin 2012, 2015; Sih 2013; Hale et al 2017). The section is organized mainly by behavioral trait type, followed by environmental change type.

It is also re-emphasized that behavioral traits as a rule co-evolve with morphology and physiology, often in the sense of a life-history syndrome, and therefore often can, but should not be treated in isolation. Thus, flight and dispersal behavior typically depend on the expression of wings and their muscles (Roff and Fairbairn 1991; Feder et al. 2010), diapause behavior on its induction and the life-history stage affected (Tauber et al. 1986), and thermoregulatory behavior on the coloration of the species (Forsman et al. 2002). Work conducted on particular species is frequently split among research disciplines mainly treating behavior versus life history, implying that the full story often cannot be gathered from single papers. This seems inevitable, but impedes our main goal of evaluating the relative importance of behavior, plasticity, and evolution in response to environmental changes.

19.3.1 Behavioral dispersal and mobility traits

As argued above, behaviors such as migration, dispersal, mobility, and (micro)habitat choice will most immediately mediate insect responses to environmental change, as well as their success, particularly if the change is drastic. To study these aspects in the field, classic mark–release–recapture studies, in combination with population genetic approaches, are the best option (see Stevens et al. 2010). In the context of environmental change such basic behavioral approaches are surprisingly rare, and generally not well integrated with studies of plastic and evolutionary responses (cf. Table 19.1). As two nice examples, Hochkirch and Damerau (2009), as well as Franzen and Nilsson (2012) found increased dispersal of grasshoppers and butterflies, respectively, during hot weather periods, although range expansion was not revealed to be specifically directed towards presumably cooler areas. Kuussaari et al. (2016) more systematically showed for the Apollo butterfly that dispersal intensified at higher temperatures and more solar radiation, i.e. on warm days, whereas it lessened when clouds, rainfall, and wind speed increased, thus demonstrating that insects can opportunistically take advantage of favourable weather conditions for their dispersal.

Rather than tracking marked individuals, recent research has focused on analysing 'big data' from repositories based on sightings and monitoring programmes over time collected by scientists as well as amateurs ('citizen scientists'), which were first set up for vertebrates (such as birds) but have now become increasingly available for insects as well. Of course, such data banks primarily cover large and charismatic insects such as butterflies, dragonflies, or grasshoppers (e.g., Hassall et al. 2007; Altermatt 2010; Phillimore et al 2012; Karlsson 2014; Penone et al. 2013), whereas other more cryptic and less popular taxa such as flies or beetles are often not well covered (but see, e.g., Maveety et al. 2013; Moret et al. 2016). However, for obvious reasons some data banks also exist for so-called pest species, which include a variety of other insect taxa (Lindström and Lehmann 2015). While data bank studies can effectively assess range shifts over time in multiple species (e.g., Hassall et al. 2007), they remain phenomenological and typically do not permit conclusions about the particular behavioral mechanisms involved. If we want to systematically assess the relative importance of the various responses of insects (and animals in general) to environmental change, several types of studies therefore need to be integrated, best also including long-term case studies of particularly well-suited single model species of the type discussed by Stoks et al. (2014) and Schilthuizen and Kellermann (2014).

Short-distance **homing behavior** can serve as a surrogate for dispersal propensity that can be assessed experimentally in the field. Several nice examples in the climate change context illustrate this approach. Short-distance dispersal, migration, or habitat selection can, for instance, be simulated and prompted by (e.g., banana) bait, which for *Drosophila* signifies both food and an oviposition site. Potentially marked flies, however previously treated in the laboratory, are released in nature (if possible at sites free of other distracting food or oviposition substrates) under various weather conditions (cool versus hot), and arrival rates at the bait serve as the main performance measure (e.g., Hoffmann and Loeschcke 2006). This approach was successfully applied also to flies from thermal selection lines (Kristensen et al. 2006), and to heat-acclimated versus untreated flies (Loeschcke and Hoffmann 2007), in both cases demonstrating expected positive effects of heat hardening or thermal background under field conditions. Thomson et al. (2001) used an analogous approach by releasing heat-hardened hymenopteran parasitoids (*Trichogramma carverae*) in the field to thereafter measure parasitism rates of host eggs, and Wennersten et al. (2012) investigated thermal microhabitat choice in an arena in the colour-polymorphic grasshopper *Tetrix subulata*. As a final example, Moya-Laraño et al. (2007) applied such techniques to investigate the temperature-dependent success of seed beetles, *Stator limbatus*, of various sizes in their scramble for mates, finding that small beetles take-off and fly more quickly towards target females, particularly at cooler temperatures. While the two latter studies were performed in a laboratory enclosure, such experiments clearly also work in field enclosures (e.g., Merckx et al. 2003).

Mobility under environmental stress has also been successfully assessed experimentally in the laboratory, often using automated activity monitoring systems originally developed to assess thermal preferences of *Drosophila* (Kjaersgaard et al. 2010). Such a system permits experimental comparisons of the short-distance walking (locomotion) activity, potentially signifying the eagerness (or appetence) to disperse or move away, of insects however treated before in various thermal or other environments. For example, Gibert et al. (2001) tested *Drosophila melanogaster* from populations raised at various temperatures, and Kjaersgaard et al. (2010) compared flies from lines selected for heat and cold shock resistance. Flight performance, as a surrogate for long-distance dispersal in nature, can be assessed in the laboratory as well using flight mills originally developed for mosquitoes (e.g., Briegel et al. 2001), which of course also work for other flying species in other contexts. In another space-for-time approach, Kjaersgaard et al. (2014) compared the temperature-dependent walking and flight performance of latitudinal European populations of the house fly *Musca domestica*, additionally assessing wing morphology. They found, as expected, that the southernmost Spanish flies showed greatest heat resistance and also flew farther under heat stress than flies from Swiss or Danish populations. On the other hand, Swiss flies were most mobile at benign temperatures in the walking test, and only some aspect of wing shape, but not size, affected flight performance. Crucially, if such 'space-for-time' population comparisons are performed under common garden conditions in the laboratory or field using insect populations held and bred in the laboratory over generations, consistent differences demonstrate that behavior is part of the evolved heritable syndrome responding to environmental change (cf. Table 19.1; Figure 19.1).

A particularly interesting case relates to a different environmental change context, that of light pollution in urban environments (Gaston et al. 2013). It has been well known for some time that nocturnal insects, such as moths, are attracted to light, presumably because they are naturally orienting and navigating according to the moon and the stars. As the outcome can be fatal when orienting towards artificial light sources (lamps or candles), this behavioral response is a good example of what has been termed an '**ecological trap**' (Kokko and Sutherland 2001; Wong and Candolin 2015). Using methods similar to those described previously in an (indoor) flight-to-light experiment, Altermatt and Ebert (2016) presented evidence that small ermine moths, *Yponomeuta cagnagella*, can apparently adapt to this problem, as individuals from 'urban' as opposed to 'rural' populations had a lower tendency to fly towards artificial light, males generally more strongly so. Such experimental approaches can and should be extended to other species and situations. For example, the tendency of some aquatic insects to confuse certain coloured or polarizing structures with the water surface would benefit from systematic experimental testing

(Kriska et al. 1998; Horváth et al. 2007), as would seemingly non-sensical colour preferences of terrestrial insects, if only for practical purposes (e.g., Vrdoljak and Samways, 2012).

To conclude this section, several laboratory and field methods are available to experimentally investigate insect mobility and dispersal behavior in an environmental change context. If combined with the aforementioned 'big data', and/or experimental field and laboratory approaches, integrated responses of behavior and morphology can be documented, which would bring us a big step closer to our goal of assessing the relative importance of behavioral, plastic, and evolutionary responses to environmental change (Table 19.1). This includes studies of morphological traits (e.g., Turlure et al 2010; San Martin y Gomes and van Dyck 2012; MacLean et al. 2016), colour polymorphisms (Wennersten et al. 2012), and entire flight syndromes or polymorphisms in some insects together with their physiological underpinnings and behavioral consequences (i.e. wing size, shape, colouration, flight muscles, etc., in heteropterans and orthopterans; Roff and Fairbairn 1991; Feder et al. 2010). It also includes the (laboratory) assessment of thermal limits in terms of cold or heat shock and the subsequent (chill or heat coma) recovery, which after all involve simple behaviors and physiology as well (e.g., Kristensen et al. 2006; Fischer and Karl 2010; Hazell and Bale 2011; MacMillan and Sinclair 2011; Santos et al. 2011; Esperk et al. 2016).

19.3.2 Diapause behavior

Winter diapause (also dormancy or quiescence) is an essential life history feature of temperate insects. Diapause may also occur in seasonal (sub)tropical environments, either in form of overwintering at higher altitudes or as aestivation during dryer periods of the season (e.g., Oostra et al. 2014). Diapause is adaptive because harsh climatic conditions with extended periods of food shortage typically preclude foraging and reproduction; particular environmental conditions regularly occurring before the inhospitable season consequently prompt profound changes in the behavior and physiology of insects and other organisms called the **diapause syndrome** (Tauber et al. 1986). Diapause is typically phenotypically plastic, i.e.

reversibly induced by specific environmental cues (mainly photoperiod and temperature) during a given life stage of an insect, and has been shown to evolve rapidly in response to climate change (Bradshaw and Holzapfel 2001; Lindström and Lehmann 2015; Ståhlhandske et al. 2017). Diapause must have evolved multiple times independently in the past as insect species of tropical origin, such as e.g., *Drosophila melanogaster*, expanded their range to colonize strongly seasonal temperate habitats at higher latitudes or altitudes. Such de novo evolution of diapause following range expansions, often mediated by climate change, is of interest but has been rarely described for any particular species (Schmidt et al. 2005).

The diapause syndrome of a given species typically encompasses multiple physiological, behavioral and sometimes morphological traits that can be experimentally assessed in various ways in the laboratory (Tauber et al. 1986) and also monitored in the field (Lindström and Lehmann 2015; Ståhlhandske et al. 2017). Laboratory investigations have repeatedly demonstrated that, once in place, diapause induction mechanisms readily evolve by natural selection when environmental conditions change (e.g., pitcher plant mosquitoes: Bradshaw and Holzapfel 2001). These are clear examples of the evolution of heritable phenotypic plasticity (Figure 19.1: V_{gxe}; see also Chapter 5). In the field, changing environmental conditions before, during, and after winter should lead to responses of insects in the induction in autumn, duration over winter, and breaking in spring of their diapause, all separate behavioral and physiological traits that can be assessed by direct observation and/or monitoring programs (e.g., Altermatt 2010; Phillimore et al. 2012; Karlsson 2014; Ståhlhandske et al. 2017). Moreover, individual insects destined to enter diapause in autumn typically exhibit particular observable behaviors to increase their overwintering chances in synchrony with their physiological changes, such as e.g., aggregation (ladybird beetles; the dung fly *Sepsis fulgens*) or digging into the ground (potato beetles; Lehmann et al. 2014). Again, therefore, an integrated research approach is required combining laboratory experiments (to assess evolutionary responses) and field observations (to assess behavioral responses) to fully understand diapause responses of insects to environmental change. Interestingly, at the southern edge of a species'

distribution, it can be expected, even predicted, that diapause may actually become lost as the winter gradually shortens and eventually disappears due to climate warming, either by random genetic drift or, more likely, by the action of natural selection. Again, this would be very interesting to observe, and to assess the concomitant presumed fitness costs associated with such 'devolution' of diapause at the southern fringes of a species distribution (e.g., Ståhlhandske et al. 2017).

19.3.3 Behavioral traits facilitating homeostasis

Thermoregulation

If animals do not or cannot move away in response to adverse environmental changes, they have to maintain their integrity (i.e. homeostasis) to continue their daily chores of foraging and reproduction in the changed, more unfavourable conditions. This can be achieved by improved thermoregulation, a physiological response, and by way of so-called behavioral thermoregulation in form of more marked microhabitat choice (Turlure et al. 2011; Huey et al. 2012; Wennersten et al. 2012). Thermoregulatory behavior and the resulting optimal body temperatures have been shown to vary even among closely related species, thus contributing to niche differentiation in terms of (micro)habitat selection (e.g., in butterflies, wild bees or damselflies: Kleckova et al. 2014; Peters et al. 2016; Rivas et al. 2016). Thermoregulation also differs among populations within species, thus showing signs of differentiation in the underlying plasticity and/or even local adaptation (e.g., in butterflies and grasshoppers: Harris et al. 2015; MacLean et al. 2016). Physiological and behavioral aspects of thermoregulation should therefore be considered when studying insects' responses to environmental change (Woods et al. 2015). Thermoregulation will of course also affect foraging and reproduction, as discussed next.

19.3.4 Foraging

Environmental changes likely lead to changes in the foraging behavior of insects in the short or medium term. To maintain their food intake and, hence, homeostasis, foragers are expected to adjust their behavior quantitatively and/or qualitatively. For its applied relevance in agriculture and forestry (e.g., Churchill et al. 2013), this topic appears to be treated almost exclusively in phytophagous insects, primarily from the plant perspective, with some additional work on insect parasitoids (for the same applied reason). Foreseeably, most studies have investigated the foraging responses of various herbivores (or parasitoids) to increased temperature (e.g., ants, Andrew et al. 2013; herbivorous beetles, Lemoine et al. 2013; aphid parasitoids, Moiroux et al. 2015; aquatic mosquito larvae, Reiskind and Janairo 2015; grasshoppers, Clissold and Simpson 2015; bees, Schürch et al. 2016). The reported reactions are detailed, often complex, and in most cases not fatal, primarily showing selectively neutral adjustments, sometimes improvements, but sometimes also responses indicating clearly negative fitness impacts (as in Tundra moth caterpillars, *Gynaephora groenlandica*, under climate warming, Barrio et al. 2016). Some responses are indirect, involving plant chemistry (e.g., Lemoine et al. 2013). Responses to environmental variables other than temperature are also researched in this realm, including barometric pressure (parasitoids, Crespo and Castelo 2012), UV light (aphids, Burdick et al. 2015), CO_2 (aphids, Sun et al. 2016), drought (butterfly larvae, Gutbrodt et al. 2011), or seasonal timing (Foster et al. 2013). Unsurprisingly in this day and age, these environmental variables again all relate to climate change. Studies further include investigations of longer-term climate effects on entire parasitoid food webs (Henri et al. 2012) and even fossil midge feeding guilds based on data banks (Luoto and Nevalainen 2015), thus providing links to the rich body of literature on community ecology and food webs. Interestingly, the comparative aspect, potentially indicating local adaptation (i.e. genetic differences) between populations and/or species in such responses seems to be missing from the literature.

19.3.5 Reproductive behavior

Maintaining homeostasis in the face of environmental change ultimately concerns all **fitness components** contributing to reproductive success. The previously discussed behaviors (dispersal, thermoregulation, and foraging) indirectly contribute to this end, too. That is, individual insects must have survived the juvenile stages in an altered environment, as assessable by specific life history assays (Schilthuizen and

Kellermann 2014; Stoks et al. 2014; e.g., for yellow dung flies, Blanckenhorn et al. 2014), and must have also survived dispersal potentially triggered by environmental changes (Travis et al. 2013; Schilthuizen and Kellermann 2014; Sánchez-Guillén et al. 2016) or altered diapause conditions (Stoks et al. 2014; e.g., Bradshaw and Holzapfel 2001), and the like.

Aside from such more general life history responses, reproduction does involve concrete behaviors, often very specifically so in terms of courtship or stimulatory mating behavior, female choice, male–male competition, oviposition site choice, etc., which all are sexually selected traits (see also Chapter 13). Researchers should therefore explicitly investigate such reproductive behaviors in the environmental change context to uncover more subtle effects and their consequences for overall reproductive success. Again, such studies investigating specific reproductive behaviors, rather than more general life history traits, are rare in the literature. One study explicitly addressing this context showed the acoustic signalling of planthoppers to be temperature dependent such that male responsiveness to female calls, and therefore mating rates, are lowered when the climate warms (Long et al. 2012). Another exemplary study demonstrated that rearing temperature influenced the choice of aphid host stages by parasitoids, thus affecting the latter's performance (Moiroux et al. 2015; see also Roux et al. 2010). A third example concerns oviposition site selection in butterflies in response to habitat fragmentation, which is mediated by behavior directly affecting reproductive success (Gibbs and Van Dyck 2009). Especially in the plant–insect interaction and applied literature, chemical (pheromone) signals frequently serve to attract herbivorous insects as well as their enemies (predators, parasites, parasitoids) in a sexual or reproductive context, so more such subtle behavioral alterations in response to environmental changes can be expected (Witzgall et al. 2010; Forrest 2015). Although no concrete examples could be found, environmental variables could also affect parental care in species that provide it, such as earwigs (e.g., Meunier and Kölliker 2012).

19.3.6 Social behavior (and learning)

For sake of completeness, environmental changes are also expected to influence social behavior, including learning, ultimately affecting the performance of insect societies. Few concrete examples involve behavioral traits, again typically in the climate (temperature) change context. Leadbeater and Florent (2014) investigated how individual bee foragers weigh the value of social (provided by the hive as a whole) versus personal foraging information in case the environment, here the availability of flowers with nectar, changes. They found that personal information has priority, meaning that the environmental change had no measurable consequence in this case. If such tests are conducted explicitly in a general environmental change context, behavioral consequences of, say, temperature shifts can be assessed in social insects; these may be subtle, but ultimately could affect the reproductive success and survival of entire colonies. Such investigations can even be conducted in the laboratory in very mechanistic settings, e.g., involving pheromone trails of foragers (Czaczkes and Heinze 2015). Field studies have shown consequences of such individual behavioral shifts, for instance that the plastic social organization of bee species can change with the climate (Schürch et al. 2016), and that such changes, positive or negative, can lead to the replacement of the ecosystem services (here pollination) of one species by another (Rader et al. 2013). Again, explicit tests of environmental changes mediating changes in social behavior are still lacking in the social insect literature, but could well complement the more common standard single-species investigations of e.g., lethal temperatures (e.g., Andrew et al. 2013).

19.4 Conclusions

There is mounting evidence that organisms in general and insects in particular can and do respond to environmental changes in the various ways outlined in Table 19.1. A priori, evolutionary (i.e. genetic) life history responses are necessarily slowest because they occur over generations, and instant behavioral responses are fastest, with physiological acclimation and plasticity being intermediate in time scale (Table 19.1). The relative importance of these various types of responses for any particular species, and in general, is more difficult to assess. We definitely do not yet have the required data, but it

can be assumed that most taxa respond by a mixture of all these possibilities.

After a general introduction and discussion of evolutionary and phenotypically plastic responses, this article focused on behavioral responses to environmental changes often caused or at least influenced by human activities. Especially climate change features prominently in the recent literature. It can be concluded that behavioral traits are under-researched relative to morphology and life history, or at least not afforded their due attention in this context, given that behavior will most immediately mediate animal responses to changing environments and this may already solve the problem to a great extent (Van Buskirk 2012; e.g., by moving away: Travis et al. 2013; Stevens et al. 2010). Therefore, evolutionary ecologists should be encouraged to incorporate more behavioral traits and responses in their work on environmental changes. It would be best if research groups, and even individuals, integrate all types of responses listed in Table 19.1 in their work, rather than belabouring merely particular elements. Such an integrative approach would also benefit doctoral dissertations by providing students with a well-rounded education combining field and laboratory work, behavior, ecology, evolution, and genetics. Only then can we ultimately and holistically understand the responses of animals to environmental change, as such knowledge depends crucially on studies of particularly well-suited model taxa. Odonata (damsel- and dragonflies) are good cases in point (see Hassall et al. 2007; Stoks et al. 2015; Bybee et al. 2016; Villalobos-Jimenez et al. 2016). To add an evolutionary and phylogenetic perspective, groups of closely related model taxa should be researched from as many angles as possible.

References

Altermatt, F. (2010). Climatic warming increases voltinism in European butterflies and moths. *Proceedings of the Royal Society London Biological Sciences*, **277**, 1281.

Altermatt, F., and Ebert, D. (2016). Reduced flight-to-light behaviour of moth populations exposed to long-term urban light pollution. *Biology Letters*, **12**, 20160111.

Andrew, N.R., Hart R.A., Jung M-P., Hemmings Z., and Terblanche, J.J. (2013). Can temperate insects take the heat? A case study of the physiological and behavioural responses in a common ant, Iridomyrmex purpureus

(Formicidae), with potential climate change. *Journal of Insect Physiology*, **59**, 870.

Angilletta Jr, M.J. (2009). *Thermal Adaptation: A Theoretical and Empirical Synthesis.* Oxford University Press, Oxford.

Atkinson, D. (1994). Temperature and organism size: a biological law for ectotherms? *Advances in Ecological Research*, **25**, 1–58.

Atkinson, D., and Sibly, R.M. (1997). Why are organisms usually bigger in colder environments? Making sense of a life history puzzle. *Trends in Ecology and Evolution*, **12**, 235.

Balanyà, J., Oller, J.M., Huey, R.B., Gilchrist, G.W., and Serra, L. (2006). Global genetic change tracks global climate warming in *Drosophila subobscura*. *Science*, **313**, 1773.

Barrio, I.C., Guillermo Bueno, C., and Hik, D.S. (2016). Warming the tundra: reciprocal responses of invertebrate herbivores and plants. *Oikos*, **125**, 20.

Berger, D., Postma, E., Blanckenhorn, W.U., and Walters, R.J. (2013). Quantitative genetic divergence and standing genetic (co)variance in thermal reaction norms along latitude. *Evolution*, **67**, 2385.

Blanckenhorn, W.U. (1991). Life history differences in adjacent water strider populations: phenotypic plasticity or heritable responses to water temperature? *Evolution*, **45**, 1520.

Blanckenhorn, W.U. (1998). Adaptive phenotypic plasticity in growth rate and diapause in the yellow dung fly. *Evolution*, **52**, 1394.

Blanckenhorn, W.U. (1999). Different growth responses to food shortage and temperature in three insect species with similar life histories. *Evolutionary Ecology*, **13**, 395.

Blanckenhorn, W.U. (2009). Causes and consequences of phenotypic plasticity in body size: the case of the yellow dung fly *Scathophaga stercoraria* (Diptera: Scathophagidae). In: D.W. Whitman and T.N. Ananthakrishnan (Eds), *Phenotypic Plasticity of Insects: Mechanism and Consequences*, pp. 369–422. Science Publishers, Enfield, NH.

Blanckenhorn, W.U. (2015). Yellow dung fly body size evolution in the field: response to climate change? *Evolution*, **69**, 2227.

Blanckenhorn, W.U., and Demont, M. (2004). Bergmann and converse Bergmann latitudinal clines in arthropods: two ends of a continuum? *Integrative and Comparative Biology*, **44**, 413.

Blanckenhorn, W.U., Gautier, R., Nick, M., and Schäfer, M.A. (2014). Stage- and sex-specific heat tolerance in the yellow dung fly. *Journal of Thermal Biology*, **46**, 1.

Bonnet, T., Wandeler, P., Camenisch, G., and Postma, E. (2016). Bigger is fitter? Quantitative genetic decomposition of selection reveals an adaptive evolutionary decline of body mass in a wild rodent population. *PLoS Biology*, **15**, e1002592.

Bradshaw, W.E., and Holzapfel, C.M. (2001). Genetic shift in photoperiodic response correlated with global warming. *Proceedings of the National Academy of Sciences USA*, **98**, 14509.

Briegel, H., Knüsel, I., and Timmermann, S.E. (2001). *Aedes aegypti*: size, reserves, survival, and flight potential. *Journal of Vector Ecology*, **26**, 21.

Burdick, S.C., Prischmann-Voldseth, D.A., and Harmon, J.P. (2015). Density and distribution of soybean aphid, *Aphis glycines* Matsumura (Hemiptera: Aphididae) in response to UV radiation. *Population Ecology*, **57**, 457.

Busso, J.P., Blanckenhorn, W.U., and González-Tokman, D. (2017). Healthier or bigger? Trade-off mediating male dimorphism in the black scavenger fly, *Sepsis thoracica* (Diptera: Sepsidae). *Ecological Entomology*, **42**(4), 517–25.

Bybee, S., Córdoba-Aguilar, A., Catherine Duryea, M., et al. (2016). Odonata (dragonflies and damselflies) as a bridge between ecology and evolutionary genomics. *Frontiers in Zoology*, **13**, 46.

Churchill, D.J., Larson, A.J., Dahlgreen, M.C., et al. (2013). Restoring forest resilience: from reference spatial patterns to silvicultural prescriptions and monitoring. *Forest Ecology and Management*, **291**, 442–57.

Clark, A.B., and Ehlinger, T.J. (1987). Pattern and adaptation in individual behavioural differences. In: P.P.G. Bateson and P. H. Klopfer (Eds) *Perspectives in Ethology*, Vol. 7, pp. 1–47. Plenum Press, New York, NY.

Clissold, F.J., and Simpson, S.J. (2015). Temperature, food quality and life history traits of herbivorous insects. *Current Opinion in Insect Science*, **11**, 63.

Conover, D.O., and Present, T.M.C. (1990). Countergradient variation in growth rate: compensation for length of the growing season among Atlantic silversides from different latitudes. *Oecologia*, **83**, 316.

Crespo, J.E., and Castelo, M.K. (2012). Barometric pressure influences host-orientation behavior in the larva of a dipteran ectoparasitoid. *Journal of Insect Physiology*, **58**, 1562.

Czaczkes, T.J., and Heinze, J. (2015). Ants adjust their pheromone deposition to a changing environment and their probability of making errors. *Proceedings of the Royal Society London Biological Sciences*, **282**, 20150679.

DeWitt, T.J., Sih, A., and Wilson, D.S. (1998). Costs and limits of phenotypic plasticity. *Trends in Ecology and Evolution*, **13**, 77.

Dingemanse, N.J., Kazem, A.J.N., Reale, D., and Wright, J. (2010). Behavioural reaction norms: animal personality meets individual plasticity. *Trends in Ecology and Evolution*, **25**, 81.

Dukas, R. (2004). Evolutionary biology of animal cognition. *Annual Reviews in Ecology Evolution and Systematics*, **35**, 347.

Esperk T., Kjaersgaard, A., Walters, R.J., Berger, D., and Blanckenhorn, W.U. (2016). Plastic and evolutionary responses to heat stress in a temperate dung fly: negative correlation between basal and induced heat tolerance? *Journal of Evolutionary Biology*, **29**, 900.

Excoffier, L., Foll, M., and Petit, R.J. (2009). Genetic consequences of range expansions. *Annual Reviews in Ecology Evolution and Systematics*, **40**, 481.

Fagen, R. (1982). Evolutionary issues in development of behavioural flexibility. In: P.P.G. Bateson and P.H. Klopfer (Eds), *Perspectives in Ethology*, Vol. 5, pp. 365–83. Plenum Press, New York, NY.

Feder, M.E., Garland Jr, T., Marden, J.H., and Zera, A.J. (2010). Locomotion in response to shifting climate zones: not so fast. *Annual Review of Physiology*, **72**, 167.

Finnay, S.C. and Edwards, L.E. (2016). The 'Anthropocene' epoch: scientific decision or political statement? *GSA Today*, **26**, 4–10.

Fischer, K., and Karl, I. (2010). Exploring plastic and genetic responses to temperature variation using copper butterflies. *Climate Research*, **43**, 17.

Forrest, J.R.F. (2015). Plant–pollinator interactions and phenological change: what can we learn about climate impacts from experiments and observations? *Oikos*, **124**, 4.

Forsman, A., Ringblom, K., Civantos, E., and Ahnesjö, J. (2002). Coevolution of color pattern and thermoregulatory behavior in polymorphic pygmy grasshoppers *Tetrix undulata*. *Evolution*, **56**, 349.

Foster, J.R., Townsend, P.A., and Mladenoff, D.J. (2013). Mapping asynchrony between gypsy moth egg-hatch and forest leaf-out: putting the phenological window hypothesis in a spatial context. *Forest Ecology and Management*, **287**, 670.

Franzén, M., and Nilsson, S.G. (2012). Climate-dependent dispersal rates in metapopulations of burnet moths. *Journal of Insect Conservation*, **16**, 941.

Gaston, K.J., Bennie, J., Davies, T.W., and Hopkins, J. (2013). The ecological impacts of night time light pollution: a mechanistic appraisal. *Biological Reviews of the Cambridge Philosophical Society*, **88**, 912.

Gibbs, M., and van Dyck, H. (2009). Reproductive plasticity, oviposition site selection, and maternal effects in fragmented landscapes. *Behavioural Ecology and Sociobiology*, **64**, 1.

Gibert, P., Huey, R.B., and Gilchrist, G.W. (2001). Locomotor performance of *Drosophila melanogaster*: interactions among developmental and adult temperatures, age, and geography. *Evolution*, **55**, 205.

Gienapp, P., Teplitsky, C., Alho, J.S., Mills, J.A., and Merilä, J. (2008). Climate change and evolution: disentangling environmental and genetic responses. *Molecular Ecology*, **17**, 167.

Gutbrodt, B., Mody, K., and Dorn, S. (2011). Drought changes plant chemistry and causes contrasting responses in lepidopteran herbivores. *Oikos*, **120**, 1732.

Hale, R., Piggo, J.J., and Swearer, S.E. (2017). Describing and understanding behavioral responses to multiple stressors and multiple stimuli. *Ecology and Evolution*, **7**(1), 38–47.

Harris, R.M.B., McQuillan, P., and Hughes, L. (2015). The effectiveness of common thermo-regulatory behaviours in a cool temperate grasshopper. *Journal of Thermal Biology*, **52**, 75.

Hassall, C., Thompson, D.J., French, G.C., and Harvey, I.F. (2007). Historical changes in the phenology of British Odonata are related to climate. *Global Change Biology*, **13**, 933.

Hazell, S.P. and Bale, J.S. (2011). Low temperature thresholds: are chill coma and CTmin synonymous? *Journal of Insect Physiology*, **57**, 1085.

Henri, D.C., Seager, D., Weller, T., and Frank van Veen, F.J. (2012). Potential for climate effects on the size-structure of host-parasitoid indirect interaction networks. *Philosophical Transactions of the Royal Society London: Biological Sciences*, **367**, 3018.

Hochkirch, A., and Damerau, M. (2009). Rapid range expansion of a wing-dimorphic bush-cricket after the 2003 climatic anomaly. *Biological Journal of the Linnean Society*, **97**, 118.

Hoffmann, A.A., and Daborn, P.J. (2007). Towards genetic markers in animal populations as biomonitors for human-induced environmental change. *Ecology Letters*, **10**, 63.

Hoffmann, A.A., and Loeschcke, V. (2006). Are fitness effects of density mediated by body size? Evidence from *Drosophila* field releases. *Evolutionary Ecology Research*, **8**, 813.

Holt, R.D. (1990). The microevolutionary consequences of climate change. *Trends in Ecology and Evolution*, **5**, 311.

Horvath, G., Malik, P., Kriska, G., and Wildermuth, H. (2007). Ecological traps for dragonflies in a cemetery: the attraction of *Sympetrum* species (Odonata, Libellulidae) by horizontally polarizing black grave-stones. *Freshwater Biology*, **52**, 1700.

Huey, R.B., Gilchrist, G.W., Carlson, M.L., Berrigan, D., and Serra, L. (2000). Rapid evolution of a geographic cline in size in an introduced fly. *Science*, **287**, 308.

Huey, R.B., Kearney, M.R., Krockenberger, A., et al. (2012). Conservation physiology: integrating physiological mechanisms with ecology and evolution to predict responses of organisms to environmental change. *Philosophical Transactions of the Royal Society London: Biological Sciences*, **367**, 1665.

Kapun, M., Schmidt, C., Durmaz, E., Schmidt, P.S., and Flatt, T. (2016). Parallel effects of the inversion In(3R) Payne on body size across the North American and Australian clines in *Drosophila melanogaster*. *Journal of Evolutionary Biology*, **29**, 1059.

Karlsson, B. (2014). Extended season for northern butterflies. *International Journal of Biometeorology*, **58**, 691.

Kjærsgaard, A., Blanckenhorn, W.U., Pertoldi, C., et al. (2014). Plasticity in behavioural responses and resistance to temperature stress in *Musca domestica*. *Animal Behaviour*, **99**, 123.

Kjærsgaard A., Demontis, D., Kristensen, T.N., et al. (2010). Locomotor activity of *Drosophila melanogaster* in high temperature environments: plastic and evolutionary responses. *Climate Research*, **43**, 127.

Kleckova, I., Konvicka, M., and Klecka, J. (2014). Thermoregulation and microhabitat use in mountain butterflies of the genus *Erebia*: importance of fine-scale habitat heterogeneity. *Journal of Thermal Biology*, **41**, 50.

Klepsatel, P., Galikova, M., Huber, C.D., and Flatt, T. (2014). Similarities and differences in altitudinal versus latitudinal variation for morphological traits in *Drosophila melanogaster*. *Evolution*, **68**, 1385.

Kokko, H., and López-Sepulcre, A. (2006). From individual dispersal to species ranges: perspectives for a changing world. *Science*, **313**, 789.

Kokko, H., and Sutherland, W.J. (2001). Ecological traps in changing environments: ecological and evolutionary consequences of a behaviourally mediated Allee effect. *Evolutionary Ecology Research*, **3**, 537.

Kriska, G., Horvath, G., and Andrikovics, S. (1998). Why do mayflies lay their eggs en masse on dry asphalt roads? Water imitating polarized light reflected from asphalt attracts Ephemeroptera. *Journal of Experimental Biology*, **201**, 2273.

Kristensen, T.N., Loeschcke, V., and Hoffmann, A.A. (2006). Can artificially selected phenotypes influence a component of field fitness? Thermal selection and fly performance under thermal extremes. *Proceedings of the Royal Society London Biological Sciences*, **274**, 771.

Kruuk, L.E.B. (2004). Estimating genetic parameters in natural populations using the 'animal model'. *Philosophical Transactions of the Royal Society London: Biological Sciences*, **359**, 873.

Kuussaari, M., Rytteri, S., Heikkinen, R.K., Heliölä, J., and von Bagh, P. (2016). Weather explains high annual variation in butterfly dispersal. *Proceedings of the Royal Society London Biological Sciences*, **283**, 20160413.

Leadbeater, E., and Florent, C. (2014). Foraging bumblebees do not rate social information above personal experience. *Behavioural Ecology and Sociobiology*, **68**, 1145.

Lehmann, P., Lyytinen, A., Piiroinen, S., and Lindström, L. (2014). Northward range expansion requires synchronization of both overwintering behaviour and physiology with photoperiod in the invasive Colorado potato beetle (*Leptinotarsa decemlineata*). *Oecologia*, **176**, 57.

Lemoine, N.P., Drews, W.A., Burkepile, D.E., and Parker, J.D. (2013). Increased temperature alters feeding behavior of a generalist herbivore. *Oikos*, **122**, 1669.

Lindström, L., and Lehmann, P. (2015). Climate change effects on agricultural insect pests in Europe. In: C. Björkman and P. Niemelä (Eds), *Climate Change and Insect Pests*, 136, pp. 136–53. CAB International, Wallingford.

Loeschcke, V., and Hoffmann, A.A. (2007). Consequences of heat hardening on a field fitness component in *Drosophila* depend on environmental temperature. *American Naturalist*, **169**, 175.

Long, Y., Chaoxing Hu, Shi Baokun, Xiao Yang, and Maolin Hou (2012). Effects of temperature on mate location in the planthopper, *Nilaparvata lugens* (Homoptera: Delphacidae). *Environmental Entomology*, **41**, 1231.

Luoto, T.P., and Nevalainen, L. (2015). Climate-forced patterns in midge feeding guilds. *Hydrobiologia*, **742**, 141.

MacLean, H.J., Higgins, J.K., Buckley, L.B., and Kingsolver, J.G. (2016). Geographic divergence in upper thermal limits across insect life stages: does behavior matter? *Oecologia*, **181**, 107.

MacMillan, H.A., and Sinclair, B.J. (2011). Mechanisms underlying insect chill-coma. *Journal of Insect Physiology*, **57**, 12.

Maveety, S.A., Browne, S.A., and Erwin, T.L. (2013). Carabid beetle diversity and community composition as related to altitude and seasonality in Andean forests. *Studies on Neotropical Fauna and Environment*, **48**, 165.

Merckx, T., van Dyck, H., Karlsson, B., and Leimar, O. (2003). The evolution of movements and behaviour at boundaries in different landscapes: a common arena experiment with butterflies. *Proceedings of the Royal Society of London Series B*, **270**, 1815.

Merilä, J., and Hendry, A.P. (2014). Climate change, adaptation, and phenotypic plasticity: the problem and the evidence. *Evolutionary Applications*, **7**, 1.

Meunier, J., and Kölliker, M. (2012). When it is costly to have a caring mother: food limitation erases the benefits of parental care in earwigs. *Biology Letters*, **8**, 547.

Moiroux, J., Boivin, G., and Brodeur, J. (2015). Temperature influences host instar selection in an aphid parasitoid: support for the relative fitness rule. *Biological Journal of the Linnean Society*, **115**, 792.

Moret, P., de los Ángeles Aráuz, M., Gobbi, M., and Barragán, A. (2016). Climate warming effects in the tropical Andes: first evidence for upslope shifts of Carabidae (Coleoptera) in Ecuador. *Insect Conservation and Diversity*, **9**, 342.

Mousseau, T.A., and Roff, D.A. (1987). Natural selection and the heritability of fitness components. *Heredity*, **59**, 181.

Moya-Laraño, J., El-Sayyid, M.E.T., and Fox, C.W. (2007). Smaller beetles are better scramble competitors at cooler temperatures. *Biology Letters*, **3**, 475.

Nussey, D.H., Postma, E., Gienapp, P., and Visser, M.E. (2005). Selection on heritable phenotypic plasticity in a wild bird population. *Science*, **310**, 304.

Öckinger, E., and van Dyck, H. (2012). Landscape structure shapes habitat finding ability in a butterfly. *PLoS One*, **7**, e41517.

Oostra, V., Mateus, A.R.A., van der Burg, K.R.L., et al. (2014). Ecdysteroid hormones link the juvenile environment to alternative adult life histories in a seasonal insect. *American Naturalist*, **184**, E79.

Paetkau, D., Slade, R., Burden, M., and Estoup, A. (2004). Genetic assignment methods for the direct, real-time estimation of migration rate: a simulation-based exploration of accuracy and power. *Molecular Ecology* **13**, 55.

Penone, C., Le Viol, I., Pellissier, V., et al. (2013). Use of large-scale acoustic monitoring to assess anthropogenic pressures on Orthoptera communities. *Conservation Biology*, **27**, 979.

Peters, M.K., Peisker, J., Steffan-Dewenter, I., and Hoiss, B. (2016). Morphological traits are linked to the cold performance and distribution of bees along elevational gradients. *Journal of Biogeography*, **43**, 2040.

Phillimore, A.B., Stålhandske, S., Smithers, R.J., and Bernard, R. (2012). Dissecting the contribution of plasticity and local adaptation to the phenology of a butterfly and its host plants. *American Naturalist*, **180**, 655.

Rader, R., Reilly, J., Bartomeus, I., and Winfree, R. (2013). Native bees buffer the negative impact of climate warming on honey bee pollination of watermelon crops. *Global Change Biology*, **19**, 3103.

Rannala, B., and Mountain, J.L. (1997). Detecting immigration by using multilocus genotypes. *Proceedings of the National Academy of Sciences of the United States of America*, **94**, 9197.

Reiskind, M.H., and Shawn Janairo, M. (2015). Late-instar behavior of *Aedes aegypti* (Diptera: Culicidae) larvae in different thermal and nutritive environments. *Journal of Medical Entomology*, **52**, 789.

Rivas, M., Martínez-Meyer, E., Muñoz, J., and Córdoba-Aguilar, A. (2016). Body temperature regulation is associated with climatic and geographical variables but not wing pigmentation in two rubyspot damselflies (Odonata: Calopterygidae). *Physiological Entomology*, **41**, 132.

Roff, D.A., and Fairbairn, D.J. (1991). Wing dimorphisms and the evolution of migratory polymorphisms among the insecta. *American Zoologist*, **31**, 243.

Roff, D.A. (1997). *Evolutionary Quantitative Genetics*. Chapman and Hall, New York, NY.

Roux, O., Le Lann, C., van Alphen, J.J., and van Baaren, J. (2010). How does heat shock affect the life history traits of adults and progeny of the aphid parasitoid *Aphidius avenae* (Hymenoptera: Aphidiidae)? *Bulletin of Entomological Research*, **100**, 543.

San Martin y Gomez, G., and van Dyck, H. (2012). Ecotypic differentiation between urban and rural populations of the grasshopper *Chorthippus brunneus* relative to climate and habitat fragmentation. *Oecologia*, **169**, 125.

Sánchez-Guillén, R., Cordoba-Aguilar, A., Hansson, B., Ott, J., and Wellenreuther, M. (2016). Evolutionary consequences of climate-induced range shifts in insects. *Biological Reviews*, **91**, 1050.

Santos, M., Castaneda, L.E., and Rezende, E.L. (2011). Making sense of heat tolerance estimates in ectotherms: lessons from *Drosophila*. *Functional Ecology*, **25**, 1169.

Scharf, I., Bauerfeind, S.S., Blanckenhorn, W.U., and Schäfer, M.A. (2010). Effects of maternal and offspring environmental conditions on growth, development and diapause in latitudinal yellow dung fly populations. *Climate Research*, **43**, 115.

Schilthuizen, M., and Kellermann, V. (2014). Contemporary climate change and terrestrial invertebrates: evolutionary versus plastic changes. *Evolutionary Applications*, **7**, 56.

Schmidt, P.S., Matzkin, L., Ippolito, M., and Eanes, W.F. (2005). Geographic variation in diapause incidence, life-history traits, and climatic adaptation in *Drosophila melanogaster*. *Evolution*, **59**, 1721.

Schürch, R., Accleton, C., and Field, J. (2016). Consequences of a warming climate for social organisation in sweat bees. *Behavioural Ecology and Sociobiology*, **70**, 1131.

Sih, A. (2013). Understanding variation in behavioural responses to human-induced rapid environmental change: a conceptual overview. *Animal Behaviour*, **85**, 1077.

Ståhlhandske, S., Gotthard, K., and Leimar, O. (2017). Winter chilling speeds spring development of temperate butterflies. *Journal of Animal Ecology*, **86**, 718–29.

Stevens, V.M., Turlure, C., and Baguette, M. (2010). A meta-analysis of dispersal in butterflies. *Biological Reviews*, **85**, 625.

Stoks, R., Geerts, A.N., and de Meester, L. (2014). Evolutionary and plastic responses of freshwater invertebrates to climate change: realized patterns and future potential. *Evolutionary Applications*, **7**, 42.

Stoks, R., Debecker, S., Khuong Dinh Van, and Janssens, L. (2015). Integrating ecology and evolution in aquatic toxicology: insights from damselflies. *Freshwater Science*, **34**, 1032.

Sun, Y., Guo, H., and Ge, F. (2016). Plant–aphid interactions under elevated CO_2: some cues from aphid feeding behavior. *Frontiers in Plant Science*, **7**, 502.

Tauber, M.J., Tauber, C.A., and Masaki, S. (1986). *Seasonal Adaptations of Insects*. Oxford University Press, Oxford.

Teder, T., and Tammaru, T. (2005). Sexual size dimorphism within species increases with body size in insects. *Oikos*, **108**, 321.

Thomson, L.J., Robinson, M., and Hoffmann, M.M. (2001). Field and laboratory evidence for acclimation without costs in an egg parasitoid. *Functional Ecology*, **15**, 217.

Travis, J.M.J., Delgado, M., Bocedi, G., et al. (2013). Dispersal and species' responses to climate change. *Oikos* **122**, p. 1532.

Travis, J.M.J., Delgado, M., Bocedi, G., et al. (2014). 'Dispersal and species' responses to climate change. *Oikos*, **122**, 1532.

Turlure, C., Schtickzelle, N., and Baguette, M. (2010). Resource grain scales mobility and adult morphology in butterflies. *Landscape Ecology*, **25**, 95.

Turlure, C., Radchuk, V., Baguette, M., van Dyck, H., and Schtickzelle, N. (2011). On the significance of structural vegetation elements for caterpillar thermoregulation in two peatbog butterflies: *Boloria eunomia* and *B. aquilonaris*. *Journal of Thermal Biology*, **36**, 173.

Van Buskirk, J. (2012). Behavioral plasticity and environmental change. In: R.B.M. Wong and U. Candolin (Eds), *Behavioural Responses to a Changing World*, pp. 144–58. Oxford University Press, Oxford.

van Noordwijk, A.J., and de Jong, G. (1986). Acquisition and allocation of resources: their influence on variation in life history tactics. *American Naturalist*, **128**, 137.

Via, S., and Lande, R. (1985). Genotype-environment interaction and the evolution of phenotypic plasticity. *Evolution*, **39**, 505.

Villalobos-Jimenez, G., Dunn, A.M., and Hassall, C. (2016). Dragonflies and damselflies (Odonata) in urban ecosystems: a review. *European Journal of Entomology*, **113**, 217.

Vrdoljak, S.M., and Samways, M.J. (2012). Optimising coloured pan traps to survey flower visiting insects. *Journal of Insect Conservation*, **16**, 345.

Wennersten, L., Karpestam, E., and Forsman, A. (2012). Phenotype manipulation influences microhabitat choice in pygmy grasshoppers. *Current Zoology*, **58**, 392.

Williams, S.E., Shoo, L.P., Isaac, J.L., Hoffmann, A.A, and Langham, G. (2008). Towards an integrated framework for assessing the vulnerability of species to climate change. *PLoS Biology*, **6**, 2621.

Witzgall, P., Kirsch, P., and Cork, A. (2010). Sex pheromones and their impact on pest management. *Chemical Ecology*, **36**, 80.

Wong, R.B.M., and Candolin, U. (2012). *Behavioural Responses to a Changing World*. Oxford University Press, Oxford.

Wong, R.B.M., and Candolin, U. (2015). Behavioural responses to changing environments. *Behavioral Ecology*, **26**, 665.

Woods, H.A., Dillon, M.E., and Pincebourde, S. (2015). The roles of microclimatic diversity and of behavior in mediating the responses of ectotherms to climate change. *Journal of Thermal Biology*, **54**, 86.

CHAPTER 20

Behavior-based control of insect crop pests

Sandra A. Allan

Center for Medical, Agricultural and Veterinary Entomology, ARS/ USDA, Gainesville, FL, USA

20.1 Introduction

The development of integrated pest management approaches has arisen from concerns of reduced efficacy of insecticides for insect control as a result of developing resistance, and from concerns of environmental effects on non-target species. Integrated pest management relies on the rational use of chemical, biological and other control methods to reduce pest levels below economic injury levels. The approach of behavioral manipulation or the use of stimuli that either stimulate or inhibit a behavior, or change its expression (Foster and Harris 1997) has provided the foundation for a range of new strategies for control of insect crop pests.

20.2 Aspects of life cycles and behavioral repertoires utilized for control

The goal of behavioral manipulation for pest control is to disrupt the life cycle at critical points to negatively impact upon insect population levels and subsequent crop damage. Through detailed knowledge of behaviors such as mating, oviposition, host-finding, and foraging, specific stimuli guiding these behaviors can be selected and deployed to target control of the pest insects. Successful application of these stimuli requires detailed knowledge of the life

histories, sensory processes, behavior, and ecology of the target insects. These behaviors are primarily guided by chemical and visual cues that vary in importance among different groups of insects. For instance, in a meta-analysis of studies involving insect behavioral manipulation and plant volatiles, insects were grouped into feeding guilds (i.e. different modes of feeding). Attraction to plant volatiles was strongest with chewing insects, followed by sap-feeders, which may reflect greater sensitivity to the volatiles produced by the more destructive chewing actions (Szendrei and Rodriguez-Saona 2010). Chewers were considered more likely to be sensitive to plant volatiles used in baits. Based on taxonomic group, Thysanoptera were least attracted to plant volatiles with Lepidoptera the most attracted and Coleoptera in the middle (Szendrei and Rodriguez-Saona 2010).

For most insect pests, adults are the most dispersive stage and the most targeted stage for behavioral disruption. Reduction of adult numbers directly corresponds to reduction of immatures that are the most destructive stage. While the challenges of life cycles vary between species, the primary points for potential intervention include mating (sex and aggregation pheromones, lekking, vibration), oviposition (plant volatiles, visual cues), and food location (plant volatiles, visual cues, aggregation pheromones; Table

Allan, S. A., *Behavior-based control of insect crop pests*. In: *Insect Behavior: From mechanisms to ecological and evolutionary consequences*. Edited by Alex Córdoba-Aguilar, Daniel González-Tokman, and Isaac González-Santoyo: Oxford University Press (2018). © Oxford University Press.
DOI: 10.1093/oso/9780198797500.003.0020

Table 20.1 Examples of behavioral components incorporated in insect control strategies

Sensory system	Component	Function	Specificity	Targets
Chemical	Host plant odour	Attractant/repellent/confusant	Low	Both sexes
	Food odour (protein, sugar)	Attractant	Low	Both sexes
	Sex pheromone	Attractant/confusant	High	One sex
	Aggregation pheromone	Attractant	High	Both sexes
	Anti-aggregation pheromones	Repellent	High	Both sexes
	Alarm pheromones	Repellent	High	Both sexes
Visual	Visible colour reflecting	Attractant/repellent	Medium	Both sexes
	UV reflecting	Repellent	Medium	Both sexes
Vibrational	Vibrational calls	Repellent/confusant	Medium	One sex

20.1). Points in the life cycle most vulnerable to manipulation are those with heavy reliance on cues that are rare or unique in the environment. While there is a multitude of excellent studies exploring the roles of stimuli affecting insect behavior in the context of potential control strategies, this chapter focuses on those close to practical implementation.

20.2.1 Overview of behavior-based control strategies

The most common application of knowledge of insect behavior has been use of sex pheromones and other attractants for the development of surveillance traps. These attractants, in some cases in conjunction with deterrents, are the basis for several control strategies based on behavioral manipulation devised to protect a resource such as a crop from insect damage. These strategies include **mass trapping**, **attract-and-kill** (including use of bait stations), **auto-dissemination**, **mating disruption**, and **push-pull** strategies and are used for control in a wide variety of species (Table 20.2). Successful deployment of these behavioral manipulation strategies relies on detailed knowledge of the basic biology, behavior, and ecology of the targeted insects. Strengths of behavior-based control include no or highly-targeted use of insecticides, fewer non-target effects as many tactics are species-specific, and low environmental impact. Weaknesses of these strategies are that they often do not kill the target pest immediately but may require longer durations for effective control. Additionally,

efficacy of behavioral controls can be density-dependent and under such conditions, these measures may be more effective for pest populations below an outbreak phase (Gut et al. 2004). These strategies can be of particular use for pest control in ecologically sensitive environments, for management of insecticide resistance and for localized control. Advantages and disadvantages, and additional references for following strategies are provided in Table 20.3.

20.3 Strategies

20.3.1 Mass trapping

The objective of mass trapping is to reduce the pest population prior to incurring damage to a crop through the large-scale use of attractant-based traps. Traps generally consist of an attractant combined with adhesives or devices that contain and kill the attracted insects (El-Sayed et al. 2006). Focus is on utilization of critical behavioral cues such as sex and aggregation pheromones, host-plant volatiles and oviposition site odours, as well as visual attractants, to outcompete the naturally occurring cues. Mass trapping is more useful for control of low-density than high-density populations as traps at high populations become saturated. The efficacy of a mass trapping method is highly dependent on both an effective lure that can attract insects from a long distance and a trap capable of capturing and killing the attracted individuals (Rodriguez-Saona and Stelinski 2009). As sex pheromones are commonly used as attractants

Table 20.2 Examples of insect control strategies based on behavioral manipulation.

Strategy	Insect	Behavioral cue	Crop	Reference
Mass trapping	Alfalfa looper, *Autographa californica*	Host plant kairomone	Alfalfa, lettuce	Camelo et al. 2007
	Ambrosia beetle, *Gnathotrichus sulcatus*	Male-produced aggregation pheromone	Conifers	Borden 1990
	Beet armyworm moth, *Spodoptera exigua*	Female sex pheromone	Onion	Park and Goh 1992
	Chinese tortrix moth, *Cydia trasias*	Female sex pheromone	Chinese scholar trees	Zhang et al. 2002
	Codling moth, *Cydia pomonella*	Pear kairomone	Pear	Light et al. 2001
	Japanese beetle, *Popillia japonica*	Female sex pheromone Floral scent kairomone	Park	Wawrzynski and Ascerno 1998
	Mountain pine beetle, *Dendroctonus ponderosae*	Aggregation, anti-aggregation pheromone	Pine	Lindgren and Borden 1993
	Palm weevil, *Rhynchophorus palmarum*	Male-produced aggregation pheromone	Palms	Oehlschlager 2016
	Sap beetles, *Carpophilus* spp.	Male-produced aggregation pheromone, fermented food	Stone fruit	James et al. 1996
	Tephritid fruit flies	Protein and sugar	Fruit	Navarro-Llopis and Vacas 2014
	West Indian sugarcane weevil *Metamasius hemipterus*	Male-produced aggregation pheromone Host plant kairomone	Sugar cane	Giblin-Davis et al. 1996
	Wireworms, *Agriotes* spp.	Female sex pheromone	Vegetables, soft fruit	Barsics et al. 2013
Attract-and-kill	Apple maggot fly, *Rhagoletis pomonella*	Red sphere Host-plant volatile kairomone Sucrose	Apple	Prokopy et al. 2000 Stelinski et al. 2001
	Agriotes obscures Melanotus wireworms	Wheat seedlings	Potato	Vernon et al. 2016
	Blueberry maggot fly, *Rhagoletis mendax*	Yellow sphere Sucrose	Blueberry	Stelinski and Liburd 2001
	Coding moth, *Cydia pomonella*	Sex pheromone	Apple	Charmillot et al. 2000
	Helicoverpa spp.	Plant kairomones	Cotton	Mensah et al. 2013 Gregg et al. 2016
	Light-brown apple moth, *Epiphyas postvittana*	Sex pheromone	Apple	Brockerhoff and Suckling 1999
	Olive fly, *Bactrocera oleae*	Light green trap Sex pheromone, food bait	Olive	Broumas et al. 2002
	Oriental fruit moth, *Cydia molesta*	Sex pheromone	Apple	Evenden and McLaughlin 2004
	Tephritid fruit flies	Protein bait spray Food bait station	Fruit	Piñero et al. 2014 Mangan 2014
Auto-dissemination	Banana weevil, *Cosmopolites sordidus*	Aggregation pheromone	Banana	Tinzaara et al. 2007
	Brown winged green bug, *Plautia crossota stali*	Aggregation pheromone	Fruit trees	Tsutsumi et al. 2003

(*continued*)

Table 20.2 Continued.

Strategy	Insect	Behavioral cue	Crop	Reference
	Damson-hop aphid, *Phorodon humuli*	Sex pheromone	Hops	Hartfield et al. 2001
	Diamondback moth, *Plutella xylostella*	Female sex pheromone	Cabbage	Furlong et al. 1995 Vickers et al. 2004
	Dusky sap beetle, *Carpophilus lugubris*	Aggregation pheromone	Corn	Dowd and Vega 2003
	Japanese beetle, *Popilla japonica*	Floral kairomone	Grass	Klein and Lacey 1999
	Mediterranean fruit fly, *Ceratitis capitata*	Trimedlure, Food kairomone Parapheromone	Citrus	Navarro-Llopis et al. 2015
	Red palm weevil, *Rhynchophorus ferrugineus*	Aggregation pheromone Palm tree kairomone	Palm	Francardi et al. 2013
	Tobacco budworm, *Heliothis virescens*	Female sex pheromone	Tobacco	Jackson et al. 1992
Disruption				
Mating	Cabbage looper moth, *Trichoplusia ni*	Female sex pheromone	Vegetable	Shorey et al. 1967
	Carpenter moth, *Cossus insularis*	Female sex pheromone	Apple	Hoshi et al. 2016
	Citrus leafminer, *Phyllocnistis citrella*	Female sex pheromone	Citrus	Stelinski et al. 2010
	Codling moth, *Cydia pomonella*	Female sex pheromone	Apple	Ebbinghaus et al. 2001 Stelinski et al. 2007
	Currant clearwing moth, *Synanthedon tipuliformis*	Female sex pheromone	Currant	Grassi et al. 2002
	Grape root borer, *Vitacea polistiformis*	Female sex pheromone	Grape	Weihman and Liburd 2006
	Lesser peachtree borer, *Synanthedon pictipes*	Female sex pheromone	Peach	Yonce 1981
	Oriental beetle, *Anomala orientalis*	Female sex pheromone	Blueberry	Rodriguez-Saona et al. 2014
	Oriental fruit moth, *Grapholita molesta*	Female sex pheromone	Apple	Stelinski et al. 2007
	Peachtree borer, *Synanthedon exitosa*	Female sex pheromone	Peach	Yonce 1981
	Tomato pinworm, *Keiferia lycopersicella*	Female sex pheromone	Tomato	Wang et al. 1997
	Vine mealybug, *Planococcus ficus*	Female sex pheromone	Grapes	Sharon et al. 2016
Vibration	Asian citrus psyllid, *Diaphorina citri*	Mating vibration	Citrus	Lujo et al. 2016
	Leafhopper, *Scaphoideus titanus*	Mating vibration	Grape	Eriksson et al. 2012
	Pine bark beetles, *Dendroctonus* spp.	Stridulation used to disrupt tunnelling and mating	Pine	Hofstetter et al. 2014

Host-plant location repellent	Asian citrus psyllid, *Diaphorina citri*	Reflective mulch Plant kairomone	Citrus	Croxton and Stansly 2013 Mafra-Neto et al. 2014
	Carrot psyllid, *Trioza apicalis*	Conifer volatile kairomone	Carrot	Nehlin et al. 1994
	Western flower thrips, *Frankliniella occidentalis*	UV reflectance	Vegetables	Antignus et al. 1996
Push-pull	California five-spines ips, *Ips paraconfusus*	Push: anti-aggregation pheromone Pull: aggregation pheromone	Pine	Shea and Neustein 1995
	Douglas fir beetle, *Dendroctonus pseudotsuga*	Push: anti-aggregation pheromone Pull: aggregation pheromone	Fir	Ross and Daterman 1994
	German cockroach, *Blatella germanica*	Push: chemical repellent Pull: faeces (aggregation pheromones)	In food bait	Nalyanya et al. 2000
	Helicoverpa moths	Push: Chemical repellent Pull: Trap crop	Pigeon pea, Maize	Pyke et al. 1987
	Mountain pine beetle, *Dendroctonus ponderosae*	Push: anti-aggregation pheromone Pull: aggregation pheromone	Pine	Lindgren and Borden 1993 Borden 1997
	Onion maggot fly, *Delia antiqua*	Push: oviposition deterrent Pull: host plant	Onion	Miller and Cowles 1990 Cowles and Miller 1992
	Pea leaf weevil, *Sitona lineatus*	Push: antifeedant Pull: aggregation pheromone	Fava beans	Smart et al. 1994
	Stemborer moth, *Chilo pertellus*	Push: non-host plants Pull: host plants	Maize	Khan et al. 2010

Table 20.3 Advantages and disadvantages of different behavioral manipulation control strategies.

Approach	Advantages	Disadvantages	Limitations	Additional references
Mass trapping	High specificity Low non-target effects No pesticide Low environmental impact	Labour intensive Lower efficiency at high density	Trap efficiency Trap saturation Less effective at high populations Best in isolation Targets one sex Best if some damage is tolerated	El-Saved et al. 2006 Gut et al. 2004 Navarro-Llopis and Vacas 2014
Attract-and-kill	Low non-target effects Low environmental impact Targeted pesticide delivery Less trap saturation	With chemical or biopesticides Moderate labour costs	Strength of attractant contrast against background May only target one sex	Foster and Harris 1997 Gut et al. 2004
Auto-dissemination	Low cost of labour/materials Low non-target effects Low environmental effects Targeted toxicant delivery No immediate kill No need to locate insect microhabitats	Moderate labour costs Slow mortality	Strength of attractant Efficiency of device Competing odours Stability of pathogen Complete control unlikely	Baverstock et al. 2010 Gut et al. 2004

(continued)

Table 20.3 Continued.

Approach	Advantages	Disadvantages	Limitations	Additional references
Disruption—mating	High specificity Low non-target effects Low environmental effects No pesticide Increasing efficiency at low population densities	Potentially high cost of developing/synthesizing attractants Labour intensive Reduced ability for population surveillance	Incomplete coverage results in some damage	Cardé and Minks 1995 Lance et al. 2016 Miller and Gut 2015 Polajnar et al.2015 Witzgall et al 2008, 2010
Disruption—host plant (chemical, visual)	No pesticides Low non-target effects Low environmental effects Sustainable	Generally not specific	Incomplete coverage results in some damage	Antignus 2000, Diaz and Fererres 2007 Glenn and Puterka 2005 Pickett and Khan 2016 Shimodo and Honda 2013
Push-pull	Often no pesticides Low non-target effects Low environmental impact Useful for resistance management	Potentially high cost of developing/synthesizing attractants	Attraction and deterrence must be strong Requires optimal timing	Cook et al. 2007 Eigenbrode et al. 2016 Khan et al. 2016 Xu et al. 2013

for this strategy (El-Sayed 2017), this method can be highly species-specific. However, if when using female-produced pheromones, only males are collected, leaving females able to produce immatures if they are mated, and if females are not attracted to the traps then they will be present in the populations where they can damage fruit while ovipositing, even if they are virgin. Of the chemicals listed by El-Sayed (2017) for use with mass trapping, 44 and 40 per cent were used for Lepidoptera and Coleoptera, respectively. For coleopteran pests, mass trapping tends to be more commonly used than attract-and-kill strategies (Suckling 2015). The likelihood of success is enhanced when using attractants effective for both sexes. For optimal efficacy, synthetic odours must compete against background levels. This method has been used for the eradication of pests, particularly of invasive species (Suckling 2015).

20.3.2 Attract-and-kill

Attract-and-kill, also known as attraction annihilation, uses an attractant in combination with a killing agent to lure an insect to a point source where it contacts a lethal treatment (Foster and Harris 1997). Attractants generally consist of olfactory or visual stimuli optimized for maximum attraction of insects and can target a wide range of behaviors (i.e. mat-ing, feeding, oviposition, aggregation). While lethal treatments often consist of an insecticide, bioinsecticides, entomopathogenic fungi, or electrocuting grids can also be used. This strategy can be delivered in the form of a treated trap, baits, sprays, or bait stations. Of the 30+ target insects and pheromones involved in this strategy and listed in the Pherobase data base (El-Sayed 2017), half belong to Lepidoptera and a quarter belong to Diptera or Coleoptera.

20.3.3 Auto-dissemination

Auto-dissemination is similar to attract-and-kill with an attractant drawing insects into a treatment device. However, instead of rapidly killing insects, they become contaminated with a slow-acting toxicant before escaping and disseminating the toxicant to immatures and/or nest-mates (Baverstock et al. 2010; Gonzalez et al. 2016). Toxicants include materials that produce slower mortality and are most often entomopathogenic fungi (Furlong et al. 1995; Vickers et al. 2004), although bacteria and viruses have been used (O'Callaghan and Jackson 1993; Vega et al. 2000; see also Chapter 21 for similar vector toxicants). By adding an attractant, which may often be a species specific pheromone, delivery of the toxicant is highly targeted in the environment and the approach requires low amounts of attractant and toxicant. One advantage is that insect microhabitats do not have to be located by the applicator, as insects will disseminate the material back to their habitats. Effects on beneficial insects (predators, parasitoids) are often minimal because of the targeted delivery. The cost of this approach is generally low due to low cost of the delivery devices and small amounts of material required. Use of this approach may be limited by the strength of the attractants, the presence of competing odours, and stability of the pathogens in the environment.

20.3.4 Disruption of resources

Mating disruption

Mating disruption interferes with the location and mating of conspecifics through inundation of the vicinity with stimuli involved in mate location and mating (Cardé and Minks 1995). Most often, sex pheromones are the basis of mating disruption programmes and El-Sayed (2017) lists over 120 species involved in mating disruption strategies, 85 per cent of which are Lepidoptera.

Pheromones

Most commonly, mating disruptants utilize volatile synthetic sex pheromones deployed in many areas so that mate location is disrupted (see also Chapter 21 for vector control). This strategy has been one of

the more successful applications of behavior modification for insect control. Because of high efficacy, species-specificity and low amounts of materials needed, this strategy has been used for area-wide control and eradication (Rodriguez-Saona and Stelinski 2009; Suckling 2015). Use of female-emitted sex pheromones presented in many point sources throughout the environment results in males having a reduced probability of finding and mating with females, leading to cessation of mating and population collapse (Witzgall et al. 2010). Disruption of mating can occur as a result of three mechanisms (Box 20.1; Miller et al. 2006a,b; Rodriguez-Sauna and Stelinski 2009). This strategy is generally very successful, in part due to potency and long-distance attraction (in the case of sex pheromones of moths), and due to the specificity of the stimuli. The efficacy of a mating disruption treatment is often evaluated as **trap shutdown** or the point at which mating is sufficiently disrupted so that sex pheromone-baited traps no longer collect males. A disadvantage is the reduced ability for population surveillance through pheromone-baited surveillance traps (Witzgall et al. 2010; Lance et al. 2016).

Sterile insect technique

This method of pest control relies on area-wide releases of sterile insects to reduce the fertility of the

---

Box 20.1 Mechanisms of mating disruption using insect sex pheromones (based on Miller et al. 2006a,b, 2010; Rodriguez-Saona and Stelinski 2009).

Competitive disruption

This occurs when there are multiple point emitters that reduce the frequency that males find calling females as they are diverted to the emitters. There is a decrease in male visitation to calling females as males respond to false plumes from synthetic pheromone from lures.

Desensitization

This occurs when sensitivity to pheromone is decreased due to continuous exposure to high background concentrations of pheromone and may be the result of adaptation (decreased sensitivity of the peripheral nervous system) or habituation (decreased sensitivity of the central nervous system).

Camouflage

This occurs when the edges of the females plume are disguised by the background levels of pheromone from lures.

Non-competitive disruption

This involves masking females by the numerous emitters and desensitization of males to the pheromone.

pest species, depends on normal mating behavior with one of the sexes compromised so that the outcome from mating is sterility and is presented in detail in Chapter 21. Traditionally, the method consists of mass-rearing, irradiation of males usually through gamma-radiation, and release of the sterile males to mate with wild females rendering them infertile (Dyke et al. 2005; Chouinard et al. 2016). This technique, effectively used for localized and widespread control or eradication of crop pests, most often those belonging to Lepidoptera and Diptera (Koyama et al. 2004; Lance et al. 2014; Vargas et al. 2015; Suckling et al. 2017), has been particularly effective for the Mediterranean fruit fly and the Mexican fruit fly. New approaches involving genetic engineering are being developed for similar mating-driven spread of infertility in insect populations (Alphey 2002; Zabalou et al. 2009). As behavioral stimuli are not directly manipulated and normal mating behavior is essential for the success of this approach, this technique will not be highlighted in this review.

Mechanical vibration

Vibrational signals, used as a basis for communication for some insects (see Chapter 12), are a relatively new approach to mating disruption as a control strategy (Polajnar et al. 2015). These cues for mating disruption through plant-borne vibrational signals are particularly promising for hemipteran pests (Ericksson et al. 2012; Lubanga et al. 2014; Lujo et al. 2016). Use of vibrational signals to deter infestations has promise with bark beetle infestations (Hofstetter et al. 2014). Investigation into the interplay between agreement and aggressive sounds, and production of aggregation and anti-aggregation pheromones, which are often related to mating in bark beetles, show promise as a behavioral manipulation approach (Liu et al. 2017).

Host-plant location disruption

Location of host plants, consisting of behaviors such as orientation in flight or walking, landing, and oviposition site selection, is mediated through chemical and/or visual cues. These cues can serve as either attractants or **repellents**, depending on pest species, context against a background, and in the terms of chemicals, concentration, mixture of compounds, and their ratios (Antignus 2000; Foster and Harris 1997). Disruption of host-plant location through manipulation of visual cues is well documented and includes use of reflective sprays and mulches, and UV-blocking materials (Antignus 2000; Shimoda and Honda 2013). Host-plant location can be manipulated by masking or confusing host-plant odours through use of synthetic chemicals or co-planting with plants that are repellent (Szendrei and Rodriguez-Saona 2010; Du et al. 2016). While use of repellent plant volatiles to protect crops has potential, it is still not in widespread use as a solo strategy (Suckling 2015; Du et al. 2016). However, this approach is important as a component of push-pull strategies. Trap crops can serve as a pull component to attract pests from a protected crop to one that serves as a sink for the insects and associated pathogens (Shelton and Badenas-Perez 2006). Herbivore-induced defensive compounds emitted from plants alter behavior of pests and natural enemies and current research is focused on how to implement these as tools in a push-pull strategy (Turlings and Ton 2006; Pickett and Khan 2016). A new avenue is the genetic manipulation or selective breeding to enhance plant volatiles to attract predators or parasitoids (Bruce et al. 2015). Host-plant resistance is considered a component of integrated pest management and this is most commonly approached through use of resistant plants (Zehnder et al. 2007).

20.3.5 Push-pull

Push-pull (also known as stimulo-deterrence) is a strategy in which control is achieved through the integration of two opposing behavioral elements to push (or displace) pest insects from a crop to be protected (e.g. onion plants) and to pull (lure) those insects to another resource of little value (onion culls; Foster and Harris 1997). For instance, this can consist of cropping systems where either repellent synthetic chemicals or companion plants that release repellent chemicals (push) are used to protect the crop. Plants that strongly attract the insect pests can be planted surrounding the crops to pull the insect pests from the crops to the trap crop. Tools for the push components include visual deterrents, synthetic repellents, plant-host volatiles and herbivore-induced plant volatiles, anti-aggregation pheromones, alarm phero-mones, **anti-feedants**, and oviposition deterrents or pheromones. Components of the pull component include visual attractants, host-plant volatiles, and herbivore-induced volatiles, sex pheromones, aggregation pheromones, gustatory, or oviposition stimulants (Foster and Harris 1997; Cook et al. 2007; Rodriguez-Saona and Stelinski 2009).

20.4 Case examples

20.4.1 Diptera

In general, flies that are most commonly associated with behaviorally-modified control are tephritid fruit flies which cause enormous loss to fruit and vegetable production world-wide due to damage from oviposition by females and development of larvae. Traditional control methods consist of insecticides applied as sprays on vegetation and fruit, combined with protein food bait as sprays or soil drenches. Control efforts using organophos-phate and synthetic pyrethroids have transitioned to reduced risk insecticides such as spinosad, which have fewer negative impacts on natural enemies. Other methods include sterile insect releases, biological control, cultural methods and reduced risk pesticides (Lance et al. 2014; Vargas et al. 2015; Chouinard et al. 2016).

Overview of behavioral/ecological traits

Tephritid flies attacking crops are diurnal with visual cues highly important for resource location. Olfaction generally plays a secondary role in location of host plants, adult food and mates. Sex phero-mones are primarily produced by tephritid males, chemically complex, multimodal in nature and used less in **control programmes** than male-specific semiochemical lures (Benelli et al. 2014; Tan et al. 2014). Although not universal, in some tephritid species males form leks where they defend territories and produce long-range sex pheromones that attract females. Females interact and mate with males at the lekking sites. Depending on species, oviposition sites may be highly specific or include a wide host range. Adults also rely on sugar and pro-tein sources, such as yeast hydrolysate for nutrition, and many species are attracted to ammonia-based

compounds and plant-associated odours (Lance et al. 2014; Tan et al. 2014).

Specific examples

Attract-and-kill: visual and food odours

Apple maggot flies, *Rhagoletis pomonella*, are serious pests of apples due to larval damage that renders fruit unmarketable (Figure 20.1a,b). This species has a resource-based mating system where sexually mature males and females gather on apple fruit for both mating and oviposition (Prokopy et al. 1971; Prokopy and Bush 1973). Based on the strong visual attraction of flies to fruit, red sticky spheres were ini-tially used for localized control as they captured both male and female flies (Duan and Prokopy 1995; Figure 20.1c). Due to the unwieldiness of this approach using sticky material, spheres were devised to contain a contact residual insecticide treatment to provide good control (Prokopy et al. 2000). To further enhance ingestion of the insecticide on the spheres and subsequent mortality, sucrose was added as a feeding stimulant (Wright et al. 2012; Figure 20.1d). Placement of these treated spheres in trees (Figure 20.1e) on the edges of a commercial orchard was nearly as effective as insecticide treat-ments for prevention of injury to fruit (Prokopy et al. 2000; Wright et al. 2012). Addition of synthetic apple volatiles enhanced attraction and mortality (Zhang et al. 1999; Morrison et al. 2016). Similarly, insecti-cide-treated visual targets developed for the blue-berry maggot fly, *Rhagoletis mendax*, incorporate a food attractant (i.e. ammonium acetate) and pro-vides control equivalent to insecticide treatments (Stelinski and Liburd 2001).

Mass trapping/attract-and-kill: visual, food, and parapheromones

World-wide, there are many other tephritid fly spe-cies that are major pests of fruits and vegetables. The importance of visual stimuli for trap attraction varies between species, with some exhibiting reli-ance on olfactory cues as well. Early traps were clear glass, although numerous coloured traps (yel-low, orange, green) have been developed to opti-mize collection of particular species (Navarro-Llopis and Vacas 2014). Traps baited with hydrolysed pro-tein, ammonium, or amine salts are effective for

Figure 20.1 Attract-and-kill approach for apple maggot fly management. (a) Adult apple maggot fly (Joseph Berger, Bugwood.org). (b) Larva and damage in apple. (c) Red sticky sphere with trapped flies (Missouri Botanical Gardens). (d) Feeding stimulant on top of the sphere with a fly feeding on the stimulant (with permission from Tracey Lesky, USDA). (e) Red sphere placed in an apple tree (Missouri Botanical Gardens).

local control (Navarro-Llopis and Vacas 2014). A three-component lure (ammonium acetate, trimethylamine, and putrescine; trimedlure) used in conjunction with a protein bait (Heath et al. 1997), is particularly effective for the Mediterranean fruit fly, *Ceratitis capitata*, and provided the basis for effective mass trapping in Spain (Navarro-Llopis and Vacas 2014), as well as mass-trapping and surveillance of medflies around the world. With semiochemicals, such methyl eugenol as attractants for male fruit flies, male-specific lures, and traps and lure combinations were developed for targeted reduction of males (Vargas et al. 2012, 2014). Commercial lures and lure/insecticide combinations are currently available and used for targeted eradication and control efforts (Mangan 2014; Piñero et al. 2014).

Attract-and-kill: visual, food, and sex pheromones

The olive fruit fly, *Bactrocera oleae*, is considered a major pest of olives due to damage to the fruit from female oviposition and larval feeding. Spraying a pro-

tein food bait and insecticide mixture on tree foliage and fruit is a standard control method for this host-specific pest (Daane and Johnson 2010). An attract-and-kill strategy was developed using a visually attractive light green trap, baited with sex pheromones, a food attractant, and a contact insecticide. Over a 4-year period, fly populations and damage were lower where the attract-and-kill method was used compared with the standard bait sprays (Broumas et al. 2002).

20.4.2 Lepidoptera

Most of the economically important pests in Lepidoptera are nocturnal moths with economic damage of crops exclusively by caterpillar feeding. Traditional control relies on contract residual insecticides effective against ovipositing females or feeding larvae, or soil drenches for root borers. Systemic insecticides protect the entire plant, but may not deter oviposition and resistance development. Insecticide resistance is widespread in many species. Transgenic plants expressing *Bacillus thuringiensis* (Bt) toxins have been very effective against feeding caterpillars; however, these are threatened by the development of resistance (Mensah et al. 2013; Gregg et al. 2016).

Overview of behavioral/ecological traits

Adult moths are mostly nocturnal with olfaction being of primary importance for resource location (see also Chapter 10). Species-specific sex pheromones are produced by females and guide long distance orientation of males to females for mating. Hundreds of sex pheromones have been identified for moths and form the basis for lure development for pest surveillance in many different crop systems (Witzgall et al. 2010). Adults of many species rely on carbohydrate feeding to enhance survival. Females locate potential host plants for oviposition through use of specific mixtures of host plant volatiles (Witzgall et al. 2008, 2010). Female-produced sex pheromones that attract males can be used for development of control strategies. However, reduction of females is critical for protection of crops. The addition of host plant odours to these control strategies utilizing sex pheromones or food odours have the potential to enhance removal of females and their potential progeny from the crop. (Gregg et al. 2016).

Specific examples

Mass trapping: sex pheromones

The fruit and shoot borer, *Leucinodes orbonalis*, causes serious damage to eggplants in India and Bangladesh. Insecticide treatment is ineffective due to insecticide resistance and difficulty delivering insecticide to caterpillars within the fruit. This control strategy is based on the use of a sex pheromone provided in lures and distributed in traps (Cork et al. 2003). Pheromone dose, blend of components and trap placement were optimized for field use. Mass trapping without insecticide resulted in over a 50 per cent increase in marketable fruit. Avoidance of insecticide use resulted in conservation of natural enemies, further enhancing control (Cork et al. 2005).

Mating disruption: sex pheromones

The citrus leafminer, *Phyllocnistis citrella*, is an important pest of citrus world-wide due to intense damage caused by larval mining in leaves, which also predisposes the trees to citrus bacterial canker infections. Chemical control is inconsistent and ineffective due to protection of caterpillars within the mines and rapid infestation of new foliar growth. A three-component pheromone blend is highly effective, but expensive due to synthesis costs. Delivery levels of pheromone were optimized under field conditions and control was determined to be through non-competitive disruption (Stelinski et al. 2008). The pheromone blend was formulated into a flowable emulsified wax compound with three dabs of treatment with retreatment in 12 weeks. Mating disruption was obtained for over 100 days (Stelinski et al. 2010). This approach demonstrates an effective release device to attain season-long control without pesticide use.

Push-pull: repellent/attractant plants

Stem borers (*Chilo partellus*, *Busseola fusca*) cause consistent and often catastrophic crop loss in maize and sorghum throughout Africa (Khan et al. 2011). As insecticides are not economically feasible for subsistence farming, a push-pull system was developed that provided good yields at low cost. Repellent non-host plants, such as molasses grass and desmodium, planted within the crop provided the push component. Attractive trap plants such as Napier and Sudan grasses planted as a border crop drew insects from the crop (Figure 20.2). The repellent plants masked the odour of the crop and effectively hid the crop from the stem borers. The molasses grass simultaneously reduced stem borer infestations and also increased parasitism by *Cotesia* wasps through their attraction to the plant volatiles (Khan et al. 2001). Stem borers were diverted from the crop to the Napier and Sudan grasses with control provided as Napier grass does not support survival of the stem borers due to plant defences. Additionally, it enhances populations of natural enemies (Khan et al. 2011; Pickett et al. 2014; Khan et al. 2016). This is a sustainable push-pull approach particularly appropriate for subsistence farming in Africa.

Attract-and-kill: host plant odours

Widespread use of transgenic Bt cotton resulted in a decline in insecticides required to control the destructive *Helicoverpa* spp. moths. In Australia, in response

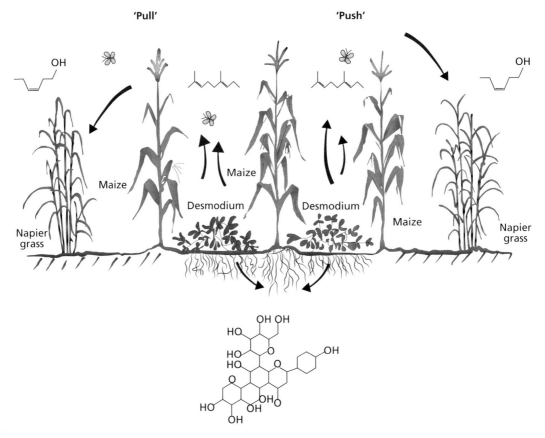

Figure 20.2 Representation of a push pull system. Volatiles from naturally repellent plants (*Desmodium*) repel moths from maize and naturally attractive plants (Napier grass) pull moths to edges of fields. Napier grass serves as a trap crop and immatures do not thrive there. (Pickett et al. 2014).

Modified from Pickett et al. (2014) with permission from Elsevier Publishing.

to increased potential for resistance by *Helicoverpa armigera and H. punctigera* (cotton bollworm) to Bt cotton and the presence of moths in conventional cotton, new attract-and-kill strategies based on attraction of moths to long-range host plant volatiles were devised. A blend of five synthetic chemicals that elicited positive pre-oviposition flight was combined in a sprayable formulation along with a non-deterrent insecticide and sucrose as a phagostimulant. Application of these treatments resulted in reductions of moth eggs and larvae on cotton (Mensah et al. 2013). As newly introduced revised strains of Bt cotton reduced the need for this approach, it has been applied to non-transgenic crops such as sweet corn (Gregg et al. 2016). In contrast to female sex pheromone-based approaches,

this approach, which incorporated host-plant volatiles resulted in mortality of both sexes.

20.4.3 Coleoptera

Over 75 per cent of beetles are phytophagous and these pose a considerable threat to agricultural crops and forests through damage from adult and larval feeding as well as transmission of pathogens causing disease. While for some species adult feeding causes damage, larvae are generally the more damaging stage with feeding on plants occurring either above or below ground. Conventional control consists of insecticide application to foliage or drenches to target control of adults and larvae feeding on foliage and roots. Systemic insecticides have been more effective, but their use is challenged with expense, development of resistance, need for precise timing to target larvae, inappropriateness for large natural areas (i.e. forests), and incompatibility with pollination and biocontrol (Barsics et al. 2013; Rodriguez-Saona and Stelinski 2009).

Overview of behavioral/ecological traits

Beetles may develop with multiple generations per year or several years per generation which poses challenges in terms of control. Critical to all, however, are the challenges of mate and host-plant location. Both visual and chemical cues are important for resource location by most species. For polyphagous feeding species, generalized plant cues can guide plant location; for those species with limited host range, mating is often associated with the preferred host plant. Male-produced aggregation or aggregation-sex pheromones mediate both mating and host-plant infestation dynamics for some wood-feeding beetle species (Byers 1989; Hanks and Millar 2016). In other species, female-produced sex pheromones attract males for mating (Hanks and Millar 2016). The presence of volatiles from host plants or fermented plant material are also known

to synergize attraction to pheromones for some species (Hanks and Millar 2016; Oehlschlager 2016).

Specific examples

Mass trapping: aggregation pheromones and visual cues

Bark beetles of the genera *Dendroctonus*, *Ips*, and *Scolytus* cause widespread destruction of coniferous trees due to direct damage from construction of maternal and larval galleries (Figure 20.3a,b) and transmission of fungal pathogens. Numerous species with specific pheromones are involved and discussed in detail by Byers (1989). Infestations initiated by beetles are often on fungi-infested trees that are more susceptible to successful infestation. Aggregation pheromones from feeding beetles attract conspecific males and females, which mount a mass attack on the trees to overcome the tree's defences. At a later stage of infestation, anti-aggregation pheromones are produced to deter additional recruits. For trapping, black log-shaped multiple funnel traps (visual cues) baited with aggregation pheromone components are used and result in large collections of beetles in traps (Borden and McLean 1981; Bentz and Munson 2000; Figure 20.3c). Components of anti-aggregation

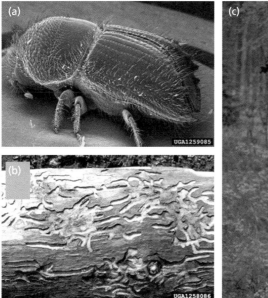

Figure 20.3 Mass trapping use visual traps and aggregation pheromones. (a) Adult bark beetles. (b) Extensive galleries under tree bark. (c) Funnel trap baited with aggregation pheromones for collection of adults (Stanislaw Kinelski, Bugwood.org).

pheromones are used to protect high value trees and are used in conjunction with the pheromones used in mass trapping (Jakus et al. 2003; Fettig and Hilszczanski 2015). This trapping strategy was particularly effective as both visual cues and aggregation pheromones attracted both sexes to the traps.

Mass trapping: aggregation pheromones

The red palm weevil, *Rhynchophorus palmarum*, is a highly damaging pest of palms, including coconut and oil palms due to larval damage on apical growing points and its role as a vector of nematodes causing red ring disease. Insecticidal control involves use of dusts, sprays, trunk injections, or drenches; however, infestations are difficult to detect and often overlooked (Faleiro 2006). Male-produced aggregation pheromones that attract both sexes are synergized by fermenting plant material (Faleiro 2006; Oehlschlager 2016). By using these attractants as bait in bucket traps in Costa Rican oil palm plantations, approximately 80 per cent reduction was obtained in the first year and in 3 years there was >93 per cent reduction in beetle numbers and >90 per cent reduction of red ring disease (Oehlschlager 2016). Currently, large scale mass trapping programmes are under way in South and Central America to reduce insect damage and disease transmission. One problem with this approach is that it draws beetles in from adjacent areas into trapping areas.

Attract-and-kill: host plant odours

Cues used by adult beetles to locate host plants can also be used in attract-and-kill strategies. An attract-and-kill study was conducted in British Columbia and Ontario, Canada against wireworm (*Agriotes obscures* and *Melanotus* spp., respectively) larvae, which cause significant damage to potatoes (Vernon et al. 2016). Wheat seed treated with fipronil and/or thiomethoxam insecticides were planted in furrows alongside potato tubers. When female beetles oviposited on the attractive emergent seedlings, larvae died after exposure to the insecticide. Reduction in wireworms relative to untreated controls was 89–100 per cent (British Columbia) and 66 per cent in Ontario. Reduction in potato yield was negligible (Vernon et al. 2016). This provided an alternate strategy for control, based on actual plant volatiles

that greatly reduced amount of insecticide needed with low environmental risk.

Mating disruption: sex pheromones

Larvae of the oriental beetle, *Anomala orientalis*, are destructive, cryptic root-feeding pests of blueberries and turf grass (Rodriguez-Saona et al. 2014). Rubber septa lures containing the female sex pheromone ((Z)-7-tetradecen-2-one, which were as effective as virgin females) were placed in capture traps at ground level. Based on mark–release captures, it was determined that males were effectively attracted and captured from a distance of 30 m. In a 3-year study in commercial blueberry fields, mating disruption provided an 87 per cent inhibition of beetle populations (trap shutdown). Trap results indicated that higher densities of baited traps placed on edges, rather than the interior of fields would optimize field-wide mating disruption (Rodriguez-Saona et al. 2014). Mating disruption appeared to work best under low to medium pest populations.

20.4.4 Hemiptera and Thysanoptera

Sap-feeding Hemiptera (i.e. aphids, whiteflies, and psyllids) and Thysanoptera (thrips) are responsible for extensive agricultural loss due to tissue damage directly from feeding, as well as indirect damage through transmission of a range of plant pathogens. Control through foliar and systemic insecticide treatments contributes to resistance development, disruption of natural enemies and negative impacts on pollinators and the environment. Horticultural oils, soaps, and insect growth regulators can be used to protect crops closer to harvest time. Transgenic Bt crops, such as cotton, which are highly effective against chewing Lepidoptera, are ineffective against suckling pests (Mensah et al. 2013).

Overview of behavioral/ecological traits

These diurnal insects are highly visually responsive. Direct damage results from sustained feeding by both adults and immatures, often co-located on host plants. This feeding provides nutritional and water needs. Adult dispersal to new plants generally involves phototactic responses to visual cues that influence orientation and landing. Host-plant odours may aid in orientation and host-plant acceptability. Mate location may be mediated by olfactory or vibrational cues (Fereres and Moreno 2009; Polajnar et al. 2016).

Specific examples

Host-plant location disruption: aphid alarm pheromones

Aphids cause devastating damage to plants worldwide through direct effects of feeding and transmission of pathogens causing disease. Over forty aphid species produce alarm pheromones, generally consisting of α-pinene, β-pinene, and (E)-β-farnesene, which result in disturbance of feeding aphids and dispersal from plants (Bushra and Tariq 2014). In contrast, these chemicals serve as kairomonal attractants for natural enemies (Bushra and Tariq 2014; Bruce et al. 2015). Wheat was genetically engineered to produce (E)-β-farnesene and, in laboratory trials,

three species of cereal aphids were repelled and foraging by a parasitoid increased (Bruce et al. 2015). In small plot field trials, however, there was no reduction in aphid numbers or increase in parasitoids, possibly due to habituation due to constant release of the pheromone. While not completely refined as a control strategy, this is a demonstration of a novel concept of genetic manipulation of the volatile organic compounds emitted from plants to manipulate pest behavior for reduced infestation, while simultaneously enhancing populations of natural enemies.

Mating disruption: vibration

Vibrational signals are utilized for communication in several species of Hemiptera. The American grapevine leafhopper, *Scaphoideus titanus* (Figure

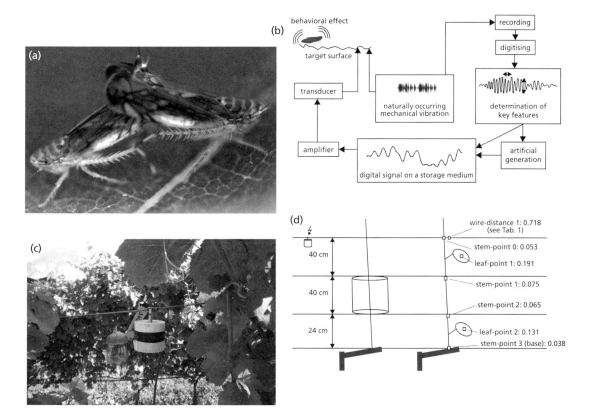

Figure 20.4 Manipulation of mating behavior of the grapevine leafhopper through use of mechanical vibrations. (a) Mating leafhoppers on a grape leaf (courtesy, Entomological Society of America). (b) Diagrammatic representation of the use of capturing natural vibrations, then generating the key components to provide disruptive vibrations on a substrate. (c) Transducer used to produce vibrations attached to wires. (d) Diagram of cage of leafhoppers tested for mating in context of the transducer, wires for vine support, and an adjacent plant.

From Polajnar et al. (2015, 2016) with permission of Wiley Publishing and Springer Publishing, respectively.

20.4a), is a host-specific pest of grapes in Europe that is a vector of the phytoplasma causing grapevine disease with control required in the European Union (Polajnar et al. 2016). This species mates solely on grape vegetation, does not use sex pheromones and only mediates mating communication through vibration. Males generally produce distinct pulses of vibration, matched by pulsed vibrations from the females (duetting). Location of the stationary female is done through interactive location pulses. Once in close proximity, courtship duetting refined identification and location of the female. Key features of the vibrational calls were identified and used for playback using a transducer (Figure 20.4b,c). In the laboratory, sexual communication was blocked. In the field, transducer-produced vibration signals were transmitted to insects along the support wires along rows between plants in a vineyard (Figure 20.4d; Polajnar et al. 2015). This approach resulted in blocking mating in 90 per cent of caged virgin females (Polajnar et al. 2016) and demonstrated that this technology was robust enough for field use, thus providing an additional tool for an integrated management programme. Similarly, mating was interrupted, by vibrational disruption in the Asian citrus psyllid, *Diaphorina citri*, a vector of bacteria causing citrus greening (Lujo et al. 2016). Male vibrational calls that elicit duetting in females were interrupted thus reducing mating.

Host-plant location disruption: visual cues

Host-plant location, particularly by dispersing aphids, whiteflies and thrips, relies heavily on visual cues. Visual and physical disguise of vegetation by sprays of whitewash or white kaolin particle films reduces attack by thrips, psyllids, and whiteflies, as well as other insects (Glenn and Puterka 2005; Larentzaki et al. 2008; Shimoda and Honda 2013). Disruption of these cues through use of coloured, white, or UV-reflecting mulch around host plants significantly reduces infestations of aphids, whiteflies, psyllids, and thrips (Polston and Lapidot 2007; Croxton and Stansly 2014; Shimoda and Honda 2013). Presumably, the reductions are related to disorientation of flight during host-finding (Shimoda and Honda 2013). Use of UV-blocking film over greenhouses or other enclosed areas also significantly reduces infestations, probably due to decreased movement of insects within or into the areas (Antignus 2000; Costa et al. 2002; Diaz and Fereres 2007; Shimoda and Honda 2013; Miranda et al. 2015).

20.5 Future approaches and concerns

20.5.1 Altered crops for anti-feedants or repellency

The approach of genetically altering crops to reduce pest pressure, as well as increase parasitoid and predator presence is a promising strategy for the future. Using a model plant, *Arabidopsis thaliana*, modifications were made so that plants would express an introduced gene (from the mint plant, *Mentha piperita*) for synthesis of the aphid alarm pheromone [sesquiterpene (E)-β-farnesene]. These genetically modified plants simultaneously repelled the peach-potato aphid, *Myzus persicae*, and increased foraging by a parasitoid wasp (Beale et al. 2006). While this effect was strong in the short term, the impact lessoned with habituation (Kunert et al. 2010). Two (E)-β-farnesene synthase genes isolated from sweet wormwood were over-expressed in tobacco and elicited repellency of *Myzus persicae* and attraction of lacewings (Yu et al. 2012). Continuing research has shown using two synthetic genes, documented strong alarm responses from cereal aphids, as well as longer foraging by parasitoids to (E)-β-farnesene emitted from wheat. Refinement of the timing of production of the alarm pheromone is anticipated to optimize this approach (Pickett et al. 2014). While not ready for implementation, such studies indicate that the approach of transgenic expression of (E)-β-farnesene or other effect compound synthases in crops may provide an avenue of pest protection of economically important crops in an environmentally benign way.

20.5.2 Role of climate change and greenhouse gas emission

While use of behavior-modifying pheromones is central to many integrated pest management strategies, there is concern about their efficacy in the face of altered climatic effects, such as increased temperatures and greenhouse gas levels (CO_2, O_3, NO_2).

A shift in temperature can potentially lessen olfactory communication through changes in pheromone biosynthesis altering ratios of compounds, emission, dispersion, perception, and behavioral response (Boullis et al. 2016). Long-range chemical signals used for insect communication could be impacted due to the extended exposure to oxidative gases during dispersal or from altered enzymatic function resulting in altered chemical ratios released (Boullis et al. 2016). Similarly, climate modifications could alter efficacy of slow-release devices for chemical release (Heuskin et al. 2011). Aspects of semiochemically based strategies prone to being impacted include mass trapping, mating disruption, push-pull, and intercropping, as well as impacting tritrophic interactions between parasitoids/predators, hosts, and plants (Boullis et al. 2015, 2016).

20.5.3 Risks of behavior-based methods

While there are many benefits of using behaviorally-based strategies, there may be unintended side effects. These risks include the potential for increasing disease transmission (Lin et al. 2016) and the potential for habitation, particularly to chemicals used as repellents or feeding deterrents (Jermy et al. 1982; Egger et al. 2014, 2015). Another concern is the potential for development of behavioral resistance that could arise from selection pressure from a particular technique, such as that reported by use of the sterile insect technique with the Mediterranean fruit fly (McInnis et al. 1996). One instance of this is the genetically-based behavioral resistance documented to food-based attractants for cockroaches (Wang et al. 2006). In terms of this example, improved bait matrices, rotational schemes between approaches, and used of integrated pest management principles can be implemented to mitigate the widespread selection of behavioral resistance by altering the selection pressure through time. Additionally, behavioral manipulation approaches can enhance other approaches with an example being the combination of sterile insect technique with transgenic corn (*Bacillus thuringiensis* (bt) toxins) to manage development of resistance to bt toxins (Wu 2010).

There are many promising avenues for pest control that are being developed using our expanding knowledge of how insect behavior is manipulated. New methods of manipulation of these behaviors provide promise for enhanced control strategies in the future. However, we should be aware of the unintended consequences and risks of these strategies, and also of unforeseen factors that may influence their success.

References

Alphey, L. (2002). Re-engineering the sterile insect technique. *Insect Biochemistry and Molecular Biology*, **32**, 1243.

Antignus, Y. (2000). Manipulation of wavelength-dependent behaviour of insects: an IPM tool to impede insects and restrict epidemics of insect-borne viruses. *Virus Research*, **71**, 213.

Antignus, Y., Mor, N., Ben Joseph, R., Lapidot, M., and Cohen, S. (1996). Ultraviolet-absorbing plastic sheets protect crops from insect pests and from virus diseases vectored by insects. *Environmental Entomology*, **25**, 219.

Barsics, F., Haubruge, E., and Verheggen, F.J. (2013). Wireworms' management: an overview of the existing methods, with particular regards to *Agriotes* spp. (Coleptera: Elateridae). *Insects*, **4**, 117.

Baverstock, J., Roy, H.E., and Pell, J.K. (2010). Entomopathogenic fungi and insect behavior: from unsuspecting hosts to targeted vectors. *BioControl*, **55**, 89.

Beale, M.H., Birkett, M.A., Bruce, T.J.A., et al. (2006). Aphid alarm pheromone produced by transgenic plants affects aphid and parasitoid behavior. *Proceedings of the National Academy of Science*, **103**, 10509.

Benelli, G., Daane, K.M., Canale, A., et al. (2014). Sexual communication and related behaviours in Tephritidae: current knowledge and potential applications for integrated pest management. *Journal of Pest Science*, **87**, 385.

Bentz, B.J., and Munson, A.S. (2000). Spruce beetle population suppression in northern Utah. *Western Journal of Applied Forestry*, **15**, 122.

Borden, J.H. (1990). Use of semiochemicals to manage coniferous tree pests in Western Canada. In: R. L. Ridgway, R.M. Silverstein, and M.N. Inscoe (Eds), *Behavior-modifying Chemicals for Insect Management*, pp. 281–315. Marcel Dekker, New York.

Borden, J.H. (1997). Disruption of semiochemical mediated aggregation in bark beetles. In: R.T. Cardé and A. K. Minks (Eds), *Insect Pheromone Research: New Directions*, pp. 421–38. Chapman & Hall, New York.

Borden, J.H., and McLean, J.A. (1981). Pheromone-based suppression of ambrosia beetles in industrial timber processing areas. In: E. R. Mitchell (Ed.), *Management of Insect Pests with Semiochemicals*, pp. 133–54. Springer, New York.

Boullis, A., Detain, C., Francis, F., and Verheggen, F.J. (2016) Will climate change affect insect pheromonal communication? *Current Opinion in Insect Science*, **17**, 87.

Boullis, A., Francis, F., and Verheggen, F.J. (2015). Climate change and tritrophic interactions: will modifications to greenhouse gas emissions increase the vulnerability of herbivorous insects to natural enemies. *Plant–Insect Interactions*, **44**, 277.

Brockerhoff, E.G., and Suckling, D.M. (1999). Development of an attracticide against light brown apple moth (Lepidoptera: Tortricidae). *Journal of Economic Entomology*, **92**, 853.

Broumas. T., Haniotakis, G., Liaropoulos, C., Tomazou, T., and Ragoussis, N. (2002). The efficacy of an improved form of the mass-trapping method, for the control for the olive fruit fly, *Bactrocera oleae* (Gmelin)(Dipt., Tephritidae): pilot-scale feasibility studies. *Journal of Applied Entomology*, **126**, 217.

Bruce, T.J.A., Aradottir, G.I., Smart, L.E., et al. (2015). The first crop plant genetically engineered to release an insect pheromone for defence. *Scientific Reports*, **5**, 11183.

Bushra, S., and Tariq, M. (2014). How aphid alarm pheromone can control aphids: a review. *Archives of Phytopathology and Plant Protection*, **47**, 1563.

Byers, J.A. (1989). Chemical ecology of bark beetles. *Experientia*, **45**, 271.

Camelo, L.D., Landolt, P.J., and Zack, R.S. (2007). A kairomone based attract-and-kill system effective against alfalfa looper (Lepidoptera: Noctuidae). *Journal of Economic Entomology*, **100**, 366.

Cardé, R.T., and Minks, A.K. (1995). Control of moth pests by mating disruption: successes and constraints. *Annual Review of Entomology*, **40**, 559.

Charmillot, P.J., Hofer, D., and Pasquier, D. (2000). Attract and kill: a new method for control of the codling moth *Cydia pomonella*. *Entomologia Experimentalis et Applicata*, **94**, 211.

Chouinard, D., Firlej, A., and Cornier, D. (2016). Going beyond sprays and killing agents: exclusion, sterilization and disruption for insect pest control in pome and stone fruit orchards. *Scientia Horticulturae*, **208**, 13.

Cook, S.M., Khan, Z.R., and Pickett, J.A. (2007). The use of push-pull strategies in integrated pest management. *Annual Review of Entomology*, **52**, 375.

Cork, A., Alam, S.N., Rouf, F.M.A., and Talekar, N.S. (2003). Female sex pheromone of brinjal fruit and shoot borer, *Leucinodes orbonalis* (Lepidoptera: Pyralidae): trap optimization and application in IPM trials. *Bulletin of Entomological Research*, **93**, 107.

Cork, A., Alam, S.N., Rouf, F.M.-A., and Telekar, N.S. (2005). Development of mass trapping technique for control of brinjal shoot and fruit borer, *Leucinodes orbonalis* (Lepidoptera: Pyralidae). *Bulletin of Entomological Research*, **95**, 589.

Costa, H.S., Robb, K.L., and Wilen C.A. (2002). Field trials measuring the effects of ultraviolet-absorbing greenhouse plastic films on insect populations. *Journal of Economic Entomology*, **95**, 113.

Cowles, R.S., and Miller, J.R. (1992). Diverting *Delia antiqua* (Diptera: Anthomyiidae) oviposition with cull onions: field studies on planting depth and a greenhouse test of stimulo-deterrent concept. *Environmental Entomology*, **21**, 453.

Croxton, S.D., and Stansly, P.A. (2014). Metalized polyethylene mulch to repel Asian citrus psyllid, slow spread of huanglongbing and improve growth of new citrus plantings. *Pest Management Science*, **70**, 318.

Daane, K.M., and Johnson, M.W. (2010). Olive fruit fly: managing an ancient pest in modern times. *Annual Review of Entomology*, **55**, 151.

Diaz, B.M., and Fereres, A. (2007). Ultraviolet-blocking materials as a physical barrier to control insect pests and plant pathogens in protected crops. *Pest Technology*, **1**, 85.

Dowd, P.F., and Vega, F.E. (2003). Autodissemination of *Beauveria bassiana* by sap beetles (Coleoptera: Nitidulidae) to overwintering sites. *Biocontrol Science and Technology*, **13**, 65.

Du, W., Han, X., Wang, Y., and Qin, Y. (2016). A primary screening and applying of plan volatiles as repellents to control whitefly *Bemisia tabaci* (Gennadius) on tomato. *Scientific Reports*, **6**, 22140.

Duan, J.J., and Prokopy, R.J. (1995). Development of pesticide-treated spheres for controlling apple maggot flies (Diptera: Tephritidae): pesticides and residue extending agents. *Journal of Economic Entomology*, **88**, 117.

Dyke, V.A., Hendrichs, J., and Robinson, A.S. (2005). *Sterile Insect Technique: Principles and Practice in Area-wide Integrated Pest Management*. Springer, Dordrecht.

Ebbinghaus, D., Losei, P.M., Romeis, J., et al. (2001). Appeal: efficacy and mode of action of attract and kill for codling moth control. *IOBC/WPRS Bulletin*, **24**, 95.

Egger, B., Spangl, B., and Koschier, E.H. (2014). Habituation in *Frankliniella occidentalis* to deterrent plant compounds and their blends. *Entomologia Experimentalis et Applicata*, **151**, 231.

Egger, B., Spangi, B., and Koschier, E.H. (2015). Continuous exposure to the deterrents cis-jasmone and methyl jasmonate does not alter the behavioural response of *Frankliniella occidentalis*. *Entomologia Experimentalis et Applicata*, **158**, 78.

Eigenbrode, S.D., Birch, A.N.-E., Lindzey, S., Meadow, R., and Snyder, W.E. (2016). A mechanistic framework to improve understanding and applications of push-pull systems in pest management. *Journal of Applied Ecology*, **53**, 202.

El-Sayed, A.M. (2017). The Pherobase: data base of pheromones and semiochemicals. Available at: http://www.pherobase.com (accessed 5 February, 2017).

El-Sayed, A.M., Suckling, D.M., and Wearing, C.H. (2006). Potential of mass trapping for long-term pest management and eradication of invasive species. *Journal of Economic Entomology*, **99**, 1550.

Ericksson, A., Anfora, G., Lucchi, A., Virant-Doberlet, M., and Mazzoni, V. (2012). Inter-plant vibrational communication in a leafhopper insect. *PlosOne*, **7**, e32954.

Evenden, M.L., and McLaughlin, J.R. (2004). Factors influencing the effectiveness of an attracticide formulation against the Oriental fruit moth, *Grapholita molesta*. *Entomologia Experimentalis et Applicata*, **112**, 89.

Faleiro, J.R. (2006). A review of the issues and management of the red palm weevil *Rhynchophorus ferrugineus* (Coleoptera: Rhynchophoridae) in coconut and date palm during the last one hundred years. *International Journal of Tropical Insect Science*, **26**, 135.

Fereres, A., and Moreno, A. (2009). Behavioural aspects influencing plan virus transmission by homopteran insects. *Virus Research*, **141**, 158.

Fettig, C., and Hilszczanski, J. (2015). Management strategies for bark beetles in conifer forests. In: F. E. Vega, and R. W. Hofstetter (Eds), *Bark Beetles: Biology and Ecology of Native and Invasive Species*, pp. 555–5 84. Springer, London.

Foster, S.P., and Harris, M.O. (1997). Behavioral manipulation methods for insect pest management. *Annual Review of Entomology*, **42**, 123.

Francardi, V., Benvenuti, C., Barzanti, G., and Roversi, P.F. (2013). Autocontamination trap with entomopathogenic fungi: a possible strategy in the control of *Phynchophorus ferrugineus* (Olivier) (Coleoptera: Curculionidae). *Redia*, **96**, 57.

Furlong, M.J., Pell, J.K., Choo, O.P., and Rahman, S.A. (1995). Field and laboratory evaluation of a sex pheromone trap for the autodissemination of a fungal entomopathogen *Zoophthora radicans* (Entomophthorales) by the diamondback moth *Plutella xylostella* (Lepidoptera: Yponomeutidae). *Bulletin of Entomological Research*, **85**, 331.

Giblin-Davis, R.M., Pena, J.E., Oehischlager, A.C., and Perexz, A.L. (1996). Optimization of semiochemical-based trapping of *Metamasius hemipterus sericeus* (Olivier) (Coleoptera: Curculionidae). *Journal of Chemical Ecology*, **22**, 1389.

Glenn, D.M., and Puterka, G.J. (2005). Particle films: a new technology for agriculture. *Horticulture Reviews*, **31**, 1–37.

Gonzalez, F., Tkaczuk, C., Dinu, M.M., et al. (2016). New opportunities for the integration of microorganisms into biological pest control systems in greenhouse crops. *Journal of Pest Science*, **89**, 295.

Grassi, A., Zini, M., and Forno, F. (2002). Mating disruption trials to control the currant clearwing moth, *Synanthedon tipuliformis*: a three-year study. *IOBC WRPS Bulletin*, **25**, 69.

Gregg, P.C., Del Socorro, A.P., Hawes, A.J., and Binns, M.R. (2016). Developing bisexual attract-and-kill for polyphagous insect: ecological rationale versus pragmatics. *Journal of Chemical Ecology*, **42**, 606.

Gut, L.J., Stelinski, L.L., Thomson, D.R., and Miller, J.R. (2004). Behaviour-modifying chemicals: prospects and constraints in IPM. In: O. Koul and G. S. Dhaliwal (Eds), *Integrated Pest Management*, pp. 73–121. CABI Publishing, Cambridge MA.

Hanks, L.M., and Millar, J.G. (2016). Sex and aggregation-sex pheromones of cerambycid beetles: basic science and practical applications. *Journal of Chemical Ecology*, **42**, 631.

Hartfield, C.M., Campbell, C.A.-M., Hardie, J., Pickett, J.A., and Wadhams, L.J. (2001). Pheromone traps for the dissemination of an entomopathogen by the Damson-hop aphid *Phorodon humuli*. *Biocontrol Science and Technology*, **11**, 401.

Heath, R.R., Epsky, N.D., Dueben, B.D., Rizzo, J., and Jeronimo, F. (1997). Adding methyl-substituted ammonia derivatives to a food-based synthetic attractant on capture of the Mediterranean and Mexican fruit flies (Diptera: Tephritidae). *Journal of Economic Entomology*, **90**, 1584.

Heuskin, S., Verheggen F.J., Haubruge, E., Wathelet, J-P., and Lognay, G. (2011). The use of semiochemicals slow-release devices in integrated pest management strategies. *Biotechnology Agronomy Society Environment*, **15**, 459.

Hofstetter, R.W., McGuire, R., and Dunn, D.D. (2014). Using acoustic technology to reduce bark beetle reproduction. *Pest Management Science*, **70**, 24.

Hoshi, J., Takabe, M., and Nakamuta, K. (2016). Mating disruption of a carpenter moth, *Cossus insularis* (Lepidoptera: Cossidae) in apple orchards with synthetic sex pheromone, and registration of the pheromone as an agrochemical. *Journal of Chemical Ecology*, **42**, 606.

Jackson, D.M., Brown, G.C., Nordin, G.L., and Johnson, D.W. (1992). Autodissemination of a baculovirus for management of tobacco budworms (Lepidoptera: Noctuidae) on tobacco. *Journal of Economic Entomology*, **85**, 710.

James, D.G., Bartelt, R.J., and Moore, C.J. (1996). Mass-trapping of *Carpophilus* spp. (Coleoptera: Nitidulidae) in stone fruit orchards using synthetic aggregation pheromones and a coattractant: development of a strategy for population suppression. *Journal of Chemical Ecology*, **22**, 1541.

Jakus, R., Schlyter, F., Zhang, Q.H., et al. (2003). Overview of development of anti-attractant based technology for spruce protection against *Ips typographus* from past failures to future success. *Journal of Pest Science*, **76**, 89.

Jermy, T., Bernays, E.A., and Szentesi, A. (1982). The effect of repeated exposure to feeding deterrents on their acceptability to phytophagous insect. *Proceedings of the 5th International Symposium of Insect-Plant Relationships Wageningen*, **1982**, 25.

Khan, Z.R., Midega, C.A.-O., Bruce, T.J.A., Hooper, A.M., and Pickett, J.A. (2010). Exploiting phytochemicals for developing a 'push-pull' crop protection strategy for cereal farmers in Africa. *Journal of Experimental Botany*, **61**, 4185.

Khan, Z., Midega, C., Pittchar, J., Pickett, J., and Bruce, T. (2011). Push-pull technology: a conservation agriculture approach for integrated management of insect pests, weeds and soil health in Africa. *International Journal of Agricultural Sustainability*, **9**, 162.

Khan, Z., Midega, C.A.O., Hooper, A., and Pickett, J. (2016). Push-pull: chemical ecology-based integrated pest management technology. *Journal of Chemical Ecology*, **42**, 689.

Khan, Z.R., Pickett, J.A., Wadhams, L., and Muyekho, F. (2001). Habitat management strategies for the control of cereal stemborers and striga in maize in Kenya. *International Journal of Tropical Insect Science*, **21**, 375.

Klein, M.C., and Lacey, L.A. (1999). An attractive trap for autodissemination of entomopathogenic fungi into populations of Japanese beetle, *Popilla japonica* (Coleoptera: Scarabaeidae). *Biocontrol Science and Technology*, **9**, 151.

Koyama, J., Kakinohana, H., and Miyatake, T. (2004). Eradication of the melon fly, *Bactrocera cucurbitae*, in Japan: importance of behavior, ecology, genetics and evolution. *Annual Review of Entomology*, **49**, 311–349.

Kunert, G., Reinhold, C., and Gershernzon, J. (2010). Constitutive emission of the aphid alarm pheromones, (*E*)-b-farnesene, from plants does not serve as a direct defense against aphids. *BMC Ecology*, **10**, 23.

Lance, D.R., Leonard, D.S., Mastro, V.C., and Walters, M.L. (2016). Mating disruption as a suppression tactic in programs targeting regulated Lepidopteran pests in the US. *Journal of Chemical Ecology*, **42**, 590.

Lance, D.R., Woods, W.M., and Stefan, M. (2014). Invasive insects in plant biosecurity; Case study—Mediterranean fruit fly. In: G. Gordh and S. McKirdy (Eds), *The Handbook of Plant Biosecurity*, pp. 447–484. Springer Dordrecht.

Larentzaki, E., Shelton, A.M., and Plate, J. (2008). Effect of kaolin particle film on *Thrips tabaci* (Thysanoptera: Thripidae), oviposition, feeding and development on onions: a lab and field case study. *Crop Protection*, **27**, 727.

Light, D.M., Knight, A.L., Henrick, C.A., et al. (2001). A pear-derived kairomone with pheromonal potency that attracts male and female codling moth, *Cydia pomonella* (L.) *Naturwissenschaften*, **88**, 333.

Lin, F., Bosquée, E., Liu, Y., et al. (2016). Impact of aphid alarm pheromone release on virus transmission efficiency: when pest control strategy could induce higher virus dispersion. *Journal of Virological Methods*, **235**, 34.

Lindgren, B.S., and Borden, J.H. (1993). Displacement and aggregation of mountain pine beetles, *Dendroctonus ponderosae* (Coleoptera: Scolytidae), in response to their antiaggrgation and aggregation pheromones. *Canadian Journal of Forest Research*, **23**, 286.

Liu, Z., Xin, Y., Xu, B., Raffa, K., and Sun, J. (2017). Sound-triggered production of antiaggregation pheromone limits overcrowding of *Dendroctonus valens* attacking pine trees. *Chemical Senses*, **42**, 59.

Lubanga, U.K., Guédot, C., Percy, D.M., and Steinbauer, M.J. (2014). Semiochemical and vibrational cues and signals mediating mate finding and courtship in Psylloidea (Hemiptera): a synthesis. *Insects*, **5**, 577.

Lujo, S., Hartman, E., Norton, K., et al. (2016). Disrupting mating behavior of *Diaphorina citri* (Liviidae). *Journal of Economic Entomology*, **109**, 2373.

Mafra-Neto, A., Fettig, C.J., Munson, S., and Stelinski, L.L. (2014). Use of repellents formulated in specialized pheromone and lure applications technology for effective insect pest management. In: M. Debboun, S. P. Frances, and D. Strickman (Eds), Insect Repellents Handbook, pp. 291–314. CRC Press, Boca Raton, FL.

Mangan, R.L. (2014). Priorities in formulation and activity of adulticidal bait sprays for fruit flies. In: T. Shelley, N. Epskey, E. B. Jang, J. Reyes-Flores, and R. Vargas (Eds), *Trapping and the Detection, Control and Regulation of Tephritid Fruit Flies*, pp. 423–456. Springer, Dordrecht.

McInnis, D.O., Lance, D.R., and Jackson, C.G. (1996). Behavioral resistance to the sterile insect technique by Mediterranean fruit fly (Diptera: Tephritidae) in Hawaii. *Annals of the Entomological Society of America*, **89**, 739.

Mensah, R.K., Gregg, P.C., Del Socorro, A.P., et al. (2013). Integrated pest management in cotton: exploiting behavior-modifying (semiochemical) compounds for managing cotton pests. *Crop and Pasture Science*, **64**, 763.

Miller, J.R., and Cowles, R.S. (1990). Stimulo-deterrent diversion: a concept and its possible application to onion maggot control. *Journal of Chemical Ecology*, **16**, 3197.

Miller, J.R., and Gut, L.J. (2015). Mating disruption for the 21st century: matching technology with mechanism. *Environmental Entomology*, **44**, 427.

Miller, J.R., Gut, L.J., De Lame, F.M., and Stelinski, L.L. (2006a). Differentiation of competitive vs. non-competitive mechanisms mediating disruption of moth sexual communication by point sources of sex pheromones: (Part 1) Theory. *Journal of Chemical Ecology*, **32**, 2089.

Miller, J.R., Gut, L.J., De Lame, F.M., and Stelinski, L.L. (2006b). Differentiation of competitive vs. non-competitive mechanisms mediating disruption of moth sexual communication by point sources of sex pheromones: (Part 2.) Case studies. *Journal of Chemical Ecology*, **32**, 2115.

Miller, J.R., McGhee, P.S., Siegert, P.Y., et al. (2010). General principles of attraction and competitive attraction as revealed by large-cage studies of moths responding to sex pheromone. *Proceedings of the National Academy of Sciences*, **107**, 22–27.

Miranda, M.P., Dos Santos, F.L., Felippe, M.R., Moreno, A., and Fereres, A. (2015). Effect of UV-blocking plastic films on take-off and host plant finding ability of *Diaphorina citri* (Hemiptera: Lividiidae). *Journal of Economic Entomology*, **108**, 245.

Morrison III, W.R., Lee, D.H., Reissig, W.H., et al. (2016). Inclusion of specialist and generalist stimuli in attract-and-kill programs: their relative efficacy in apple maggot fly (Diptera: Tephritidae) pest management. *Environmental Entomology*, **45**, 974.

Nalyanya, G., Moore, C.B., and Schal, C. (2000). Integration of repellents, attractants, and insecticides in a 'Push-Pull' strategy for managing German cockroach (Dictyoptera: Blatellidae) populations. *Journal of Medical Entomology*, **37**, 427.

Navarro-Llopis, V., Ayala, I., Sanchis, J., Primo, J., and Moya, P. (2015). Field efficacy of a *Metarhizium anisopliae*-based attractant-contamination device to control *Ceratitis capitata (Diptera: Tephritidae)*. *Journal of Economic Entomology*, **108**, 1570.

Navarro-Llopis, V., and Vacas, S. (2014). Mass trapping for fruit fly control. In: T. Shelley, N. Epskey, E. B. Jang, J. Reyes-Flores, and R. Vargas (Eds), *Trapping and the Detection, Control, and Regulation of Tephritid Fruit Flies*, pp. 513–547. Springer, New York.

Nehlin, G., Valterova, I., and Borg-Karlson, A. (1994). Use of conifer volatiles to reduce injury caused by carrot psyllid, *Trioza apicalis*, Förster (Homoptera: Psyllioidea). *Journal of Chemical Ecology*, **20**, 771.

O'Callaghan, M., and Jackson, T.A. (1993). Adult grass grub dispersal of *Serratia entomophila*. *Proceedings of the 46th New Zealand Plant Protection Conference*, p. 235–236. Christchurch, New Zealand.

Oehlschlager, A.C. (2016). Palm weevil pheromones—discovery and use. *Journal of Chemical Ecology*, **42**, 617.

Park, J.D., and Goh, H.G. (1992). Control of beet armyworm, *Spodoptera exigua* Hubner (Lepidoptera: Noctuidae), using synthetic sex pheromone. I. Control by mass trapping in *Allium fistulosum* field. *Korean Journal of Applied Entomology*, **31**, 45.

Pickett, J.A., and Khan, Z.R. (2016). Plant volatile-mediated signaling and its application in agriculture: successes and challenges. *New Phytologist*, **212**, 856.

Pickett, J.A., Woodcock, C.M., Midega, C.A.-O., and Khan, Z.R. (2014). Push-pull farming systems. *Current Opinion in Biotechnology*, **26**, 125.

Piñero, J.C., Enkerlin, W., and Epsky, N.D. (2014). Recent developments and applications of bait stations for integrated pest management of tephritid fruit flies. In: T. Shelley, N. Epsky, E.B. Jang, J. Reyes-Flores, and R. Vargas (Eds), *Trapping and the Detection, Control, and Regulation of Tephritid Fruit Flies*, pp. 457–511. Springer, New York.

Polajnar, J., Eriksson, A., Lucchi, A., et al. (2015). Manipulating behaviour with substrate-borne vibrations—potential for insect pest control. *Pest Management Science*, **71**, 15.

Polajnar, J., Eriksson, A., Virant-Doberlet, M., and Mazzoni, V. (2016). Mating disruption of a grapevine pests using mechanical vibrations: from laboratory to the field. *Journal of Pest Science*, **89**, 909.

Polston, J.E., and Lapidot, M. (2007). Management of tomato yellow leaf curl virus: US and Israel perspectives. In: H. Czosnek (Ed.), *Tomato Yellow Leaf Curl Virus Disease*, pp. 251–262. Springer, New York.

Prokopy, R.L., Bennett, E.W., and Bush, G.L. (1971). Mating behavior in *Rhagoletis pomonella* (Diptera; Tephritidae). I. Site of assembly. *Canadian Entomologist*, **103**, 1405.

Prokopy, R.L., and Bush, G.L. (1973). Mating behavior of *Rhagoletis pomonella* (Diptera: Tephritidae): IV. Courtship. *Canadian Entomologist*, **105**, 873.

Prokopy, R.J., Wright, S.E., Black, J.L., Hu, X.P., and McGuire, M.R. (2000). Attracticidal spheres for controlling apple maggot flies: commercial-orchard trials. *Entomologia Experimentalis et Applicata*, **97**, 293.

Pyke, B., Rice, M.J., Sabine, B., and Zalucki, M.P. (1987). The push-pull strategy-behavioral control of *Heliothis*. *Australian Cotton Growers*, **8**, 7.

Rodriguez-Saona, C.R., Polk, D., Holdcraft, R., and Koppenhöfer. (2014). Long-term evaluation of field-wide oriental beetle (Col., Scrabaeidae) mating disruption in blueberries using female-mimic pheromone lures. *Journal of Applied Entomology*, **138**, 120.

Rodriguez-Saona, C.R., and Stelinski, L.L. (2009). Behavior-modifying strategies in IPM: theory and practice. In: R. Pechin and A.K. Dhawan (Eds), *Integrated Pest Management: Innovation-development Process*, pp. 263–315. Springer, Dordrecht.

Ross, D.W., and Daterman, G.E. (1994). Reduction of Douglas-fir beetle infestation of high-risk stands by antiaggregation and aggregation pheromones. *Canadian Journal of Forest Research*, **24**, 2184.

Sharon, R., Zahavi, T., Sokolsky, T., et al. (2016). Mating disruption method against the vine mealybug, *Planococcus ficus*; effect of sequential treatment on infested vines. *Entomologia Experimentalis et Applicata*, **161**, 65.

Shea, P.J., and Neustein, M. (1995). Protection of a rare stand of Torrey pine from *Ips paraconfusus*. General Technical Report—Internmountain Research Service, USDA Forest Servid No. INT-318, pp. 39–43.

Shelton, A.M., and Badenes-Perez, F.R. (2006). Concepts and applications of trap cropping in pest management. *Annual Review of Entomology*, **51**, 285.

Shorey, H.H., Gaston, L.K., and Saario, C.A. (1967). Sex pheromones of noctuid moth. XIV. Feasibility of behavioural control by disrupting pheromone communication in cabbage loopers. *Journal of Economic Entomology*, **60**, 1541.

Shimodo, M., and Honda, K. (2013). Insect reactions to light and its application to pest management. *Applied Entomology and Zoology*, **48**, 413.

Smart, L.E., Blight, M.M. Pickett, J.A., and Pye, B.J. (1994). Development of field strategies incorporating semiochemicals for the control of the pea and bean weevil, *Sitona lineatus* L. *Crop Protection*, **51**, 1366.

Stelinski, L.L., Gut, L.J., Haas, M., McGhee, P., and Epstein, D. (2007). Evaluation of aerosol devices for simultaneous disruption of sex pheromones communication in *Cydia pomonella* and *Grapholita molesta* (Lepidoptera: Tortricidae). *Journal of Pest Science*, **80**, 225.

Stelinski, L.L., Lapointe, S.L., and Meyer, W.L. (2010). Season-long mating disruption of citrus leafminer, *Phyllocnistis citrella* Stanton, with an emulsified wax formulation of pheromone. *Journal of Applied Entomology*, **134**, 512.

Stelinski, L.L., and Liburd, O.E. (2001). Evaluation of various deployment strategies of imidacloprid-treated spheres in highland blueberries for control of *Rhagoletis mendax* (Diptera: Tephritidae). *Journal of Economic Entomology*, **94**, 905.

Stelinski, L.L., Liburd, O.E., Wright, S., et al. (2001). Comparison of neonicotinoid insecticides for use with biodegradable and wooden spheres for control of *Rhagoletis* species (Diptera: Tephritidae). *Journal of Economic Entomology*, **94**, 1142.

Stelinski, L.L., Miller, J.R., and Rogers, M.E. (2008). Mating disruption of citrus leafminer mediated by a non-competitive mechanism at a remarkably low pheromone release rate. *Journal of Chemical Ecology*, **34**, 1107.

Suckling, D.M. (2015). Can we replace toxicants, achieve biosecurity, and generate market position with semiochemicals? *Frontiers in Ecology and Evolution* **3**, 17.

Suckling, D.M., Conlong, D.E., Carpenter, J.E., et al. (2017). Global range expansion of pest Lepidoptera requires socially acceptable solutions. *Biological Invasions*, **19**, 1107.

Szendrei, Z., and Rodriguez-Saona, C. (2010). A meta-analysis of insect pest behavioral manipulation with plant volatiles. *Entomologia Experimentalis et Applicata*, **134**, 201.

Tan, K.H., Nishida, R., Jang, E.B., and Shelly, T.E. (2014). Pheromones, male lures, and trapping of Tephritid fruit flies. In T. Shelley, N. Epskey, E.B. Jang, J. Reyes-Flores, and R. Vargas (Eds), *Trapping and the Detection, Control and Regulation of Tephritid Fruit Flies*, pp. 15–74. Springer, Dordrecht.

Tinzarra, W., Gold, C.S., Dicke, M., et al. (2007). The use of aggregation pheromone to enhance dissemination of *Beauveria bassiana* for the control of the banana weevil in Uganda. *Biocontrol Science and Technology*, **17**, 111.

Tsutsumi, T., Teshiba, M., Yamanka, M., Ohira, Y., and Higucki, T. (2003). An autodissemination system for the control of brown winged green bug, *Plautia crossota stali* Scott (Heteroptera: Pentatomidae) by an emtomopathogenic fungus, *Beauveria bassiana* E-9102 combined with aggregation pheromone. *Japanese Journal of Applied Entomology and Zoology*, **47**, 159.

Turlings, T.C., and Ton, J. (2006). Exploiting scents of distress: the prospect of manipulating herbivore-induced plant odours to enhance the control of agricultural pests. *Current Opinion in Plant Biology*, **9**, 421.

Vargas, R.I., Leblanc, L., Piñero, J.C., and Hoffman, K.M. (2014). Male annihilation, past, and future. In: T. Shelley, N. Epskey, E.B. Jang, J. Reyes-Flores, and R. Vargas (Eds), *Trapping and the Detection, Control and Regulation of Tephritid Fruit Flies*, pp. 493–511. Springer, New York.

Vargas, R. I., Piñero, J.C., and Leblanc, L. (2015). An overview of pest species of *Bactrocera* fruit flies (Diptera: Tephritidae. and the integration of biopesticides with other biological approaches for their management with a focus on the Pacific region. *Insects*, **6**, 297.

Vargas, R.I., Souder, S.K., Mackey, B., et al. (2012). Field trials of solid triple lure (Trimedlure, methyl eugenol, raspberry ketone, and DDVP) dispensers for detection and male annihilation of *Ceratitis capitata*, *Bactrocera dorsalis*, and *Bactrocera cucurbitae* (Diptera: Tephritidae) in Hawaii. *Journal of Economic Entomology*, **105**, 1557.

Vega, F., Dowd, P.F., Lacey, L.A., et al. (2000). Dissemination of beneficial microbial agents by insects. In: L.A. Lacey and H.K. Kaya (Eds), *Field Manual of Techniques in Invertebrate Pathology*, pp 153–177. Kluwer Academic Publishers, New York.

Vernon, R.S., van Herk, W., Clodius, M., and Tolman, J. (2016). Companion planting attract-and-kill method for wireworm management in potatoes. *Journal of Pest Science*, **89**, 375.

Vickers, R.A., Furlong, M.J., White, A., and Pell, J.K. (2004). Initiation of fungal epizootics in diamondback moth populations within a large field cage: proof of concept for auto-dissemination. *Entomologia Experimentalis et Applicata*, **111**, 7.

Wang, C., Scharf, M.E., and Bennett, G.W. (2006). Genetic basis for resistance to gel baits, fipronil, and sugar-based attractants in German cockroaches (Dictyoptera: Blatellidae). *Journal of Economic Entomology*, **99**, 1761.

Wang, K., Ferguson, G., and Shipp, J.L. (1997). Incidence of tomato pinworm, *Keiferia lycopersicella* (Walsingham), (Lepidoptera: Gelechiidae) on greenhouse tomatoes in Southern Ontario and its control using mating disruption. *Proceedings of the Entomological Society of Ontario*, **128**, 93.

Wawrzynski, R.P., and Ascerno, M.E. (1998). Mass trapping for Japanese beetle (Coleoptera: Scarabaeidae) suppression in isolated areas. *Journal of Arboriculture*, **24**, 303.

Weihman, S.W., and Liburd, O.E. (2006). Mating disruption and attract-and-kill as reduced-risk strategies for control of grape root borer *Vitacea polistiformis* (Lepidoptera: Sesiidae) in Florida vineyards. *Florida Entomologist*, **89**, 245.

Witzgall, P., Kirsch, P., and Cork, A. (2010). Sex pheromones and their impact on pest management. *Journal of Chemical Ecology*, **36**, 80.

Witzgall, P., Stelinski, L., Gut, L., and Thomson, D. (2008). Codling moth management and chemical ecology. *Annual Review of Entomology*, **53**, 503.

Wright, S.E., Leskey, T.C., Jacome, I., Piñero, J.C., and Prokopy, R.J. (2012). Integration of insecticidal, phagostimulatory, and visual elements of an attract and kill system for apple maggot fly (Diptera: Tephritidae). *Journal of Economic Entomology*, **105**, 1548.

Wu, K. (2010). No refuge for insect pests. *Nature Biotechnology* **28**, 1273.

Xu, Q.X., Hatt, S., Lopes, T., et al. (2018). A push-pull strategy to control aphids combines intercropping with semiochemical releases. *Journal of Pest Science*, **91**, 93–103.

Yonce, C.E. (1981). Mating disruption of the lesser peachtree borer, *Synanthedon pictipes* (Grote and Robinson), and the peachtree borer *S. exitosa* (Say), with a hollow fiber formation. *Miscellaneous Publications of the Entomological Society of America*, **12**, 21.

Yu, X., Jones, H.D., Ma, Y., et al. (2012). (*E*)-β-Farnesene synthase genes affect aphid (*Myzus persicae*) infestation in tobacco (*Nicotiana tabacum*). *Functional Integrative Genomics*, **12**, 207.

Zabalou, S., Apostolake, A., Livadaras, I., et al. (2009). Incompatible insect technique: incompatible males from a *Ceratitis capitata* genetic sexing strain. *Entomologia Experimentalis et Applicata*, **132**, 232.

Zehnder, G., Gurr, G.M., Köhne, S., et al. (2007). Arthropod pest management in organic crops. *Annual Review of Entomology*, **52**, 57.

Zhang, A. Linn Jr., C., Wright, S., et al. (1999). Identification of a new blend of apple volatiles attractive to the apple maggot, *Rhagoletis pomonella*. *Journal of Chemical Ecology*, **25**, 1221.

Zhang, G-F., Meng, S-Z., Han, Y., and Sheng, C-E. (2002). Chinese tortrix *Cydia trasisas* (Lepidoptera: Olethreutidae): suppression on street-planting trees by mass trapping with sex pheromone traps. *Environmental Entomology*, **31**, 602.

Zhang, Z., Sun, X., Luo, Z., Gao, Y., and Chen, Z. (2013). The manipulation mechanism of 'push-pull' habitat management strategy and advances in its application. *Acta Ecologia Sinica*, **33**, 94.

CHAPTER 21

Behavior-based control of arthropod vectors: the case of mosquitoes, ticks, and Chagasic bugs

Ana E. Gutiérrez-Cabrera[1], Giovanni Benelli[2], Thomas Walker[3], José Antonio De Fuentes-Vicente[4], and Alex Córdoba-Aguilar[5]

[1] CONACyT-Centro de Investigación sobre Enfermedades Infecciosas, Instituto Nacional de Salud Pública 62100, Cuernavaca, Morelos. México

[2] Department of Agriculture, Food and Environment, University of Pisa, via del Borghetto 80, 56124 Pisa, Italy

[3] Department of Disease Control, London School of Hygiene and Tropical Medicine, London, WC1E 7HT, United Kingdom

[4] Universidad Pablo Guardado Chávez, Libramiento Norte Oriente 3450, Fracc. Residencial Las Palmas, 29040, Tuxtla Gutiérrez, Chiapas, México

[5] Departamento de Ecología Evolutiva, Instituto de Ecología, Universidad Nacional Autónoma de México, Apdo. Postal 70-275, Ciudad Universitaria, 04510 México D.F., México

21.1 Arthropod disease vectors

21.1.1 Patterns and occurrence of major diseases transmitted by arthropod vectors

Among arthropods, mosquitoes, ticks, and Chagasic bugs, represent major vectors of high public health relevance. The World Health Organization (WHO) estimates more than one billion people are infected, resulting in more than one million deaths per year. Vector-borne diseases impose the biggest effect on the least-developed countries in tropical regions, particularly communities associated with poverty. The development of synthetic residual pesticides in the 1940s was followed by large-scale spraying in the 1950s and 1960s. This has had significant effects on many vector-borne diseases (Brathwaite-Dick et al. 2012). Vector control programmes emphasized elimination of breeding sites (source reduction) with limited use of insecticides. By the 1960s, only Africa was considered to have significant transmission of vector-borne diseases (Gubler 1998). However, vector-borne diseases re-emerged in the 1970s due to factors such as climate change, urbanization, and a sustained reduction in the intensity of vector control programmes. One consequence of these factors is the spread of key mosquito species which are responsible for transmission of important diseases including malaria, dengue, yellow fever, filariasis, Japanese encephalitis, and Zika virus (Benelli 2015a; Benelli et al. 2016a). Mosquitoes also transmit key pathogens and parasites that dogs, horses, and birds are very susceptible to. Unfortunately, no treatment is available for most of the **arboviruses** vectored by mosquitoes. Another major vector are ticks as they are behind the widest array of disease-causing organisms of all hematophagous arthropods, second only to mosquitoes in their capacity to transmit disease agents of importance to human and veterinary health (Mehlhorn 2015). Finally, Chagas disease is recognized by WHO as one of the world's thirteen most neglected tropical diseases (Hotez et al. 2007). Chagas is of great concern in many Latin American countries, mainly because it has no cure. This disease, also known as American trypanosomiasis, is caused by the protozoan parasite *Trypanosoma cruzi*, whose vectors are more than 130 species of the triatomine bug family.

The aim in this chapter is to describe how behavior can influence control strategies to manage arthropods that transmit diseases. Please note that this is not a scholastic review, but a student-orientated guide, which should serve students of insect behavior how to use this area for an applied topic.

21.1.2 The need of behavior-based strategies of control

The absence of vaccines or effective drugs for the treatment of most vector-borne diseases has resulted in vector control being the only effective form of protection. For a number of insect vectors, control strategies have historically involved the use of chemical insecticides such as insecticide treated bednets for malaria control. However, bednets are predominantly effective only against mosquito species that bite during the night (*Anopheles* and *Culex* spp.), but not in aedine species, such as *Aedes albopictus*. Alternative control strategies have been under consideration, since insecticide resistance is widespread in many mosquito species (e.g. Ranson and Lissenden 2016). The widespread use of insecticides is also financially demanding in developing countries so various biological control strategies have been explored (e.g. Benelli 2015a; Benelli et al. 2016a). These strategies include bio-friendly and self-sustained methods that either kill mosquitoes or replace them with mosquitoes that are unable to transmit pathogens through mass releases. In addition, special attention has been paid to promising novel mosquito control strategies, such as *Wolbachia* bacteria (see Section 2.3 for the effects of this bacteria on mosquitoes and insects in general) and the release of sterile male mosquitoes

through genetic engineering (Wilke et al. 2018). Nevertheless, the success of these strategies is highly dependent on mosquito behavior. For example, the mating competitiveness of males, either sterile or infected with *Wolbachia*, will have a direct impact on the effectiveness of mass released mosquitoes. A greater understanding of behavior that underpins mosquito mate searching, recognition and swarming (including sexual chemical ecology) will improve vector control programs to reduce mosquito-borne transmission (Benelli and Mehlhorn 2016).

In the search of alternative control methods, a similar strategic path has taken place for other insect vectors. For the case of tick-borne diseases, several outbreaks indicate that such diseases are becoming more prominent in different countries (e.g. Baltic States, The Russian Federation and Slovenia; Mehlhorn 2015). A number of novel routes have been attempted such as the development of vaccines against viruses vectored by ticks, pheromone-based control tools, 'lure and kill' techniques, biological control programmes relying on tick natural enemies and pathogens, and integrated pest management practices aimed at reducing tick interactions with livestock (see Benelli et al. 2016b). Other vector example is the case of triatomine bugs, vectors of the Chagas disease agent, *Trypanosoma cruzi*. The use of insecticide to control triatomines is becoming less extensive given the increase in insecticide resistance (Vassena et al. 2000). Chagas disease has been increasing quite extensively in recent years, especially in North America. Estimated number of infected people are up to seven million just for this region, while for the entire American continent are up to eighteen million (Hotez et al. 2013). Bugs are the main source of infection and this has been exacerbated by the new climatic opportunity windows that global warming is providing (Ibarra-Cerdeña et al. 2009). Several strategies have emerged including prophylactic and therapeutic vaccines to prevent and cure people, respectively. The former is recommended in highly **endemic** zones, while the latter is recommended for infected patients and can be used in combination with chemotherapy (Beaumier et al. 2016). However, the fact that Chagas disease is linked to poverty, implies a reduced market incentive to produce an effective drug-based treatment. This renders drug-based treatment as a very inefficient strategy.

However, there have been recent drug developments possibly as a response to the spread of Chagas in the USA in the last decade (Ribeiro et al. 2009).

21.2 Mosquitoes

21.2.1 Traditional mosquito control strategies

Current prevention tools are represented by the employment of mosquito repellents [e.g. N,N-diethyl-meta-toluamide (DEET), dimethyl phthalate (DMP), N,N-diethyl mendelic acid amide (DEM), N, N-diethyl-3-methylbenzamide (DEET), ethyl 3-[acetyl(butyl) amino] propanoate (IR3535), para-menthane-3,8-diol (PMD), picaridin as well as natural-based products (Mehlhorn 2015), light-coloured clothes covering as much of the body as possible, and sleeping under mosquito nets (Benelli 2015a; see Figure 21.1). These measures should be used along with Culicidae breeding site control in areas with endemic mosquito borne diseases, as well as with chemical or microbiological ovicide, larvicide, and pupicide treatments (e.g. Benelli 2015b). Concerning the employment of synthetic pesticides, attention should be given to the development of resistant mosquito strains, as well as to environmental concerns (e.g. Naqqash et al. 2016). Indeed, in the past, Culicidae larvae and pupae have been massively targeted using organophosphates, carbamates, and pyrethroids, with huge negative effects on human health and the environment. Later on, insect growth regulators and microbial control agents have been introduced, and *Bacillus thuringiensis* var. *israelensis* is currently the most common mosquito larvicide employed in European countries. However, also these tools induce resistance in a number of mosquito species (e.g. Tabashnik, 1994), pointing out the need for novel, cheap, and reliable mosquito control programmes (Benelli 2015a).

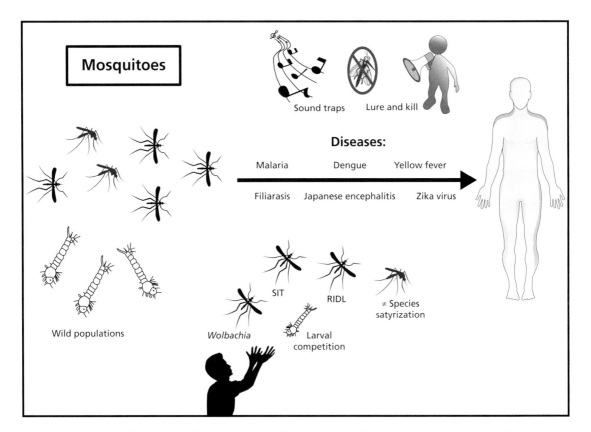

Figure 21.1 A summary of diseases vectored by mosquitoes, and behavioral-linked control strategies based on non-manipulated animals (sound traps, lure and kill) and release of manipulated animals (*Wolbachia*, SIT, satyrization, larval competition).

21.2.2 How mosquito behavioral ecology boost vector control?

Control strategies that are influenced by mosquito behavior require basic knowledge regarding the mosquito mating ecology (particularly sexual chemical ecology) (Benelli 2015a; Benelli and Mehlhorn 2016). Detailed information on the mating ecology of mosquito vectors may help to improve the success of mosquito control programmes in several ways. Indeed, this information is crucial for any control programme against mosquito vectors of medical and veterinary importance, since it is widely known that the success of the sterile insect technique (SIT; a genetic suppression strategy that involves rearing large numbers of male insects that have been sterilized through either irradiation or treatment with chemosterilants), is largely dependent on the ability of sterile males to compete for mates with wild ones in the field (e.g. Oliva et al. 2014). Moreover, behavioral knowledge on mosquito swarming and mating behavior may be used to perform comparisons of courtship and mating **ethograms** among different vector species and strains, allowing monitoring and optimization of quality of mass-reared males (i.e. sexual competitiveness and mating success) over time in SIT and 'boosted SIT' programmes, as well as to monitor the mating performances in mass-reared *Wolbachia*-induced phenotypes (Benelli 2015a,c).

It is worthy to note that the quantitative analyses of mating ethograms in Culicidae are rare. The majority of the studies focused on the sexual behavior of medically important *Aedes* species and compared insemination ability in sterilized and wild males (e.g. Madakacherry et al. 2014), without behavioral quantification of courtship and mating events (but see, for example, Oliva et al. 2013a; Benelli 2015c). Notably, limited information is also available on the potential molecules mediating aggregation and mating dynamics in mosquitoes (e.g. Cabrera and Jaffe 2007). Finally, there is the potential for genetic engineering to modify reproductive traits, based on *Anopheles* evidence in traits that include male accessory glands (Dottorini et al. 2007), sperm storage organ enzyme activity (Shaw et al. 2014), and male fertility (Pompon and Levashina 2015) to cite some examples.

Sound traps

The first attempts to control mosquito populations by using sound traps was in Cuba in 1949 against *Anopheles albimanus*, where the use of a sound trap resulted in an increase in the number of males collected (Kahn and Offenhauser 1949). Further similar approaches led to the successful control of mosquitoes from the genera *Aedes*, *Anopheles*, *Culex*, and *Mansonia* (Kanda et al. 1988, 1990; Benelli and Mehlhorn 2016). However, even if it has been claimed that the partial overlapping of wing-beat frequencies may be advantageous, attracting more than a single mosquito species, there are at least two major shortcomings for the use of sound traps in the field (Diabate and Tripet 2015; see also Cator and Harrington 2010). The first one is that the design and production of sound traps able to attract mosquitoes from long distances, with proper amplification features, is quite difficult. The second shortcoming is that traps should be placed in sites that are close to those selected by mosquito vectors for swarming. Unfortunately, these sites are frequently unknown (Diabate and Tripet 2015).

'Lure and kill' technique

The 'lure and kill' approach (see also Chapter 20) has been used successfully against a number of arthropod pests (Kanda et al. 1988, 1990; Benelli and Mehlhorn 2016). Consequently, recent research has highlighted the possibility to disrupt or enhance swarms by manipulating artificial markers. These efforts may lead to create 'kill zones' (e.g. within or in close proximity of villages) where high numbers of mosquitoes can be attracted and killed. Further research on the exact nature of chemical cues routing mate searching and choice dynamics in mosquito vectors is required, as well as behavioral studies dissecting the relative importance of visual (with special reference to swarming landmarks), vibrational, olfactory, and tactile cues perceived during swarming and mate recognition (Benelli and Mehlhorn 2016).

The sterile insect technique (SIT)

Insect behavior plays a pivotal role in the potential success (and often failure) of a number of novel mosquito control strategies that involve releasing mosquitoes into wild populations. These control strategies can be divided into **population suppression** (significant reduction or elimination of the target population) or population replacement with mosquitoes that are unable to transmit disease. The SIT involves the generation of sterile males through either irradiation or treatment with chemosterilants followed by mass releases with the aim of successfully mating with wild females and producing no progeny (see also Chapter 20). A successful SIT programme results in an increasing sterile:wild male ratio with a decreasing overall insect population that ultimately can result in eradication of the population. Mass rearing of the target species is required with sex separation prior to extensive releases of male only insects (Benedict and Robinson 2003). Indeed, over the past 50 years SIT has been particularly successful against agricultural insects (see Chapter 14). The use of SIT for mosquitoes that transmit diseases of public health importance has been limited and this is partly down to mating success of sterilized males. Key male mosquito fitness parameters need to be assessed prior to release. These parameters include longevity, body size, and sperm production (Oliva et al. 2013a,b). In addition, male mosquito performance in the field can be very different from that in the laboratory with unpredictable behaviors occurring in the wild such as flight ability and mating behavior in swarms (Benelli et al. 2015). With this information, the level of radiation must balance the induced sterility of mosquitoes and the effects on male mating competitiveness. Furthermore, for SIT programmes targeted towards *Aedes* mosquitoes, there are logistical difficulties in initially reducing the wild population densities prior to the release of sterile males. Such difficulties include population fluctuation properties (dispersal, density, and mortality rate), sexual competitiveness of target animals, and number of sterile males to be released (Ito and Yamamura 2005).

Alternative strategies to supress mosquito populations include the use of juvenile hormones analogs such as pyriproxygen, which regulate insect development by delaying metamorphosis until larvae have attained an appropriate stage and size (for effect of hormones on insect development, see Chapter 4). Auto-dissemination by adult

female *Aedes* mosquitoes contaminated with dissemination stations of juvenile hormone have been used to treat breeding habitats (Caputo et al. 2012). Using this method, female mosquitoes contaminated with juvenile hormones analogs would lay eggs and the insect growth regulators would inhibit further development to the adult stage. With this methodology, pyriproxygen led to successful suppression of *Ae. aegypti* in Peru (Devine et al. 2009), and *Ae. albopictus* in Italy (Caputo et al. 2012). The release of sterile males in combination with juvenile hormone control methods could also allow contamination of females during mating to 'boost SIT' (Bouyer and Lefrancois 2014). The use of species-specific pathogens such as mosquito **densoviruses** (MDVs) could also be exploited to enhance SIT (Bouyer et al. 2016). Dispersion of MDVs by sterile males to wild females could results in detrimental effects on *Aedes* larval habitats as a result of their 'skipping' oviposition behavior (Bouyer et al. 2016). MDVs kill mosquito larvae, prevent development to the adult stage and can be transmitted vertically to progeny. Therefore, further research is needed to determine if MDVs could play an important role in supressing wild mosquito populations, particularly in combination with SIT strategies.

Genetically modified mosquitoes

The detrimental effects of radiation for SIT compromises the mating competitiveness of sterilized males when released into wild populations. Genetics has been exploited to develop alternative ways to generate sterility in mosquito species, with the aim of reducing effects on male mosquito fitness. Unlike SIT, genetically modified strategies imply alteration of gene material to reduce mosquito fitness (including sterilization). This approach, pioneered by the British biotech company Oxitec (www.oxitec.com), aims to introduce a self-limiting lethal gene into mosquito populations to supress and ultimately eradicate local populations. The release of insects with a dominant lethal (*RIDL*) gene is species-specific and the lethal gene can be repressed using an antidote (tetracycline) during the rearing phase to achieve adult mosquitoes prior to release of males into wild populations (Thomas et al. 2000). RIDL technology was used to produce a strain of *Ae. aegypti*, OX513A, to carry a dominant, repressible, non-sex-specific, late-acting lethal genetic system (Phuc et al. 2007). In the absence of tetracycline, larvae carrying one or more copies of the OX513A insertion develop normally, but die at pupation. Field trials in the Cayman Islands with *Ae. aegypti* OX513A was shown to suppress a wild population of *Ae. aegypti* (Harris et al. 2012). In Malaysia, OX513A males were shown to have similar longevity and dispersal capabilities (Lacroix et al. 2009) and a recent release in Brazil led to strong suppression of the target wild population (Carvalho et al. 2015). Genetics can also be used to supress or eliminate mosquito populations through induction of an extreme male-biased sex ratio (Schliekelman et al. 2005). Indeed, naturally occurring **sex ratio distorters** have been found in *Aedes* and *Culex* mosquitoes, but population suppression was not achieved in cage experiments (Robinson 1983). Still, genetic modification can be used to induce a male gamete production bias by inducing preferential breakdown of the X chromosome during male meiosis. Breakdown of paternal X chromosomes in *An. gambiae* prevents it from being transmitted to the next generation, resulting in fully fertile mosquito strains that produce >95 per cent male offspring (Galizi et al. 2014). On the other hand, synthetic distorter male mosquitoes suppress caged wild type mosquito populations providing evidence for potential new control strategies. Transgenic mosquitoes are also being developed to create both anti-pathogen effector genes in combination with gene-drive systems that will allow **introgression** into wild mosquito populations (Gantz et al. 2015). Finally, highly effective **clustered regularly interspaced short palindromic repeats (CRISPR)**-associated protein 9 (Cas9)-mediated gene-drive systems have been developed in *Anopheles* species precise editing of mosquito DNA sequences (e.g. Gantz et al. 2015).

21.2.3 *Wolbachia* endosymbiotic bacteria

Wolbachia are endosymbiotic bacteria that infect more than 65 per cent of insect species and, in mosquitoes, induce a reproductive phenotype termed **cytoplasmic incompatibility** (CI). Unviable offspring result from matings between uninfected females and *Wolbachia*-infected males. However, *Wolbachia*-infected females can produce viable progeny when they mate with both infected and uninfected males. These crossing patterns provide a reproductive advantage to *Wolbachia* infections allowing this maternally transmitted bacteria to invade insect host populations. Natural *Wolbachia* infections are present in some mosquito vectors, such as *Cx. quinquefasciatus* and *Ae. albopictus* but no natural infections are present in *Ae. aegypti*. The first experiments that used *Wolbachia* for mosquito control utilized CI to eradicate *Culex quinquefasciatus* mosquito populations from Myanmar in the late 1960s (Laven 1967). This incompatible insect technique (IIT) relies on the release of a large number of *Wolbachia*-infected males that compete with wild type males to induce sterility and suppress the mosquito population (outlined in Atyame et al. 2015). IIT is currently being used in trials with *Ae. albopictus* through the generation of a triple *Wolbachia*-infected strain (wAlbA, wAlbB, and wPip; Zhang et al. 2015) and *Aedes polynesiensis*, a vector of lymphatic filariasis in the South Pacific (O'Connor et al. 2012). Trials for IIT are being undertaken by the biotech company MosquitoMate (www.mosquitomate.com) using *Ae. albopictus* male releases. The long-term success of IIT is likely dependent on the environment and ecology of the target mosquito population.

Wolbachia strains in their native *Drosophila* fruit fly hosts provide strong protection against infection by pathogenic RNA viruses (Hedges et al. 2008). This discovery has been used for mosquito biocontrol strategies that use *Wolbachia* to prevent human pathogens from replicating within mosquitoes (Iturbe-Ormaetxe et al. 2011). In this sense, the 'World mosquito program' project (https://www.world-mosquitoprogram.org) has been able to demonstrate that *Wolbachia* bacteria can prevent dengue virus (DENV) transmission in mosquitoes without significant fitness costs. *Wolbachia*-infected *Ae. aegypti* lines were generated using embryo microinjection and significantly reduced the vector competence of *Ae. aegypti* for DENV under laboratory conditions (Walker et al. 2011). Interestingly, *Wolbachia* bacteria in transinfected *Ae. aegypti* are found at high levels in mosquito salivary glands, which is likely the principle reason behind their ability to inhibit DENV transmission (shown through the absence of infectious virus in the saliva; Walker et al. 2011). In addition to a 'virus-blocking' phenotype, *Wolbachia* strains in

Ae. aegypti show maternal transmission rates close to 100 per cent and induce high levels of CI (Walker et al. 2011). To assess fitness costs and the ability of *Wolbachia* strains to invade mosquito populations, semi-field cage experiments were undertaken with two *Wolbachia* strains (*w*Mel and *w*MelPop). The fecundity of *w*MelPop-infected female mosquitoes was reduced by ~60 per cent relative to uninfected wildtype and *w*Mel-infected mosquitoes and *w*Mel-Pop strain invaded at a slower rate when compared with *w*Mel (Walker et al. 2011). In field conditions, *Ae. aegypti* mosquitoes infected with the *w*Mel strain were introduced in the wild through open releases in two locations near Cairns in north Queensland, Australia and reached near fixation within a few months (Hoffmann et al. 2011). The long-term success of any *Wolbachia*-based release programme will be dependent on maintenance and stability of the inhibitory effects on DENV replication in wild mosquito populations. This was tested with *Wolbachia*-infected mosquitoes 1 year following field release and vector competence experiments revealed strong inhibition of DENV replication and dissemination (Frentiu et al. 2014). Recently, a *Wolbachia*-superinfected *Ae. aegypti* line was generated with the *w*Mel and *w*AlbB strains and this combination resulted in greater inhibitory effects on DENV replication than the single *w*Mel strain when challenged with blood meals from viraemic dengue patients (Joubert et al. 2016). *Wolbachia* superinfections could be used to replace single infections in wild populations and may provide an effective strategy to help manage potential resistance by DENV to field deployments of single infected strains. As with all novel mosquito control strategies at the preliminary stage of development, *Wolbachia*-based biocontrol requires further research for long-term implementation in the field. An important decision will be which is the best *Wolbachia* strain/combination of strains for applied use given the balance required between inhibition of pathogen transmission and vector fitness costs. What impact will *Wolbachia* strains have on dengue transmission? A potential answer for this came via mathematical modelling of DENV transmission that predicted that *w*Mel would reduce the basic reproduction number, R0, of DENV transmission by 66–75 per cent (Ferguson et al. 2015), but the predicted effect may be different to the observed effect. *Wolbachia* has

also been shown to inhibit the transmission of other arboviruses and parasites (reviewed in Benelli et al. 2016c). Japanese encephalitis virus is predominantly transmitted by *Cx. tritaeniorhynchus* mosquitoes and a *Wolbachia*-based biocontrol strategy has the potential to reduce transmission if stably-infected lines can be generated (Jeffries and Walker 2015). Additional mosquito species, such as *Cx. quinquefasciatus* and *Ae. albopictus* that contain resident *Wolbachia* strains, are also potential targets for introducing 'transinfected' strains that are likely to grow to higher densities and, therefore, impact pathogen transmission (Jeffries and Walker 2016).

21.2.4 Satyrization

Frequently, two species whose distribution ranges do not overlap and are therefore allopatric can, for different reasons, become sympatric. The range of many invasive mosquito species that can transmit arboviruses, particularly of the genus *Aedes*, are increasing (Medley 2010). This can lead to new shared geographical areas, where sympatric species can have sexual encounters (Sánchez-Guillén et al. 2016). Assuming that species are closely related and isolation barriers do not prevent interspecific mating, different species can give raise to effective matings, but no hybrids. This phenomenon has been observed in insects including mosquitoes and has been labelled as satyrization. This implies reproductive costs for sexes of the different species but especially for females. These asymmetric costs is precisely the case for the observed sympatric areas in Florida where *Ae. aegypti* and *Ae. albopictus* inhabit (Tripet et al. 2011). Interestingly, only females of *Ae. aegypti* become refractory to further matings, which serves as a sterilizing mechanism for this species (Tripet et al. 2011). Although it seems that *Ae. aegypti* has evolved more effective premating isolation in sympatry to reduce the fitness costs of satyrization (Bargielowski et al. 2013), this phenomenon has been proposed as a way to control invasion of mosquitoes of other species such as *Aedes* spp. Recent cage experiments looking at populations of *Ae. aegypti* and *Ae. albopictus* outlined that they were relatively susceptible to satyrization (Lounibos et al. 2016). Further work on satyrization of invasive mosquito species could result in opportunities to influence control strategies.

21.2.5 Larval competition

Following the same logic that mosquito species may share habitats, another areas in which mosquito species can interact is through competition at the larval stage. Since mosquito larvae may compete for food and such competition is density-dependent (e.g. Novak et al. 1993), carry-over effects have been observed at the adult stage in terms of body size and viral infection (Alto et al. 2008). Basically, high competition leads to a reduced body size and higher viral infection rates (Alto et al. 2008). This phenomenon may explain species-specific distributions in the USA (Omeara et al. 1995) and may be used to control virus spread. The effect of interspecific competition on important vectors of arboviruses must be taken into consideration for control strategies. For example, temporal shifts in the distribution of *Ae. albopictus* and *Cx. pipiens* in North-eastern Italy were due to competition (Marini et al. 2017) and these two species have very different behavioral traits that require different control strategies.

21.3 Ticks

21.3.1 Traditional tick control strategies

Arsenic dips were the first effective method for controlling ticks and tick-borne diseases, and were used for 50 years before resistance to the chemical became a problem. Since the discovery of organochlorines, virtually every chemical group of pesticides developed for the control of arthropods is represented among the list of products employed for the control of ticks on cattle (George 2000). In recent years, effective improvements in the development of acaricides with low mammalian toxicity (e.g. pyrethroids and avermectins) enhanced the efficacy of treatments against ticks, but at greatly increased cost (Sonenshine 2006; Sonenshine and Roe 2014). Furthermore, the evolution of tick resistance to acaricides has been a major determinant of the need for new products (George 2000).

To prevent tick-borne diseases, the best strategy remains preventing tick attachment and contraction of tick-vectored disease organisms. This can be done

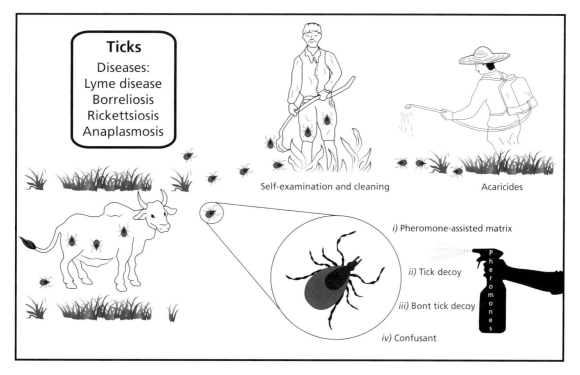

Figure 21.2 A summary of diseases vectored by ticks and some traditional preventive measures (self-examination and acaricides), as well as behavioral-based control measures (pheromone-assisted matrix, tick decoy, bont tick decoy and confusant).

by limiting exposure to tick habitat, through self-examination after contact with tick habitat, and use of personal arthropod repellents onto naked skin and clothes (Lupi et al. 2013). In contrast to mosquitoes and other arthropods of public health importance, there is less research conducted on tick behavior and repellence mechanisms (Lupi et al. 2013).

Finally, about two decades ago, the first vaccine against the tick *Rhipicephalus microplus*, based on a recombinant antigen (Bm86), was released (Willadsen et al. 1995; de la Fuente et al. 1999). However, vaccines protect only against infections by viruses vectored by ticks (see Willadsen 2006 for a review), which may induce meningoencephalitis, while for bacterial diseases such as borreliosis, rickettsiosis or anaplasmosis, the best strategy remains the avoidance of tick bites (Lupi et al. 2013).

21.3.2 Behavioral knowledge helps: pheromone-assisted tick control

Tick control strategies that focus on tick chemical ecology (Sonenshine 2006) have shown promise through the development of pheromone-assisted tick control. They include:

- pheromone-assisted matrix for application to vegetation(see Figure 21.2);
- tick decoy;
- bont tick decoy;
- confusants.

The pheromone-assisted matrix for application to vegetation relies on the employment of arrestment nymphal and adult pheromone components (i.e. guanine, xanthine, and hematin) of blacklegged ticks (also known as deer ticks), incorporating these molecules and an acaricide (e.g. permethrin) into oily droplets or microfibers (e.g. Sonenshine and Roe 2014).

Tick decoy allows disrupting tick reproduction, relying to a pheromone-acaricide-impregnated device, which attract and kill male ticks before they could mate with feeding females (e.g. Abdel-Rahman et al. 1998). In addition, a peculiar modification of the tick decoy concept was adapted to control *Amblyomma hebraeum* and *A. variegatum* ticks on cattle. These ticks are major vectors of *Anaplasma ruminantium*, responsible of the deadly heartwater disease. *A. hebraeum* and *A. variegatum* form aggrega-

tions on their hosts in response to a male-originated attraction-aggregation-attachment pheromone. Methyl salicylate, *o*-nitrophenol, 2,6-DCP, and phenylacetaldehyde, inducing attraction-aggregation-attachment in these tick vectors, were incorporated, with a pyrethroid acaricide, into a plastic strip attached to the animal's tail (Norval et al. 1996). Besides direct application on livestock, the bont tick decoy may be also useful to attract large quantities of ticks in a treated site, killing them before they can infest livestock (Maranga et al. 2003; Sonenshine 2006). Notably both methods lead to a strong reduction in the quantity of applied acaricides.

Lastly, a further way of disrupting tick mating is to confuse the males, releasing multiple pheromone plumes in the environment. In this way, males are unable to discriminate differences in pheromone concentration on a given host, locating the feeding females (Sonenshine et al. 1985). Sonenshine (2006) pointed out that the longer they move in search of potential mates, the greater the likelihood that they will acquire a lethal dose of acaricide.

21.4 Chagasic bugs

21.4.1 Traditional triatomine control strategies

Despite some claims in the beginning of this century, Chagas disease has recently spread in Latin America (Dias et al. 2008; de Fuentes-Vicente et al. 2018). In contrast, Latin American government have increasingly invested less funds towards any type of control (Bonney 2014) or have stuck to traditional insecticide-based strategies (e.g. Carbajal de la Fuente et al. 2017; see Figure 21.3). Insectides against triatomines include organophosphates, carbamates and pyrethroids (e.g. bifenthrin, cyfluthrin, and delamethrin; Mougabure-Cueto and Picollo 2015). Insecticide use, however, has induced resistance in triatomines, which partly explains bug reinfestation in many cases (Dumonteil et al. 2004). Some other explanations for reinfestations are that insecticide use was not carried out properly (i.e. not reaching triatomine refuges) and reinfestation from neighbouring, insecticide-free areas (Mougabure-Cueto and Picollo 2015). However, the different resistance mechanisms (i.e. detoxifying enzymes), in different regions in Latin America indicate that resistance is indeed an acute problem now (Mougabure-Cueto and Picollo 2015).

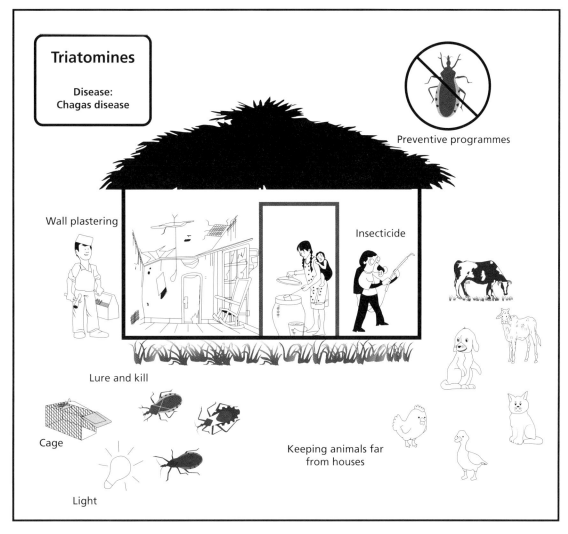

Figure 21.3 A summary of traditional control strategies (wall plastering, insecticides, and keeping domestic animals far from houses) of triatomine bugs (Chagas disease vectors) as well as alternative behavioral-based control measures (lure and kill based on odour and light attractants).

As an alternative to insecticides, preventive measures are the most effective for a number of insect vectors. In the case of Chagasic bugs, such measures aim to prevent bugs getting close to people or domestic animals that surround people. Currently, such measures are the following:

- plastering of house walls or clearing log and block piles;
- keeping domestic animals away from people;
- 'lure and kill'.

Right from the onset, not all measures can be applied to all situations, which in general is the case for many vector species. Next the mode of action of each of these measures will be described. Given that triatomines use cavities to hide, plastering these places has been shown to be an effective methodology. For this to be possible, plastering material has to be available, which is not necessarily the case given the poor socio-economical condition of affected people. In Guatemala (Monroy et al. 2009), the use of plastering material was viable as this was made from the debris left by volcano emissions. This debris had no cost as it had to be collected from the surroundings, but this is not possible in other locations. Given the habits of triatomines of hiding in dark places, bugs hide in wood, and block piles. These piles have to be cleared or placed distant from people. On the other hand, having domestic animals away involved keeping mainly chickens in cages (Waleckx et al. 2015). For this methodology to be successful, chickens had to be relatively far away from houses, a practice that is not necessarily well admitted by people. Also, building cages implies an economical cost and not-so-healthy practice given the value for free-range chickens. For 'lure and kill', bugs have to be attracted to traps and the ways to do this is via the use of odour-based attractants. The attraction approach was used for two species of *Triatoma* in Argentina and Paraguay by using hexanal, octanal, and benzaldehyde semiochemicals from polyethylene vials which was fairly successful (Rojas de Arias et al. 2012) These attractants can be those similar or analogous to bug's faeces or epicuticular lipid-based compounds, human odours or a combination of all (Lazzari et al. 2013). Since different species of bugs may vary in the odours they are attracted to (Lazzari et al. 2013), this control strategy has to be species-specific. The second step for this strategy is that of trapping, for which species-specific cages have to be designed too (for one example, see Rojas de Arias et al. 2012). In relation to killing after trapping, one strategy is that of using fungicides (Flores-Villegas et al. 2016). Ideally, the above measures have to be implemented along with an education-based programme to communities for a more integral approach of prevention. In fact, such programmes were quite successful up until a few decades ago (Dias et al. 2008).

21.4.2 Behavioral ecology to meet triatomine control

Despite its importance in vectoring a terrible disease, we know surprisingly little of triatomine behavior. A number of triatomine species exhibit strong or some degree of domiciliation, meaning that they tend to live within or around houses. Triatomine bugs move around urbanized surroundings via flight or walking, and these behaviors are expressed more inten-

sively from 02.00 to 04.00 hours. This circadian rhythm seems adaptive as dark time is used mainly for finding blood sources, but also mates and oviposition, which is when mammal hosts are asleep, while day time is used to hide, presumably to evade predators (Lazzari 1992). Interestingly, active predators of triatomines are not known, which can be looked for at such times. Triatomine locomotion activity is propelled by a number of factors. One first key factor that induces flight dispersal is artificial light, whose proper identification has allowed bugs to be so successful in finding domestic animals and humans as blood sources (Ramírez-Sierra et al. 2010). However, only adults, but not nymphs, show artificial light attraction, which is highly intense during night and suddenly stops before day breaks (Pacheco-Tucuch et al. 2012; Erazo and Cordovez 2016). Why nymphal stages are not attracted by artificial light is unclear and one possible reason is that such ability may depend on learning, although this possibility has not been explored. Furthermore, low intensity white light elicits a stronger phototaxic response, while high intensity red light does not produce such an effect (Pacheco-Tucuch et al. 2012). Such orientation is genetically based, given the existence of gene encoding for a protein that resembles those of opsin photodetectors in *Drosophila* with maximum sensitivity occurring at 508, 420, and 478 nm (Pacheco-Tucuch et al. 2012). Being phototaxis essential during the evolution of Triatominae, it is unclear how such sensitivity evolved as low intensity white light is being used just recently in human history. Possibly, Triatominae relied on other attractants (see later) before artificial light was invented, but the light-detecting ability was positively selected with human progress. However, one way of driving away triatomines from human dwellings is via a different source of artificial light and/or modifying human life style, such as keeping low intensity light relatively distant from buildings (see also Longcore et al. 2015).

A second key factor that allows triatomine dispersal, in this case via walking, is carbon bioxide (CO_2) as these are emitted by the hosts. Experiments with sub-adults that were exposed to different intensities of CO_2 indicated that 400 ppm (i.e. 0.04 per cent of CO_2 present in an airflow) was enough to elicit a positive response (Castillo-Neyra et al. 2015).

This response was more intense with upper CO_2 levels. One would suppose that such response would be exacerbated if the bug has spent some starvation time, but this expectation was not supported. It has been shown that CO_2 induce electrophysiological responses from the triatomine antennae and a similar response has been observed for other mammal-derived attractants such as L-lactic acid, pyruvic acid, short-chain carboxylic acids, aldehydes, pyridine, furan, terpenes, alcohols and amines (e.g. Guerenstein and Guerin 2001; Diehl et al. 2003). As a matter of fact, odours emitted by mammals are a complex cocktail of different combinations of the above. In support of this, CO_2 along with L-lactic acid induced a strong positive triatomine response in subadults (Castillo-Neyra et al. 2015). Starting from knowledge given previously, it has been proposed that baits that include the these attractants can be used to trap and kill triatomine bugs (e.g. Botto-Mahan et al. 2002). However, a full 'lure and kill' programme has not been developed yet. Such a programme, has to consider that kissing bugs bearing *T. cruzi* parasites are far more responsive, which may be a behavioral adaptation, induced by the parasite so that the bug can track potential blood sources more intensively (Fellet et al. 2014).

21.5 Concluding remarks

It is our hope that, after reading this chapter, the reader should have learned that the behavior of arthropod vectors underlies our ability to implement control strategies to reduce disease transmission. Fundamentally, we require knowledge of how arthropods attract con- and heterospecifics, and transmit pathogens to humans has a significant influence on how to design novel control strategies. This information is needed now more than ever given the rise of insecticide resistance. This knowledge implies gathering pieces related to different levels of organization around behavior such as genetics, physiology, endocrinology, and ecology. Only then, mechanisms for control can be established. For mosquitoes, some of the more promising recent strategies involve the release of mosquitoes for which behavior and ability to compete with wild type populations is critical to their success. For many other vectors, such as ticks and triatomines, a synergistic approach that includes odour-related control programmes seem more promising. One can learn from mosquito research as this has paved the way in terms of combining behavioral knowledge with other methodologies. Even in mosquitoes, Integrated Vector Management involving multiple strategies require further behavioral findings to abate the currently overwhelming threats to humanity.

References

Abdel-Rahman, M.S., Fahmy. M.M., and Aggour, M. (1998). Trials for control of ixodid ticks using pheromone acaricide tick decoys. *Journal of the Egyptian Society of Parasitology*, **28**, 551.

Alto, B.W., Lounibos, L.P., Mores, C.N., and Reiskind, M.H. (2008). Larval competition alters susceptibility of *Aedes* mosquitoes to dengue infection. *Proceedings of the Royal Society, B: Biological Sciences*, **275**, 463–71.

Atyame, C.M., Cattel, J., Lebon, C., et al. (2015). *Wolbachia*-based population control strategy targeting *Culex quinquefasciatus* mosquitoes proves efficient under semi-field conditions. *PLoS One*, **10**, e0119288.

Bargielowski, I.E., Lounibos, L.P., and Carrasquilla, M.C. (2013). Evolution of resistance to satyrization through reproductive character displacement in populations of invasive dengue vectors. *Proceedings of the National Academy of Sciences*, **110**, 2888–92.

Beaumier, C.M., Gillespie, P.M., Strych, U., et al. (2016). Status of vaccine research and development of vaccines for Chagas disease. *Vaccine*, **34**, 2996.

Benedict, M.Q., and Robinson, A.S. (2003). The first releases of transgenic mosquitoes: an argument for the sterile insect technique. *Trends in Parasitology*, **19**, 349.

Benelli, G. (2015a). Research in mosquito control: current challenges for a brighter future. *Parasitology Research*, **114**, 2801.

Benelli, G. (2015b). Plant-borne ovicides in the fight against mosquito vectors of medical and veterinary importance: a systematic review. *Parasitology Research*, **114**, 3201.

Benelli, G. (2015c). The best time to have sex: mating behaviour and effect of daylight time on male sexual competitiveness in the Asian tiger mosquito, *Aedes albopictus* Diptera: Culicidae. *Parasitology Research*, **114**, 887.

Benelli, G., Jeffries, C.L., and Walker, T. (2016c). Biological control of mosquito vectors: past, present and future. *Insects*, **7**, E52.

Benelli, G., Lo Iacono, A., Canale, A., and Mehlhorn, H. (2016a). Mosquito vectors and the spread of cancer: an overlooked connection? *Parasitology Research*, **115**, 2131.

Benelli, G., and Mehlhorn, H. (2016). Declining malaria, rising dengue and Zika virus: insights for mosquito vector control. *Parasitology Research*, **115**, 1747.

Benelli, G., Pavela, R., Canale, A., and Mehlhorn, H. (2016b). Tick repellents and acaricides of botanical origin: a green roadmap to control tick-borne diseases? *Parasitology Research*, **115**, 2545.

Benelli, G., Romano, D., Messing, R.H., and Canale, A. (2015). First report of behavioural lateralisation in mosquitoes: right-biased kicking behaviour against males in females of the Asian tiger mosquito, *Aedes albopictus*. *Parasitology Research*, **114**, 1613.

Bonney, K.M. (2014). Chagas disease in the 21st century: a public health success or an emerging threat? *Parasitei*, **21**, 11.

Botto-Mahan, C., Cattan, P.E., and Canals, M. (2002). Field tests of carbon dioxide and conspecifics as baits for *Mepraia spinolai*, wild vector of Chagas disease. *Acta Tropica*, **82**, 377.

Bouyer, J., Chandre, F., Gilles, J., and Baldet, T. (2016). Alternative vector control methods to manage the Zika virus outbreak: more haste, less speed. *Lancet Global Health*, **4**, e364.

Bouyer, J., and Lefrançois, T. (2014). Boosting the sterile insect technique to control mosquitoes. *Trends in Parasitology*, **30**, 271.

Brathwaite-Dick, O., San Martín, J.L., Montoya, R.H., et al. (2012). The history of dengue outbreaks in the Americas. *American Journal of Tropical Medicine and Hygiene*, **87**, 584.

Cabrera, M., and Jaffe, K. (2007). An aggregation pheromone modulates lekking behavior in the vector mosquito *Aedes aegypti* Diptera: Culicidae. *Journal of the American Mosquito Control Association*, **23**, 1.

Caputo, B., Ienco, A., Cianci, D., et al. (2012). The 'auto-dissemination' approach: a novel concept to fight *Aedes albopictus* in urban areas. *PLoS Neglected Tropical Diseases*, **6**, e1793.

Carbajal de la Fuente, A.L., Lencina, P., Spillmann, C., and Gürtler, R.E. (2017). A motorized vehicle-mounted sprayer as a new tool for Chagas disease vector control. *Cad Saúde Pública*, **33**, 1.

Carvalho, D.O., McKemey, A.R., Garziera, L., et al. (2015). Suppression of a field population of *Aedes aegypti* in Brazil by sustained release of transgenic male mosquitoes. *PLoS Neglected Tropical Diseases*, **9**, e0003864.

Castillo-Neyra, R., Barbu, C.M., Salazar, R., et al. (2015). Host-seeking behavior and dispersal of *Triatoma infestans*, a vector of Chagas disease, under semi-field conditions. *PLoS Neglected Tropical Diseases*, **9**, e3433.

Cator, L.J., and Harrington, L.C. (2010). The harmonic convergence of fathers predicts the mating success of sons in *Aedes aegypti*. *Animal Behaviour*, **82**, 627.

De Fuentes-Vicente JA, Gutiérrez-Cabrera AE, Flores-Villegas LA, Lowenberger C, Benelli G, Salazar-Schettino PM, Córdoba-Aguilar A (2018) What makes an effective Chagas disease vector? Factors underlying Trypanosoma cruzi-triatomine interactions. *Acta Tropica*, **183**, 23, doi: 10.1016/j.actatropica.2018.04.008.

de la Fuente, J., Rodriguez, M., Montero, C., et al. (1999). Vaccination against ticks *Boophilus* spp.: the experience with the Bm86-based vaccine Gavac TM.). *Genetic Analysis and Biomolecular Engineering*, **15**, 143.

Devine, G.J., Perea, E.Z., Killeen, G.F., et al. (2009). Using adult mosquitoes to transfer insecticides to *Aedes aegypti* larval habitats. *Proceedings of the National Academy of Sciences, USA*, **106**, 11530.

Diabate, A., and Tripet, F. (2015). Targeting male mosquito mating behaviour for malaria control. *Parasites & Vectors*, **8**, 347.

Dias, J.C.-P., Prata, A., and Correia, D. (2008). Problems and perspectives for Chagas disease control: in search of a realistic analysis. *Revista da Sociedade Brasileira de Medicina Tropical*, **41**, 193.

Diehl, P.A, Vlimant, M., Guerenstein, P., and Guerin, P.M. (2003). Ultrastructure and receptor cell responses of the antennal grooved peg sensilla of *Triatoma infestans* Hemiptera: Reduviidae.). *Arthropod Structure and Development*, **31**, 271.

Dottorini, T., Nicolaides, L., Ranson, H., et al. (2007). A genome-wide analysis in *Anopheles gambiae* mosquitoes reveals 46 male accessory gland genes, possible modulators of female behavior. *Proceedings of the National Academy of Sciences, USA*, **104**, 16215.

Dumonteil, E., Ruiz-Piña, H., Rodríguez-Félix, E., et al. (2004). Re-infestation of houses by *Triatoma dimidiata* after intra-domicile insecticide application in the Yucatan peninsula. *Memorias do Instituto Oswaldo Cruz*, **99**, 253.

Erazo, D., and Cordovez, J. (2016). The role of light in Chagas disease infection risk in Colombia. *Parasites & Vectors*, **9**, 1.

Fellet, M.R., Lorenzo, M.G., Elliot, S.L., Carrasco, D., and Guarneri, A.A. (2014). Effects of infection by *Trypanosoma cruzi* and *Trypanosoma rangeli* on the reproductive performance of the vector *Rhodnius prolixus*. *PloS One*, **9**, e105255.

Ferguson, N.M., Kien, D.T., Clapham, H., et al. (2015). Modeling the impact on virus transmission of *Wolbachia*-mediated blocking of dengue virus infection of *Aedes aegypti*. *Science Translational Medicine*, **7**, 279ra37.

Flores-Villegas, A.L., Cabrera-Bravo, M., Toriello, C., et al. (2016). Survival and immune response of the Chagas disease vector *Meccus pallidipennis* Hemiptera: Reduviidae. against two entomopathogenic fungi, *Metarhizium anisopliae* and *Isaria fumosorosea*. *Parasites & Vectors*, **9**, 176.

Frentiu, F.D., Zakir, T., Walker, T., et al. (2014). Limited dengue virus replication in field-collected *Aedes aegypti* mosquitoes infected with *Wolbachia*. *PLoS Neglected Tropical Diseases*, **8**, e2688.

Galizi, R., Doyle, L.A., Menichelli, M., et al. (2014). A synthetic sex ratio distortion system for the control of the human malaria mosquito. *Nature Communications*, **5**, 3977.

Gantz, V.M., Jasinskiene, N., Tatarenkova, O., et al. (2015). Highly efficient Cas9-mediated gene drive for population modification of the malaria vector mosquito *Anopheles stephensi*. *Proceedings of the National Academy of Sciences, USA*, **112**, e6736–43.

George, D.R. (2000). Present and future technologies for tick control. *Annals of the New York Academy of Sciences*, **916**, 583.

Gubler, D.J. (1998). Resurgent vector-borne diseases as a global health problem. *Emerging Infectious Diseases*, **4**, 442.

Guerenstein, P.G., and Guerin, P.M. (2001). Olfactory and behavioural responses of the blood-sucking bug *Triatoma infestans* to odours of vertebrate hosts. *Journal of Experimental Biology*, **204**, 587.

Harris, A.F., McKemey, A.R., Nimmo, D., et al. (2012). Successful suppression of a field mosquito population by sustained release of engineered male mosquitoes. *Nature Biotechnology*, **30**, 828.

Hedges, L.M., Brownlie, J.C., O'Neill, S.L., and Johnson, K.N. (2008). *Wolbachia* and virus protection in insects. *Science*, **322**, 702.

Hoffmann, A.A., Montgomery, B.L., Popovici, J., et al. (2011). Successful establishment of *Wolbachia* in *Aedes* populations to suppress dengue transmission. *Nature*, **476**, 454.

Hotez, P.J., Dumonteil, E., Cravioto, M.B., et al. (2013). An unfolding tragedy of Chagas disease in North America. *PLoS Neglected Tropical Diseases*, **7**, e2300.

Hotez, P.J., Molyneux, D.H., Fenwick, A., et al. (2007). Control of neglected tropical diseases. *New England Journal of Medicine*, **357**, 1018.

Ibarra-Cerdeña, C.N., Sánchez-Cordero, V., Peterson, A.T., and Ramsey, J.M. (2009). Ecology of North American triatominae. *Acta Tropica*, **110**, 178.

Ito, Y., and Yamamura, K. (2005). Role of population and behavioural ecology in the sterile technique. In: V.A. Dyck, J. Hendrichs, and A.S. Robinson (Eds), *Sterile Insect Technique Principles and Practice in Area-wild Integrated Pest Management*, p. 177. Springer, Dordrecht.

Iturbe-Ormaetxe, I., Walker, T., and O' Neill, S.L. (2011). *Wolbachia* and the biological control of mosquito-borne disease. *EMBO Reports*, **12**, 508.

Jeffries, C.L., and Walker, T. (2015). The potential use of *Wolbachia*-based mosquito biocontrol strategies for Japanese encephalitis. *PLoS Neglected Tropical Diseases*, **9**, e0003576.

Jeffries, C.L., and Walker, T. (2016). *Wolbachia* biocontrol strategies for arboviral diseases and the potential influence of resident strains in mosquitoes. *Current Tropical Medicine Reports*, **3**, 20.

Joubert, D.A., Walker, T., Carrington, L.B., et al. (2016). Establishment of a *Wolbachia* superinfection in *Aedes aegypti* mosquitoes as a potential approach for future resistance management. *PLoS Pathology*, **12**, e1005434.

Kahn, M.C., and Offenhauser, Jr. W. (1949). The first field tests of recorded mosquito sounds used for mosquito destruction. *American Journal of Tropical Medicine and Hygiene*, **29**, 811.

Kanda, T., Kerdpibule, V., Deesin, T., et al. L. (1990). Strategies for mosquito control by using a sound trap system and an insect growth regulator pyriproxyfen—a review. *Tropical Biomedicine*, **7**, 159.

Kanda, T., Loong, K.P., Chiang, G.L., Cheong, W.H., and Lim, T.W. (1988). Field study on sound trapping and the development of trapping method for both sexes of *Mansonia* in Malaysia. *Tropical Biomedicine*, **5**, 37.

Lacroix, R., Delatte, H., Hue, T., and Reiter, P. (2009). Dispersal and survival of male and female *Aedes albopictus* Diptera: Culicidae. on Reunion Island. *Journal of Medical Entomology*, **46**, 1117.

Laven, H. (1967). Eradication of *Culex pipiens fatigans* through cytoplasmic incompatibility. *Nature*, **216**, 383.

Lazzari, C.R. (1992). Circadian organization of locomotion activity in the haematophagous bug *Triatoma infestans*. *Journal of Insect Physiology*, **38**, 895.

Lazzari, C.R., Pereira, M.H., and Lorenzo, M.G. (2013). Behavioural biology of Chagas disease vectors. *Memorias do Instituto Oswaldo Cruz*, **108**, 34.

Longcore, T., Aldern, H.L., Eggers, J.F., et al. (2015). Tuning the white light spectrum of light emitting diode lamps to reduce attraction of nocturnal arthropods. *Philosophical Transactions of the Royal Society, B*, **370**, 20140125.

Lounibos, L.P., Bargielowski, I., Carrasquilla, M.C., and Nishimura, N. (2016). Coexistence of *Aedes aegypti* and *Aedes albopictus* Diptera: Culicidae. in peninsular Florida two decades after competitive displacements. *Journal of Medical Entomology*, **53**, 1385.

Lupi, E., Hatz, C., and Schlagenhauf, P. (2013). The efficacy of repellents against *Aedes, Anopheles, Culex* and *Ixodes* spp.—a literature review. *Travel Medicine and Infectious Disease*, **11**, 374.

Madakacherry, O., Lees, R.S., and Gilles, J.R.-L. (2014). *Aedes albopictus* Skuse males in laboratory and semi-field cages: release ratios and mating competitiveness. *Acta Tropica*, **132**, 124.

Maranga, R., Hassanali, A., Kaaya, G.P., and Mueke, J.M. (2003). Attraction of *Amblyomma variegatum* ticks. to the attraction-aggregation-attachment pheromone with or without carbon dioxide. *Experimental and Applied Acarology*, **29**, 121.

Marini, G., Guzetta, G., Baldacchino, F., et al. (2017). The effect of interspecific competition on the temporal dynamics of *Aedes albopictus* and *Culex pipiens*. *Parasites & Vectors*, **23**, 10.

Medley, K.A. (2010). Niche shifts during the global invasion of the Asian tiger mosquito, *Aedes albopictus Skuse* (Culicidae), revealed by reciprocal distribution models. *Global Ecology and Biogeography*, **19**, 122.

Mehlhorn, H. (2015). *Encyclopedia of Parasitology*. Springer, New York.

Monroy, C., Bustamante, D.M., Pineda, S., et al. (2009). House improvements and community participation in the control of *Triatoma dimidiata* re-infestation in Jutiapa, Guatemala. *Cd. Saúde Pública*, **25**, 168.

Mougabure-Cueto, G., and Picollo, M.I. (2015). Insecticide resistance in vector Chagas disease: evolution, mechanisms and management. *Acta Tropica* 149, 70.

Naqqash, M.N., Gökçe, A., Bakhsh, A., and Salim, M. (2016). Insecticide resistance and its molecular basis in urban insect pests. *Parasitology Research*, **115**, 1363.

Norval, R.A., Sonenshine, D.E., Allan, S.A., and Burridge, M.J. (1996). Efficacy of pheromone-acaricide-impregnated tail-tag decoys for controlling the bont tick, Amblyomma hebraeum (Acari: Ixodidae.), on cattle in Zimbabwe. *Experimental and Applied Acarology*, **20**, 31–46.

Novak, M.G., Higley, L.G., Christianssen, C.A., and Rowley, W.A. (1993). Evaluating larval competition between *Aedes albopictus* and *A. triseriatus* (Diptera: Culicidae.) through replacement series experiments. *Environmental Entomology*, **22**, 311.

O'Connor, L., Plichart, C., Sang, A.C., et al. (2012). Open release of male mosquitoes infected with a *Wolbachia* biopesticide: field performance and infection containment. *PLoS Neglected Tropical Diseases*, **6**, e1797.

Oliva, C.F., Damiens, D., and Benedict, M.Q. (2014). Male reproductive biology of *Aedes* mosquitoes. *Acta Tropica*, **132**, 512.

Oliva, C.F., Damiens, D., Vreysen, M.J., Lemperière, B., and Gilles, J. (2013a). Reproductive strategies of *Aedes albopictus* (Diptera: Culicidae) and implications for the sterile insect technique. *PLoS One*, **8**, e78884.

Oliva, C.F., Vreysen, M.J., Dupe, S., et al. (2013b). Current status and future challenges for controlling malaria with the sterile insect technique: technical and social perspectives. *Acta Tropica*, **132**, 130.

Omeara, G.F., Evans, L.F, Gettman, A.D., and Cuda, J.P. (1995). Spread of *Aedes albopictus* and decline of *Aedes aegypti* (Diptera, Culicidae) in Florida. *Journal of Medical Entomology*, **32**, 554.

Pacheco-Tucuch, F.S., Ramirez-Sierra, M.J., Gourbière, S., and Dumonteil, E. (2012). Public street lights increase house infestation by the Chagas disease vector *Triatoma dimidiata*. *PLoS One*, **7**, e36207.

Phuc, H.K., Andreasen, M.H., Burton, R.S., et al. (2007). Late-acting dominant lethal genetic systems and mosquito control. *BMC Biology*, **5**, 11.

Pompon, J., and Levashina, E.A. (2015). A new role of the mosquito complement-like cascade in male fertility in Anopheles gambiae. *PLOS Biology*, **13**, e1002255.

Ramirez-Sierra, M.J., Herrera-Aguilar, M., Gourbière, S., and Dumonteil, E. (2010). Patterns of house infestation dynamics by non-domiciliated *Triatoma dimidiata* reveal a spatial gradient of infestation in rural villages and potential insect manipulation by *Trypanosoma cruzi*. *Tropical Medicine and International Health*, **15**, 77.

Ranson, H. and Lissenden, N. (2016). Insecticide resistance in African *Anopheles* mosquitoes: a worsening situation that needs urgent action to maintain malaria control. *Trends in Parasitology*, **32**, 187.

Ribeiro, I., Sevcsik, A.M., Alves, F., et al. (2009). New improved treatments for Chagas disease: from the R&D pipeline to the patients. *PLoS Neglected Tropical Diseases*, **3**, e484.

Robinson, A.S. (1983). Sex-ratio manipulation in relation to insect pest control. *Annual Review of Genetics*, **17**, 191.

Rojas de Arias, A., Abad-Franch, F., Acosta, N., et al. (2012). Post-control surveillance of *Triatoma infestans* and *Triatoma sordida* with chemically-baited sticky traps. *PloS Neglected Tropical Diseases*, **6**, e1822.

Sánchez-Guillén, R.A., Córdoba-Aguilar, A., Hansson, B., Ott, J., and Wellenreuther, M. (2016). Evolutionary consequences of climate-induced range shifts in insects. *Biological Reviews*, **91**, 1050.

Schliekelman, P., Ellner, S., and Gould, F. (2005). Pest control by genetic manipulation of sex ratio. *Journal of Economic Entomology*, **98**, 18.

Shaw, W.R., Teodori, E., Mitchell, S.N., et al. (2014). Mating activates the heme peroxidase HPX15 in the sperm storage organ to ensure fertility in Anopheles gambiae. *Proceedings of the National Academy of Sciences, USA*, **111**, 5854.

Sonenshine, D.E. (2006). Tick pheromones and their use in tick control. *Annual Review of Entomology*, **51**, 557.

Sonenshine, D.E. and Roe, R.M. (2014). *Biology of Ticks*. Oxford University Press, New York, NY.

Sonenshine, D., Taylor, D., and Corrigan, G. (1985). Studies to evaluate the effectiveness of sex pheromone impregnated formulations for control of populations of the American dog tick, *Dermacentor variabilis* (Say) (Acari: Ixodidae). *Experimental and Applied Acarology*, **1**, 23.

Tabashnik, B.E. (1994). Evolution of resistance to *Bacillus thuringiensis*. *Annual Review of Entomology*, **39**, 47.

Thomas, D.D., Donnelly, C.A., Wood, R.J., and Alphey, L.S. (2000). Insect population control using a dominant, repressible, lethal genetic system. *Science*, **287**, 2474.

Tripet, F., Lounibos, L.P., Robbins, D., et al. (2011). Competitive reduction by satyrization? Evidence for interspecific mating in nature and asymmetric reproductive competition between invasive mosquito vectors. *American Journal of Tropical Medicine and Hygiene*, **85**, 265.

Vassena, C.V., Picollo, M.I., and Zerba, E.N. (2000). Insecticide resistance in Brazilian *Triatoma infestans* and Venezuelan *Rhodnius prolixus*. *Medical and Veterinary Entomology*, **14**, 51.

Waleckx, E., Camara-Mejia, J., Ramírez-Sierra, M.J., et al. (2015). An innovative ecohealth intervention for Chagas disease vector control in Yucatan, Mexico. *Transactions of the Royal Society of Tropical Medicine and Hygiene*, **109**, 143.

Walker, T., Johnson, P.H., Moreira, L.A., et al. (2011). The *w*Mel *Wolbachia* strain blocks dengue and invades caged *Aedes aegypti* populations. *Nature*, **476**, 450.

Willadsen, P. (2006). Tick control: thoughts on a research agenda. *Veterinary Parasitology*, **138**, 161.

Willadsen, P., Bird, P., Cobon, G.S., and Hungerford, J. (1995). Commercialization of a recombinant vaccine against *Boophilus microplus*. *Parasitology*, **110**, 43.

Wilke ABB, Beier JC, Benelli G (2018) Transgenic mosquitoes– fact or fiction? Trends in Parasitology, doi: 10.1016/j.pt.2018.02.003.

Zhang, D., Zheng, X., Xi, Z., Bourtzis, K., and Gilles, J.R. (2015). Combining the sterile insect technique with the incompatible insect technique: I-impact of *Wolbachia* infection on the fitness of triple- and double-infected strains of *Aedes albopictus*. *PLoS One*, **10**, e0121126.

CHAPTER 22

Insect behavior in conservation

Tim R. New

Department of Ecology, Environment and Evolution, La Trobe University, Victoria 3086, Australia

22.1 Introduction

Practical insect conservation is founded, at the species level, in the ecological understanding of what a target species needs, how it finds, exploits, and benefits from those needs, its access to resources such as habitat, food, and mates, and its capability to withstand or defend against environmental changes—including invasive competitors, predators, parasitoids, and pathogens. For the great majority of individual species selected for conservation attention, those needs must be satisfied for two very distinct life forms within a complex life cycle and limitations set by a diversity of behaviors and sensory responses. Larvae and adults of holometabolous insects usually have dramatically different opportunities and capabilities, based on their morphological traits, feeding habits and life styles—the sedentary herbivorous caterpillar is a very different functional entity from the corresponding mobile nectar-feeding adult butterfly, for example, and the larval aquatic dragonfly has very different needs from the strongly flying terrestrial adult stage. Needs of both phases, spanning two very different environments and resource sets, are central planks of a conservation platform—and, for adults, behavior of the two sexes may differ considerably, and change with age or maturity.

Behavioral variety ('ethodiversity') can allow for an individual, or a species, to respond in different ways to environmental features and changes. The extent of this plasticity differs widely and reflects the relative allocation of species as 'specialists' or 'generalists'. These form a continuum, but attributes of the former—such as narrow food requirements, restricted microclimate regimes or other resource need—contrast with the broader ecological flexibility and tolerances of relatively generalized species. Recognition of ethodiversity led Cordero-Rivera (2017) to claim that 'behavior has profound ecological consequences, particularly in species interactions... and is a crucial element in the adaptability of animals to new environments', and also that 'maintaining ethodiversity is essential in conservation'. Enabling conditions for 'normal' behavior (which, in conjunction with the ecological specializations of many threatened species, may have very limited flexibility) is an implicit component of effective conservation, but commonly not signalled specifically for attention (Cordero-Rivera 2017), in contrast to the ecology of processes such as dispersal (both locally and across landscapes), resource discovery and selection, seasonal adaptations, and the impacts of anthropogenic change and natural processes such as succession.

The subtleties of behavior are displayed by the idiosyncrasies of individual species, imposed on general principles of wider application and relevance, so that this essay deals mainly with the single-species focus of insect conservation, rather than the increasingly advocated focus of wider entities such as

New, T. R., *Insect behavior in conservation*. In: *Insect Behavior: From mechanisms to ecological and evolutionary consequences*. Edited by Alex Córdoba-Aguilar, Daniel González-Tokman, and Isaac González-Santoyo: Oxford University Press (2018). © Oxford University Press.
DOI: 10.1093/oso/9780198797500.003.0022

communities. I draw on conservation management of selected species to illustrate the seemingly endless array of behaviors that have contributed vital practical benefits and knowledge, against the obvious background that all fundamental ecological interactions between insects and their wider environments are mediated in some way through behavioral traits. My 'Box examples' are of Australian Lepidoptera, as cases with which I am familiar, but can be paralleled for many other insect groups across the world, most notably in the temperate regions where interest, support, and faunal knowledge is greatest (Stewart and New 2007). The history of insect conservation reveals many cases in which either awareness of behavioral subtleties has benefited conservation programmes or, conversely, ignorance of those intricacies has accelerated declines and **extirpations**. Many behavior themes link strongly with the other major activity in applied entomology, pest management, and in some contexts very similar behavioral understanding has been applied to similar contexts, but with the diametrically opposed aims of suppressing and killing (pests; see for example, Chapters 20 and 21), or enhancing and sustaining (conservation) the focal taxa. Across both disciplines, the parallels are instructive. The vast spectrum of sensory responses, activity patterns, and mechanisms that enable insects to prosper may all, at times, be amenable to manipulation in some way; specific visual, olfactory, or auditory responses to environmental cues, for example, are all parts of normal behavior repertoires. Their relevance and applications in conservation are very varied, and the topics below draw attention to contexts in which understanding behavioral responses relates directly to practical insect conservation assessment or management. As Berger-Tal and Saltz (2016) elegantly remarked 'The behavior of an organism is, in a sense, the mediator between the organism and its environment'. Their book is a valuable expansion and synthesis of that paradigm and, for vertebrates, develops support for behavioral awareness in the three main conservation roles of:

- helping to predict impacts of rapid anthropogenic changes;
- formulating better management interventions;
- enabling use of behavioral changes as an 'early warning system' from monitoring changes and declines detected.

The principles are equally relevant for insects, but in general are less prominent in many conservation agendas. Many insect conservation initiatives are founded in the knowledge that a particular ecological association (such as a specific host plant for an insect herbivore) is necessary, but without understanding the nuances and mechanisms by which that resource is discovered and used.

Some of the many ways in which behavioral knowledge of an insect species may contribute to the design and implementation of conservation planning and management are outlined below, in a broader context of some of the key ecological themes that are more conventionally given priority in such programs.

22.2 Insect habitats

Restriction of many insects to particular environments that are susceptible to changes by people or natural processes, is a major conservation concern. Most species given conservation priority are, in some way, ecological 'specialists', as well as being distributed narrowly (many as narrow range endemics), found only in low numbers, and now threatened from loss or change to their environments. Their conservation necessitates understanding of why and how they persist, or can be preserved or restored in some way. The formerly widespread definition of 'habitat' as 'a place to live' is progressively giving way to that of an environment in which a species' critical resources and conditions co-occur, and are accessible as the basic requirements for survival. This concept, known as the 'resource–based definition of habitat', advanced elegantly by Dennis et al. (2003, 2007) and flowing largely from studies on European butterflies (Dennis 2010), marked a pivotal paradigm shift in insect conservation biology in emphasizing that the distribution and abundance of critical resources in the landscape are fundamental determinants of ability for a species to occupy an area—a bounded space, not knowing what that space includes, or does not include, in relation to a focal insect's needs, is clearly an oversimplification, in itself insufficient for practical conservation.

The resources needed, which must be sought and exploited by the various life history stages are separable broadly into two categories as 'consumables' and 'utilities', for both of which scale of appraisal is important—appraisal of resources on a small site, for example, may differ markedly from that on the wider landscape. The dispersion and accessibility of those critical resources in relation to:

- the dispersal prowess of the insect seeking them;
- their distribution and the structure of the landscape.

Finding isolated clumps of a specific food plant separated by inhospitable terrain in a highly altered urban or agricultural landscape, for example, are both important considerations in assessing conservation needs.

The first category, essentially 'foods', are easier to define, but the range of food types and how these are discovered and used, and how they are distributed all pose behavioral questions. Particular plant species needed by specialist insect herbivores, or prey or hosts for specialist predators or parasitoids, may each be distributed unevenly, sparsely, or available only seasonally within a site—and many may have chemical or other defences to deter attack by the very species we seek to conserve. Many insects initially believed to be monophagous may also be able exploit other food species in their environment if these are available or more easily accessible.

'Utilities' are the conditions needed for survival and persistence—those that enable a species to 'behave normally' and pursue an unstressed life style. For example, particular butterflies may need open sites, such as bare ground for basking (thermoregulation), display areas, or territorial perches (mate encounters), rests or roosts (including long-term retreats for aestivation or hibernation), oviposition sites (some associated with selection of sites for larval feeding), larval retreats, pupation sites, and other specific features, in addition to more general and widespread needs, such as enemy-free space. Access to any of these reflects availability, the distance and form of dispersal, and local 'conditioners', such as climate, topography, and exposure. It also indicates the range of background behavioral information that is useful in planning species' management for conservation and without which that management may be deficient. They are the clues by which an optimal environment—the insect's true habitat—can be defined, and their patterns relate to insect behavior in practical conservation. They also warn that the historically common appraisal of landscapes as divisible into 'patch' (suitable habitat areas) and 'matrix' (commonly assumed to be inhospitable intermediate areas) is also over simplistic. Thus, matrix is more simply the contrasting milieu separating obviously suitable areas, and may support the species to varying extents, rather than being wholly alien.

Dennis (2010) further noted the three distinct variable features of resource distribution. **'Composition'** is the presence/absence and abundance of specific resources, and how these vary in space and time. Particular foods, such as nectar or pollen for particular pollinators, for example, may reflect a restricted flowering period to which the insect's foraging pattern and seasonal development has become attuned, or a larval host plant may occur only in particular vegetation communities governed by other factors such as edaphic and plant nutritional conditions, and hydrology. **'Physiognomy'** refers to the nature and distribution of the resource, described by features such as geographical location, size, orientation, shape, and extent of fragmentation. **'Connectivity'**, a critical theme in insect conservation, refers to the extent that resources are functionally linked within a wider landscape, so linked with a species' (and life stage) mobility and ability to track those resources, and variable also with landscape features such as the presence/absence of functional 'barriers' that can halt or impede movement or of **'corridors'** that can enhance dispersal. Behavior of flying insects at any such landscape feature may be characteristic; open spaces, linear features, such as roadways or power line easements, hedgerows, riparian vegetation, or other contrasting vegetation may effectively impede movement (causing insects to turn back, rather than enter or cross such areas) or, conversely, enable them to use such features to increase rates and efficiency of dispersal and, in some cases, as additional breeding areas. Connectivity may relate also to any characteristic population or metapopulation structure through which dispersal need or propensity can be indicated.

Elucidating the cues by which resources are discovered and used has direct relevance in defining conservation management of insect species. In conjunction with ecological influences, longer-term evolutionary heritage may have effectively moulded and limited an insect's behavioral options to cope with changes imposed by human activity.

22.3 Status evaluation and monitoring

Gaining sound information on a species' distribution, presence/absence at putatively occupied or occupiable sites, and abundance is the template for according it conservation significance or priority. The World Conservation Union's categories of threat (IUCN 2001), for example, used globally as guidelines for conservation status assessment, draw heavily on these parameters. Monitoring subsequent conservation management involves detecting and quantifying changes in these, to determine the success (or otherwise) of the measures taken (New 2009). Many standard techniques for sampling or appraising insect species or diversity (such as species richness and relative abundance as signals of wider community structure and integrity) draw on behavior both for initial investigation, and later

---

Box 22.1 Understanding behavior in reliably sampling a threatened insect, the Golden sun-moth (*Synemon plana*, Castniidae) in Australia

The Golden sun-moth is categorized nationally as 'critically endangered', and is an important flagship species for remnant native grasslands threatened with imminent development for peri-urban residential and industrial expansion in south eastern Australia (Figure 22.1). The larvae feed underground on grass roots and are difficult to sample non-destructively, so that the diurnal adult is the only stage easily available for detection and monitoring. Until recently, ignorance of adult behavior severely hampered attempts to find the moth and assess its abundance. It is still unknown whether the moth is univoltine or whether it takes 2 or even 3 years to develop, so that an annual cohort on a site may be only an unknown fraction of the entire resident population. Quantitative surveys are very difficult to undertake. Relevant behavioral features are noted below.

1. Adult moths live for only 3–5 days, and do not feed. On any site the entire flight period is of the order of 6–8 weeks in early summer, with continuous emergences from underground pupae over this time. 'Hotspots' of emergences may reflect topography, and soil temperatures affecting duration of development, with time of any single inspection giving misleading impressions of distribution.

2. Adults are strongly sexually dimorphic in appearance and behavior. Males fly about 1 m above ground, 'patrolling' in search of females, but not undertaking longer dispersals than, at most, a few hundred metres. Females fly little, and rest on open ground. They expose their brightly coloured hind wings in response to overflying males, which are attracted to land, and mate.

3. Males fly for only a few hours each day, limiting the 'sampling window' to ca 10.00–15.00 hours. They are unlikely to be detected at other times, so that surveys must be scheduled for that interval, and opportunities for cross-site surveys on the same day are limited to it.

4. Male flight is also strongly weather-dependent. Activity is favoured by sunny, dry, calm conditions at temperatures within the approximate range of 16–28° C, and flight does not occur in cloudy, rainy, or windy conditions or when temperatures are unsuitable.

Recognition of these observational constraints in seeking the moth has led to more effective surveys and discovery of numerous additional populations.

Figure 22.1 A female golden sun-moth in a characteristic pose, with her brightly coloured hind wings partially exposed. The forewings are repeatedly raised and lowered to create a 'flash pattern' of colour that attracts overflying male moths as mates.

Photo: L. A. Gibson.

Further information: Gibson and New (2007); New (2015).

periodic measurements (such as site-based intergenerational counts of adult butterflies using standard methods on each occasion) by which monitoring is undertaken. Compendia of insect sampling methods for conservation (such as Samways et al. 2010) summarize numerous approaches and contexts, but for designated rare or threatened species found only in low numbers, and highly localized methods that kill or otherwise harm the insects should be avoided. For

all, the limitations of methodology in relation to the focal insect's behavior must be appreciated as hypothetical contexts that counter conventional 'wisdom'.

- many moths do not come to lights (phototaxis);
- many edaphic beetles are not sufficiently active to enter pitfall traps;
- yellow or green pan traps (colour attractants) are often highly selective in catches;

Box 22.2 Monitoring nocturnal butterfly larvae, the Eltham copper, *Paralucia pyrodiscus lucida*, Lycaenidae, in Australia

The Eltham copper is one of rather few threatened insect taxa in which monitoring has incorporated regular surveys of both adults and larvae, as twin indicators of intergenerational changes (Figure 22.2). Surveys have focused on tiny isolated peri-urban remnant sites near Melbourne, Victoria, as remaining fragments of the butterfly's formerly more extensive range now swallowed by urban expansion. The butterfly is now conservation-dependent, with individual site management needed to sustain it, and those sites differing in the intensity and variety of urban pressures and successional patterns: the small sites also present problems with surveys, in order to minimize physical disturbances.

Adult butterflies are monitored during direct visual surveys based on transect walks and point inspections, as standard

Figure 22.2 A male Eltham copper butterfly, perched on the highest point of the larval food plant, from where short venturing flights to patrol or interact with other males, or potential mates take place, with the individual returning to the same 'sentry point' in a form of territoriality.

Photo: A.A. Canzano.

methods for butterfly surveys, and their activity, although weather dependent, extends over much of the day. Butterflies live for several weeks.

Larvae present greater constraints and problems, reflecting their behavior, including obligate mutualisms with specific endemic ants, as follows:

1. They are nocturnal, and forage on the sole larval food plant (*Bursaria spinosa*, Pittosporaceae) for several hours after dusk, at temperatures above about 15° C. In inclement conditions they are not active.
2. Although long-lived, larvae feed little during the cooler winter months, when they rest within the subterranean nest chambers of the host ants (*Notoncus* spp., with different species of this small genus hosting *Paralucia* larvae in different parts of the butterfly's range).
3. Larvae also rest by day in ant nest chambers around the roots of the host plants during their feeding season, and are not then accessible for counting.
4. At night, larvae are 'herded' up the *Bursaria* by the ants, which are glossy, active, and easily detected by torchlight. Their presence signals the nearby presence of the more cryptic and slow-moving larva. Several ants may accompany a single larva and remain with it whilst it feeds. If disturbed, larvae retreat to the ground, and usually do not re-emerge on the same night.
5. Quantitative assessment is limited through individual larvae not feeding every night, so that non-sighting equates to both chance and non-activity. They can be marked individually with coloured pigments for recognition.
6. Larvae remain on the same individual host plant during development, and pupate within the ant nest chamber.

Further information: New (2011).

- observable territorial behavior or mating displays may occur only in particular sexes, for individual age classes of adults, or under limited weather conditions; acoustic monitoring (sonotaxis) may have only very limited use for sparse populations of Orthoptera;
- many chemical baits (olfactory stimuli) need careful evaluation before use.

Each can be used effectively only against known phenology of the focal species, and validity and success depends on the behavioral idiosyncrasies of that species. Many insects also exhibit well-defined patterns of circadian, seasonal, or weather-related activity (Box 22.1), and monitoring or inspections may target either or both of adult and larval stages (Box 22.2).

In planning any monitoring of insect responses in conservation management programmes, the initial aims and needs must be formulated very carefully, and any/all known behavioral idiosyncrasies are an important component of the ecological framework that guides the wider management plan. Important decisions include:

- whether detection of presence of the insect is sufficient, or whether measures of abundance, sex ratio, seasonal activity or development, or other parameters are needed;
- whether changes across generations are required to demonstrate longer-term changes in response to management and, if so, the level of precision expected;
- whether across-site comparisons, perhaps in relation to environmental conditions or changes, are needed.

Behavioral and ecological variables are legion, but underlie the need to sustain and protect a viable population of the target species.

Reflecting the high costs and ecological intricacies of single species conservation, the impracticality of managing large numbers of needy candidate species individually, and each within a template of far greater and largely undocumented insect richness, alternatives such as focusing on sites or communities to conserve large numbers of co-occurring native insect species are highly attractive. Naturally evolved behavior patterns may thus remain largely undisturbed.

22.4 Perspective

If environmental conditions are changed, so that resource quality or supply is diminished or disrupted, an insect species or population has three main 'options', each reflecting its evolutionary history and current capability:

- it moves away, to occupy another site on which the requisite resources are suitable;
- it adapts to the changed conditions, and persists without moving away;
- it becomes extinct.

The major conservation aims for threatened insects are:

- to facilitate the first option, if needed, perhaps using **translocation** or assisted migration, or changing landscape features;
- to acknowledge the realistically low potential of many ecologically specialized species to adjust substantially, and so to compensate by measures that continue to provide its key needs on the site, or elsewhere to enable translocation;
- to prevent extinction, while recognizing that this is the natural fate of most species over evolutionary time, and consider a variety of ex situ options (such as captive breeding for later release), threat reduction, and **ecological restoration**.

'Changes' vary in intensity and impact, leading to different levels of threat, and a species' (and conservation manager's) response may differ accordingly—in much policy the most severely threatened species are given highest priority, and also pose the greatest challenges for success. Elucidating the behavioral and ecological actions involved is difficult.

Most of this chapter emphasizes conservation of individual threatened species, those insects conventionally selected for priority attention. The other 'face' of insect conservation is to prevent currently non-threatened species from becoming so—declines of pollinating insects in both local diversity and richness and single species' distribution and abundance, for example, continue to cause concerns.

22.5 A caveat!

Some problems of obtaining the requisite behavioral information for optimal conservation and decisions are difficult to address, but must be considered when status assessment and management is contemplated. Unlike abundant pests, as near-ideal study models for acquiring basic knowledge (with studies likely to be well-funded because of their economic implications), many insect species in need of conservation are (as noted earlier) in some way 'rare' or elusive, and believed to be declining further as vulnerable to continuing environmental changes. A consequence of scarcity is that the species are hard to study. Detection may be uncertain and quantitative data on abundance are usually difficult to obtain, populations are not amenable to experimental manipulations, and loss of even a few individuals (for example, by accidental deaths or inadvertent destructive sampling) might pose additional threat. All these endorse the reality that conservation management may necessarily be limited in scope by basic ignorance of the species' true resource needs beyond, usually, some basic and often incomplete knowledge of consumables. A related frustration is that the species may be known from single or few sites, sometimes newly discovered there as a stimulus for 'crisis management', which are perhaps degraded or remnant fragments, but which are also the only models for resource assessment and emulation. It is usually unclear whether the site conditions are indeed suitable/optimal, or if our focal species is 'clinging to the edge of existence' on only marginally suitable sites.

Also unfamiliar to many non-entomologists is that even closely related species of insects may have very different biologies and behaviors, so that uncritical transfer of observations from even near relatives may be misleading. Nevertheless, all such background information must be assessed carefully, as even fragmentary information on activity patterns, dispersal, and resource spectrum and use by the focal species must be appraised as ably as possible. With suitable caution, data from common and well-studied relatives can be incorporated into that appraisal. For 'popular' insect groups, such as butterflies, dragonflies, and larger beetles, much of that information may not be published formally, but rests in the accumulated folklore of hobbyists and collectors, whose knowledge of rare and desirable species by far exceeds that of most 'professionals'. Learning where and when to look, and what to look for, in seeking information on rare insect species may need wide consultation.

22.6 Behavior and vulnerability

Imposed impacts on natural environments can lead to changes in insect behavior that sometimes have severe consequences. Ecological traps, syndromes important in some insect conservation programmes, have been implicated in vulnerability of both insect species and larger groups. Two examples are:

- the Richmond birdwing butterfly, *Ornithoptera richmondia*, Papilionidae, in Australia (Box 22.3) (Sands and New 2013);
- the impacts of artificial lighting ('light pollution') on nocturnal moths and others, recognized as a contributor to the widespread declines of moths (Macgregor et al. 2015) as the predominant group of nocturnal insect pollinators, and so affecting pollination networks and contributing to the widespread global concerns over pollinator declines.

Impacts include diverting moths from normal behavior and resources by superstimuli, and concentrating them around lights where they become more susceptible to vertebrate predators. A further urban example is the attraction of some aquatic insects (notably some mayflies and beetles) to the polarized light reflections of asphalt road surfaces, on which futile oviposition is a behaviorally induced population drain, with potential to cause threat to localized species (Horvath et al. 2010).

Understanding how behavior may initiate or enhance threats to an endangered insect is important in planning to mitigate or avoid those threats. Prevention of mortality from traffic impacts during insect migration or other dispersal, for example, includes measures such as:

- controlling traffic speed, as recommended in the United States FWS Recovery Plan for Hine's emerald dragonfly, where rail lines run across two of the few major breeding sites (Soluk et al. 1998) and speed reductions occur during the dragonfly flight season;

Box 22.3 Recognition of an ecological trap and its conservation significance; the Richmond birdwing butterfly, *Ornithoptera richmondia*, Papilionidae, in Australia

One of Australia's most spectacular butterflies, the strongly-flying Richmond birdwing (Figure 22.3) was lost from much of its former range in central eastern Australia during the first half of the twentieth century, due to a combination of forest clearance (destroying its primary habitat) and widespread ornamental planting and subsequent naturalization of an alien South American vine, *Aristolochia elegans*, Aristolochiaceae, that is superattractive to the ovipositing female butterflies, but which is also toxic to hatchling larvae, and kills them after they feed. The vine is, therefore, an ecological trap, and could accelerate the declines of a diminishing butterfly population, as a population sink.

The conservation programme sought to build on the butterfly's strong flight behavior, to focus on restoring it to its entire recorded historical range by the twin strategies of progressive removal of the alien vine from natural and urban areas and extensive plantings of the native food plant vine, *Pararistolochia praevenosa* (itself having become threatened by the same forest clearances), from specially produced nursery stocks. With this change in resource balance, it was hoped that the butterfly would progressively recolonize its former range, if necessary supported by selected releases. The area-wide programme, from the late 1980s on, has been strongly supported by networks of community volunteers, whose activities are coordinated and publicized continually. Replanting of *Pararistolochia* has concentrated on extensive patch plantings as local food plant enrichment, and corridor plantings between forest patches to promote connectivity through use of gardens, school grounds, parks and other open spaces, and the campaign has met with considerable success.

Further information: Sands and New (2013).

Figure 22.3 A male Richmond birdwing butterfly maturing after eclosing from the pupa. The bright red lateral thorax patch, found on both sexes of this highly dimorphic species, may be some form of aposematic or counter-predator sign, with the larvae having fed on food plants likely to be toxic.

Photo: D. P. A. Sands.

• placement of tall nets along highway lengths in Taiwan, to force migrating *Euploea* butterflies to fly above traffic when crossing the road (Wu 2009).

Effectiveness of any such measure reflects detailed knowledge of the seasons, heights, and speeds of flight. Likewise, in agricultural landscapes, knowledge of seasonal or diurnal dispersal activity patterns may be used to time pesticide applications on crops to avoid harming the insect of concern.

A high proportion of insect species of conservation concern are threatened primarily by one or both of two major causes. Loss or change of resource supply through some form of 'habitat change' is the most widespread, almost universal, concern and behavioral issues also come to the fore in considering the impacts of invasive alien species. The behavioral features of invasive animals or plants may render them aggressive competitors and colonizers that displace native species and disrupt ecological interactions. Many invasive ants, for example, are highly aggressive and can rapidly eliminate sensitive native species and intricate native food webs. In many contexts, behavioral differences between species such as red imported fire ant, *Solenopsis invicta*, and Argentine ant, *Linepithema humile*, from native ants in their invaded environments gives then a strong competitive advantage and increases their threat to those natives (New 2012, for background). As for some other social Hymenoptera, their very large numbers, ease of undetected transport, and rapid spread cause massive ecological and economic concerns. Conservation measures to counter their impacts may necessitate knowledge of the behavior of an invasive insect or plant weed in defining the severity of impacts and how (and if) they should be countered. Impacts may be intricate. Alien bumble bees and other Apidae, some introduced deliberately for pollination services, provide contexts such as:

- severe decline of the endemic *Bombus dahlbomi*, a keystone specialized pollinator species in some temperate forests in South America, from displacement by native bees and likely spill-over of parasites introduced in the alien *Bombus terrestris* (Goulson and Hughes 2015);
- impacts of *Bombus terrestris* in Tasmania, where its feeding methods, including chewing holes to gain nectar from the base of flowers with a deep corolla, is associated with considerable disruption to intricate pollination associations amongst native species (Hingston 2007).

Many such impacts were summarized by New (2016), where the numerous practical lessons ensuing from continued studies of behavior of deliberately introduced species, notably of classical biological control agents for control of insect or plant pests are also treated. Attempts to assure the 'safety' of such agents and minimizing their non-target impacts on native taxa in the receiving environments, necessarily incorporate considerable behavioral understanding in designing and undertaking convincing screening trials, and predicting outcomes. Knowledge of many aspects of dispersal, host discovery, and selection, and resource needs, contributes to the acceptance and outcomes of such strategies.

22.7 Dispersal and population structure

Ability to track patchily distributed resources across local areas or wider landscapes varies considerably amongst different insects and, whereas most winged adult insects can fly, many only rarely do so. Species with closed populations may be functionally restricted to small patches—some of the isolated peri-urban sites occupied by the Eltham copper near Melbourne, Australia (Box 22.2) are only 1–3 hectares in extent, and conservation management must focus on maintaining key resources there, because butterfly movement across the urbanized inhospitable intervening area is unlikely and would almost certainly be suicidal—the butterfly there is conservation-dependent. Also site-dependent, but on a very different scale, the long distance migratory flights of the North American continental populations of the Monarch butterfly, *Danaus plexippus*, to reach their restricted overwintering sites in California and Mexico, have emphasized the roles of such seasonal sites and the vulnerability their loss can confer. Should the oyamel forests in Mexico be lost, catastrophic declines of the monarch are seemingly inevitable. Such concentrations of vast numbers of butterflies also create vulnerability to stochastic events—in Mexico, an estimated 467.5 million monarchs were killed in a single winter storm in 2002 (Brower et al. 2004). Such transgenerational site fidelity provides fixed points in a species' life cycle.

Large insect aggregations occur in other seasonally resting taxa, such as the aestivating Bogong moth, *Agrotis infusa*, Noctuidae, in Australia, whose cool alpine aestivation sites may become vulnerable to climate warming. Any such sites regularly used by any seasonally diapausing insect may need active protection, but most insects with open populations and that move freely over landscapes are not considered threatened at present, and—beyond remarking the unusual spectacle of such migrations and concentrations—behavioral observations have focused largely on more restricted taxa.

The wider concern for many insect species is landscape quality in relation to dispersal prowess and need, whether patches of key resources in largely inhospitable matrices are accessible as potential harbours for metapopulation units or whether connectivity between them can be facilitated. In practice, unless this population structure is recognized, resource evaluation can become uncertain because currently unoccupied patches may differ very little from those supporting apparently viable populations. Fragmented habitats, such as forest remnants may simply be too far apart for insects to move between them; a recent study of a Neotropical damselfly (Odonata), for example, using tethered adults, showed that they could fly across a 25-m gap, but larger distances (50–100 m) provided far greater barrier effects (Khazan 2014). Assuring success of dispersal may entail constructing 'stepping stones' between dispersed patches or wider landscape restoration to provide corridors of resources between them. Patch size and shape, together with the nature of their surroundings, may influence colonization success and later population performance. Wide natural corridors that do not expose the users to external influences or edge effects from changed external areas may benefit ecological specialists with little tolerance for alternative conditions. Drawing on behavioral awareness, ecological networks for insect conservation can now be designed on the twin premises of maintaining natural land (nodes) and instigating landscape linkages between these, following what Samways (2007) referred to as 'the golden thread of the metapopulation trio of large patch size, good patch quality and reduced patch isolation'.

Behavior in relation to landscape topography is also relevant in conservation assessment and planning. In southern England, some butterflies in the decades before climate warming became obvious were at the northernmost fringe of their predominantly European distributions, and needed south-facing slopes in the landscape on which to bask for efficient thermoregulation. Similar observations have been made on grasshoppers and beetles elsewhere. Small scale movements are important in tracking diurnal changes of this sort, and conservation measures proposed for the Adonis blue, *Lysandra bellargus*, Lycaenidae, included the seemingly drastic measure of bulldozing to create south-facing slopes as additional utility areas (Thomas 1983). Many scarce butterflies and other insects normally widely dispersed in the landscape can 'hilltop', a behavior whereby they may move from up to several kilometres away to concentrate on hills or other prominences. Males often arrive first, and may either establish territories or patrol to await females. The

behavior is a means of increasing chances of finding a mate, and females then disperse to oviposit. Some species that are extremely difficult to detect in open landscapes may be detected by concentrating searches on local hills (Britton et al. 1995), while discovery may not indicate where the species breeds, it at least confirms incidence in the local environment—a hilltop is simply a 'facilitating utility' used regularly and predictably in the insect's life.

Unexpected features of dispersal can confuse interpretation. Biased sex ratios found in surveys may indicate patterns such as protandry. In some butterflies, males emerge first and can either establish territories or await female emergence—in extreme cases, they congregate and wait near pupae, and rape the females as they emerge! Female pupae of the Australian Imperial blue, *Jalmenus evagoras*, emit volatile chemical attractants for males shortly before they hatch, so that dozens of males can surround a pupa (Pierce and Nash 1999). The pupae also produce noises that recruit ants, in this context believed to protect pupae from predators (Travassos and Pierce 2000).

Differential movements of the two sexes are not uncommon in insects, as exemplified by the endangered North American Hines' emerald dragonfly, *Somatochlora hineana*, in which newly emerged females move into areas well away from waterside sites for several weeks to mature, then return to mate and oviposit. Males stay close to water and sampling only there shows a strongly biased sex ratio. Those additional areas used by females—perhaps of many species—are a critical resource that is often unheeded in conservation management unless the background behavior has been clarified, and female movement has been suggested to avoid peak male activity and harassment for pre-reproductive females (Foster and Soluk 2006).

22.8 Mutualisms

The interactions between threatened insects and their ecological partners involve some of the most intricate of all their behavioral adaptations, with a full repertoire of sensory mechanisms involved in honing and sustaining those dependencies, thwarting antagonisms (such as by defence against natural enemies and avoiding their attacks). They also pose central challenges in practical conservation, perhaps nowhere more so than amongst mutualisms, in which the behavior and ecology of the participating species can become strongly intertwined. Mutualisms range from rather casual (as in many pollinator-plant associations) to highly specific and obligatory, but are unified by the partners gaining mutual benefit, in contrast to the antagonistic relationships that typify herbivore–plant or predator–prey interactions in which one partner benefits at the expense of the other.

Understanding mutualisms between ants and lycaenid butterflies (the 'blues' and 'coppers') has been a critical aspect of conserving the latter, notably in campaigns for notable flagship species such as the large blues (*Maculinea* spp.) in Europe, including the re-introduction of *M. arion* to Britain after it had become extinct there, in part due to failure to appreciate the attending ants needs (Thomas 1999; Thomas et al. 2009). Larvae of these butterflies are initially herbivorous, but are later taken into the nests of specific ant hosts (of *Myrmica* spp.), and complete their development there. The behavioral sequence therefore includes:

- the mated female butterfly finding a specific food plant within a diverse community, and ovipositing there on influorescences;
- the hatchling larvae adopting that plant for food;
- their later discovery and acceptance by a suitable host ant species, with the adoption process and ensuing care enabled in part by chemical features of the ant cuticular hydrocarbons;
- the butterfly emerging successfully and escaping from the ant nest without being eaten.

The specific interactions reflect a balance between the benefits of the butterfly larvae producing sugar-rich secretions sought by the ants as enticing food, and the ants tending the larvae and protecting them from insect predators and parasitoids. The larval secretions are common as mediators of associations between the ants and their mutualistic species, and a recent study (Hojo et al. 2015) described them as 'manipulable drugs that can enforce the ants' cooperative behavior'. That study revealed the lycaenid secretions to have a variety of roles, such as moderating the locomotory and aggressive behavior of the ants, and also altered the brain dopamine levels of the

ants. The apparent specificity of many ant-butterfly mutualisms is a key area for clarification in conservation programmes, with habitat restoration needs to provide for the wellbeing of both ants and butterflies. Many high profile lycaenid conservation programmes are on ecologically complex myrmecophilous species, for which conserving such mutualisms is vital. Some have, in the past, been hampered severely by lack of awareness of behavioral idiosyncrasies and the scope of the 'conservation module' that must be addressed. Thus, for this context the focus must be on the tripartite association of 'butterfly-specific larval food plant-specific host ant', but conservation may need to also heed needs of a specific pollinator of the plant, and specific predators or parasitoids of the butterfly larvae. In passing, it should be noted that conservation of the latter, involving detailed specific host-finding behavior by tiny wasps or other parasitoids, is very rarely considered, although the parasitoids may be far rarer and far more threatened than their host species, they are viewed more commonly as threats for mitigation, rather than as deserving conservation targets in their own right.

Searching behavior by insect herbivores for specific plants and oviposition sites, or by predators or parasitoids for specific insect prey or hosts (usually of particular growth stages), may build on very specific cues, and also be opposed by 'defences' from the target species. Most detailed studies have been on species evaluated as biological control agents against pests, where many details of searching behavior and modes of specificity have been clarified. For the ecological specialists of major concern in conservation programmes, elucidating those clues and incorporating behavioral knowledge into management can be highly beneficial—and the key to a successful outcome. With somewhat different emphasis, pollinators may depend on very specific cues to discover their target plants, and maintain associations without which both partners may suffer declines in fitness.

22.9 Discussion

Flexibility in behavior provides for possible variations in responses and in interactions between species, with options for alternatives and adaptations, such as adopting secondary food plants, prey, or hosts. These may all contribute to the scope of insect species conservation plans. Behavioral considerations in insect conservation span variety and scale—from the intricacies of interactions between rare and highly specialized taxa, and how to assure these on tiny sites supporting small populations, to landscape-level planning involving supplies of critical resources across wide regions to facilitate insect movements, or to enlarge existing habitat patches or promote development of additional areas as residential sites. Knowledge of a species' behavior can hone ecological information to contribute meaningfully to the design of insect conservation programmes at all scales, and enhance subsequent adaptive management. The latter depends largely on understanding how an ecologically specialized insect species thrives, and the patterns of its resource supply and need in the face of environmental changes and threats. However, designation and definition of relevant behavior is currently an under-valued component of species management. Individual species' peculiarities, many apparently unique, can have substantial 'novelty interest', but are not always directly relevant to conservation management. The Sandhill rustic moth, *Luperina nickerlii leechi*, Noctuidae, is known only from a single small coastal shingle bar in southern England, and its breeding area can be inundated by high tides. The moths rarely fly, so stay within the restricted area—a trend that is common amongst insect species characteristic of small isolated habitats—but can survive submersion in sea water, and can crawl into the water and grasp underwater vegetation (Spalding 2015). Insects do strange and unexpected things, and assessing the importance of these in conservation can be difficult and emphasizes the need for careful study.

Fostering interest in insects is a key need for the future of insect conservation (Cheesman and Key 2007) and, although rarely acknowledged as such, insect behavior and activity are powerful tools in this endeavour. Butterfly houses and similar displays involving free-flying conspicuous and popular insects can help convey significant conservation messages. Open areas, such as butterfly gardens and ponds constructed as 'dragonfly trails' can (with adequate signage) convey behavioral idiosyncrasies to young people and others, and illustrate features of territoriality, courtship, and spatial distributions 'in real life' (Suh and Samways 2001).

Displays associated with captive breeding of threatened insects also provide educational opportunities. A key purpose is to rear such insects under conditions that avoid normal field mortality (for example, by avoiding natural enemies in those confined conditions) and eventually provide stocks for release, perhaps to rehabilitated and protected sites, or to augment tiny field populations. Success of such operations must heed behavior. One impediment to captive breeding of the strongly flying birdwing butterflies (Papilionidae), a group of major conservation interest, has been the need for large, expensive, flight cages in which adults can pursue their mate-seeking rituals in conditions that more resemble their normal environment (Parsons 1992). One cage erected for the Richmond birdwing (Box 22.3) measured 15 × 4 × 2.5 m, for example. Such extremes clearly do not apply to smaller, less mobile insects, but key utilities are still needed. The wingless and nocturnal Lord Howe Island stick insect, *Dryococelus australis*, rests by day in hollows in trees, and provision of such suitable retreats was a key need for captive maintenance of this highly endangered insect (Honan 2008).

Referring to pollinators, but expressing a sentiment also of very wide relevance in other contexts, Waser (2006) wrote 'a modern understanding of insect sensory and cognitive abilities suggests more subtlety than imagined earlier'. Exploring those subtleties and their relevance in practical conservation, recognizing that the maintenance of ecological interactions is a major key to understanding biodiversity, is a major challenge for the future. For any threatened insect species for which conservation intervention is contemplated, listing behavioral idiosyncrasies of both immature and adult stages of that species and deliberately considering their relevance in management schemes, as exemplified in the three boxes in this chapter, could have important benefits.

Acknowledgements

I am very grateful to Andrea Canzano, Lucy Gibson, and Don Sands for generously allowing me to use their photographs in this chapter.

References

Berger-Tal, O., and Saltz, D. (Eds) (2016). *Conservation behavior: applying behavioral ecology to wildlife conservation and management.* Cambridge University Press, Cambridge.

Britton, D.R., New, T.R., and Jelinek, A. (1995). Rare Lepidoptera at Mount Piper, Victoria: the role of a threatened butterfly community in advancing understanding of insect conservation. *Journal of the Lepidopterists' Society*, **49**, 97.

Brower, L.P., Kust, D.R., Salimas, E.T., et al. (2004). Catastrophic winter storm mortality of monarch butterflies in Mexico during January 2002. In: K. S. Oberhauser and M. J. Solensky (Eds), *The Monarch Butterfly. Biology and Conservation*, pp. 151–66. Cornell University Press, New York, NY.

Cheesman, O.D., and Key, R.S. (2007). The extinction of experience: a threat to insect conservation? In: A.J.A. Stewart, T.R. New, and O.T. Lewis (Eds), *Insect Conservation Biology*, pp. 322–48. CAB International, Wallingford.

Cordero-Rivera, A. (2017). Behavioral diversity (Ethodiversity): a neglected level in the study of biodiversity. *Frontiers in Ecology and Evolution*, **5**, 7.

Dennis, R.L.H. (2010). *A Resource-based Habitat View for Conservation. Butterflies in the British Landscape.* Wiley-Blackwell, Oxford.

Dennis, R.L.H., Shreeve, T.G., and Sheppard, D.A. (2007). Species conservation and landscape management: a habitat perspective. In: A.J.A. Stewart, T.R. New, and O.T. Lewis (Eds), *Insect conservation biology*, pp. 92–126. CAB International, Wallingford.

Dennis, R.L.H., Shreeve T.G., and Van Dyck H. (2003). Towards a functional resource-based concept of habitat: a butterfly biology viewpoint. *Oikos*, **102**, 417.

Foster, S.E., and Soluk, D.A. (2006). Protecting more than the wetland: the importance of biased sex ratios and habitat segregation for conservation of Hine's emerald dragonfly, *Somatochlora hineana* Williamson. *Biological Conservation*, **127**, 158.

Gibson, L.A., and New, T.R. (2007). Problems in studying populations of the Golden sun-moth, *Synemon plana* (Lepidoptera: Castniidae) in south-eastern Australia. *Journal of Insect Conservation*, **11**, 309.

Goulson, D., and Hughes, W.O.H. (2015). Mitigating the anthropogenic spread of bee parasites to protect wild pollinators. *Biological Conservation*, **191**, 10.

Hingston, A. (2007). The potential impact of the large earth bumblebee *Bombus terrestris* (Apidae) on the Australian mainland: lessons from Tasmania. *Victorian Naturalist*, **124**, 110.

Hojo, M.K., Pierce, N.E., and Tsuji, K. (2015). Lycaenid caterpillar secretions manipulate attendant ant behavior. *Current Biology*, **25**, 2260.

Honan, P. (2008). Notes on the biology, captive management and conservation status of the Lord Howe Island Stick Insect (*Dryococelus australis*) (Phasmatodea). *Journal of Insect Conservation*, **12**, 399.

Horvath, G., Kriska G, Malik, P., et al. (2010). *Asphalt Surfaces as Ecological Traps for Water-seeking Polarotactic Insects: How can the Polarized Light Pollution of Asphalt Surfaces be Reduced.* Nova, New York, NY.

IUCN (World Conservation Union) (2001). *The IUCN Red List of Threatened Species. Categories and Criteria* (Version 3.1). Gland, IUCN.

Khazan, E.S. (2014). Tests of biological corridor efficacy for conservation of a neotropical giant damselfly. *Biological Conservation*, **177**, 117.

Macgregor, C.J., Pocock, M.J.O., Fox, R., and Evans, D.M. (2015). Pollination by nocturnal Lepidoptera, and the effects of light pollution: a review. *Ecological Entomology*, **40**, 187.

New, T.R. (2009). *Insect Species Conservation.* Cambridge University Press, Cambridge.

New, T.R. (2011). *Butterfly Conservation in South-eastern Australia: Progress and Prospects.* Springer, Dordrecht.

New, T.R. (2012). *Hymenoptera and Conservation.* Wiley-Blackwell, Oxford.

New, T.R. (2015). *Insect Conservation and Urban Environments.* Springer International Publishing, Basel.

New, T.R. (2016). *Alien Species and Insect Conservation.* Springer International Publishing, Basel.

Parsons, M.J. (1992). The world's largest butterfly endangered: the ecology, status and conservation of *Ornithoptera alexandrae* (Lepidoptera: Papilionidae). *Tropical Lepidoptera 3*, **1**, 33.

Pierce, N.E., and Nash, D.R. (1999). The Imperial blue: *Jalmenus evagoras* (Lycaenidae). In: R.L. Kitching, E. Scheermeyer, R.E. Jones, and N.E. Pierce (Eds) *The Biology of Australian Butterflies*, p. 279. CSIRO Publications, Melbourne.

Samways, M.J. (2007). Implementing ecological networks for conserving insect and other biodiversity. In: A.J.A. Stewart, T.R. New, and O.T. Lewis (Eds), *Insect Conservation Biology*, pp. 127–43. CAB International, Wallingford.

Samways, M.J., McGeoch, M.A., and New, T.R. (2010). *Insect Conservation. A Handbook of Approaches and Methods.* Oxford University Press, Oxford.

Sands, D.P.A., and New, T.R. (2013). *Conservation of the Richmond Birdwing Butterfly in Australia.* Springer, Dordrecht.

Soluk, D.A., Zercher D.S., and Swisher, B.J. (1998). *Preliminary Assessment of Somatochlora hineana Larval Habitat and Patterns of Adult Flight over Railway Lines near Lochport and Lemont, Illinois.* Illinois Natural History Survey, Champaign, IL.

Spalding, A. (2015). *Loe Bar and the Sandhill Rustic Moth: the Biogeography, Ecology and History of a Coastal Shingle Bar.* E.J. Brill, Leiden.

Stewart, A.J.A., and New, T.R. (2007). Insect conservation in temperate biomes: issues, progress and prospects. In: A.J.A. Stewart, T.R. New, and O.T. Lewis (Eds), *Insect conservation biology*, pp. 1–33. CAB International, Wallingford.

Suh, A.N., and Samways, M.J. (2001). Development of a dragonfly awareness trail in an African botanical garden. *Biological Conservation*, **100**, 345.

Thomas, J.A. (1983). The ecology and conservation of *Lysandra bellargus* (Lepidoptera: Lycaenidae) in Britain. *Journal of Applied Ecology*, **20**, 59.

Thomas, J.A. (1999). The large blue butterfly: a decade of progress. *British Wildlife*, **11**, 22.

Thomas, J.A., Simcox, D.J., and Clarke, R.T. (2009). Successful conservation of a threatened *Maculinea* butterfly. *Science*, **325**, 80.

Travassos, M.A., and Pierce, N.E. (2000). Acoustics, context and function of vibrational signaling in a lycaenid butterfly-ant mutualism. *Animal Behaviour*, **60**, 13.

Waser, N.M. (2006). Specialization and generalization in plant-pollinator interactions: a historical perspective. In N.M. Waser and J. Ollerton (Eds), *Plant-pollinator Interactions: From Specialization to Generalization*, pp. 3–18. Chicago University Press, Chicago, IL.

Wu, S. (2009). Taiwan´s purple crow butterfly conservation, a story of progress. *China Post*, 2 April 2009. Available at: http://www.chinapost.com.tw/print/202723.htm (accessed 5 September 2016)

General Glossary

Acclimation (also acclimatization) a physiological adjustment by an organism to a novel environment, thus maintaining its performance (i.e. homeostasis) across a range of environmental conditions (e.g. altitude, temperature, humidity, photoperiod, or pH).

Acoustic niche (related to the acoustic niche hypothesis) it is analogous to the classical ecological concept of niche partitioning. This suggests that acoustic signals of different signallers should display partitioning in their frequency (calling at frequencies not occupied by others), time (calling when other signallers are silent), or space (calling at places free of other signallers) domain. All mechanisms would reduce the range of masking, and increase the effectiveness of intraspecific communication.

Additive genetic variance (V_A) the proportion of the variance expressed in the phenotype that is due to the additive effects of genes. Additive gene effects are transmitted directly from parents to offspring and, therefore, are the major contributor to evolutionary change.

Additive genetic variance-covariance matrix (G) a square, symmetrical matrix with additive genetic variance for traits along the diagonal and additive genetic covariances in the off-diagonal positions.

Afferent nerve is the axon of a sensory neuron in the peripheral nervous system towards the central nervous system.

Age-related division of labour see 'polyethism'.

Aggregation pheromones a secretion or excretion of chemical attractants that result in assemblages of individuals of the same species. These attractants can function to choose a mate, to overcome host-plant resistance through mass attack or to defend against predators.

Aggregation a group of individuals that forms without distinct displays of organized cooperative behaviors. Animal herds are one example.

Aggressive spillover hypothesis it refers to the idea that when positive selection on aggressiveness in one context (e.g. foraging) inadvertently causes increases in aggressiveness in other contexts (e.g. mating).

Ageing an age-dependent decline in the survival probability caused by an inevitable gradual loss of physiological function.

Airspeed the speed of an airborne animal (or any airborne object) relative to the air through which it is flying, i.e. it is equivalent to the self-powered flight speed (c.f. ground speed).

Alarm pheromone a chemical substance produced and released by an organism, that warns a conspecific of an impending danger.

Allometric the change in size of a body part or organ that does not change proportional to the change in size of the body.

Alloparental care a type of parental care observed in groups where individuals other than the parents take care of the offspring.

Altricial helpless at birth or hatching.

Altruism a behavior that benefits another individual in which the direct fitness cost to the actor are greater than the benefits.

Amphisexual care (or flexible parental care compensation) a behavior that occurs when individuals of the non-caring sex assume parental activities if individuals of the caring sex die or desert the offspring.

Animal personality a metric of an individual's behavioral tendencies that is temporally consistent across time and context. Typical measurements estimate individuals' aggressiveness, boldness, sociability, exploration, and activity using repeated behavioral assays.

Animal signal any stimulus produced by a sender that modifies the behavior of one or more receivers.

Anisogamy sexual reproduction as achieved by the fusion of morphologically different gametes. It typically refers to difference between sperm cells (of males) and egg cells (of females) thought to have been favoured by disruptive selection early in the evolution of life on earth.

Anti-feedant any chemical substance that inhibits feeding. An example would be the cardiac glycosides that deter some insects from feeding on milkweed plants.

Anxiety a negative emotion associated with a general anticipation of loss at some point in the future.

Aposematism a defensive strategy characterized by the co-occurrence of a conspicuous signal and a defence (often chemical) in prey, such that predators may ultimately learn to avoid prey bearing the signal.

Appetite   willingness to eat.

Apposition eye   an eye structured such that light reaches the photoreceptors of each ommatidium as a single spot, and these point samples are subsequently combined for image formation.

Arbovirus   An acronym for ARtropod-Borne VIRUSes that refers to any virus transmitted by arthropods.

Assortativity   a phenomenon that describes when individuals prefer to associate or mate with others that share their phenotypes (positive assortativity). Conversely, disassortativity refers to individuals that tend to interact with unlike phenotypes.

Atmospheric boundary layer (or ABL)   a meteorological term referring to the lowest part of the atmosphere that is directly influenced by its contact with the planetary surface. The ABL corresponds to the altitude range in which migrant insects fly.

Attract-and-kill   pest management strategy that combines an attractant (e.g. pheromone) with a killing agent (e.g. insecticide) to reduce the pest population.

Auto-dissemination   the spread of a material (generally a toxicant) through the environment by the insect. For example, an autodissemination trap uses an attractant to draw the pest to the trap where the pest is contaminated with the material, which it disseminates to its environment (e.g. next oviposition site).

Batesian mimicry   an evolved resemblance of a species that is profitable to attack to a species that is unprofitable to attack by predators, thereby protecting the mimic from predation.

Behavioral fever   An acute change in thermal preference following infection or pathogen recognition.

Behavioral hypervolumes   akin to the 'niche hypervolume' concept of ecology. It refers to a multivariate measurement of variation within or among individuals in multidimensional behavioral trait space. The metric is calculated as an n-dimensional hypervolume, where each axis represents a different personality trait.

Behavioral immunity   mechanisms that allow hosts to detect and avoid contact with infectious environments or conspecifics, and behaviors that lower the severity of infection.

Behavioral manipulation   the use of stimuli that either stimulates or inhibits a behavior.

Behavioral reaction norms   the behavioral response of an individual expressed across an environmental gradient.

Behavioral response thresholds   the minimum stimulus amount necessary to elicit a behavioral response from an individual.

Behavioral syndromes   correlations between personality metrics; typically assessed at the population level.

Binaural cues   for an animal equipped with two ears, for a sound broadcast from one side there will be a propagation delay between two ears, generating interaural time differences. At the same time, the head or body where the ears are located may produce a shadowing effect predominantly on high frequencies due to diffraction, which will generate differences in the intensity of sound at the two ears.

Biogenic amines   a set of neurotransmitters (e.g. dopamine, octopamine, serotonin) that mediate many behaviors in insects and other arthropods.

Bioluminescence   production and emission of light by living organisms (such as fireflies and deep-sea fishes).

Bourgeois/anti-bourgeois   each refers to hypotheses conceived in the application of game theory to territorial conflict in biological systems. The bourgeois hypothesis (also framed as an uncorrelated asymmetry) describes a situation where territorial contests are settled according to the simple rule of 'resident always wins'. Anti-bourgeois describes the opposite scenario. Both are theoretically predicted to minimize individual-level costs under select circumstances in territorial insect competition.

BT × BT interactions   an instance where the outcome of an ecological interaction depends on the behavioral types of two or more interactors.

Camouflage   the use of any combination of material, colouration, scent, or illumination, that make an animal hard to detect by members of the same or different species.

Caste   behaviorally and/or physiologically distinct group of workers in a colony of social insects.

Central pattern generator   a neuronal circuit, which produces rhythmic output in relative isolation, i.e. the absence of rhythmic input or sensory feedback (though sensory feedback may influence it in the intact animal). Examples are networks for locomotion or respiration rhythms.

Chemical ecology   or chemoecology, is the study of the chemical interactions between living organisms and their environment.

Chill coma recovery   the measurable time needed for an organism (typically insects) to recover after being exposed to a specified sub-freezing temperatures.

Chromophore   a part of a molecule—typically a protein—that gives colour to the entire complex.

Clustered regularly interspaced short palindromic repeats (CRISPR)   a new genome editing method that is based on a bacterial CRISPR-associated protein-9 nuclease (Cas9) from *Streptococcus pyogenes*. *CRISPR* and *Cas9* genes are essential in adaptive immunity in bacteria in response to eliminating invading genetic material. In the acquisition phase, foreign DNA is incorporated at the CRISPR loci and Cas9 endonucleases cleave foreign DNA.

Cognition the processes concerned with the acquisition, retention, and use of information.

Cognitive bias the interpretation of an ambiguous cue based on an individual's emotional state. A positive state leads to an optimistic bias, while a negative state causes a pessimistic bias.

Colour vision the ability to distinguish stimuli based on differences in their spectral shape, independent of luminance (subjective brightness).

Common garden experiment an experimental design in which the offspring of individuals from different origins (various populations, environments, families, etc.) are reared in a common laboratory or field environment to control for environmental influences on phenotypic traits. The term derives from agriculture and horticulture, where plants are literally reared in a common garden.

Composition refers to presence/absence and abundance of specific resources, and how they vary in space and time.

Communication the overt transfer of signals from one individual to another.

Conflict the phenomenon wherein two individuals have opposing interests leading to different optima. This could occur at the intra- (e.g. between genes or organs) or inter-individual level (e.g. between sexes).

Connectivity the facility for a species to disperse between habitat patches, so reflecting aspects of the species' dispersal capacity in relation to features of the intervening area that can enhance or impede the likelihood or success of movement.

Connectomics a field of neuroscience dedicated to the analysis and interpretation of comprehensive maps of the neuronal connections within nervous systems.

Control programme a programme aimed at reducing or eliminating a given vector or pest from a selected area.

Cooperation (behavior) a behavior that generates mutually beneficial direct fitness benefits. Often equated with social behavior.

Corpora allata endocrine glands located underneath the *corpora cardiaca* that secrete juvenile hormone. Sometimes fused with *corpora cardiaca* in dipteran species to form Weismann's ring gland.

Corpora cardiaca endocrine glands located behind the insect brain that secrete a prothoracicotropic hormone synthesized by neurosecretory cells in the brain.

Corridor a linear feature in the landscape that can be used for dispersal or as a breeding area, and that links larger patches of suitable habitat across relatively unsuitable intermediate areas.

CRISPR/Cas9 technique an efficient and versatile method for genome editing using a modified version of a virus defence system in bacteria. Cas9 is an enzyme which cuts DNA at a sequence which is specified by so-called guide RNA.

Critical thermal maximum/minimum the maximal or minimal temperature above or below which, respectively, a given organism (animal), typically estimated using multiple individuals, responds with unorganized locomotion and ultimately death, setting the extreme margins of the thermal performance curve.

Crypsis a form of camouflage that works by matching the background on which the organism appears, thereby reducing its likelihood of detection.

Cryptic female choice female influences on fertilization that occur during or after insemination and that are usually not detected by a human observer.

Cue an act or structure that conveys information to a receiver, but that have not been selectively favoured for that purpose, and which may not have positive fitness consequences.

Cumulative assessment a hypothesized model that describes how information regarding competitive ability is signalled/received as the sum of behaviors that accumulate during an agonistic interaction.

Cytoplasmic incompatibility reproductive disorder that results in sperm and eggs being unable to form viable offspring (commonly resulting from a *Wolbachia* bacteria-infected male mating an uninfected female).

Dear enemy effect a hypothesized explanation for the reduced level of aggression sometimes observed among residents of adjacent territories. This predisposes the ability to recognize competitors across repeat interactions, and is thought to be beneficial by circumventing the costs inherent with negotiating territorial boundaries.

Death feigning adoption of a fixed posture or immobility in potential prey, which arises in response to the presence of predators. Despite its name, it may not always involve mimicking death.

Deimatic display the sudden unleashing unexpected defences or conspicuous signals. These signals tend to elicit a reflexive response from the predator that inhibits their attack.

Deimatism a defensive strategy characterized by the combined use of multiple defences, such as camouflage and aposematism, with a rapid transition between the two as a deimatic (or 'startle') component.

Densoviruses invertebrate, single-stranded DNA viruses that can specifically infect some species of mosquitoes causing mortality.

Desensitization reduced response to the presence of a semiochemical as a result of either peripheral sensory adaptation of olfactory receptors or habituation affecting the processing of information to the central nervous system.

Diapause syndrome the ensemble of traits making up the physiological and behavioral state of dormancy of an animal (primarily referring to insects or arthropods), allowing the animal to survive predictable and longer unfavourable conditions (primarily the winter, but also periods of drought).

Diapause an altered developmental state generating a developmental delay or reproductive arrest. Typically, diapause occurs in response to environmental triggers that herald oncoming adverse conditions, such as photoperiod or temperature.

Direct benefits material resources that increase the fitness of individuals practicing mate choice.

Direct fitness fitness derived from personal reproduction.

Dispersal the more or less directed movement of an organism from one place to another (as opposed to undirected, random movement). Often referring to natal dispersal—the directed movement of an animal individual away from its site of birth. Also, loosely applied to plant seeds or spores.

Dispersive propagation the speed of bending wave propagation in the substrate; depends on the square root of the wave frequency, with high frequencies travelling faster than low frequencies. Thus, the time-frequency structure of a vibrational signal containing high and low frequencies gradually changes with distance from the source.

Disruptive colouration a form of camouflage that works by breaking up the characteristic outlines and/or body patterns of the organism, thereby preventing it from being recognized by a predator.

Disruptive selection type of natural selection that favours extreme trait values and simultaneously acts against intermediate values of a trait.

Division of labour a group level metric, conveying the degree to which individuals specialize on different tasks within a social group.

Dominance variance (V_D) the proportion of the variance in phenotype that is due to the interaction between alleles at a single locus. Dominant gene effects are not transmitted directly because only one allele at each locus is inherited from each parent and it is the combination of parental alleles at each locus that determines the dominance relationships in offspring.

Eavesdropping the secret and intentional interception of signal(s) intended for a different recipient.

Ecdysis-triggering hormone (ETH) hormone that is secreted by inka cells, triggers ecdysis and increases secretion of eclosion hormone.

Ecdysis the process of moulting from one instar to another.

Ecdysone steroid prohormone that triggers moulting, secreted by prothoracic glands, activated by fat body and cuticle to the active 20-hydroxyecdysone.

Eclosion hormone (EH) secreted by brain, triggers preprogrammed eclosion behavior, increases production of ETH.

Ecological (or evolutionary) trap the phenomenon by which organisms may orientate towards and eventually settle in unsuitable, sub-optimal, or poor quality habitats or sites. This mismatch can result if artificial, often man-made structures by chance exploit the naturally evolved sensory, neural, or physiological responses of organisms to environmental cues that, under natural circumstances, lead the organism correctly to suitable places or habitats conferring high fitness. For example, artificial light sources being mistaken by the moon.

Ecological restoration management undertaken to restore lost or degraded areas to conditions that more closely resemble their former undisturbed state and are deemed ecologically more desirable.

Economic defendability principle it refers to the idea that a defence of a resource has costs, such as energy expenditure or risk of injury, and benefits such as priority access to resources. Territorial behavior is said to arise when the benefits of territory defence outweigh the costs.

ECR and USP proteins that form the ecdysone receptor.

Efferent nerve a nerve sending information from neurons in the central nervous system to muscles or sensory organs in the peripheral nervous system.

Electrophysiology measurement and analysis of the electrical activity patterns (voltage and/or current across the cell membrane) of neurons by intra- or extracellular recordings with electrodes.

E-methylgeranate pheromone secreted by a breeding pair of burying beetles, structurally similar to juvenile hormone, enables mate identification.

Emotion an internal state elicited by rewards and punishers, where rewards are entities an individual is willing to work for and punishers are things an individual will work to avoid.

Endemic referring to a species or higher level taxon, locally restricted to a given area, with the connotation that it does not occur naturally elsewhere. Scale ranges from highly localized sites to continental, with qualifications such as 'narrowly endemic' implying a very restricted distribution range.

Environmental variance (V_E) the proportion of the variance in phenotype that is due to the environment. Environmental variance can take two main forms: general environmental variance and special environmental variance.

Epigenetic acting above the genome; not due to a change in DNA sequence, but usually due to methylation.

Ethogram a graph depicting a selected behavioral sequence.

Eusocial a level of sociality defined by cooperative brood care, overlapping adult generations, and a division of reproductive and non-reproductive labour. It may or may not involve morphologically distinguishable or sterile castes.

Evolutionarily stable strategy a population-level strategy that is incapable of being invaded by any potential (biologically-realistic) alternative.

Evolutionary game theory a mathematical basis for predicting optimality when the net pay-off (benefit–cost) of a particular biological strategy depends upon those adopted by competitors.

Extirpation a local extinction, usually applied to loss of a species from a particular site or area, or to loss of individual isolated populations, but not implying wider or global extinction of the taxon involved.

Eyespot circular or quasi-circular patterns on an organism's body, which superficially resemble eyes and inhibit predators from attacking on encounter. Eyespots may have evolved for several reasons, but they are widely thought to deter would-be predators through their similarity to their own predators' eyes.

Fear an emotion associated with impending loss.

Fibre tract/nerve tract A bundle containing the long extensions (axons) of neurons, connecting different areas of the nervous system without making synaptic connections on the way.

Fitness component the biological concept of fitness is estimated in various ways, depending on the field of biological inquiry. Examples are the expected or realized time to extinction of a species (in paleontology), the population growth rate r in the Euler–Lotka growth equation (in population biology). Behavioral ecologist usually focus on the fitness of individuals, whose lifetime reproductive success depends on survival to reproduction and the number of offspring produced (fecundity). For males, the latter in turn depends on the number of mates acquired and their offspring inseminated. Survival, fecundity, and mating success are thus the major *fitness components* of an individual.

Flight boundary layer the layer of the atmosphere, extending a variable distance up from the ground, where the wind speed is less than an insect's (self-propelled) flight speed. Within this layer, insects can make some progress upwind.

Flight trajectory (flight path) the path through the air taken by an insect, from initiation (take-off) to termination (landing).

Fluorescence the absorption and subsequent reemission of light, typically at longer wavelengths, i.e. at lower energy.

General environmental variance (V_{Eg}) the proportion of the variance in phenotype that is due to general environmental effects. Examples of the general environment include temperature and diet.

Genetic correlation a measure of the degree to which two traits are affected by the same genes (pleiotropy) or pairs of genes (linkage disequilibrium). A genetic correlation can be measured specifically for additive genes

(r_A) or for the total genetic contribution (r_G), which does not distinguish additive and non-additive gene effects. In both cases, values range between 1 and −1, as estimates of genetic covariances are standardized by the sum of the variance of the two traits.

Genetic variance (V_G) the proportion of the variance in phenotype that is due to genes, both with additive and non-additive effects.

Genome editing the permanent modification of DNA in an organism, i.e. the generation of a transgenic organism. In contrast to random mutagenesis, DNA sequence and gene expression are changed in a controlled fashion.

Genome-wide association study (GWAS) an observational study of a genome-wide set of genetic variants in different individuals to determine if any variant is associated with a phenotypic trait.

Genomics A field of science devoted to the characterization and quantification of genes.

Genotype-by-environment interaction (GEI) the phenomenon where different genotypes respond differently to variation in the abiotic environment. GEIs represent an important source of phenotypic variation (V_{GEI}).

Genotype-by-social environment interaction (GSEI) the phenomenon where different genotypes respond differently to variation in the social environment. GSEIs represent an important source of phenotypic variation (V_{GSEI}) that can be very different that GEIs.

Geometric framework a model of nutrition that considers the interaction between all components of food as predictors of individual preference for food or any fitness-related trait.

Glomerulus a compacted, demarcated neuropil structure, often indicating a spatial processing unit in the nervous system.

Golgi method a staining technique for nervous system tissue published by Camillo Golgi in 1873, which allowed for the first time to analyse the cellular composition and circuit architecture of nervous systems. In the procedure, a few cells are randomly labelled in black by precipitation of silver chromate, allowing for visualization of single neurons.

Ground speed the horizontal speed of an airborne insect (or any airborne object) relative to the ground.

Ground track (or track) a line connecting an animal's consecutive positions on the ground. The track direction is therefore the direction of displacement over the ground. The track and ground speed of a flying insect will be the vector sum of its airspeed and heading direction, and the ambient wind velocity.

Group-living when individuals of the same species live together and interact with each other more than they do with others over a period of time.

Habitat selection or choice most animals have preferred habitats in which they live and function best. These need to be found, unless they are already born into them, in which case the mother has made the choice.

Habitat traditionally, 'a place to live', but with recent greater functional emphasis on habitat as a milieu in which the key resources and other demands of a species co-occur and are accessible. Those resources and needs are divisible into 'consumables' (food), 'utilities' (wider factors that enable a species to behave and pursue life activities), and 'conditioners' (factors such as climate that enable or impede relationships between a species and its key resource needs).

Haplo-diploid a sex determination mechanism where (haploid) males emerge from unfertilized eggs and (diploid) females emerge from fertilized eggs.

Hemimetabolous A type of metamorphosis, often called 'incomplete metamorphosis', in which an insect develops through three stages as egg, nymph, and adult. The nymph resembles the final adult stage in many ways, but lacks wings. Termites, thrips, and aphids are examples of hemimetabolous insects.

Heritability (h^2) the proportion of the total phenotypic variance that is due to genetic causes. h^2 can be estimated in the broad sense as V_G/V_P or in the narrow-sense as V_A/V_P.

Heritable variation variation that will respond to artificial or natural selection resulting in an evolutionary change in mean behavior phenotype in a population.

Holometabolous a type of metamorphosis, often called 'complete metamorphosis', in which an insect develops through four stages as egg, larva, pupa, and adult. The larva differs dramatically from the final adult stage, which is typically winged. Ants, bees, and wasps are examples of holometabolous insects.

Homeostasis maintenance of a dynamic steady state by regulatory mechanisms that compensate for changes in external conditions.

Homing behavior describes the often innate ability of many animals to navigate, through unknown territory, to the location of their home or origin. Such homing behavior may occur over short distances, e.g. finding a nest, or long distance, e.g. finding the overwintering place.

Host manipulation the expression in an infected host of a behavior which is a product of infection and which increases the fitness of the infecting parasite (e.g. increases transmission); an adaptive trait of the parasite expressed in the phenotype of the host.

Hue a property of colours described by the dominant wavelengths of light in a given spectrum.

Identified neuron a neuron with characterized anatomical and functional properties, which can be identified across different animals and even species.

Immune priming a decrease in disease susceptibility following a prior pathogenic challenge.

Inclusive fitness the sum of direct and indirect fitness, and takes into account the effects of individuals on each other, weighted by relatedness.

Indirect benefits genetic benefits that increase the fitness of individuals practicing mate choice.

Indirect fitness fitness derived from helping related individuals.

Indirect genetic effects these occur when the genotype of one individual affects the phenotypic traits of a conspecific.

Infection avoidance behavior that prevents hosts from becoming infected.

Infection tolerance the ability of a host to maintain health during infection, independently of changes in infection loads, usually mediated by mechanisms that prevent or repair infection-derived tissue damage.

Inka cells endocrine cells located near the peripheral trachea system, secrete ETH.

Inquiline a form of social association where an organism lives within the nest of a heterospecific social group. Inquilines may be commensal or parasitic.

Insemination direct or indirect movement of sperm from males to females.

Integrated pest management a process used to solve pest problems, while minimizing risks to people and the environment. It is an ecosystem-based strategy that focuses on long-term prevention of pests or their damage through a combination of techniques, such as biological control, habitat manipulation, modification of cultural practices, and use of resistant varieties.

Interneuron a neuron that connects other neurons, i.e. every neuron that is neither a sensory nor a motor neuron.

Introgression the transfer of genetic information from repeated backcrossing through mating.

Iteroparous species or populations with more than one opportunity for reproduction in a breeding season.

Juvenile hormone (also JH) The name given to acyclic sesquiterpenoids produced by the *corpora allata* glands located behind the insect brain. Juvenile hormone regulates development, reproduction, diapause, polyphenisms, and social behavior.

Kairomone a semiochemical emitted by an organism that mediates interspecific interactions that benefit an individual of another species that perceives it, but detrimental to the emitter. An example would be a chemical emitted by a plant than a phytophagous insect uses to locate and feed on the plant.

Keystone individual akin to the keystone species concept in community ecology; individuals within a social group or population that exhibit an inordinately large influence over group or population dynamics relative to other individuals.

Kin selection a form of natural selection where apparent selection against an individual's own survivorship or reproduction can best be explained by its conferring survivorship or reproductive advantage to related individuals.

Latitudinal cline a systematic increase or decrease in a life history trait, e.g. body, size, diapause incidence, development time, along latitude. The most prominent example is Bergmann's rule, the phenomenon that mammals or birds, but also some insects, become larger towards the poles. Analogous altitudinal clines exist.

Learning the process of internally representing new information.

Life history traits standard traits that describe the life cycle of an organism, the sequence of events/milestones in its life: birth, growth, reproduction, death. Examples are propagule (egg or seed) size, growth rate, development time, age/size at first reproduction, fecundity, offspring number, longevity, body size, and the like. These traits are more or less directly linked to fitness, and typically involve together in a syndrome. For example, large animals tend to have large birth size, long development times, and late ages at first reproduction, but they are long lived.

Linkage disequilibrium the non-random association of alleles at different loci in a population.

Lipopolysaccharides (LPS) large molecules found on the outer membrane of Gram-negative bacteria that elicit strong immune responses.

Local adaptation an evolutionary process that leads populations to evolve traits that render them more well-suited to their local environment compared with other populations of the same species.

Locomotor cross-over hypothesis originally formulated to describe trophic patterns at the species level, this hypothesis states that more actively foraging predators will encounter and consume inactive prey, whereas inactive (sit-and-wait) predators are more likely to encounter active prey.

Macronutrients chemical compounds, including carbohydrates, proteins, and lipids that provide nutritional value for living organisms.

Masquerade a defensive strategy in which prey species impair their recognition by predators through physical resemblance to an inanimate object, such as a twig or leaf.

Mass trapping the use of a semiochemical with traps to capture enough of the pest population before mating, oviposition, or feeding to prevent crop damage.

Mate guarding the defence of a member of the opposite sex in which mating and/or fertilization is already realized, dependent upon, or guaranteed by such behavior.

Mating disruption method disrupting normal mating behavior with the aim of overall population reduction. Generally, involves release of large amounts of synthetic sex pheromone into a crop over a period of time to suppress pest reproduction by interfering with mate finding.

Mating system the type of association between males and females during a breeding season. This encompasses the details of mate encounter, competition and mating rates.

Mating a general term encompassing the reproductive events leading up to fertilization, including courtship.

Mechanoreceptors all sensory receptors with excitatory ion channels directly gated by mechanical forces. These ion channels are involved in the conversion of mechanical forces into biological signals. Senses of touch, mechanical pain, hearing, or balance depend on such mechanically-activated channels.

Memory internal storage of information. Newly learned information is consolidated into short-, then middle-term memory. There are two independent pathways, one leading to long-term memory, which can last a lifetime, the other leading to anaesthesia-resistant memory, which unlike long-term memory, is resistant to cold shock, requires no protein synthesis, and decays within a few days.

Methoprene artificial juvenile hormone analogue.

Migration syndrome a suite of co-adapted traits (behavioral, physiological, morphological, and life-history), which enable and promote migratory activity. The syndrome incorporates, for example, the locomotory apparatus and capabilities, the responses to environmental cues that determine when migration is initiated and how long it lasts, the orientation behavior that steers the locomotor activity, and sets of other characteristics that determine the fitness of the migrants.

Migration system a way of describing and integrating various components and processes involved in the migration of a species—the movements themselves, population processes, the physical and biotic environments, the suite of traits expressed in the migratory phenotype, and the underlying genetic package.

Migration typically recurring seasonal long-distance movement of animals from one place (e.g. where they breed) to another (e.g. where they overwinter). The persistent and undistracted nature of migratory movements is enabled by the temporary inhibition of an individual's responses to resource cues.

Mobility a generic term describing any type of short- or longer-distance movement behavior of animals.

Monandrous (monandry) a mating system in which females mate only once during their lifetime.

Monogamy a mating system in which individuals of both sexes have only one mate during a single breeding season.

Motor neuron a neuron-forming synapses with a muscle and controlling muscle contraction with its activity.

Müllerian mimicry an evolved resemblance of warning signals among two or more species that are unprofitable (i.e. distasteful) to reduce probability of attack by common predators.

Multi-dimensional care parental care involving two or more significant forms of care.

Nasty neighbour effect a hypothesized explanation for the increased level of aggression sometimes observed among residents of adjacent territories.

Nest structure where eggs are deposited and food for the resulting nymphs or larvae may be brought from outside by the parent(s).

Neurite a thin and elongated process formed by a neuron, which makes synapses with other neurons. Neurites can receive information (then also called dendrite) or send out information (then also called axon). On some insect neurites, input and output synapses can also be intermingled.

Neuroblast progenitor cells of neurons, which divide under tight spatial and temporal control to generate the adult cells of the nervous system.

Neuromodulator similar to neurotransmitters, neuromodulators are molecules mediating the chemical communication between neurons and neurons and other cells. Neuromodulators are not only released at synapses, but also other sites. They can affect large groups of cells and have longer lasting, global effects.

Neuropil area of the nervous system consisting of neuronal arboriz\ations and synapses, with no cell bodies or fibre tracts.

Neurotransmitter small molecule transmitting information across chemical synapses. Neurotransmitters are stored in vesicles in the presynaptic neuron and released into the synaptic cleft upon activation of the presynaptic neuron. They bind to receptors on the postsynaptic neuron and by this change its activity.

Non-additive genetic variance the collective term used to describe the variance in a phenotypic trait that is due to dominance variance and epistasis variance (V_D and V_I).

Non-competitive disruption mechanism of mating disruption that occurs when the locations of sex pheromone-emitting females are masked by synthetic sex pheromone emitted from numerous lures throughout the environment, as well as by desensitization of males to the sex pheromone.

Nuptial gift items presented to females by males before or during copulation.

Nutrient any component of food that is necessary for completion of the life cycle of an organism.

Obesity accumulation of fat reserves that has a negative effect on health and causes a decline in fitness.

Ommatidium a single functional unit of the insect compound eye that broadly consists of two lenses overlying several photoreceptors.

Oogenesis-flight syndrome the tendency in many (but not all) insect species for the adult life history (particularly in females) to be partitioned into a migration phase followed by a reproductive phase. In other words, in the young adult, energy and nutrients are first allocated to the migration apparatus and capabilities, and sexual maturation is characteristically delayed. After migration, resources are switched to reproductive activities (and, in particular, egg production). It is an aspect of the migration syndrome.

Operational sex ratio the number of fertilizable females compared with the number of sexually active males in a population at a given time.

Opsin a type of G protein-coupled receptor that binds to a retinoid chromophore to become photosensitive.

Optogenetics technique for using light to control neuronal activity. This is achieved by transgenic expression of light sensitive membrane channels, which open upon absorbance of photons.

Pain a negative emotion associated with actual or potential tissue damage.

Parasites organisms requiring a living host to achieve positive fitness and whose presence has negative fitness consequences for the host.

Parasitoid a parasitic organism that not only must infect a host individual of another species to complete its life cycle, but also requires host death as a result of successful infection.

Parasocial a term that describes groups composed of adults usually from the same generation and their offspring, encompassing communal, quasisocial, and semisocial types.

Parental compensation the caregiving response made by one parent to a change in parental effort by its partner.

Parental investment any investment by the parent in an individual offspring that increases the offspring's survival and reproductive success at the cost of the parent's ability to invest in other current or future offspring.

Parental negotiation the rules for responding to a partner's parental effort over a series of behavioral interactions.

Parental-care parasitism (or social parasitism) interaction in which an individual (the parasite) obtains reproductive benefits, while reducing or completely eliminating the costs of parenting by exploiting any type of parental care provided by other individuals (the hosts).

Parthenogenesis the production of genetically identical offspring without meiosis and without sexual reproduction.

Perception the process of capturing information from the external environment and translating it into internal representations.

Phagostimulants chemical compounds that promote the initiation and/or continuation of feeding (contrasted with phagodeterrents, which have the opposite effect).

Phenology the evolved fit of the life cycle of organisms (plants, animals) to periodic (daily, monthly, yearly, seasonal) variation in climatic events (winter, summer, reproductive season, etc.).

Phenotypic plasticity In narrow terms, it refers to the response of a genetically homogeneous group of organisms (clone) or to a single organism's response to a changing environment. A broader definition refers to the average change in response of a population of genetically variable individuals in response to environmental change, in which case genetic variation is not taken into account. Phenotypic plasticity, behavioral flexibility, and modifiability are often used interchangeably in the behavioral literature.

Phenotypic variance (V_p) the variation in phenotype expressed by members of the population.

Pheromone semiochemical emitted from one organism that stimulates a response in another individual of the same species.

Phonotaxis the act of moving in response to sound. Positive phonotaxis refers to movement towards sound, such as when female crickets walk towards a singing male, whereas negative phonotaxis refers to movement away from sound, such as when a flying cricket flies away from an echolocating bat.

Phoresy a relationship which entails one animal (the phoretic) 'hitchhiking' on the body of another (usually larger and more mobile) host animal. The phoretic animal receives an ecological or evolutionary advantage by migrating from its current habitat while superficially attached to a selected interspecific host. Notice that the term phoresy is not applied to relationships where the individual being transported is directly parasitizing the host individual.

Photoperiod hours of light in a 24-hour period.

Physiognomy refers to the nature and distribution of a resource, given by features such as geographical location, size, orientation, shape, and extent of fragmentation.

Pigment any substance, although typically an organic molecule, which selectively absorbs a portion of the visible spectrum, e.g. melanin and carotenoids.

Pleiotropy the situation where one locus affects more than one phenotypic trait.

Plesiomorphic state refers to the ancestral form of a trait.

Polyandry a form of polygamy in which some females mate with multiple males during a single breeding season.

Polyethism the sequential variation in behavioral tasks or jobs that an insect does, over developmental time.

Polygyne or monogyne colony forms of social organization, for example in ants, characterized by the number of queens a colony will tolerate as reproductive: many queens = 'polygyny' or only one queen = 'monogyny'.

Polygyny a form of polygamy in which some males mate with multiple females during a single breeding season.

Polyphenism a particular case of phenotypic plasticity in which morphological and behaviorally distinct morphs develop depending on environmental cues experienced as larvae.

Population suppression control strategies that reduce or eliminate the target vector population.

Population trajectory (or its long-term average, the population pathway) it refers to the spatio-temporal demography of a population or subpopulation as it changes its geographical locations over succeeding generations due to migration.

Precocene a juvenile hormone inhibitor that acts by destroying *corpora allata*.

Precocial active and able to move freely at birth or hatching.

Precopulatory sexual cannibalism a behavior where individuals (usually females) attack and consume their potential mates before copulation occurs.

Prophylactic a medicine or behavior intended to prevent disease.

Proprioreceptors and exteroreceptors proprioreceptors involve all sensory receptors in internal organs, such as stretch receptors, chemoreceptors monitoring carbon dioxide and oxygen levels in the blood, thermoceptors in the brain providing feedback on internal body temperature, or mechanoreceptors responding to changes in the relative position of body parts. By contrast, exteroreceptors include all senses responding to physical stimuli in the environment, such as sound, light, or odours.

Prothoracicotropic hormone (PTTH) a hormone synthesized by brain neurosecretory cells and released from *corpora cardiaca* that stimulates prothoracic glands to produce ecdysone.

Push-pull technique a behavioral manipulation strategy for pest management that incorporates behavior-modifying stimuli to push the pest from the resource to be protected and pull the pest to a location where control strategies can be applied.

Quantitative trait loci a genomic region that is responsible for the variation in a quantitative trait. Also known also as QTL.

Quasisocial parasocial group with alloparental care.

Range shift the shift of the whereabouts or limits of occurrence/existence of species in response to environmental, e.g. climate, changes.

Ranging a type of movement that, like migration, takes an animal away from its home range, but in contrast to migration, responses to sensory inputs from resources are not inhibited, and movement ceases when the appropriate resource is found. Thus, ranging movements are continuously exploratory, and equate to 'natal dispersal' in vertebrates, where young animals leave their natal site in order to find a new home range.

Reaction norm a figure showing the mean phenotype of a series of different genotypes in multiple environments.

Repellents synthetic or natural products leading to an innate negative chemotaxis in selected arthropod vectors.

Resource-holding power (potential; RHP) the property of being able to dominate or monopolize limiting resources. In animal contest theory, RHP refers specifically to variation in the physical or physiological properties that engender success in pairwise battles over receptive females or the resources that determine access to them.

Resource value in animal contest behavior, is conceptualized as the increment of reproductive success at stake for each contestant.

Retina the light-sensitive layer of cells lining the inner surface of an eye.

Rhabdom the photoreceptive element of insect eyes, formed by a set of fused or closely-space ommatidia.

Secondary metabolites chemical compounds that are found in plants and define their adaptation to the environment, but are not part of the primary pathways of growth and reproduction. They include alkaloids, terpenoids, and phenols, and act as defence toxins, attractors for pollinators and feeding deterrents to herbivores, among other functions.

Self-medication any behavior by which infected hosts choose to exploit food sources or other compounds to reduce or clear infections.

Semelparous species or populations constrained by phylogeny, season, and/or resources to a single reproductive event.

Semiochemical a chemical that carries a message to another organism, often triggering a behavioral response.

Semisocial a parasocial group with alloparental care and reproductive division of labour, but no overlap between adult generations.

Sensory neuron a neuron in the periphery that detects sensory stimuli and transforms them into electrical activity. Examples are photoreceptors for vision, olfactory sensory neurons for olfaction, and gustatory neurons for taste.

Sequential assessment a hypothesized basis for how information on competitive ability is revealed across discrete stages of escalation during an agonistic interaction.

Sex pheromone a pheromone that is emitted by one sex of a species that elicits a positive response by the other sex of the species in terms of mating behavior.

Sex ratio distorter a genetic factor that changes the normal sex ratio of 1:1 that can be exploited to reduce the number of female insects in a population.

Sexual conflict an evolutionary phenomenon that occurs when males and females have different optimal fitness-accruing strategies concerning the mode and frequency of mating, leading to antagonistic co-evolution in which one sex evolves a favourable trait that is offset by a countering trait in the other sex.

Sexual deception the act wherein an individual releases misinformation regarding its quality as a potential mate. For example, in deceptive orchids, flowers provide no gain in fitness from the attempted mating act by a pollinator.

Sexual selection a form of natural selection that arises when individuals compete for priority access to mates.

Shared ancestry a relationship among species from their most recent common ancestor.

Sickness behaviors a set of behavioral changes that occur in sick individuals (usually following immune challenge), including increased lethargy, somnolence, and anorexia.

Signal an act or structure that codes and conveys information to a receiver, and that affects change in the receiver with positive fitness consequences, on average, for both signaller and receiver.

Single-nucleotide polymorphisms (SNPs) a variation in a single nucleotide that occurs at a specific region in the genome.

Social environment the environment that is provided by other individuals in the population.

Social heterosis when groups perform better by containing individuals of a diversity of phenotypes (behavioral, morphological, etc.) compared with less diverse groups.

Social learning the process of internally representing new information acquired through watching other individuals or their products.

Social niche construction the action by which individuals actively shape their social environment.

Social niche specialization an individual specialized at a specific social role within its social environment.

Special environmental variance (V_{Es}) the proportion of the variance in phenotype that is due to special environmental effects. Examples of the special environment include parental care or competitive interactions.

Sperm competition competition between sperm of two or more males for fertilization of a particular set of ova.

Station-keeping movements a variety of movements, e.g. foraging, territorial behavior, commuting, which are resource-directed and tend to keep an individual within its home range or local habitat patch, in marked contrast to migratory movements or to 'ranging', which result in displacement away from the home range.

Sterile insect technique (STI) a method frequently utilized in pest or vector control, whereby male sterile insects are used to compete with non-sterile males for mates. Females that mate with a sterile male produce no offspring.

Stridulation the production of sound by rubbing together certain body parts, such as in grasshoppers and other insects, where a hind leg scraper is rubbed against the adjacent forewing, or in crickets and katydids, where a file on one wing is rubbed by a scraper on the other wing.

Structural colouration a colour based on the wavelength-dependant reflection of light from photonic crystals, rather than via the selective absorption of light by pigments.

Subgenual organ an organ in insects involved in the perception of substrate vibrations, sometimes with enormous sensitivity (threshold of less than 1 nm of displacement). The organ is located just below the knee in the tibia of all legs in most insects.

Subsocial a term that describes groups of parents and their offspring with extended parental care.

Superadditive effects a term understood as the efforts of cooperating partners being complementary and synergistic.

Superposition eye an eye in which all light rays except those entering the central facet of a group are intercepted. This is typically achieved through the separation of the lens and rhabdom, and is common among nocturnal insects as it is generally capable of producing brighter images.

Swarming behavior a collective behavior exhibited by animals that aggregate together, perhaps milling about the same spot or moving *en masse* or migrating in some direction.

Synapomorphic trait it refers to a derived character present in a group of species that share a common ancestor.

Synomone an interspecific chemical message that benefits both the signaller and the recipient.

Temporal polyethism organisms that alter their task participation across development. For example, honeybees move from being brood care specialists to foragers across development. See also 'polyethism'.

Territoriality the aggressive defence of an object and/or area, as engendered by a sense of ownership (or motivation thereof). Insect territoriality is commonly studied in the contexts of resource and/or mate acquisition.

Thermal performance curve a non-linear relationship between any measure of organism performance or fitness (on the y-axis) and temperature (on the x-axis).

Thermoregulation the ability to regulate and maintain their body temperature within certain non-lethal limits (cf. homeostasis). Endothermic (= homeothermic) animals (birds and mammals) do this rather precisely by physiological means. Ectothermic (= poikilothermic) animals (fish, reptiles, invertebrates) have limited capacity to do so usually via physiological heat generation (e.g. bumblebees), evaporative cooling or behavioral means (sun basking, shade seeking, etc.).

Trade-offs negative genetic correlations between at least two traits such that selection for one trait results in an antagonistic outcome in the other so that fitness cannot be maximized.

Trail pheromone an intraspecific, chemical message that elicits directional movement (e.g. toward food resources) by conspecifics.

Transcriptome the total set of mRNA transcripts produced by the organism often assayed as the expressed genes in one body part, under a particular environmental condition.

Translocation in conservation practice, the deliberate movement and release of individuals to a site, variously as a reintroduction to places from where it has been lost, as an expansion of range, or to augment small field small populations.

Trap shutdown a measurement used in mating disruption studies to characterize success of the treatment. It is calculated as the percentage of moths collected in a sex pheromone-treated area compared with an untreated area.

Traumatic insemination mating practice in which a male pierces a female with his genitals and directly injects sperm into her body cavity.

Tremulation the behavior involves rocking of the entire body or the abdomen; the subsequent vibrations are transferred through the legs to the substrate.

Trophallaxis the exchange of body fluids between individuals, typically for nutrients, transfer of beneficial microbes, or communication. Common in many social insects.

Ultraviolet the region of the electromagnetic spectrum with wavelengths ranging from 100–400 nm.

Uni-dimensional care Parental care that consists primarily of one form.

Urbanization the gradual change of natural (e.g. forests) or rural (e.g. grasslands) to urban landscapes (e.g. villages, cities) by the action and activity of humankind (buildings, parks, etc.).

Vector an agent carrying and transmitting pathogens or parasites to another organism.

Virulence the harm experienced by a host during infection. In evolutionary ecology, virulence refers broadly to any fitness costs of infection.

Voltinism transition zones environmental locales where populations of a given species shift from predominantly univoltine (one mating generation per year) to multivoltine (one mating generation per year) or vice versa.

Wind-related orientation a situation where the heading direction of a migrant insect is influenced by variations in wind velocity. For example, a group of migrants attempting to orientate in a downwind direction should show changes in their heading distribution if the wind direction changes.

Index

(Bt) toxin, 319, 325
(E)-β-farnesene, 154, 323–324
(Z)-7-tetradecen-2-one, 322
20E, 49–51
3D stereoscopic vision, 162
α-pinene, 323
β-pinene, 323

A

Abedus, 207
Abiotic, 21, 102, 178–179, 181, 186, 261–262
   condition, 204–205, 216
   environment, 4, 20–21, 23, 25, 103, 367
   feature, 80
Abundance, 24, 33, 124, 126, 248, 349–351, 353–354
Acanthocephalan, 283, 286
Acanthos celides obtectus, 19–20
Accessory gland, 55, 190, 335
Acclimation, 293–294, 303, 363
Acetylcholine, 37
Acheta domesticus, 43, 89, 197
Achroia grisella, 10, 15, 21–22
Aconitum napellus, 261
Acoustic, 141, 174–178, 181–183, 302, 353
   aposematism, 140–141
   moth, 15, 17
   niche, 180, 363
   signal, 4, 17, 22, 43, 138, 145, 175–176, 181–182, 194, 215, 243, 363
Acoustically orientating parasitoid fly, 24
Acridid grasshopper, 138
Acrididae, 264
Acripeza reticulate, 138, 167
Acromyrmex, 230
   insinuator, 279
Aculeata, 222
Acyrthosiphon pisum, 12, 237

Adaptation, 2–3, 13, 39, 41, 53, 81, 105, 126, 130, 134, 139–140, 159–160, 165, 180, 182, 198–199, 210, 231, 244, 248, 258–259, 262, 273, 278–279, 281, 283–284, 292, 294–295, 297, 301–302, 315, 343, 348, 357, 359, 365, 369, 372
Adaptive, 5, 32, 36, 83–84, 100, 134, 140, 145, 151, 180, 195, 216, 242, 247, 250, 258, 260, 262, 264, 281, 284, 286, 298, 300, 342, 359, 364
   genetic, 294
   mating preference, 195
   sickness behavior, 274–275, 281
   significance, 133, 200, 257–258, 264, 268, 275
Additive gene effect, 7, 25, 363, 367
Additive genetic, 13–14, 18–20, 363
   variance (V_A), 5, 7, 13–14, 24–25, 363
   variance-covariance matrix, 24, 363
Adipokinetic hormone, 280
Adonis blue, 357
Adrenalin, 37
Adult, 34, 40–43, 49–57, 59, 65–67, 69–70, 80, 83, 99, 102, 119, 204, 208–209, 221–223, 225–228, 230–231, 242–244, 284, 287, 292, 296, 309, 317–322, 336, 339–342, 348, 351–353, 356–357, 359, 366, 368, 370, 372
   emergence, 42, 53, 82
   foraging area, 82
   mating aggregation, 223–224
   nervous system, 37, 42
   olfactory response, 70
   sucrose response, 70
Advanced eusociality, 223–224
Advertisement, 22, 84
   call structure, 8
   signal, 194
Adzuki bean beetle, 139
Aedeagus, 198
Aedes, 333, 335–338
Aedes aegypti, 53, 67

Aerial, 37, 88–89, 99–100, 105, 108, 112
   manoeuvre, 32
   pursuit, 89
Afferent nerve, 36, 363
Africa, 98–99, 109–110, 204, 228, 319, 332
Africanized honey bee, 230
Age, 6, 15, 43, 58, 65, 70, 85, 88–89, 93, 98, 209, 245, 265, 298, 302, 348, 353, 369
Ageing, 15, 17, 363
Age-matched forager, 43
Age-related division of labor (labour), 363
Aggregation, 65, 82, 124, 134–135, 137, 175–176, 220, 222–224, 226, 229, 231–232, 276, 282, 301, 311, 314, 316, 336, 340, 356, 363
   pheromone, 268, 309–313, 317, 321–322, 363
Aggression, 10, 22, 37, 52, 57–58, 65, 69–70, 90, 92, 175, 239, 246, 276, 365, 370
Aggressive, 58–59, 65–66, 71, 81, 90, 93, 155, 226, 237–238, 242–243, 245–248, 264, 316, 355–356, 373
   behavior, 10, 23, 71, 358
   encounter, 39
   spillover hypothesis, 247, 363
Agriculture, 98, 302, 365
Agriotes, 311
   obscurus, 311, 322
Agrotis infusa, 107, 356
Air turbulence, 146–147
Airborne chemical, 34
Airspeed, 104, 108, 363, 367
Alarm pheromone, 140, 146–147, 154–155, 310, 317, 323–324, 363
Aldehydes, 148, 343
Alfalfa looper, 311
Alkaloid toxin, 283
Allatostatin, 122
Allele frequency, 297
Allocation, 58, 67, 73, 147, 214, 348

Allogrooming, 279
Allometric, 363
  relationship, 148
  slope, 190
Allonemobius socius, 15, 17
Alloparental care, 227–229, 363,
  371–372
Alograpta oblicua, 136
Alternative reproductive tactic, 195
Altitudinal, 104, 294, 369,
Altricial, 212–213, 215, 363
Altriciality, 213, 214
Altruism, 219–220, 222, 227,
  231–232, 363
Altruistic, 220, 222
Ambrosia beetle, 209, 223, 226–227, 311
Amegilla dawsoni, 89
American rubyspot damselfly, 166
Amfor, 70
Aminesalt, 317
Amino acids, 118, 121, 261
Ammonia-based compound, 317
Ammonium acetate, 317, 318
Amniotic egg, 210
Amphisexual care, 213, 363
Anabrus simplex, 124
Anachoresis, 133
Anaesthesia-resistant memory, 260, 369
Analysis of covariance (ANCOVA),
  14, 21
Anax parthanopes, 161
Andrenidae, 224
Anechura harmandi, 207, 209
Angle, 41, 106, 138, 161–162, 164,
  250–251, 266, 303
Animal, 1–2, 4, 18, 32, 36, 38–41, 43,
  51–54, 70, 81, 86, 90, 92, 100–101,
  112, 116–118, 121–123, 126, 153,
  159–160, 163, 165, 168–169,
  180–181, 186, 189, 191, 193, 196,
  200, 210, 219–220, 222, 236,
  239–243, 249, 257–260, 262, 264,
  274–276, 280–282, 293, 297–299,
  301, 303, 334, 336, 340–342, 348,
  355, 363–373
  personality, 2, 236–237, 239, 241,
    244, 248, 250–251, 363
  signal, 194, 363
Anisogamy, 80–82, 191–192, 363
Anomala orientalis, 312, 322
ANOVA, 6, 7, 9, 18–19, 21, 64
Anorexia, 124, 126, 281–282, 372
Anosmic, 39–40
Ant, 12, 32, 40, 50–51, 57–58, 66–67,
  70, 72, 118, 125, 134–135, 140, 147,
  149, 169, 191, 203–204, 207–208,
  215, 219–220, 222–225, 228–232,
  238, 244–246, 250, 257, 268,

277–280, 282, 284–285, 302, 353,
  356–358, 368, 371
  queen, 190
  repellent, 204
  trail pheromones, 149
Ant-like jumping spider, 134
Antagonistic sexual conflict, 192, 196
Antenna, 39–40, 45, 68, 102, 106, 130,
  174, 177, 182, 184, 343
Antennal lobe (AL), 36, 39, 243
  glomerulus, 39, 45
Anterior wing margin, 39
Anthropocene, 293
Anthropogenic changes, 111,
  348–349
Anti-aggregation pheromone,
  310–311, 313, 316–317, 321
Anti-aphrodisiac pheromone, 71
Anti-bourgeois, 87, 364
Anti-feedant, 313, 317, 324, 363
Antimicrobial secretion, 279
Anti-predator, 132, 137, 139–141
  behavior, 130–133, 137,
    142, 260
  defense (defence), 130, 132–133,
    140–141
Anti-predatory, 130
  behavior, 260–261
  benefit, 124
  regurgitation, 140
Antlions, 169
Anuran, 210–211
Anurogryllus muticus, 279
Anxiety, 269, 363
Aphaenogaster senilis, 149
Aphid clonal migrant, 106
Aphid-farming ant, 279
Aphididae, 223
Aphidius matricariae, 154–155
Aphis fabae, 68, 101
Apidae, 224, 356
Apis, 231
  cerana japonica, 140
  mellifera, 3, 11, 15, 40, 43, 51, 64,
    160, 167, 177, 194, 225, 229, 258,
    266, 281
Apocrita, 223
Apoidea, 222, 224
Apolipophorin, 124
Apollo butterfly, 299
Aposematic, 135, 137–138, 141,
  168–169, 355
Aposematism, 135, 137, 140–141, 169,
  363, 365
Appetite, 116–117, 121–122, 124–125,
  282, 364
Appetitive learning, 70

Apple maggot fly, 311,
  317–318
Applied entomology, 349
Apposition eye, 258, 364
Aquarius remigis, 296
Aquatic beetle, 40
Arabidopsis thaliana, 324
Arachnid, 208
Arborization, 36–37, 42–43, 45
Arbovirus, 332, 338–339, 364
Argentine ant, 356
Arista, 184
Aristolochia elegans, 355
Aristolochiaceae, 355
Armyworm, 98
Arthropod, 32, 49, 88, 99–100, 105,
  145–146, 148, 154–155, 159, 203,
  207–208, 210, 213, 215, 239, 241,
  243–244, 247–248, 251, 280,
  332–333, 335, 339–340, 343,
  364, 366, 372
Artificial, 141, 259, 266, 300, 335, 342,
  354, 366, 369
  light environment, 298
  selection, 3, 7, 8, 14–17, 21, 26,
    63–65, 139, 368
Ascia monuste, 107
Asexual reproduction, 190
Asian
  citrus psyllid, 312, 324
  corn-borer moths, 195
Assessment, 73, 86–87, 92, 166, 168,
  287, 300, 349, 351–352, 354, 357,
  365, 372
Assisted migration, 111, 353
Assortative, 247
Assortativity, 246, 364
Assassin bug, 169
  nymphs, 152
Astrocyte, 37
Atlantic Ocean, 106
Atmospheric boundary
  layer, 104–105, 364
Atta, 230
  cephalotes, 278
  sextans, 225, 228
Attack, 130–135, 137, 140–141, 147,
  155, 213, 230, 238–239, 246–247,
  321, 324, 350, 357, 363–365,
  370–371
Attention, 4, 90, 98, 134, 137, 139,
  163, 177, 210, 215–216, 236, 238,
  248, 257, 280, 303, 333–334,
  348–349, 353
Attract-and-kill, 310–311, 313–314,
  317–320, 322, 364
Attractant-based trap, 310
Attraction annihilation, 314

Attractiveness, 15, 17, 85, 93, 191, 199, 213,
Auditory display, 82
Aulacorthum solani, 155
Australia, 24–25, 103, 107, 152, 338, 351–352, 354–356
Australian, 177, 349
   field cricket, 12, 22, 199
   imperial blue, 357
   plague locust, 109
   shield bug, 207
   tessaratomid, 210
Auto-dissemination, 310, 311, 314, 336, 364
Autographa
   californica, 311
   gamma, 108
Autogrooming, 279
Autosomal, 18
Aversive, 40, 70, 238–239
Axonal, 37, 45
Azuki bean weevil, 15

B

Bacillus thuringiensis, 319, 334
Bacteria, 261–262, 277, 279, 292, 314, 324, 333, 337, 364–365, 369
Bactericera cockerelli, 15
Bactrocera
   cucurbitae, 15
   oleae, 311, 318
Balamara gidya, 197
Banana weevil, 311
Bark beetle, 312, 316, 321
Bark-resting cryptic moths, 133
Bat echolocation, 33
Batesian
   mimicry, 134, 364
   mimics, 134, 169
Bed bug, 150, 197
Bee, 3, 11–12, 15, 32–33, 37–41, 43, 57–58, 64–67, 70, 72, 83, 118, 126, 130, 134, 140, 152, 154, 158–161, 167–168, 177, 191, 195–196, 204–205, 220, 222–223, 225, 227–232, 243–244, 258–259, 261–263, 266–269, 276–279, 281–282, 301–303, 356, 368
Beet armyworm moth, 311
Beetle, 8, 15, 17, 19–20, 22, 32, 37–38, 40, 43, 54–58, 65, 67, 71, 85, 99, 118, 125–126, 134, 139–140, 159, 164, 166–167, 169, 178, 183, 191, 193, 198–199, 204–205, 207–209, 214–215, 223, 225–227, 229–230, 232, 237, 240, 244, 258, 279, 284–285, 299, 301–301,

311–313, 316, 320–322, 352, 354, 357, 366
Behavioral, 2, 3, 32, 34, 36–43, 49, 51, 57–59, 64, 66–67, 69–70, 72–73, 80, 100–101, 119, 124, 130–131, 133–136, 138–140, 142, 145, 149–150, 164, 174, 182, 198, 205, 216, 220–221, 228–229, 231–232, 236–239, 251, 260–262, 264, 269, 273–277, 283–284, 286, 293–296, 298–303, 309–311, 313, 316–317, 321–322, 323, 325, 335, 339–340, 342–343, 348–351, 353–359, 363–367, 369–372
   change, 68, 221, 249, 262, 274–275, 280–285, 349, 372
   disruption, 309
   epigenetics, 72
   fever, 283, 364
   flexibility, 45, 63, 72, 87, 138
   hypervolume, 249, 364
   immunity, 274, 364
   manipulation, 285–287, 309–311, 313, 316, 325, 364, 371
   mimicry, 131, 134–135, 140
   modifiability, 63, 371
   plasticity, 56, 66, 67–68, 73, 84, 241–242, 278
   reaction norm, 241–242, 364
   response, 64–65, 70–71, 111, 125, 131–132, 141, 262, 273–276, 283, 286–287, 292–293, 298, 300–301, 303, 325, 349, 364, 372
   response threshold, 245, 364
   syndrome, 236, 238, 240, 242–244, 246–247, 250, 364
   variability, 63, 72, 241
Belenois aurota, 98
Belostomatid, 207, 209, 213
Bergmann cline, 295
Bertholdia trigona, 138
Binaural, 147
   cue, 181, 264
   information, 147
Binocular vision, 162
Biocontrol, 250, 321, 337–338
Biodiversity, 80, 251, 293, 359
Biogenic amines, 37, 243, 364
Bioinsecticide, 314
Biological control, 155, 250, 317, 333, 356, 358, 369
Bioluminescence, 137, 141, 163, 166, 364
Bioluminescent, 164, 166
   glow-warm larvae, 137
Biomass, 99, 154
Biotic, 21, 23, 103, 179, 181, 186, 207–208, 210, 261, 369
   condition, 207, 216

factor, 80, 102, 261–262
Biparental, 212, 214–215
   care, 210 216, 229
   family group, 223
Birdwing butterfly, 354–355, 359
Birth, 100, 153, 363, 366, 369, 371
Biston betularia, 133, 167
Biting, 55, 139–140, 142, 286
Blaberid cockroaches, 205
Blaberus discoidalis, 41
Black, 91, 107, 110, 117, 133, 140, 159, 161, 162, 164, 169, 238, 267, 295–297, 321, 367
   bean aphid, 101
   box, 4, 9, 86, 112, 199
   field cricket, 8, 16, 199
   melanic form, 133
Blatella germanica, 313
Blattellidae, 209
Blattidae, 209
Blattodea, 182, 193, 206, 209, 223, 231
Blood-feeding species, 116
Blowfly, 33, 36, 116, 118, 121, 126
Blue jays, 133
Blueberry, 311, 312, 322
   maggot fly, 311, 317
Body, 33–34, 36–37, 38, 40–41, 50, 53, 67, 89, 99, 105, 118, 122, 131–133, 140, 148–149, 159, 166, 168–169, 178, 182–185, 190, 196–197, 199, 205, 207, 209–210, 213, 242, 244, 266, 283, 296, 301–302, 304, 363–364, 366–367, 369, 371–373
   condition, 71, 192
   size, 15, 17, 86, 88, 91, 139, 148, 166, 181, 192, 204, 211, 213, 292–297, 336, 339, 369
Bogong moth, 107, 356
Bombardier beetle, 140
Bombus
   bohemicus, 279
   dahlbomi, 356
   fervidus, 261
   impatiens, 266
   terrestris, 58, 287, 356
   vagans, 261, 266
Bombycoidea moths, 141
Bombyx mori, 3, 12
Boundary layer, 104, 105, 364, 367
Bourgeois, 87–88, 90, 364
Brain, 32–34, 36–41, 43–44, 50–53, 55, 59, 66, 68, 70, 106, 121–122, 126, 161–163, 231, 245, 257–260, 264, 284, 287, 358, 365–366, 368, 371
Bristletail, 196
Britain, 107–108, 358
British Columbia, 322

Broad-horned flour beetle, 15, 17
Broad-sense estimate, 13
Brood care, 54, 66, 209, 222, 227–228, 230, 232, 244–245, 366, 373
Brown, 160, 164, 168, 311
  planthopper, 12
  winged green bug, 311
BT × BT interaction, 248, 364
Buffering neurotransmitter, 37
Bug, 42–43, 53–54, 56–57, 102, 125, 150, 152, 154, 169, 180–181, 197, 199, 203, 207, 209, 211, 213, 223, 286, 311, 332–333, 340–343
Bumblebee, 38, 57–58, 72, 134, 222, 261, 266, 283, 286–287, 356, 373
Burrow, 24, 53, 82, 89, 131, 213
  system, 215
Bursa copulatrix, 190
Bursaria spinosa, 352
Burying beetle, 15, 17, 54, 56–58, 65, 71, 134, 193, 209, 214, 223, 225–226, 229, 279, 366
Busseola fusca, 319
Butterfly, 8, 15, 17, 38, 41, 51, 53–54, 68, 82–85, 88, 90–93, 99, 102, 106–107, 109–112, 126, 133–135, 138–139, 145, 152, 164–165, 166, 169, 196, 198, 204, 224, 247, 260, 277, 283, 294–295, 299, 301–302, 348–350, 352, 354–359
  migration, 98–99
Buzz courtship behavior, 20

C

Cabbage looper, 148
  moth, 312
Cabbage white butterflies, 60, 260
Calcium indicator, 34
California, 356
  five-spines ips, 313
Callosobruchus
  chinensis, 15, 139
  maculatus, 8, 15, 19–20, 195, 197
Calopteryx, 83, 90
  haemorrhoidalis asturica, 195
  maculata, 89–90
Calyces, 33, 39–40
Camouflage, 131, 134, 139, 141, 152, 168–169, 315, 364–366
Camponotus
  fellah, 279
  floridanus, 43, 72
Canada, 69, 269, 322
Cannibalism, 124–125, 196–197, 238, 241, 247, 371
Canopy, 106, 164, 181

Captive breeding, 353, 359
Carabidae, 140
Carbohydrate, 119–121, 123, 125–126, 264, 282, 319, 369
Carcass, 54, 56–57, 193, 209, 214, 229, 279
Cariana abbreviata, 136
Caribbean, 99
Carotenoid, 164, 371
Carpenter
  ant, 43, 279
  bee, 222
  moth, 312
Carpophilus, 311, 312
Carrion, 109, 210, 212, 229
  beetle, 225
Carrot psyllid, 312
Caste, 34, 43, 51, 57–59, 220–222, 228, 230, 246, 278–279, 364, 366
Castniidae, 351
Cataglyphis ants, 67
Caterpillar, 53, 55, 116, 119, 121, 125, 131–134, 137, 139–141, 149, 221, 224, 227, 229–231, 260, 283, 285, 298, 302, 318–319, 348
Catocala, 130, 133, 142
Catopsilia florella, 98
Causation, 213, 274–275
Cave, 107, 190, 216
Celestial compass orientation, 106
Cell, 10, 32–34, 36–38, 40, 42–44, 49–50, 52, 108, 119, 121, 160–162, 183–185, 189–190, 204, 221, 261, 280, 363, 365–368, 370–372
Centipede, 169, 210
Central, 3, 5, 33, 36–37, 39–44, 68, 99, 104, 106, 121, 123, 137, 207, 229–230, 232, 241, 243, 246–247, 257, 268, 279, 281, 287, 296, 324, 355, 358, 373
  America, 125, 286, 322
  complex, 36, 38, 40, 106
  excitatory site state, 118
  nervous system, 34, 52, 121, 181, 275, 280, 315, 363, 365–366
  pattern generator, 33, 364
  plank, 348
Centris pallida, 83, 89
Cephalopod, 159
Cessation of mating, 315
cGMP dependent protein kinase (PKG), 69–70
Ceratina australensis, 222
Ceratitis capitata, 312, 318
Cerebral ganglia, 34–35, 257

Chagas, 286, 333–334, 340–341
Chalcosyrphus, 136
Chameleon, 159
Charis cadytis, 89
Cheilosia pontiaca, 136
Chemical, 34, 37, 42, 55, 58, 105, 118, 135, 139–141, 146–150, 152–153, 161, 164, 194–195, 203, 302, 309–310, 313–314, 316–317, 320, 323, 325, 333–335, 339–340, 357–358, 363, 364, 368–370, 372–373
  bait, 353
  communication, 66, 145–146, 150, 153–155, 260, 370
  control, 319
  cue, 118, 168, 309, 316, 321, 335
  defense, 139, 141, 350
  deterrent, 203
  diversity, 148
  ecology, 149, 364
  ecology of fear, 153
Chemosensory sensilla, 39
Chemotaxis, 261, 372
Chewing insect, 309
Chill coma recovery, 294, 364
Chilo partellus, 313, 319
Chinese
  scholar tree, 311
  tortrix moth, 311
Chlorophanus viridis, 38
Choosy sex, 194, 195
Choristoneura fumiferana, 102, 109
Chortoicetes terminifera, 103
Chorus, 175–176, 180
Chroma, 163, 166
Chromophore, 161, 364, 370
Chromosome, 9–12, 19, 69, 337
  inversion, 297
Chrysomelid, 125, 207, 209
Chrysoperla carnea, 184, 285
Chrysopidae, 203, 204
Cicada, 40, 175, 177–178, 180, 182, 185
Cichlid fish, 210
Cimex lectularius, 150
Circadian, 13, 73, 342, 353
  clock, 68, 102, 106
Circuit, 32–34, 36–38, 40–45, 49, 86, 99, 110, 119, 121–122, 126, 281, 364, 367
Cis-vaccenyl acetate, 36, 45
Citizen science, 111, 294, 297
Citrus, 39, 312, 319, 324
  leafminer, 312, 319
Clasper, 196

Climate, 111, 208, 302–303, 325, 350, 368
  adaptation, 295, 297
  change, 2, 111, 250, 294–295, 298–299, 301–303, 324, 332, 371
  shift, 297
  warming, 294, 301–302, 356–357
Climatic factor, 295
Clock-and-compass navigation, 106
Clonal, 5, 106
  offspring, 190
  raider ant, 66
Clustered regularly interspaced short palindromic repeats (CRISPR), 337, 364
CO_2, 302, 324, 342–343
Coccinella semtempunctata, 167
Coccinellidae, 204
Coccoidea, 209
Cockroach, 33, 39, 41–42, 58–59, 119, 185, 196, 205, 209, 269, 283, 313, 325
Codling moth, 311–312
Co-evolution, 165, 298, 372
Co-evolutionary, 151, 163
  arms race, 153
  relationships, 116
Co-evolve, 86, 103, 298
Co-foundress ant, 215
Cognition, 2, 70, 257–258, 268–269, 365
Cognitive, 72, 92, 257–258, 268–269, 359
  bias, 269, 365
  specialization, 268
  trait, 257–258, 267–269
Co-gradient adaptational pattern, 295
Coleomegila maculate, 285
Coleoptera, 8, 15, 19, 22, 99, 146, 169, 182, 206–207, 209, 222–223, 225, 231, 277, 309, 314, 320
Cold place, 216
Colletidae, 224
Colony, 51, 58, 64, 66, 70, 72, 92, 125, 140, 149, 220, 225–226, 228–229, 231–232, 243–246, 248, 250, 266, 268, 277–279, 364, 371
Color(our), 17, 38, 120, 131, 133, 137, 158–159, 161, 163–166, 169, 259, 262, 264, 266, 268, 299–300, 310, 351–352, 364, 368, 373
  attractant, 352
  patterns, 17, 130, 133, 135, 166, 168–169
  polymorphism, 300
  vision, 38, 365

Color (colour)-polymorphic aposematism, 169
Commensal, 116, 368
Common silkmoth caterpillar, 140
Common garden, 22, 297, 300, 365
  approach, 23
  experiment, 6, 21, 287, 365
  laboratory, 294
Communal, 98, 220, 223–224, 227, 229, 232, 370
  aestivation sites, 107
  group, 223, 227, 230
  sweat bee, 227
Communally cooperative, 223–224, 227, 229
Communication, 17, 39, 45, 66, 110, 145–148, 150, 153–155, 157, 160, 162–163, 165–170, 174–186, 194, 231, 260–261, 265–266, 316, 323–325, 363, 365, 370, 373
Community ecology, 250, 302, 368
Comparative, 32–35, 59, 92, 142, 211–212, 302, 372
  approach, 45, 140, 215
  method, 130
Compensatory feeding, 124
Competition, 57–58, 80, 82, 84–86, 92, 124, 125, 148, 151, 301, 334, 339, 364
Competitive disruption, 315
Competitor, 21, 22, 80–82, 84, 90, 149, 150–151, 198, 204, 210, 250, 262, 348, 355, 365, 367
Complex life cycle, 348
Computational ethology, 119
Composition, 38, 117, 119, 121, 125, 245, 246, 249–250, 264, 269, 350, 367
Compound eye, 38, 160–162, 258, 370
Condition, 6, 42, 53–54, 56, 64–65, 67–69, 71, 73, 84, 87, 104–106, 126, 132, 138–139, 150–151, 160, 163–166, 178–181, 192, 194, 199, 203–205, 207–208, 216, 229, 239, 242, 246, 248, 266, 293, 297, 299–302, 310, 319, 337–338, 341, 347–354, 357, 359, 363, 366, 368, 373
Conditioning, 42, 268
Conflict, 25, 57–59, 122, 133, 149–150, 166, 192, 196, 229, 239, 247, 364, 365, 372
Confusion over residency, 90
Conifer, 311
  volatile kairomone, 312
Connectivity, 32, 34, 40, 350, 355–356, 365

Connectomics, 34, 365
Conopidae, 283
Conservation, 2, 68, 100–111, 125, 319, 348–350, 359, 373
  measure, 356–357
Conspecific, 21, 38–39, 45, 57–58, 80, 82–83, 86, 89–90, 124, 146, 150–151, 163, 166, 168, 176, 178, 205, 207, 210, 215, 220, 238–240, 243, 263, 269, 278, 281–282, 315, 321, 363–364, 368, 373
Conspicuous appearance, 135
Conspicuousness, 137, 141, 166
Consumptive effects, 153
Contact chemical, 34
Contamination, 126, 336
Contest, 58–59, 81–82, 84–93, 243, 248, 364, 372
  competition, 86
  duration, 81, 85
  success, 88
Control, 2–3, 7, 11, 32–34, 36–43, 45, 49–53, 55–58, 72, 88–89, 102, 104, 106, 117, 119, 121–122, 124–126, 135, 140, 141, 147, 155, 176, 194, 196, 208, 250, 261–262, 264, 267, 275, 284, 309, 310, 313, 315, 316–322, 324–325, 332–343, 356, 358, 365, 368, 370, 372
  program, 317, 332–335, 343, 365
  strategy, 310–311, 313, 316, 319, 323, 325, 333–343, 371
Cooperation, 149, 168, 214–215, 219–222, 227, 229–230, 232, 244, 365
Cooperative, 215, 220–224, 226–232, 366
  behavior, 195, 220–221, 229–230, 358, 363
  sociality, 220, 223–224
Coordinated behavior, 220
Copulation, 8, 10, 43, 80, 83–84, 168, 176, 189–190, 192, 194–199, 209, 373
  duration, 8, 15, 19–20
  latency, 10
  occurrence, 10
Corbiculate, 224
  Bee, 220, 223–224, 228
Corn, 12, 106, 195, 285, 312, 320, 325
Cornea, 160
Corpora
  allata, 50–51, 53, 55, 58, 356, 368, 371
  cardiaca, 50, 122, 280, 365, 371

Corridor, 83, 350, 355, 357, 366
Cosmopolites sordidus, 311
Cossus insularis, 312
Cost, 9, 14, 71, 81, 84–86, 92, 111, 122,
    124, 132, 133, 138, 140, 146–147,
    150–151, 153, 175, 181–182, 189,
    191, 197, 205, 210–211, 214–216,
    221–222, 229, 232, 237, 245, 260,
    265, 273–274, 277, 278–279,
    281–283, 285, 287, 301, 313–315,
    319, 337 339, 342, 353, 363–367,
    370, 373
Costa Rica, 322
Cotesia, 319
    glomerata, 260
Cotton, 311, 319–320, 322
    bollworm, 320
Counter-gradient adaptational
    pattern, 295
Courtship, 4, 8, 10, 12, 19–20, 43–45,
    49, 55, 70, 164, 176–177, 195,
    199–200, 212, 263, 276, 278, 281,
    282, 302, 324, 335, 359, 369
    display, 4, 22–23, 43, 176
    duration, 8, 19–20
    latency, 10
    occurrence, 10
    signals, 194–195
    song, 8, 10–12, 15–17, 22–23,
        43–44, 176
    song structure, 8
    wing display, 10
Covariance, 4, 13–14, 21, 24, 102, 221,
    363, 367
Cowpea
    seed, 20
    seed beetle, 8
    weevil, 195, 197
Crematogaster, 134
Cricket, 7, 8, 10, 12, 15–17, 21–24,
    31, 33, 38, 42–43, 64, 84, 102,
    124–125, 148, 150, 174–178,
    180–182, 192, 197–199, 224, 237,
    239, 243–244, 249, 279–280,
    285–286, 371, 373
Criotettix japonicas, 139
CRISPR/Cas9, 34
    system, 66
    technique, 365
Crithidia bombi, 287
Critical thermal
    maximum, 294, 365
    minimum, 365
Crop pest, 2, 309, 316
Cross, 7, 18–19, 24, 34, 110, 135, 137,
    141, 236, 238–239, 247–248, 293,
    350, 351, 369

Cross-generational, 293–294
Cross-sectional sightings, 294
Crustacean, 208, 280
Crypsis, 130, 132–134, 137–138, 168, 365
Cryptic, 88, 111, 130–131, 133–134, 137,
    159, 168, 199, 251, 299, 322, 352,
    female choice, 198–200, 365
Cryptocercus punctulatus, 209
Cryptotermes secundus, 226
Cuckoo bee, 279
Cue, 23, 33, 39, 41, 43, 57, 64, 67–68,
    87–88, 101, 106–107, 108, 118,
    120–121, 145–146, 148, 154–155,
    160, 165–166, 168, 175–176, 178,
    181, 239, 241, 243, 258, 264, 266,
    268, 274, 281, 301, 309–311, 313,
    316–317, 321–324, 335, 349–350,
    358, 364–366, 369, 371
Culex pipiens, 67
Cumulative assessment, 87, 365
Curculionidae, 209, 214, 225, 230
Currant, 312
    clearwing moth, 312
Cuterebra austeni, 89
Cuticle, 50–52, 71, 130, 176, 183, 197,
    279, 366
Cuticular hydrocarbon, 4, 8, 22, 24,
    150, 279, 358
    production, 10
Cutter bee, 224
Cyanocitta cristata, 133
Cycloptiloides canariensis, 197
Cydia molesta, 311
    pomonella, 311–312
    trasias, 311
Cyphoderris strepitans, 150
Cyrtodiopsis dalmanni, 89, 167
Cytoplasmic incompatibility, 337, 365

D

Damselfly, 83, 85, 90, 159, 166, 169, 195,
    199, 243–244, 260, 277, 301, 357
Damson-hop aphid, 312
Danaus plexippus, 41, 53–54, 68, 99,
    169, 277, 356
Darwin, 139, 191, 257
Dasysyrphus venustus, 136
Dead leaf mantis, 167–168
Dear enemy effect, 86, 92, 365
Death feigning, 131, 139, 365
Death-grip, 230
Deception, 152, 153, 159, 165, 372
Deceptive, 153, 160, 372
    signal, 150, 152
Decision, 42, 64, 80–82, 93, 121–122,
    140, 151, 210, 212, 257, 260,
    268–269, 338, 353–354

making, 81, 231, 241, 258
Decorated cricket, 8, 15, 17, 21, 22
Defense (defence), 56–57, 71, 80, 82,
    84–85, 89, 92, 130–133, 135,
    137–142, 160, 169, 194, 197, 205,
    208, 210, 215, 227, 229–230, 232,
    239, 245, 246, 274, 281, 319, 321,
    350, 357–358, 363, 365–366, 369,
    372–373
Deilephila elpenor, 259
Deimatic, 365
    display, 138, 365
    signal, 169
Delia antiqua, 313
Dendritic region, 37
Dendroctonus, 312, 321
    ponderosae, 311, 313
    pseudotsuga, 313
Density, 53, 65–66, 72, 84, 105, 108,
    153, 179, 204, 211–213, 297, 310,
    313, 336, 339
Densoviruses, 336, 365
Depreciable care, 214
Dermaptera, 206, 210
Deroplatys, 168, 169
    dessicata, 167
Desensitization, 315, 365, 370
Desert, 85, 110, 112, 214, 216, 363
    ant, 38, 118
    arthropod, 215
    locust, 35–36, 41, 99, 106, 109, 283
Desertion, 211
Desiccation, 24, 203, 205, 208, 210, 213
Desmodium, 319, 320
Detectability, 130, 134, 137, 140, 146,
    151, 163
Detection, 38, 90, 131, 133–134, 138,
    148, 149, 165, 168–169, 179–180,
    182–183, 186, 243, 278, 286, 297,
    351, 353, 354, 365
Deterrent, 118, 123, 140, 203, 310, 313,
    316–317, 320, 325, 372
Development, 1, 7, 9, 17, 22, 33–34, 37,
    42, 45, 49, 50, 53–54, 56–58, 65–68,
    70–73, 81–83, 87, 98, 109, 116–117,
    121–123, 132, 159, 166, 169, 183,
    193, 204, 209, 212, 222, 229, 231,
    241, 243–244, 258, 262, 269, 276,
    279, 280, 294, 296, 309–310,
    317–319, 321–322, 325, 332–334,
    336, 339–340, 350–353, 358, 359,
    368–369, 373
    in biology, 80
    rate, 295, 296
Developmental rate, 67, 230
Deviance, 18
Diallel, 3, 7, 18

Diamondback moth, 312
Diapause, 54, 56, 65, 67–68, 80, 109,
    121, 292, 294, 300–302, 366,
    368–369
  behavior, 65, 298, 300
  syndrome, 300–301, 365
Diapausing gynes, 68
Diaphorina citri, 312, 324
Diastatops obscura, 89
Dicrocoelium dendriticum, 284
Dictyoptera, 169
Diet, 21–23, 25, 117, 119, 122–126,
    141, 230–231, 240, 250, 262,
    264–265, 282, 367
  choice, 121
  formulations, 124
  niche, 124
Dietary self-medication, 282–283
Digestion, 117, 124, 230
Diminutive moth, 160
Dinocampus coccinellae, 285
Diploid, 5, 149, 153, 368
  insect, 80
Diploptera punctate, 196
Diplura, 206
Diptera, 8, 10–12, 15, 19, 22, 34, 38, 89,
    99, 146, 177, 182, 206, 223, 226,
    277, 314, 316–317
Direct, 14, 21, 41, 49, 102, 133, 148,
    153, 164, 166, 175, 178, 182, 185,
    195–196, 198, 213, 220, 229, 232,
    264, 266, 280–281, 283, 293,
    298, 301, 321–323, 333, 340,
    350, 352, 368
  benefit, 191–192, 195, 221, 366
  sperm transfer, 196
Direct fitness, 221–222, 229, 363,
    366, 368
  benefit, 221, 229, 365
Directionality, 55, 107, 146–147, 175
Disease, 2, 43, 59, 126, 273, 276, 282,
    286, 320, 322–325, 332–334, 336,
    339, 340–343, 368, 371
  resistance, 229
Dispersal, 17, 53–56, 80, 82, 111, 155,
    221–222, 225, 227–228, 232,
    247–248, 250, 292–293, 298–300,
    302, 322–323, 325, 336–337,
    342, 348, 350–351, 354–357, 365,
    366, 371
  behavior, 64, 82, 298, 300, 302
Dispersive
  propagation, 180, 366
  stage, 309
Display, 4, 10, 12, 22–23, 43, 59, 64, 82,
    112, 131, 135, 137–138, 141–142,
    163–164, 166, 168–169, 175, 185,

192–194, 241, 285, 350, 353, 359,
    363, 365
Disruptive, 131, 133, 323
  coloration (colouration),
    140, 366
  selection, 81, 214, 363, 366
Distribution, 3–4, 8, 24, 81–83, 86,
    104, 107, 124, 126, 141, 176, 205,
    211–212, 215, 219, 230, 245, 246,
    248–249, 261–262, 277, 284, 301,
    338–339, 349–351, 353, 357, 359,
    366, 371, 373
Division of labor (labour), 4, 43, 51,
    57, 125, 214–215, 219–220, 225,
    228–229, 232, 243–246, 363,
    366, 372
DNA methylation, 72
DNA methyltransferase, 72
Dominance, 4, 6, 14, 18–20, 25, 49,
    57–59, 69, 228–229, 366
  variance (V_D), 5, 6, 366, 370
Dopamine, 37, 59, 122, 243,
    358, 364
Dormancy, 65, 300, 365
Dorsal rim area, 38
doublesex(dsx), 12, 45, 71
Douglas fir beetle, 313
Downwind direction,
    107–108, 373
Dragonfly, 33, 83, 85, 88, 98–99,
    133–134, 160–162, 169, 260,
    293, 299, 303, 348, 354,
    357, 359
Drosophila, 1, 7, 11–12, 21, 33–34,
    36–37, 39–41, 45, 55, 67, 69,
    71–72, 118–119, 121–122,
    125–126, 150, 160, 176–177,
    184, 192, 199, 262, 276, 281,
    294, 299, 337, 342
  C Virus, 281
  bifurca, 190
  elegans, 10, 12
  gunungcola, 10, 12
  littoralis, 12
  melanogaster, 3, 8, 10–12, 14–15, 22,
    33–36, 43–44, 65, 67, 70, 72, 117,
    123, 177, 184, 195–196, 237, 242,
    245, 281, 299, 301
  mojavensis, 23
  montana, 192
  nigrospiracula, 15
  pseudoobscura, 263
  sechellia, 10
  serrata, 8, 24
  simulans, 8, 10, 22
  tripunctata, 19–20
  virilise, 10, 12

Drosulfakinin, 122
Drumming, 178
Dryococelus australis, 359
Dung beetle, 8, 71, 126, 198–199,
    204–205, 208
Dusky sap beetle, 312
Dyadic contest, 81, 86
Dynamic-state model, 85

E

Ear, 174–175, 181–186, 364
Early life, 8, 15, 69
Earth's magnetic field, 108
Earwig, 57, 207, 209, 269, 302
East Asia, 109
Eastern, 98–99, 109, 260
  Australia, 25, 107, 351, 355
  spruce budworm moth, 109
  sword-bearing katydid, 194
  tent caterpillar, 140
Eavesdropping, 146, 147, 151, 154,
    181, 366
Ecdysis, 52, 244, 366
  triggering hormone (ETH),
    52–53, 366
Ecdysone, 49–51, 53, 366, 371
Ecdysteroids, 49, 51, 58–59, 71
Echolocation, 33, 179, 183
Eclosion, 52–55, 83, 102, 151, 196,
    212, 266
  hormone (EH), 52–53, 266
Ecological, 2, 38, 63, 86, 100, 123,
    140, 150–151, 155, 169, 180,
    185, 191, 194, 219–220, 222,
    229, 231–232, 236, 240, 241,
    247, 249–251, 277, 280, 286,
    317, 319, 321–322, 348–349,
    350, 353, 356–359, 363, 371
  interaction, 349, 355, 359, 364
  niche, 160, 180
  restoration, 353, 366
  trap, 300, 354–355, 366
Economic defendability
    (defensibility) principle, 84,
    189, 366
Ecosystem, 116–117, 123, 125–126,
    368
  dynamics, 125
  service, 250, 303
Ectoparasite, 17, 248, 276
Ecotoxicology, 250
ECR, 49, 366
Ectoparasite, 17, 248, 275
Ectotherm, 283
Edaphic beetle, 352
Edith's checkerspot butterfly, 8
Efferent nerve, 36, 366

Egg, 8, 19–20, 42, 50, 53–58, 68, 71, 81, 100, 121–123, 125, 149–152, 189, 190–195, 198–199, 203–213, 215, 221, 227–229, 246, 260, 267–268, 277, 280, 284, 296, 299, 320, 336, 363, 365, 368–370
　coating, 208
　stage, 212–213, 215
Egg-laying, 58, 100, 122, 211
Eggfly, 164
Egyptian leafworm, 282
Ejaculate, 150, 198–199
Elaphrothrips, 212
Electrical, 37, 119
　activity, 34, 366, 372
　properties, 42
Electrophysiology, 32, 366
Electroshock, 42
Elementary motion detector, 38
Elevated glycogen store, 67
Ellipsoid body, 40–41
Eltham cooper, 352, 356
Embiidina, 204
Embryo, 190, 205, 209, 337
E-methylgeranate, 57, 366
Emotion, 98, 269, 363, 366–367, 370
Empoasca fabae, 106
Enallagma, 260
Enchenopa binotata, 22, 194
Encounter, 39, 43, 54, 82–84, 86, 89–91, 131–133, 135, 147, 150–151, 175, 243, 248–249, 259, 261, 274, 338, 350, 367, 369
Endemic, 333–334, 349, 352, 356, 366
Endurance test, 85
Energy reserve, 88
England, 107, 229, 294, 357, 359
Enteroendocrine cell, 121
Entomopathogenic fungi, 283, 314
Environmental, 5–6, 11, 23, 25, 43, 49, 51, 54, 58, 64–65, 68–70, 72–73, 102–103, 105, 109, 111–112, 117, 125, 126, 146–147, 160, 165, 194, 203, 205, 210, 215, 241, 243, 247, 257–258, 262, 284, 293–297, 299–302, 309–310, 313–314, 322, 334, 348–349, 353, 363–366, 369, 371–373
　change, 49, 73, 111, 262, 264, 292–293, 297–303, 348, 354, 359, 371
　variance (V$_E$), 5, 21, 366–367, 372
　variation, 5, 20, 63
Enzyme, 68–69, 71–72, 153, 164, 335, 340, 365
Ephemeral, 109
　aggregation, 220, 223–224
　resources, 210

Ephemeroptera, 206, 223
Epigenetic, 64, 72, 73, 366
Epiphyas postvittana, 311
Epistasis, 4, 14, 18–20, 25, 370
Epistatic variance, 5, 18–20, 25
Epistrophe
　emarginata, 136
　grossulariae, 136
Eradication, 314–316, 318, 336
Erectile organ, 190
Eristalis, 136
Ermine moth, 300
Escape response, 33, 37
Escherichia coli, 261, 287
Ethodiversity, 348
Ethogram, 335, 366
Euchromatin histone methyl transferase, 72
Euglossini, 195
Eulerian approach, 109
Eunica bechina, 204
Eupeodes, 136
Euphydryas edita, 8
Euploea, 355
Eurema, 82
European corn borer, 12
Eurosta solidaginis, 19–20
Eusocial, 57, 67, 72, 162, 205–206, 219–220, 222–225, 227–232, 246, 366
　insect, 32, 43, 66, 70, 72, 125, 140, 215, 219–220, 222, 229–231, 244, 250, 278–279, 282
　insect species, 39, 58
　wasp, 65, 72, 222
　system, 229, 245
Eusociality, 40, 219, 223–224, 227–231, 244, 246
Evapotranspiration, 208
Evolutionarily stable strategy, 81, 367
Evolutionary, 1–3, 5, 13, 21, 23–26, 43, 49, 58, 68, 70, 80–81, 87, 92, 99–100, 116, 118, 124, 126, 141, 148, 151, 153, 163, 165, 169, 182, 189, 193, 197, 199–200, 209, 211–215, 227, 236, 250–251, 258, 260, 269, 274–276, 280–281, 286–287, 292–294, 298, 303, 351, 353, 363, 368–369, 371–373
　game theory, 80–81, 367
　radiation, 80
　response (R), 7, 14, 24–26, 292–294, 298–301, 303
　time, 86, 153, 180, 214, 275, 298, 353
　time scale, 64
　trap, 366
Excitability, 42, 101
Exoskeleton, 52, 159–160, 204

Exploratory, 101, 250, 371
　climbing, 41
　walking, 41
Explosive penis, 190
Extatosoma tiaratum, 134
Extended parental care, 222–223, 227, 229–232, 373
External, 32, 45, 117–118, 206, 209, 212, 258, 262, 280, 357, 368, 370
　fertilization, 211
　mouthparts, 118
Exteroreceptor, 371
Extinction, 100, 246, 353, 367
Extirpation, 349, 367
Eyespot, 51, 132, 137–139, 141, 152, 169, 367

F

Facultative, 135, 137, 228
　sex-role reversal, 82
Facultatively eusocial, 220
Fall armyworm moth, 109
False eyespot, 152
Family sums, 18
Fat body, 50, 53, 122, 366
Fava bean, 313
Fear, 98, 125, 153, 248, 269, 367
Featherwing, 37
Fecundity, 3–4, 8, 14–17, 132, 192, 197, 211, 213, 338, 367, 369
Feeding, 11, 33, 39–40, 42, 53, 55, 57, 66, 67, 100, 116–119, 121–126, 154, 180, 197, 204–205, 210, 212–213, 220, 223, 226–228, 264–265, 276, 282, 286, 309, 314, 317–323, 325, 340, 348, 350, 352, 356, 363, 369, 370–372
　decision, 121, 122
　guild, 302, 309
Female, 3–4, 6–8, 10, 12, 14–26, 43–45, 53–57, 67, 71, 80–84, 88–89, 102, 121–123, 125, 134, 147, 149–151, 153, 155, 164–166, 168, 174–178, 182, 185, 189–200, 203–215, 219, 221, 226–227, 229, 237–239, 243–244, 247, 263–264, 267–268, 277–278, 283, 285, 296–297, 299, 302, 314–319, 321–322, 324, 336–338, 340, 351, 355, 357–358, 363, 365, 368–373
　care, 210–211, 213–216
　choice, 198–200, 212, 302, 365
　choosiness, 22, 200
　desertion, 211
　experience, 247, 263
　fecundity, 192, 197, 213
　genitalia, 190, 196
　mate discrimination, 10

preference, 10, 22, 122, 211
reproductive tract, 122, 192, 195, 197–199, 205
sensory system, 195
sex pheromone, 311–312, 319–322
sperm storage, 190
Fermented food, 311
Fertility, 315, 335
Fertilization, 82, 84, 126, 153, 190–191, 195, 197–199, 203, 211, 365, 369, 372
Fibre tract, 32, 367, 370
Fidelity, 84–85, 87, 135, 356
Field cricket, 7–8, 12, 15–17, 22–24, 176, 181–182, 199, 237, 239, 243
Fight, 32, 58, 71, 166, 168, 191, 229–230, 280, 283
Fighting, 15, 17, 81, 86, 93, 193, 281–282
ability, 8, 86, 166, 192
back, 131, 133, 139
Fir, 99, 313
Fire ant, 66, 125, 250, 278, 356
Firefly, 163–164, 166–167, 192, 364
Fish, 133–134, 210–212, 215, 260, 286, 364, 373
Fission-fusion society, 227
Fitness, 3, 18, 23, 39, 65–67, 71–72, 80–81, 86, 93, 103, 126, 132, 149, 153–154, 159, 182, 192, 196–197, 209, 220–222, 228–229, 239, 242, 248, 250, 257–259, 261, 264–265, 268–269, 274, 278, 281–284, 301–302, 336–338, 358, 363, 365–370, 372–373
benefit, 181, 191, 197, 220–222, 229, 281, 284, 365
component, 302, 367
Flagella, 262
Flagship species, 351, 358
Flatwing, 12, 24
mutation, 24
Fleeing, 130–131, 133, 135, 137–139
Flexible parental care
compensation, 363
FLIC, 119
Flight, 17, 33, 37–38, 42–43, 53–56, 68, 85, 89, 99–103, 105, 107, 109, 111–112, 134, 139–140, 158, 163–164, 183, 229, 231, 247, 266, 277, 280, 298–300, 316, 320, 324, 342, 351–352, 354–356, 359, 363, 367, 370
ability, 293, 336
boundary layer, 104, 367
fuel, 102, 108
initiation distance, 138
muscle, 105, 140, 300

network, 42
path, 109, 134, 159, 367
phenology, 111
trajectory, 109, 367
Flightless, 102, 124
Floral, 38, 165, 266, 268
kairomone, 312
scent kairomone, 311
Florida, 107, 338
carpenter ant, 43
Flour beetle, 15, 17, 22, 139, 199, 237, 244, 284
Flower, 153, 158–159, 165, 195, 248, 259, 261–262, 266–267, 284, 287, 303, 313, 356, 372
Fluorescence, 358, 367
Fluorescent protein, 33, 34
Fly, 1, 8, 10, 12, 15, 19–20, 22–24, 32–33, 35, 37, 40–41, 43, 45, 54–55, 65, 67, 69–70, 72, 83, 85, 99, 106–107, 116, 118–119, 121–124, 126, 130, 133, 136, 145, 151–152, 158–160, 165–168, 176–177, 181–184, 186, 195, 199, 205, 209, 223, 237, 242, 262–263, 267–269, 276–278, 281–283, 286, 294–297, 299–302, 311–313, 316–318, 325, 337, 351, 355–357, 359, 364, 371
Fly dance, 118
FlyPAD, 119
Follicle, 50, 190
Food, 40–41, 53–54, 56–57, 65, 67–71, 116–125, 147, 152, 158, 165–166, 190–191, 194–195, 203–206, 208–210, 212, 222, 227–232, 240, 245, 248–249, 261, 264–265, 267–268, 278–279, 281–283, 286, 297–300, 311–312, 317–318, 325, 339, 348, 350, 352, 355, 358–359, 367–368, 370, 372–373
availability, 65–67, 69, 71, 247, 296
bait, 311, 313, 317–318
bait station, 311
chain length, 125
choice behavior, 124
choice, 116–117, 121, 154, 267
collection, 125
deprivation, 65, 69
intake, 69–70, 116, 118, 121–122, 281–282, 301
items, 86, 119, 192, 278, 298
location, 309
odor, 310, 317, 319
quality, 54, 230, 266
sharing, 229–230
stimuli, 117
web, 248, 302, 356
web dynamics, 125

Foot shock, 40
for gene, 12, 242
Forager, 43, 57–58, 64–66, 70, 72, 116, 151, 230, 258, 261, 266, 268, 301, 303, 373
Foraging, 4, 15, 32, 37, 39–41, 58, 65, 69–70, 82, 99, 101, 118, 121, 137, 139, 147, 149, 151, 155, 160, 168, 194–195, 223–224, 227, 229, 230, 232, 236, 238–239, 240, 242, 244–247, 249–250, 261, 266, 278, 281, 286–287, 294, 300–303, 309, 323, 324, 350, 363, 369, 372
bee, 259, 267
behavior, 12, 58, 70, 124–125, 197, 248, 281, 301
decision, 64, 151
efficiency, 192, 278
gene, 69–70
Forest tent caterpillar, 125
Forestry, 302
Forficulid earwig, 209
Formica podzolica, 279
Formicidae, 222, 224
Fortress defense (defence), 230
Foundress behavior, 68
Foxglove aphid, 154, 155
Fragmentation, 298, 302, 350, 371
Frankliniella occidentalis, 313
Frequency-dependent, 81
Freshwater habitats, 80
Fruit, 109, 124, 151–152, 190, 311–312, 314, 317–319
fly, 8, 10, 12, 15, 19, 22, 24, 33, 35, 70, 72, 119, 124, 159–160, 177, 183–185, 195, 237, 242, 262–263, 267–269, 276, 281–283, 311–312, 316–318, 325, 337
tree, 311
fruitless(fru), 12, 43, 45, 70
sex determining gene, 12
Full-sibling, 6, 14
analysis, 6, 14
design, 6
Fungal, 225, 280, 283, 321
attack, 213
infection, 207–208, 279–280
Fungi, 205, 208–209, 212, 228, 279, 283, 284, 314, 321
Fungus, 225, 227, 230, 277–278, 280, 282
Fungus-infected grape, 278
Fungus-tending ambrosia
beetle, 225

G

G9a/EHMT, 72
GABA, 37

GAL4/UAS transcriptional system, 34
Gall-forming aphid, 222–223, 230
Game theory, 80–81, 364, 367
Gamergate, 58, 229
Gamete size, 81
Game-theoretic modeling (modeling), 86, 91
Ganglion, 34–36, 39, 51–52, 55, 257, 284
Gap-climbing, 39
Gape-limited predator, 139
Gas exchange, 209
Gene, 2–5, 7, 9–13, 15, 17–20, 23–25, 34, 43, 45, 50–51, 63–64, 66–73, 102–103, 121–122, 161, 192, 200, 242, 245–246, 262, 273, 280, 284, 297–298, 324, 336–337, 342, 363–367, 373
    expression, 19, 34, 50–51, 64–66, 68, 70, 72–73, 367
    frequency, 19, 293
    network, 73
    transcription, 68
Gene-by-environment interaction (G×E or GEI), 4–5, 11, 20–23, 25, 64–65, 69–70, 72, 294, 367
General environmental variance (V_{Eg}), 5, 21, 366–367
Generalist, 66, 230, 250, 287, 348
Generation, 2, 3, 5–7, 9, 13–14, 18, 26, 64–65, 82, 99, 110, 130, 139, 162–164, 178, 198–199, 220, 228, 244, 247, 265–267, 279, 292–294, 296, 300, 303, 321, 323, 336–337, 353, 366–367, 370–373
    time, 3, 7, 72, 189
Genetic, 1–9, 11, 13–14, 17–18, 21, 23–25, 33–34, 39, 51, 63–71, 100, 102–103, 117, 121, 125–126, 149, 200, 221, 229, 241–242, 245, 273, 281, 284–285, 293–294, 295, 297–298, 301–303, 335–337, 343, 363–364, 367–369, 371–372
    architecture, 3–4, 6, 11, 23–25, 213, 262
    basis, 2–4, 6–7, 18, 21, 26, 245, 261
    correlation, 13–15, 17, 25, 71, 367, 373
    engineering, 39, 316, 333, 335
    manipulation, 316, 323
    marker, 9, 11, 297
    network, 262
    response, 292, 294
    variance (V_G), 3–5, 7, 13–14, 18–20, 24–25, 363, 367, 370
Genetically homogenous organism, 63

Genetically-modified crop, 111
Genetic-linkage map, 9
Genital, 8, 190, 195–199, 373
    organ, 190, 199
    shape, 15, 17
Genome, 3–4, 9–12, 25, 34, 41, 45, 64, 66, 72–73, 100, 103, 281, 366, 367, 372
    editing, 34, 364–365, 367
Genome-wide association study (GWAS), 9, 11–12, 367
Genomic, 3–4, 9, 11, 17, 23–25, 297–298, 367, 371
Genotype-by-social environment, 22
    interaction (GSEI), 4, 21, 23, 25–26, 367
Geometric framework, 117, 122–125, 367
German cockroach, 269, 313
Ghost of Lamarck, 72
Giant sperm, 199
Glia cell, 34, 37
Global change, 2, 126
Glomerulus, 36, 39, 367
Glucose, 10, 262
Glutamate, 37
Glyptapanteles, 285
Gnathal ganglia, 34–35, 39
Gnathotrichus sulcatus, 311
Gnatocerus cornutus, 15, 17
Golden egg bug, 199
Golden sun-moth, 351
Goldenrod, 20
Golgi method, 32–33, 367
Good gene, 17
    model, 192, 200
Gp-9, 66
Grammia incorrupta, 283
Gram-negative bacteria, 279, 369
Gram-positive bacteria, 279
Grape, 278, 312, 323–324
    root borer, 312
Grapevine, 324
    leafhopper, 323
    moth, 278
Graphium sarpedon, 38
Grapholita molesta, 312
Grass, 125, 154, 237, 284, 312, 319–320, 322, 351
Grasshopper, 33, 53–54, 106, 125, 134, 138–139, 147, 154–155, 159 162, 168, 175–176, 185, 224, 264, 264, 283, 286, 293, 299, 301–302, 357, 372
Gravity current, 105
Grazer, 155
Great southern white butterfly, 107
Green-headed ant, 125

Greenhouse gas, 324
Gregarious, 66, 72, 137, 159, 266–267, 282
Gregariousness, 137
Grooming, 4, 228, 275, 277, 279, 282
Ground, 1–2, 80, 99, 104–108, 111, 133, 168, 227, 230, 301, 320, 322, 350, 351, 352, 355, 367
    cricket, 15, 192
    plan, 34–35, 222
    speed, 104–105, 363, 367
    track, 105, 367
Group defense (defence), 227, 230, 232
Group-living, 220, 222, 225–226, 367
Growth, 43, 52, 68, 71, 100, 101–102, 194, 201, 211, 232, 244, 262, 264–265, 279, 280–281, 283, 294–295, 319, 322, 334, 336, 358, 367, 369, 372
Gryllodes sigillatus, 8, 15, 17, 21–22, 197
Gryllus
    bimaculatus, 8, 176, 197
    firmus, 8, 15, 17
    integer, 237, 239, 243
    texensis, 280
Gustatory, 121, 317
    neuron, 44–45, 372
    receptor, 121
    response, 197
    responsiveness, 121
    sensilla, 39
Gut-brain neural circuit, 121
Gryllodes sigillatus, 8, 15, 17, 21–22, 197
Gynaephora groenlandica, 302
Gynosome, 190
Gypsy moth, 278

H
Habitat, 41, 54–55, 80, 82–84, 100–103, 109–112, 124, 131, 134, 142, 163–164, 166, 179–181, 210, 216, 219, 238, 246, 260, 293, 301, 315, 336, 339–340, 348–350, 355, 357–359, 365–368, 371–372
    change, 101, 355
    choice, 80, 292–293, 298, 367
    fragmentation, 298, 302
    loss, 126
    selection, 80, 92, 133, 299, 301, 367
    specificity, 82
Habituation, 315, 323–324, 365
Haemolymph (hemolymph), 42, 50–51, 54, 56, 121, 185, 209, 280
Half-sibling, 3, 6, 14, 17–18, 20–21
Halictidae, 222, 224

Halticoptera rosae, 151–152
Handicap hypothesis, 166
Haplodiploid, 149, 153, 212, 213, 219, 229, 368
Haploid DNA, 81
Haplothrips, 212
Harassment, 357
Harlequin bug, 169
Harmful microbe, 281
Harpegnathos, 43
    saltator, 39, 58
Harsh environmental condition, 205, 293
Hasarius adansoni, 139
Hawaii, 24
Hawaiian island, 24
Hawk-dove game, 81
Hawkmoth, 14, 38, 52, 259
    caterpillar, 141
Hearing, 133, 174–175, 179, 181–186, 258, 369
Heat,140, 166, 264, 299–300, 373
    hardening, 299
    shock response, 294
Heat-seeking behavior, 283
Hedgehog limb developmental pathway, 71
Heliconiines, 83
Heliconius, 17, 83, 135, 196
    charithonia, 83
    cydno, 17, 135
    erato, 135
    melpomene, 135
    pachinus, 17
    sapho, 135
Helicoverpa, 311, 319
    moth, 313
    armígera, 103, 320
    punctigera, 103, 320
    zea, 106
Heliothis subflexa, 12
    virescens, 312
Helophilus fasciatus, 136
Helper, 221–222, 228, 231
Hemideina maori, 194
Hemileuca lucina, 229
Hemimetabola, 50
Hemimetabolous, 83, 211–213, 215, 231, 368
    eusocial insect, 231
    insect, 42, 211, 213, 222, 230–231, 244, 368
Hemipepsis ustulata, 85, 89, 91
Hemiptera, 15, 22, 102, 169, 182, 205–207, 209, 212, 215, 222–223, 226, 231, 322–323
Herbicide, 111, 250

Herbivore, 148, 150, 155, 302, 349, 350, 358, 372
Herbivore-induced, 317
    defensive compound, 316
    plant volatile, 316
Heritability (h^2), 5, 7–8, 13, 14, 20, 69, 368
Heritable, 111, 139, 247, 300–301
    component, 292, 294
    variation, 63, 258, 269, 368
Hermaphroditism, 190
Hetaerina americana, 166
Heterospecific, 38, 80, 205, 239, 263, 343, 368
Hexamerin protein, 68
High reproductive rate, 189
High-cellulose diet, 230
High-frequency sound, 148
High-order interaction, 25
Hilltop, 83–85, 89, 357
Hindgut fluid, 209
Hippoboscidae, 205
Histone modification, 72
Hive, 58, 64, 66, 70, 140, 158, 177, 266, 267, 279, 283, 303
Hold courtship behavior, 20
Holometabolous, 41–42, 50–51, 196, 210–211, 213, 215, 230–231, 244, 278, 348, 368
    group, 212–213
Home range, 100–101, 371, 372
Homeostasis, 119, 125, 301–302, 363, 368, 373
Homing behavior, 299, 368
Homozygous, 6
Homoptera, 12, 119, 178
Honest signal, 57, 150–152, 191
    of unprofitability, 135
Honeybee, 3, 11–12, 15, 39–40, 43, 57–58, 64–67, 70, 72, 140, 154, 159–161, 165, 167, 177, 194, 220, 222, 225, 228, 229–231, 243, 245, 258, 266, 268, 281–283, 373
    drone, 190
Hopkins host-selection principle, 42
Hop, 312
Hormone, 2, 42, 49–59, 67, 68, 71, 122, 153, 242, 280, 336, 365–366, 368–369, 371
Horn, 39–40, 44–45, 71, 191
    development, 71
Hornet, 140, 250
Hornless male, 71
Horsehair, 285, 286
Host, 19, 20, 23–24, 42, 53, 55, 71, 82, 101, 108–109, 111, 116, 124, 140, 146, 148–152, 182, 190, 248, 260, 273–275, 277–287, 299, 302,

316–318, 320–321, 324–325, 337, 340, 342, 350, 352, 356, 358–359, 364, 368, 370, 371, 372, 373
    behavior, 273–275, 278, 281, 283–284
    ecology, 274, 278
    manipulation, 274–275, 281–286, 368
Host-finding, 309, 324, 358
Host plant, 19, 22, 39, 54, 83, 105, 108, 111, 118, 154–155, 180, 203–205, 209, 260, 310, 313, 314, 316–317, 319, 321–322, 324, 349–350, 352, 363
    choice, 68
    kairomone, 311
    location, 312, 316, 321, 323, 324
    odour, 310, 316, 319, 322
    volatile kairomone, 311
Hot weather, 299
Hotspot, 351
Hours of light, 65, 371
Housefly, 19–20
Housing photoreceptor neuron, 38
Hoverfly, 38, 130, 135–136
Hue, 158, 163–164, 368
Human disturbance, 126
Hydrocarbon, 4, 8, 10, 22, 24, 148, 150, 279, 358
Hydrogen, 140
Hydrolysed protein, 317
Hydrophilidae, 209
Hydroquinone, 140
Hymenoptera, 8, 10, 15, 19, 34, 89, 92, 99, 136, 140, 146, 150, 177, 182, 191, 205, 206, 207, 219, 222–225, 228–231, 269, 277, 356
Hymenopteran parasitoids, 190, 299
Hymenopus coronatus, 159, 165
Hypolimnas bolina, 84, 89, 92, 164
Hypomecis roboraria, 134

I

Idolothips, 212
Illness-induced anorexia, 124, 126
Immobile, 139, 166, 176, 293
Immune, 71, 197, 273–274, 280–281
    deployment, 274
    challenge, 72, 280, 372
    function, 17, 124, 280
    mechanism, 274
    priming, 280, 368
    response, 124, 165, 168, 280–282, 369
Immunity, 14–15, 17, 24, 124, 274, 280–282, 287, 364
Immunocompetence, 278
Immunological, 273

Immunology, 273
Immunopathology, 274
In situ adaptation, 292
Inachis io, 138
Inbred line, 6–9, 14–15, 18, 21–22
Inclusive fitness, 149, 154,
    220–222, 368
Indirect, 49, 120, 125, 166, 196,
    286–287, 293, 297, 302, 322, 368
  benefit, 191–192, 194–195, 368
  fitness, 221–222, 228–229, 281, 368
  genetic effect, 25, 221, 368
  sperm transfer, 193, 196
Individual learning, 265–268
Infanticidal intruder, 215
Infanticide, 193
Infection, 15–16, 71, 124, 197, 207–208,
    273–274, 276, 278–287, 319, 333,
    337–340, 364, 368, 370, 372–373
  avoidance, 274, 277–280, 287, 368
  risk, 273, 277, 279, 287
  tolerance, 282, 368
Influorescence, 358
Informationalist, 145
Infrared, 276
Inhibitory signal, 37, 58, 118
Inka cell, 52, 366, 368
Innate behavior, 264
Inquiline, 248, 368
Insect, 1–4, 7–26, 32–39, 41–43, 49–53,
    55–59, 63–64, 66–68, 70–73, 80–93,
    98–109, 111–112, 116–119,
    121–122, 124–126, 130, 132–135,
    137–142, 146–151, 155, 158–170,
    174–186, 189–191, 193–200,
    203–206, 208–216, 219–220,
    222–223, 225–232, 236–237, 240,
    242–244, 246–247, 248, 250–251,
    257–258, 261–262, 264–269,
    273–287, 292–296, 298–303,
    309–311, 313–317, 319, 322,
    324–325, 333–338, 341, 348–359,
    363–373
  cognitive abilities, 268
  control, 309–311, 315
  crop pest, 309
  damage, 310, 322
  development, 66, 336
  egg, 210
  learning, 63, 261, 264, 268
  life history, 231
  pest, 98, 309, 316
Insecticide, 250, 309–310, 314,
    317–320, 322, 332–333, 340–341,
    343, 364
Insemination, 55, 150, 195–198, 335,
    365, 368, 373

Instinct, 257
Insulin, 67
  pathway, 68
  signaling, 67–68, 71, 73
Insulin-like peptides, 122
Insulin-releasing neurons, 121
Integrated pest management, 250,
    309, 316, 324–325, 333, 368
Interaction, 2, 4, 7, 9, 18–21, 23, 25,
    39, 56, 63–64, 66, 69, 72, 80–81,
    85–88, 91, 103, 108, 116, 119,
    123–126, 150, 153, 163–164,
    175, 177, 192, 207–208, 214–215,
    220–221, 238–240, 243–245,
    247–250, 261, 266–267, 269,
    277–280, 282, 298, 302, 325,
    333, 348–349, 356–359, 364–368,
    370, 372
Intergenerational inheritance, 72
Interloper, 57, 85, 90
Internal, 45, 112, 117–118, 196,
    261–262, 269, 278, 366, 369–371
  egg carrying, 205–206, 210
  fertilization, 199, 211
  map, 106
  orientation sense, 41
Interneuron, 34, 37–39, 42, 45, 368
Interommatidial angle, 160
Intersexual selection, 191
Interval mapping, 9
Intralocus sexual conflict, 25
Intrasexual selection, 191, 196
Intraspecific communication, 39,
    182–184, 186, 363
Introgression, 337, 368
Intruder, 57, 81, 84, 85, 90–91, 100,
    140, 149, 215
Invasive, 66, 250, 338, 355–356
  competitor, 250, 348
  species, 125, 250, 314
Ion chanel(channel), 34, 369
Ips, 321
  paraconfusus, 313
Iridescent, 164, 169
  butterfly scale, 38
  signal, 164
Iso-female line, 3, 6–8, 14, 21–22
Isolation, 13, 33, 168, 176, 282, 298,
    313, 338, 357, 364
Isopod, 210
Isoptera, 146, 222–223, 277
Iteroparous, 132
IUCN, 351

J

Jalmenus evagoras, 357
Jankowskia fuscaria, 134

Japanese, 207, 332, 334, 338
  beetle, 311–312
  giant hornet, 140
  honey bee, 140
  redbug, 190
Jumping spider, 134, 139, 250
Juvenile, 49, 80, 82–83, 165, 213, 222,
    225, 231, 243–244, 247, 302
  growth, 203
  hormone (JH), 49–51, 53–57, 59,
    67–68, 336, 365–366, 368–369, 371

K

Kairomonal attractants, 323
Kairomone, 148, 311–312, 368
Katydid, 138–139, 159, 167, 169,
    175–177, 180, 182–185, 190,
    194, 373
Kauai, 24
Keiferia lycopersicella, 312
Kenyon cell, 33, 40
Ketones, 148
Keystone individual, 246, 368
Kicking, 139
Kin, 168, 219, 266
  aggregation hypothesis, 135
  selection, 135, 219, 278, 282, 369
Kin-based relatedness, 219
Kinship, 219, 266
Kirkaldyia, 213
Kissing bug, 286, 343

L

Lacewing, 184–185, 203–204,
    285, 324
Ladybird beetle, 164, 167, 169,
    285, 301
Ladybug, 130, 204
Lagrangian approach, 109
Lamina, 36, 38
Lampyridae, 166–167
Lampyris noctiluca, 137
Landing, 101, 108–109, 134, 267, 316,
    322, 367
Landscape topography, 357
Large cabbage white butterfly, 260
Larvae, 22, 24, 33, 37, 41–42, 49–51,
    53–54, 56–57, 65–72, 112, 125,
    133–134, 137, 151, 182, 204–205,
    208–209, 212, 227, 229–230, 244,
    260, 278, 282, 283–285, 287, 292,
    302, 317–318, 320–322, 334,
    336–337, 339, 348, 351–352, 355,
    358, 368, 370–371
Larval, 34, 42, 49–51, 54, 56, 57, 67, 69,
    70–71, 111, 209, 223–224, 226,
    229, 231, 242–244, 260, 280,

283–285, 297, 317–320, 322, 334, 336, 339, 348, 350, 352–353, 358
flatworm, 284
gallery, 321
Lasiocampidae, 224
Lasioglossum, 230
zephyrum, 227
Lasiorhynchus barbicornis, 167, 168
Lasius neglectus, 280
Lateral horn, 39–40, 44–45
Latitude, 99, 105, 107, 295, 296, 301
Latitudinal, 110, 247, 294–295, 300
cline, 24, 369
Laupala, 10, 11
Leaf miner parasitoid, 151
Leafcutter ant, 228, 230
Leafhopper, 106, 312, 323
Learning, 2, 4, 33, 37, 39–41, 43, 59, 63, 69–70, 72, 120–121, 137, 158, 168, 257–258, 260, 261–269, 277, 302–303, 342, 354, 369, 372
ability, 40
Leishmanias, 286
Lejops lunulatus, 136
Lek, 83, 85, 309, 317
Lens, 160, 161, 169, 189, 200, 258, 275, 370, 373
Lepidoptera, 8, 10, 12, 15, 22, 42, 53, 55, 89, 111, 124, 134, 137, 146, 164, 169, 182–183, 203, 206, 222, 224–226, 231, 277, 309, 314–316, 318, 322, 349
Leptinotarsa decemlineata, 65, 67, 125
Leptopilina
boulardi, 71
heterotoma, 71, 281
Lesser, 260
peachtree borer, 312
wax moth, 10, 21–22
Lestes sponsa, 277
Lethal treatment, 314
Lethargy, 276, 281, 372
Lethocerus, 213
Lettuce, 311
Leucinodes orbonalis, 319
Libellula pulchella, 89
Libythea labdaca, 98
Life-history, 4, 18, 23, 59, 67–68, 71, 85, 100–102, 123, 219, 221–222, 227, 230–231, 244, 293, 294, 298, 300, 302–303, 369–370
stage, 80, 274, 298, 349
theory, 93
trade-off, 220
trait, 14, 18, 24, 295, 302, 369
Lifespan, 12, 14–17, 24, 85, 123, 124, 197, 220, 229, 258, 260, 269

Lifetime monogamy, 229
Lift wing courtship behavior, 20
Light, 32, 34, 36, 38, 41, 65, 68, 105–106, 108, 117–118, 137, 147, 158–161, 163–166, 169, 195, 250, 258–259, 268, 278, 295, 298, 300, 302, 334, 341–342, 352, 354, 364, 366–368, 370–373
green trap, 311, 318
pollution, 300, 354
sensitivity, 68
Light-brown apple moth, 311
Ligurotettix coquilletti, 89
Limenitis archippus, 169
Line-cross technique, 18
Linepithema humile, 356
Linkage disequilibrium, 6, 11, 13, 25, 242, 367, 369
Lipid transport, 124
Lipopolysaccharide (LPS), 281–282, 369
Listeria monocytogenes, 282
Liver fluke, 284–285
Livestock, 26, 125, 279, 333, 340
Lobesia botrana, 278
Lobula, 36, 38–39
Plate, 36, 38
Local adaptation, 248, 294–295, 301–302, 369
Local search behavior, 118
Locomotion, 37, 39, 41–43, 55, 69, 101, 112, 125, 276, 299, 342, 364–365
Locomotor activity, 100–101, 276, 281, 295, 369
Locomotor cross-over hypothesis, 248, 369
Locus, 4–7, 9–10, 12–13, 17, 25, 364, 366, 369, 371, 373
Locust, 33, 35–38, 41–43, 51, 64–66, 72, 98–99, 103, 106, 109, 116, 119–121, 123, 125–126, 159, 185, 264, 267, 280, 283
invasions, 98
swarms, 98–99
Locusta, 118–120, 267, 280
Long distance, 41, 53, 108–109, 147, 299, 310, 315, 319, 335, 356, 368–369
communication, 194
migration, 53, 68
Long-lived, 93, 111, 258, 269, 352
Long-range migration, 98, 111
Long-term, 25, 103, 110, 262, 269, 294, 298–299, 337–338, 350, 368, 371
local adaptation, 295
memory, 72, 257, 260, 369
Long wing, 17

Lord Howe Island stick insect, 359
Lovebug, 198
Low temperature, 25, 65, 67
Luciferase, 164, 166
Luciferin, 164, 166
Luminance, 38, 159, 163, 165, 365
Lungecourtship behavior, 19–20
Luperina nickerlii leechi, 359
Lycaenidae, 352, 357
Lycorma delicatula, 138
Lymantria dispar, 278
Lysandra bellargus, 357

M

Macroglossinae, 141
Macro-parasite, 124
Macronutrient, 124, 269, 282
intake, 119, 125
Macroptery, 17
Macrotermes bellicosus, 228
Maculinea, 358
Maghreb, 110
Maize, 313, 319–320
Malacosoma
americanum, 140
disstria, 125
Male, 3–4, 6–8, 10, 12, 14–20, 22–25, 32, 43–46, 53–58, 71, 80–93, 122, 147, 149–150, 153, 155, 164–166, 168, 174–178, 181–182, 185, 189–200, 206–215, 227, 229, 238–239, 242–243, 245, 247, 263–264, 278, 285, 296, 300, 302, 314–319, 321–322, 324, 333, 335–337, 340, 351–352, 355, 357, 363, 365, 367–373
care, 210–216
courtship behavior, 8, 10, 12, 19, 45, 195, 212
courtship song, 8, 10, 12, 22
pheromone production, 10
Male-derived protein, 192
Male-male competition, 302
Male-produced aggregation pheromone, 311, 321–322
Mallota, 136
Malpighian tubule function, 70
Mammal, 37, 71–72, 99, 168, 214, 227, 231, 283–285, 342–343, 365, 369, 373
Mammalian homologue norepinephrine, 280
Manduca, 42, 52, 131
Mantid, 138, 181, 183, 197
Mantis, 41, 43, 139, 158–159, 162, 165, 167–169, 183, 207
Mantodea, 182, 193, 206

Marching behavior, 124
Masquerade, 134, 141, 168, 369
Mass, 17, 67, 88–89, 108–109, 178, 190, 209, 246, 249, 257, 267–268, 284, 321, 333, 335–336, 363
  landing, 108
  migration, 98, 107, 124–125
  provisioning, 203–204, 205, 208
  trapping, 310–311, 313–314, 317–319, 321–322, 325, 369
Massachusetts, 133
Mass-rearing, 316, 336
Mate, 4, 10, 17, 22, 43, 53, 55, 57, 66, 68, 80, 82–84, 86, 88, 93, 99, 150, 153–154, 161–166, 168, 174–176, 178, 181, 189, 191–199, 211, 213, 216, 220–221, 226, 228–229, 239–240, 242–243, 245–247, 262, 264, 266, 268, 278–281, 285, 287, 299, 314, 316–317, 321, 333, 335, 337, 340, 342, 348, 350–352, 357, 359, 363–364, 366–367, 369, 371–373
  choice, 22–24, 26, 82, 84, 160, 191–192, 195, 246–247, 262–263, 269, 278, 292, 366, 368
  competition, 191, 198
  guarding, 82, 84, 369
  location, 40, 82–84, 194, 315, 322
  quality, 150, 163, 166
  recognition, 38, 40, 57, 335
Mate-encounter site, 82–83
Mate-guarding, 196–198
Mate-locating male, 83
Mate-seeking behavior, 84
Maternal, 6, 18–20, 57, 68, 205, 210–211, 321, 338
  behavior, 68
  provisioning, 80
Mating, 4, 6, 8–9, 17, 20, 22, 53, 55–57, 67–68, 71, 80, 82–84, 88, 92, 100, 121–122, 149–150, 153, 166, 176, 185, 189–192, 194–199, 209–213, 220, 223–224, 229, 236, 239, 241–242, 245–247, 258, 281–282, 287, 294, 302, 309, 312, 314–317, 319, 321, 323–324, 333, 335–338, 340, 353, 363, 365, 368–369, 372–373
  disruption, 310, 312, 314–316, 319, 322–323, 325, 369–370, 373
  frequency, 8, 22–23, 66, 150, 153
  habitat, 82, 84
  opportunities, 71, 84, 210, 211, 212
  plug, 82, 190, 198–199
  position, 193, 196
  rate, 15, 17, 22, 150, 302, 369
  strategy, 43, 197

success, 23, 93, 176, 193–195, 335–336, 367
  system, 82–83, 192–194, 212, 242, 250, 317, 369
  vibration, 312
Maturity, 102, 348
Mauritania, 110
Mayfly, 40, 165, 223, 354
Mechanistic, 32, 70, 159, 242–243, 250, 258, 274, 280, 287, 298, 303
Mechanoreceptor, 182–183, 185, 369, 371
Medication, 124, 277, 279, 280–283, 372
Mediterranean fruit fly, 312, 316, 318, 325
Medulla, 36, 38
Megachilidae, 224
Megalopta genalis, 41
Megapodes, 211
Megoura viciae, 68
Melanotus, 322
  wireworm, 311
Melangyna umbellatarum, 136
Melanostoma mellinum, 136
Melbourne, 352, 356
Meliponini, 231
Melittidae, 224
Melon fly, 15
Membracid, 205
Memory, 37, 39–42, 63, 69–70, 72, 257–260, 268, 369
  formation, 40
Mentha piperita, 324
Meta-analysis, 88, 92, 293, 309
Meta-regression, 208
Metabolite, 116, 118, 121, 123, 147, 372
Metamasius hemipterus, 311
Metamorphosis, 41–43, 45, 49–53, 55–56, 196, 222, 231, 243–244, 336, 368
Metapopulation unit, 356–357
Metarhizium
  anisopliae, 280
  robertsii, 283
Methoprene, 57–58, 369
Methyl eugenol, 318
Mexican fruit fly, 316
Mexico, 68, 106, 356
Microbial association, 124
Microbiota, 278
Micro-environment, 231
Microhabitat, 109, 134, 142, 313–314
  choice, 134, 299, 301
  selection, 131, 133–134
microRNA, 72
Microstructure of feeding, 118–119
Microvilli, 160–161

Middle-term memory, 260
Migrant, 41, 98–99, 101–112, 364, 369, 373
Migration, 49, 52–56, 66, 68, 80, 98–103, 105–109, 111–112, 124–125, 220, 227, 298–299, 353–354, 356, 369–371
  arena, 102
  syndrome, 100–103, 369–370
  system, 100, 102–103, 369
Migratory, 51, 53–54, 68, 98, 101–103, 106, 108–109, 111, 224, 369, 372
  behavior, 53, 64, 100–101, 103
  flight, 54, 56, 101–102, 105, 108–109, 356
  insect pest, 98
  locust, 66, 280
  pest moth, 103
Milkweed bug, 43, 53, 56
Millipede, 141
Mimicry, 131, 134–135, 139, 141, 152, 169, 364, 369
Mining bee, 224
Mint plant, 324
MN5, 42
Mobile, 81, 210, 231, 261, 293, 300, 348, 359, 371
Mobility, 231, 298–300, 350, 369
Modern ratite, 211
Molasses grass, 319
Molecular, 1, 7, 9–10, 17, 37, 50, 52, 67–69, 71–72, 106, 112, 117, 121, 125–126, 183, 261, 281
  readout, 65
  weight, 148
Mollusk, 280
Monandrous, 84, 369
Monandry, 369
Monarch, 68, 102, 106, 169, 356
  butterfly, 41, 53–54, 68, 99, 106, 109, 111, 169, 277, 283, 356
Monkshood, 361
Monogamous, 219
Monogamy, 55, 193–194, 198, 199, 229, 369
Monogyne colony, 66, 371
Monophagous, 350
Monophyletic lineage, 222
Mormon cricket, 124
Morph, 15, 70
Morpho butterfly, 164
Morphological trait, 17–18, 102, 246, 294, 300–301, 348
Morphology, 4, 17, 25, 42, 49, 112, 148, 150, 165, 189, 247, 298, 300, 303
Mortality, 4, 85, 111, 123, 208, 210, 213, 250, 313, 314, 317, 320, 336, 354, 359, 365

Mosquito, 53, 55, 67, 183–186, 276, 286, 300–302, 332–340, 343, 365

Moth, 3, 8, 10, 12, 15, 17, 21–22, 43, 52, 53, 71, 102, 106, 108–109, 111, 121, 131, 133, 138, 146, 160, 167–168, 229, 278, 285, 302, 311–313, 320, 351, 356, 359

Motion, 36–38, 101, 104, 139, 159, 161, 163, 164–165, 174, 177, 184

Motivation, 86, 88–89, 93, 262, 373

Motor, 37, 39–41, 43, 52, 99, 174, 261–262, 264, 268

 control, 33, 36, 40–41

 neuron, 34, 36, 38, 42, 44–45, 368–369

 pattern, 33, 43, 118, 257, 262

 skill, 261, 264, 268

Motyxia, 141

Moult, 42, 50–51, 53, 82–83, 210

Mound-building termite, 228

Mountain, 99, 106, 107

 katydid, 138, 159, 167

 pine beetle, 311, 313

Müllerian mimicry, 134–135, 169, 369

Multi-dimensional care, 212–216, 370

Multiple, 6, 9, 24, 37, 39, 59, 64, 66, 69–70, 82–83, 109, 117, 119, 122–124, 141, 163, 166, 190, 194, 197, 211–213, 220, 228–229, 236, 242, 244–245, 249, 251, 258, 277, 279, 281, 283, 294, 299, 301, 315, 321, 340, 343, 365, 371–372

 generations, 3, 26, 64, 199, 294, 321

 mating, 150, 194, 211

 QTL, 9, 11

 trait mapping, 9

Mung bean seed, 20

Musca domestica, 19, 300

Muscle, 32, 34, 36, 42, 49, 52, 105, 140, 148, 176–177, 181, 298, 300, 366, 369

 power, 88

Mushroom, 20, 39

 body, 32–33, 36, 39, 243

Mutated protein, 34

Mutual cooperation, 219

Mutually cooperative, 220

Myrmarachne melanotarsa, 134

Myrmecophilous, 358

Myrmica, 358

Mythimna separata, 103

Myzus persicae, 324

N

Napier grass, 319, 320

Narrow-sense estimate, 5, 13

Nash equilibrium, 81

Nasonia

 giraulti, 10

 oneida, 10

 vitripennis, 8, 19–20

Nasty neighbor, 86, 92, 370

Nasutitermes, 229–230

Natural, 6, 11, 40, 63, 67, 81, 89–90, 98–100, 126, 132, 134, 140, 165, 168, 208, 213, 219, 239, 244, 248, 273, 276–277, 284, 297, 321, 323, 334, 337, 348–349, 353–355, 357, 366, 372–373

 enemy, 153, 155, 204–205, 207–208, 316–317, 319, 322–323, 357, 359

 pest control, 126

 selection, 63–64, 100, 102–103, 112, 148, 162, 165, 191, 236, 244, 258, 294, 298, 301, 366, 368–369, 372

Navigation, 38, 40–41, 68, 106

Nearest-neighbor (neighbor) distance, 276

Nectar rewards, 158

Negative genetic correlations, 14, 17, 71, 273

Neighbors (neighbours), 85–86, 90, 92, 221, 262, 276, 370

Nematode, 265, 277, 284, 322

Neoconocephalus ensiger, 194

Neodohrniphora curvinervis, 278

Neotrogla, 190

Neriid fly, 199

Nerve tract, 367

Nervous system, 2, 32–35, 37–38, 41–43, 52, 121, 181–182, 186, 257, 275, 280, 363, 365–367, 370

 plasticity, 41

Nest, 43, 56–58, 70, 118, 125, 140, 161, 168, 190, 203–205, 208–209, 212–215, 221–222, 225, 227–230, 232, 245, 261–262, 266, 268–269, 278–280, 282, 314, 352, 358, 368, 370

 architecture, 231, 244, 279

Nested ANOVA, 6

Network, 40, 42, 44, 73, 111, 161, 169, 176, 231, 240, 262, 354–355, 357, 364

Neurite, 34, 37, 370

Neuroanatomy, 32–33

Neurobiology, 32–34, 43, 280–281

Neuroblast, 42, 49, 370

Neuroendocrine-immune axis, 274, 280

Neurogenesis, 42–43

Neuromere, 34

Neuromodulator, 37, 66, 121–122, 370

Neuromodulatory neuron, 37, 39–40

Neuron, 32, 33–34, 36–45, 49, 118, 121–122, 159, 183–184, 257–259, 262, 280, 363, 366–367, 370, 372

silencing, 121

Neuropil, 32, 34, 36–43, 367, 370

 glia, 37

Neuroptera, 182, 185

Neurosecretory cell, 50, 365, 371

Neurotransmitter, 37, 59, 243, 364, 370

New England buck moth caterpillar, 229

New Zealand giraffe weevil, 167–168

Niche, 13, 136, 172, 192, 198, 200, 257, 261, 263–265, 313, 357, 375, 376, 384

 construction, 257–258, 265, 267, 289, 291, 384

Nicrophorus, 56, 193, 225–226

 orbicollis, 54

 tomentosus, 134

 vespilloides, 15, 17, 71, 214, 279

Night vision, 258–259, 268

Nilaparvata lugens, 12, 106

Nitrogen-rich food, 230

NO_2, 324

Noctuidae, 356, 359

Nocturnal, 104, 107–108, 112, 133, 146, 175, 180, 258, 318–319, 352, 354, 359

 boundary layer, 104

 hawkmoth, 259

 insect, 106, 258, 268, 300, 354, 373

 mammalian predator, 141

 migrant, 104, 106, 109

 surface inversion, 104

Noise, 146, 148, 160, 163–165, 178–182, 236, 277, 357

Non-additive genetic, 4, 18–20, 23, 25 variance, 3, 5, 7, 20, 370

Non-additive gene effect, 7, 18, 367

Non-competitive disruption, 315, 319, 370

Non-consumptive effect, 153

Non-genetic maternal effect, 18, 20

Non-learned behavior, 261, 264

Non-resident, 85, 87–88

Non-target species, 309

Noradrenaline, 37

Norm of reaction, 62

North, 68, 98–99, 106, 108, 112, 124, 295–296, 338–339, 356

 America, 41, 68, 99, 106, 109, 247, 250, 260, 277, 333, 356

 American Hines' emerald dragonfly, 357

North Carolina Design III, 7, 18–19

Notoncus, 352

Nuptial gift, 17, 190–192, 370

 quality, 199

Nutrient-rich spermatophore, 190

Nutrient, 67, 70, 116–117, 119–126, 190–192, 203, 205, 261–262, 264, 282, 369–370, 373
  availability, 71, 121
  cycling, 125
  deprivation, 120
Nutrition, 66–67, 69, 71–72, 116–117, 122–126, 282, 367
Nutritional, 58, 63, 66, 69, 73, 80, 116–117, 119, 221–226, 210, 233, 249, 250, 264, 282–283, 369
  ecology, 116–117, 249
  independence, 203, 205, 209, 215
  vector, 123
Nutritive supplement, 190
Noctuid moth, 12, 109
Nurse, 43, 58, 64–66, 70, 279
Nursing tasks, 68
Nycteribiidae, 205
Nymphalidae, 152, 204

O

O$_3$, 324
Oahu, 24
Obesity, 125, 270
Octopamine, 36, 59, 243, 280, 364
Octopaminergic, 37
Odonata, 89, 206, 244, 277, 303, 357
Odorant, 39
  binding protein, 66
Odor (odour), 39–40, 42, 66, 106, 147–149, 153–154, 259, 262–263, 266, 268, 310, 319, 341–343
Oedaleus asiaticus, 125
Offspring
  carrying, 203, 206, 209
  feeding, 210
  provisioning, 8, 80
  survival, 205, 209–210, 279
Olfaction, 39–40, 70, 317, 319, 372
Olfactory, 39–40, 42–43, 45, 66, 70, 107, 121, 176, 266, 281, 314, 317, 323, 325, 335, 349, 372
  memory, 40
  receptor, 39, 66, 365
  stimuli, 36, 353
Olfactory-based aversive memory, 70
Olive, 311
  fly, 311
  fruit fly, 318
Omaspides brunneosignata, 207
Ommatidia, 38, 160–162, 372
Ommatidium, 160–161, 364, 370
Onion, 311, 313, 316
  maggot fly, 313
Ontario, 269, 322
Onthophagus, 65, 71

taurus, 8, 198
Ontogeny, 243, 276
Ooceraea biroi, 39, 66
Oocyte, 53
Oogenesis-flight syndrome, 102, 370
Ootheca, 209
Open box model, 86, 87
Operational sex ratio (OSR), 82, 194, 370
Ophiocordyceps, 284
Ophryocystis elektroscirrha, 283
Opsin, 161, 342, 370
  protein, 161
Optic, 160
  flow, 38, 161
  glomeruli, 39
  lobe, 36, 38
  nerve, 160
Optical sensitivity, 258
Optima, 149–150, 192, 250, 365
Optimal, 80, 81, 84, 93, 112, 123–124, 126, 137, 147, 149, 151, 192, 239, 246, 264, 268, 301, 314, 350, 354, 366, 372
  internal environment, 262
  mating rate, 150
  timing, 314
Optimization, 81, 335
Opius dimidiatus, 151
Optogenetic activation, 39–40, 45
Optogenetics, 34, 370
Orchid, 152–153, 372
  bee, 195
  mantis, 158–159, 165
orco gene, 66
Organophosphate pyrethroid, 317
Oriental
  armyworm, 183
  beetle, 312, 322
  fruit moth, 311–312
Orientation, 38–39, 41, 84, 98, 102, 105–106, 108, 112, 131, 133–134, 142, 161, 164–165, 196, 264, 316, 319, 322, 342, 350, 369, 371, 373
Ormia ochracea, 24, 182
Ornament, 190–191, 199
Ornithoptera richmondia, 354, 355
Orthoptera, 8–12, 15, 22, 72, 89, 146, 169, 182, 194, 206, 224, 226, 286, 300, 352
Ostrinia
  furnacalis, 195
  nubilalis, 12
Ova production, 80
Ovary, 58, 66–67
Overlapping generation, 244, 266–267
Overwinter, 99, 221, 369

Oviparity, 212
Oviposition, 2, 4, 8, 39, 54, 56–57, 80, 83–84, 89, 147, 150–151, 198–199, 204, 213, 260, 277, 279, 281, 283, 299, 309, 314, 317, 318–320, 336, 342, 354, 369
  decision, 42, 80
  deterrent, 313, 317
Oviposition site, 19–20, 82–83, 89, 203–204, 212–214, 299, 310, 317, 350, 358, 364
  choice, 302
  selection, 204, 302, 316
Ovipositor, 39, 196, 205, 210, 212
Ovoviviparity, 205, 212
Ovoviviparous species, 205, 209
Ovulation, 190
Owl butterfly, 152
Ownership, 85, 88, 90–91, 373
Oxygen, 148, 210, 371
  flow, 209
Oxygenation, 164, 209
Oyamel forest, 99, 356

P

Pachliopta aristolochiae, 134
Pachydiplax longipennis, 134
Painted lady butterfly, 99, 109–112
Paired noduli, 41
Pairwise, 81, 85–88, 92, 238, 249, 372
  assessment, 86, 92
  epistatic effect, 25
Palatable prey species, 132
Palm, 312, 322
  tree kairomone, 312
  weevil, 311–312, 322
Panamanian tortoise beetle, 159
Panorpa vulgaris, 8
Pantala flavescens, 99
Paperwasp, 67, 161, 167–168, 220, 225, 244
Papilio, 141
  machaon, 139
  polytes, 134
Paralucia pyrodiscus lucida, 352
Pararge aegeria, 89, 91
Parapheromone, 312, 317
Pararistolochia, 355
Parasite
  establishment, 284
  evasion, 278
  load, 88–89
Parasitic Hymenoptera, 99
Parasitism, 205, 208, 273, 275, 277–279, 286–287, 299, 319, 370
Parasitoid, 71, 72, 124, 146, 148–152, 154–155, 181, 182, 190, 204, 207,

210, 213, 223, 250, 273, 277, 280, 283, 285–286, 299, 302, 315–316, 323–325, 348, 350, 358, 370
  fly, 24, 181–182, 277–278
  wasp, 8, 19–20, 71, 111, 152, 208, 250, 260, 277, 279, 280–281, 285, 324
Parasocial, 226, 370–372
Parastrachia japonensis, 207, 209
Parental, 5–8, 18–19, 57, 208, 210, 211, 214–216, 363, 366
  care, 26, 49, 52, 56–57, 80, 190–192, 194, 203, 205–206, 210, 212, 214–216, 222–223, 225, 227–232, 266, 269, 281, 302, 363, 370, 372–373
  compensation, 214, 370
  effort, 211, 214, 370
  investment, 82, 191, 200, 370
  negotiation, 214, 370
Parental–care parasitism, 205, 370
Parent-daughter, 20
  coefficient, 18
  regression, 18
Parent-offspring, 21, 227
  prophylaxis, 279
  regression, 3, 6, 8, 14
  relationship, 226–227
Parent-son, 20
  coefficient, 18
  regression, 18
Parhelophilus laetus, 136
Park, 311, 355, 373
Parthenogenesis, 190–191, 370
Parthenogenetic, 51, 68, 102
Partner effort, 214
Passalid, 209, 215, 223, 225, 230
Paternal, 6, 337
  care, 211–213
Paternity, 192, 198–199, 211, 213
  assurance, 213
Pathogen, 111, 273, 274, 280–285, 313, 315–316, 320–323, 333, 336–338, 343, 348, 364, 368, 373
Pathogenicity, 274
Pathology, 274, 281, 284
Pattern recognition, 41
Pea, 11, 313
  aphid, 237
  leaf weevil, 313
Peach, 312
Peach-potato aphid, 324
Peachtree borer, 312
Peacock butterfly, 138
Peak male activity, 357
Pear, 311
  kairomone, 311

Pedestrian migration, 100
Pedigree, 297
Pellucopia crassiventris, 207
Pentatomid bug, 213
Peppered moth, 167
Perception, 23, 84, 90, 108, 159, 163, 165, 181, 184, 257–258, 325, 370, 373
Peripheral nervous system, 34, 315, 363, 366
Perithemis tenera, 88
Persistence, 81, 85, 88, 146–147, 210, 350
Personality, 2, 4, 236–239, 241–251, 363–364
Pest management, 59, 250, 309, 316, 324–325, 333, 349, 364, 368, 371
Phagostimulant, 118, 320, 370
Pharynx, 39
Phasmatodea, 169
Phasmid, 134
Pheidole, 230
  bicarinata, 51
  pallidula, 70
Phenological strategies, 80
Phenology, 11, 126, 295, 298, 353, 370
Phenoloxidase, 280
Phenotype, 4–7, 9, 18–19, 21, 24, 64, 68, 69–70, 83–84, 88, 102, 133, 141, 153, 199, 221, 236, 239, 247, 249–250, 262, 284, 292, 298, 335, 337, 363–364, 366–369, 371–372
Phenotype-environment matching, 133
Phenotypic, 5–7, 13–15, 17–18, 20–21, 24–26, 103, 191, 244, 261–262, 273, 294, 297, 305, 365, 367–368, 370–371
  plasticity, 51, 63–64, 69–73, 126, 262, 294, 298, 301, 371
  variance (V_p), 5, 294, 368, 371
Pherobase data base, 314
Pheromone
  biosynthesis, 55, 325
  production, 10, 12, 55, 71
  signal, 84, 302
Phlaeothripidae, 204, 212, 224
Phloeaquadrata, 209
Phloeidae, 209
Phonotaxis, 175, 371
Phosphatidylinositol 3 kinase, 67
Phoresy, 100, 371
Phoretic attachment, 101
Phorid fly, 277–278
Phorodon humuli, 312
Photinus, 192

Photoperiod, 54, 65, 67–68, 101, 296, 361, 363, 366, 371
Photoreceptor, 38, 160–161, 165
  cell, 160–161
Phototransduction, 161
Phthiraptera, 206
Phyllocnistis citrella, 312, 319
Phyllomorpha laciniata, 199
Phylogenetic, 136, 140, 166, 183, 193, 213, 275, 286, 303
  analyses, 141, 227
  history, 219
  meta-regression, 208
Phylogeny, 92, 130, 140–141, 182, 275, 372
Physical, 58, 66, 83, 85, 88–89, 148, 150–151, 164, 166, 169, 175, 178, 182–183, 191, 194, 195–196, 198, 242, 261, 324, 352, 369, 371–372
  defense, 130–131
  fighting, 86
  reinforcement, 165
Physiognomy, 350, 371
Physiological response, 52, 183, 280, 292, 301, 343, 366
Physiology, 1, 55, 68, 73, 102, 112, 121–125, 160, 183, 189, 198, 200, 262, 273, 293, 297, 298, 300, 343
Phytophagous stick insect, 134
Pieris
  brassica, 247, 260
  napi, 15, 17
  rapae, 260
Pigeon pea, 313
Pigment, 150–164, 352, 371, 373
  absorbance, 163
  cell, 160
  molecule, 161
Pine, 204, 311–313
  bark beetle, 312
Pipiza femoralis, 136
Pittosporaceae, 352
Plague locust, 103, 109
Planococcus ficus, 312
Plant
  kairomone, 311–312
  volatile, 309–311, 316–317, 319–320, 322
Planthopper, 12, 106, 109, 302
Plant-host volatile, 316
Plastic, 42, 63, 70–71, 86, 153, 261–262, 287, 292, 294–295, 298, 300, 303, 340
  infection avoidance response, 287
  response, 69–70, 274, 292–295, 298–300, 303

Plasticity, 41–43, 51, 56, 63–64, 66–73, 84, 126, 241–242, 261–262, 278, 287, 293–295, 297–298, 301, 303, 348, 371
Platycheirus
  confuses, 136
  hyperboreus, 136
  nearcticus, 136
  obscurus, 136
Platypotidae, 223
Plautia crossotastali, 311
Plecia nearctica, 198
Plecoptera, 206
Pleiotropy, 13, 69–70, 73, 242, 367, 371
Plesiomorphic
  behavior, 213
  state, 203, 371
Plutella xylostella, 71, 312
Poecilimon thessalicus, 190
Polarization, 36, 38, 106, 159, 161, 163–165
  plane, 41, 159
Polarized color(colour), 165
Polistes, 58, 67–68, 220, 223
  fuscatus, 167
  metricus, 67, 220–221
Polistine wasp, 228
Pollen basket bee, 222
Pollination, 2, 59, 126, 287, 303, 321, 354, 356
Pollinator, 153, 248, 322, 350, 354, 356, 358–359, 372
Pollinator-plant association, 358
Pollution, 298, 300, 354
Polyandrous species, 83
Polyandry, 8, 19–20, 24, 194, 229, 371
Polyethism, 245, 363, 371, 373
Polygamy, 193–194, 198–199, 371
Polygyne colony, 66, 371
Polygyny, 84, 194, 212–213, 215, 229, 271
Polymodal information, 40
Polymorphic, 169, 222, 299
  geometrid species, 133
  swallowtail butterfly, 134
Polynesia, 24
Polynesian field cricket, 8, 12, 15, 22
Polyphagous feeding species, 321
Polyphenism, 51, 64, 66–68, 71–73, 102, 368, 371
Ponerine, 43, 224
Popillia japonica, 311–312
Population
  collapse, 315
  structure, 356, 357
  suppression, 336–337, 371
  trajectory, 100, 103, 109
Post-copulatory fertilization, 84

Post-infection, 275
Post-migratory appetitive flights, 109
Post-ovipositional
  care, 203, 205, 207–210, 215–216
  parental care, 203, 205–206, 216
Post-pharyngeal gland, 149
Post-synaptic neuron, 37
Potato, 67, 311, 322, 324
  beetle, 301
  leafhopper, 106
  psyllid, 15
Potential mate, 163–166, 176, 194–195, 240, 247, 262, 340, 352, 371–372
Poultry, 26
Praying mantis, 43, 139, 159, 162, 169, 207
Precocene, 58, 371
Precocial, 212, 215, 371
Precopulatory, 83, 198–199, 241
  behavior, 194
  mate-guarding, 196
  sexual cannibalism, 238, 247, 371
Predation, 2, 52, 80, 100, 125, 130–133, 138, 153, 168, 180, 182, 189, 203, 207–208, 210, 248, 265, 281, 284, 286, 364
  risk, 141, 203, 243
  sequence, 131–132
Predator
  avoidance, 40, 182–183
  protection, 220
Pre-infection, 275, 286
Pre-ovipositional
  care, 203, 215
  parental care, 203
Pre-reproductive female, 357
Presynaptic neuron, 37, 370
Primary memory phase, 260
Primate, 159, 215, 241
Primitive eusociality, 223–224
Primitively eusocial, 58, 65, 67, 72, 222–223, 227–229
Production cost, 146
Prophylactic, 333
  medication, 279–281, 371
Proprioceptive neuron, 37
Proprioreceptor, 182, 371
Protandry, 82, 357
Protective signaling, 168
Protein, 10, 33–34, 49–50, 52–53, 66–69, 119–120, 126, 161, 192, 198–199, 282, 310–311, 317, 337, 342, 364, 366, 369–370
  depleted food, 119
  food bait, 317–318
Protein bait, 318
  spray, 311
Protein: carbohydrate ratio, 123, 282

Protein-rich, 119, 122
  shell, 190
Prothoracic gland, 50–51, 366, 371
Prothoracicotropic hormone (PTTH), 50–51, 365, 771
Protospirura muricola, 284
Protozoan, 277, 283, 286, 333
Pseudoplusia includens, 71
Pseudo-pupil, 162
Pseudoxycheila tarsalis, 135
Psinidia fenestralis, 138
Psocoptera, 204, 206
Psyllid, 15, 312, 322, 324
Ptiliidae, 37
Public information, 147, 151, 154
Punishment, 40
Pupa, 42, 50–52, 82–83, 197, 209, 285, 334, 351, 355, 357, 368
Pupal guarding, 209
Push-pull, 310, 313–314, 316, 319, 325
  technique, 371
Putrescine, 318
Pygmy grasshopper, 134, 139
Pyramid moth, 8

Q

QTL, 9–11, 23, 271
  genetic linkage mapping, 17
  mapping, 9, 12
Quantitative genetic, 4–5, 7–9, 11, 13–14, 18, 21, 23, 241, 294, 297
  analyses, 69
  breeding design, 3, 13
Quantitative trait loci (see QTL), 7
Quasisocial, 277, 370–371
Queen, 43, 51, 58, 65–66, 71, 149, 168, 190–191, 222, 228–229, 231, 244–245, 371
Quiescence, 80, 101, 240, 300

R

Radar, 99, 105–109, 111–112, 138, 266
Rana nigromaculata, 139
Range shift, 111, 293, 299, 371
Ranging, 7, 52, 100–101, 130, 133, 142, 371–373
Rape, 357
Rapid evolution, 24
Raptorial forelegs, 159
Raspberry viruses, 286
Rate of copulation, 80
Reaction norm, 21, 23, 25, 241–242, 294, 364, 372
Realized h^2, 7
Receiver, 162, 166, 174–175, 177–182, 184, 194, 363, 365, 372
Receptive female, 82–84, 372

Receptor, 10, 39, 49, 51–52, 163, 174, 182–185, 264, 366, 370
Recipient, 52, 145 117, 150, 152 153, 221, 366, 373
Reciprocal transplant field approach, 294
Recruitment, 149, 177, 278
Red, 159, 164, 166, 169, 322, 342, 355
    flour beetles, 139, 199
    imported fire ant, 250, 356
    palm weevil, 312, 322
    sphere, 311, 318
    sticky sphere, 318
Reduviidae, 203, 211–212
Reflected light, 158
Reflective mulch, 312, 316
Regurgitation, 139–140
Relatedness, 5, 199, 219–221, 229, 275, 368
Relocate, 99, 118, 205, 293
Remote sensing system, 159
Rendezvous location, 83
Repellent, 140, 204, 310, 312–313, 316, 319–320, 325, 334, 340, 372
Reproduction, 4, 14, 15, 17, 43, 49, 53–54, 56, 58, 67–68, 71, 80, 100–102, 106, 111, 116, 121–122, 124–125, 146, 189–193, 195, 222, 226, 228, 246–247, 262, 264, 279, 281, 283, 300–302, 338, 340, 363, 366–370, 372
Reproductive
    conflict, 229
    division of labor (labour), 219, 225, 228, 232, 244, 372
    success, 17, 71, 124, 153, 191–192, 199, 221, 302, 303, 367, 370, 372
Residency, 85, 86, 87, 88, 89, 90, 91, 92
Resident, 84–85, 87–88, 90–91, 215, 247, 338, 351, 364–365, 370
Resident-intruder role, 81
Resin bug, 203, 205
Resistance, 24, 124, 196, 229, 299–300, 309, 310, 314, 316, 318–322, 325, 333–334, 338–340, 343, 363
Resolution, 86, 149, 161, 181, 276
Resource
    acquisition, 80, 83
    availability, 109, 126, 230–231
    holding power (potential; RHP), 86–88, 90–93, 372
    need, 348, 354, 356, 368
    value, 86, 89, 372
Resource-based, 82, 317, 349
Response surface, 123–124
Reticulitermes fukienensis, 279
Retina, 38, 372
Retinula, 160

Retroactive interference in memory, 70
Reward, 40, 84–86, 152–153, 158, 232, 260–261, 266, 296, 366
Rhabdom, 160–161, 258, 372–373
Rhabdomere, 160–161
Rhagoletis
    basiola, 151–152
    mendax, 311, 317
    pomonella, 311, 317
Richmond birdwing butterfly, 354–355
Rhingia nasica, 136
Rhinocoris, 213
    bug, 211
    carmelita, 211
    tristis, 211
Rhodopsin, 34, 161
Rhynchophorus
    ferrugineus, 312
    palmarum, 311, 322
Rhytidoponera metallica, 125
Ritualized flight, 163, 164
RNA, 37, 67, 337, 365
    expresion pattern, 66
    interference, 121
RNA-seq analysis, 24
Roach, 223
Robotics, 159
Rocky Mountains, 106
Rodent, 119
Ropalidia marginata, 228
Round dance, 168
Rove beetle, 279
Rose hip fly, 151
Rothamsted Insect Survey, 111

S
s-tim allele, 65
Sagebrush cricket, 150
Sahara, 110, 112
Salient signal, 162
Salinity change, 298
Salmonella typhimurium, 282
Sand
    cricket, 8, 15
    flies, 286
Sandhill rustic moth, 359
Sap, 210
    beetle, 311–312
    feeder, 125
    feeding, 322
Sarcophagidae, 205
Satellite behavior, 84
Saturnidae, 141
Sawfly, 204, 224
Scalability, 276
Scaphoideus titanus, 312, 323
Scarabaeidae, 204, 205

Schistocerca
    alutacea, 138
    americana, 264
    gregaria, 35–36, 41, 106, 267
Schoepfia jasminodora, 207
Scolytidae, 223
Scolytinae, 214
Scolytus, 321
Scorpionfly, 8, 169, 192
Scramble competition, 84
Screening pigment, 163
Season length, 295–296
Seasonal, 51, 53, 63, 67–68, 73, 80, 98, 99, 108–112, 232, 261, 295, 300–302, 348, 350, 353, 355–356, 369, 371
    adaptation, 348
    clock, 102
Secondary metabolites, 116, 118, 121, 123, 147, 372
Sedentary, 51, 56, 111, 240, 248, 348
Seed beetle, 8, 15, 19–20, 299
Selenia dentaria, 134
Self-grooming, 282
Self-medication, 124, 282–283, 372
Semelparous, 55, 132, 212
Seminal fluid, 55, 122, 199
    protein, 198–199
Semiochemical, 155, 317–318, 325, 342, 365, 368–369, 371–372
Semisocial, 223–224, 227–228, 370, 372
Sender, 145, 150, 154, 162, 174–175, 178, 180, 194, 363
Senescence, 54
Sense, 5, 39, 41, 82, 87, 90, 108, 112, 121, 126, 138, 146, 158, 169, 240, 244, 278, 284, 298, 337, 349, 368–369, 371, 373
Sensory, 32–33, 36, 38–39, 42–43, 45, 59, 80, 86, 108, 118, 121, 134, 165, 175, 179, 181–184, 185, 194–195, 259, 357, 359, 364–366, 369, 371, 373
    bias, 160, 195, 197
    ecology, 40
    exploitation, 166
    information, 45, 257, 274
    neuron, 34, 36, 39, 43, 183–184, 363, 368, 372
    process, 33, 43, 309
    response, 121, 348–349
    system, 33, 158, 160, 163, 166, 179–180, 195, 274, 278, 281, 310
Sepsis
    cynipsea, 296
    fulgens, 301
    punctum, 296
    thoracica, 297

Sequential assessment, 87, 372
Sericomyia, 136
Serotonin, 39, 66, 243, 364
    (5-hydroxytryptamine), 37
Sesquiterpene (E)-β-farnesene, 324
Sex
    determination cascade, 43
    pheromone, 55, 146, 148–149, 153,
        166, 310–312, 315, 318–321,
        323–324, 369–370, 372–373
    ratio, 82, 153, 194, 337, 353, 357,
        370, 372
    ratio distorter, 337, 372
    role, 80–82, 212
    role reversal, 82, 212
Sex-linked single gene, 24
Sexual
    antagonism, 150
    cannibalism, 196–197, 238, 247, 371
    coercion, 196
    conflict, 25, 122, 192, 196, 229,
        247, 372
    deception, 152–153, 372
    ornament, 190, 199
    selection, 17, 82, 84, 153, 165, 189,
        191–192, 194, 196–200, 213, 216,
        262, 372
    signal, 189, 191–192, 194, 243
Shade, 158, 165, 221, 259, 373
Shape, 15, 17, 32, 38, 71, 82, 84, 90,
        116, 125, 137, 147, 153, 158–159,
        163, 165–166, 168–169, 198, 219,
        221, 244–245, 248, 259–260, 261,
        268, 296, 300, 350, 357, 365,
        371–372
Shared ancestry, 141, 372
Sharer, 90, 266
Short day length, 54, 67, 68, 71
Short-lived, 101, 153, 242, 258, 278
Short-range, 99
    courtship signal, 194
    navigational tool, 41
    signal, 194
Short-term memory, 70, 260
Sibling-based group, 222
Sibling-sibling prophylactic, 280
Sickness, 286–287
    behavior, 274–275, 281–282, 285,
        287, 372
Signal
    generation, 163
    transmission, 164
Signal-to-noise ratio, 146, 148
Signaling, 10, 52, 59, 67–69, 71, 73, 92,
        121, 141, 145, 148, 150, 158–159,
        163–166, 168, 174–176, 179–182,
        189, 194–195, 302

environment, 162
    sex, 189
Silk, 3, 52–53, 196, 204, 208
    nest, 121
Silphidae, 134, 209, 223, 225
Silver Y moth, 108, 111
Silverfish, 196
Single-marker, 9
Single-nucleotide polymorphism
        (SNPs), 11, 24, 372
Siphonaptera, 226
Site
    defence (defense), 82, 84, 88–89, 92
    fidelity, 84–85, 87, 356
    selection, 82, 131, 133–134, 204,
        302, 316
    value, 84–85
Sitona lineatus, 313
Sitter, 65, 69–70
Sky, 36, 105–106, 158, 165, 266
    compass, 36, 41
    polarization vision, 106
Skylight cue, 68
Sleep, 33, 37, 41, 286, 334, 342
    deprivation, 70
    pattern, 281
Slope, 6, 21, 106, 190, 208, 295, 357
Slow-acting toxicant, 314
Slow-moving larva, 352
Small fly, 99
Smell, 39, 118, 146, 258
Sneaker behavior, 84
Snowshoe hare, 153
Social
    beetle, 230
    behavior, 12, 49, 57, 59, 66, 69,
        220–221, 232, 269, 282, 285,
        302–303, 365, 368
    caterpillar, 224, 227
    cognition, 70
    contact, 277–278, 282
    cricket, 279
    environment, 4–5, 21–23, 25, 66, 68,
        242, 245, 367, 372
    heterosis, 246, 373
    insect, 54, 64, 66, 70, 72, 86, 92, 149,
        168, 215, 219–220, 222, 225–226,
        229, 231–232, 243–244, 265–269,
        279, 303, 364
    interaction, 39, 124, 177, 215, 220,
        238–240, 243, 245, 278, 282
    learning, 257, 265–269, 372
    living, 219–220
    niche, 245–246, 372
    parasitism, 370
    phenotype, 221
    spider, 237, 246, 248

Sociality, 160, 219–220, 222–225,
        225–232, 244, 246, 366
Society, 98, 116, 119–222, 226–232,
        244–246, 303
Soil fertilization, 126
Solar radiation, 261, 299
Solenopsis, 125
    geminata, 278
    invicta, 66, 250, 356
Solidaginis, 20
Solitary, 40, 51, 54, 56, 58, 66, 72, 92,
        159, 168, 219–220, 222–225,
        227–228, 230, 248, 284
    insect, 72
    living, 221
    wasp, 83, 85, 285
Soldier caste, 223–224, 228, 230
Somatochlora hineana, 357
Somnolence, 276, 281, 372
Sound mimicry, 152
Song, 8, 10–12, 15–17, 22–24, 43–45,
        174–177, 179, 180, 195, 195
Sonotaxis, 353
South America, 99, 125, 152, 355–356
Space-for time approach, 300
Spatial, 34, 42, 70, 73, 83, 85, 90, 164,
        181, 258–259, 261–262, 264, 266,
        268, 276, 294–295, 367, 370
    avoidance, 277
    configuration, 257
    differentiation, 295
    distribution, 83, 261, 277, 359
    learning, 40
    memory, 41
Special environmental variance (V_{Es}),
        5, 21, 366, 372
Specialist, 2, 287, 294, 348–350,
        357–358, 373
Speciation, 3, 17, 23–24, 262–263
Specificity, 82, 146, 152, 310, 313–315
Speckled wood butterfly, 88, 90–91
Spectral, 106, 164, 365
    reflectance, 163
    sensitivity, 38
Speed, 41, 104–105, 108, 146–147,
        180–181, 240, 260, 299, 354, 355,
        363, 366–367
Sperm, 81, 150, 153, 189, 190–199, 336,
        363, 365, 372–373
    competition, 82, 198–199, 372
    gigantism, 190
    length, 8
    number, 199
    package, 190–191
    size, 190, 199
    storage, 190, 199, 211, 335
Spermatheca, 190

Spermatophore, 15, 17, 190, 192, 196–199
Spermatophylax, 15, 17, 190, 192
 investment, 8
Spermatozoon, 190
Sphaerophoria, 136
Sphecidae, 209
Sphingidae, 141
Spider predation, 125
Spilomyia, 135–136
Spine, 130, 135, 139, 159, 176, 190, 195, 197, 220, 225, 228, 313
Spinosad, 317
Spirurid nematode, 284
Split brood design, 21
Spodoptera
 exigua, 311
  frugiperda, 109
  littoralis, 282
Sporothrips, 212
Spotted lanternfly, 138
Spreadwing damselfly, 277
Spring phenology, 295, 298
Springtail, 196
Spruce budworm moth, 102, 109
Stalk-eyed fly, 166–168
Standardized ground trapping networks, 111
Starlight, 259
Startle display, 131, 138, 141
Starvation, 67, 98, 343
Station-keeping, 100–102
 movement, 372
Stator limbatus, 299
Stem borer, 319
 moth, 313
Stenaptinus insignis, 140
Stenotritidae, 224
Stenus comma, 279
Sterile insect technique, 315, 325, 335, 372
Stick insect, 33, 134, 359
Stigmella microtheriella, 160
Stimuli, 33–34, 36, 38–40, 43–45, 53–54, 101, 105, 117–118, 161, 182, 183, 215, 257, 259, 262, 309–310, 314–317, 353, 364–365, 371–372
Stimulo-deterrence, 316
Sting, 130, 135–136, 140
Stinging, 11, 130, 135, 139
Stingless bee, 140, 222, 230–231
Stink, 154, 181
Stone fruit, 311
Strepsipteran, 205
Stress hormone, 153, 280
Stridulation, 33, 148, 176–178, 312, 372

Structural coloration (colouration), 164, 373
Struggling, 131, 139
Subgenual organ, 174, 184–185, 373
Subjugation, 131
Subsocial, 205, 222, 227–228, 373
 species, 205, 228
Subsociality, 205, 222, 228–229
Sucrose, 70, 122, 311, 317, 320
Sudan grass, 319
Sugar, 40, 67, 118, 121, 126, 261–262, 310–311, 317, 358
 cane, 311
Suicidal drowning behavior, 287
Sun compass orientation, 102
Superadditive effect, 215, 373
Supernormal stimulus, 166
Superorganism, 125, 228, 244
Superposition eye, 258, 373
Suppression pheromone, 149
Surface: volume ratio, 209
Survival value, 138, 275
Susceptibility to eavesdropping, 146–147
Swallowtail, 141
 butterfly, 134, 139
Swarming behavior, 32, 335, 373
Sweat bee, 222, 224, 227, 230
Sweden, 294
Swiss army knife, 132
Symbiont, 116
Symbiotic microbe, 230
Synanthedon, 312
Synapomorphic trait, 205, 373
Synapse, 37, 40, 45, 369–370
Syndrome, 100–103, 112, 236, 238, 240, 242–244, 246–247, 250, 298, 300–301, 354, 364–365, 369–370
Synemon plana, 351
Synomone, 373
Synthetic, 153, 203, 241, 264, 314–317, 320, 324, 332, 334, 337, 369, 370, 372
 pyrethroid, 317
 repellent, 316
Syritta pipiens, 136
Syrphid, 152
Syrphus, 136

T

Tachinidae, 205
Tactic, 81–84, 86–87, 93, 139, 151, 155, 195, 236, 239, 242, 250, 310
Taiwan, 355
Take-off, 103, 105, 299, 367
Tarachodes, 207
Target-tracking, 41

Task specialization, 214, 215
Tasmania, 356
Taste, 39, 118, 121–122, 140, 146, 258, 264, 372
Taste-receptor, 264
Tectocoris diopthalmus, 169
Teleogryllus, 8, 12, 15, 16, 22–24, 197, 199
Telemetric, 109, 111
Telostylinus angusticollis, 199
Temnostoma, 135–136
Temnothorax, 238, 282
Temperate, 107, 111, 208, 231, 300, 301, 349, 356
Temperature-size-rule, 295–296
Temporal
 avoidance, 277, 278
 polyethism, 245, 373
Temporary interlopers, 85
Tenacity, 84
Tenebrio molitor, 55
Teneral mating, 196
Tent caterpillar, 125, 133, 140, 221, 229, 231
Tephritid, 19–20, 311, 317
Termite, 68, 140, 149, 219–220, 222–223, 225–226, 228–232
Terrestrial arthropod, 100, 105, 208, 210, 239, 241
Territorial, 71, 80–82, 84–93, 100, 166, 281, 350, 352–353, 359, 364–366, 372–373
 contest, 85, 364
 male, 81, 86, 93
 perch, 350
Territory, 58, 71, 84–86, 88, 90–92, 149, 166, 168, 194, 212, 237, 239, 317, 357, 365–366, 368, 370
Territory-based polygyny, 213
Tessaratomid, 207, 209–210
Testes, 65, 71, 190
Tetrix, 134, 299
Thanatosis, 139
Thermal
 background, 299
 benefit, 229
 microclimate, 80
 performance curve, 295–296, 365, 373
Thermoregulation, 2, 165, 283, 292, 301–302, 350, 357, 373
Thermoregulatory behavior, 301
Threatened, 126, 137, 319, 348, 349, 351–353, 355–359
Thrip, 154, 204, 212–213, 222, 224, 228, 230, 232, 313, 322, 324, 368
Thyrinteina leucocerae, 285

Thysanoptera, 206, 210, 222, 224, 231, 309, 322
Tiarothrips, 212
Tiger
    beetle, 135
    moth, 121, 138
Time in residency, 85, 90–91
Time-compensated sun compass, 68
Timeless, 65, 67
Tinbergen, 257, 274–276, 283
Tiny sperm, 199
Toad, 137
Tobacco, 52, 312, 324
Tomato, 20, 312
Tomentose burying beetle, 134
Tonic immobility, 139, 142
Topography, 261, 350–351, 357
Touch, 19, 43, 45, 57, 66, 139, 142, 159, 163, 219, 258, 369
Toxin, 34, 130, 283, 292, 319, 325, 372
Toxomerus, 136
Tracheal system, 52
Track, 104–105, 107, 109, 111, 118, 119, 121, 164, 238, 240–241, 249, 276, 277, 297, 299, 343, 350, 356–357, 367
Trade-off, 65, 71, 73, 111, 124, 137, 139, 149, 166, 210, 214, 220, 236, 238–239, 244, 247, 373
Trail pheromone, 149, 373
Training, 40, 120, 259
Transcriptome, 66, 373
Transcript, 10, 24, 69–70
Transgenerational, 72, 356
Transgenic, 33–34, 69, 319–320, 322, 324–325, 337, 367, 370
Translocation, 105, 353, 373
Transmigration phase, 106–107
Trap
    crop, 313, 316, 320
    efficiency, 313
    saturation, 313
    shutdown, 315, 322, 373
Traumatic insemination, 150, 195–198, 373
Treehopper, 22–23, 194
Trematode, 284–285, 287
Tremulation, 178, 373
Tribolium, 22, 139, 199, 237, 284
Trichogramma carverae, 299
Trichoplusia ni, 148, 312
Trichoptera, 206
Triglyceride, 70
Trigona, 138, 140
Trikinetic, 276
Trimedlure, 312, 318
Trimethylamine, 318

Trioza apicalis, 312
Triple test cross, 7, 19
Trophallaxis, 230, 373
Trophic
    avoidance, 277–278
    interaction, 2, 125, 325
Tropical, 41, 51, 110, 135, 180, 207–208, 225, 228, 278, 300–301, 332–333
Tropidia quadrata, 136
True
    bug, 223
    navigator, 106
Turf grass, 322
Twig-mimicking caterpillar, 134
Tympanum, 184–185
Tyramine, 37

U
Ultrasound hearing, 133
Ultraviolet, 159, 373
Uni-dimensional
    care, 210, 214, 373
    parental care, 214
Uniparental
    female care, 210, 212, 214–216
    male care, 210–213, 215–216
Unipolar cell, 37
United States, 66, 204, 354
Univariate breeder's equation, 7, 24
Univoltine, 247, 351, 373
Ultraspriacle (USP), 49, 366
Utilities, 350, 359, 368
Ultraviolet (UV), 38, 159, 161, 302, 310, 313, 316, 324, 373

V
Vaccenyl acetate, 36, 45, 268
V_D, 5, 7, 18–19, 366
Vector navigation, 106
Vegetable, 311–313, 317
Velvet worm, 169
Ventilation, 33
Ventral nerve chord (VNC), 36, 38, 39, 42, 45, 52
Vespa mandarinia, 140
Vespid, 92, 152, 204, 222–223
Vespine, 220, 225, 228
V_I, 5, 7, 370
Vibration, 8, 174–186, 309–310, 312, 316, 323–324, 335, 366, 373
Vibratory signal, 176, 181, 194
Viceroy butterfly, 169
Victoria, 352
Video-based tracking, 276
Vine mealybug, 312
Virulence, 274, 285, 287, 373

Virus, 71, 278, 281–282, 285–286, 314, 332–334, 336–340, 364–365
Visible color reflecting, 310
Vision, 37–38, 41, 106, 162–164, 258–259, 268, 365, 372
Visual
    adaptation, 258
    communication, 160, 163, 165–170
    cue, 120–121, 166, 266, 281, 309, 316–317, 321–322, 324
    neuron, 258
    neuropil, 38
    place learning, 41
    stimuli, 44, 105, 314, 317
    system, 43, 158–159, 162–163, 165
Vitacea polistiformis, 312
Vitamin, 126
Vitellogenesis, 57
Viviparity, 203, 205, 211–212
Vocal calling, 84
Volatile chemical attractant, 357
Voltinism, 247, 373

W
Wader, 211
Waggle dance, 168, 266–268
War of attrition, 81, 86–87, 89
Warning
    display, 131, 135, 137
    signal, 131, 134–135, 137, 140–141, 159, 164, 166, 369
Wasp, 8, 10, 19, 20, 40, 57–58, 65, 67–68, 71–72, 83, 85, 91–92, 111, 130, 135, 140, 151–153, 161, 167–169, 191, 204–205, 208–209, 220, 222–223, 225, 228–229, 231–232, 244, 250, 260, 276–277, 279–281, 285, 319, 324, 358, 368
Water
    strider, 178, 196, 237, 239, 242, 247, 296
    surface, 38, 178, 183, 207, 209, 300
Waterborne migration, 100
Wavelength, 105, 161, 164, 166, 181, 367–368, 373
Wavelength-specific
    photoreceptors, 161
Wax moth, 10, 21–23
Weapon, 71, 82, 86, 88–89, 135, 139, 166, 168, 191
Web spinner, 204
Weevil, 15, 167–168, 195, 197, 311–313, 322
Western
    flower thrip, 313
    Sahara, 110
Wheat, 311, 322–324

White butterfly, 15, 17, 107, 260
Wind
  convergence, 105, 112
  speed, 104–105, 108, 299, 367
Wind-related orientation, 108, 373
Wing
  flapping, 139
  polymorphism, 64, 102
  size, 300
  spot, 10
Wing out courtship behavior, 20
Winter
  diapause, 294, 300

  survival, 194
Wireworm, 311, 322
Wolbuchiu, 278, 333–335, 337–338, 365
Wood-boring beetle, 214
Wood-dwelling, 223, 226–228
Worker, 15, 51, 57–58, 64–68, 70, 72,
    116, 125, 140, 149, 158, 168,
    219–222, 225, 228–231,
    243–245, 250, 278–280,
    282, 364
  honeybee, 57–58, 64
Working memory, 70, 257
World Conservation Union, 351

X
X chromosome, 10–11,
    69, 337
X-linked, 25
Xylota confuse, 136

Y
Yeast hydrolysate, 317
Yellow
  dung flies, 297, 302
  sphere, 311
Yolk, 50, 53, 205
Yponomeuta cagnagella, 300